Microbiology

a clinical approach

SECOND EDITION

Microbiology

a clinical approach

SECOND EDITION

Anthony Strelkauskas

Angela Edwards

Beatrix Fahnert

Gregory Pryor

Jennifer Strelkauskas

Garland Science
Taylor & Francis Group

Garland Science
Vice President: Denise Schanck
Senior Editor: Elizabeth Owen
Editorial Assistants: Louise Dawnay and Deepa Divakaran
Production Assistant: Deepa Divakaran
Senior Production Editor: Georgina Lucas
Media Production Editor: Natasha Wolfe
Illustrator and Cover Designer: Matthew McClements,
Blink Studio, Ltd
Director of Digital Publishing: Michael Morales
Editorial Assistant for Multimedia: Jasmine Ribeaux
Layout: EJ Publishing
Copyeditors: Bruce Goatly and Jo Clayton
Proofreader: Sally Huish
Indexer: Medical Indexing Ltd

About the Authors
Anthony Strelkauskas, formerly Lead Instructor, Trident
Technical College, South Carolina, USA
Angela Edwards, Instructor, Trident Technical College,
South Carolina, USA
Beatrix Fahnert, Senior Lecturer, Cardiff University, UK
Gregory Pryor, Associate Professor, Francis Marion University,
South Carolina, USA
Jennifer Strelkauskas, Doctor of Veterinary Medicine in Oregon
and Washington, USA

ISBN 978-0-8153-4513-8

Library of Congress Cataloging-in-Publication Data
Strelkauskas, Anthony J., 1944- , author.
  Microbiology : a clinical approach / Anthony Strelkauskas, Angela Edwards,
Beatrix Fahnert, Gregory Pryor, Jennifer Strelkauskas. -- Second edition.
      p. ; cm.
  Includes index.
  ISBN 978-0-8153-4513-8 (paperback) -- ISBN 978-0-8153-4544-2 (loose leaf)
  I. Edwards, Angela, 1964- , author. II. Fahnert, Beatrix, author. III. Pryor, Gregory,
1973- , author. IV. Strelkauskas, Jennifer, author. V. Title.
  [DNLM: 1. Communicable Diseases--microbiology. 2. Infection--microbiology.
3. Microbiological Phenomena. QW 700]
  QR46
  616.9'041--dc23
                              2015017351

Published by Garland Science, Taylor & Francis Group, LLC,
an informa business,
711 Third Avenue, New York, NY, 10017, USA, and
3 Park Square, Milton Park, Abingdon, OX14 4RN, UK.

Printed in Great Britain by Ashford Colour Press Ltd
15 14 13 12 11 10 9 8 7 6 5 4 3 2

GS **Garland Science**
   Taylor & Francis Group

Visit our web site at http://www.garlandscience.com

Preface

Health care professionals see the results of human interaction with microbes every day, usually in the form of infectious diseases. New diseases are emerging and old ones are re-emerging due to changes in the environment and the rise of antibiotic resistance. It is essential that anyone working in health care understands the fundamental principles of microbiology.

When Anthony Strelkauskas began thinking about writing the first edition of this textbook, he had been teaching microbiology to pre-nursing and health care professionals for years and continually had to rewrite material so that it answered the first question his students asked: "Why is this important for me?" His experience in the classroom showed that the way to teach microbiology was to make the answers timely, interesting, and personally relevant. He used to say to his students: "Someday, I may be laying on a gurney looking up at one of you and I want to KNOW you know what you are doing." Believe it or not, that situation actually happened and he was so proud of the professionals that his students had become. Dr. Strelkauskas decided he would write a textbook in which everything was designed to be an effective tool for teaching microbiology to a clinically oriented audience and the outcome was the first edition of *Microbiology: A Clinical Approach*. His efforts were well-received but, unfortunately, he died before he could start work on a second edition. A new team of teachers, scientists, and health care professionals has come together to continue the work.

The authors past and present have put all their experience into writing a textbook and online resources that will help you, the student, succeed in your studies. The text begins with a special Learning Skills section that gives practical strategies for improving reading comprehension and retention of information. The first chapter presents a series of case studies designed to demonstrate the vital importance of studying microbiology. All chapters address that most important question: "Why do I need to know this?" The opening page of each chapter has a human story, to emphasize that microbiology is a real-world experience, and a chapter map to help orientation. Short summaries of individual sections of each chapter, called "Keep in Mind," are interspersed throughout, to help you digest important facts and concepts before continuing to the next section. Line drawings, photographs, and micrographs beautifully illustrate important concepts. The Pathogen List at the end of the book is a quick-and-easy reference to the pathogens mentioned throughout the book, their characteristics, and the diseases they cause. There are three types of end-of-chapter questions. Self Evaluation and Chapter Confidence questions test basic comprehension. Depth of Understanding questions require you to integrate important concepts in a more challenging way. Clinical Corner questions ask you to apply what you have just learned to a specific clinical setting or problem. Answers and explanations for the correct answers are available on the Student Resource Website. The website also has a robust set of animations specifically developed to make some of the challenging topics more accessible. Why is this information important to you? Continue reading and find out!

To do justice to the whole of microbiology would take a book much longer than this one, and as it is written specifically for students interested in health care, it focuses on the area most relevant to human health: the relationship between microbes and human disease. The presence of microorganisms in the body is now known to be beneficial and even essential to good health, and this aspect of microbiology has been expanded and updated in the new edition. All information on pathogens has been updated, especially with respect to emerging and re-emerging infections, and a section on clinical diagnosis added. One of the major challenges to health care today, antibiotic resistance, is emphasized throughout and, as with the first edition, an entire chapter is dedicated to it. Clinical themes are interwoven throughout the individual chapters, and the book as a whole, by emphasis and repetition of key clinical concepts. Particular emphasis is placed on the five requirements for infection: get in, stay in, defeat the host defenses, damage the host, and be transmissible. These requirements are introduced in the first chapter and revisited throughout the text.

We wish all students and instructors the best success. It is important to us that the book is accurate, authoritative, and interesting. Despite our best efforts, there may be errors. We encourage readers who find errors to email us at science@garland.com in order that we may continue to refine the text in subsequent printings and editions. Your input is very important and we thank you in advance for your comments and questions.

The authors would like to thank all the staff at Garland Science for their persistence, professionalism, humor, and friendship throughout.

AE, BF, GP, JS

ACKNOWLEDGMENTS

The authors and publishers have benefitted greatly from the many microbiologists, immunologists, and instructors who have reviewed the text and figures and made suggestions for improvements. We would like to thank the following for their contribution:

Maryam Ahmed (Appalachian State University); Simon Allard (Madison Area Technical College); Yassine Amrani (University of Leicester); Justin Anderson (Radford University); Graham Avery (University of Essex); Mamatha Ballal (Manipal University); David Beck (Tennessee Technological University); Christine Bezotte (Elmira College); Wilbert Bitter (VU Amsterdam); Damien Brady (Athlone Institute of Technology); Alan Bruce (University of Abertay, Dundee); Claire Carpenter (Yakima Valley Community College); Shaqil Chaudary (Liverpool John Moores University); Philip Cheetham (Nottingham Trent University); Mary Ellen Clark (St. John Fisher College); Leslie Dafoe (Sault College); Giulia De Falco (University of London); Natasha Dean (La Sierra University); Alex Dent (Indiana University School of Medicine); Kathleen DiCaprio (Touro College of Osteopathic Medicine); Jeff Duerr (George Fox University); Jessica Elliott (Holmes Community College); Shankar Esakimuthu (University of Malaya); Barbara Fenner (King's College); Sherry Fuller-Espie (Cabrini College); Kamal Gandhi (United States University); Heather Garcia (Troy University); David Glick (King's College); Brinda Govindan (San Francisco State University); Linda Harris-Young (Motlow State Community College); Anne Heise (Washtenaw Community College); Cheryl Hertz (Loyola Marymount

University); Arthur Hosie (University of Bedfordshire); Carmel Kealey (Athlone Institute of Technology); George Keller (Samford University); Martin Khechara (University of Wolverhampton); Sue Lang (Glasgow Caledonian University); Jessica Lea (Central Carolina Technical College); Blaine Legaree (Keyano College); Julie Letchford (University of Bath); Stacey Lettini (Gwynedd Mercy University); Stephen Longdo (McDaniel College); Sarah Maddocks (Cardiff Metropolitan University); Bernardino Madsen (University of Wyoming); Prabir Mandal (Edward Waters College); Michelle McEliece (Gwynedd Mercy University); John McEvoy (North Dakota State University); Preeti Mehta (Seth GS Medical College & KEM Hospital, Mumbai); Nicola Milner (Anglia Ruskin University); Veronica Mittak (New York Chiropractic College); Lorena Navarro (University of California, Davis); Jamee Nixon (Northwest Nazarene University); Maja Nowakowski (Touro College of Osteopathic Medicine); Fernando Ontiveros-Llamas (St. John Fisher College); Cameron Perkins (South Georgia State College); Mark Pilgrim (Lander University); Rachel Polando (Manchester College); David Ponsonby (Canterbury Christ Church University); Edith Porter (California State University, Los Angeles); Kumar Rajakumar (University of Leicester); Nadia Rajsz (Orange County Community College); Sarah Richart (Azusa Pacific University); Marie-Claire Rioux (John Abbott College); Jennifer Robbins (Xavier University); Beverly Roe (Erie Community College); Mary Russell (Kent State University, Trumbull); Lauri Sammartano (Roane State University); Suchitra Shenoy Mala (Kasturba Medical College); Stuart Siddell (University of Bristol); Bharathi Srinivasan (The Oxford College of Scicnce); Munir Syed (Hartwick College); Milly van Dijk (VU University); Kerstin Voelz (University of Birmingham); Margaret Wexler (University of East Anglia); Karl Wooldridge (University of Nottingham); Jun Yu (University of Strathclyde).

I would like to thank my husband Michael and my children Anthony and Rachel for all their love and support. I would like to dedicate my efforts to my mother who always said I should write something someday. And I would like to thank Tony Strelkauskas for having the vision, skill, and determination to write this book. It was a pleasure to know and work with him and an honor to try to carry on a small part of what he created.

AE

I am grateful for the thousands of students I have taught, who in turn have taught me to be a more engaging, confident, and compassionate teacher. I am especially appreciative of my wife, Dr. Tamatha Barbeau, who played so many supportive roles while I worked on this book.

GP

I'd like to thank my dad first and foremost for allowing me to be a part of this project. He truly believed that knowledge is power and that spreading knowledge to students was his greatest achievement. Out of the many years of schooling I have been through, he was the greatest teacher I ever had. I miss him every day. I would also like to thank my mom for being the rock we all relied on for support, for help, and for generally making sure we could make it all work. She showed me strength can bc formidable when paired with flexibility. Thank you to my husband and my boys for their love and support and for giving me the reasons to keep pushing!

JS

All the teaching and learning resources for instructors and students are available online. Resources for instructors are password protected and available only to qualifying instructors. Resources for students are available to everyone.

ONLINE HOMEWORK PLATFORM WITH INSTRUCTOR DASHBOARD

The online homework platform for *Microbiology: A Clinical Approach* is designed to improve and track student performance. It allows instructors to select homework assignments on specific topics and review the performance of the entire class, as well as individual students, via the instructor dashboard. The user-friendly system provides a convenient way to gauge student progress, and tailor classroom discussion, activities, and lectures to areas that require specific remediation. The assignments have been written by one of the authors, Gregory Pryor (Francis Marion University).

Instructors can obtain access to the online homework platform from their sales representative or by emailing science@garland.com. Students who wish to use the platform must purchase access and, if required for class, obtain a course link from their instructor.

INSTRUCTOR RESOURCES

Instructor resources are available at www.garland science.com/instructors. This website provides access not only to the teaching resources for this book, but to the resources for all Garland books. Qualifying instructors can obtain access from their sales representative, or by emailing science@garland.com.

The Art of Microbiology: A Clinical Approach

The images from the book are available in two convenient formats, PowerPoint® and JPEG, and have been optimized for display. Figures are searchable by figure number, figure name, or by keywords used in the figure legend.

Instructor's Manual

The authors provide a guide to each section that highlights the most important topics to be covered in class with suggestions for assignments outside of class. For each chapter, the authors provide detailed guidance on presentation strategies, clinical connections, instructional goals, and potential problem areas for students.

Lecture Outlines

Two PowerPoint presentations are provided for each chapter. Both can be used "as is" in the classroom or edited for personal preference. One presentation is concise and based on the "Keep in Mind" points from the chapter; the other is longer and has been written for each chapter by the authors with illustrations and tables integrated. The Lecture Notes provided to students are based upon these presentations.

Question Bank

Written by Arnold Kaplan, Washington University in St. Louis, the Question Bank contains over 1000 questions in a variety of formats: multiple-choice, true-false, fill-in-the-blank, matching, and depth-of-understanding question. Each question is graded according to Bloom's taxonomy and answers are also given. The questions test basic retention of scientific facts and the ability to understand and apply concepts. They are designed for quizzes and examinations, and the multiple-choice questions are suitable for use with personal response systems (clickers).

Diploma® Test Generator Software

The questions from the *Microbiology: A Clinical Approach* Question Bank have been loaded into the Diploma Test Generator Software. The software is easy to use and can scramble questions to create multiple tests. Questions are organized by chapter and type, and can be additionally categorized by the instructor according to difficulty or subject. Existing questions can be edited and new ones added. It is compatible with several course management systems, including Blackboard®.

MicroMovies

Short movies have been developed to complement material in a select number of chapters, with a special emphasis on molecular genetics, virology, and immunology. Each movie has a voice-over narration, and the text for this voice over is located in the Media Guide. The movies can easily be imported into PowerPoint presentations, and are available in two handy formats: WMV and QuickTime®. The WMV versions are suitable for importing movies into PowerPoint for Windows®.

The QuickTime versions are suitable for importing the movies into PowerPoint for Macintosh®. Students are provided with the movies and movie scripts on the Student Resource Website.

Media Guide

This document overviews the multimedia package available for students and instructors. It also contains the text of the voice-over narration for all of the MicroMovies.

STUDENT RESOURCES

All resources for students are freely available from www.garlandscience.com/micro2.

E-Tutor

The E-Tutor provides guidance for answering all of the questions found at the end of each chapter. The E-Tutor not only provides the right answer, but also explains in detail exactly why a particular answer is correct. The E-Tutor covers the Self Evaluation and Chapter Confidence questions, the Depth of Understanding questions, and the Clinical Corner questions.

MicroMovies

The short movies (animations and videos) concentrate on topics in molecular genetics, virology, and immunology. The transcripts of the voice-over narration for each movie are available in the Media Guide.

Bug Parade

The Bug Parade is an enhanced version of the Pathogen List found at the end of the book. Clicking on the name of the bug gives the correct pronunciation.

Flashcards

Each chapter contains a set of flashcards that allows students to review key terms from the text.

Glossary

The complete glossary is available online and can be searched or browsed.

Lecture Notes

These outlines are based on the lecture slides and designed to help students follow a lecture on a particular chapter. They can be easily printed out and brought to class.

Contents

Learning Skills   L1

PART I   Foundations   **1**
Chapter 1   What Is Microbiology and Why Does It Matter?   3
Chapter 2   Fundamental Chemistry for Microbiology   15
Chapter 3   Essentials of Metabolism   31
Chapter 4   An Introduction to Cell Structure and Host–Pathogen Relationships   47

PART II   Disease Mechanisms   **73**
Chapter 5   Requirements for Infection   75
Chapter 6   Transmission of Infection, the Compromised Host, Epidemiology, and
               Diagnosing Infections   97
Chapter 7   Principles of Disease   131
Chapter 8   Emerging and Re-Emerging Infectious Diseases   147

PART III   Characteristics of Disease-Causing Microorganisms   **183**
Chapter 9   The Clinical Significance of Bacterial Anatomy   185
Chapter 10   Bacterial Growth   213
Chapter 11   Microbial Genetics and Infectious Disease   235
Chapter 12   The Structure and Infection Cycle of Viruses   271
Chapter 13   Viral Infection   293
Chapter 14   Parasitic and Fungal Infections   317

PART IV   Host Defense   **349**
Chapter 15   The Innate Immune Response   351
Chapter 16   The Adaptive Immune Response   383
Chapter 17   Failures of the Immune Response   413

PART V   Control and Treatment   **433**
Chapter 18   Control of Microbial Growth With Disinfectants and Antiseptics   435
Chapter 19   Antibiotics and Antimicrobial Drugs   459
Chapter 20   Antibiotic Resistance   485

PART VI   Microbial Infections   **507**
Chapter 21   Infections of the Respiratory System   509
Chapter 22   Infections of the Digestive System   535
Chapter 23   Infections of the Genitourinary System   571
Chapter 24   Infections of the Central Nervous System   593
Chapter 25   Infections of the Blood and Lymph   615
Chapter 26   Infections of the Skin and Eyes   641

Epilogue   661
Multiple Choice Answers   662
Glossary   663
Pathogen List   687
Figure Acknowledgments   697
Index   700

Detailed Contents

Learning Skills   L1

PART I   Foundations   1

**Chapter 1   What Is Microbiology and Why Does
It Matter?**   3
HUMAN STORIES   4
Ivan Goes to Chicago…   4
It's for the Birds…   4
Special Delivery…   5
Hamburger Havoc…   5
Did You Wash Your Hands?…   6
The Hospital Can Be Dangerous…   7
THE RELEVANCE OF MICROBIOLOGY TO HEALTH CARE   8
Infectious Disease   9
    Epidemiology   10
    Pathogenesis   10
    Host Defense   10
Treatment of Infectious Diseases   11

**Chapter 2   Fundamental Chemistry for
Microbiology**   15
CHEMICAL BONDING   16
Ionic Bonds   16
Covalent Bonds   17
Hydrogen Bonds   18
WATER   19
ACIDS, BASES, AND pH   19
BIOLOGICAL MOLECULES   20
Carbohydrates   20
Lipids   21
    Fats   21
    Phospholipids and Glycolipids   22
    Steroids   23
Proteins   23
    Properties of Proteins   23
    Protein Structure   24
    Types of Protein   24
Nucleic Acids   25
    Structure of Nucleic Acids   25
ATP   27

Chapter 3   Essentials of Metabolism   31
BASIC CONCEPTS OF METABOLISM   32
Oxidation and Reduction Reactions   32
Respiration   33
Metabolic Pathways   33
ENZYMES   34
Properties of Enzymes   34
Coenzymes and Cofactors   35
Enzyme Inhibition   36
Factors That Affect Enzyme Reactions   38
CATABOLISM   38
Glycolysis   38
The Krebs Cycle   40
The Electron Transport Chain   40
Chemiosmosis   41
Looking at the Numbers   41
Fermentation   42
Homolactic Fermentation   42
Alcoholic Fermentation   43
ANABOLISM   43

**Chapter 4   An Introduction to Cell Structure
and Host–Pathogen Relationships**   47
CLASSIFICATION OF ORGANISMS   48
BACTERIA   48
Size, Shape, and Multicell Arrangement   48
Staining   50
    The Gram Stain   51
    The Negative (Capsule) Stain   52
    The Flagella Stain   53
    The Ziehl–Neelsen Acid-Fast Stain   53
    The Endospore Stain   53
HOST–PATHOGEN RELATIONSHIPS   54
Pathogenicity: A Matter of Perspective   55
Opportunistic Pathogens and Primary Pathogens   55
Disease and Transmissibility   55
BACTERIAL PATHOGENICITY AND VIRULENCE   56
Quorum Sensing   56
Biofilms   57
THE HOST CELL   59
Prokaryotic Versus Eukaryotic Cell Structure   59
Plasma Membrane   60
    The Role of the Plasma Membrane in Infection   62
Cytoplasm   62
    The Role of the Cytoplasm in Infection   62
Cytoplasmic Structures Not Enclosed by a Membrane   62
    Cytoskeleton   63
    The Role of the Cytoskeleton in Infection   63
    Cilia   63
    The Role of Cilia in Infection   64
Flagella   64
Ribosomes   64
    The Role of the Eukaryotic Ribosome in Infection   64
Cytoplasmic Structures Enclosed by a Membrane   65
    Mitochondria   65
    Endoplasmic Reticulum and Golgi Apparatus   65

The Role of the Endoplasmic Reticulum in Infection 66
Lysosomes 66
The Role of the Lysosome in Infection 66
Proteasomes 66
The Role of the Proteasome in Infection 66
Peroxisomes 66
Nucleus 66
The Role of the Nucleus in Infection 67
Endocytosis and Exocytosis 67
The Role of Endocytosis and Exocytosis in Infection 69

PART II   Disease Mechanisms   73

Chapter 5   Requirements for Infection   75

PORTALS OF ENTRY (GETTING IN) 76
Mucous Membranes 76
The Respiratory Tract 76
The Gastrointestinal Tract 77
The Genitourinary Tract 78
Skin 79
The Parenteral Route 80

ESTABLISHMENT (STAYING IN) 81
Increasing the Numbers 83

AVOIDING, EVADING, OR COMPROMISING HOST DEFENSES (DEFEAT THE HOST'S DEFENSES) 83
Capsule and Cell Wall (Passive) Protection Strategies 84
Enzyme (Active) Protection Strategies 84

DAMAGING THE HOST 87
Exotoxins 87
Anthrax Toxin 88
Diphtheria Toxin 88
Botulinum Toxin 89
Tetanus Toxin 90
Cholera Toxin 90
Other Exotoxins 90
Endotoxins 91
Viral Pathogenic Effects 93

Chapter 6   Transmission of Infection, the Compromised Host, Epidemiology, and Diagnosing Infections   97

TRANSMISSION OF INFECTION 98
Reservoirs of Pathogens 98
Mechanisms of Transmission 99
Contact Transmission 99
Vehicle Transmission 100
Vector Transmission 100
Factors Affecting Disease Transmission 102
Portals of Exit 102

THE COMPROMISED HOST 102
Neutropenia 103
Organ Transplantation 104
Burn Patients 104
Opportunistic Infections 105
Nosocomial Infections 105
Universal Precautions 106
Preventing and Controlling Nosocomial Infections 107

EPIDEMIOLOGY 108

Incidence and Prevalence 108
Morbidity and Mortality Rates 108
Herd Immunity 109
Types of Epidemiological Study 110

DIAGNOSING INFECTIONS 112
Growth-Based Diagnostics 112
Serology-Based Diagnostics 113
Biotechnology and Health 113
Recombinant DNA Technology 115
Monoclonal Antibodies 116
Serology-Based Diagnostic Methods 119
Precipitation 119
Agglutination 120
Immunoblotting 120
ELISA 121
Diagnostic Methods Based on Nucleic Acids 121
PCR 121
Hybridization 123
Sequence Identification 123
The Future Is Here: New Approaches and Opportunities 124

Chapter 7   Principles of Disease   131

THE RELATIONSHIP BETWEEN THE HUMAN HOST AND MICROORGANISMS 132

THE ETIOLOGY OF DISEASE 136

DEVELOPMENT OF DISEASE 137
Communicable and Contagious Disease 138
Duration of Disease 139
Persistent Bacterial Infections 139

THE SCOPE OF INFECTIONS 142
Toxic Shock and Sepsis 142

Chapter 8   Emerging and Re-Emerging Infectious Diseases   147

EMERGING INFECTIOUS DISEASES 148
Environment and Infectious Disease 151
Foodborne Infections 152
Globalization and Transmission 153
Hurdles to Interspecies Transfer 154
Emerging Viral Diseases 154
SARS (Severe Acute Respiratory Syndrome) 154
West Nile Fever 156
Dengue Fever 157
Viral Hemorrhagic Fever (VHF) 158

RE-EMERGING INFECTIOUS DISEASES 161
Tuberculosis 161
Immune Status and Fitness of the Host 162
Virulence/Fitness of the Pathogen 162
Influenza 163

AVIAN INFLUENZA 165

PRIONS AND PRION DISEASES (TRANSMISSIBLE SPONGIFORM ENCEPHALOPATHY) 167
The Prion Hypothesis 167
The Biology of Prion Diseases 167
Creutzfeldt–Jakob Disease 168
Variant CJD (vCJD) 169

BIOTERRORISM 170

Anthrax	172
Botulism	173
Plague	173
Smallpox	175
Tularemia	176
Viral Hemorrhagic Fevers	176
The Danger of Bioweapons	177
Warning Signs	177

PART III   Characteristics of Disease-Causing Microorganisms — 183

Chapter 9   The Clinical Significance of Bacterial Anatomy — 185

THE BACTERIAL CELL WALL	186
Building the Cell Wall	187
The Cytoplasmic Phase	187
The Membrane-Associated Phase	187
The Extracytoplasmic Phase	187
Additional Cell Wall Components	188
Wall Components in Gram-Positive Bacteria	189
Wall Components in Gram-Negative Bacteria	189
Clinical Significance of the Bacterial Cell Wall	191
STRUCTURES OUTSIDE THE BACTERIAL CELL WALL	193
The Glycocalyx	193
Clinical Significance of the Glycocalyx, Slime Layer, and Capsule	193
Fimbriae and Pili	194
Clinical Significance of Fimbriae and Pili	194
Axial Filaments	195
Clinical Significance of Axial Filaments	196
Flagella	196
Structure of the Bacterial Flagellum	196
Flagellar Configurations	198
Clinical Significance of Flagella	198
Intracellular Movement	199
STRUCTURES INSIDE THE BACTERIAL CELL WALL	199
The Plasma Membrane	199
Structure of the Plasma Membrane	200
Energy Production	200
Membrane Transport	201
Secretion	204
Clinical Significance of the Plasma Membrane	204
The Nuclear Region	205
Plasmids	205
Clinical Significance of DNA and Plasmids	205
Ribosomes	206
Clinical Significance of Ribosomes	206
Inclusion Bodies	206
Endospores	206
Clinical Significance of Sporogenesis	207

Chapter 10 Bacterial Growth — 213

REQUIREMENTS FOR BACTERIAL GROWTH	214
Physical Requirements for Growth	214
Temperature	214
pH	216
Osmotic Pressure	217
Chemical Requirements for Growth	218
Carbon	218
Nitrogen	218

Sulfur	218
Phosphorus	218
Organic Growth Factors	219
Oxygen	219
Growth of Anaerobic Organisms	220
GROWTH MEDIA	222
Complex Media	222
Chemically Defined Media	222
Media as a Tool for Identifying Pathogens	224
CHARACTERISTICS OF BACTERIAL GROWTH	226
The Bacterial Growth Curve	227
Measurement of Bacterial Growth	228
CLINICAL IMPLICATIONS OF BACTERIAL GROWTH	229
Specimen Collection	229

Chapter 11 Microbial Genetics and Infectious Disease — 235

THE STRUCTURE OF DNA AND RNA	236
Deoxyribonucleic Acid (DNA)	236
Ribonucleic Acid (RNA)	238
DNA REPLICATION	239
DNA Separation and Supercoiling	239
DNA Polymerase	240
Proofreading By DNA Polymerase	241
The Replication Fork	242
Initiation and Termination of Replication	243
THE GENETIC CODE	244
GENE EXPRESSION	245
Transcription	245
Translation	246
Messenger RNA in Translation	247
Transfer RNA in Translation	247
The Ribosome in Translation	248
Formation of Peptide Bonds in Translation	250
Translation Initiation	250
Translation Elongation	251
Translation Termination	251
REGULATION OF GENE EXPRESSION	252
Induction	253
Repression	254
MUTATION AND REPAIR OF DNA	256
Replication Errors	256
How DNA Damage Occurs	257
Repair of DNA Damage	257
TRANSFER OF GENETIC INFORMATION	259
Transposition	259
Transformation	260
Transduction	260
Conjugation	262
GENETICS AND PATHOGENICITY	265

Chapter 12 The Structure and Infection Cycle of Viruses — 271

VIRUS STRUCTURE	272
The Virion	272
Viral Nomenclature	272
The Capsid	273

Helical Viruses 273
Icosahedral Viruses 274
Complex Viruses 274
Viral Envelopes 274
Envelope Glycoproteins 275
THE INFECTION CYCLE 275
Attachment 276
When Virus Meets Host 277
Receptor Binding 277
Penetration and Uncoating 278
Penetration and Uncoating by Non-enveloped
Viruses 278
Penetration and Uncoating by Enveloped Viruses 279
Cytoplasmic Transport of Viral Components 279
Transport of the Viral Genome into the Nucleus 280
Biosynthesis 281
DNA Viruses—Replication 281
DNA Viruses—Transcription 282
RNA Viruses—Replication and Transcription 283
Retroviral Transcription and Integration 284
Viral Control of Translation 284
Maturation 285
Intracellular Trafficking 285
Assembly 287
Release 287
Budding from the Plasma Membrane 288
SPREAD OF VIRUSES 289

Chapter 13 Viral Infection 293

PATTERNS OF VIRAL INFECTION 294
Acute Infections 295
Antigenic Variation 295
Persistent Infections 296
Chronic Infections 297
Latent Infections 297
Slow Infections 298
DISSEMINATION AND TRANSMISSION OF VIRAL
INFECTION 299
Portals of Entry 299
Respiratory Tract 299
Gastrointestinal Tract 300
Genitourinary Tract 300
Eyes 301
Skin 301
Dissemination Pathways 301
Bloodstream 302
Nervous System 303
Internal Organs 303
Viral Transmission 304
Transmission via the Respiratory Tract 304
Transmission via the Epidermis 305
Transmission via Bodily Fluids 305
Transmission via the Fecal–Oral Route 306
Fetal Infection 306
VIRULENCE 306
Virulence and Host Susceptibility 307
VACCINE DEVELOPMENT 307
Recombinant DNA Vaccines 310
Viral Culture 310

VIRUSES AND CANCER 311
Oncogenic Viruses 311
HOST DEFENSE AGAINST VIRAL INFECTION 312

Chapter 14 Parasitic and Fungal Infections 317

PARASITES AND THEIR INFECTIONS 318
Significance of Parasitic Infections 318
Protozoan Morphology and Pathogenesis 319
Helminth Morphology and Pathogenesis 320
Life Cycles and Transmission Pathways of Protozoans
and Helminths 321
Parasites That Use a Single Host 321
Parasites That Use Multiple Hosts 321
EXAMPLES OF PROTOZOAN INFECTIONS 323
Malaria (*Plasmodium* species) 323
Toxoplasmosis (*Toxoplasma gondii*) 326
Amebiasis (*Entamoeba histolytica*) 327
Trichomoniasis (*Trichomonas vaginalis*) 328
Trypanosomiasis (*Trypanosoma* species) 329
EXAMPLES OF HELMINTHIC INFECTIONS 330
Intestinal Nematodes 330
Enterobiasis 331
Ascariasis 332
Tissue Nematodes 333
Trichinosis 333
Cestodes 334
Trematodes 335
Paragonimiasis 335
Clonorchiasis 336
Schistosomiasis 337
FUNGAL INFECTIONS 338
Fungal Structure and Growth 338
Yeasts and Molds 338
Dimorphism 340
Fungal Infections 340
Superficial Mycoses 340
Cutaneous and Mucocutaneous Mycoses 341
Subcutaneous Mycoses 342
Deep Mycoses 342
Pathogenesis of Fungal Infections 343
Adherence 343
Invasion 343
Tissue Injury 344

PART IV Host Defense 349

Chapter 15 The Innate Immune Response 351

BARRIER DEFENSES 352
Skin 352
Mucous Membranes 353
The Lacrimal Apparatus 354
Saliva 354
Epiglottis 354
Sebum 355
Perspiration 355
Gastric Juice 355
Urine and Vaginal Secretions 355
Transferrins 355
MOLECULES OF INNATE IMMUNITY 356

Toll-Like Receptors 357
Cytokines 357
CELLS INVOLVED IN INNATE IMMUNITY 358
Neutrophils 358
Basophils 360
Eosinophils 361
Macrophages and Monocytes 361
Dendritic Cells 363
Mast Cells 364
Natural Killer Cells 365
THE FIVE MECHANISMS OF INNATE IMMUNITY 367
Phagocytosis 367
Adherence 367
Ingestion 368
Digestion 368
Excretion 368
Avoiding Phagocytosis 369
Inflammation 369
Vasodilation and Vascular Permeability 369
Phagocyte Migration 370
Acute-Phase Response 370
Fever 371
The Complement System 372
Activation of the Classical Pathway 372
Activation of the Alternative Pathway 372
Activation of the Lectin-Binding Pathway 373
C3 and Beyond 374
Evasion of the Complement System 375
Interferon 375
IFN-α and IFN-β 376
IFN-γ 376
Therapeutic Use of Interferon 376
GENETIC SUSCEPTIBILITY TO INFECTION 378

Chapter 16 The Adaptive Immune Response 383

COMPONENTS OF ADAPTIVE IMMUNITY 384
Strategic Lymphoid Structures 384
Cells of the Adaptive Immune Response 386
Dendritic Cells 386
Lymphocytes 387
DEVELOPMENT OF LYMPHOCYTE POPULATIONS 388
Clonal Selection of Lymphocytes 388
Survival of Lymphocyte Populations 389
Lymphoid Tissues 390
ANTIGEN PRESENTATION 391
Class I MHC 392
Class II MHC 392
T-Cell Response to Superantigens 392
CELLULAR (T CELL) RESPONSE 393
Production of Activated Effector Helper T Cells 393
Activation by Dendritic Cells 393
Activation by Macrophages 394
Activation by B Cells 394
Functions of Activated Effector T Cells 394
Cytotoxic T Cells 394
Helper T Cells 395
Regulatory T cells 396
HUMORAL (B CELL) RESPONSE 396

The Immunoglobulin Molecule 397
Distribution and Function of Immunoglobulins 399
IgM 399
IgG 399
IgA 399
IgE 400
IgD 401
Timing of Immunoglobulin Release 401
Antibody Activation of Immune Cells 401
B-Cell Activation and Cooperation with T Cells 402
COURSE OF THE ADAPTIVE RESPONSE 403
IMMUNOLOGICAL MEMORY 404
Natural and Artificial Immunity 405
OVERALL IMMUNE RESPONSE 406

Chapter 17 Failures of the Immune Response 413

IMMUNODEFICIENCIES CAUSED BY INFECTION 414
Acquired Immune Deficiency Syndrome 414
Cellular Targets of HIV 415
Modes of HIV Transmission 415
HIV Genome 416
Dynamics of HIV Replication in Infected Patients 416
Course of Infection 417
Response to HIV Infection 418
Major Tissue Effects of HIV Infection 419
Other Infections 420
PRIMARY IMMUNODEFICIENCY DISEASES 421
B-Cell Defects 421
T-Cell Defects 422
Defects in Accessory Cells 422
AUTOIMMUNE DISEASE 424
Causes of Autoimmunity 424
Mechanisms of Autoimmunity 425
HYPERSENSITIVITY (ALLERGIC REACTIONS) 426
Effector Mechanisms in Allergic Reactions 427
Phases of Allergic Reactions 428
Clinical Effects of Allergic Reactions 428

PART V Control and Treatment 433

Chapter 18 Control of Microbial Growth With Disinfectants and Antiseptics 435

CONTROLLING MICROBIAL GROWTH 436
TARGETS FOR DISINFECTANTS AND ANTISEPTICS 437
The Cell Wall 437
The Plasma Membrane 437
Protein Structure and Function 438
Nucleic Acids 439
MICROBIAL DEATH 439
Factors That Affect the Rate of Microbial Death 439
CHEMICAL METHODS FOR CONTROLLING MICROBIAL GROWTH 441
The Potency of Disinfectants and Antiseptics 441
Evaluation of Disinfectants and Antiseptics 441
The Phenol Coefficient 441
The Disk Diffusion Method 442
The Use Dilution Method 442

Selecting an Antimicrobial Agent 442
Types of Chemical Agent 444
   Phenol and Phenolic Compounds 445
   Alcohols 445
   Halogens 445
   Oxidizing Agents 446
   Surfactants 446
   Heavy Metals 447
   Aldehydes 447
   Gaseous Agents 447

PHYSICAL METHODS FOR CONTROLLING MICROBIAL
GROWTH 448
Heat 448
Moist Heat Methods 449
Refrigeration, Freezing, and Freeze-Drying 451
Filtration 452
Osmotic Pressure 452
Radiation 452
A Word About Hand Washing 454

Chapter 19 Antibiotics and Antimicrobial Drugs 459

SOURCE OF ANTIBIOTICS 460
Antibiotics Are Part of Bacterial Self-Protection 460

ANTIBIOTIC SPECTRA 461
Antibiotic Structure 462

ANTIBIOTIC TARGETS 465
The Bacterial Cell Wall 466
   Penicillins 466
   Cephalosporins 466
   Carbapenems 467
   Monobactams 467
   Glycopeptide Antibiotics 467
   Peptide Antibiotics Effective Against Mycobacteria 467
   Polypeptide Antibiotics 468
The Bacterial Plasma Membrane 468
Synthesis of Bacterial Proteins 469
   Macrolides 469
   Tetracyclines 469
   Aminoglycosides 470
Bacterial Nucleic Acids 470
   Quinolones 471
   Rifamycins 471
Bacterial Metabolism 471

ANTIVIRAL DRUGS 472
Nucleoside Analogs 474
   Acyclovir 474
   AZT 474
Nucleotide Analogs 475
Non-nucleoside Inhibitors 475
Inhibiting Viral Assembly and Release 475
   Protease Inhibitors 475
   Neuraminidase Inhibitors 475
Inhibiting Viral Uncoating 476
Inhibiting Viral Entry 476
DRACO 476

ANTIFUNGAL DRUGS 476
Antifungals That Affect the Plasma Membrane 477
   Polyenes 477
   Azoles 477

Antifungals That Affect the Cell Wall 477
Antifungals That Affect Nucleic Acid Synthesis 478
DRUGS FOR PARASITIC INFECTIONS 478
Anti-protozoan Drugs 478
   Antimalarials 478
   Folate Agonists 479
   Drugs for Anaerobic Protozoa 479
   Drugs for *Trypanosoma* and *Leishmania* 480
Anti-helminthic Drugs 480

Chapter 20 Antibiotic Resistance 485

DEVELOPMENT OF ANTIBIOTIC RESISTANCE 486
Bacterial Growth and Mutation Rates 486
Plasmids and Conjugation 486
Inappropriate Clinical Use of Antibiotics 487
Use of Antibiotics in the Food Chain 488
Immunocompromised Population 488
Health Care Facilities 488
Lifestyle 488

TIMELINE OF ANTIBIOTIC RESISTANCE 489
Susceptibility Testing 490

MECHANISMS OF RESISTANCE 492
Inactivation of Antibiotic 492
   AmpC β-lactamase 492
   Aminoglycoside-Inactivating Enzymes 493
Efflux Pumping of Antibiotic 493
   Mechanism of Efflux Pumps 494
   Tetracycline Efflux Pumps 494
Modification of Antibiotic Target 495
   Penicillin-Binding Proteins 495
   Modification of Target Ribosomes 495
Alteration of a Metabolic Pathway 496

CLINICALLY DANGEROUS RESISTANCE 496
MRSA 496
VREs 498
E. coli 498
Re-emerging Diseases 498
Superinfections 499

RESISTANCE TO ANTIVIRALS AND ANTIPARASITICS 500
Antiviral Resistance 500
Parasitic Resistance 500

HOPE FOR THE FUTURE — DEVELOPMENT OF
NEW ANTIBIOTICS 501
DNA and RNA Analysis 501
Structural Analysis 501
Auxiliary Targets 501
Automated Synthesis and Screening 502
Virulence Factors 502
Further Investigation of Known Antibiotic Compounds 502
Testing of Antibiotics 502

PART VI Microbial Infections 507

Chapter 21 Infections of the Respiratory System 509

ANATOMY OF THE RESPIRATORY SYSTEM 510

PATHOGENS OF THE RESPIRATORY SYSTEM 510
Bacteria That Infect the Respiratory System 512

BACTERIAL INFECTIONS OF THE UPPER RESPIRATORY TRACT | 512
Otitis Media, Mastoiditis, and Sinusitis | 512
Pharyngitis | 512
Scarlet Fever | 513
Diphtheria | 514
VIRAL INFECTIONS OF THE UPPER RESPIRATORY TRACT | 515
Rhinovirus Infection (the Common Cold) | 515
Parainfluenza | 515
BACTERIAL INFECTIONS OF THE LOWER RESPIRATORY TRACT | 516
Bacterial Pneumonia | 516
Chlamydophila Pneumonia | 518
Mycoplasma Pneumonia | 518
Tuberculosis | 519
Pertussis (Whooping Cough) | 521
Inhalation Anthrax | 522
Legionella Pneumonia (Legionnaires' Disease) | 523
Q Fever | 524
Psittacosis (Ornithosis) | 524
VIRAL INFECTIONS OF THE LOWER RESPIRATORY TRACT | 525
Influenza | 525
Respiratory Syncytial Virus (RSV) | 527
Hantavirus Pulmonary Syndrome (HPS) | 528
FUNGAL INFECTIONS OF THE RESPIRATORY SYSTEM | 528
Pneumocystis Pneumonia (PCP) | 529
Blastomycosis | 530
Histoplasmosis | 530
Coccidioidomycosis | 531
Aspergillosis | 532

Chapter 22 Infections of the Digestive System | 535

CLINICAL SYMPTOMS AND EPIDEMIOLOGY | 536
Watery Diarrhea | 536
Dysentery | 536
Enteric Fever | 536
Treatment and Management Options for Gastrointestinal Infections | 536
Endemic and Pandemic Gastrointestinal Infections | 537
Traveler's Diarrhea | 537
Nosocomial Infections | 537
Food Poisoning | 537
DENTAL AND PERIODONTAL INFECTIONS | 540
Formation of Dental Plaque | 540
Dental Caries | 540
Gingivitis and Periodontitis | 541
Necrotizing Periodontal Disease | 541
BACTERIAL INFECTIONS OF THE DIGESTIVE SYSTEM | 542
Escherichia coli | 545
Enterotoxigenic *E. coli* | 546
Enteropathogenic *E. coli* | 546
Enterohemorrhagic *E. coli* | 547
Treatment of all five forms of *E. coli* infection | 547
Shigella | 547
Salmonella | 549
Salmonella Gastroenteritis | 550
Typhoid Fever | 551
General Treatment of *Salmonella* Infections | 552

Vibrio | 552
Campylobacter Enteritis | 553
Helicobacter pylori | 554
VIRAL INFECTIONS OF THE DIGESTIVE SYSTEM | 555
Rotavirus | 556
Norovirus | 557
Enterovirus | 558
Hepatitis Viruses | 558
Hepatitis A | 559
Hepatitis B | 560
Hepatitis C | 561
Hepatitis D | 562
Hepatitis E | 562
Hepatitis G | 562
PARASITIC INFECTIONS OF THE DIGESTIVE SYSTEM | 563
Giardiasis (*Giardia duodenalis*) | 563
Cryptosporidiosis (*Cryptosporidium*) | 564
Whipworm (*Trichuris trichiura*) | 565
Hookworms | 566

Chapter 23 Infections of the Genitourinary System | 571

URINARY TRACT INFECTIONS | 572
BACTERIAL INFECTIONS OF THE URINARY SYSTEM | 573
Pathogenesis of Bacterial UTIs | 573
Symptoms and Diagnosis of Bacterial UTIs | 574
Urethritis and Cystitis | 575
Nephritis | 575
Prostatitis | 575
Diagnosis | 575
Treatment of Urinary System Bacterial Infections | 575
INFECTIONS OF THE REPRODUCTIVE SYSTEM | 576
BACTERIAL INFECTIONS OF THE REPRODUCTIVE SYSTEM | 577
Common Clinical Conditions Associated With STIs | 578
Syphilis (*Treponema pallidum*) | 579
Gonorrhea (*Neisseria gonorrhoeae*) | 581
Non-gonococcal Urethritis (*Chlamydia trachomatis*) | 583
VIRAL INFECTIONS OF THE GENITOURINARY SYSTEM | 586
Herpes Simplex Virus Type 2 | 586
Human Papillomavirus | 588
FUNGAL INFECTIONS OF THE GENITOURINARY SYSTEM | 589
Vaginal Candidiasis (*Candida albicans*) | 589

Chapter 24 Infections of the Central Nervous System | 593

ANATOMY OF THE CENTRAL NERVOUS SYSTEM | 594
COMMON PATHOGENS AND ROUTES FOR INFECTIONS OF THE CNS | 596
General Treatment of Infections of the CNS | 598
MENINGITIS | 599
Bacterial Meningitis | 599
Viral Meningitis | 600
Chronic Meningitis | 600
Symptoms of Meningitis | 600
Diagnosis and Treatment of Meningitis | 600
NON-MENINGITIS BACTERIAL INFECTIONS OF THE CNS | 601

Tetanus (*Clostridium tetani*) 601
Botulism (*Clostridium botulinum*) 602
VIRAL INFECTIONS OF THE CNS 603
Rabies 603
Polio 605
Viral Encephalitis 606
Persistent Viral Infections of the CNS 606
PRIONS 607
FUNGAL INFECTIONS OF THE CNS 609
Cryptococcosis 609
PARASITIC INFECTIONS OF THE CNS 610
Primary Amebic Meningoencephalitis 610

Chapter 25 Infections of the Blood and Lymph 615
CIRCULATING PATHOGENS 616
Bacteremia 616
Intravenous Line and Catheter Bacteremia 616
Sepsis and Septic Shock 618
Intravascular Infections 618
    Infectious Endocarditis 618
BACTERIAL INFECTIONS OF THE BLOOD AND LYMPH 620
Plague 620
Tularemia 621
Brucellosis 622
Lyme Disease 623
Relapsing Fever 625
RICKETTSIAL INFECTIONS OF THE BLOOD 626
Rocky Mountain Spotted Fever 627
Epidemic Typhus 627
Endemic Typhus 628
VIRAL INFECTIONS OF THE BLOOD 629
Cytomegalovirus 629
Epstein–Barr Virus 630
Arbovirus Infections 631
Filovirus Fevers 632
PARASITIC INFECTIONS OF THE BLOOD AND LYMPH 633
Chagas' Disease 633
Filariasis 635

Chapter 26 Infections of the Skin and Eyes 641
ANATOMY OF THE SKIN 642
BACTERIAL INFECTIONS OF THE SKIN 644
Fasciitis 644
Erysipelas 645
Folliculitis 645
Acne 645
Impetigo 646
Scalded Skin Syndrome 646
Gas Gangrene (*Clostridium perfringens*) 647
Cutaneous Anthrax 647
VIRAL INFECTIONS OF THE SKIN 648
Measles 648
Rubella (German Measles) 649
Smallpox (Variola) 649
Chickenpox and Shingles 650
Herpes Simplex Type 1 651
Warts 652
FUNGAL INFECTIONS OF THE SKIN 653
Cutaneous Candidiasis 653
Dermatophytosis 653
PARASITIC INFECTIONS OF THE SKIN 655
Cutaneous Leishmaniasis 655
Pediculosis (Lice) 655
INFECTIONS OF THE EYES 656
Neonatal Eye Infections 658

Epilogue 661

Multiple Choice Answers 662

Glossary 663

Pathogen List 687

Figure Acknowledgments 697

Index 700

Learning Skills:
Using Your Brain Effectively

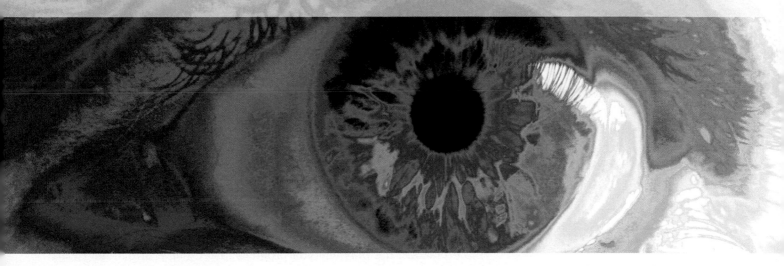

TURNING YOUR BRAIN ON

You are equipped with one of the most powerful learning tools known to exist: the human brain. Unlike other tools, your brain has its own will—your will. If you do not desire to learn, your brain simply stays turned OFF. To learn, you must want to learn. In addition, your brain avoids wasting effort. So if you believe that you cannot learn something, your brain will not learn. If you think you cannot learn a subject, then learning that subject will be impossible or, at best, extremely difficult. The ON switch for your brain requires you to believe that you want to learn and can learn.

Making the Subject Familiar

Once your brain turns ON, it operates by connecting new ideas to ideas that you have already mastered. If you fail to relate new ideas to your past experience, fail to make sense of it, then learning slows. This can cause a capable person to falsely think: "I cannot do this," and your brain switches OFF. You must connect new ideas to what is familiar to keep your brain turned ON to mastering the new knowledge or skills. As you connect ideas together, they form a growing network of knowledge. We experience this as greater understanding and skills in a subject. As the network of ideas for a subject grows, related topics become easier, faster, more interesting, and more fun to learn. When you learn enough, you take interest in a subject and become more powerful at turning ON new learning in that subject. You can foster interest in anything: desire to learn it, believe in yourself, and actively connect with the new ideas.

It is more difficult to connect new unfamiliar ideas. Again, the brain will tend to turn OFF and you will get frustrated, lost, bored, or will think, "I can't." In this case, you can achieve success by seeking knowledge and skills from other resources such as lower-level books, films, websites, classmates, tutors, and so on. Using such resources takes time and effort, but they are well invested as you will develop the necessary network of fundamental ideas more rapidly. Once the necessary stepping stones are mastered, your brain can stay turned ON to more unfamiliar and complex ideas in a subject. Persistence in connecting new ideas to your familiar ideas will make the new ideas more familiar. The newly achieved familiarity then makes future learning in the subject even easier, because now there are more familiar ideas available for making connections.

Planning a Learning Strategy

You can direct your brain to turn ON to any subject. To start learning, you should honestly evaluate yourself and identify weaknesses in your background for a subject. Then you should develop a learning strategy that includes mastering the more fundamental prerequisite skills first. Obviously, you cannot instantly increase your background to reduce the time needed for study. However, proper planning can help. If your background in a subject is low, you must factor enough study time into your learning strategy. Choose your commitments carefully to avoid being overwhelmed. Time management and planning skills are essential for learning. As a first step you need to break down your overall task into necessary subtasks. Then you estimate the time each of those subtasks will take you. You fit this suitably into your available time to achieve everything in good time for any deadline. It is difficult to identify all subtasks and their timings if you have not done some of the tasks before. Add additional time here as a backup, but also always in general. With more practice, not only will you progress faster, but your planning will be more accurate. It is important not to split your task into big chunks, because they appear insurmountable. This can lead to procrastination and frustration. Smaller tasks are more likely to be approached and accomplished. Progress and completion are also easier to monitor. Put in enough breaks; otherwise your time plan is unrealistic and not supportive of your learning. Good action and time planning also permit a good work–life balance.

In summary, to switch your brain ON to learning, you must:

- Believe you can learn

- Want to learn

- Have enough background knowledge

- Make enough time to learn

- Spend your study time personally connecting to the new ideas

If you believe you cannot do it, you should recruit a support network of people to encourage you and foster positive expectations. This will most probably require face-to-face support from friends, family, advisors, counselors, therapists, and so on—a human touch. **Figure 1** shows a flow chart to start your learning.

The next section of this chapter addresses motivation: why you want to learn. The remaining sections then address learning methods for understanding concepts and developing communication skills.

MOTIVATIONS FOR LEARNING

During learning, your brain wants to know "Why am I learning all this?" If you lack interest or sufficient reason to learn something, your brain turns OFF to the new ideas. If you repetitively memorize something without knowing how it fits with your personal knowledge, you will quickly forget it. You must have motivation to keep your brain turned ON and ideally develop new connections that build your knowledge and skills. When properly motivated, you will learn. All possible drivers are from two fundamental sources: internal (driven from within) and external (driven from outside).

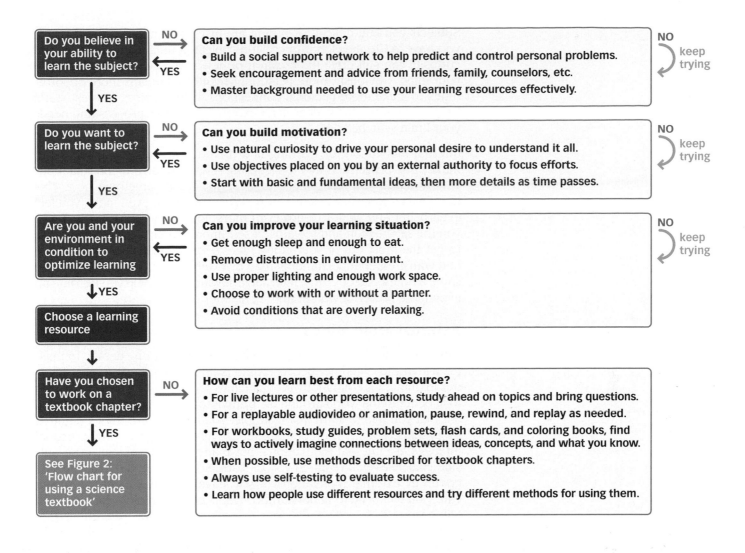

Internal Motivations

Internal motivations are the most powerful, and humans are born with an internal motivation for learning. As an infant, you learned instinctively. When as an infant you were faced with something unfamiliar (most things, at that time), you automatically learned as much as you could. Adults can have child-like curiosity, too! This natural curiosity is your personal motivation for self-improvement. Tap into your powerful natural curiosity, to make new learning faster, more satisfying, and more fun. New ideas change from being something you have to learn to something you want to learn. Accept a new subject as a new hobby.

As you gain experience when growing up, your models of the world (real and imaginary) grow and became interconnected. New ideas that more easily interconnect with your existing models are learned more easily. The models that you created in your past have developed into your present interests and skills in certain subjects. Yet, existing interests tend to limit your naturally broad curiosity. You tend to seek ideas that connect more easily to what is already familiar. Thus, as you gain experience, you tend to narrow your broad curiosity and focus on interests that are already more developed. In turn, interests help us to focus more on particular sets of ideas for a deeper understanding of those subjects. Internal motivation is driven by your naturally broad curiosity and your more focused interests that continue to develop over time.

Figure 1 Flow chart for getting started on learning. This chart lists some issues that you face each time you start studying (whether you think about them or not). This takes you up to choosing a learning resource. Note that there are many different learning resources that are used in a wide variety of ways. If you want to develop strategies for using learning resources, then seek books or other resources that focus on learning skills. This chart (and this book) cannot detail the many possibilities. However, this might help get you started.

Subjects you find difficult or boring are just less familiar ones in which you have yet to develop interest. If a subject is unfamiliar, your natural curiosity can still drive the creation of new stories and explanations in your mind. Obviously, more personal background means more opportunities to make more personal connections to the subject. Interest starts to grow. Learning new ideas is an extremely creative process and needs your brain switched ON.

Internal motivation for learning is a drive for self-improvement. What makes it so powerful is that the learning is its own reward. Internal motivation makes you feel good whenever you learn something new. This then makes you want to learn more. When you learn even more, you feel even better and want to learn even more. Internal motivations will give you the satisfaction and rewards of self-improvement. This will help you to get the most out of your study time, tie more ideas together, faster, and with greater understanding than by external motivation alone. In the same amount of study time, internal motivations will help you become an even more powerful worker in or consumer of health care.

External Motivations

Obviously, there is too much possible knowledge for one person to master within their lifetime. Because time is limited, each of us must focus our learning. External motivations help us focus on subjects that help us to fit better in the world around us. Undoubtedly at times in your past you felt that you had to learn. For example, you had to learn things required by parents, teachers, or perhaps employers. Currently, your personal interests are internal motivations, but many were developed from outside influences that required you to learn. For these, initial learning was driven by external motivations. If you have chosen a career in health care, you will need to understand concepts of microbiology because you will encounter infectious diseases in your work. If not, you still cannot avoid your own health care needs, not to mention those of your family, friends, and other people who will inevitably get infectious diseases.

For learning, external motivation is vulnerable. Humans make mistakes. If you are only externally motivated, then your mistakes tend to make you dwell on the resulting losses, pains, or punishments. Obviously, this leads to discomfort, promotes negative attitudes, and slows the development of curiosity and interest. External motivation alone cannot maximize your learning. When you persist long enough in a subject, you make enough sense of it (connections to yourself) to eventually acquire a self-motivated interest in the subject. Thus, starting from external motivation, you gain internal motivation for a subject that otherwise might never have developed.

MOTIVATIONS OF GOOD LEARNERS

Don't limit your learning potential by relying mostly on external motivations. If you primarily learn for purposes such as just getting that job, just getting that paycheck, or just getting through the task at hand, then you have probably limited yourself to external motivation. Again, this makes you vulnerable to negative attitudes. If you truly learn because you want to understand more, to have greater abilities, or to be a better person, then you gain internal motivation. Internal motivation recognizes errors and mistakes as opportunities for greater learning. Internal motivations maximize learning and make you resistant to discouragement when external motivations try to correct mistakes.

Internal motivations will make you want to know it all. However, so much is known about microbiology that your limited study time will never allow you to master it all. Use external motivations to focus on fundamental concepts first and to choose details that you need most.

You will learn best with a combination of internal and external motivations. Internal motivation is your best protection against the vulnerabilities of external motivation. Poor learners stay reliant on external authority or circumstance to force them to learn. When poor learners finish a course of study, they do not want to return to concepts through which they suffered. Good learners develop a personal drive to understand a subject. When good learners finish a course of study, they continue to think spontaneously about things they learned; they enjoy returning to the concepts they mastered, and want to share what they learned with others.

The best learners continue to imagine new and different ways to make connections between all the different ideas that they ever achieved. Thus, after a learning requirement ends, learning continues in good learners but ends in poor learners once the threat of punishment or promise of reward ends. When a course of study ends, a weak learner is glad it's over, but a strong learner is glad to have begun!

The feelings you use to motivate your learning are your choice. For better motivation, an internal wanting to learn works much better than an external having to learn. Good learners use both external and internal motivations for their different strengths. Internal motivations make learning about anything enjoyable and continuous; external motivations focus learning on the needs of your circumstances. All of us have had and will get infections. This is strong external motivation to learn about microbiology, especially as it applies to health care. If you have chosen a career in health care, then you must already have some personal interest in understanding, preventing, and treating diseases. Your interest and desire for personal growth in health care provide strong internal motivations for learning new and unfamiliar concepts. Allow your motivations to help you become a good learner of microbiology.

QUESTION NEW CONCEPTS TO HELP YOU LEARN

If you want to learn, believe you can learn, and have planned enough time for a subject, you are ready to do the actual learning. Education systems and this textbook are designed to help you understand concepts, connect them together, communicate them correctly, and finally apply concepts to solve problems.

Understanding Concepts

A concept is a general idea about something, often a set of related things. Concepts can be about anything: physical objects, properties of objects, processes of change, abstract ideas, and even things that are completely imaginary. Even the idea of concepts is itself a concept! Concepts are units of knowledge that your brain uses to deal with information and interrelate different ideas. A concept is how an idea can connect to other ideas. As you imagine how a new idea relates to familiar ideas and experiences, you build connections to that new idea. Eventually, the new idea becomes more familiar, it makes more sense, and you begin to understand its concept. If you learn enough to recognize a new, never before experienced, presentation of an idea, then you have gained an understanding of the concept. Now the idea is familiar, and you can more easily understand additional new ideas that relate to this concept.

Concepts help us organize relationships. They allow us to answer fundamental questions of understanding: who, what, when, where, why, and how? Actively answering questions about an idea will help you master its concept:

- How is the concept defined?
  - Collect different definitions for the same concept.
  - Create a definition in your own words.

- How many different examples of the same concept can you list?
  - List some examples.
  - Explain why these examples belong with this concept.

- How is this concept distinguished from different closely related concepts?
  - List closely related concepts that might be confused with this one.
  - Explain why this concept is different from the closely related concepts.

- Does this concept have an opposite? If so, what is its opposite?
  - Describe the opposing relationship.

- How does this concept interact with other concepts to form more complex concepts?
  - Describe how it is part of something bigger.

- How do smaller concepts interconnect to form this concept?
  - Describe how it is composed of smaller things.

Connecting Concepts

As you increase your repertoire of concepts, you also want to connect different concepts together. More complex concepts are just ideas that tie together multiple simpler concepts. Thus, as you understand and tie more concepts together, you begin to master even more complex concepts. Your learning becomes easier, because you gain concepts in your brain, which you can use to connect to even more new ideas and experiences. You get smarter, faster. If you memorize a separate new fact or idea, then you still have not mastered a concept. A common error by students is to repeat the same idea, in the same way, over and over. Repetition is good for motor skills (muscle coordination) but fails to develop concepts. Furthermore, your brain cannot recognize different versions of the same idea that are too different from a memorized one.

As someone interested in health care, you have already become motivated to learn microbiology. A good learner will want to use productive learning techniques, such as asking questions about the ideas you are trying to master. As you study microbiology, remember to ask questions related to infectious organisms, how they reproduce, how they get through the body's defenses, and how they produce disease. How does each concept relate to the other concepts? Is it an organism? Is it a part of an organism? How do the structures of an infectious organism interact with structures of the human body? How are smaller structures combined to form larger structures? How do they work? Is the concept a step-by-step process of change? What are the necessary steps of the process? Does this process contribute to larger, more complex processes? As you proceed to study, constantly ask and find answers to such questions. This will allow you to forge new connections between ideas and develop your understanding of microbiology concepts.

Communication

We all depend on precise and clear communication. Unless you can efficiently communicate a concept, you may thoroughly understand it but its use is limited and you cannot prove that you know it. If you understand a concept and can communicate it, then you can go one step farther: you can apply the concept to solve problems.

ALWAYS USE THE POWER OF IMAGINATION TO HELP YOUR BRAIN

Statements such as "What do I have to know?" or "What's the least I've got to study?" or "What's the minimum to pass the course?" indicate a lack of motivation and a predisposition to memorize, not understand. Poor motivation will block efficient learning and can only be addressed by you. Once sufficiently motivated, you can choose creative learning experiences and use your imagination to master concepts.

Only after you have mastered a concept can you use it to learn faster, enrich your knowledge, gain skills, solve problems, and make life better. Some concepts can be learned by going out and physically experiencing related circumstances. However, your brain has a much more powerful ability: imagination. Imagination frees you from the constraints of direct experience. You learn most concepts by imagination, because many concepts are too inefficient or are impossible to experience physically. For example, we can imagine that the Sun is 93 million miles away, without traveling those 93 million miles. Imagination also allows us to learn concepts communicated from other people.

Many concepts are developed over centuries of thinking and learning, through the experiences of many people. A complex concept can require millions of person-hours to figure out. This is much more than one human lifetime, for just a single concept! However, our ancestors recorded ideas about concepts and developed explanations or stories to help us understand the concept. Now we can read and rapidly imagine what took a long time and many people to figure out. Within seconds, we can imagine things that no one has ever physically experienced: incredibly long distances, incredibly large objects, extremely small objects, or very abstract ideas. Without imagination, many concepts are impossible to learn. With imagination, you can learn from other people's past work and master a concept that took many lifetimes to build—but you can do it in just minutes or hours, instead of years, decades, or centuries!

The power of modern health care derives from an understanding of concepts down to the molecular level: the molecules of chemicals and how they interact. Microbiology involves understanding organs of the human body, how they are composed of tissues, which themselves are composed of cells, which are composed of organelles, which are composed of molecules. These structures provide our normal functions, protection from infections, and the routes by which organisms infect and move through our bodies. Most infectious organisms are cells or particles much smaller than cells. Like all stable matter, they are composed of the molecules of chemicals. The prefix micro- in microbiology literally means small, so small that phenomena are invisible to the naked eye. Although we can observe changes in patients during infectious diseases, our best understanding must reach down to microscopic structures and the chemicals that compose them. This requires imagination to build the worlds of microbiology in your mind. Micrographs, drawings, or animations of these worlds will be examples to help you develop your imagination.

Once you imagine enough of these microscopic worlds, you will explore how they connect to tell the stories of microbiology. Ultimately, reading or hearing descriptions will allow you to form pictures in your mind, and then to change and improve the world of microbiology that you build in your own imagination. As you more accurately imagine microbiology down to the molecular level, you will gain ability to contribute to health care.

USE LEARNING RESOURCES EFFECTIVELY

When ready to learn, the most important thing you must study is yourself. After building confidence and motivation, you next choose your learning circumstances, including your learning resources such as this textbook. Of course, you are your most important tool for learning. You must be in as good a condition as possible (for example fed and rested) to optimize learning. Then choose an environment that best supports your ability to concentrate and imagine. Resources that explain concepts come in many forms: texts, diagrams, figures, lectures, videos, experts, and other learning resources. As you face a new learning resource (lecturer, book, video, and so on), ask yourself:

- Do I have the necessary communication skills to interpret this resource?

- Do I have enough background concepts to comprehend this resource?

Using an Appropriate Resource

Communication skills include not only the ability to send information to other people but also your ability to receive and understand information. The latter is essential for you to use a learning resource. For example, if you cannot understand the French language, then you will never understand an explanation in French, no matter how simple the explanation. This is obvious, but less obvious is when you attempt to use a resource that assumes a higher level of background knowledge and skills. You might be able to read the words, but you might not be able to interpret the intended message. Can you imagine what the speaker or writer is trying to say to you? Different disciplines often have different specialized ways of communicating—their own languages. For example, a high-level resource for a science will assume that you have mastered some basic scientific terminology used in that discipline. Most post-high-school science assumes some ability to use tables, graphs, charts, figures, and standard symbols. For your sake, you should quickly identify which assumed skills you still need for using a resource. Then, before proceeding, you should do whatever is necessary to master those skills. Otherwise, you could get lost, become frustrated, and fail to learn, even though you are making an effort. Successfully identifying and mastering the necessary communication skills and background concepts can save large amounts of wasted effort.

Resources come in a wide range of difficulties from low-level to high-level. Higher-level resources make more assumptions about what you know (and can do) than lower-level ones. If you find explanations or presentations difficult or impossible to understand, they probably assume knowledge of basic concepts or communication skills that you must master first. Rather than continuing with a resource that is incompatible with you, your best strategy is to identify what the presenter (author or lecturer) assumes that you (the audience) should know. Seek resources that present concepts at a lower, more fundamental level, but be careful because simpler resources sometimes use over-simplifications that are incorrect at higher levels of understanding. As you work your way up to higher-level resources, over-simplifications and misconceptions will appear as contradictions. Do not ignore the contradictions. By resolving contradictions, you will further increase your higher-level understanding. As you master enough background, you will find greater success with using the higher-level resources. For example, if you do not understand how to read a graph, it will be impossible to learn about a concept presented on a graph. However, you certainly can find a resource that teaches about graphs. Once you have learned how to use graphs, you can go back to the graph of a concept and learn about the concept. Formal courses of study often include required resources. Keep in mind that you might need additional resources to master basic concepts and skills to use the course's required resources effectively.

If you have many concepts to master, start with the more fundamental or basic ones. Then decide on a strategy for mastering the concepts. There are many ways to learn. Will you read part of a textbook? Will you watch an animation or short video? Will you discuss concepts with someone who already knows them? Will you create a diagram, a table, or some other sketch of the ideas? Will you create a story, an explanation, a joke, a song, or a dance? Will you get up and try to act out the concept? Studying is a creative process, because you must create your own understanding of the concept in your mind. This will require new experiences, real or imagined. Whenever possible, choose a method that gets you actively involved with the learning. Once you select target concepts and a learning strategy, it's time to begin actual learning. Limit yourself to about 20 minutes.

Using a Science Textbook

If you read a chapter of a science textbook like you would read a novel (straight through, uninterrupted, from beginning to end), you are approaching the science text incorrectly. Your study of this book will be more productive after mastering science textbook reading skills, because it assumes you have these skills. If an explanation uses unfamiliar vocabulary to explain a new concept you might need to use an earlier part of the book, look terms up in the glossary or dictionary, or seek simpler presentations in lower-level biology or chemistry books. Once you have mastered the assumed knowledge and skills in your book, you will more effectively use the text to continue building your understanding. **Figure 2** summarizes a commonly used approach to using a science textbook.

ALWAYS USE SELF-TESTING TO HELP YOUR BRAIN

A common cause of unproductive studying is a failure to self-test. If you self-test, you can discover how much you have learned and confirm what you have achieved. If you fail to self-test, you could waste large amounts of time by using a study method that is not working. For example, self-testing could help you identify how well a particular learning resource is working. Learning cannot be rushed, but if your progress is slower than it could be, you can focus on identifying and fixing the problem that holds you back. Self-testing is essential to make studying as efficient as possible. By definition, the more efficiently you study, the more you can learn in the same amount of time. Who wouldn't want to do that?

Here is a general approach to self-testing. First, select a concept or group of related concepts that you think you can make sense of within

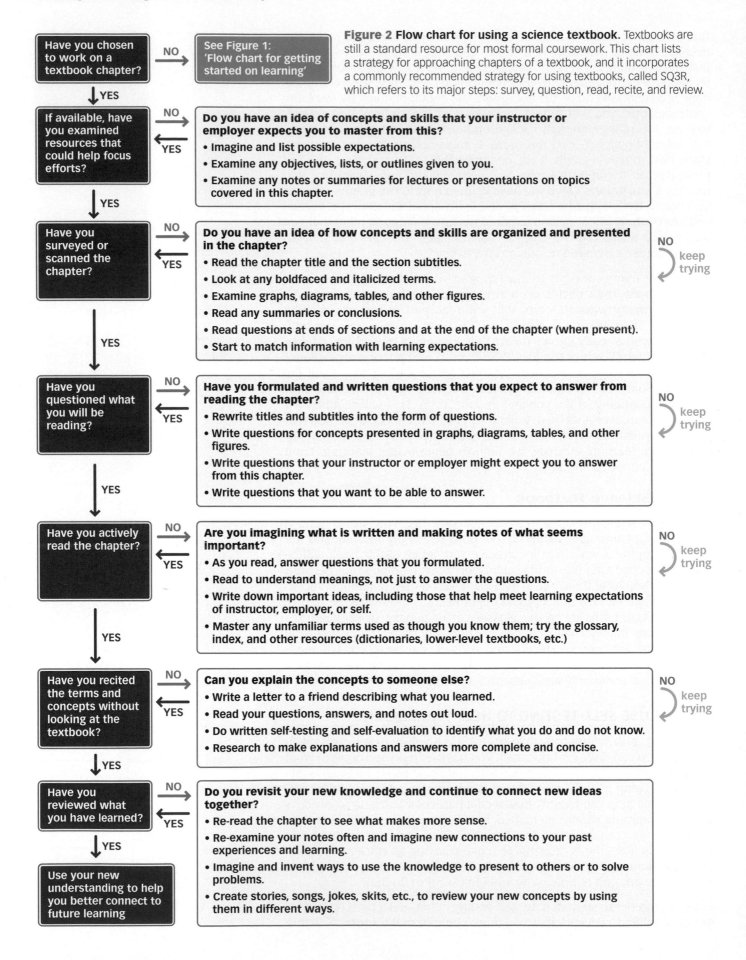

Figure 2 Flow chart for using a science textbook. Textbooks are still a standard resource for most formal coursework. This chart lists a strategy for approaching chapters of a textbook, and it incorporates a commonly recommended strategy for using textbooks, called SQ3R, which refers to its major steps: survey, question, read, recite, and review.

Have you chosen to work on a textbook chapter? — NO → See Figure 1: 'Flow chart for getting started on learning'

↓ YES

If available, have you examined resources that could help focus efforts? — NO → / ← YES

Do you have an idea of concepts and skills that your instructor or employer expects you to master from this?
- Imagine and list possible expectations.
- Examine any objectives, lists, or outlines given to you.
- Examine any notes or summaries for lectures or presentations on topics covered in this chapter.

↓ YES

Have you surveyed or scanned the chapter? — NO → / ← YES

Do you have an idea of how concepts and skills are organized and presented in the chapter?
- Read the chapter title and the section subtitles.
- Look at any boldfaced and italicized terms.
- Examine graphs, diagrams, tables, and other figures.
- Read any summaries or conclusions.
- Read questions at ends of sections and at the end of the chapter (when present).
- Start to match information with learning expectations.

NO keep trying

↓ YES

Have you questioned what you will be reading? — NO → / ← YES

Have you formulated and written questions that you expect to answer from reading the chapter?
- Rewrite titles and subtitles into the form of questions.
- Write questions for concepts presented in graphs, diagrams, tables, and other figures.
- Write questions that your instructor or employer might expect you to answer from this chapter.
- Write questions that you want to be able to answer.

NO keep trying

↓ YES

Have you actively read the chapter? — NO → / ← YES

Are you imagining what is written and making notes of what seems important?
- As you read, answer questions that you formulated.
- Read to understand meanings, not just to answer the questions.
- Write down important ideas, including those that help meet learning expectations of instructor, employer, or self.
- Master any unfamiliar terms used as though you know them; try the glossary, index, and other resources (dictionaries, lower-level textbooks, etc.)

NO keep trying

↓ YES

Have you recited the terms and concepts without looking at the textbook? — NO → / ← YES

Can you explain the concepts to someone else?
- Write a letter to a friend describing what you learned.
- Read your questions, answers, and notes out loud.
- Do written self-testing and self-evaluation to identify what you do and do not know.
- Research to make explanations and answers more complete and concise.

NO keep trying

↓ YES

Have you reviewed what you have learned? — NO → / ← YES

Do you revisit your new knowledge and continue to connect new ideas together?
- Re-read the chapter to see what makes more sense.
- Re-examine your notes often and imagine new connections to your past experiences and learning.
- Imagine and invent ways to use the knowledge to present to others or to solve problems.
- Create stories, songs, jokes, skits, etc., to review your new concepts by using them in different ways.

↓ YES

Use your new understanding to help you better connect to future learning

20 minutes. Try setting a timer with an alarm, then study the concepts as you planned. When the alarm sounds, stop the learning, and start self-testing. Put all your books, notes, and practice away. Take out a blank sheet of paper and write an explanation of what you just tried to figure out. For example, write a letter to a friend explaining the way you connected ideas together to make sense of the concept. Your written answers record what you have learned, providing data about yourself that you can evaluate.

Write Your Answers Down and Evaluate Them

Far too often, students limit self-testing to thinking answers, instead of writing answers. If you write the answer, you will check your resources to decide whether the ideas have stuck in your head. When you only have thoughts to check, you face a fundamental flaw in self-evaluation. Your brain automatically fills in gaps in stories and explanations. In addition, your brain likes to feel progress. As you look at resources to check your knowledge, you can easily think you know things that in fact you still do not know well enough. Furthermore, this in-your-head evaluation will not catch things that you know but still cannot recall. Thus, as you check your books and notes, your brain will recognize the presentations and can make you believe, "Oh, I thought that—I know that," when in fact you still cannot explain the concept correctly and completely. In this case, you mistakenly think that you mastered the concept, and move on, without making sufficient progress.

A written answer for self-testing is a permanent record of what you accomplished during your study time. Your brain cannot magically make words or drawings appear that you did not write. Now you can evaluate your written answer objectively. Did you leave any important ideas out? Did you make sense? How can you improve and expand on your recorded understanding? Furthermore, you can take your answer to a friend, classmate, tutor, or teacher to get help in evaluating your answer. Your written answer is data about what you could do at that time, and you can use it to make better plans for future learning. Now you can spend more time making connections to things that you forgot to include in your answer. Learn something new about the forgotten ideas to make them more a part of you and easier to recall. Make sure you revise what you learned to reinforce the connection. Finally, you can save your results and compare them with future efforts, allowing you to gauge your progress. In addition, you can record the date, time, and study method to help improve your choices for future learning experiences.

Are You Learning?

What if you study for 20 minutes and the timer goes off, but you can write only little or nothing about what you tried to learn? Was that 20 minutes a waste of time? Not yet. However, if you repeat the same study method, you should expect a similar result: little or nothing. If you do this, you will waste 40 minutes: the first 20 minutes and another 20 minutes in repeating a learning method that already proved unproductive. Do not ignore your self data. Your self-test is only useful if you evaluate and act on the results. At the end of an unproductive 20-minute study interval, instead of repeating a mistake, try to figure out what went wrong. Ask yourself, "What must I change to improve studying?" It might not be the study method itself. Do you understand all the background concepts and communications in the resources you used? Are there too many distractions in your learning environment? Have you eaten enough? Have you slept enough? Identify the problems and fix them. If the study method is the problem, then try something different. Develop a variety of ways that you can use to study. Different types of concepts and skills can require different learning methods to optimize efficiency. In addition, people have personal learning styles, so what works for someone else might not be the most efficient for you. Try various study techniques, study their effectiveness for you, and learn which work best for different learning objectives.

In all endeavors, including health care, people refer to ideas that are parts of larger stories and explanations. People attempt to communicate more efficiently by assuming that you already know the stories and explanations. Of course, to understand what your co-workers mean, you must already have those stories and explanations in your mind. When problems arise, you often will not have time to seek resources and learn what you need to know. As a health care professional, you are expected to have that understanding in your head already. Self-testing is the only way to make sure that you get the knowledge and skills that you must master, before a demand is placed on you. Otherwise, when a need arises, you might not be able to understand what is happening, and you will not be able to solve the health care problems that you face.

Effective Learning

Ineffective studying might not just waste time and effort: it can convince your brain that you cannot be successful. Remember: if you believe that you cannot learn, then your brain turns OFF. Failure to make necessary changes to improve learning can lead to frustration, and frustration can lead to anger or sadness. If anger is aimed inward at your own unproductive behaviors, then you might attack them, eliminate them, and turn yourself around. If your anger is aimed outward at your learning resources, then your attacks might eliminate valuable resources. If you get angry at the subject matter, then your brain will dislike it, maybe even hate it, and refuse to learn it.

To be effective, study time must be as active and creative as possible. Studying cannot be done by just putting in the time. If you are passive about learning, then you are just looking at information or listening to information. If you fail to really imagine what ideas mean and connect them to other ideas, you will not master the concepts. For this course of study, as your abilities grow, new microbiology concepts that you encounter will more quickly become familiar, and you will be able to solve more problems that you face as a health care worker or consumer.

COMMUNICATION

Learning concepts requires you to know yourself, gain your own experience, and connect to ideas that are in your own mind. Learning to communicate requires you to know other people, share experiences with them, and connect to ideas that are in their minds. To show what you know, you must correctly interpret messages from other people and concisely communicate your ideas to other people, using the standard language established for each discipline of study. Errors in interpreting messages or constructing messages will indicate a lack of knowledge and skills, even ones that you actually have.

Human beings are social animals with advanced abilities to communicate. Our communication skills allow us to coordinate activities, create societies, and reach achievements greater than one person could ever do alone. If you cannot communicate, then as far as other people are concerned, your knowledge and skills do not exist. Lack of communication skills will make your good ideas of limited or no use to others.

Communication allows you to integrate with society effectively. First, you must understand concepts, and then you can develop skills at communicating those concepts. Avoid reversing the process. If you do not understand a concept and attempt to communicate it, then you could send someone the wrong idea. This also puts you at risk of sounding ignorant, incompetent, or foolish. Such appearances undermine confidence in you, and reduce or eliminate your ability to work with other people. When you work well with other people, your efforts can be added to a society, to reach your greatest potential.

Communication skills fall into two major categories: receiving messages and sending messages. To receive and understand messages, you must know enough background concepts and standard terminology. You must also be able to interpret specialized communication forms that are used to express relationships, such as diagrams, figures, tables, graphs, and equations. Messages that you receive can teach you new concepts. Learning resources contain messages specialized to help you master new concepts.

Sending messages is more challenging than receiving them, because you must know how the message would get received. You can understand the message but still might not be able to send it. For sending, you must create a message that can be received and interpreted by someone else's brain. You must predict how the targets of your message would comprehend them. To send an efficient message, you must not only understand all of the concepts, but also know your audience and use language that they will understand. Is the message to your boss, to an instructor, to a friend, to a colleague, or to a more general audience? What can you assume about the language skills of your audience? If the message is for a less educated audience, it must use simpler language. If the message is for a more educated audience, then it should use more advanced terminology. If your message uses terms that are too simple, too general, or unnecessarily wordy, your message might communicate that you lack an understanding of the concepts. Every time you communicate, endeavor to choose the most precise terms that still include everything that must be addressed in the message.

We have already seen that learning involves evaluating yourself. To learn better communication, you must learn more terminology and language skills, practice communicating, and evaluate the practice messages that you create. Recorded messages are best for effective self-evaluation. For verbal messages, try audio recordings that can be played back. If you understand and can communicate microbiology concepts, then you will more readily understand what health care personnel say to you about infectious disease. Furthermore, you will express knowledge that will gain respect and increase the confidence of your patients, co-workers, and other health care professionals. An important part of communicating at higher levels is to achieve sufficient vocabulary. Vocabulary allows you to communicate more concisely and avoid appearing ignorant to educated people.

VOCABULARY AND TERMINOLOGY

Vocabulary refers to the use of terminology to communicate a subject. Terminology refers to the set of words, symbols, and phrases that are used to communicate ideas and their concepts. Every discipline of study has specialized terminology used in the vocabulary for that discipline. The ability to define terms is necessary but it does not mean that you have mastered vocabulary. To have a good vocabulary, you must understand the terms, speak the terms, write the terms, and use the terms correctly in messages you create. Your vocabulary is only as good as the terms that you can use to interpret ideas from messages and send ideas in your own created messages. As your vocabulary increases, you will understand higher-level messages and can use terminology to create more concise and precise messages. Learning the terminology generally begins as you master the concepts. When you learn how to use a concept, your brain naturally wants to learn how to communicate it. Mastering vocabulary can provide even more conceptual connections in your brain.

Understanding New Terminology

A good place to start with new terms is to recall other terms that look or sound similar. A useful way to make connections to new terminology is to break a term into its parts. What is the meaning for each word in a phrase? For larger words, what are the meanings for each word part: prefix, suffix, and word root? If you encounter new terminology that you have never seen before, you still might be able to predict meaning from examining the word parts. When you master a large enough vocabulary of prefixes, suffixes, and word roots, you will be able to begin identifying possible meanings of new terminology. Your vocabulary plus the context in which the new terminology is used will sometimes be sufficient for you to figure out the meaning or a close approximation. Try to learn meanings for the parts of each term. Better dictionaries show the origins of words and define word parts. If the word you are learning is not in your dictionary, look for words that use the same or similar word parts.

Can you identify shared meanings between new terms and terms you already know? If you find words that you already know, these can help you connect a new word to your personal background, making the new term become familiar faster. Keep in mind that words and word parts occasionally have more than one meaning in different contexts. However, for a new term, each time you recognize a word or word part, your curiosity can help you wonder whether it is used in a way that is similar to your background knowledge. This active thinking helps you work with the new concept. Often the same or similar word parts in different terms will have a shared meaning. When this happens, it makes conceptual connections that help you remember all the related terms better, and you more rapidly become familiar with the new terminology.

Learning New Terminology

The abstract nature of new terminology poses a challenge: how do you learn a foreign-looking new term? The short answer is: practice. For terminology that you receive in messages, the more messages that you read or hear, the better you will imagine the meanings. Obviously, you first must thoroughly understand the idea to which the term refers. Then each time you hear or see the term, imagine its meaning. If the word sounds or looks like something familiar to you, then create a fantasy story in your mind in which the familiar idea interacts with the new concept. Now the unfamiliar term can make you think of a familiar idea, and when you recall your fantasy story it will help you remember the meaning of the unfamiliar term. When you want to use the unfamiliar term (but cannot remember it), your fantasy story can help you recall your familiar term that sounds or looks like the unfamiliar one. This might then be enough to trigger recall of the unfamiliar term. Now you can use the unfamiliar term to send a message. As you use a term more, it will become more familiar, and even easier to recall. Eventually, you will not need memory tricks, because when you imagine the idea or concept, the vocabulary term will automatically come to mind.

The easier part of vocabulary is recognizing the meaning of a term that you are given. However, you have not mastered a term's vocabulary until you can recall it and use it correctly to send messages. Your brain has a specialized language center dedicated to motor skills for producing terms. You must practice language motor skills to get good at them. After learning a new concept, each time you imagine its meaning also think of the term. Imagine how the term is pronounced and say it correctly. Imagine how the term is written, and write it correctly. The more you practice the term, the better you will produce it.

Using Correct Terminology

Of course, to send good messages, you must be able to speak or write terms correctly. If you use incorrect pronunciations or spellings, then you can send incorrect information or make your audience believe that you are ignorant and do not understand the concepts behind the terms. Be careful: language centers in human brains have good memories. Avoid misspelling and mispronunciation whenever possible. Unfortunately, some terms are not pronounced the same way that they are spelled. Each time you misspell or mispronounce a term, you must overwhelm the incorrect motor memory with the correct skill. Achieve this by repeating the correct spelling or pronunciation multiple times. For newly acquired terms, each incorrect attempt must be accompanied by at least seven correct repeats to override it successfully. However, if you commit to writing and speaking correctly the first time and every time, then your language motor skills will improve more rapidly with the least amount of work.

Repetitive memorization works well for the spelling and pronunciation of terminology, because these are mostly motor skills. However, speaking and writing about a concept cannot be done unless you connect to and recall those motor skills. Again, these connections can be made by practice, but now the practice needs to show your brain the usefulness of the concept and its corresponding term. You must imagine using a concept in a variety of different ways, and then create different messages using the correct terminology. Each different correct use will improve your understanding of the concept, and each different message will improve your ability to recall the proper term. As you practice more uses, your brain will value the concept more and will recall the correct vocabulary more easily. As your understanding grows, your skill at recognizing new, different, novel presentations of the concept will grow.

BEYOND MEMORIZATION: INTEGRATION OF CONCEPTS

Integration of concepts involves actively relating concepts to each other. Memorization does not improve your ability to integrate concepts, it only gives you the ability to do simple recognition. It is seductive, because you can rapidly learn to recognize and recall many terms in a short time. If the usefulness of a concept is limited, then your brain rapidly forgets it (Figure 3). When memorization stops, forgetting happens equally quickly and most memorized concepts are lost. Furthermore, you can only solve simple recognition problems, and must do so within the limited time before you've forgotten the concept.

Conceptual integration is much more powerful than memorization. Conceptual integration means adding ideas together, relating different concepts to each other, and making more connections between different ideas in your brain. This increases your understanding of a concept, which increases the usefulness of a concept to you. As understanding and usefulness increase, your brain values the concept and keeps it for

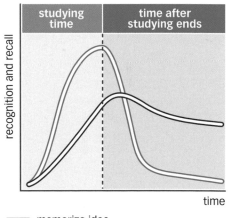

Figure 3 Effect of study method on learning. The blue curve shows repetitive memorization of concepts, and the red curve shows conceptual integration and using ideas. Increases in the graphs indicate learning curves, whereas decreases indicate forgetting curves. The vertical dashed line indicates when studying stops. Note that the learning curve for conceptual integration and using ideas continues past the dashed line, because these activities often keep the mind thinking about concepts even after studying ends.

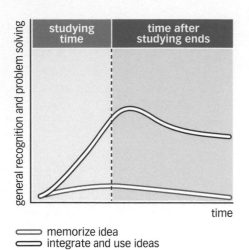

memorize idea
integrate and use ideas

Figure 4 Effect of study method on general recognition and problem solving. The blue curve shows repetitive memorization of concepts, and the red curve shows conceptual integration and using ideas. Increases in the graphs indicate learning curves, whereas decreases indicate forgetting curves. The vertical dashed line indicates when studying stops. Note that the learning curve for conceptual integration and using ideas continues past the dashed line, because these activities often keep the mind thinking about concepts even after studying ends.

future use. Making connections for each concept takes time, so recognition and recall of concepts increase more slowly than for rote memorization. In the same study time using integration of concepts, you learn to recognize less than with memorization. However, forgetting is significantly inhibited by integration of concepts. After studying stops, integration allows you to remember many times more concepts for much longer periods than rote memorization. Furthermore, conceptual integration allows you to solve more complex problems, which involve more than just rote recognition.

General Recognition and Problem Solving

To identify concepts even when you experience them in ways that you have never seen before requires a firm understanding of the concepts, which does not occur much during memorization (**Figure 4**). If you repeat the same thoughts about the same concept, without making new connections to other ideas, you will only be able to recognize that concept in that context. You will not develop the ability to generalize. Using memorization alone, it takes more time to do general recognition and problem solving, because you need much more time to think through connections that you failed to make during studying. How can you tell whether you rely too heavily on memorization? General recognition and problem solving must often be done quickly within a limited time. If you rely too much on memorization during learning, then when problem-solving time is limited, you will feel rushed for time (or run out of time), feel like you do not know what you are doing, and have little or no confidence in your answers.

As you make more connections to concepts, you increase your understanding, which is essential for these more complex skills. Thus, as you integrate ideas and use them in different ways, you will more easily be able to do general recognition and problem solving. Conceptual integration and practice using concepts is the only way to achieve your greatest ability to understand and solve the most difficult problems. If you use study time to gain deeper understanding, you will be able to do more general recognition, solve more problems, and perform skills in less time and with more confidence than someone who studied too much by rote memorization.

As a health care consumer or worker, you will face problems addressed by microbiology concepts. The problems will rarely, if ever, present themselves in the same way in which you learned the concepts. Thus, it is incredibly important to reach for the highest levels of understanding so that you can generalize and recognize concepts in the novel ways in which they will show up in health care. Greater breadth and depth of understanding of how the concepts relate to each other will provide greater ability for you to solve health care problems.

SUMMARY

Re-read this chapter or seek other resources to develop your learning skills as necessary. To optimize your brain's work for learning, review these reminders:

- Always study yourself first.

- Want to learn, believe you can learn, and prepare a learning strategy.

- Foster your curiosity and interest in new concepts—learn them as though you will always remember them.

- Enjoy learning and reward yourself with feelings of self-accomplishment each time you master another concept.

- Plan ahead and give yourself the time you need to reach your learning objectives.

- Actively learn new concepts and test yourself often—use recorded answers to evaluate your learning, what you master, and what you still need to master.

- Start with convenient or assigned resources—if these resources expect skills that you lack, then seek simpler resources to master more basic concepts and skills. Afterward, return to higher-level resources for greater success.

- There are many ways to study—don't limit yourself to one method, try a greater variety of approaches for more difficult concepts.

- Examine different presentations of a concept and different uses—imagine common examples, compare and contrast closely related concepts that might lead to confusion.

- Analyze smaller parts of a new concept that combine to make the idea complete.

- Mastering a concept requires much more than just memorizing—you must relate it to other concepts and integrate them into more complex ideas.

- Imagine concepts in your mind and practice using them.

- Invent questions and answers about who, what, where, when, why, and how (the questions can also be used for self-testing).

- Create stories, jokes, songs, drawings, tables, graphs, classification schemes, and so on, that relate concepts to each other and to your personal background.

- Develop your vocabulary—use your prior knowledge and messages sent to you to recognize and interpret terminology.

- Practice sending messages using your new vocabulary, and correct any mispronunciations and misspellings with enough repetition.

- Strive to understand concepts enough to be able to generalize, solve complex problems, and reach goals using the concepts.

PART I

Foundations

The study of microbiology is an important and fascinating course for health care professionals. To understand the information we will be discussing as we travel through the chapters of this book, we have to prepare a sound foundation. **Part I** of this book contains chapters that will work toward establishing that foundation.

The opening chapter gives us some good examples of the involvement of microbiology in our everyday lives. It also gives us our initial look at infectious diseases and the concepts of virulence and pathogenicity, which we will see repeatedly as we proceed.

In **Chapter 2** we review and re-establish the principles of chemistry that are required for understanding many of the topics to be discussed in later sections of the book. Basic concepts of bonding, pH, and types of biological molecule will be very important as we discuss molecular structures and infectious disease mechanisms.

Chapter 3 gives us a working knowledge of the basic mechanisms associated with metabolism. This is important because it is metabolism that controls the growth of microorganisms. When we use the term 'growth' in microbiology, we mean an increase in the number of organisms. For pathogens, which are our main interest, growth is an integral part of the infectious process.

The final chapter in this part (**Chapter 4**) builds on what was introduced in Chapter 1 by reinforcing our fundamental understanding of pathogenicity. We accomplish this by focusing on the relationship between the host and the pathogen. This chapter also provides a preliminary look at the differences between prokaryotes (bacteria) and eukaryotes. Finally, the chapter describes the structure of the eukaryotic host cell, which is the target of infectious disease. Here, in addition to refreshing your memory about the structures found in the eukaryotic cell, we examine the role of many of these host cell structures during the infectious process.

When you have finished **Part I** you will:

- Have developed a foundation for understanding pathogenicity and virulence and the relationship between the pathogen and the host.

- Have a foundation in the chemistry required for the study of microbiology.

- Be able to relate how bacteria need and acquire the energy needed for growth.

- Have reviewed the structures of the host cell that are involved in the process of infection.

This will make the chapters in the rest of the book easier to understand and will establish a thread of continuity that will continue throughout the book. Enjoy the journey!

What Is Microbiology and Why Does It Matter?

Chapter 1

Why Is This Important?

Microbiology is more relevant than ever in today's world. Infectious diseases are a leading health-related issue, especially in societies in which the elderly population is increasing, and in developing countries.

OVERVIEW

In this chapter we will look at some of the reasons why it is important for you to understand microbiology and infectious diseases as a health care professional. As we examine some modern examples of pathogenic, or disease-causing, microbes and how they affect us every day, we will briefly tie them into some historical perspectives.

We will divide our discussions into the following topics:

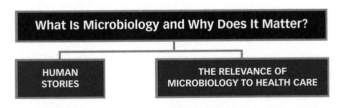

"C'est les microbes qui auront le dernier mot." (It is the microbes who will have the last word.)

Louis Pasteur (1822–1895)

The above quote fits very well with today's concept of microbiology and infectious disease. Although Pasteur—one of the early pioneers in microbiology—made this remark in the 1800s, his emphasis on the importance of microbiology in human health still holds true today. It is surprising how many times microbiology and infectious diseases are mentioned in the news and in everyday life.

For example, infectious diseases often wreak havoc in the aftermath of natural disasters, such as the devastating hurricane Katrina in New Orleans (USA, 2005), the earthquake in Port-au-Prince (Haiti, 2010), the earthquake and tsunami in Fukushima (Japan, 2011), and typhoon Haiyan/Yolanda (Philippines, 2013). The lack of sanitary conditions after these disasters led to the rapid spread of infectious diseases among survivors. Let's look at a few other examples of how microbiology affects everyday life, and how these present-day examples of infectious diseases are linked to the past.

Figure 1.1 A crowded hospital waiting room is a perfect environment for the spread of infectious diseases such as tuberculosis. Many of the people here are already debilitated from illness, which increases the chances that they could become infected.

HUMAN STORIES

Ivan Goes to Chicago...

Ivan has been a prisoner in Russia for 5 years. During this time he has contracted multi-drug resistant **tuberculosis**. He has served his time and after being released, he decides to visit relatives in Chicago. He travels on a full flight to New York's Kennedy International airport. While on the plane he experiences several severe coughing spasms, which he thinks little of because he has active disease and has experienced them for several years. In New York he changes to another full flight and lands safely in Chicago. After several days with relatives, he feels poorly and goes to the county hospital for treatment. Because Ivan is not bleeding or experiencing chest pain he is told to take a seat in the waiting room, which is full (**Figure 1.1**). While waiting, Ivan once again begins to cough. In point of fact, Ivan is a one-man epidemic.

He is contagious, and with each cough he is expelling infectious *Mycobacterium tuberculosis* organisms in an aerosol. If the droplets are small enough, they can be expelled a considerable distance and hang in the air for a relatively long period. During Ivan's travels, he has exposed all 300 people on the flight from Russia to New York, many of whom have changed planes in New York and exposed their fellow passengers to the pathogens they got from Ivan. In addition, while waiting in the hospital, he has exposed everyone else in the waiting room. More importantly, he has managed to bring a deadly pathogen into an environment (the hospital) where people are debilitated from being ill or **immunocompromised**, a condition in which their immune system is not functioning properly. If these people become infected, they will have difficulty dealing with the tuberculosis pathogen that Ivan is carrying. Because of today's rapid methods of travel (as discussed in Chapter 8), this kind of scenario is not that far-fetched. Experts estimate that some respiratory diseases could be moved around the world in less than 48 hours.

Just as the speed with which people travel the globe today was inconceivable before the development of airplanes, before microscopes it must have been unimaginable that diminutive microorganisms, too small to see with the unaided eye, could be transmitted among people via coughed-up droplets. We credit the ability to observe tuberculosis-causing bacteria and other microbes to Antonie van Leeuwenhoek. In the 1600s, he was the first to see bacteria through his simple yet superbly fashioned microscopes (**Figure 1.2**). Although he did not implicate microorganisms and infectious disease, his microscopes and early descriptions of microbes opened the doors for other microbiologists to proceed in that direction.

It's for the Birds...

Many people get a flu shot (vaccine) each year to protect them from the influenza virus. One of the most compelling problems in microbiology and infectious disease today is **influenza**, which is a viral disease. As we will see in later chapters, there are numerous strains of the influenza virus and some have been very deadly. In fact, there have been three severe epidemics of influenza, the worst of which occurred in 1918 (the Spanish flu) and killed 30–50 million people in one year (**Figure 1.3**) with considerable global social and economic impact. Fortunately, it has not been seen since, but there is the potential for genes from the virus that infects humans to combine with genes of a type of influenza virus that is currently seen in birds (avian influenza, or bird flu). The transmission of bird flu to humans in the 1990s suggested that a pandemic similar to

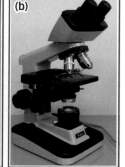

Figure 1.2 Early and recent microscopes. Panel a: An original handcrafted microscope used by Leeuwenhoek in the seventeenth century to observe bacteria. Panel b: A modern-day microscope, similar to those used by researchers, teachers, and students today.

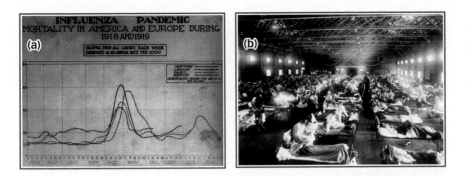

Figure 1.3 The 1918 influenza pandemic. Panel a: The Spanish flu pandemic of 1918 was a worldwide infection that killed an estimated 50 million people. Panel b: In hospitals of that era it was difficult to limit the spread of infection, especially when so many became so sick, so quickly. Could we handle this type of epidemic today?

that seen in 1918 may be imminent. Similar concerns accompanied the appearance of swine flu, as seen in 2009. If or when these flu outbreaks occur, will a flu vaccine help and will there be enough of the vaccine to go around, or will decisions have to be made about who gets it? Even more importantly, how many people will die?

Whereas crude inoculations were used to immunize against viral diseases such as smallpox for centuries in the Far East, it wasn't until Edward Jenner developed a much safer vaccine in the late 1700s that immunization became well established and routine in Western medicine. We will learn more about how vaccines work in conjunction with the immune system in Chapter 16. New vaccines for emerging flu viruses and other infectious diseases are being developed as you read this, and although they are not without risk, they promise to save unknown numbers of lives.

Special Delivery...

You go to the mail box and take out the mail. While looking through it you notice that one of the letters has a fine white powder on it that blows off as you walk back to your house. As you open the letter, more of the powdery substance fills the air and you can't help breathing some of it in. Without knowing it, you have inhaled spores of the bacteria *Bacillus anthracis* and contracted **inhalation anthrax**. You will be dead in a matter of days, and there is little anyone can do to stop it. Does this sound like a sci-fi movie? It's not. In fact, it happened in 2001 and could happen again because **anthrax** is a potential weapon for bioterrorists.

The fears surrounding anthrax are nothing new. Robert Koch, a microbiologist who helped establish the **germ theory of disease** (which states that microbes are the cause of infectious diseases), studied the history and transmission of anthrax in the 1800s. His research led to the postulates, or principles, by which modern microbiologists isolate pathogenic microbes and demonstrate their ability to cause disease. We will examine Koch's postulates in Chapter 7.

Hamburger Havoc...

Your 6-year-old daughter loves hamburgers, so you decide to take her to her favorite hamburger restaurant. She delightedly gobbles up a huge burger but that night she experiences a bout of **diarrhea** (the leading cause of death in children worldwide) and the next day begins to vomit. The diarrhea and vomiting continue for 2 days and are accompanied by severe abdominal cramping. You suspect food poisoning and continue giving her fluids, but you notice that her stool now has blood in it. By the time you take her to the doctor, she is anemic and in renal failure and is immediately put on dialysis. The doctor finally tells you that she has been infected by enterohemorrhagic bacteria called *Escherichia coli* (**Figure 1.4**), which are found in improperly processed ground beef, and had she not been seen in time she would have died. This kind of **bacterial hemorrhagic disease** is a problem that is caused by the high volume of food production and has resulted in strong recommendations by the beef industry that all restaurants and consumers thoroughly cook ground beef (**Figure 1.5**).

Most of the varieties, or strains, of *E. coli* are not pathogenic. However, some strains, such as O157:H7 (the enterohemorrhagic strain mentioned above) and O104:H4 cause serious food poisoning in humans. These names may seem confusing at first, but there is a good reason for such detail. In this case, the O and H refer to outer-membrane lipopolysaccharides and **flagella** (whip-like structures used for locomotion) proteins on the bacteria. The numbers refer to specific polysaccharides (O antigens) and proteins (H antigens) within these categories. Harmless and harmful strains can thus be differentiated by carbohydrate and protein analysis, and the origins of harmful strains can be traced by these means. We will discuss the pathogenic strains of *E. coli* in Chapter 22. Incidentally, the scientific name *Escherichia* is much easier to understand when you know that it was named after Theodor Escherich, who discovered it in the late 1800s, and *coli* designates its habitat (the colon).

Did You Wash Your Hands?...

"Employees Must Wash Hands Before Returning To Work." You can see this sign in the bathrooms of many restaurants throughout the world today. Most people know that fecal bacteria on the hands of employees can contaminate food if their hands go unwashed after using the

Figure 1.4 The enterohemorrhagic bacterium *E. coli* O157:H7 can cause life-threatening illness.

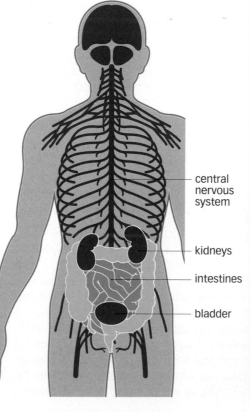

life-threatening illnesses

disease of the central nervous system

in adults, infection may progress to this stage, causing seizures and coma

after-effects: blood clots in the brain, death

urinary tract infection

most commonly strikes children under 5 years and the elderly
• destroys red blood cells
• acute kidney failure

treatment: blood transfusions, kidney dialysis

after-effects: chronic kidney failure, bowel disorders, blindness, stroke, or seizures

non-life-threatening illness

bloody diarrhea

bacteria colonize the intestines, producing a powerful toxin, causing up to 10 days of:
• severe abdominal cramps
• watery diarrhea, often bloody
• vomiting and nausea
• inflammation

treatment: usually resolves itself

after-effects: usually none

central nervous system

kidneys

intestines

bladder

the path of *E. coli* O157:H7

ground meat:

unsanitary slaughtering causes meat to be contaminated by fecal material; the most common source of *E. coli* O157

bacteria multiply in the intestinal tracts of healthy cattle and in contaminated water

milk: bacteria from cows' udders get into milk

water: inadequate sewage treatment can cause water to be contaminated by feces

how to prevent infection

avoid raw, rare, or undercooked ground meat; make sure the meat is brown, not pink, throughout, the juices run clear and the inside is hot

infected persons should wash hands carefully with soap to reduce risk of spreading infection

avoid raw, unpasteurized milk products

drink bottled water if traveling to places where water quality is uncertain

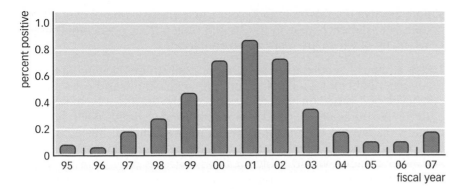

Figure 1.5 Results of the examination of raw ground beef for the presence of *E. coli* O157:H7. Notice the increases in positive samples between 1999 and 2003. The meat-processing industry has incorporated new safeguards to protect the public.

bathroom. Still, a 2012 survey showed more than half of people have witnessed others leaving a public restroom without washing their hands, and about half of people questioned do not handwash after using public transportation or shared exercise equipment, handling money, or sneezing, coughing, or blowing their noses. It is critical for health care professionals to wash their hands frequently, especially after working with a patient.

In the 1800s, Ignaz Semmelweis investigated the high death rate of new mothers in his obstetric clinic. An epidemic of 'childbed fever' in this hospital, caused by Group A streptococcal pathogens normally found on human skin, killed hundreds of new mothers each year (a death rate of 10%). When the records were carefully examined, they showed that in the ward where these deaths were most prevalent, only doctors and medical students delivered the babies. In the other delivery ward, the babies were delivered by midwives and there was a very low incidence of these infections. It was found that many of the doctors and medical students had moved directly from performing autopsies to the delivery room without properly washing their hands. Semmelweis deduced that the infections were caused by the transfer of contaminated materials from the sick or dead to healthy women. When his staff began to wash their hands thoroughly, the number of deaths caused by this infection dropped to less than 1%. Although he was not a microbiologist and wasn't familiar with the link between microbes and infectious disease, the work of Semmelweis was instrumental in developing the standards for hospital (and restaurant) hygiene today.

The Hospital Can Be Dangerous...

Uncle Harry went to the hospital for a hip operation. The surgery went well and the family expects Harry home in a matter of days. You have volunteered to get Uncle Harry and bring him home from the hospital. When you arrive, Harry is not in his room, and you find that he has been transferred to intensive care (**Figure 1.6**). Upon arriving at the intensive care unit you see that Uncle Harry is in a room with a sign outside that reads "**MRSA**—Authorized Personnel Only." The doctor tells you that Uncle Harry has developed a **nosocomial**, or hospital-acquired, infection with a bacterium called *Staphylococcus aureus* that is resistant to the **antibiotic** methicillin (MRSA stands for methicillin-resistant *Staphylococcus aureus*). Harry stays in intensive care for three more days and dies at 8 a.m. on the fourth day. You and the rest of the family are in shock because Harry was not sick and was going into the hospital for a

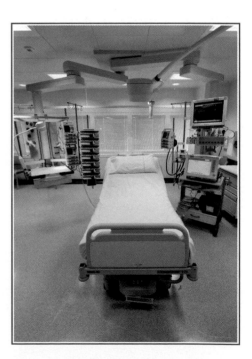

Figure 1.6 An intensive care unit (ICU) that is closed to all except authorized personnel as a result of infection with MRSA (methicillin-resistant *Staphylococcus aureus*). The same ICU conditions are used for patients with VRSA (vancomycin-resistant *Staphylococcus aureus*) and other antibiotic-resistant bacteria, which are on the rise in hospitals.

simple elective operation. In fact, your Uncle Harry is one of over 20,000 people who die each year in the United States from nosocomial infections, many of which are resistant to antibiotics.

Today, strains of *S. aureus* are resistant not only to methicillin but also to many other antibiotics, including penicillin. Penicillin was the first antibiotic developed, based on research conducted by Alexander Fleming in the early 1900s. Ironically, Fleming was researching *S. aureus* when he accidentally discovered the inhibitory effect of penicillin on these bacteria. One might wonder if he ever imagined the true extent to which *S. aureus* and other bacteria would evolve resistance to his miracle drug, or the impressive diversity of antibiotics used today. We will discuss Fleming, penicillin, and several other antibiotics in Chapter 19.

Keep in Mind

- Microbiology has relevance to everyday life.
- We do not live in a sterile environment; therefore we interact with microbes all the time.
- Travel allows the movement of infectious diseases around the world in a relatively short time.

THE RELEVANCE OF MICROBIOLOGY TO HEALTH CARE

The above examples are just a small sample of the type of microbiological problems that we have to face. The better that health care professionals understand how a pathogen causes a disease, the more they can help patients, both in accessing suitable treatment and by preventing disease in the first place. As a health care professional you may be involved in cases like these. How will you react? What will you communicate to the patients and their relatives? The chapters that follow are designed to help you if and when that time comes.

Infectious disease is a major concern for health care providers. Rough estimations attribute 50% of health problems to infectious diseases and their secondary complications. The majority are caused by viruses, which limits our treatment options. If we look back through history, there has always been disease. In fact, spores of bacteria have been found that are over 25 million years old. In the nineteenth century, tuberculosis and other pulmonary infections were the scourges of the world and were the leading cause of premature death. Although the causes of some of these infections were being discovered, there was little that could be done to stop or prevent them. The links between several pieces of evidence had to be made first: microbes exist, microbes are not formed by spontaneous generation, and microbes have the potential to cause disease. Then microbial features had to be discovered before they could be used for prevention (for example vaccine development; see Chapter 16), in diagnosis (see Chapter 6), control and treatment (see Part V). Many of those links were established in the first golden age of microbiology (1875–1910), when microbes were directly associated with the diseases they caused. This was followed by advances in public health that decreased the number of deaths resulting from infection. Before, infectious diseases were believed to be caused by supernatural forces, poisonous vapors (miasmas), or an imbalance of bodily fluids.

In the first half of the twentieth century, scientists studied the structure, physiology, and genetics of microbes and found links between microbial properties and the development and symptoms of infectious disease.

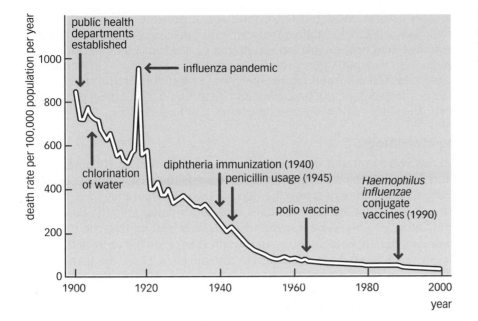

Figure 1.7 Death rates for infectious disease in the United States during the twentieth century. Notice the decline after the introduction of public health measures, immunization, and antibiotics.

During this period, the discovery of penicillin in 1929 and sulfonamide drugs in 1936 opened the door for effective treatments. By the end of the twentieth century, better sanitation methods and advances in the fields of molecular biology and genetics had increased our knowledge at the molecular level, contributing to the development of vaccines and antibiotics that all changed the face of infectious disease (**Figure 1.7**).

Now in the twenty-first century, infectious diseases that we thought were conquered are reappearing, and resistance to treatments that were previously completely effective is beginning to grow. New diseases are emerging and organisms that were thought to be harmless are now being found to cause disease in certain circumstances. The potential for bioterrorism has gone from fiction to fact. For students in the health sciences, the fundamental understanding of microbiology has never been more relevant.

Infectious Disease

It is important to remember that of the thousands of species of bacteria, viruses, fungi, and parasites, only a tiny fraction are involved in disease of any kind. The majority of microbes are actually essential for human existence because of their ecological and industrial relevance, and their role in Earth's history and evolution. Furthermore, of the small number involved in infections, only an even smaller fraction cause disease in humans; the rest infect other organisms. Among the human pathogens, the potential to cause disease is referred to as **pathogenicity** and we can categorize pathogenic microorganisms based on their degree of fitness to overcome the body's defenses and establish themselves, which is called **virulence**.

Many microbes are part of what we call the normal microbial flora of the body, where they naturally colonize the skin, mucosal surfaces, and large intestine. Most of the time these organisms are harmless and in some cases they actually provide us with important products such as vitamins. In certain circumstances, such as an increased host susceptibility, the organisms can become pathogenic (opportunistic pathogens), but when they do they are only mildly virulent. An example is *Candida albicans*, causing thrush and vaginal yeast infections. In contrast, there are pathogens, such as *Yersinia pestis* (causing the plague), that are always associated with disease (primary or obligate pathogens). Some microbes (accidental pathogens) can cause disease but are actually not usually found near people; an example is *Clostridium tetani* (causing tetanus). For most pathogens, the basic aspects of how they cause disease can be looked at from three perspectives: epidemiology (the study of the factors determining the frequency and distribution of disease), pathogenesis (the study of how disease develops), and host defense.

Epidemiology

As we will see in our discussions of the disease process (Chapters 5, 6, and 7), pathogens must accomplish five steps to cause disease successfully. First they must get into the body or the cell. Then they must be able to stay in, defeat the host defenses, and cause damage to the host. Last, they must be able to be transmitted to a new host so that the infection will continue. In epidemiology, each pathogen is looked at from this perspective and classified with regard to transmission. Some are transmitted by air, some by food and water, some by insect vectors, and still others through person-to-person contact. In addition, geography has a role in infectious disease: some pathogens are found worldwide, whereas others are restricted to certain geographic areas. Knowing how an organism gains entry and how it spreads are vital to our efforts to care for infected individuals and to protect those around them.

Outbreaks of disease are fostered by factors such as poor socioeconomic conditions and hygiene, ignorance of the cause of infection, and natural disasters. The possibility of recurrence of old pandemics such as the Spanish flu of 1918 remains a problem today, and we are already seeing an extended pandemic with HIV infection and **AIDS**.

Pathogenesis

It is not easy for pathogens to produce disease in humans. This is the primary reason that so few organisms are successful pathogens. Multiple factors known as **virulence factors** are required for the organism to persist in the host, cause disease, and escape the host defenses so that the infection can continue.

Pathogens can use a variety of methods to damage host cells and tissues. Bacterial pathogens can produce digestive enzymes and toxins, while the intracellular multiplication of viruses seen as part of viral infections produces massive numbers of viral particles, which cause the death of the infected cell. In fact, some of the damage associated with infection is due to an overcompensating host defense. For example, redness, fever, pain, swelling, and loss of function are universal symptoms of inflammation that result from host defense reactions designed to fight infection. If these reactions are too severe, they can damage the host. Symptoms can also be associated with the particular organs that are involved in the infection. For example, coughing, diarrhea, and nervous system dysfunction are symptoms seen in the respiratory, digestive, and central nervous systems and often allow transmission to a new host.

Host Defense

We will see throughout our discussions that infection is essentially a complex competition. The rules and conditions of the battle change all the time. The pathogen uses multiple methods to survive and thrive, while the host defense (immune system) becomes involved immediately in an attempt to kill or expel the pathogens. Much of the outcome depends on the success or failure of the host defense. The best example of host defense failure is the disease called acquired immune deficiency syndrome (AIDS), in which the defense mechanism of the host is compromised and infections (such as **pneumonia**) that would have been easily defeated end up overwhelming and killing the host.

There are two basic types of defense against infection. The first is the nonspecific defense referred to as the **innate immune response**, the one you are born with and representing the first line of defense by the host (see Chapter 15). These responses involve a variety of cellular mechanisms coupled with chemical and mechanical factors. In many cases these are all that are needed to defeat the infection. The second line of defense is called the **adaptive immune response** (see Chapter 16), which you acquire through being exposed to your environment. The adaptive immune response is specific and has the benefit of memory, which means that if you see that same infection later, the response to it will be immediate and powerful. It is the adaptive defenses that keep us safe in a hostile atmosphere. The critical role of these systems can be seen in burn patients who have lost part of an important innate barrier (the skin), and in patients with AIDS, who lose the adaptive immune defense.

Consequently, for a pathogen to be successful it must find a way to defeat, evade, or hide from these potent host defenses, and many pathogens have developed intriguing ways of doing so. Some will attack the defending cells, while others will change their looks (a form of camouflage) such that the adaptive defense response is fooled. Some pathogens will use the defending cells as a place to hide from the host defense.

Treatment of Infectious Diseases

During the course of our discussions we will focus on some of the treatments that are used for specific infections. Many potent and successful therapeutic tools to treat us and protect us from infection have been developed over the past 75 years. These include antibiotics as well as **disinfectants** and **antiseptics**. As we will see later, antibiotics can have adverse side effects and they therefore have to be selective. They must kill the disease-causing microorganisms but not harm the patient. The first antibiotic, penicillin, also happens to be naturally selective. It prevents the formation of one of the components of the bacterial cell wall called **peptidoglycan**, which is not found in human cells. Unfortunately, there are very few drugs that are selective for bacteria and even fewer for fungi and parasites, which have close structural and metabolic similarities to our cells.

Treatment becomes even more difficult for viral infections because viruses are, by definition, intracellular parasites. If treatment does not negate the virus before it invades the host cell, the only defense is to kill the infected cell. Fortunately, our defense mechanisms can accomplish this task, but there is still the collateral damage of losing the cell.

The best treatment for any disease is prevention, and for infectious disease this involves public health measures and immunization. To take effective public health measures, we must understand the transmission mechanisms of infection so that we can interfere with those mechanisms. Public health measures involve activities such as the disinfection of water supplies, increased care in the preparation of food supplies, insect control, proper hygiene and sanitation, proper care in waste removal and treatment, and many other measures that prevent contact with infectious agents. Immunization requires that we understand the immune mechanisms and that we design vaccines that will successfully stimulate protection.

> **Keep in Mind**
>
> - When we understand how microbial pathogens cause disease, we are better able to treat and defeat these infections.
> - There are thousands of species of bacteria, viruses, fungi, and parasites, but only a small fraction are involved in disease.
> - The potential to cause disease is referred to as pathogenicity, and pathogens can be categorized on the basis of their degree of virulence.
> - Part of the disease process is based on the host's ability to defend itself against infection by using the innate and adaptive immune responses.

SUMMARY

- Microbiology is very relevant to our everyday lives.
- We are exposed to potentially dangerous pathogens on a daily basis.
- These pathogens possess virulence factors that allow them to persist in the host, evade host defenses, and cause disease.
- Pathogens must accomplish five tasks to be successful in causing disease. They must get in, stay in, defeat the host defenses, damage the host, and be transmissible.

We live among infectious organisms that could be potentially lethal to us. Consequently, our understanding of how microbes cause disease and how we defend ourselves is basic to our survival, and the following chapters are intended to increase that understanding.

Q SELF EVALUATION AND CHAPTER CONFIDENCE

Multiple Choice

Answers are given in the back of the book and help can be found in the student resources at:
www.garlandscience.com/micro2

1. Which of the following is true about the relationship between humans and pathogens?

   A. The human body has its own population of microbes

   B. Humans do not come into contact with many microbes

   C. Most microbes are pathogenic and cause disease

   D. Non-pathogenic organisms can never be pathogenic

2. In the past 10 years the Earth has become warmer. This has probably caused which of the following?

   A. More respiratory disease (transmitted by droplets)

   B. Less respiratory disease

   C. Less digestive disease

   D. Spread of insect-transmitted disease

3. If the availability of clean drinking water decreases, then we can expect

   A. More respiratory disease

   B. More insect-borne disease

   C. More digestive disease

   D. Less respiratory disease

4. A term reserved for diseases acquired in a clinical setting is a(n)_____ infection.

   A. Nosocomial

   B. Enterohemorrhagic

   C. Contagious

   D. Multi-drug resistant

   E. Acute

5. Which type of microorganism causes the majority of infectious diseases?

   A. Bacteria
   B. Viruses
   C. Fungi
   D. Parasites

6. Which of the following are important in guarding against the spread of infectious disease?

   A. Handwashing between working with different patients in health care situations
   B. Handwashing after using restrooms
   C. Handwashing before and after food preparation
   D. None of the above
   E. All of the above

7. The general term for features of microorganisms that enhance pathogenesis is

   A. Toxins
   B. Inflammatory agents
   C. Antibiotics
   D. Virulence factors

8. Redness, swelling and pain indicate which of the following?

   A. Blush reflex
   B. Presence of nonpathogenic organisms
   C. Inflammation
   D. Adaptive immune response

9. Why are infectious diseases a serious healthcare problem in the developed world again?

   A. The human immune system is becoming less effective
   B. There are more pathogenic organisms than there used to be
   C. There is increasing resistance to treatments that were once completely effective
   D. Pathogens have a wider variety of mechanisms to damage host cells

 DEPTH OF UNDERSTANDING

Questions listed here require you to bring together the concepts you have learned in this chapter into a discussion format. This helps you to increase your depth of understanding of the material you have learned. Help can be found in the student resources at: www.garlandscience.com/micro2

1. Because microorganisms are ubiquitous and live in, on, and around us, why is it that most people are free from infectious disease?

2. Why is it that new viral diseases and epidemics occur frequently?

3. Seven thousand years ago the measles virus was found only in cattle. Now it is seen only in humans and never in cattle. Explain the reason for this change.

Fundamental Chemistry for Microbiology

Chapter 2

Why Is This Important?

An understanding of chemistry is essential for the topics we will be covering and will help you better understand cellular structure and function, which are paramount for your study of microbiology.

Sabrina is a typical 12 year-old girl who, like many of us, dreads going to the dentist. She brushes her teeth once a day and gets her teeth cleaned every six months. Now she has five new caries (dental cavities) that have caused her pain and discomfort when chewing food. Her dentist suspects she has an unusually high density of cariogenic (cavity-causing) *Streptococcus mutans* bacteria residing in her mouth, and collects a saliva sample to test this prediction. The dentist inoculates a Snyder test (a test tube filled with culture medium that encourages the growth of *S. mutans* bacteria) with Sabrina's saliva and incubates it overnight. During that time, the medium in the test tube changes from dark green to yellow, confirming the dentist's suspicions.

The Snyder test is based on the chemical reactions that occur in Sabrina's mouth. Cariogenic bacteria, including *S. mutans*, create a biofilm on the teeth (we will learn more about biofilms in Chapter 5, and about dental caries in Chapter 22). These bacteria chemically break apart sugar molecules and release acid; the acid, in turn, erodes the enamel and causes caries. The Snyder test contains sugar for the bacteria to feed on, and a pH indicator that turns from green to yellow when acid is present. If there is a large population of cariogenic bacteria in the saliva sample, the pH is reduced and the color change is rapid and obvious as the bacteria proliferate in the medium. Using this elegant, simple test, Sabrina's predisposition to caries is confirmed and the dentist gives her advice on better oral hygiene that should reduce the amount of leftover sugars and acid-secreting bacteria in her mouth.

OVERVIEW

This chapter is not intended to be a thorough examination of chemistry. Instead, it is designed to provide a basic understanding of chemical bonding, and the structure of biological molecules. It is important to understand these concepts, because the structures and functions of microorganisms are molecular and many of the pathogenic effects of infection occur at the molecular level. Therefore, to understand the infectious process we need to understand basic chemistry.

We will divide our discussions into the following topics:

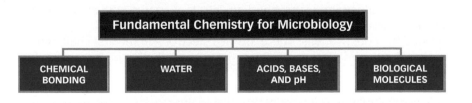

Fundamental Chemistry for Microbiology — CHEMICAL BONDING — WATER — ACIDS, BASES, AND pH — BIOLOGICAL MOLECULES

"She was puzzled as to why learned people didn't adopt chemistry as a religion."

Betty Smith (1896–1972)

As seen in the above quote from her book *A Tree Grows In Brooklyn*, Smith's character Francie marvels at the basic principles of chemistry: nothing is ever lost or destroyed, even if it burns or rots away. Life is based on the predictable bond-forming and bond-breaking properties of atoms, a series of chemical reactions. Chemistry is the underlying foundation of many of the things that we will study in later chapters of this book. We will see that many microbial pathogens infect and damage host cells, tissues, organs, and organ systems, which are organized in the following sequence:

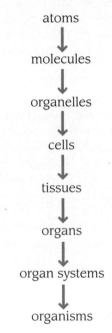

atoms

↓

molecules

↓

organelles

↓

cells

↓

tissues

↓

organs

↓

organ systems

↓

organisms

CHEMICAL BONDING

Atoms bond together to make molecules, and molecules join together to make organelles and cells. Cells associate to make tissues, which form the organs and organ systems of host organisms. It should be pointed out that the infectious processes that we will be looking at are usually found at the level of cells and tissues, so we will focus on these. We will begin by looking at the three main ways in which atoms bond together. The properties of protons, neutrons, and electrons, as well as atomic structures and electron shells, should have already been covered in your introductory biology and chemistry classes.

Ionic Bonds

As a way of illustrating ionic bonding, let's look at two elements. The first is the element sodium, which has 11 protons in the core and 11 electrons around the core in shells. There are two electrons in the first shell, which makes it full, and there are eight in the second shell, which makes it full too—but there is one electron left, which occupies the third (outermost) shell. Because this shell is not full, the atom is chemically unstable and this instability makes it possible for the sodium atom to donate the electron to another atom.

Now let's look at the element chlorine, which has 17 protons and 17 electrons. If we fit them into their shells we will have two in the first, eight in the second, and seven in the third. As we saw in our example of sodium, the outer shell is not filled with electrons and is therefore unstable.

If sodium and chlorine were to meet, they could solve each other's problem because sodium has a single electron and wants to donate it. Chlorine has only seven electrons in its outer shell and needs eight to be stable, so it is ready to accept an electron. That is precisely what happens (**Figure 2.1**): the lone electron from sodium moves to the outer shell of chlorine, thereby filling the outer shell and making the chlorine atom stable. When this giving and receiving takes place, the overall electrical charges of both sodium and chlorine also change and they are classified as **ions**. Because sodium has lost an electron it becomes more positive and is referred to as a **cation**. In contrast, chlorine has gained an electron, making it more negative, and it is now classified as a chloride **anion**. The opposing charges of these cations and anions are attracted, and the two ions form an **ionic bond**.

Covalent Bonds

Another major type of chemical bond is the **covalent bond**. Covalent bonds are the type of bond seen in biological molecules. In contrast to the ionic bond, the covalent bond is formed when electrons in the outermost shells of the atoms are shared. If one pair of electrons is shared, a single covalent bond will form; if two pairs of electrons are shared, there will be a double covalent bond. In addition, covalent bonds can have polarity, which can be viewed as the equality of sharing that occurs. For example, in a **nonpolar covalent bond**, the sharing of the electrons between two atoms is equal, and in this case the bonds are electrically neutral.

However, if the sharing is unequal (one side is pulling more than the other) there can be a polarity, which results in a weak electrical charge. The best example of this polar covalent bond is seen with water, which consists of two hydrogen atoms that are bound covalently to one oxygen (**Figure 2.2**). In this sharing, the oxygen pulls on the hydrogen's bonding electrons more than the hydrogen pulls on the oxygen's bonding electrons. This results in a slightly positive charge on the two hydrogens and a slightly negative charge on the oxygen.

When we look at the mechanism of covalent bonding, atomic structure also has a role. We can use carbon to demonstrate this bonding. In fact, biological life on this planet is based on carbon, which has six protons in the core and six electrons in shells around the core. Because it takes two electrons to fill the first shell, there will be four left in the second. Carbon therefore needs four more electrons to fill the shell. This can be accomplished by bonding to other atoms.

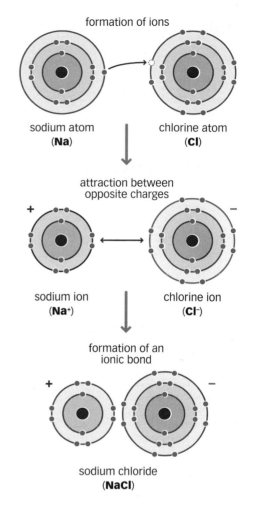

Figure 2.1 The formation of an ionic bond. In this sequence of events sodium and chloride form an ionic bond. The first step is the donation of an electron from the sodium atom to the chlorine atom. This ionizes the atoms, allowing the ionic bond to form.

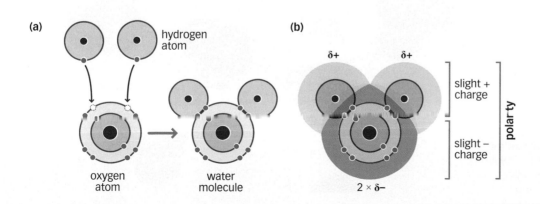

Figure 2.2 The formation of a covalent bond. Panel a: The water molecule is formed through covalent bonds in which each hydrogen shares its electrons with oxygen. Panel b: This is a polar covalent bond in which the water molecule has a slight negative charge (δ–) on the oxygen side of the molecule and a slightly positive charge (δ+) on the hydrogen side of the molecule.

(a)

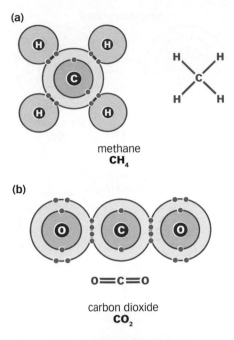

methane
CH₄

(b)

O=C=O

carbon dioxide
CO₂

Figure 2.3 The binding properties of carbon. Panel a shows the ability of carbon, which has only four electrons in the outer orbital, to bind covalently to other atoms (in this case hydrogens). Panel b: The formation of CO₂ takes advantage of the same binding properties of carbon. However, in this case two pairs of electrons are shared with each oxygen, causing the formation of double covalent bonds.

Because covalent bonding involves sharing, it is not necessary to move electrons from one atom to another. All that needs to be done to create a covalent bond with a carbon atom is to bring other atoms near enough so that their outer shells are close to its outer shell. We can illustrate this with hydrogen atoms, which have only one electron in the outer shell and need two electrons to be stable (**Figure 2.3**). Because carbon has only four electrons in its outer shell, four hydrogens can bond covalently to carbon by sharing their electrons. When the four hydrogens are covalently bonded to it, the carbon will have eight electrons in its outer shell, which makes it stable. Each of the hydrogens will have its own electron plus one shared from carbon, giving the hydrogens two in their outer shell and making them stable as well.

Carbon's ability to bond covalently with other atoms makes it useful for building biological molecules such as carbohydrates, lipids, proteins, and nucleic acids. In addition, each of these biological molecules will have a chemical shape that will be directly related to the function of the molecule. Part of achieving and maintaining that shape depends on the last type of bond, the **hydrogen bond**.

Hydrogen Bonds

These are chemical bonds that are found between and within molecular structures, and they help to give molecules their shape. Hydrogen bonds form because of the polarity of molecules seen in covalent bonding. Hydrogen is slightly positively charged and is attracted to negatively charged atoms, such as oxygen, nitrogen, or fluorine. If we consider water, this becomes easy to understand. In **Figure 2.4** we see that the polarity of water allows one molecule to be attracted to another through the positive hydrogen side of one water molecule and the negative oxygen side of another.

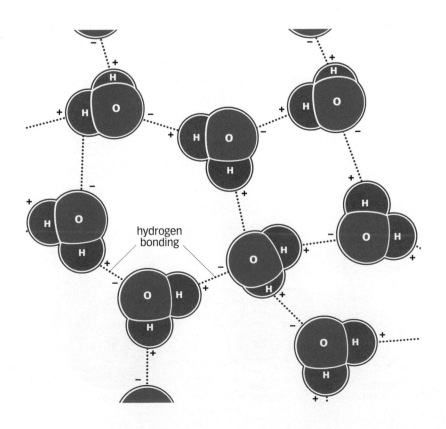

Figure 2.4 Hydrogen bonding between water molecules occurs because of the slight positive and negative sides of the water molecules. These hydrogen bonds are the weakest of the chemical bonds and can rapidly break and reform.

Hydrogen bonding gives water a fluid property when these weak bonds between water molecules are breaking and reforming quickly. If we lower the temperature, the breaking and rejoining slows down and the water becomes ice. If we heat the water, the bonds break so fast that there is little reforming and the water molecules dissipate as steam.

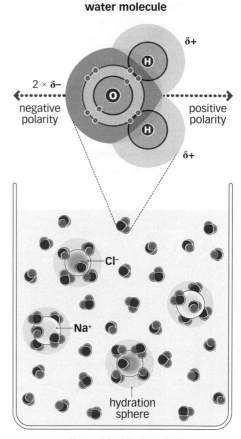

water molecule

sodium chloride in solution

Figure 2.5 The formation of spheres of hydration, which surround compounds that dissolve in water. In this example, the polarity of the water molecules causes the dissociation of the sodium and chloride ions and the formation of spheres of hydration.

| Keep in Mind |

- Atoms can bind together to make molecules, which can join together to make cells.

- There are three main types of chemical bonding: ionic, covalent, and hydrogen bonding.

- Ionic bonds form when electrons are donated to or received by atoms.

- In covalent bonds, electrons are shared by atoms.

- In hydrogen bonds, positively charged hydrogen is attracted to negatively charged atoms.

WATER

Water may be the most important component for life because it has properties that are critical for physiological functions. We will look at three of them: solubility, reactivity, and heat capacity.

1. **Solubility**—There are a remarkable number of molecules that can dissolve in water. In any solution there is the solvent (which in this case is water) and the solute (which is the material that is dissolved). Ionic bonds and the polarity of water are responsible for how water can dissolve other compounds. For example, let's use sodium chloride as the solute. We already know that it is a compound made up of positive sodium (Na^+) ions and negatively charged chloride (Cl^-) ions. Water molecules, which are polar, can come between these elements and essentially surround them, forming **spheres of hydration** (**Figure 2.5**). This dissociates (separates) them and keeps them separated. Consequently, when you put salt in water it dissolves. However, if you continue to add salt, it stops dissolving. This is because there are no longer enough water molecules to surround the ions, and some of the sodium will stay bound to the chloride.

2. **Reactivity**—Chemical reactions normally occur in water, and water can also participate in reactions. When dehydration (the removal of water) is used to build molecules, it is referred to as **dehydration synthesis**, and when water is used to break down molecules, it is called **hydrolysis**.

3. **Heat capacity**—Heat capacity is the ability to absorb and retain heat. Many chemical reactions give off heat as a by-product; water, which has a high heat capacity, can absorb this heat.

ACIDS, BASES, AND pH

From a chemical point of view, most of life on Earth lives in a neutral environment. However, there are microorganisms that can live in highly acidic or alkaline environments. When we look at an acid, we can view it as a donor that donates hydrogen ions (H^+) to a solution. These hydrogen ions can also be donated as part of a chemical reaction. In contrast, the hydroxyl ion (OH^-) can neutralize H^+ ions. A solution that contains an excess of hydroxyl ions is more alkaline, or basic. The concept of pH is

| Fast Fact | Bacteria that thrive in low-pH solutions are called **acidophiles**. Some acidophilic bacteria can even survive the acidity of the stomach.

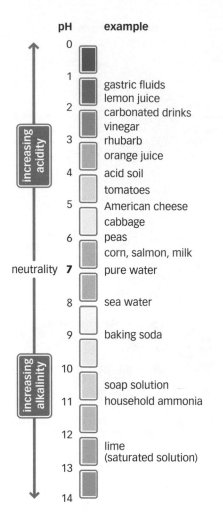

Figure 2.6 The pH scale shows the acidity or alkalinity of common substances.

basically a definition of the acidity or alkalinity of a solution. On the pH scale (**Figure 2.6**), high H⁺ ion concentration is related to low pH, and vice versa. The pH scale goes from 0, which is the most acidic, to 14, which is the most alkaline, with 7.0 being a neutral pH. Pure water has a pH of 7.0 and is completely neutral.

Keep in Mind

- Water has several properties that are important for physiological functions, including solubility, reactivity, and heat capacity.

- pH measures the acidity or alkalinity of a solution.

- Acidity can be considered as the amount of free hydrogen ions (H⁺) in a solution.

BIOLOGICAL MOLECULES

Now that we have examined bonding, water, and pH, we can look at biological molecules, which are also referred to as organic molecules. There are four major categories of biological molecule: carbohydrates, lipids, proteins, and nucleic acids. All of these molecules are naturally occurring and all use carbon as the primary building block for their structure. Recall from our discussions above that carbon has the ability to form covalent bonds and link up in long chains that can form a wide variety of structural types of molecule. Carbon-based biological molecules occur in all living things. Carbons can form long chains and these chains can have hydrogens and other atoms such as oxygen and nitrogen attached to them. We can also subdivide these molecules into parts such as the functional group, which is that part of the molecule that participates in chemical reactions and gives the molecule its particular chemical properties. Molecules can possess more than one functional group.

All living organisms require energy, and large organic molecules can provide this energy. Organisms harvest the energy in organic molecules by breaking the chemical bonds that hold the molecules together. In fact, the major biological molecules can be viewed within the context of the energy they contain and the ease with which that energy can be obtained. This energy is used to synthesize **ATP** (adenosine triphosphate), which is the form of energy that cells use to do cellular work such as active transport or protein synthesis. We discuss these principles in Chapter 3.

Carbohydrates

Carbohydrates can be viewed as the most easily used source of energy. Microorganisms break down carbohydrates from their environment and they can also synthesize carbohydrates. For example, bacteria produce a special type of carbohydrate called peptidoglycan that they use to produce cell walls.

All carbohydrates contain carbon, oxygen, and hydrogen, with hydrogen usually being found in a 2:1 ratio to the other atoms. We can group carbohydrates into three major categories based on how many building blocks are involved. A monosaccharide is the smallest type of carbohydrate and is the building block used to make larger carbohydrates. A disaccharide has two monosaccharide building blocks joined together. In some cases, we can join many of these monosaccharides together to form polysaccharides.

A monosaccharide consists of a carbon chain with several functional groups attached to it; a good example is the glucose molecule (**Figure 2.7**). It has the formula $C_6H_{12}O_6$, meaning that there are 6 atoms of carbon,

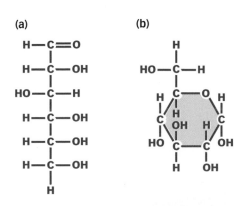

Figure 2.7 The structure of glucose. Panel a shows the structural formula in the straight-chain form. Panel b shows the structure in the ring form, which is the most common form in nature.

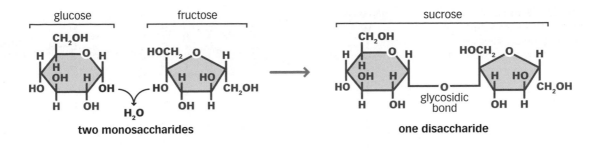

glucose   fructose   sucrose

two monosaccharides   one disaccharide

12 hydrogens, and 6 oxygens bonded into the structure. This molecule can also be represented in a ring form, which is the form it would fold into in a solution. Disaccharides are formed when two monosaccharides are linked together. This occurs by dehydration synthesis in which the removal of water results in the joining of two monomers into a dimer (Figure 2.8). Polysaccharides (Figure 2.9) are formed in the same way except that the molecule can become very large, with monomers repeatedly added on to the end of the growing polysaccharide molecule. The polysaccharide form is the storage molecule for carbohydrates and is a tremendous source of energy because single monomers can be broken off and metabolized individually.

Lipids

Lipids are a chemically diverse group of substances that includes fats, phospholipids, and steroids. These substances are relatively insoluble in water, which makes them very useful as elements of cellular structure, in particular the plasma membrane. Lipids can also be used as energy sources, and some lipids actually contain more energy than carbohydrates. In contrast with carbohydrates, lipids contain more hydrogen and less oxygen.

Fats

Fats are lipids that contain the three-carbon molecule glycerol and one or more **fatty acids**. These fatty acids are long chains of carbons with bound hydrogens. As with other organic molecules, dehydration synthesis is used to synthesize fat. During this process, bonds form between the glycerol and one end of each of the fatty acids (Figure 2.10).

Fatty acids can be saturated or unsaturated, depending on how many hydrogen atoms are attached to pairs of carbons in the tails. Saturated fatty acids contain all of the hydrogens that can possibly be bound. In contrast, the unsaturated forms have lost hydrogens and formed double bonds at the locations of the missing hydrogens (Figure 2.11).

Figure 2.8 The formation of a disaccharide is accomplished through a dehydration synthesis reaction in which water is removed and two monosaccharides are joined together.

Fast Fact Carbohydrate-based molecules, such as peptidoglycan, are found in the cell walls of many bacteria and are destroyed by lysozymes (enzymes) in mucus, tears, and chicken egg whites. Penicillin antibiotics work by interfering with peptidoglycan synthesis.

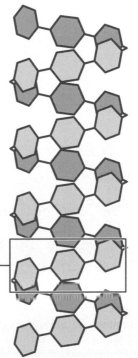

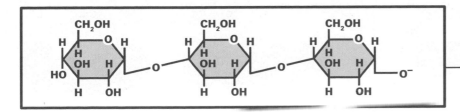
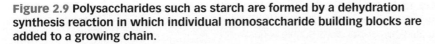

Figure 2.9 Polysaccharides such as starch are formed by a dehydration synthesis reaction in which individual monosaccharide building blocks are added to a growing chain.

Figure 2.10 The formation of a triglyceride lipid. This involves the combination of one glycerol molecule and three fatty acids through dehydration synthesis.

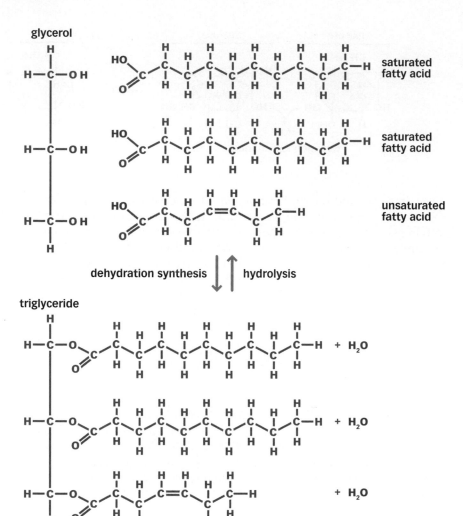

Phospholipids and Glycolipids

Some lipids have other molecules attached to them, such as carbohydrate (to form a **glycolipid**) or a phosphate (to form a **phospholipid**) (**Figure 2.12**). The phospholipid molecule is the foundation of a cell's plasma membrane structure, giving the membrane unique properties. The phosphate end of the molecule is **hydrophilic**, which means it can interface with water, but the opposing end of the molecule is **hydrophobic** and cannot interface with water. These phospholipids are used to form a barrier between the water outside the cell and the water inside the cell.

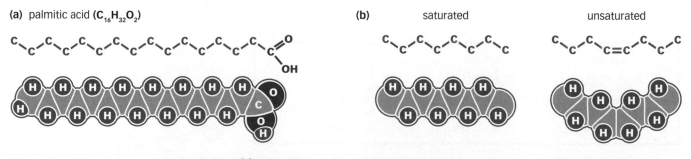

(a) palmitic acid ($C_{16}H_{32}O_2$)

(b) saturated   unsaturated

Figure 2.11 The common characteristics of fatty acids. Panel a: Palmitic acid demonstrates the two common characteristics of fatty acid molecules: the long chain of carbons and the reactive carboxyl group (COOH). Panel b: A fatty acid is either saturated (contains the maximum number of hydrogen atoms) or unsaturated (has one or more double bonds).

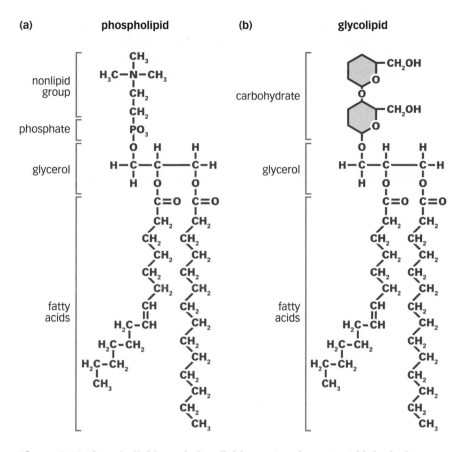

(a) phospholipid

(b) glycolipid

Figure 2.12 Phospholipids and glycolipids are two important biological molecules. Panel a: Phosopholipids have a phosphate group attached to the lipid molecule. Panel b: Glycolipids have a carbohydrate molecule attached.

Steroids

Steroids are four-ring carbon structures that are different from other types of lipid. One of the most important steroids is cholesterol, which is insoluble in water and is found in the cell membrane of some eukaryotic cells and also in the bacterium *Mycoplasma*. Another steroid, ergosterol, is found in the cell membranes of fungi and is the target for many anti-fungal drugs.

Proteins

Proteins are one of the most important of the biological molecules. They are very diverse in both structure and function, and each has a specific three-dimensional configuration (shape) that is directly related to function.

Properties of Proteins

Proteins are made up of amino acid building blocks, which are carbon structures that have at least one amino group (NH_2) and one carboxyl group (COOH) (**Figure 2.13**). These two parts of the amino acid structure are important because they are involved in the formation of a **peptide bond**. The peptide bond is formed by dehydration synthesis in which the amino group of one amino acid bonds to the carboxyl group of another amino acid (**Figure 2.14**).

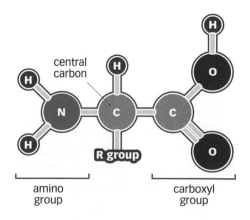

Figure 2.13 The structure of an amino acid. Each amino acid is composed of a central carbon atom to which four different groups are attached. The amino group, the carboxyl group, and a hydrogen atom are found on all 20 amino acids, but each amino acid differs at the R group.

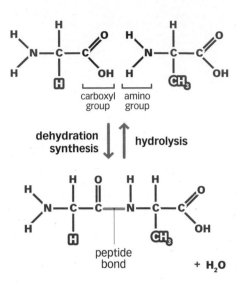

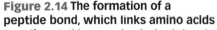

Figure 2.14 The formation of a peptide bond, which links amino acids together. In this example, alanine is bonded to glycine. Notice that the bond is caused by dehydration synthesis and always involves the amino group of one amino acid and the carboxyl group of the other. This linking of amino acids can continue and form a polypeptide chain.

Proteins are made up of long sequences of these linked amino acids, which are called peptides. These peptides can vary in length from two amino acids (dipeptide) to many amino acids (polypeptide). Some amino acids contain sulfur atoms, and these can bond to sulfur atoms in other amino acids to form **disulfide bridges**, which are important in the folding of the protein (**Figure 2.15**).

Protein Structure

As mentioned above, proteins are three-dimensional molecules, and the structure of the molecule is critical to the function of the molecule. This three-dimensional structure can be broken down into four levels (**Figure 2.16**).

1. The primary level is the sequence of the amino acids in the polypeptide chain.

2. The secondary level consists of a folding or coiling of the polypeptide that is brought about by the sequence of the amino acids and hydrogen bonds that form between amino acids. This secondary folding is usually seen as either a helix or a pleated sheet form.

3. The tertiary level involves the folding of the chain upon itself. This folding confers the major three-dimensional structure of the polypeptide and is held in place in part by hydrogen bonding and by the formation of disulfide bridges. This folding can result in several shapes, including globular and fibrous (threadlike) structures.

4. Quaternary structure occurs in very large proteins for which more than one polypeptide is joined together. This joining together of individual polypeptide chains also occurs through the formation of bonds and between amino acids in the individual polypeptide chains.

It is important to understand that the three-dimensional structure of a protein is susceptible to disruption by chemical changes that affect the hydrogen bonds and other weak bonds that hold the structure in place. Factors such as pH and temperature can change the structure of the protein and thereby the function of the protein. This is referred to as protein denaturation, which can be a lethal event for microorganisms.

Types of Protein

There are a variety of proteins, but structural proteins and enzymes are among the most important. Structural proteins contribute to cellular structures such as the cell membranes and the cytoskeleton that maintains the shape of the cell. Structural proteins are also involved in the motility seen with flagella (structures that allow an organism to move). In

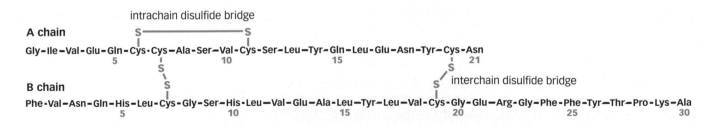

Figure 2.15 The formation of disulfide bridges occurs between amino acids that contain sulfur atoms. These bridges contribute to the three-dimensional shape of polypeptides, and they can be intrachain (within a polypeptide chain) or interchain (linking two chains together).

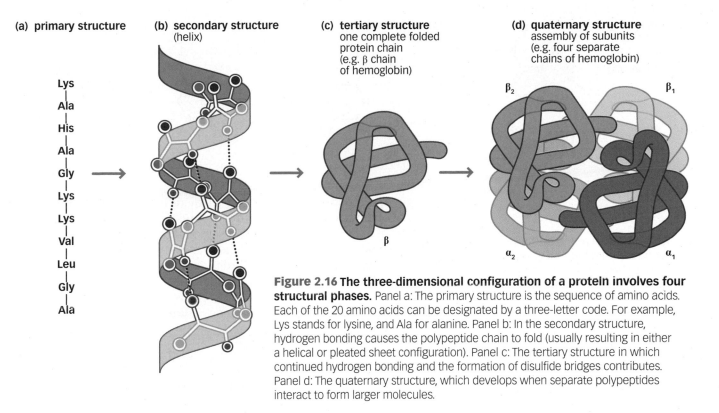

(a) primary structure

Lys
|
Ala
|
His
|
Ala
|
Gly
|
Lys
|
Lys
|
Val
|
Leu
|
Gly
|
Ala

(b) secondary structure
(helix)

(c) tertiary structure
one complete folded
protein chain
(e.g. β chain
of hemoglobin)

β

(d) quaternary structure
assembly of subunits
(e.g. four separate
chains of hemoglobin)

β₂   β₁

α₂   α₁

Figure 2.16 The three-dimensional configuration of a protein involves four structural phases. Panel a: The primary structure is the sequence of amino acids. Each of the 20 amino acids can be designated by a three-letter code. For example, Lys stands for lysine, and Ala for alanine. Panel b: In the secondary structure, hydrogen bonding causes the polypeptide chain to fold (usually resulting in either a helical or pleated sheet configuration). Panel c: The tertiary structure in which continued hydrogen bonding and the formation of disulfide bridges contributes. Panel d: The quaternary structure, which develops when separate polypeptides interact to form larger molecules.

contrast, the enzymatic proteins are involved in many cellular functions such as metabolism. We discuss enzyme activity in further detail when we look at metabolism in Chapter 3.

Nucleic Acids

Nucleic acids are organic molecules that carry genetic information. There are two types of informational molecule: deoxyribonucleic acid (DNA) and ribonucleic acid (RNA). The building blocks of nucleic acids are called nucleotides. Each DNA nucleotide consists of a nitrogenous base (adenine (A), thymine (T), guanine (G), or cytosine (C)), the sugar deoxyribose, and a phosphate (**Figure 2.17**). Adenine and guanine are purines, which are double-ring structures. In contrast, thymine and cytosine are pyrimidines, which are single-ring structures (**Figure 2.18**). The differences between the purine and pyrimidine structures are important for the overall structure of DNA. The structure of RNA nucleotides is similar except that the sugar is **ribose**, and the nitrogenous base uracil (U) is used instead of thymine.

Structure of Nucleic Acids

These are long polymeric structures in which nucleotides are linked together in a special way. Each nucleotide contains the nitrogenous base,

(a)

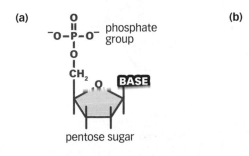

phosphate group

pentose sugar

(b)

deoxyribose

ribose

Figure 2.17 The structure of nucleotides. Panel a: The basic structure of nucleotides, which are composed of a phosphate, a pentose sugar, and a nitrogenous base. Panel b: The pentose sugar deoxyribose is found in DNA, whereas the sugar ribose is part of the RNA molecule.

(a) purines

adenine (**A**)

guanine (**G**)

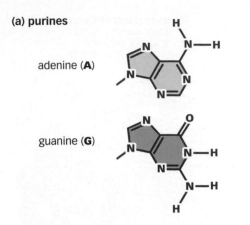

(b) pyrimidines

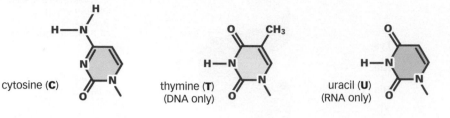

cytosine (**C**)

thymine (**T**)
(DNA only)

uracil (**U**)
(RNA only)

Figure 2.18 Panel a: The structure of the purine bases adenine and guanine. Panel b: The structure of the pyrimidine bases cytosine, thymine, and uracil.

a sugar, and a phosphate. There are covalent bonds between the phosphate of one nucleotide and the sugar of the next, and we can view this chaining of nucleotides together as a spine made up of alternating sugars and phosphates with the nitrogenous bases extending inward from it. The ends of the spine are chemically different: one end is called the 5′ end, in which the 5′ carbon of the sugar molecule is attached to the phosphate. At the other end, the 3′ carbon of the sugar is unattached (**Figure 2.19**). It is here that additional nucleotides can be linked.

For cells and most viruses that use DNA as the genetic material, the molecule is double-stranded, meaning that there are two strands that go in opposite directions (antiparallel, as shown in **Figure 2.20**). The antiparallel nature of the strands has a significant role in the replication of DNA, as we will see in Chapter 11. In DNA, the bases face each other,

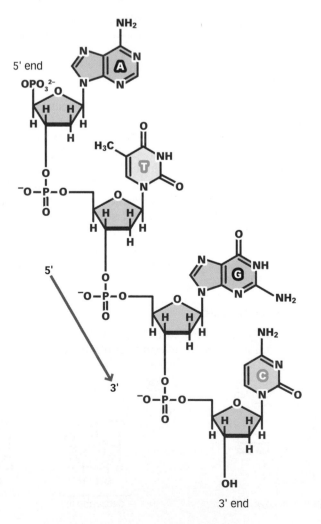

Figure 2.19 The structure of the DNA strand, which is composed of a phosphate and sugar backbone linked together in a 5′ to 3′ direction. Notice that the nitrogenous bases project inward from the backbone.

and because of the chemical nature of these bases they can hydrogen bond to each other. This occurs in a specific way, in that A always bonds to T and G always bonds to C. In addition, the bonding of A to T involves two hydrogen bonds, whereas the bonding of C to G involves three hydrogen bonds. It is important to remember that these hydrogen bonds are weak and that they are easily broken during the replication process, during which the chains separate. The bonding of A to T and G to C is referred to as complementary base pairing and is very important for both replication and gene expression. Because of the hydrogen bonding that occurs within other parts of the structure, the DNA molecule is twisted into a double helix (**Figure 2.21**).

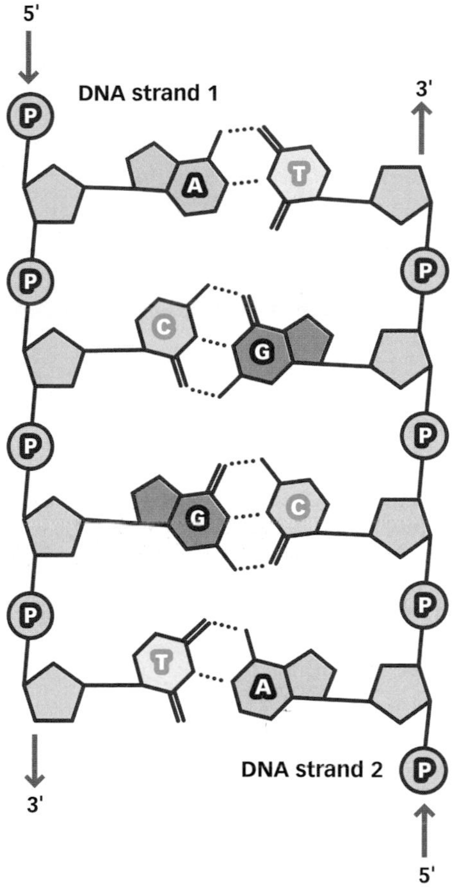

DNA is the genetic material for all life on this planet—including animals, plants, and bacteria. It carries the instructions for producing all of the structures found in the cell and also for the genetic regulation that is required for successful growth of the organism. DNA is faithfully replicated and carefully regulated so that there are no mistakes and the information that is handed down from generation to generation is correct. However, mutations that can change genetic information can occur.

Ribonucleic acid is a single-chain molecule that incorporates ribose sugar and is made by the same bonding mechanisms that link DNA nucleotides together, with uracil used instead of thymine. It is important to note that complementary base pairing occurs between DNA and RNA and is the basis for the transcription of DNA into messenger RNA, which is used to encode a protein sequence.

ATP

As we mentioned above, energy is required for all the processes in the cell, and some of these processes, such as protein synthesis, can be very energy-consuming. The major energy molecule in biological systems is ATP (adenosine triphosphate), which contains the nitrogenous base adenosine, a ribose sugar, and a chain of three phosphates bonded to the sugar (**Figure 2.22**). The bonds between these phosphates are high-energy bonds that yield energy when they are broken. When the outermost phosphate is removed from ATP it yields ADP (adenosine diphosphate). In most cases, this ADP can be recycled back into ATP if a phosphate is added, and the ATP can then be used for energy again. The recycling between ATP and ADP is important to the overall physiology of cells, but making ATP from ADP requires energy. In some cases, the phosphate from ADP can be removed to yield AMP (adenosine monophosphate). The breaking of this bond also yields energy, but less than when ATP is converted to ADP. Energy derived from breaking the bonds of ATP can be used for synthesis reactions, for motility, and also for the active

Figure 2.20 The DNA strands are antiparallel, with the strands running in opposite directions. Notice the bonding between the bases, in which there are two hydrogen bonds between the bases A and T and three hydrogen bonds between the bases G and C.

Fast Fact Some bacteria, such as *Corynebacterium diphtheriae* (the cause of diphtheria, a respiratory disease), store excess phosphates in granules called volutin. These phosphates can later be used for ATP production. Volutin can be stained and viewed under the microscope to help discern *C. diphtheriae* from other bacteria.

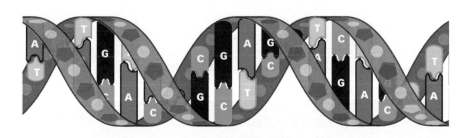

Figure 2.21 An illustration of the double-helical structure of DNA.

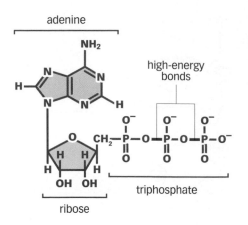

Figure 2.22 The nucleotide adenosine triphosphate (ATP), which is the immediate source of energy for most active living cells. Breaking the high-energy bonds between the phosphates releases energy that can be used for cellular functions.

transport of materials across the plasma membranes of cells. ATP must be continually replenished; this is carried out through metabolism, which we will cover in Chapter 3.

Keep in Mind

- Biological molecules use carbon as the primary building block of their structures.
- There are four types of biological molecule: carbohydrates, lipids, proteins, and nucleic acids.
- The three-dimensional structure of a protein is directly related to the function of the protein.
- ATP is the major energy molecule in biological systems.

SUMMARY

- Atoms join together to make molecules, molecules join to make cells, cells join to make tissues, and tissues join to make organs.
- There are three main types of chemical bond: ionic, covalent, and hydrogen bond.
- Ionic bonds form when electrons are donated or accepted.
- Covalent bonds form when electrons are shared.
- Hydrogen bonds form between hydrogen and negatively charged atoms and are the weakest form of chemical bond.
- Biological molecules use carbon atoms as their building blocks.
- There are four types of biological molecule: carbohydrates, lipids, proteins, and nucleic acids.
- ATP is the major energy molecule of the cell.

In this chapter we have explored some of the chemical principles and structures that will help us understand the information to come. We saw how bonds are formed and how molecules are constructed. These molecules are the foundation of cells and tissues, which are the targets for infectious disease. In the next chapter, we will apply this information as we begin to examine metabolism.

(Q) SELF EVALUATION AND CHAPTER CONFIDENCE

Multiple Choice

Answers are given in the back of the book and help can be found in the student resources at:
www.garlandscience.com/micro2

1. An atom has 17 electrons. It is most likely to

    A. Be inert
    B. Form an anion
    C. Form a cation
    D. React with a Cl⁻ ion

2. Electron sharing leads to covalent bonds. Which compound contains covalent bonds?

    A. H_2O
    B. Glucose
    C. Fatty acids
    D. Amino acids
    E. All of the above

3–7. Order the substances based on acidity, most acidic first:

Acidity	Substance
3. Most acidic	**A.** Pure water
4. Second most acidic	**B.** Gastric fluids
5. Third most acidic	**C.** Baking soda
6. Next to the least acidic	**D.** Soap solutions
7. Least acidic	**E.** Tomatoes

8. Which of the following has just two reactive groups?

- **A.** Amino acids
- **B.** Fatty acids
- **C.** Glucose
- **D.** Proteins
- **E.** Nucleic acids

9. Which of these is not a component of the plasma membrane?

- **A.** Glucose
- **B.** Triglyceride
- **C.** Phospholipid
- **D.** Cholesterol
- **E.** Glycolipid

10. Which is not a component of DNA?

- **A.** Cytosine
- **B.** Deoxyribose
- **C.** Phosphate
- **D.** Thymine
- **E.** Ribose

 DEPTH OF UNDERSTANDING

Questions listed here require you to bring together the concepts you have learned in this chapter into a discussion format. This helps you to increase your depth of understanding of the material you have learned. Help can be found in the student resources at: www.garlandscience.com/micro2

1. Describe the binding that can occur between element X, which has an atomic number of 11, and element Y, which has an atomic number of 9.

2. Describe how protein building blocks are bound together to make polypeptides.

3. Discuss the formation of spheres of hydration and why they are so important in biological systems.

Essentials of Metabolism   Chapter 3

Why Is This Important?

It is important to have a basic understanding of metabolism because it governs the survival and growth of microorganisms. Growth of microorganisms can have a direct effect on infectious disease.

Cedric suffers from pain in his stomach and has tried taking over-the-counter antacids, to no avail. His condition is common: at any given time, approximately 5 million people in the United States suffer from peptic (stomach and small intestine) ulcers. Whereas spicy foods, alcohol, and physical and emotional stress are often blamed for causing ulcers, microbes such as the bacterium *Helicobacter pylori* can cause ulcers (see Chapter 22). Diagnosis of *H. pylori* infection includes the urea breath test, in which a sample of isotope-labeled urea is swallowed, and the patient's breath is tested for isotope-labeled carbon dioxide. How does this test work?

The bacteria secrete the enzyme urease, which converts the reactants urea and water into the products carbon dioxide and ammonia. The urea consumed in this test has a unique isotope of carbon added to it. If this carbon shows up in the exhaled carbon dioxide, the bacterial enzyme urease must be present, and the *H. pylori* bacteria must also be present.

Treatment for ulcers is straightforward and includes antibiotics coupled with antacids and acid secretion inhibitors. After getting the proper medications, Cedric's ulcer problems go away and he is relieved. As shown in this example, and as we will see in this chapter, enzymes orchestrate biochemical reactions in all organisms and have a critical role in microbiology.

OVERVIEW

Before we look at the effects that infectious diseases have on humans, it is important to get a basic idea of how metabolism works. In many cases, infectious agents must be present in the body in great numbers to do harm, and attaining such high numbers requires an orderly progression of metabolic functions followed by cell division. Metabolism involves catabolism, in which molecules are broken down and energy is released, and anabolism, in which energy is used to build molecules. These mechanisms are subject to regulation, which coordinates the myriad of biochemical processes involved in metabolism.

We will divide our discussions into the following topics:

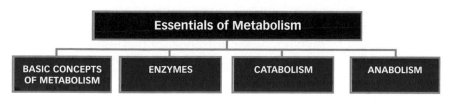

"Every year, 98 percent of the atoms of your body are replaced. This non-stop chemical replacement, metabolism, is a sure sign of life."

Lynn Margulis (1938–2011)

As Margulis sharply pointed out in the above quote, if there is no metabolism, there is no life. **Metabolism**, which is defined as the chemical processes that go on inside any living organism, also provides energy to organisms. Because all organisms require energy to live, if there is no energy, there is no life. The better an organism can metabolize, the better it grows and reproduces. For pathogens, good metabolic function makes the pathogens more successful at causing disease. In this chapter, we look at the basics of metabolism that allow microorganisms to flourish. In later chapters, we will see how the severity of an infection can be linked to increases in the numbers of pathogens in the infected host.

BASIC CONCEPTS OF METABOLISM

Microorganisms have two ways of obtaining the energy and carbon they need to live: autotrophy and heterotrophy. **Autotrophy** can be defined as 'self-feeding,' because autotrophic organisms obtain carbon atoms (the building blocks of organic molecules) from inorganic sources such as carbon dioxide (CO_2) in the environment. Green plants are one familiar example of autotrophic organisms. Autotrophs can be further divided into photoautotrophs and chemoautotrophs. **Photoautotrophs** are organisms that use sunlight as a source of energy, and carbon from carbon dioxide. **Chemoautotrophs** are autotrophic organisms that obtain energy from chemical reactions involving inorganic substances such as nitrates and sulfates.

Heterotrophy can be looked at as 'other feeding,' because heterotrophic organisms get carbon atoms from the organic molecules present in other organisms. Humans are heterotrophs because we get our carbon atoms from the food we eat. **Photoheterotrophs** obtain energy from sunlight and obtain carbon from organic compounds and **chemoheterotrophs** obtain energy by breaking down organic compounds.

Because nearly all infectious organisms are chemoheterotrophs, they will be our main focus in this chapter. Now let's look at some of the basic concepts of metabolism more closely.

Oxidation and Reduction Reactions

Metabolism can be broken down into two parts, catabolism and anabolism (**Figure 3.1**). **Catabolism** is the collective term for all the metabolic processes in which molecules are broken down to release the energy stored in their chemical bonds. **Anabolism** is the collective term for all the metabolic processes in which the energy derived from catabolism is used to build large organic molecules from smaller ones. All catabolic pathways involve electron transfer.

Electron transfer is part of the oxidation and reduction reactions. An **oxidation** reaction is a chemical reaction in which an atom, ion, or molecule loses one or more electrons. A **reduction** reaction is a chemical reaction in which an atom, ion, or molecule gains one or more electrons. (You can easily remember this as **ox**idized molecules have **lost** electrons, and **red**uced molecules have **re**ceived electrons.) Oxidation and reduction reactions always occur together, and for this reason the combination of an oxidation reaction and its reduction partner is referred to as a **redox reaction**. In many cases, oxygen is the acceptor for electrons in a redox reaction, but other substances can also accept them.

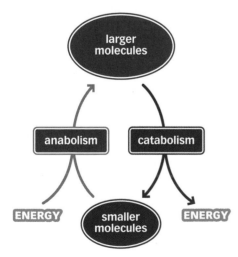

Figure 3.1 Metabolism is composed of catabolism and anabolism. In catabolism, large molecules are broken down and energy is released. In anabolism, large molecules are created from smaller ones, with energy consumed in the process.

The classic example of this is seen in the chemical reaction

$$Na^+ + Cl^- \longrightarrow NaCl$$
$$\text{sodium}$$
$$\text{chloride}$$

In this reaction, the sodium atom loses an electron and so becomes oxidized, while the chlorine atom gains the electron lost by the sodium atom and so becomes reduced.

Respiration

Respiration means two different things in the biological sciences. At the macroscopic level, respiration is the exchange of carbon dioxide and oxygen in the lungs. At the cellular level, which is the level we are interested in, the term respiration is used to describe catabolic processes, and these can be found in two forms, **aerobic respiration** (meaning metabolism that uses oxygen) and **anaerobic respiration** (metabolism that occurs without the use of oxygen). Whether a microorganism uses aerobic respiration or anaerobic respiration for catabolism determines the amount of ATP that the microorganism produces. We will see below that anaerobic respiration through glycolysis yields only 2 ATP molecules for every glucose molecule consumed, whereas coupling glycolysis with aerobic respiration using the Krebs cycle and electron transport yields a total of 38 ATP molecules for every glucose molecule consumed by bacteria. Because ATP is vitally important, organisms that use aerobic respiration are able to increase in number more quickly than those that use the anaerobic form.

It is important to note that some organisms can carry out metabolic processes either way, aerobically or anaerobically. These organisms are called **facultative anaerobes**. They grow well in the presence of oxygen, but when oxygen is absent they can still grow, although not as well as with oxygen. Many pathogens (for example, *Shigella*, *Salmonella*, *E. coli*, and other members of the family Enterobacteriaceae, which cause a variety of gastrointestinal diseases) are facultative anaerobes.

The chemical reaction for the complete oxidation of glucose by aerobic respiration is

$$C_6H_{12}O_6 \ + \ 6\,O_2 \longrightarrow 6\,CO_2 \ + \ 6\,H_2O \ + \ \text{energy}$$
$$\text{glucose} \quad \text{oxygen} \qquad \text{carbon} \quad \text{water}$$
$$\text{dioxide}$$

All the substances to the left of the reaction arrow are called either the **reactants** or the **substrates** of the reaction, and all the substances to the right of the reaction arrow are called the **products** of the reaction.

Metabolic Pathways

Nearly all chemical processes in living organisms consist of a series of chemical reactions called pathways. In these pathways the product of one reaction serves as the substrate for the next reaction. An example of a pathway is

$$\text{Enzyme 1} \quad \text{Enzyme 2} \quad \text{Enzyme 3} \quad \text{Enzyme 4}$$
$$A \longrightarrow B \longrightarrow C \longrightarrow D \longrightarrow E$$

In this example, A is the initial substrate and E is the final product of the pathway, with B, C, and D being intermediates. Notice that each step in the pathway is mediated by an enzyme. These enzymes facilitate, or promote the running of, certain reactions under physiological conditions and are required for the reactions to proceed at an appropriate speed.

Before we move on to our discussion of metabolic pathways, let's take a closer look at enzymes.

> ### Keep in Mind
>
> - Metabolism is the chemical process that provides or stores energy for organisms.
> - Metabolism can be broken down into two parts: catabolism, in which molecules are broken down and energy is released, and anabolism, in which molecules are constructed and energy is consumed.
> - Nearly all infectious organisms are chemoheterotrophs, obtaining energy by breaking down organic molecules.
> - Oxidation–reduction (redox) reactions involve the transfer of electrons. When a molecule gives up an electron it is oxidized, and when a molecule obtains an electron it is reduced.
> - Cellular respiration can be aerobic, in which oxygen is the final acceptor of electrons, or anaerobic, in which oxygen is not involved.
> - Nearly all chemical processes consist of a series of chemical reactions known as pathways.

ENZYMES

Recall from Chapter 2 that enzymes are protein molecules and therefore have a distinctive three-dimensional shape. The shape of a protein is directly related to its function. If the shape is changed, the function can be inhibited or even stopped. Enzymes are found in all living organisms, and most cells contain hundreds of types of enzymes, which are constantly being manufactured and replaced.

In chemistry, a **catalyst** is a substance that speeds up a chemical reaction but is not itself changed in any way by the reactions. Enzymes act as catalysts for the reactions that take place in organisms. This means that enzymes help to get reactions started and help them to proceed.

Properties of Enzymes

Enzymes work in a chemical reaction by lowering the **energy of activation**, which is the energy required to start the reaction (**Figure 3.2**). In

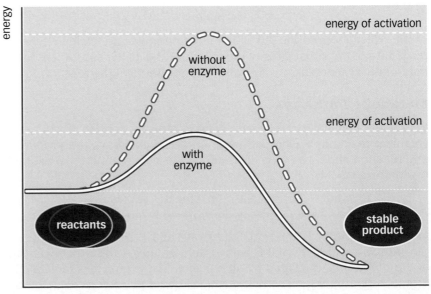

Figure 3.2 Energy of activation.
A chemical reaction cannot take place unless there is enough activation energy available to start the reaction. Enzymes lower the amount of activation energy necessary to start the reaction.

any metabolic reaction, the energy of activation is required for the substrate to be converted to product. The amount of energy needed is greater when no enzyme is present and less when enzyme is present.

To conceptualize the role of enzymes, imagine there is a boulder poised to roll down a hill, but a small mound at the edge of the hill keeps the boulder from moving. The mound represents the activation energy needed for the reaction, or the effort it would take to push the boulder to start it rolling. This mound is high when no enzyme is present, but when we reduce the height of the mound (by adding an enzyme) it allows the boulder to more easily roll over the mound and down the hill.

For a more realistic example of enzyme-mediated reaction, suppose you dissolve a tablespoon of table sugar (sucrose) in a glass of water. At room temperature, the sucrose will never be broken into glucose and fructose unless the enzyme sucrase is present. This is because the enzyme reduces the amount of activation energy required for the reaction to take place.

Because enzymes are proteins, each enzyme molecule has a three-dimensional shape that results from the folding of the molecule. The shape of the molecule provides a site where the enzymatic activity takes place, called the **active site**, and it is here that the substrate fits into the enzyme (**Figure 3.3**). This is a precise fit based on the shape of the active site, which means that only one shape of substrate can fit at each active site. It is at the active site that the enzyme and substrate interact to form what is referred to as an **enzyme–substrate complex**. As a result of this binding to the active site, some of the chemical bonds in the substrate molecule undergo changes. These changes lead to the formation of the product of the reaction. In the hypothetical pathway shown in the last section, substrate A binds to enzyme 1, and this binding changes A to product B. After the change has taken place, A is released from the active site of enzyme 1 as product B. Product B now becomes the substrate of enzyme 2, and the reaction between B and enzyme 2 forms product C. This continues until the final product, E, is formed.

Enzymes are generally highly specific. This means that a given enzyme catalyzes only one type of reaction. In addition, most enzymes react with only one particular substrate. Therefore, in our hypothetical pathway, the only reaction that enzyme 1 catalyzes is the one in which substrate A is converted to product B. This specificity is important because it allows a cell to function in an orderly way and use its energy stores efficiently. The shape of an enzyme molecule and the electrical charges found at the active site are responsible for the enzyme's specificity.

Coenzymes and Cofactors

Many enzymes can catalyze a reaction only if other substances are present. When the helper substance is an inorganic ion (mineral), such as magnesium, zinc, manganese, or iron, it is called a **cofactor**. When the helper substance is a nonprotein organic molecule, it is called a **coenzyme**. Just as in humans, microbes have specific cofactor and coenzyme requirements for growth. Some antimicrobial chemicals work by binding to cofactors and coenzymes. For example, the lactoferrin in tears, saliva, and breast milk bind iron tightly and make it unavailable to microbes, thus acting as an antimicrobial compound.

Metabolic reactions may require the presence of **carrier molecules** to carry hydrogen atoms or electrons in redox reactions. Coenzymes and cofactors can be used as carrier molecules. When a carrier molecule receives either hydrogen atoms or electrons, it becomes reduced. When

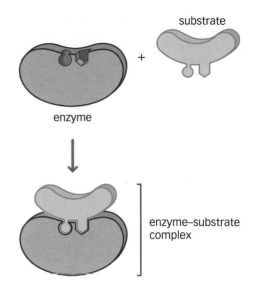

Figure 3.3 Active site on an enzyme. Only a substrate that fits exactly in the active site can combine with the enzyme. When the substrate binds to the active site, an enzyme–substrate complex is formed.

Fast Fact Minerals such as iron are critical for the growth of some pathogenic microbes, including strains of *Salmonella* and *E. coli*. Microbes that thrive in high iron concentrations are called siderophilic. Patients with excess iron in their blood, resulting from too many blood transfusions or from the genetic condition hemochromatosis, are very susceptible to infections from siderophilic microbes.

Figure 3.4 Coenzymes in redox reactions. The coenzyme FAD can carry two hydrogen atoms and two electrons. The coenzyme NAD^+ can carry one hydrogen atom and one electron. When coenzymes are reduced, they gain electrons, which increases their energy level. When the reduced forms of the coenzymes are oxidized, they lose electrons, which decreases their energy level.

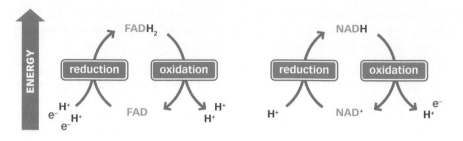

the carrier molecule releases them, it becomes oxidized. Two coenzyme carrier molecules frequently encountered in biological reactions are **FAD (flavin adenine dinucleotide)** and NAD^+ **(nicotinamide adenine dinucleotide)**. Both are vitamins and are represented in **Figure 3.4**. FAD can carry two hydrogen atoms and two electrons, and when it does it is called $FADH_2$, which is the reduced form of FAD. The coenzyme NAD is positively charged in its oxidized state (NAD^+) because it is lacking one electron. When it is reduced, it has gained a hydrogen atom and an electron and so becomes NADH. When these two carrier molecules are in their reduced state—$FADH_2$ and NADH—it is the electron that carries the energy that is being transferred from one molecule to another.

Enzyme Inhibition

Organisms cannot allow continuous enzyme activity to occur because this requires a great deal of energy and chemical components that may be hard to come by for the organism. In addition, continuous enzyme activity may result in the buildup of potentially harmful products. It is therefore necessary to be able to regulate enzyme activity. This is accomplished in two major ways: competitive inhibition and allosteric inhibition.

In **competitive inhibition**, an inhibitor molecule, which is a molecule similar in structure to the substrate for a given enzyme, competes with the substrate to bind to the enzyme's active site (**Figure 3.5**). Once the inhibitor molecule has bound to the active site, the substrate cannot bind. Because the inhibitor cannot react with the enzyme, the metabolic pathway is stopped. This competition is reversible and depends on the relative numbers of inhibitor molecules and substrate molecules present. If the number of substrate molecules is high and the number of inhibitor molecules is low, the active sites of only a few enzyme molecules are blocked, which leads to only a slight decrease in enzyme activity. If the number of inhibitor molecules is high and the number of substrate molecules is low, the active sites of most of the enzyme molecules are blocked, leading to a steep decrease in enzyme activity.

Fast Fact Competitive inhibition is an example of how sulfonamide ('sulfa') drugs work in inhibiting bacterial growth. The sulfonamide acts as a competitive inhibitor of the enzyme dihydropteroate synthase, thus blocking folate (vitamin B_9) production in the microorganism. Without the essential nutrient folate, the bacteria quickly die.

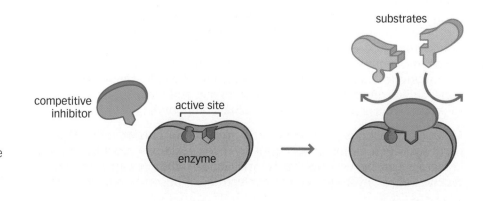

Figure 3.5 Competitive inhibition. The competitive inhibitor fits into the active site of the enzyme, preventing the substrate from binding to the enzyme.

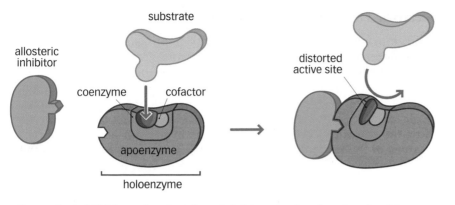

Figure 3.6 Allosteric inhibition. The allosteric inhibitor binds to a location on the enzyme that is different from the enzyme's active site. This binding changes the shape of the active site, and as a result enzymatic activity is stopped.

Allosteric inhibition also involves inhibitor molecules, but in this case the molecules do not block the active site. Instead they bind to another part of the enzyme, at a site called the allosteric site (**Figure 3.6**). Before an allosteric inhibitor binds, the enzyme's active site is the right shape to accept substrate. The binding of the inhibitor at the allosteric site changes the three-dimensional structure of the enzyme in such a way that the active site changes and can no longer fit properly with the substrate. As a result, there is no enzymatic activity. Some allosteric inhibitors bind reversibly and so can unbind and allow activity to resume. Others do not, however. When the binding is irreversible, the enzyme is permanently disabled.

Feedback inhibition also regulates enzymatic activity and is used for many of the metabolic pathways found in the cell. With this type of inhibition, the final product in a pathway accumulates and begins to bind to and inactivate the enzyme that catalyzes the first reaction of the pathway (**Figure 3.7**). This inactivation is once again associated with a change in the three-dimensional structure of the enzyme molecule such that it no longer functions properly. Feedback inhibition is reversible, and when the level of end product decreases, the inhibition stops, and the pathway begins to function again.

> **Fast Fact** Some examples of irreversible allosteric inhibitors are lead, mercury, and other heavy metals. This is one of the reasons that we don't use lead-based paint any more.

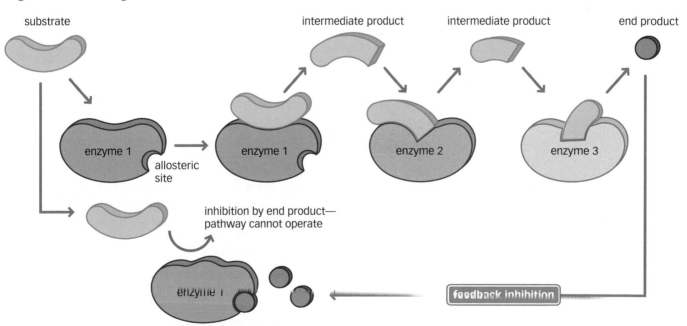

Figure 3.7 Feedback inhibition. The inhibitor molecule is one of the end products of the metabolic pathway. If the inhibitor levels build up sufficiently, inhibitor molecules bind to an allosteric site on the enzyme, changing the shape of the active site and preventing the reactions of the pathway from continuing.

Fast Fact Even though refrigerator temperatures slow down microbial enzymes, it does not stop them completely, and eventually microbial growth spoils the leftover food (often it looks as if it had grown hair!). At freezing temperatures, however, microbial metabolism stops. Upon thawing, some microbes resume their metabolism and food spoilage occurs.

Factors That Affect Enzyme Reactions

Three major factors affect enzyme activity: temperature, pH, and the concentrations of substrate, product, and enzyme. Extremely high temperatures can break the hydrogen bonds that hold the enzyme in its three-dimensional shape, and this change in structure can inhibit enzyme activity. However, minor increases in temperature can help reactions occur more rapidly. Lower temperatures slow metabolic reactions and slow the growth of organisms. Changes in pH can alter the electrical charges in the enzyme molecule, and these changes in charges inhibit the molecule's ability to bind to its substrate. Very high or very low pH can denature the protein-based enzyme by causing it to precipitate out of solution. Microbial enzymes function best at an optimal temperature and pH, which are specific to that microbe.

The concentrations of enzyme, substrate, and product also affect enzyme reactions. The job of an enzyme in a reaction is to lower the activation energy needed, and thereby stimulate the reaction to occur. Obviously, the smaller the number of enzyme molecules available, the smaller the number of substrate molecules that can bind to enzyme at any instant and so the slower the reaction will be. If the substrate concentration is too low, this limits the number of enzyme–substrate complexes that can form at any instant and so also affects reaction rate. In some cases, once an enzyme reaction has run long enough to build up a quantity of product, the product can begin feedback inhibition, and so again the reaction rate is changed.

Keep in Mind

- Enzymes are proteins that work in metabolism by lowering the energy of activation.

- These proteins have a specific three-dimensional shape and bind with the substrate they act upon at a place on the molecule called the active site.

- Enzymes are highly specific and in some cases require cofactors or coenzymes to function.

- Enzyme function can be regulated by competitive inhibition or allosteric inhibition.

- Temperature, pH, and the concentration of substrate, product, and enzymes all affect the function of enzymes.

CATABOLISM

Now let's turn to the details of the various types of metabolic reaction seen in microorganisms. Here we will concentrate on **catabolic** processes, which are those that break down large molecules into smaller ones. After a detailed look at several types of catabolic reaction, we'll close the chapter with a brief look at **anabolic** reactions, which are those that assemble smaller molecules into larger ones.

Nutrient molecules can be processed in several ways to release energy: glycolysis, the Krebs cycle, and the electron transport chain. Each of these pathways involves a sequence of chemical reactions.

Glycolysis

In **glycolysis**, the catabolic pathway used by most organisms, glucose is broken down through a series of steps that ultimately result in the

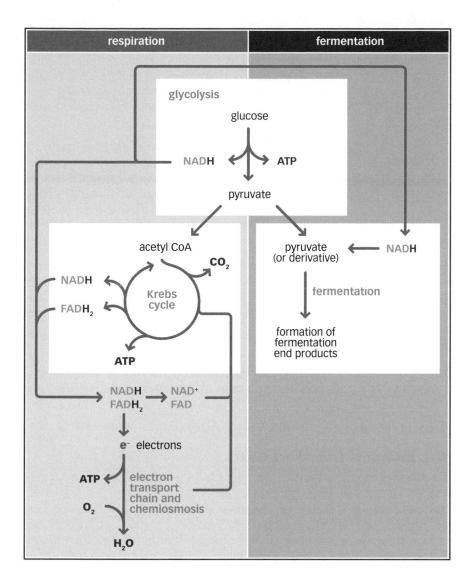

production of two molecules of ATP for each molecule of glucose involved. This process occurs in the cytoplasm and does not require oxygen.

During glycolysis:

- Phosphate groups (PO_4^{3-}) are removed from ATP and transferred to substrates. This process is called **phosphorylation** and makes the substrates more energetic.

- After a series of steps, the six-carbon glucose molecule is broken in half, yielding two three-carbon molecules called **pyruvate**. These pyruvate molecules can be processed and moved into the Krebs cycle or into fermentation pathways (**Figure 3.8**).

- During glycolysis, two electrons are transferred to the carrier molecule NAD^+, converting it to NADH. The NADH then carries the electrons to electron-acceptor molecules.

- Although four ATP molecules are produced in glycolysis, there is a net gain of only two ATP molecules because the first steps of the pathway consume two ATPs.

Figure 3.8 The fate of pyruvate. The pyruvate produced in glycolysis can be used in either respiration or fermentation. Respiration involves both the Krebs cycle and the electron transport chain, with oxygen as the final electron acceptor.

The Krebs Cycle

The Krebs cycle is a catabolic pathway seen in aerobic cellular respiration in which the pyruvate produced in glycolysis is metabolized further (see Figure 3.8). The cycle was named for the German biochemist Hans Krebs, who identified the steps in the 1930s. It is also referred to as the tricarboxylic acid cycle (TCA) or the citric acid cycle.

The cycle accepts only two-carbon molecules, and because pyruvate contains three carbons it must be modified before it can enter the cycle. This modification involves the conversion of the three-carbon molecule to a two-carbon molecule through the removal of one carbon in the form of CO_2. This is a complex reaction involving the carrier molecule NAD^+ and the addition of a molecule called coenzyme A (CoA) to form the complex acetyl-CoA.

The Krebs cycle is a sequence of reactions in which hydrogen atoms are removed and their electrons are transferred to coenzyme carrier molecules. CO_2 is also given off during this cycle, and each step in the cycle is controlled by a specific enzyme. During the Krebs cycle, three important things occur:

1. Carbon is oxidized to CO_2.

2. Electrons are transferred to coenzyme carrier molecules (NADH and $FADH_2$) that take the electrons to the electron transport chain (described below).

3. Energy is captured and stored when ADP is converted to ATP.

The Electron Transport Chain

Electron transport is the cellular process in which electrons are transferred to a final electron acceptor. In aerobic metabolism, that final electron acceptor is oxygen. In anaerobic metabolism, the final electron acceptor is an inorganic molecule such as nitrate (NO_3^-), nitrite (NO_2^-), or sulfate (SO_4^{2-}). During electron transport, hydrogen atoms are transferred to NAD^+ carrier molecules, which transfer the hydrogen atoms to proteins; these proteins are found in a precise arrangement in the microbial cell membrane or in the inner membrane of the mitochondria in eukaryotic cells (**Figure 3.9**).

The transfer of hydrogen atoms from one molecule to another in the electron transport chain involves oxidation and reduction reactions. Each member of the chain becomes reduced as it picks up electrons. When it gives the electrons to the next molecule in the chain, it becomes oxidized and the accepting molecule becomes reduced (see Figure 3.9). From the catabolism of a single molecule of glucose, there are 10 pairs of electrons transported to the electron transport chain by NAD^+ and an additional 2 pairs transported by FAD. When oxygen is the final electron acceptor in the chain it becomes reduced to the final form found in the H_2O molecule.

Electron transport differs from organism to organism, and some organisms can use more than one type. Keep in mind that during aerobic metabolism the electron transport chain also keeps the Krebs cycle turning by being a place for the reduced carrier molecule NADH to drop off electrons. If the electron transport chain were blocked, the Krebs cycle would shut down.

We can think of the electron transport chain as a series of steps in a stairway. NADH enters at the top step and passes the electrons it is carrying to the next step. As these electrons descend the stairs, energy is released via a process called **chemiosmosis**.

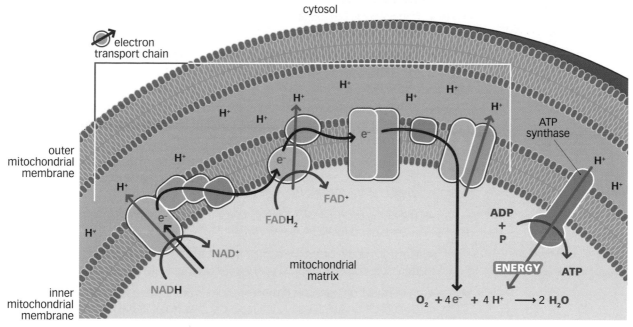

Figure 3.9 An illustration of the electron transport chain with oxygen as the final electron acceptor. This process is where the majority of ATP is produced. In this series of steps, electrons and protons (H⁺ ions) are passed between the intermediates in a sequence of redox reactions. In this case, the final electron acceptor is molecular oxygen, which then uses the acquired electrons and protons to form water, H_2O. P, phosphate.

Chemiosmosis

As electrons are transferred along the electron transport chain, protons (which carry a positive electrical charge) are pumped out of the cell. In eukaryotic cells, chemiosmosis occurs in the mitochondria, and these protons are pumped into the intermembrane space. Prokaryotes do not have mitochondria, so their electron transport chain is located in the plasma membrane. Thus, in prokaryotes, the electron transport chain pumps protons across the plasma membrane. This causes the proton concentration outside the cell to be higher than the proton concentration inside the cell, causing a **concentration gradient** to form. Because it is a difference in proton concentrations, this particular concentration gradient is called the **proton motive force**. Whenever a concentration gradient exists, the particles tend to move from the region of high concentration to the region of low concentration. So, because the proton concentration is higher outside the cell, the proton motive force moves protons back into the cell. As the protons move through specialized proteins embedded in the membrane, energy is released and used to bind inorganic phosphate to ADP and form the high-energy molecule ATP.

Looking at the Numbers

During the aerobic catabolism of one glucose molecule, 10 pairs of electrons are transported by NAD⁺, and each pair releases from the glucose molecule enough energy for 3 ATPs, yielding 30 ATPs. Each pair of electrons carried by FAD generates only enough energy for 2 ATPs. Because there are 2 pairs of electrons carried by FAD, there is a gain of 4 ATPs by this carrier molecule. Include the net gain of 2 ATPs obtained from glycolysis and the 2 ATPs from the Krebs cycle and we see that aerobic respiration yields a total of 38 ATP molecules from one molecule of glucose.

Fast Fact Because prokaryotes do not have mitochondria, they do not waste energy transporting molecules across a mitochondrial membrane. During aerobic catabolism, prokaryotes gain 38 ATPs from each glucose, whereas eukaryotes, having 'paid a toll' of 2 ATPs to move molecules into the mitochondrion, actually gain only 36 ATPs from each glucose.

> **Keep in Mind**
>
> - Catabolism is the metabolic process in which organic molecules are broken down to release energy.
> - Catabolism can involve glycolysis, which occurs in the cytoplasm, as well as the Krebs cycle and electron transport, which occur at the bacterial plasma membrane.
> - When glycolysis is linked to the Krebs cycle and electron transport, and oxygen is the final electron acceptor, 38 ATP molecules are produced for every molecule of glucose that is broken down.
> - Aerobic cellular respiration uses oxygen as the final electron acceptor, whereas anaerobic cellular respiration uses an inorganic molecule other than elemental oxygen as the final electron acceptor.
> - The Krebs cycle involves a series of chemical changes that generate the release of protons and electrons.
> - Electrons and protons are carried from the Krebs cycle to the electron transport chain by reduced coenzymes such as NADH and $FADH_2$.
> - In the electron transport chain, the protons and electrons are moved through a series of oxidation–reduction reactions that result in the formation of a concentration gradient called the proton motive force.
> - Protons are moved by the proton motive force across the plasma membrane of the bacterial cell, and energy is released.
> - The energy produced from protons moving across the plasma membrane is used to form ATP.

Fermentation

Fermentation is defined as the anaerobic enzymatic breakdown of carbohydrates in which an organic molecule serves as electron donor and final electron acceptor. ATP is synthesized by substrate-level phosphorylation, which is not linked to the electron transport chain, and oxygen is not required. In fermentation, there is no Krebs cycle or electron transport chain. In most cases, fermentation does not increase the yield of ATP from what it is in glycolysis: two ATPs for every glucose molecule. This means that a huge amount of glucose must be consumed by organisms that use fermentation if they are to satisfy their energy requirements.

Different microorganisms use different fermentation pathways, and in some cases these differences can be used diagnostically to aid in identifying microorganisms (**Figure 3.10**). Microbial fermentation end products are also critical in industry and food manufacturing.

Two of the most important and common fermentation pathways are **homolactic** fermentation and **alcoholic** fermentation. Neither pathway stores energy in ATP, but both fermentation pathways recycle the carrier molecule NAD^+ so that it can be used to continue glycolysis. This recycling is accomplished by removing the electrons from the reduced form, NADH, so that the resulting NAD^+ can go back to the glycolytic pathway and acquire more electrons.

Homolactic Fermentation

This is the simplest fermentation pathway for metabolizing pyruvate. The pyruvate molecule is converted directly to lactate with the reactions driven by the energy of the electrons from NADH (**Figure 3.11**). Note that the homolactic pathway does not produce CO_2. Unlike most

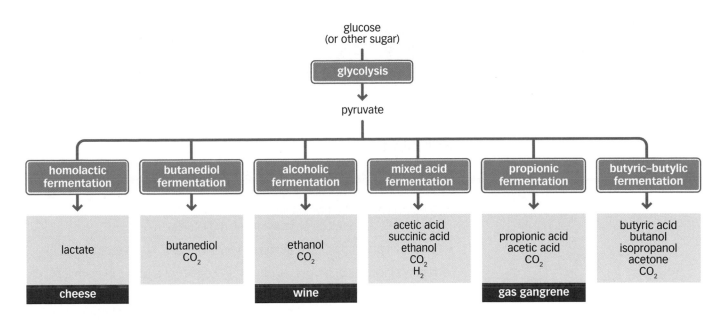

Figure 3.10 **There are a variety of fermentation pathways possible for the pyruvate formed in glycolysis.** Which pathway is used depends on which microorganism is doing the fermenting.

other fermentation pathways, this pathway does not produce gas as a by-product. Lactic fermentation occurs in several forms of bacteria and is also seen in mammalian muscle cells. Also, as Figure 3.10 indicates, it is a major method used in the making of cheese.

Alcoholic Fermentation

In alcoholic fermentation, CO_2 is released from pyruvate to form the intermediate acetaldehyde (**Figure 3.12**). This intermediate is quickly reduced to ethanol by electrons from NADH, and the NADH is oxidized to NAD^+. Alcoholic fermentation is rarely seen in bacteria but is very common in yeasts, as seen in the production of beer, wine, and other alcoholic beverages.

ANABOLISM

Anabolic reactions are used to synthesize all the biological molecules needed by the cells of living organisms. Biosynthetic reactions form the network of pathways that produce the components required by the cell for growth and survival. These reactions are fueled by the energy stored in high-energy bonds in ATP. Using anabolic reactions, cells construct carbohydrates, lipids, proteins, nucleic acids, and all the other molecular biological building blocks they need. These anabolic pathways serve as a target for treatment strategies. As we will see in Chapter 19, the sulfonamide and trimethoprim antibiotics act by disrupting anabolic biosynthetic pathways.

Anabolic reactions cost energy and therefore must proceed in an orderly, efficient, coordinated fashion. In general, microorganisms cannot control their environment, and so any environmental change (in temperature or availability of nutrients) can disrupt or inhibit metabolic reactions, and this could be lethal for the microorganism.

Control of enzyme activity is the most common means by which organisms modulate the flow of materials through the catabolic and anabolic pathways. This ensures that the flow of carbons from the major substrates through the various pathways is adjusted to meet the biosynthetic demands of the cell. In biosynthetic pathways, reactions are also controlled through feedback inhibition.

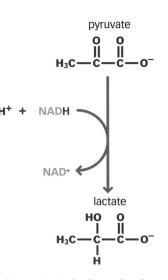

Figure 3.11 **The homolactic fermentation pathway.** Pyruvate obtained from glycolysis is reduced to lactate through a redox reaction with the carrier molecule NAD^+.

Figure 3.12 The alcoholic fermentation pathway. In this two-step process, a molecule of CO_2 is first removed from the pyruvate to form acetaldehyde. NADH then reduces the acetaldehyde to ethanol. At the same time, the NADH is oxidized to NAD^+.

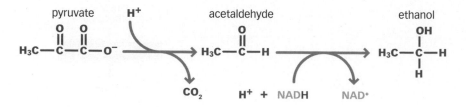

SUMMARY

- Metabolism is the chemical process that provides or stores energy for the organism.

- Metabolism can be divided into two parts, catabolism (breaking down molecules) and anabolism (building up molecules).

- Oxidation and reduction reactions involve the transfer of electrons.

- Nearly all chemical processes of the cell consist of a series of chemical reactions known as a pathway.

- Enzymes are proteins that encourage reactions to proceed by lowering the energy of activation.

- Enzymes work on the basis of their three-dimensional shape. They are specific and in some cases require cofactors or coenzymes to function.

- Temperature, pH, and concentration of substrate all affect enzyme function.

- When oxygen is involved, catabolism occurs through glycolysis, the Krebs cycle, and electron transport.

- Aerobic metabolism requires oxygen and yields 38 ATPs from the breakdown of one molecule of glucose.

- Breakdown of one molecule of glucose without oxygen (fermentative metabolism) yields only two molecules of ATP.

In this chapter we have briefly reviewed the essentials of metabolism. Metabolism is all about building molecules (anabolism) and breaking them down (catabolism). The energy released from catabolic processes is first stored in high-energy ATP bonds and then released to be used in anabolic processes. Keep in mind the connection between metabolism and infectious disease. Organisms that metabolize well increase their numbers more quickly, and increased numbers can enhance the infectious process.

(Q) SELF EVALUATION AND CHAPTER CONFIDENCE

Multiple Choice

Answers are given in the back of the book and help can be found in the student resources at:

www.garlandscience.com/micro2

For questions **1–5** choose the best matches among **A** to **E**.

1. Enzyme	**A.** Precursor of protein
2. Alcohol	**B.** Inorganic compound
3. Acetaldehyde	**C.** Precursor of ethanol
4. CO_2	**D.** Catalyst
5. Amino acid	**E.** Ethanol

For questions **6–10** choose the best matches among **A** to **E**.

6. Energy of activation	**A.** Energy required to start a reaction
7. Energy (fuel)	**B.** Stabilize water and protein structure
8. Gradient	**C.** Where enzyme catalyzed reactions occur
9. Hydrogen bond	**D.** Uneven distribution of a chemical in solution or across a membrane
10. Active site	**E.** Released by catabolic reactions

11. Which of the following is the product of catabolism?

   A. DNA
   B. RNA
   C. Protein
   D. Phospholipid
   E. Pyruvic acid

12. Which of the following is a product of anabolism?

   A. CO_2 generated during oxidative phosphorylation
   B. Lipoproteins generated in membranes
   C. H^+ generated during oxidative phosphorylation
   D. Lactic acid generated during fermentation
   E. Acetic acid generated during fermentation

13. Autotrophic organisms obtain carbon from

   A. Organic molecules
   B. CO_2
   C. Photosynthesis
   D. The soil
   E. Water

14. Heterotrophic organisms obtain carbon from

   A. Organic molecules
   B. CO_2
   C. Photosynthesis
   D. The air
   E. Water

15. Anabolism is the process in which

   A. Molecules are broken down
   B. Molecules are transported out of the cell
   C. Energy in the form of ATP is released
   D. Energy in the form of ATP is used up

16. Catabolism is the process in which

   A. Molecules are broken down
   B. Molecules are transferred out of the cell
   C. Molecules are built up
   D. Energy is used up

17. Oxidation is defined as

   A. Gaining an electron
   B. Utilizing CO_2 for metabolism
   C. Destroying CO_2 during metabolism
   D. Losing an electron
   E. None of the above

18. During a redox reaction

   A. Protons are lost
   B. Electrons are generated
   C. Protons are gained
   D. Electrons are gained and lost
   E. Electrons are not used at all

19. During a reduction reaction a substance

   A. Gains an electron and becomes more positively charged
   B. Loses an electron and becomes more negatively charged
   C. Gains an electron and becomes more negatively or less positively charged
   D. Loses an electron and becomes positively charged
   E. Neither gains nor loses an electron

20. In prokaryotes, the greatest amount of ATP per glucose molecule is produced through

   A. Fermentation
   B. Anaerobic respiration
   C. Aerobic cellular respiration
   D. Both A and B
   E. None of the above

21. Organisms that can metabolize either in the presence or absence of oxygen are called

   A. Obligate anaerobes
   B. Micro-aerobes
   C. Obligate aerobes
   D. Facultative anaerobes

22. Complete oxidation of glucose yields____.

   A. CO_2, water, and energy
   B. Only CO_2 and water
   C. Only CO_2 and energy
   D. Only CO_2
   E. None of the above

23. Enzymes

   A. Decrease the energy of activation
   B. Cause a reaction to happen
   C. Cause a reaction to happen more slowly
   D. Increase the activation energy
   E. Have no effect on the energy of activation

24. Which of the following can be required for an enzyme to function

   A. Cofactors
   B. Coenzymes
   C. Prosthetic groups
   D. Metal ions
   E. All of the above

25. Allosteric inhibition occurs

   A. At the active site
   B. Because a compound binds at a site which reduces the maximum rate at which the enzyme works
   C. Because of conformational changes in the enzyme
   D. When inhibitors bind close to the active site
   E. Because of a molecule identical to the substrate

26. Which of the following affect enzyme activity?

   A. Temperature
   B. pH
   C. Concentration of substrate
   D. Concentration of product
   E. All of the above

27. The net gain of molecules of ATP solely from glycolysis is

   A. 4
   B. 0
   C. 19
   D. 36
   E. 2

28. Fermentation can produce which of the following?

   A. Ethanol
   B. Propionic acid
   C. Butyric acid
   D. CO_2
   E. All the above

29. Homolactic fermentation

   A. Results in a loss of ATP
   B. Produces lactic acid
   C. Is not involved in anaerobic respiration
   D. Is a form of aerobic respiration
   E. Produces carbon dioxide

30. The Krebs cycle only accepts

   A. Two-carbon molecules
   B. Three-carbon molecules
   C. Pyruvate from the glycolytic pathway
   D. Single carbons
   E. Products derived from fermentation

31. The final electron acceptor in aerobic respiration is

   A. Oxygen
   B. Water
   C. Hydrogen
   D. Both A and B

32. The majority of electrons are carried to the electron transport chain by

   A. FAD
   B. NAD^+
   C. Glycolysis
   D. The Krebs cycle
   E. None of the above

33. In anaerobic metabolism, the final electron acceptor can be any of the following except

   A. Nitrite
   B. Sulfate
   C. Nitrate
   D. Oxygen

34. The total number of ATP molecules that result from the breakdown of one molecule of glucose via glycolysis and oxidative phosphorylation by a prokaryote is

   A. 4
   B. 38
   C. 16
   D. 34
   E. 32

For questions 35–37, answer true (**A**) or false (**B**) to the following statements.

35. Pathogens with good metabolic function are more successful at causing disease.

36. A gastric biopsy is required to determine whether a patient with an ulcer infected with *Helicobacter*.

37. Iron is critical for the growth of some pathogenic microbes.

 DEPTH OF UNDERSTANDING

Questions listed here require you to bring together the concepts you have learned in this chapter into a discussion format. This helps you to increase your depth of understanding of the material you have learned. Help can be found in the student resources at: www.garlandscience.com/micro2

1. Using your knowledge of metabolism, describe the relationship between anabolism and catabolism. Why is this relationship important?

2. Describe the relationship between glycolysis and the Krebs cycle.

3. What are the advantages of aerobic cellular respiration over anaerobic respiration?

An Introduction to Cell Structure and Host–Pathogen Relationships

Chapter 4

Why Is This Important?

This chapter provides fundamental information on cell structure required for success in studying microbiology in general and the processes of infection in particular.

Sydney was a 4-year-old girl who spent most weekdays in daycare. One evening, she had a fever and told her parent that her head hurt. By the next morning, she was comatose and was rushed to the hospital. The doctor immediately took a sample of cerebrospinal fluid (the normally sterile, colorless fluid around and inside the brain and spinal cord) and sent it to the laboratory for analysis. Several antibiotics were also administered, but it was too late. Tragically, Sydney died later that day.

The lab results showed she had contracted an infection by the Gram-negative diplococcus *Neisseria meningitidis*, also known as meningococcus. This meningitis-causing bacterium is spread through saliva, and could have been acquired by Sydney at her daycare facility through activities such as chewing on toys. It infects the host cell by sticking to it using specialized proteins. That is why different strains of the bacterium are dangerous to different age groups (for example children or young adults). Once infected, the host presents general symptoms such as fatigue, fever, headache, and neck stiffness, and the disease can quickly result in coma and death. Infections with this particular strain are most common in children under 5 years of age, and vaccines are available. In this case, she had not been vaccinated, and neither were many of the children in her daycare facility. Therefore, the other children and everyone who was exposed to Sydney in the previous week were given the antibiotic rifampicin to prevent them from getting the disease.

OVERVIEW

In the preceding two chapters, we have reviewed first the fundamental chemical principles applicable to microbiology and then the basics of metabolism in microorganisms. These topics are important to our understanding of many of the subjects we discuss as we explore microbiology and infectious disease. In this chapter we look at the basic characteristics of bacterial cells, the concepts of pathogenicity and virulence, and the relationship between a pathogen and its host. The last section of the chapter is devoted to a review of the eukaryotic host cell, which is the target for infectious disease. Although this chapter focuses on bacteria, it is important to remember that viruses, fungi, and parasites can also cause infections. Each of these groups of microorganisms is discussed in detail in subsequent chapters.

We will divide our discussions into the following topics:

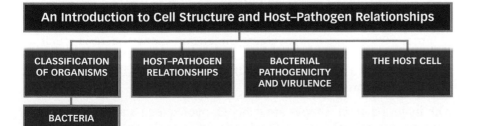

"The man who catches a meningococcus is in considerably less danger for his life, even without chemotherapy, than meningococci with the bad luck to catch a man."

Lewis Thomas (1913–1993)

The quote above, from *The Lives of a Cell*, poignantly describes the host–pathogen relationship: pathogens and their host cells are in a constant battle, and in many cases the host cell wins. In fact, some of the most successful pathogens are not **highly virulent** (immediately lethal), because if the initial host died quickly, it would prevent the spread of the pathogen to other hosts. In this chapter, the nature of host–pathogen relationships is discussed at the cellular level, and the defensive features of eukaryotic human cells are described.

CLASSIFICATION OF ORGANISMS

All living organisms can be classified as either prokaryotes or eukaryotes. **Prokaryotes** (Bacteria and Archaea) are less complex cells that lack a true nucleus and any other structures enclosed by a membrane. **Eukaryotes** are organisms made up of cells that do contain a membrane-enclosed nucleus as well as other membrane-enclosed structures outside the nucleus. All the membrane-enclosed structures are collectively referred to as **organelles**. We will look at the major differences between prokaryotic (bacterial) and eukaryotic cells later in this chapter.

Biologists classify organisms by their **genus** name and species name. For example, humans are referred to as *Homo sapiens* (genus *Homo*, species *sapiens*), the bacterium that causes **tetanus** is called *Clostridium tetani* (genus *Clostridium*, species *tetani*), and the bacterium that causes botulism is *Clostridium botulinum* (genus *Clostridium*, species *botulinum*).

A genus (plural genera) can include one species or multiple species. For example, the genus *Homo* now has only one species, *sapiens*, whereas the genus *Clostridium* has several species, including *tetani* and *botulinum*. The convention for writing the genus and species names of organisms is to put the entire name in italics with the genus name capitalized and the species name in lower case.

BACTERIA

Bacteria are the product of about 3 billion years of natural selection and have emerged as immensely diverse and very successful organisms that colonize all parts of the world and its inhabitants. Our initial discussion of bacteria begins with a look at their size and their shape, and how clusters of them form identifiable multicell arrangements. We then examine the procedures used to stain bacteria so that they can be seen microscopically.

Size, Shape, and Multicell Arrangement

Bacteria are the smallest living organisms. They are microscopic (hence the term microbiology) and we cannot see them with the naked eye. Those that either colonize or infect humans range from 0.1 to 10 μm (1 μm = 10^{-6} meter) in the largest dimension. Some bacteria, such as those belonging to the genera *Rickettsia*, *Chlamydia*, and *Mycoplasma*, are almost as small as some viruses, as the overlap of the bars for viruses and bacteria in **Figure 4.1** shows.

The major bacterial shapes are spheres, ovoids (egg-shaped rather than true sphere), straight rods, curved rods, and spirals. Spherical and ovoid

Fast Fact Bacteria that cause infections in humans are referred to as pathogens.

Fast Fact Some bacteria are referred to as pleomorphic, meaning that they can change their shape.

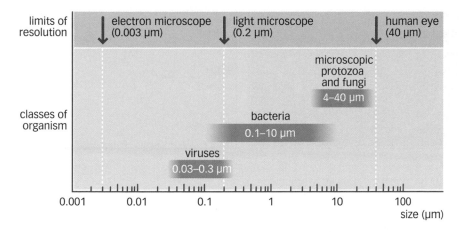

Figure 4.1 **The relative average sizes of microorganisms.**

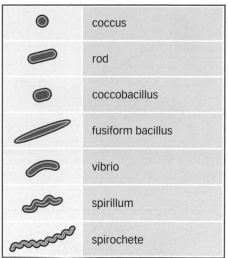

Figure 4.2 **Major bacterial shapes.**

bacteria are called cocci (singular **coccus**). Straight rods were previously called bacilli (singular **bacillus**). Very short rods, which can be mistaken for cocci, are referred to as coccobacilli. Rod-shaped bacteria that have tapered ends are called fusiform bacilli. Spiral bacteria are called spirilla (singular **spirillum**) if the cell is rigid, and **spirochetes** if it is flexible and undulating (**Figure 4.2**).

In addition to their distinctive shapes, bacteria can form distinctive multicell arrangements that are easily identified (**Figure 4.3**). One factor determining the shape of these multicell arrangements is the degree of stickiness of the organisms (which can vary depending on growth conditions). Cocci can form two-cell arrangements called **diplococci** (**Figure 4.4**). This arrangement is seen in bacteria such as *Streptococcus pneumoniae*, which causes respiratory infections, and *Neisseria gonorrhoeae*, which causes the sexually transmitted disease we call **gonorrhea**. In addition, cocci can form chains (**Figure 4.5**), as in *Streptococcus pyogenes*, and clusters (**Figure 4.6**), as occurs with staphylococcal organisms. These arrangements do not only apply to cocci. Rods can, for instance, also form chains, as in the case of *Bacillus anthracis*.

Although some bacteria are named for their distinctive shapes or multicell arrangements, these criteria cannot be used alone to identify particular bacteria. For an additional degree of morphological classification, we use differential staining.

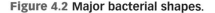

Figure 4.3 **Multicell arrangements seen with spherical bacteria (cocci).**

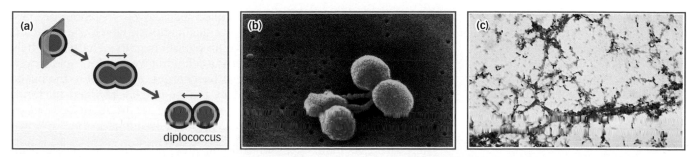

Figure 4.4 **Arrangement of diplococci.** Panel a: Cocci can arrange in pairs based on the plane of division. Panel b: A scanning electron micrograph of *Streptococcus pneumoniae*. Panel c: A photomicrograph of *Streptococcus pneumoniae* grown from a blood culture.

Figure 4.5 Arrangement of streptococci. Panel a: Streptococci can grow in chains. Panel b: A scanning electron micrograph of chains of streptococci. Panel c: A sputum smear showing the typical chaining arrangement.

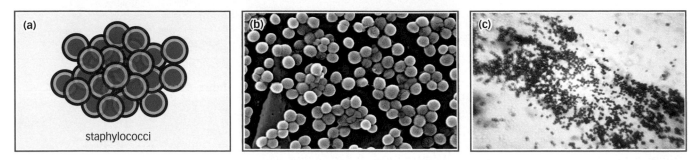

Figure 4.6 Arrangement of staphylococci. Panel a: Staphylococci arrange in grape-like clusters. Panel b: A scanning electron micrograph of *Staphylococcus aureus*. Panel c: *Staphylococcus aureus* in a sputum sample from a patient with staphylococcal pneumonia.

Keep in Mind

- All living organisms can be divided into prokaryotes and eukaryotes.

- Prokaryotes do not contain a nucleus or membrane-enclosed structures like those found in eukaryotes.

- Bacteria can be classified by genus and species, just like all other organisms.

- Further classification of bacteria can be based on size, shape, and arrangement.

Staining

In addition to being very small, microorganisms are essentially colorless under the light microscope.

We therefore use stains to see them under a microscope. Staining is caused by dye molecules that bind to the microorganisms and give them color so that they are visible. Microorganisms can be categorized according to which stains color them and which do not.

The most commonly used stains (called 'basic dyes' in laboratory jargon) contain positively charged molecules (cations), which are attracted to bacterial cells because they have an overall negative charge on their surface. These basic dyes can enter the cell and give it color. Acidic dyes are less common. Their anions (negative charges) give color to the background because they are repelled by the negatively charged bacterial cells. Procedures using basic dyes are known as positive staining; procedures using acidic dyes are known as negative staining.

There are two main types of staining procedure that use basic dyes (**Table 4.1**). **Simple stains** consist of only one dye, and these stains are used to identify the shape and multicell arrangement of bacteria. Methylene blue, carbolfuchsin, safranin, and crystal violet are some of the

Fast Fact If bacteria are grown for long enough on solid media for us to see colonies, we can see the color of the colonies, which are an indication of the color of the cells (based on components in or on the cell wall and membrane). *Staphylococcus aureus* is golden in color, and thus is called 'aureus' (based on the Latin word for gold). The scientific names of bacteria or fungi often relate to their shape or other features.

Table 4.1 **A comparison of staining procedures for bacteria.**

Example	Outcome of the staining	Use
Simple stain (single dye)		
Methylene blue	Cells are stained uniformly blue	For analyzing size, shape, and multicell arrangement
Safranin	Cells are stained uniformly red	
Crystal violet	Cells are stained uniformly purple	
Differential stain (two or more dyes)		
Gram	Gram-positive cells appear purple Gram-negative cells appear pink Gram-variable specimens show intermediate or mixed colors Gram-nonreactive specimens show poor staining or no stain at all	For distinguishing the main cell wall structure types
Ziehl–Neelsen acid-fast	Acid-fast organisms retain red of carbolfuchsin; non-acid-fast organisms are counterstained blue with methylene blue	For distinguishing members of genera *Mycobacterium* and *Nocardia* based on the presence of mycolic acid in the cell wall
Negative	Stains background of the sample, not the cells	Allows bacterial capsules to be seen
Flagella	Dyes or silver used to build up layers on flagella	Allows flagella to be seen
Endospore	Endospores retain malachite green color while cell counterstained with safranin is red	Used to identify endospores present in bacterial cells

most commonly used simple stains. **Differential stains** use two or more dyes to distinguish either between two or more organisms or between different parts of the same organism. The format of a typical differential stain is first the addition of the primary stain, followed by the decolorizing agent, and last the counterstain. The major differential stains are Gram (cell wall), negative (capsule), flagella, Ziehl–Neelsen acid-fast, and endospore stain.

Simple stains let us see only the shape, size, and arrangement of the organisms, but differential stains allow us to begin classifying organisms and also show us many of the structures associated with bacteria. Let's take a look at the individual types of differential stain one by one.

The Gram Stain

The **Gram stain**, developed by the Danish physician Hans Christian Gram in 1884, can be used to separate most bacteria into four major groups: Gram-positive, Gram-negative, Gram-variable, and Gram-nonreactive.

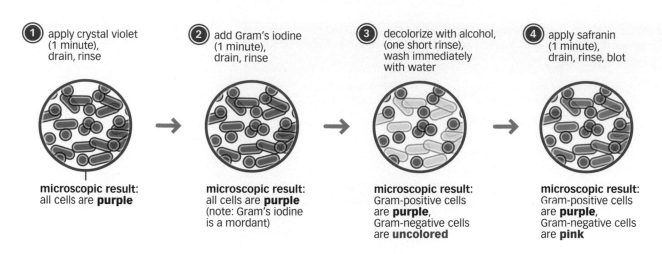

Figure 4.7 The steps involved in the Gram stain. Notice the change of color of the Gram-negative rods after the decolorizing and counterstaining steps.

The Gram stain takes advantage of the differences in the cell walls of these groups of bacteria, which we will discuss in detail in Chapter 9. Most bacteria are either Gram-positive or Gram-negative.

The Gram staining procedure (**Figure 4.7**) initially uses the dye crystal violet, which is taken up by all bacteria. The cells are then treated with iodine as a **mordant** (a substance that sets the color inside the cell and makes it permanent), which helps the Gram-positive cells retain the crystal violet because it cannot get out of the cell again owing to the structure of the Gram-positive cell wall. When alcohol is added in the next step, the Gram-negative cells lose their color but the Gram-positive cells retain the violet dye. Because the now-colorless Gram-negative cells would be invisible under a microscope at this point, they are counterstained with the red dye safranin. Any Gram-variable bacteria in the mix are recognizable by the way in which they stain unevenly. Gram-nonreactive bacteria do not stain and must be stained by a different method so as to be seen. Cell age is also a factor with old cells not staining properly. The cell wall becomes leaky in old cells, and Gram-positive cells lose the dye during the alcohol wash step.

The Negative (Capsule) Stain

The negative stain can be used to identify bacterial shapes, and in particular spirochetes. It can also be used to identify the presence of a **capsule**, which is a structure that surrounds certain bacterial cells. The capsule is important in bacteria that infect humans because it can facilitate adherence and undermines the host's defense mechanisms by inhibiting phagocytosis. It also limits the access of antiseptics, disinfectants, and even antibiotics, thereby protecting the infecting bacteria. In fact, some bacteria can infect humans only if the bacteria are surrounded by a capsule.

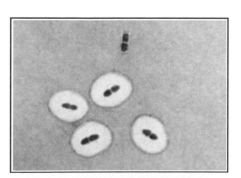

Figure 4.8 The negative stain. Staining the background makes the bacterial capsules visible. The bacteria shown here are encapsulated *Streptococcus pneumoniae*.

The negative staining procedure uses dyes such as nigrosin and India ink to color the background surrounding encapsulated bacteria in a sample being tested, making the capsule very visible. The stain colors only the background and not the capsule itself. It is the coloring of the background that allows us to see the capsule. A second dye can be added to color the bacterium inside each capsule (**Figure 4.8**).

The Flagella Stain

Some types of bacteria have flagella, a property that allows the bacteria to be motile (capable of moving). Motility is an important part of the infectious process because it allows the invading organisms to move from the initial site of the infection. Because bacterial flagella are too thin to be seen under the microscope, a **flagella stain** is used to coat the surface of the flagella either with layers of dye or with metals such as silver. The coating makes the flagella visible (Figure 4.9). This staining procedure is very difficult to carry out and is very time-consuming, so it is not routinely done.

The Ziehl–Neelsen Acid-Fast Stain

The Ziehl–Neelsen acid-fast staining procedure is used to detect *Mycobacterium tuberculosis*, the causative agent of tuberculosis, and *Mycobacterium leprae*, which causes leprosy. These bacteria have a cell wall that contains mycolic acid (a waxlike molecule) and lipids, and the presence of these substances in the cell wall makes the wall difficult to penetrate. Therefore, heat is used as part of the staining procedure to break down the mycolic acid and permit the entry of stain. Acid-fast bacteria stain poorly or not at all with the Gram stain.

The term acid-fast refers to whether or not a bacterium that is first stained and then washed with acid retains the staining dye. Dyed cells that do not lose their color when washed with acid are classified as acid-fast. An everyday use of fast in this sense is the "colorfast" label you sometimes see on clothing and linens. Products made from colorfast fabrics retain their color through numerous launderings (or so the manufacturers would have us believe).

The presence or absence of a cell wall containing mycolic acid is the underlying principle of Ziehl–Neelsen acid-fast staining because only this type of cell wall is acid-fast. A sample is treated first with heated carbolfuchsin, a red dye that can penetrate all cell walls, regardless of whether or not they contain mycolic acid. The next step is the addition of an acid–alcohol solution that removes the red color only from those cells in the sample in which the cell wall does not contain mycolic acid—in other words, only from cells that are not acid-fast. The acid washing does not disturb the red color of carbolfuchsin in any acid-fast cells. Because any cells in the sample that are not acid-fast are now colorless, the procedure is finished by counterstaining with methylene blue to give color to these cells (Figure 4.10).

The Endospore Stain

An **endospore** is a small, tough, dormant structure that forms in bacterial cells and several types of bacteria can undergo **sporulation** (the process in which endospores are formed). Endospore formation involves gathering a copy of the genetic information of the bacterium together with other important chemicals and enclosing the collection inside a tough coating. Bacteria that can undergo sporulation are particularly difficult to neutralize because the endospores are extremely resistant to antiseptics, disinfectants, radiation, and antibiotics.

The endospore stain is a differential stain in which a sample is colored by heating with malachite green for 5 minutes. The endospore walls are so thick and resistant that extensive heating is required to make them permeable to the stain. The sample is then washed thoroughly with water for 30 seconds, and in this washing the dye is removed from all of each cell except the endospore, which stays green. A final step of counterstaining

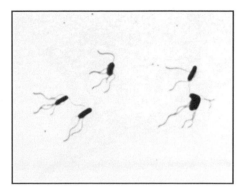

Figure 4.9 The flagella stain. A photomicrograph showing the flagella found on *Bordetella bronchiseptica*.

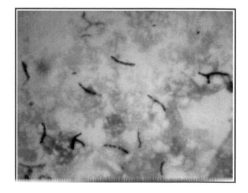

Figure 4.10 The Ziehl–Neelsen acid-fast stain. The vivid red shown here is typical of acid-fast bacteria, such as *Mycobacterium tuberculosis* and *Mycobacterium leprae*.

Figure 4.11 The endospore stain. This is a differential stain that shows green-dyed endospores inside red-dyed rod-shaped bacteria.

with safranin turns the non-spore part of each cell red (**Figure 4.11**). If the cells were stained with safranin only, the endospores would appear as colorless regions in the cell interior. This is also the case for Gram-stained cells.

In the above discussions we have seen that bacterial cells have distinct sizes, shapes, and multicell arrangements and that the various types of bacterium can be differentiated by staining. In later chapters we will focus on the structures associated with bacteria in great detail. Now, however, let's examine the relationship between bacteria and their host cells.

| Keep in Mind |

- Staining is used to make organisms visible under the microscope.

- There are two major types of stain: simple stains, which use one dye, and differential stains, which use more than one dye.

- The Gram stain is used to classify bacteria on the basis of their cell wall structure.

- Several stains, such as the capsule, flagella, and endospore stains, are used to identify structures associated with the bacterial cell.

- The acid-fast stain is used to confirm the identity of *Mycobacterium* species.

- Cell morphology and staining are often used for diagnostic testing.

HOST–PATHOGEN RELATIONSHIPS

Pathogens are organisms that can infect humans and cause illness in them. Before we move into our examination of pathogens and the infectious diseases they cause, it's important to look at the relationship that we as hosts have with pathogens. Infectious diseases have been the major cause of human death and suffering throughout history, up to and including the present day.

Even though *Homo sapiens* has been around since about 300,000 years ago, only over the past several centuries do we have detailed enough records for us to be able to appreciate the effect of disease on our evolution. Consider, for example, the bubonic plague of the fourteenth century (also referred to as the **Black Death**), which led to the death of one-third to one-half of the population of Europe at that time. It is interesting to speculate on the effect of this epidemic in terms of what may have been lost from the human gene pool. For example, people with blood type B were more susceptible and thus likely to die from plague. We know of many important individuals who survived the plague and went on to affect the course of history, but it is likely that just as many potentially influential individuals died, and we will never know what they might have gone on to do. The same can be said in general with regard to epidemics such as AIDS and also the newly emerging infectious diseases.

Fast Fact Poverty, crowding, unsanitary conditions, and malnutrition are leading contributors to increased susceptibility to infection and disease, while war, famine, and civil unrest also increase the level of disease.

Infectious diseases are complex and involve a series of shifting interactions between pathogen and host. These interactions can vary, depending on the pathogen in question; relevant factors include the pathogen's ability to either overcome or evade host defenses, to increase in number, and to establish the infection. In addition, the pathogen must find a way to transmit the infection to new hosts. At the same time, the host uses its defenses to control or eliminate the pathogen. It is essential to keep in mind that the ability of a pathogen to infect or cause disease depends in large part on the host's susceptibility to infection. This susceptibility is usually associated with a diminished host defense.

Pathogenicity: A Matter of Perspective

We have defined a pathogen as an organism capable of causing **disease**, which means damaging the host and thus causing symptoms. From the perspective of the pathogenic organism, however, being pathogenic is simply a strategy for survival. In fact, the human body is host to a myriad of microorganisms, both externally and internally: from mouth to anus and from head to toe. Every cell exposed to the outside world interacts with a variety of microorganisms. Most of these microorganisms are harmless and in some cases perform useful functions in our lives by providing protection from pathogens, producing vitamins, and helping us to digest food. These microorganisms are classified as **mutualistic**, meaning that they depend on us for their survival and we benefit from their presence. If given the opportunity, however, some mutualistic organisms become opportunistic pathogens, as we discuss in the next section (also see Chapter 1).

Opportunistic Pathogens and Primary Pathogens

Organisms that cause disease by taking advantage of a host's increased susceptibility to infection are called **opportunistic pathogens**. Under normal conditions, when the host is not especially susceptible, these organisms are not pathogenic. In contrast to opportunistic pathogens, **primary** (**obligate**) pathogens are those that can cause disease in individuals who are healthy. Primary pathogens include the viruses that cause diseases such as colds and **mumps** and the bacteria that cause diseases such as **typhoid fever**, gonorrhea, tuberculosis, and **syphilis**. Primary pathogens have evolved mechanisms that allow them to overcome the defenses of the host and, once inside a host, to multiply greatly. Some primary pathogens are restricted to humans, whereas others infect both humans and other animal hosts.

Disease and Transmissibility

Infection by any pathogen, whether opportunistic or primary, requires that the pathogen (1) be able to multiply in sufficient numbers to secure establishment in the host and (2) be transmissible to new hosts. An interesting point to note is that some signs of a disease are reflective of pathogens accomplishing the objective of transmission. For example, coughing promotes the transmission of diseases such as influenza and tuberculosis from the respiratory system, and diarrhea spreads pathogenic bacteria, viruses, and **protozoan** parasites from the digestive tract of an infected host.

From the pathogen's standpoint, we can look at the death of an infected host as an inadvertent, unfavorable outcome. In fact, pathogens that are well adapted to humans usually spare the majority of those infected. The pathogens benefit from the illness produced because it permits easier transmission from one host to another. However, if a large number of hosts die, spreading the infection is no longer possible. For example, Marburg and HIV are both deadly viruses, but HIV is transmitted far more effectively, and is far more successful, because it progresses slowly. This allows infected hosts to survive long enough to transmit the disease. In contrast, Marburg virus kills its hosts so rapidly that outbreaks usually die out fairly quickly. So in most cases there is a subclinical resolution to the infection in which the pathogen causes damage but no disease. In this case there can be continued transmission, which is critical to the pathogen's survival.

Keep in Mind

- Pathogens are organisms that cause disease in humans.

- Infectious disease is a complex process that involves both the pathogen and the host.

- Pathogens can be classified as opportunistic (causing disease if the host's defenses are compromised in some way) or primary/obligate (causing disease even if the host's defenses are intact).

BACTERIAL PATHOGENICITY AND VIRULENCE

We know that only a small fraction of all the bacteria that exist are associated with humans either as normal flora or pathogens. Therefore, we can ask the question: What makes a bacterium pathogenic? As we touched on in Chapter 1, pathogens must be able to accomplish the following:

- A potential pathogen must be able to adhere to, penetrate, and persist in the host cell (the get-in-and-stay-in rule).

- It must be able to avoid, evade, or compromise the host defense mechanisms.

- It must damage the host and permit the spread of the infection.

- It must be able to exit from one host and infect another host.

Virulence refers to just how fit a pathogen is when it comes to fighting the host, and thus how harmful a given pathogen is to a host. How virulent a given pathogen is depends on genetically encoded factors (virulence factors) of the pathogen. Most bacterial pathogens have to survive in two very different environments: the environment external to the host and the environment either on or in the host. Depending on which of these two environments a pathogen is in, some of its genes are active and others are not. For example, it would be a waste of resources for a bacterium to produce substances harmful to a host when it is not in or on the host. The genes for producing these substances are therefore turned off when the pathogen is in the external environment. The question is how the pathogen knows where it is, and which genes to turn on and which to turn off.

It is now well-established that pathogens often carry clusters of genes whose activity results in the production of factors that increase its virulence. Bacteria typically have only one chromosome, and these virulence genes are carried either on that chromosome or on mobile genetic elements such as **plasmids**. Plasmids can move from one bacterium to another and can cause previously harmless types to become pathogenic, and previously mildly virulent ones to become highly virulent. Certain clusters of virulence genes are called **pathogenicity islands** because they occupy distinct regions on the chromosome, and all genes on the islands are involved with pathogenicity. Some virulence genes are regulated by quorum sensing, an environment-sensing mechanism that we discuss next.

Quorum Sensing

During **quorum sensing**, specialized proteins in a pathogenic cell called sensing proteins relay information about the cell's environment (in this case, the concentration of small diffusible molecules, which is related to the cell density) to other proteins that regulate genes controlling the transcription of virulence genes. This environment-sensing mechanism is referred to as quorum sensing because it is based on cell population densities; that is, certain genes are expressed only when there is a sufficient population density such as an infectious dose. (A quorum is the minimum number of things that must be present for something to happen. For instance, a vote cannot take place in a committee meeting unless a quorum of members is present in the room.) As an example, *Salmonella* bacteria, which cause food poisoning, do not secrete the toxins responsible for the signs of food poisoning until there are sufficient numbers—a quorum—of *Salmonella* present in the host. It is as if the bacteria do not want to tip their hand before they are present in large enough numbers to make the host sick. The scientific explanation is that the bacteria have evolved to delay the production of toxins as a means of hiding from the host defenses, which would have no trouble in dealing with small numbers of pathogens.

Quorum sensing occurs when bacteria in a host secrete into the host circulation small diffusible molecules (*N*-acyl homoserine lactones in the case of many Gram-negative bacteria; peptides in the case of Gram-positive bacteria) that can be sensed by other bacteria to then start the communication chain via the sensing proteins. In this process, called auto-induction, these other bacteria then wait before starting the transcription of toxin genes until a sufficient increase in population has occurred.

There is a difference in the genetic involvement of quorum sensing between opportunistic pathogens and primary pathogens. For example, in *Pseudomonas aeruginosa*, which is an opportunistic pathogen, only 5–20% of its virulence genes are subject to quorum-sensing regulation. In a primary pathogen, such as *Staphylococcus aureus*, quorum sensing is used to control all of its virulence genes.

Biofilms

Whether in natural conditions such as soil or marine environments or on artificial surfaces such as the surfaces of prosthetic devices, catheters, or internal pacemakers, bacteria adhere and grow as aggregated assemblies of cells referred to as a **biofilm**. The regulation of biofilm formation is also subject to quorum sensing. A biofilm is important for health care because it can either impede or totally prevent the entry of antimicrobial agents and other molecules that are potentially toxic to bacteria. Moreover, biofilms are very difficult to remove and thus are a challenge when cleaning and maintaining health care equipment. At the same time, a biofilm is able to capture and retain nutrients, thereby allowing the bacterial population to increase in number. The adhesion of a biofilm can also affect the host's inflammatory response by attracting host defensive cells that attempt to engulf the biofilm. This results in what is referred to as frustrated phagocytosis, causing the formation of gigantic cells that eventually form a tough collagen capsule that inhibits new blood vessel development and interferes with wound healing. On the other hand, some biofilm formation actually facilitates wound healing. The timely and balanced development of microbial normal flora communities in wounds triggers some aspects of the host immune response in wound healing. However, an impaired normal microbial progression in the wound is also detrimental for healing.

The body usually reacts to the implantation of a medical device by first coating it with a protein film (we will see in Chapter 22 that the same coating process occurs on our teeth). This protein film is composed of

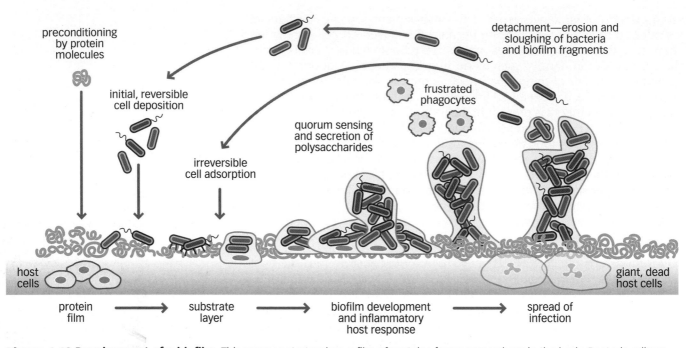

Figure 4.12 Development of a biofilm. This process starts when a film of proteins forms somewhere in the body. Bacteria adhere to these proteins and then aggregate with other bacteria to form the substrate layer. More bacteria (of different species) join and use quorum sensing to sense their environment and neighboring bacteria. When suitable conditions are sensed, polysaccharides are secreted and a mature, robust biofilm is formed. Phagocytes recognize the bacteria in the biofilm but cannot engulf them, forming giant host cells and resulting in cell death. Bacteria and fragments of biofilm can detach and spread the infection.

Fast Fact The production of medical devices and tissue-engineered materials is an industry generating $170 billion per year, with 5 million prosthetic devices such as heart valves, orthopedic implants, and catheters implanted each year in the United States alone. Unfortunately, many nosocomial infections are associated with implants and indwelling catheters.

fibronectin, fibrinogen, albumin, immunoglobulins, and other proteins that serve as binding sites and receptors for bacteria. Through a series of steps, bacteria begin to accumulate on this protein film (which is then referred to as a substrate layer) and form a biofilm (**Figure 4.12**). Several nonexclusive things can happen at the level of the biofilm:

1. The bacteria can accumulate on the device.

2. The bacteria can detach from the device surface and move into the liquid surrounding the biofilm.

3. The biofilm can form a tenacious gelatinous mass that is resistant to host defenses.

4. Large pieces of the biofilm can detach from the device surface and spread to other locations. In some cases, these pieces can also be large enough to cause fatal thromboembolisms (clots).

Keep in Mind

- Infection by a pathogen requires the pathogen to get in, stay in, defeat the host defense, damage the host, and be transmissible.

- Virulence refers to how fit a pathogen is to survive in the host and thus how harmful it is.

- There are virulence genes that are carried on the bacterial chromosome or on extrachromosomal pieces of DNA such as plasmids.

- Virulence gene expression can be regulated by quorum sensing.

- A biofilm can assist in the infectious process by inhibiting the exposure of bacteria to the host defense, antibiotics, and other molecules toxic to bacteria.

THE HOST CELL

In this section, we compare prokaryotic and eukaryotic cell structures. In addition, we examine the structures found in the eukaryotic cell in detail because the eukaryotic cell is the primary target for infectious diseases. Because many structures of the eukaryotic cell have important roles in the infectious process, we include a brief description of how some of these structures are involved in infection.

Prokaryotic Versus Eukaryotic Cell Structure

Bacteria and Archaea are prokaryotic cells (having no membrane-enclosed nucleus), whereas the cells of humans, other animals, and fungi are eukaryotic (having a membrane-enclosed nucleus). Table 4.2 lists the major differences between the two cell types. The eukaryotic cell is bigger and far more complex than the prokaryotic cell in almost

Table 4.2 A comparison of bacterial prokaryotic and eukaryotic cells.

Characteristic	Bacterial prokaryotic cells	Eukaryotic cells
Genetic structures		
Genetic material	Usually found in single circular chromosome	Usually found in paired chromosomes
Location of genetic material	Nuclear region (nucleoid)	Membrane-enclosed nucleus
Nucleolus	Absent	Present
Histones	Absent	Present
Extrachromosomal DNA	In plasmids	In mitochondria and plasmids
Intracellular structures		
Mitotic spindle	Absent	Present
Plasma membrane	Lacks sterols	Contains sterols
Internal membranes	Only in photosynthetic organisms	Numerous membrane-enclosed organelles
Endoplasmic reticulum	Absent	Present
Respiration (ATP)	At cell membrane	In mitochondria
Golgi	Absent	Present
Lysosomes	Absent	Present
Peroxisomes	Absent	Present
Ribosomes	70S	80S in cytoplasm and on endoplasmic reticulum, 70S in mitochondria
Cytoskeleton	Absent	Present
Extracellular structures		
Cell wall	Peptidoglycan, lipopolysaccharide, and teichoic acid	None in animal cells; chitin in fungal cells
External layer	Capsule or slime layer	None in most eukaryotic cells; pellicle or shell in certain parasites
Cilia	Absent	Present in certain cell types
Pili	Present	Absent
Flagella	Present	Present in certain cell types
Reproduction		
Cell division	Binary fission (primarily)	Mitosis or meiosis
Reproduction mode	Asexual	Sexual or asexual

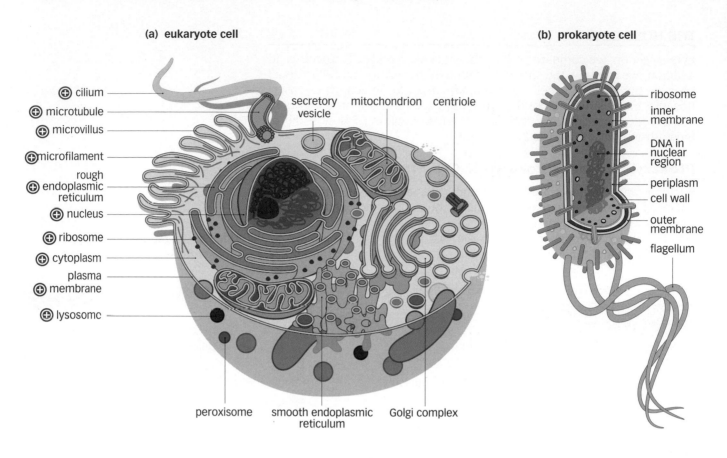

(a) eukaryote cell

cilium
microtubule
microvillus
microfilament
rough endoplasmic reticulum
nucleus
ribosome
cytoplasm
plasma membrane
lysosomc

secretory vesicle   mitochondrion   centriole

peroxisome   smooth endoplasmic reticulum   Golgi complex

(b) prokaryote cell

ribosome
inner membrane
DNA in nuclear region
periplasm
cell wall
outer membrane
flagellum

Figure 4.13 A eukaryotic host (panel a) cell is much more complex than the prokaryotic cell (panel b). The internal structures of the host cell that are involved in the infectious process are indicated by ⊕.

every way (**Figure 4.13**). During the following discussions, we will look at eukaryotic cell structures by using the human cell as the example. However, keep in mind that there are many other types of eukaryotic cell.

Plasma Membrane

The plasma membrane is the outer layer of the eukaryotic cell. It is structurally similar to the plasma membrane of bacteria but with significant differences in the amounts of lipid and certain other components. For example, the human cell plasma membrane contains cholesterol, and the fungal cell plasma membrane contains ergosterol, both of which confer support and strength. The bacterial plasma membrane does not contain cholesterol because in prokaryotes this membrane is surrounded, supported, and protected by a strong outer cell wall.

The eukaryotic plasma membrane is made up of a phospholipid bilayer (**Figure 4.14**). Recall from Chapter 2 that lipids are not soluble in water and for this reason they are said to be hydrophobic (water-fearing). Lipids therefore provide a perfect barrier between water on the outside of the cell and water on the inside of the cell. However, the cell must be able to interact with its environment. To accomplish this, the lipid is bound with phosphate ions, which are hydrophilic (water-loving). In the plasma membrane, lipid molecules with their bound phosphate groups align in two rows, with the lipid chains pointing toward the center of the membrane and the phosphate groups facing outward and forming the two surfaces of the membrane, as Figure 4.14 shows.

The plasma membrane contains a variety of other molecules (**Figure 4.15**). In particular there are proteins involved in communication and transport and also in structural roles, connecting to the cytoskeleton

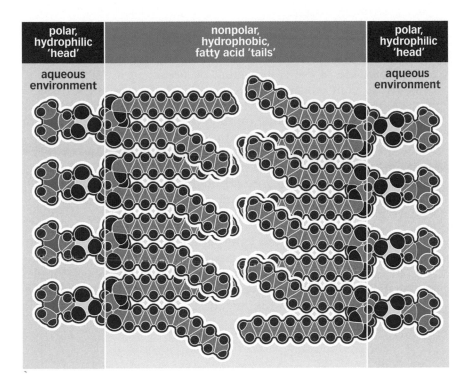

Figure 4.14 The plasma membrane of a eukaryotic cell. The structure is a phospholipid bilayer that has the polar heads of the bilayer facing the inside and outside of the cell. These polar heads are hydrophilic, a property that allows them to interact with the aqueous environments on both sides of the plasma membrane. The inner portion of the bilayer consists of the nonpolar, hydrophobic fatty acid tails that separate the aqueous environments inside and outside the cell.

of the eukaryotic cell. Some of these proteins act as receptors for signals sent by other cells; they also can serve as a site for virus attachment. These proteins are not stationary in the membrane but instead move freely through the lipid bilayer. (Because of this mobility, the plasma membrane represented in Figure 4.15 is referred to as the **fluid mosaic model of the membrane**.) It is important to remember that the phospholipid bilayer structure is also seen on membrane-enclosed structures inside the cell. This structural characteristic allows the plasma membrane to interact with these internal structures during certain cellular functions.

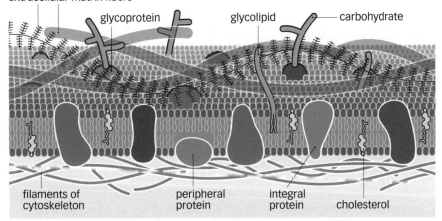

Figure 4.15 Components of the eukaryotic plasma membrane. The proteins and carbohydrate molecules are free-floating in the lipid bilayer. This model of the structure of the plasma membrane is called the fluid mosaic model.

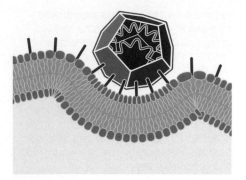

Figure 4.16 Diagram of a rhinovirus (which causes the common cold) binding to the plasma membrane of a host cell. Viruses infect host cells by binding to the receptors on the plasma membrane.

The Role of the Plasma Membrane in Infection

Because the plasma membrane is the barrier between the inside and the outside of the cell, it must be breached if microorganisms are to gain entrance. This is particularly true for viruses. Viruses are able to infect host cells that have a specific receptor for viral particles located on their plasma membrane. A viral particle attaches to these receptors on a host cell and thereby gains entry into the cell (Figure 4.16). A good example of this is the influenza virus, which binds to receptors called ICAM-1 receptors. These receptors are normally found on the surface of cells of the respiratory system, and therefore these cells are very susceptible to infection by the influenza virus.

In many viral infections, the viral particle acquires a part of the host's plasma membrane as the particle leaves the infected host cell. This piece of plasma membrane wraps around the virus and is referred to as the **viral envelope**. We will see that this envelope is important both in the infection of new host cells and in protection from the host defense.

Cytoplasm

The **cytoplasm** of the eukaryotic cell is all the volume of the cell that is inside the plasma membrane but outside the nucleus. It is made up of (1) a semifluid material consisting mainly of water plus a variety of dissolved substances, referred to as the cytosol, (2) membrane-enclosed structures called organelles, and (3) structures that are not enclosed by a membrane.

Cytoplasm is found in both prokaryotic cells and eukaryotic cells, but membrane-enclosed organelles are found only in the latter. Even though eukaryotic cells are much larger than prokaryotic cells, there is actually less cytosol in eukaryotic cells. This is because the cytosol of eukaryotic cells must share the extranuclear volume with many organelles and other structures.

The Role of the Cytoplasm in Infection

The cytoplasm is involved in a variety of infections, particularly viral infections. During a viral infection, the structures in the host cell are taken over and used by the virus. It is in the host cell's cytoplasm that the individual viral particles are manufactured and assembled.

To study the interior of the eukaryotic cell, we can separate the structures found in the cytoplasm into two groups: those not enclosed by a membrane and those enclosed by a membrane. Recall in what we discuss next that the structures named all float in the semifluid cytosol.

Cytoplasmic Structures Not Enclosed by a Membrane

In the eukaryotic cell, there are three major categories of cytoplasmic structure that are not enclosed by any membrane: cytoskeleton structures, cilia, and ribosomes.

Figure 4.17 Intercellular infection by *Shigella*. Panel a: Infection of the intestinal lining by the bacterium *Shigella*. After entering a cell that lines the intestinal lumen, *Shigella* uses the actin molecules of microfilaments in the host cell to move from one cell to the next. Panel b: A colorized photomicrograph of *Shigella*. Here *Shigella* organisms (stained red) are propelling themselves through the cytoplasm with their actin tails (stained green).

Cytoskeleton

The cytoskeleton of the cell is much like that of the skeleton of the body in that it gives the cell structural integrity. There are three cytoskeletal components: microfilaments, intermediate filaments, and microtubules. In addition to maintaining the shape of the cell, the filaments are also involved in determining how cells are joined together to form tissue.

Microfilaments are thin structures made up of molecules of the protein actin. They are solid structures and anchor the cytoskeleton to proteins in the plasma membrane (much like scaffolding). These small filaments give the cytoplasm a gel-like consistency.

Intermediate filaments are larger than microfilaments and are composed of a variety of different molecular subunits belonging to the family of proteins called keratins. These structures provide additional strength and stability to the cytoskeleton. They are also involved with positioning cells alongside one another in tissues.

Microtubules are hollow tubes made up of the protein tubulin. They are the largest of the cytoskeletal structures and are found in the cilia and flagella seen on some types of eukaryotic cells. Microtubules are involved in the movement of other structures that reside in the cytoplasm, in particular chromosomes during mitosis and meiosis.

The Role of the Cytoskeleton in Infection

Many pathogens take advantage of the cytoskeleton as part of the infectious process. For example, *Shigella* bacteria, which cause serious infections in the digestive system, use the cytoskeleton as part of the infectious process. These organisms destroy the lining of the intestinal tract when they infect the cells forming that lining. As the infection spreads, the *Shigella* bacteria move laterally from one cell to the next by protruding a finger-like structure made up of host-cell microfilaments (Figure 4.17). The movement of *Listeria monocytogenes* using host cell actin can be seen in Movie 4.1. The cytoskeleton also has an integral role in phagocytosis, which is one of the most important host defense mechanisms.

Cilia

Found only on eukaryotic cells, **cilia** (singular cilium) are made up of microtubules arrayed in an arrangement in which nine pairs of microtubules form a ring encircling two central microtubules (a 9 + 2 arrangement; Figure 4.18). The cilia on a cell project from the surface of the cell and are anchored to the plasma membrane. By moving in unison, cilia move liquids and secretions across the surface of the membrane.

There are many ciliated cells in the human body. A good example is the lower respiratory tract, which is lined with ciliated cells that work together with mucus-producing cells to move trapped particles upward and out of the respiratory tract.

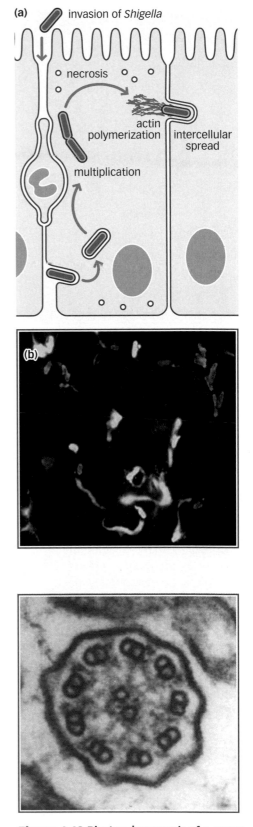

(a) invasion of *Shigella*

necrosis

actin polymerization

intercellular spread

multiplication

(b)

Figure 4.18 Photomicrograph of a cross section of a cilium associated with eukaryotic cells. Notice the 9 + 2 arrangement of the microtubules: a ring of nine groups surrounding two central ones. These cytoskeleton structures give cilia support and strength.

Figure 4.19 Scanning electron micrograph of ciliated cells of the respiratory tract being infected by *Bordetella pertussis* (colorized lime green), the bacterium that causes whooping cough.

Fast Fact The bacterium *Neisseria gonorrhoeae*, which causes gonorrhea, can sometimes attach to sperm cells and ride along from one partner to the other to spread this sexually transmitted disease. However, this is not the primary method of infection with other *Neisseria* pathogens.

The Role of Cilia in Infection

Cilia are involved both in the infection strategy of invading pathogens and in the host's defense against the invasion. Ciliated cells work with mucus-secreting cells to remove foreign materials, including microorganisms, from the respiratory tract, thereby preventing the staying-in requirement for successful infection. In some respiratory diseases, such as pertussis (**whooping cough**) or diphtheria, pathogens attach to host ciliated cells as an initial part of the infection (Figure 4.19).

Flagella

Flagella are responsible for cell motility. They are composed of the globular protein tubulin. In humans, the only cells that contain flagella are sperm cells. The flagellum is anchored in the plasma membrane of the cell and uses a complex sequence of reactions, almost like motorized turning, at the anchor point. This action moves the flagellum and permits the whip-like motion seen in the flagella found on eukaryotic cells.

We will see in Chapter 9 that flagella are commonly found in the microbial world and have a significant role in the infectious disease process by allowing pathogens to move from one location to another in a host body.

Ribosomes

The third category of cytoplasmic structures not enclosed by a membrane is the **ribosome**, found in both prokaryotic and eukaryotic cells and responsible for the production of proteins. Recall from your introductory biology course that ribosomes are found either floating free in the cytosol or, in eukaryotes, attached to the endoplasmic reticulum.

The ribosomes in eukaryotic cells differ from those found in prokaryotic organisms, although both types of ribosome are made up of protein and a specific form of RNA (called ribosomal RNA). We will discuss the ribosomes of bacteria in Chapter 9.

The Role of the Eukaryotic Ribosome in Infection

Eukaryotic ribosomes are actively involved in viral infections. They are part of the cellular mechanisms that are taken over by the virus. All of the protein components of new viral particles are made by the host cell's ribosomes. Although ribosomes are not directly involved in the bacterial infectious process, they do have a role in the treatment of bacterial infections because the bacterial ribosome is one of the targets attacked by certain antibiotics. For example, erythromycin and streptomycin target the subunit proteins of bacterial ribosomes, but not those of eukaryotic ribosomes. When this happens, the invading bacterium can no longer make protein and therefore dies. Because bacterial ribosomes are structurally different from those in human cells, this targeting is referred to as selective toxicity. This means that the antibiotic kills the bacteria but not the host cell. We will see in Chapter 19 that selective toxicity is an important consideration when developing and prescribing antibiotics.

Cytoplasmic Structures Enclosed by a Membrane

Now let's turn to those components of the cytoplasm that are enclosed by a membrane—the cell's **organelles**. It is important to keep in mind that the membrane enclosing any organelle is of the same type as the plasma membrane surrounding the whole cell—a phospholipid bilayer. This two-layer configuration makes it possible for organelles to fuse with one another and also with the plasma membrane.

Mitochondria

Mitochondria (singular mitochondrion) are the energy-producing elements of the eukaryotic cell. They are the organelles where most ATP is produced. There are large numbers of mitochondria in cells that are working and fewer in cells that are resting. A mitochondrion has two membranes enclosing it, and ATP is made on the folds (called cristae) of the inner membrane (**Figure 4.20**). Mitochondria also contain their own ribosomes and their own DNA, which replicates independently of the host cell.

Mitochondria have many characteristics that are similar to those seen in bacteria: they replicate independently just as bacteria do, the mitochondrial chromosome is single and circular, and the ribosomes in mitochondria are different from those in the rest of the eukaryotic cell but the same as those seen in bacteria. The RNA and DNA polymerase molecules found in mitochondria are also structurally similar to those seen in bacteria. The mechanism of ATP production on the inner membrane of mitochondria is similar to that seen in bacteria. All these similarities between mitochondria and bacteria have fostered the **endosymbiotic theory**, which describes a process whereby early in evolution, bacteria and eukaryotic organisms had a symbiotic relationship—a relationship in which two organisms lived as one unit. Over time, the bacteria were integrated into the eukaryotic cells as mitochondria.

Endoplasmic Reticulum and Golgi Apparatus

Both the **endoplasmic reticulum** (ER) and the **Golgi apparatus** are systems of membranes that form numerous flattened sacs and platelike structures in the cytoplasm of the eukaryotic cell (**Figure 4.21**). These structures are not found in prokaryotes. The ER is the site where various cellular components are synthesized. It is sometimes associated with ribosomes. If ribosomes are attached to the ER, it is referred to as rough endoplasmic reticulum. ER to which no ribosomes are attached is called smooth endoplasmic reticulum. Rough ER is where proteins are produced; smooth ER is the site where lipids and other nonprotein components are produced. The ER moves synthesized materials either to the Golgi apparatus, where additional finishing steps are completed, or directly to the plasma membrane for transport out of the cell.

The Golgi apparatus has three functions: (1) modifying and packaging products that come from the ER, (2) renewing the plasma membrane, and (3) producing lysosomes. Because both the ER and the Golgi apparatus are surrounded by the same type of phospholipid bilayer, the two organelles can interact with each other by fusing together. It is this mechanism that is used to move newly synthesized components from the ER to the Golgi apparatus.

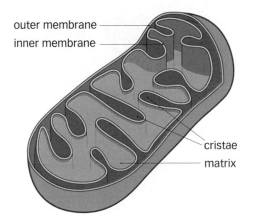

Figure 4.20 The mitochondrion has a double-membrane structure.

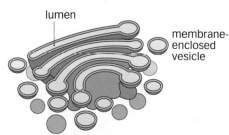

Figure 4.21 The Golgi apparatus is the cellular organelle in which molecules produced in the ER are completed. Notice the membrane-enclosed vesicles that have 'pinched' off from the flat membrane sacs. Lysosomes are produced in the Golgi apparatus.

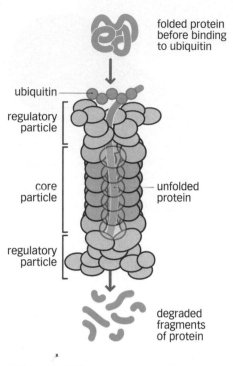

Figure 4.22 The proteasome is an organelle involved in protein degradation. When the proteins being degraded are from invading pathogens, the proteasome participates in the host's defense reactions.

The Role of the Endoplasmic Reticulum in Infection

In viral infections, the ER is the site of the biosynthesis and assembly of viruses. The ER is also associated with the adaptive immune response, a major host defense against infection.

Lysosomes

Lysosomes are the organelles responsible for destroying invading microorganisms and any other foreign materials that get inside the cell. They are produced by the Golgi apparatus and bud off as vesicles containing powerful enzymes that destroy the invaders. Lysosomes are also responsible for recycling any host-cell components that are no longer needed or are no longer functioning properly. The enzymes in lysosomes break down these components and get rid of anything that is not recyclable. Any inhibition of lysosomal function can be a lethal event for the eukaryotic cell.

The Role of the Lysosome in Infection

Lysosomes have a pivotal role in phagocytosis, which is part of the innate immune response. In this process, invading pathogens are enclosed by the host phagocytic cell's plasma membrane to form a vesicle. Once this membrane has fused with the lysosome membrane, the lysosomal enzymes destroy the pathogens.

Proteasomes

Proteasomes are organelles composed of ring structures stacked together (**Figure 4.22**). They function in the degradation of proteins. Proteins tagged with molecules of a small protein called **ubiquitin** are recognized and bound by the regulatory proteins of the proteasome and enter the core of the rings, where they are degraded into fragments. In many cases, these fragments are further degraded, and the useful components are reused by the cell.

The Role of the Proteasome in Infection

Proteasomes degrade proteins associated with pathogens and move them into the endoplasmic reticulum. Here the degraded proteins are combined with self-proteins (identifying the cell as belonging to the host) and then moved to the cell surface. The combination of the self-protein and the protein fragment from the pathogen serve to trigger the host's immune system to attack the pathogens.

Peroxisomes

Peroxisomes are the organelles responsible for the breakdown of fatty acids in the eukaryotic cell. Many of the by-products of this breakdown are poisonous to the cell, and one role of the peroxisomes is to get rid of these products. For example, hydrogen peroxide (H_2O_2), a substance extremely toxic to cells, is one by-product of fatty acid oxidation. Peroxisomes contain catalase, an enzyme that breaks H_2O_2 down to oxygen (O_2) and water (H_2O).

Nucleus

The nucleus is a structure unique to eukaryotic cells. It is not found in prokaryotic cells. It is an organelle and so is enclosed by a phospholipid membrane, in this case called the **nuclear membrane**, and contains a unique form of cytoplasm called **nucleoplasm**. There are one or more structures called **nucleoli** (singular nucleolus) in the nucleus, and it is in the nucleoli that ribosomal RNA is made.

The nucleus is where the DNA of the eukaryotic cell is stored. When the cell is not dividing to create new cells, the DNA is in the form of **chromatin**, which has a hairlike structure. When the cell is in the process of dividing, the chromatin condenses to form pairs of **chromosomes**.

The nuclear membrane is a double phospholipid bilayer (**Figure 4.23**). Pores in this membrane allow material to move into and out of the nucleus. The nucleus is the location of **transcription**, the process in which DNA is used as a template to produce RNA. Newly formed RNA uses the nuclear membrane pores to move from the nucleus to the cytoplasm.

The Role of the Nucleus in Infection

The nucleus of the host cell is important in many infections, particularly those caused by viruses that contain DNA. For this type of virus to infect a host, the viral DNA must enter the host nucleus. This involves a fascinating series of steps that move the viral DNA from the host cytoplasm through the nuclear pores into the nucleus. In some cases, the viral DNA that enters the nucleus becomes incorporated into the host's DNA, and the virus becomes **latent** in the host.

Endocytosis and Exocytosis

Before we finish looking at the eukaryotic host cell, it will be helpful to review the processes of **endocytosis**, the process by which a cell takes in materials, and **exocytosis**, the process by which a cell expels unneeded materials. Endocytosis can occur in three ways: pinocytosis, phagocytosis, and receptor-mediated endocytosis.

When a cell takes in material from the extracellular fluid via **pinocytosis** (**Figure 4.24a**), the plasma membrane of the cell will roll over anything

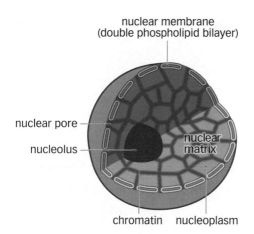

Figure 4.23 The nucleus of the eukaryotic cell, showing the double phospholipid bilayer and the nuclear pores. Notice the nucleolus, the region where ribosomal RNA is produced.

(a) pinocytosis

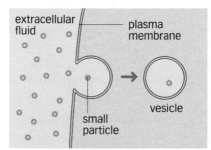

(b) phagocytosis

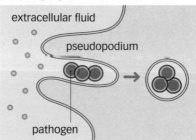

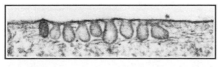

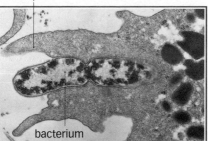

Figure 4.24 Pinocytosis and phagocytosis. Panel a: Pinocytosis involves the formation of a concave region in the membrane that flaps over the material to be taken in and creates a vesicle. Pinocytosis is seen when small molecules are brought into the cell. Panel b: Phagocytosis is the process by which large items are taken into the cell. This process involves the formation of pseudopodia that attach to and enclose the organism, as shown in the electron micrograph.

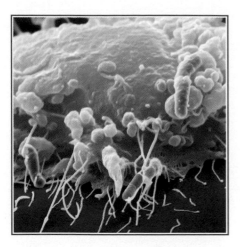

Figure 4.25 Phagocytosis. A scanning electron micrograph of phagocytosis of bacteria (colorized red).

that is in contact with it. This is seen with small molecules and results in the formation of a small vesicle that is like a bag surrounded by plasma membrane. This vesicle is then moved into the cytoplasm. **Phagocytosis (Figure 4.24b)** also involves the formation of a vesicle, but in this process the cell membrane is pushed outward to form pseudopodia (singular pseudopodium, false foot) as shown in Figure 4.24b and **Figure 4.25**. The pseudopodia surround the material to be brought into the cell, forming a membrane-enclosed vesicle. As noted earlier, phagocytosis is part of the host defense mechanism, and so materials imported into the cell by this process are usually pathogens that the cell must destroy.

The third form of endocytosis is **receptor-mediated endocytosis**. In this process, receptors on the surface of the cell bind with the extracellular material that is to be brought into the cell. Then the plasma membrane begins to sink into the cell interior, as in pinocytosis, and eventually a vesicle forms **(Figure 4.26)**.

Exocytosis is essentially the reverse of endocytosis. Vesicles are formed by organelles such as the endoplasmic reticulum and the Golgi apparatus. These vesicles then move to the plasma membrane and, because of their phospholipid bilayer structure, fuse with the plasma membrane and release their components to the extracellular fluid **(Figure 4.27)**.

Figure 4.26 An illustration of receptor-mediated endocytosis. Notice the formation of the vesicle and the fusion of the vesicle with the lysosome. This fusion of vesicle and lysosome is possible because of the phospholipid bilayer structure of the membrane.

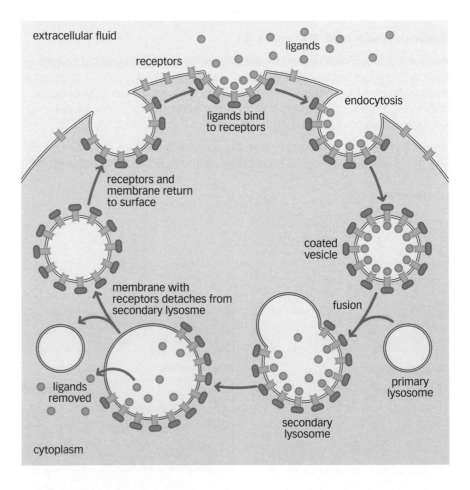

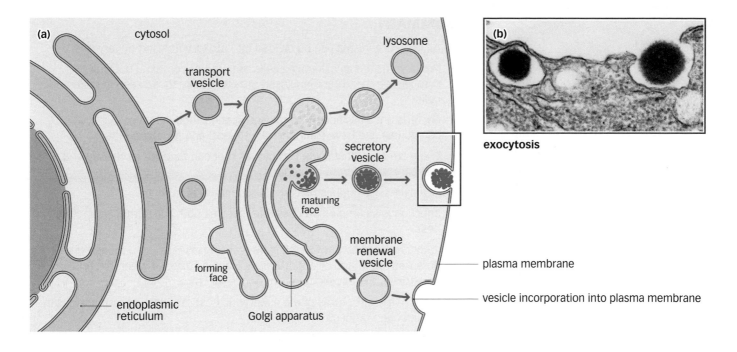

(a) cytosol

lysosome

transport vesicle

secretory vesicle

maturing face

membrane renewal vesicle

forming face

endoplasmic reticulum

Golgi apparatus

(b)

exocytosis

plasma membrane

vesicle incorporation into plasma membrane

The Role of Endocytosis and Exocytosis in Infection

For many pathogens, entry into a host cell is the ultimate goal because once the pathogens are inside, they are protected from the host's immune response. Some pathogens enter by binding to receptors on the cell surface and then entering through receptor-mediated endocytosis. This is particularly true of viruses.

For other pathogens, the host's defensive mechanism of phagocytosis is used for entry. As noted in our discussion of lysosomes, in phagocytosis the plasma membrane of a host cell encloses a pathogen, forming a vesicle that fuses with a lysosome in the cell. The lysosome is filled with enzymes that destroy the pathogen. Some pathogens, however, are resistant to the enzymes and survive, and even grow, inside the phagocytic vesicle. Others destroy the vesicle and move into the cytoplasm of the cell to continue the infection.

Viruses that have envelopes use the envelope, which is essentially plasma membrane from a previously infected host cell, to fuse with the plasma membrane of a new host cell. In this way viral particles are released into the cytoplasm of the new host, and viral reproduction begins.

Figure 4.27 Exocytosis. Panel a: An illustration of exocytosis, in which vesicles pinch off from the Golgi apparatus and move to the surface of the cell. Here they become incorporated into the cell membrane. Panel b: A transmission electron micrograph of exocytosis.

Keep in Mind

- The structure of the prokaryotic cell is distinctly different from that of the eukaryotic cell.

- Many of the structures of the eukaryotic cell have important roles in the infectious disease process.

SUMMARY

- All living organisms can be divided into prokaryotes and eukaryotes.
- Prokaryotes are less complex cells that do not contain a nucleus or cytoplasmic membrane-enclosed organelles like those seen in eukaryotic cells.
- Bacteria are classified by genus and species (and in higher taxonomic categories) and have distinct sizes, shapes, and arrangements.
- There are several staining techniques that can be used to classify and characterize different bacteria.
- Pathogens are organisms that cause disease in humans.
- Infection is a complex process that involves both the pathogen and the host.
- Pathogens can be primary/obligate (causing disease even though the host's defenses are intact) or opportunistic (causing disease when the host's defenses are diminished).
- Many of the structures of the eukaryotic host cell have important roles in the infection process.

In this chapter you have been introduced to bacterial sizes, shapes, and multi-cell arrangements. We also looked at the structures associated with the eukaryotic cell, which is the target of pathogenic microorganisms. Pathogen–host relationships were also discussed and these help us understand the events associated with infectious disease that are explored in the next two chapters. Keep in mind that in addition to bacteria, there are also viral pathogens, fungal pathogens, and parasitic pathogens that can infect humans, and each of these topics is discussed in detail in later chapters of this book.

(Q) SELF EVALUATION AND CHAPTER CONFIDENCE

Multiple Choice

Answers are given in the back of the book and help can be found in the student resources at:
www.garlandscience.com/micro2

1. Which of the following pairs is mismatched?
   A. Gram-negative bacteria—*Staphylococcus*
   B. Alcohol—decolorizer
   C. Acid–alcohol—decolorizer
   D. Iodine—mordant
   E. None of the above

2. Place the steps of the Gram stain in the correct order, using this code: 1 alcohol wash, 2 crystal violet stain, 3 safranin stain, 4 iodine.
   A. 4-3-2-1
   B. 1-2-3-4
   C. 1-3-2-4
   D. 2-1-4-3
   E. 2-4-1-3

3. Name the genus of *Neisseria meningitidis*
   A. *Neisseria*
   B. *Meningitidis*
   C. Bacjk
   D. ib-1/2
   E. None of the above

4. Which of the following stains allows you to see the capsule of an encapsulated bacterium under the microscope?
   A. Negative
   B. Ziehl–Neelsen acid-fast
   C. Endospore
   D. Gram

5. Virulence genes are carried on
   A. The chromosome
   B. The ribosomes
   C. The endoplasmic reticulum
   D. Plasmids
   E. Only A and D above

6. Virulence genes are often arranged into
   A. Reservoirs
   B. Pathogenicity islands
   C. Clusters
   D. Plasmids
   E. Individual chromosomes

7. Properties of quorum sensing include which of the following?

   A. Relaying information about the cell's environment to other proteins
   B. Regulation genes controlling the transcription of virulence genes
   C. Detection of pathogen population density
   D. Involvement of small diffusible molecules
   E. All of the above

8. Biofilms

   A. Are aggregations of many bacterial cells
   B. Often contain polysaccharide matrices
   C. Protect bacteria
   D. All of the above

9. Organisms that can cause disease even in healthy individuals are

   A. Opportunistic pathogens
   B. Variable pathogens
   C. Directed pathogens
   D. Primary pathogens
   E. None of the above

10. A pathogen must be able to

    A. Adhere to, penetrate, and persist in the host cell
    B. Avoid, evade, or compromise the host defense mechanisms
    C. Damage the host and permit the spread of the infection
    D. Accomplish all of the above

11. Which of the following is false regarding the Gram stain and the acid-fast stain?

    A. In one the primary stain is crystal violet; in the other the stain is carbolfuchsin
    B. Acid-fast-positive bacteria stain poorly or not at all with the Gram stain
    C. In the Gram stain a negative result is red, whereas in the acid-fast stain a positive result is red
    D. Mycobacteria stain negative in the acid-fast stain, but positive in the Gram stain
    E. One uses an alcohol wash, the other an acid–alcohol wash

In questions **12–15**, starting from the site of entry of a protein into a proteasome, indicate the order of the proteasome parts it will encounter using the following choices:

12. First part encountered	**A.** First regulatory particle
13. Second part encountered	**B.** Second regulatory particle
14. Third part encountered	**C.** Core particle
15. Fourth part encountered	**D.** Ubiquitin

 DEPTH OF UNDERSTANDING

Questions listed here require you to bring together the concepts you have learned in this chapter into a discussion format. This helps you to increase your depth of understanding of the material you have learned. Help can be found in the student resources at: **www.garlandscience.com/micro2**

1. Describe the host cell structures that are involved in the infection process.

2. Describe the relationship between quorum sensing and infectious disease.

Part I gave us a basic understanding of the infectious process and a good foundation in chemistry, which allows us to understand concepts on a molecular level. In **Part II**, we look at the disease process, which is a fundamental topic for health care professionals who study microbiology.

Chapter 5 gives you a basic understanding of the disease process, including detailed information on four of the five requirements for a successful infection that were introduced in Part I (getting in, staying in, defeating the host defense, and damaging the host).

In **Chapter 6** we discuss the fifth requirement, transmissibility. Equally importantly, this chapter also describes the compromised host, which is a major target for infection. Throughout these discussions the concepts you learned in Chapter 5 will help in your understanding. In Chapter 6 we also take a look at epidemiology and its importance in how we view infectious disease. Because this also requires detecting infections, we discuss traditional and modern diagnostics. Understanding core principles and methods in biotechnology enables future health care workers to operate at the forefront of the profession.

Chapter 7 provides you with discussions of the etiology or cause of disease as well as how diseases develop. This includes topics such as communicable and contagious diseases and persistent infections. The chapter concludes by discussing the scope of infections, including topics such as toxic shock.

Chapter 8 gives you a look at emerging infectious diseases that are now becoming major problems for public health. In addition, this chapter talks about re-emerging diseases, which are those that were once controlled but are now becoming a threat to public health again. We also discuss the potential intentional use of pathogens as bioweapons.

When you have finished **Part II** you will:

- Understand the tactics used by microbial pathogens to get into the host, remain in the host, defeat the defenses of the host, and damage the host.

- See how infection can be spread from one person to another. Be aware of how individuals can be immunocompromised and at greater risk for infection.

- Have an understanding of the link between diagnosis and epidemiology and how it can help us understand, cure, and prevent infection better.

- Understand the principles of disease including the etiology and classifications used to describe diseases.

- Understand the principles of traditional and modern diagnostic methods, and the underlying core concepts of biotechnology.

- Have a better understanding of the importance of emerging infectious diseases and also how some diseases that were controlled are now re-emerging as public health problems, including potential bioweapons.

Requirements for Infection   Chapter 5

Why Is This Important?

This chapter introduces you to the mechanisms involved in the infectious disease process. What you learn here will be the foundation for the rest of your studies in medical microbiology.

Nicky is eating breakfast at the kitchen table, overhearing her mother having a chat with one of her friends over the phone. Her mother happily describes the dress she bought the other day. She is obviously looking forward to wearing it for a big school reunion. After hanging up she joins Nicky, and mentions how curious she is about seeing everyone after so many years. She wonders what her class mates look like now, and whether she will recognize them all. She sighs and says that she wishes there were a miracle to look young for much longer. Nicky cannot believe her mother does not know! Just last week Nicky had visited her classmate Penny's house to do homework together. Penny's mother had returned from a youth treatment in the afternoon. Nicky did not understand how it works and cannot remember all details, but she tells her mother about some injections in the face, which makes people look young. She now recalls the name: Botox®!

Her mother nods and laughs in amusement, also smiling even more to herself, when it strikes her that those smiles both light up her face and cause the lines and wrinkles that show her age. She then explains that this 'youth serum' is derived from the most potent poison on the planet. Botox® is the commercial brand of botulinum toxin, a neurotoxin produced by the bacterium *Clostridium botulinum*. In 2002 its use for cosmetic reduction of forehead frown lines was approved.

The toxin causes a blockage of the release of acetylcholine (a neurotransmitter). Nicky hears from her mother that this means that the muscles depending on the function of nerves are essentially paralyzed. The muscles cannot contract any longer, which over time reduces lines in the skin. The body compensates within 2–3 months for the localized poisoning by forming new synapses. People then have a new dose injected. Nicky finds this rather gross, which is very obvious in her facial expression. And she rather does not want to think about the needles! Her mother smiles again. She would never volunteer to impair such essential nonverbal communication. Not even for "eternal youth"!

OVERVIEW

This is a very important chapter because it is here that we begin to look at the fundamentals of the infectious disease process. Armed with the concepts from previous chapters, we are well prepared for these discussions about the specifics of what is involved. We will then build on this information as we move through future chapters. In this chapter, we examine four of the five requirements necessary for a successful infection, namely portals of entry (getting in), establishment (staying in), defeating the host defenses, and damaging the host.

We will divide our discussions into the following topics:

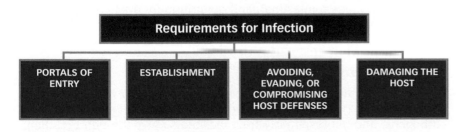

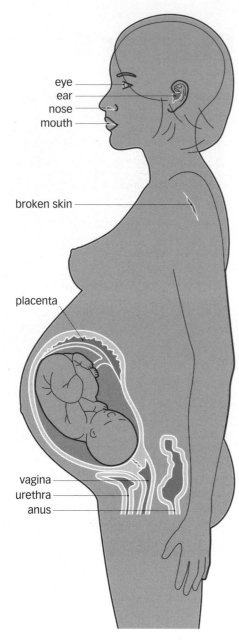

eye
ear
nose
mouth

broken skin

placenta

vagina
urethra
anus

Figure 5.1 Portals of entry.

As we saw in Part I of this book, there are five fundamental requirements for a successful infection. Pathogens must be able to do the following:

1. Enter the host (get in).

2. Have the ability to remain in a location while the infection gets established. We call this process establishment (staying in), and this can include growth (increasing the number of pathogenic organisms).

3. Avoid, evade, or compromise the host defenses (defeat the host defenses).

4. Damage thc host.

5. Exit from the host and survive long enough to be transmitted to another host (transmissibility, which is covered in Chapter 6).

For pathogens to accomplish this, they use their **virulence factors**. These are the characteristics that the pathogens possess that allow them to thrive and survive in the host's environment. In other words, how virulent a pathogen is depends largely on its being equipped with virulence factors. Their mode of action is related to many of the symptoms in the host. As we discuss the requirements for a successful infection in this chapter, you will see the important role of these virulence factors in the process. Those factors are not all continuously used by the microbe (it often depends on the stage of the infection or disease): they can be lost or obtained.

PORTALS OF ENTRY (GETTING IN)

The entry of pathogens is a primary requirement for infection and it is relatively easy in humans and other animals, because so much of the body is open to the outside world (**Figure 5.1**). Any point at which organisms can enter the body is called a **portal of entry**, and we can divide these portals into three categories: mucous membranes, skin, and parenteral routes (**Table 5.1**). The skin and mucous membranes are in direct contact with the exterior environment and are therefore in close proximity to potential pathogens. In contrast, pathogens that enter the body via the **parenteral route** take advantage of breaks in the body's barriers to gain access.

Mucous Membranes

Recall from your studies of anatomy that mucous membranes are located in areas of the body that are adjacent to the outside world. These membranes are found in the respiratory tract, the gastrointestinal tract, and the genitourinary tract.

You can think of the enormous surface area of the mucous membranes as a border, analogous to the border between two countries. Fortunately, as part of the immune system the body has a variety of powerful border defenses that prevent entry. Consequently, even though the surfaces of the respiratory, gastrointestinal, and genitourinary tracts are in contact with potential pathogens, the surfaces have means of protecting the body against the entry of microorganisms.

The Respiratory Tract

Of all of the portals of entry, this is probably the most favorable to pathogens (**Figure 5.2**). We live in a cloud of potentially dangerous microbial pathogens, and the respiratory tract facilitates entry through breathing.

Portal of entry	Pathogen	Disease
Respiratory-tract mucous membranes	*Streptococcus* species	Pneumonia
	Mycobacterium tuberculosis	Tuberculosis
	Bordetella pertussis	Whooping cough
	Influenza virus	Influenza
	Measles virus	Measles (rubeola)
	Rubella virus	German measles (rubella)
	Varicella-zoster virus	Chickenpox
Gastrointestinal-tract mucous membranes	*Shigella* species	Shigellosis (bacillary dysentery)
	Escherichia coli	Enterohemorrhagic disease
	Vibrio cholerae	Cholera
	Salmonella enterica	Salmonellosis
	Salmonella typhi	Typhoid fever
	Hepatitis A virus	Hepatitis A
	Mumps virus	Mumps
Genitourinary-tract mucous membranes	*Neisseria gonorrhoeae*	Gonorrhea
	Treponema pallidum	Syphilis
	Chlamydia trachomatis	Nongonococcal urethritis
	Herpes simplex virus	Herpes
	Human immunodeficiency virus	Acquired immune deficiency syndrome
Skin or parenteral route	*Clostridium perfringens*	Gas gangrene
	Clostridium tetani	Tetanus
	Rickettsia rickettsii	Rocky Mountain spotted fever
	Hepatitis B and C virus	Hepatitis
	Rabies virus	Rabies
	Plasmodium species	Malaria

Table 5.1 Portals of entry for some common pathogens.

We cannot just stop breathing if we are in a dangerous area (for example on a train with many people coughing around us). It is much easier to just not eat certain food if we are not sure it is safe to eat. Organisms that can go on to cause many diseases (such as colds, pneumonia, tuberculosis, influenza, **measles**, and even **smallpox**) can be found on droplets of moisture in the air and even on dust particles, and use this portal of entry. As we will see later in this chapter, the respiratory tract is also a very productive **portal of exit** that can be used to transmit pathogens through coughing or sneezing. A portal of exit is any point at which microorganisms can leave the body.

The Gastrointestinal Tract

This system is also open to the outside world, and organisms can enter the body via the foods and liquids we eat and drink. The gastrointestinal

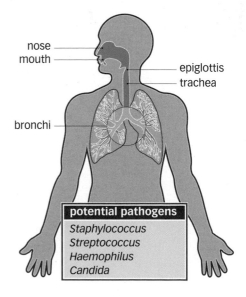

Figure 5.2 Mucous membranes of the upper respiratory tract are portals of entry to the body for potential pathogens. The upper respiratory tract is the body's most accessible portal of entry because organisms are brought in through the breathing process.

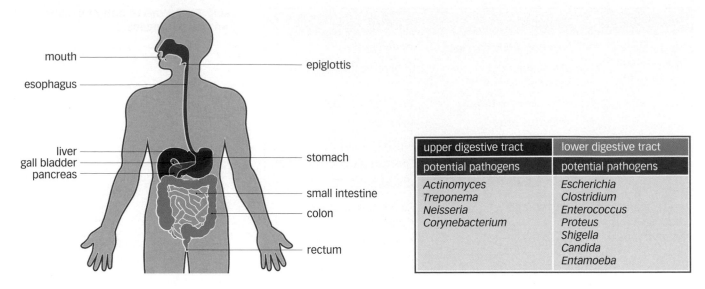

upper digestive tract	lower digestive tract
potential pathogens	potential pathogens
Actinomyces *Treponema* *Neisseria* *Corynebacterium*	*Escherichia* *Clostridium* *Enterococcus* *Proteus* *Shigella* *Candida* *Entamoeba*

Figure 5.3 Mucous membranes of the gastrointestinal tract are portals of entry to the body for potential pathogens. Microorganisms enter the gastrointestinal system with food or water. The gastrointestinal system is largely an inhospitable environment. It is the preferred portal of entry for some pathogens such as *Salmonella*.

tract has many protective barriers against pathogens, the most obvious of which is the production of stomach acid and bile. These substances are required for normal digestion but produce hostile environments that limit the survival of most pathogens. Still, there are many organisms that not only use this portal of entry but actually prefer or require it. For example, the polio virus uses the gastrointestinal tract as a required part of its infectious cycle. In addition, the tract is a preferred entry point for hepatitis A virus, the parasite *Giardia*, the bacterium *Vibrio cholerae*, and bacteria that cause **dysentery** and typhoid fever (**Figure 5.3**).

A very interesting pathogen to use this portal of entry is *Helicobacter pylori*. This organism is carried in the gastric mucosa of one out of every two people in the world, and infection with this organism is a known risk factor for the development of gastroduodenal **ulcers** and associated with cancers of the stomach. For many years, it was thought that the acidity of the stomach (about pH 1.0) would preclude bacterial survival but it turns out that *Helicobacter* can survive by making its environment (microhabitat) more hospitable. One of its virulence factors is the enzyme urease, which hydrolyzes urea into carbon dioxide and ammonia, protecting *H. pylori* during its journey in the stomach by neutralizing the acidic pH around itself. The organism eventually makes its way to the mucus that lines the wall of the stomach and duodenum of the small intestine. Nestled in this mucus, it is protected and can begin the process of infection that results in the destruction of tissue and the formation of an ulcer. Eventually even cancer can develop, depending on other factors such as diet.

The gastrointestinal tract is also a leading portal of exit for pathogens in feces. In fact, we will see throughout our discussions of infectious disease that the **fecal–oral route of contamination** has a major role in many infections, especially with Gram-negative bacteria, viruses, protozoa, and other parasites.

The Genitourinary Tract

The urinary and reproductive tracts are also open to the outside world, but unlike the respiratory and gastrointestinal tracts, they are more complicated with respect to entry.

Urinary tract infections are more common in women than in men because of the anatomical relationship between the anus and urethra,

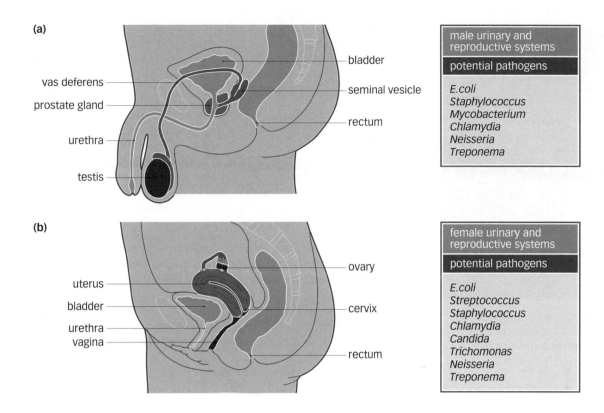

(a)

vas deferens
prostate gland
urethra
testis

bladder
seminal vesicle
rectum

male urinary and reproductive systems
potential pathogens
E.coli   *Staphylococcus*   *Mycobacterium*   *Chlamydia*   *Neisseria*   *Treponema*

(b)

uterus
bladder
urethra
vagina

ovary
cervix
rectum

female urinary and reproductive systems
potential pathogens
E.coli   *Streptococcus*   *Staphylococcus*   *Chlamydia*   *Candida*   *Trichomonas*   *Neisseria*   *Treponema*

which is much closer in women than in men (**Figure 5.4**). Because fecal material contains bacteria, it is easy for these organisms to find their way to the urinary tract. As the urethra in women is shorter, it is also easier for pathogens to ascend to the bladder. One bacterium that can cause urinary tract infections is *Proteus mirabilis*. It also has urease as one virulence factor, which ultimately causes its environment to become more alkaline. This presents more advantageous conditions for *Proteus* to grow than for competing organisms in that habitat.

Diseases of the reproductive tract are usually sexually transmitted, and either abrasions or tiny tears in the tissues that routinely occur during sexual activity serve as additional portals. Once the mucous membrane barrier is broken, pathogens gain entry. Conditions such as syphilis, gonorrhea, chlamydia, herpes, genital warts, and HIV infections are caused by pathogens that can use this portal of entry. The genitourinary tract can also be used as a portal of exit through which these infections can be transmitted.

Skin

The skin is the largest organ in the body, and like the mucous membranes it has a vast surface area through which microorganisms may enter the body. However, unlike the case with the mucous membranes, the association between the skin and microorganisms does not depend on their being taken in through breathing, eating, or sexual activity. The microorganisms are already there. In fact, the skin is literally covered with many types of microorganism (**Figure 5.5**) and is easily accessible to many other types, including pathogens. Fortunately, the skin is impenetrable to most microorganisms. In fact, many bacteria, fungi, and some parasites that live on the skin are completely harmless to the host. To initiate an infection, these organisms must find an opening, such as a hair follicle, a sweat gland, or a break in the skin, so as to gain entry.

Figure 5.4 Genitourinary portals of entry to the body for potential pathogens. Panel a: The male urinary and reproductive tract. Panel b: The female reproductive and urinary tract. In both sexes, both tracts are portals of entry for pathogens. Urinary tract infections occur more frequently in females than in males because of the length of the urethra and the anatomical relationship between the anus and the urethra.

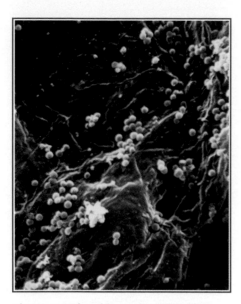

Figure 5.5 The skin and conjunctiva of the eye are in constant contact with microorganisms but present an impenetrable barrier to entry. These barriers must be compromised in some way if an organism is to enter. This scanning electron micrograph shows microorganisms that inhabit the skin.

We will see that these potential entry points are very well guarded. Some pathogens are able to open doors to the body, for example by disintegrating cells or tissue.

The Parenteral Route

Movement of organisms past the barrier of the skin requires a break in the barrier, and the portal of entry referred to as the parenteral route depends on such breaks. Things such as injections, which are routinely used in clinical applications, can easily become parenteral portals of entry any time that microorganisms are present in close proximity to the site of injection. Entry can also occur via insect bites (referred to as **vector transmission**), and many organisms, such as *Plasmodium* (the causative agent of malaria), use this as a way to enter the host.

Obviously, any cut or wound is going to allow the entry of skin-dwelling organisms, but the extent of the trauma can also have a role in the severity of the subsequent infection. Recall that the skin is made up of two basic layers, the epidermis and the dermis (**Figure 5.6**). Because the epidermis is made up of dead and dying cells, there is no access of blood to this layer. Therefore, cuts or wounds limited to this layer are less likely to spread beyond the site of entry. In contrast, the dermis is associated with blood vessels, and cuts or wounds that involve this layer or go deeper are far more likely to cause more serious systemic infections. This is even more apparent when we look at surgical procedures, in which contaminating organisms can gain access deep into internal tissues. Equipment used in general and intensive health care can also ease the access of pathogens to the body: **cannulas** used for injections provide a parenteral portal, catheters allow entry into the urinary tract, and intubation tubes give access to the respiratory tract. They are not only a convenient portal for pathogens: normal flora can colonize the surface of the equipment and eventually cause disease.

Some organisms have preferred portals of entry, and only entry through the preferred portals will result in infection, because the pathogens find a suitable environment necessary for growth. For example, *Salmonella enterica* serovar Typhi must be swallowed if it is to cause intestinal infections, whereas *Streptococcus pneumoniae* must be inhaled to cause

Figure 5.6
A diagrammatic representation of a cross section of the skin. Pathogens that gain access to the epidermis usually result in localized infections, whereas those that enter the dermis can cause systemic infection when entering the blood vessels in this layer of the skin.

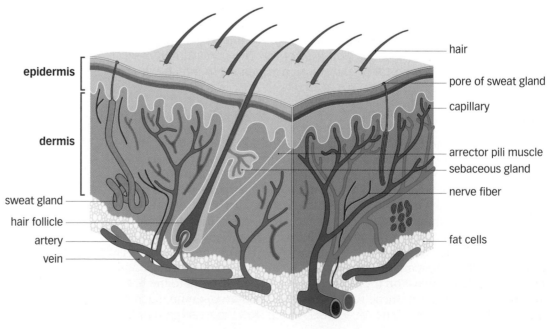

epidermis

dermis

sweat gland
hair follicle
artery
vein

hair
pore of sweat gland
capillary
arrector pili muscle
sebaceous gland
nerve fiber
fat cells

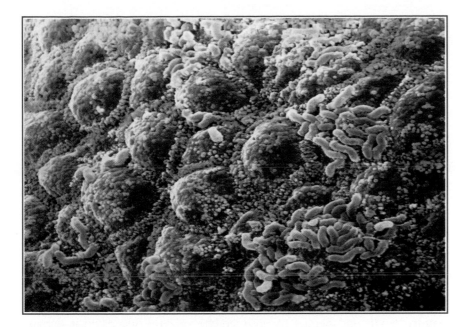

Figure 5.7 **A false-color scanning electron micrograph of the surface of the colon mucosa with clusters of bacteria (possibly** *E. coli***) attached.**

pneumonia. However, many organisms can cause infection no matter what entry point is used. *Yersinia pestis*, the organism that causes bubonic plague, uses multiple entry routes (for example parenteral and respiratory). In the Middle Ages it wiped out one-third to one-half of the population of Europe.

ESTABLISHMENT (STAYING IN)

Gaining entry into the host is just the beginning of the challenges that most pathogens face. After entry, a pathogen must find a way to stay in the host if it is to establish the focus of the infection. This task is very difficult for a variety of reasons. For instance, there can be physical obstacles to overcome. Let's use *Neisseria gonorrhoeae* as an example. If, after gaining access to the genitourinary tract, this microorganism does not have a way of adhering to the tissue, it might be flushed back out during urination. On top of this is the vast array of defenses that the body has in place to destroy the organism.

In Chapter 9 we will look at the anatomical structures of bacteria and see how these structures can have a role in clinical pathogenesis. During these discussions, we will see that organisms such as *Streptococcus pneumoniae* are not infectious without a surrounding capsule that allows them to adhere to the body's tissue and inhibit the host defense. In point of fact, almost all pathogens have some means of attachment. Some Gram-negative organisms—for instance, *Escherichia coli*—use structures called **fimbriae** to attach to certain receptors on cells of the small intestine, colon, and bladder (**Figures 5.7, 5.8,** and **5.9**).

In many cases, pathogens use molecules called **adhesins** (surface proteins) as a means of adhering to tissue (**Table 5.2**). For example, *N. gonorrhoeae* can use fimbriae coated with adhesins to adhere to tissue in the genitourinary tract. Pathogens can also take advantage of sticky glycoproteins found on the surface of host cells.

Let's look at a good example of adherence. The organism *Streptococcus mutans* has for a long time been accused of causing tooth decay. In fact, this is not strictly true because tooth decay actually starts with fluids produced by your oral tissues. These fluids form a dental **pellicle**, which

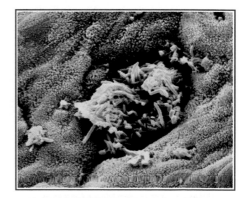

Figure 5.8 **Rod-shaped bacteria, probably** *E. coli* **(colored green in a scanning electron micrograph) filling a gland opening in the wall of the colon.**

Figure 5.9 A colorized scanning electron micrograph of cells of the human bladder infected with bacteria. Rod-shaped *E. coli* (colored yellow) are seen attached to the epithelial cells of the bladder (colored blue). Note the purple-colored epithelial cells, which have swelled and developed a rough surface as a result of this chronic bladder infection.

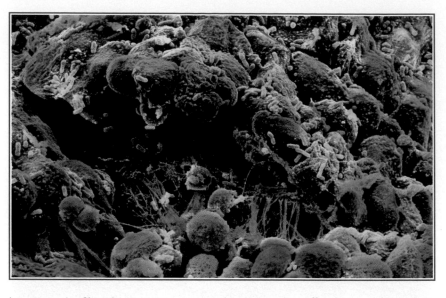

is a protein film that coats your teeth. *S. mutans* adheres to this pellicle (**Figure 5.10**) and begins to produce the enzyme glycosyltransferase, which acts on the sugar we eat (sucrose), resulting in a habitat that allows other organisms to adhere to *S. mutans*, forming the plaque, a biofilm (see Chapter 4), which is essentially a living coating on the teeth. This combination of organisms and the enzymes they produce causes the

Table 5.2 Adherence factors associated with infection.

Location	Bacterium	Disease	Examples of mechanisms of adherence
Upper respiratory tract	*Mycoplasma pneumoniae*	Atypical pneumonia	Cell surface adhesion molecules bind to receptors on cells lining respiratory tract
	Streptococcus pneumoniae	Pneumonia	Adhesion molecules attach to carbohydrates on respiratory cells
	Neisseria meningitidis	Meningitis	Adhesion molecules on bacterial cell attach to respiratory cells
Genitourinary tract	*Treponema pallidum*	Syphilis	Bacterial proteins attach to cells in reproductive tract
	Neisseria gonorrhoeae	Gonorrhea	Adhesion molecules on bacterium attach to cells in reproductive tract
Gastrointestinal tract	*Shigella dysenteriae*	Dysentery	Adhesion molecules on bacterium bind to colonic mucin and glycolipids
	Escherichia coli	Diarrhea	Adhesion molecules on bacterial pili attach to gastrointestinal cells
	Vibrio cholerae	Cholera	Adhesion molecules on bacterial flagella attach to gastrointestinal cells

destruction of the tooth enamel (acids as a metabolic by-product lower the pH and cause demineralization), resulting in the formation of a cavity. The plaque the dentist removes from your teeth is made up of this complex of organisms. When you consider how hard the dentist has to work to pry this plaque from your teeth, you get a good idea of establishment.

Some organisms, such as *Treponema pallidum* (which causes syphilis), avoid the need for protracted periods of adherence by boring into the tissues (**Figure 5.11**).

Increasing the Numbers

For pathogens, there is safety and success (that is, infectivity) in numbers. In fact, the doubling time for some bacteria (such as *E. coli*) can be as little as 20 minutes. This extremely high reproductive rate requires that the growth environment be satisfactory, and for most pathogens the tissues and fluids of the human body are an ideal environment. So the number of organisms required for successful infection can easily be achieved.

There is considerable variability among organisms with regard to the number required for success, and we can run specific experiments to establish the criteria for virulence for any given pathogen. The **lethal dose 50% (LD$_{50}$)** is the number of organisms required to kill 50% of the hosts, and the **infectious dose 50% (ID$_{50}$)** is the number of organisms required for 50% of the population to show signs of infection. Pathogens having the lowest LD$_{50}$ and ID$_{50}$ values are the most virulent. Using these types of information, we can categorize organisms according to virulence, as shown in **Figure 5.12**.

It is important to remember that most bacteria divide by **binary fission** (one bacterium splits into two) and that a pathogen that has a low LD$_{50}$ and a short doubling time could be extraordinarily dangerous, with the rapid increase in organisms quickly overwhelming a patient. Fortunately, attacking only rapidly growing bacteria is the way in which many antibiotics work, thereby preventing negative outcomes from infection. Unfortunately, this story is changing, as we will see when we discuss the rapidly expanding resistance to antibiotics.

In viral infections, the number requirement is easier to understand. Upon virus production, host cells can be severely damaged and die. The more cycles of infection and virus production that the host tissue or organ has to go through, the fewer viable host cells are available. This can lead to organ failure and death of the host.

AVOIDING, EVADING, OR COMPROMISING HOST DEFENSES (DEFEAT THE HOST'S DEFENSES)

If a pathogen has managed to get into a host, stay in, and rapidly increase its numbers, it is on the way to causing a successful infection. Unfortunately for the pathogen, these steps are usually not enough because there are many ways in which the host can defend itself. Thus, the pathogen has to deal with the host's defenses. There are two basic ways in which the pathogen can be successful; one involves the structure of the pathogen cells, which is a built-in (passive) strategy, and the other involves attacking the host's defenses (an active strategy).

The main strategic structural components of pathogenic bacteria are capsules and cell wall components. In fact, as we noted earlier, encapsulation is required for some organisms to cause certain symptoms. For example, unless they are encapsulated, *S. pneumoniae* will not cause

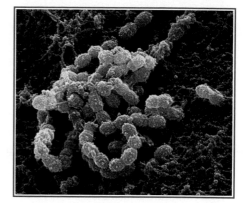

Figure 5.10 A colorized scanning electron micrograph of dental plaque. The yellow-green structures are *Streptococcus mutans* cells adhering to the pellicle that covers the teeth. Other organisms adhere to *S. mutans* and cause further development of the biofilm. The organisms in this biofilm produce the enzymes and then metabolic by-products that can eventually cause destruction of the tooth, resulting in the formation of a cavity.

| Fast Fact | Drastic increases in pathogen numbers could be a cause for worry except for the fact that humans have developed tremendous defense mechanisms. Consequently, even though a pathogen could potentially proliferate from a single cell to 10^{21} cells in 24 hours, each antibody-secreting plasma cell in the human immune system can produce antibodies against that pathogen at the rate of 2000 antibody molecules per second. More importantly, there can be millions of plasma cells!

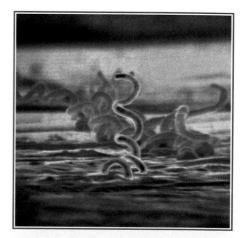

Figure 5.11 The spirochete *Treponema pallidum* 'corkscrewing' into tissue.

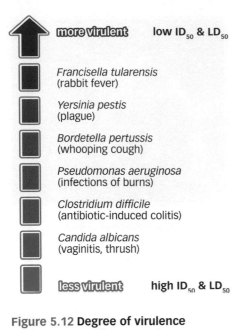

more virulent low ID$_{50}$ & LD$_{50}$

Francisella tularensis
(rabbit fever)

Yersinia pestis
(plague)

Bordetella pertussis
(whooping cough)

Pseudomonas aeruginosa
(infections of burns)

Clostridium difficile
(antibiotic-induced colitis)

Candida albicans
(vaginitis, thrush)

less virulent high ID$_{50}$ & LD$_{50}$

Figure 5.12 Degree of virulence attributed to different pathogens. This type of appraisal can be made after determining the LD$_{50}$ and ID$_{50}$ of the pathogens. Remember, the lower the LD$_{50}$ and ID$_{50}$, the more virulent the pathogen.

pneumonia and *Klebsiella pneumoniae* will not cause Gram-negative bacterial pneumonia.

Capsule and Cell Wall (Passive) Protection Strategies

In humans, a first line of defense for the innate immune response is **phagocytosis**. In this process, cells known as **phagocytes** ingest pathogens and then destroy them. Pathogens can encapsulate themselves, covering their entire surface in a slimy capsule, which protects against phagocytosis (**Figure 5.13**). The capsular material seems to prevent a phagocyte from adhering to the surface of the bacterium. This adherence is required for the phagocyte to develop pseudopodia (false feet) that then surround the bacteria.

You might think that capsule protection would be all that the pathogen required to overcome the host defense, but that is not quite true. The host can use the adaptive immune response to produce antibodies against the capsule. When the antibody molecules bind to the capsule (a process known as **opsonization**, which we will discuss in detail in Chapter 16), phagocytic cells can use the antibody molecules as bridges to adhere to the organism and eventually phagocytose it.

A second structural defense available to bacteria is the bacterial cell wall, which is a complex structure that protects the bacteria from environmental pressure. The components of the wall can help to increase virulence by protecting against host defenses. For instance, *Streptococcus pyogenes* incorporates **M proteins** into its cell wall. These proteins increase the virulence of this pathogen by increasing adherence to host target cells and by making the pathogen resistant to heat and to acidic environments. In addition, M proteins also inhibit phagocytosis and are intimately involved in the condition known as toxic shock. Fortunately, the host's immune response can provide antibodies and also specific serum proteins (fragments of **complement proteins** that opsonize these bacteria).

Another structural protection is seen in *Mycobacterium tuberculosis*, in which the cell wall is infused with **mycolic acid**, a waxy substance that protects the bacterium so it can live within unactivated phagocytes and provides a defense against antiseptics, disinfectants, and antibiotics.

Enzyme (Active) Protection Strategies

Capsules and cell wall components are passive measures used to defeat the host defenses. Bacteria also use active measures against a host by producing extracellular enzymes that enable the infection to spread,

Figure 5.13 A diagrammatic representation of the inhibition of phagocytosis by an encapsulated bacterium. Encapsulation is a defense mechanism that pathogens have against the defenses of the host. Panel a: The capsule keeps the surface of the engulfing phagocyte from 'sticking' to the bacterium, and the bacterium is more likely to go free. Panel b: Some bacteria are phagocytosed but then multiply once inside the phagocyte.

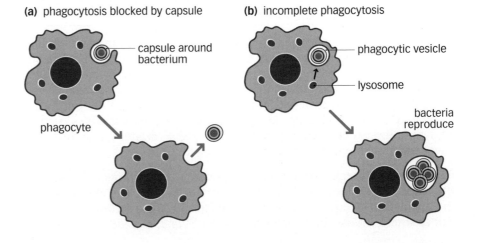

(a) phagocytosis blocked by capsule

capsule around bacterium

phagocyte

(b) incomplete phagocytosis

phagocytic vesicle

lysosome

bacteria reproduce

and by killing off host defensive cells. The following enzymes are useful against the host defenses. They are technically exotoxins (see below).

Leukocidins are enzymes that destroy white blood cells in the host. White blood cells are an important part of both the innate and the adaptive host defense systems. Two types of white blood cell are neutrophils and macrophages, which are powerful phagocytic cells. In addition, the white blood cells known as lymphocytes, which are responsible for the adaptive immune response to infection, are destroyed by leukocidins. Leukocidins are produced by staphylococcal and streptococcal pathogens, and it is easy to see how the ability to kill off host defenders can make these organisms dangerous.

Hemolysins are membrane-damaging toxins that disrupt the plasma membrane of host cells and cause the cells to lyse. These toxins can damage the plasma membrane of both red blood cells and white blood cells and are produced by a variety of bacteria, including staphylococcal species, *Clostridium perfringens* (which causes **gas gangrene**), and streptococcal species. Hemolysin produced by streptococcal bacteria is referred to as streptolysin and can be divided into different types, such as types S and O. The various types of streptolysin differ from one another in the cell destruction that they cause. For example, streptolysin O is associated with β-hemolysis (complete destruction) of red blood cells.

Coagulase (**Figure 5.14a**) is a pathogen-produced enzyme that causes fibrin clots to form in the blood of a host. Clot formation can be used by

Fast Fact Knowledge about streptococcal bacteria and their kinase enzymes has saved the lives of many cardiac patients! Scientists have been able to make use of streptococcal bacteria in the lab to produce these enzymes (such as streptokinase), which are then used as a medical treatment to destroy blood clots.

Figure 5.14 Some of the enzymes produced by pathogenic bacteria. Panel a: Pathogens can wall themselves off from host defenses by using coagulase but can also use enzymes (for example streptokinase) to dissolve clots. Panel b: The enzyme hyaluronidase allows pathogens to invade deep tissues.

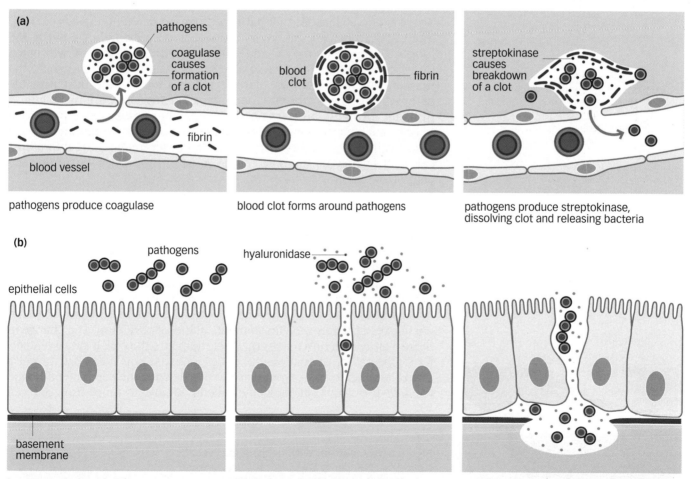

(a)

pathogens
coagulase causes formation of a clot
fibrin
blood vessel
pathogens produce coagulase

blood clot
fibrin
blood clot forms around pathogens

streptokinase causes breakdown of a clot
pathogens produce streptokinase, dissolving clot and releasing bacteria

(b)

pathogens
epithelial cells
basement membrane
invasive pathogens reach epithelial surface

hyaluronidase
pathogens produce hyaluronidase

pathogens now able to invade deeper tissues

both host and pathogen during an infection. The host can use clotting as part of the defense against infection. This clotting occurs in blood vessels around the site of the infection, thereby closing in the pathogens and preventing the spread of the infection. An example of bacterial use of clotting is a **boil** resulting from a localized staphylococcal infection. Here the organism will wall itself off to avoid the host defenses.

Staphylokinase is produced by staphylococcal species, and streptokinase is produced by streptococcal species. Those enzymes (see Figure 5.14a) start a cascade that eventually breaks down fibrin and dissolves clots. These enzymes can be used by a pathogen to overcome attempts by the host to wall off the infection, thereby ensuring its spread.

Hyaluronidase and collagenase are enzymes that break down connective tissue and collagen in a host, allowing infections to spread (**Figure 5.14b**). Both are active, for instance, in the infection gas gangrene (caused by *C. perfringens*), with the result that the infection is usually widespread, involving the massive destruction of connective tissue.

The process of spreading out is fundamental for the increased virulence of many pathogens. Group A streptococci are a good example. These organisms do not infect any mammals other than humans. They use enzymes as described above to inhibit the clotting machinery or to break down clots formed by the host, allowing the pathogens to spread. There may be a genetic predisposition in humans that has a role in whether streptococcal infections are minor, such as boils, or major, such as **necrotizing fasciitis** (flesh-eating; **Figure 5.15**).

There is one final tactic that pathogens use to defend themselves. They hide! Pathogens can use any of the tactics described above, but in the long run the host's innate and adaptive immune response will catch up with them. Therefore, being able to get inside a host cell, where the humoral immune response cannot detect them, is the best possible defense. This is easy for viral pathogens because they are by definition **obligate intracellular parasites**. (Obligate here means able to survive in only one environment, and for viruses that one environment is inside a host cell.) Bacteria, in contrast, have a harder time getting into a host cell, and so they let the host cell do most of the work by usurping the cell cytoskeleton.

Recall from Chapter 4 that the eukaryotic cell has a variety of fibers—microfilaments, intermediate filaments, and microtubules—that are part of the cellular cytoskeleton and are responsible for cellular support and intracellular movement. Bacteria use these filaments and tubules to penetrate and move around inside the infected cell. As an example, let's look at *Salmonella*. When this pathogen makes contact with a host cell, the contact causes the plasma membrane of the host cell to change its configuration. *Salmonella* produces a molecule called **invasin** that can change the structure of actin filaments in the cytoskeleton. The change in these filaments in turn moves the bacterium into the cell. It gets even better for the pathogen once inside the cell, because now it can use the cell's actin filaments to move from place to place inside the cell. The pathogen can also use a host cell molecule called **cadherin** to move from one cell

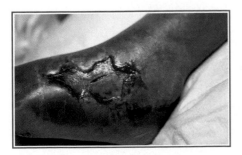

Figure 5.15 A patient with necrotizing fasciitis. This infection is caused for example by *Streptococcus pyogenes* using extracellular enzymes that enable the infection to spread, and by killing off host defensive cells.

to another without ever exposing itself to the host's immune defenses searching for it. We will see this mechanism again during our discussion of infections of the digestive tract.

| Keep in Mind |

- There are five requirements for a successful infection: get in, stay in, defeat the host defenses, damage the host, and be transmissible.

- Places at which pathogens enter the body are called portals of entry.

- The major portals of entry are the mucous membranes, the skin, and parenteral routes.

- Mucous membrane portals of entry are associated with the respiratory, digestive, and genitourinary tracts of the body.

- Establishment (the requirement of staying in) can be accomplished using adhesin molecules, which are surface proteins. In addition, some pathogens take advantage of structures such as fimbriae to adhere to tissues.

- Virulence for a given pathogen can be gauged by the ID_{50} and LD_{50} of that organism.

- Pathogens can defeat a host's defenses in two ways: passively (by using structures such as the capsule) and actively (by attacking the host defense directly through the production of enzymes).

DAMAGING THE HOST

In the above discussions, we talked about some of the requirements for a successful infection. Now let's look at the damage that occurs to host cells during the disease process. Most of the damage associated with infection can be divided into two parts: damage that occurs because the bacteria are present (direct or indirect damage), and damage that is a by-product of the host response.

Direct damage is the obvious destruction of host cells and tissues and is usually localized to the site of the infection. In direct damage, the host defense response is timely and potent, usually limiting the damage done. Indirect damage is seen in most serious infections and is much more dangerous to the host because it involves systemic disease. This type of distal pathology results from the production of bacterial toxins. These toxins are soluble proteins and easily diffuse into, and move through, the blood and lymph systems. Thus, they can travel throughout the body quickly. This causes pathogenic changes far away from the initial site of infection.

Toxins can produce fatal outcomes in patients. They also have some common characteristics associated with them, such as fever, shock, diarrhea, cardiac and neurological trauma, and the destruction of blood vessels. There are two types of toxin—exotoxins and endotoxins—and they are very different from each other.

Exotoxins

Exotoxins are toxins that are specific to a certain pathogen and secreted by the pathogen. They can therefore act remotely without the presence of the pathogen, and can even enter host cells (**Figure 5.16**). Exotoxins are among the most lethal substances known: some exotoxins are a million times more potent than the poison strychnine. Many of the genes that encode these toxins are carried on prophages or plasmids in the

Figure 5.16 Exotoxins are produced by living pathogenic bacteria. These toxins then enter cells of a host and prevent those cells from functioning properly.

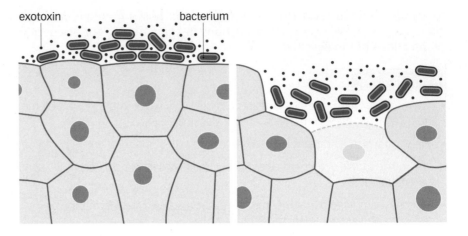

pathogens, which make them even more dangerous because these are mobile genetic elements that can transfer genetic information from one bacterium to another.

There are several types of exotoxin. They are often grouped via their structure or mode of action. Examples are **cytotoxins**, which kill cells that they come in contact with; **neurotoxins**, which interfere with neurological signal transmission; and **enterotoxins**, which affect the lining of the digestive system (**Table 5.3**). Let's look at some of the more dangerous exotoxins.

Anthrax Toxin

Anthrax toxin is a cytotoxin produced by the bacterium *Bacillus anthracis*, a Gram-positive rod commonly found in the soil of pastures. The toxin is made up of three parts: an edema factor (EF), a protective antigen (PA, which is a transmembrane factor also used for vaccine production), and a lethal factor (LF). It is so dangerous that each bacterium produces the parts separately and assembles them on the bacterial cell surface in such a way that the complex is not yet toxic. The complex then leaves the bacterium and attaches to a host cell. Once attached to its target cell, the complex is endocytosed into a vesicle, where the low-pH environment causes a conformational change in the complex, converting it to the toxic form. In this state, the protective antigen forms a pore in the vesicle membrane, and the edema and lethal factors temporarily change their shape so that they can squeeze out into the host cell's cytoplasm.

Anthrax toxin interrupts the signaling capability of host macrophages and causes their death. It also interrupts the signaling capability of dendritic cells but does not kill them. However, even though they are still alive, the infected dendritic cells are no longer able to participate in the host defense against the infection. The importance of the loss of macrophage and dendritic cell function will become obvious in Chapters 15 and 16.

 Fast Fact As with many other toxins, anthrax toxin has become important in discussions of potential biological weaponry.

Diphtheria Toxin

Diphtheria toxin is a cytotoxin that was discovered in the 1880s and is produced by the Gram-positive rod *Corynebacterium diphtheriae*. It works by inhibiting protein synthesis and is produced in an inactive form, just as anthrax toxin is. After secretion from the bacterium, the diphtheria toxin is changed enzymatically into an active form. This toxin is very potent, and a single molecule of it is sufficient to kill a cell.

The general structure of the diphtheria toxin seems to be the same as that seen in many other exotoxins. It is an A–B toxin composed of two

Exotoxin	Organism producing toxin	Action of toxin	Symptoms
Anthrax toxin (cytotoxin)	*Bacillus anthracis*	Increases vascular permeability	Hemorrhage and pulmonary edema
Enterotoxin (enterotoxin)	*Bacillus cereus*	Causes host body to lose electrolytes	Diarrhea
Botulinum toxin (neurotoxin)	*Clostridium botulinum*	Blocks release of acetylcholine	Respiratory paralysis
Alpha toxin (cytotoxin)	*Clostridium perfringens*	Breaks down cell membranes	Extensive cell and tissue destruction; gas gangrene
Tetanus toxin (neurotoxin)	*Clostridium tetani*	Impairs function of inhibitory synapses	Violent skeletal muscle spasms (lockjaw), respiratory failure
Diphtheria toxin (cytotoxin)	*Corynebacterium diphtheriae*	Inhibits protein synthesis	Heart damage, possible death weeks after apparent recovery, throat epithelial damage, suffocation
Shiga-like toxin (enterotoxin)	*Escherichia coli* O157:H7	Inhibits protein synthesis	Destruction of intestinal lining, kidney failure, hemolytic uremic syndrome
Shiga toxin (cytotoxin)	*Shigella dysenteriae*	Inhibits protein synthesis	Severe diarrhea
Cholera toxin (enterotoxin)	*Vibrio cholerae*	Causes excessive loss of water and electrolytes	Severe diarrhea; death can occur within hours
Erythrogenic toxin (cytotoxin)	*Streptococcus pyogenes*	Causes vasodilation	Maculopapular lesions, as seen in scarlet fever

Table 5.3 Effects of exotoxins.

polypeptide chains, A and B. The B chain is responsible for binding to the target cell. This occurs through receptor molecules on the target cell and allows transport of the A chain across the cell membrane. Once inside the host cell, the A chain inhibits protein synthesis. Without the B chain, the toxin is harmless because it cannot gain entry into the host cell.

Botulinum Toxin

Botulinum toxin is a neurotoxin produced by the bacterium *Clostridium botulinum*, a Gram-positive anaerobic rod. There are seven types of botulinum toxin, and all of them inhibit the release of the neurotransmitter acetylcholine. This disrupts the neurological signaling of skeletal muscles and results in paralysis, because muscles cannot contract. The mechanism of action is fascinating and affects specialized proteins known as SNARE proteins to eventually block neurotransmitter release. SNARE proteins are found on vesicle and cytoplasmic membranes in the host cell and normally facilitate the fusion of vesicles to the cell membrane so that routine exocytosis can occur. The botulinum toxin clips these SNARE proteins off the vesicle membranes, thereby disrupting fusion and inhibiting the exocytosis of the neurotransmitter. Because it also affects muscles required for respiration, the resulting paralysis can lead to the death of the patient. Botulinum toxin has become a favorite candidate for potential use as a biological weapon.

Tetanus Toxin

Tetanus toxin is a neurotoxin produced by the bacterium *Clostridium tetani*, a Gram-positive, obligate anaerobic rod commonly found in soil. This toxin is closely related to botulinum toxin and causes a loss of skeletal muscle control by blocking relaxation. This loss of control results in convulsive muscle contractions. The condition known as **lockjaw**, in which the facial muscles contract uncontrollably, is a symptom of infection with *C. tetani*.

Cholera Toxin

Cholera toxin is an enterotoxin produced by the bacterium *Vibrio cholerae*. This toxin also consists of two polypeptide chains, and binding of the B chain to receptors on the target cell also allows the A chain to enter the cell. Once inside, the A chain induces the epithelial cells of the digestive system to release large quantities of electrolytes, followed by an excessive passive loss of water (up to 20 l a day). The action of cholera toxin can be seen in **Movie 5.1**. The result is severe diarrhea and vomiting that can be lethal. One symptom of **cholera** seen in patients with advanced disease is what is called rice-water stool, in which the fecal material is mainly liquid with bits of mucus in it. Cholera is transmitted through the fecal–oral route of contamination and is an endemic problem in many parts of the world that are socioeconomically depressed. Treatment involves proper sanitation, killing of the organisms, and large-scale infusion of electrolyte solution.

Other Exotoxins

Exotoxin effects are also seen in scarlet fever. Here, *S. pyogenes* produces a cytotoxin that destroys blood capillaries, and the result of this capillary destruction is the characteristic rash seen with this disease. If not treated, this exotoxin can cause occult heart damage that may lead to death. The exotoxin is classed as a **superantigen**. Superantigens cause a massive immune response, which damages the host itself.

The bacterium *Staphylococcus aureus* also produces a superantigen. It is called **toxic shock syndrome** toxin and can cause toxic shock. This condition causes significant loss of liquids from the body, a loss that can lower the blood volume and blood pressure to the point at which the patient goes into shock and then dies.

Fortunately, exotoxins are very antigenic, which means they are substances that stimulate a host to produce antibodies. It is therefore relatively easy to generate antibodies for vaccine production against these toxins. In fact, the DTaP vaccination that children receive is derived from the exotoxins from *C. diphtheriae* and *C. tetani* coupled with

Fast Fact Staphylococcal organisms such as *S. aureus* can also cause a form of toxic shock that has been found to be associated with the use of tampons leading to staphylococcal toxic shock syndrome. The exact connection remains unclear, but researchers suspect that certain types of high-absorbency tampon provide a moist, warm environment where these organisms can thrive.

antigenic components of *Bordetella pertussis* (the organism that causes **whooping cough**). So that they can serve as vaccines, these toxins have been chemically treated to inactivate them. The treatment causes them to lose their toxicity but not their antigenicity (the ability to induce an antibody response). After this inactivation treatment, toxins are referred to as **toxoids**.

Endotoxins

Endotoxins are bacterial toxins that are part of the outer membrane of the cell wall of Gram-negative bacteria. They are active only after release into the bloodstream from the bacterium during cell division or once the bacterial cell is dead.

Endotoxins are very different from exotoxins (Table 5.4). For a start, endotoxins are a component of Gram-negative organisms, whereas most exotoxins are produced by Gram-positive organisms. Endotoxins are not nearly as toxic, do not have specific targets in the host, and are a normal part of the bacterial cell wall. We will see in Chapter 9 that Gram-negative

Property	Exotoxins	Endotoxins
Organism	Mainly in Gram-positives	In almost all Gram-negatives
Location	Excreted by living pathogen into its environment	Part of pathogen cell wall outer membrane, released when cell dies or during cell division
Chemistry	Polypeptide	Lipopolysaccharide complex
Stability	Unstable; mostly denatured above 60°C	Stable; can withstand 60°C for hours
Toxicity	Among the most potent toxins known (100 to 1 million times more lethal than strychnine)	Weak, but fatal at high doses
Effects	Highly specific, several modes of action	Nonspecific; local reactions, such as fever, aches, and possible shock
Fever production	No	Yes, rapid rise to very high fever
Usefulness as antigen	Very good, long-lasting immunity conferred	Weak, no immunity conferred
Conversion to toxoid form	Yes, by chemical treatment	No
Lethal dose	Small	Large
Examples of related diseases	Botulism, gas gangrene, tetanus, diphtheria, cholera, plague, scarlet fever, staphylococcal food poisoning	Salmonellosis, typhoid fever, tularemia, meningococcal meningitis, endotoxic shock

Table 5.4 A comparison of exotoxins and endotoxins.

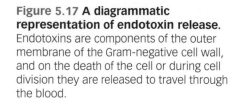

Figure 5.17 A diagrammatic representation of endotoxin release. Endotoxins are components of the outer membrane of the Gram-negative cell wall, and on the death of the cell or during cell division they are released to travel through the blood.

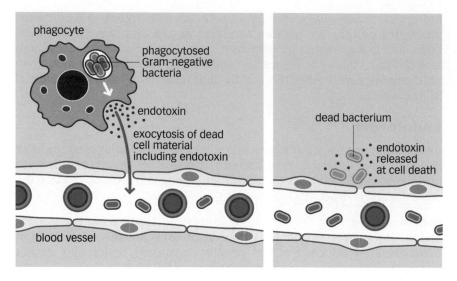

bacteria have an outer layer made up of lipoproteins, phospholipids, and lipopolysaccharides (LPS). When the organism dies, the layer falls apart (**Figure 5.17**). The lipopolysaccharides of this layer contain a particular lipid called lipid A, which has endotoxin properties.

All endotoxins cause essentially the same symptoms: chills, fever, muscle weakness, and aches. However, large amounts of endotoxin can cause more serious problems, such as shock and **disseminated intravascular coagulation** (**DIC**), a condition in which minor clotting occurs throughout the body and can lead to multiple organ failure. The minor clotting also uses up the clotting elements so that they are not available when needed for serious blood loss situations.

Endotoxins can sometimes elicit antibody production in the host, but in general this immune response is extremely weak because endotoxins are not very antigenic. As a result, the antibody response to them is usually poor. One of the problems with endotoxins is that because they are fairly stable and independent of viable cells, they can contaminate materials and equipment used in clinics and hospitals, and remain toxic for long periods. They can still cause damage to the host after successful treatment with antibiotics. The potential for long-term contamination makes the chance of endotoxins being transferred to patients a problem. Therefore, a test to determine whether there is endotoxin contamination of supplies or equipment was developed. This test is referred to as the *Limulus* **amebocyte lysate assay** (**LAL**) and takes advantage of the white blood cells of the horseshoe crab (*Limulus*), which are very different from human white blood cells. In the presence of endotoxin, *Limulus* white blood cells clot into a gel-like matrix that becomes turbid (cloudy). The degree of turbidity can be used as a measure of the amount of endotoxin contamination present.

Keep in Mind

• Damage to the host can be either direct or indirect. Direct damage is usually localized, and indirect damage is usually systemic through the production of soluble toxins.

• Exotoxins are extremely lethal substances produced by living cells (usually Gram-positive bacteria) and in most cases are proteins.

- Exotoxins can be cytotoxins (which kill cells), neurotoxins (which interfere with neurological signaling), or enterotoxins (which affect the lining of the digestive system).

- Exotoxins can cause the production of antibodies.

- Endotoxins are part of the bacterial cell wall and are released on the death of the organism or during cell division.

- Endotoxins are products of Gram-negative bacteria, do not effectively cause the generation of antibodies, and are less toxic than exotoxins.

- Lipid A, which is part of the Gram-negative phospholipid outer membrane of the bacterial cell wall, has endotoxin properties.

Viral Pathogenic Effects

The pathogenic effects caused by viruses are discussed in detail in Chapters 12 and 13. For now, we can say that we classify virally caused cell damage or death as a **cytopathic effect** (CPE) and that the cytopathology associated with viral infections can occur in three ways.

The most obvious way is viral overload, a condition that causes the virus to explode out of the host cell and infect and lyse surrounding host cells (**Figure 5.18**). A second type of CPE occurs when host defense mechanisms identify and kill virally infected cells. This is categorized as a **cytocidal effect**. The third type of viral cytopathology occurs when a virus shuts down the host DNA, RNA, and protein synthesis and thereby forces the host cell to devote all its efforts to virus production. This is classified as a **non-cytocidal effect**.

Viral cytopathology can be identified microscopically. In some cases, **inclusion bodies** filled with virus become visible inside the cell. For example, **Negri bodies** are inclusion bodies seen in rabies viral infections (**Figure 5.19**) and can be diagnostic for the disease. Some viral infections cause breaks in the host chromosomes, and these breaks are easily identifiable. The formation and appearance of **syncytia** (gigantic cells formed as several infected host cells merge) can also be a visual indication of viral infection (**Figure 5.20**). More subtle pathogenic effects, such as the production of hormones or interferon, are also signs of virally infected cells.

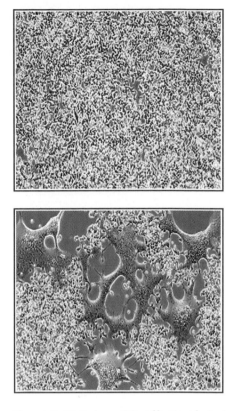

Figure 5.18 Cytopathic effects of viral infection. The left panel shows uninfected cells, which grow to confluence (meaning that they completely cover the surface). The right panel shows the same cells after viral infection. The host cells are destroyed as the viral particles burst out and enter the surrounding cells, causing their destruction.

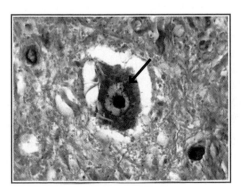

Figure 5.19 The cytopathology of the rabies virus, which produces large Negri bodies in the cell. These bodies contain newly formed viral particles.

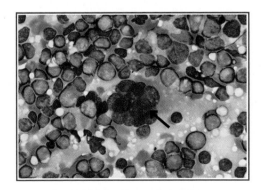

Figure 5.20 Formation of syncytia (arrow) in cells infected with virus. Formation of these structures is one type of cytopathic effect seen in viral infections.

The pathogenic effects seen in diseases caused by fungi and parasites will be discussed in Chapter 14. Parasites usually cause damage to the host as a result of moving in the body, taking up space and causing host immune responses.

SUMMARY

- The requirements for infection include entry, establishment, avoiding host defenses, damaging the host, and exiting from the host.

- Portals of entry for pathogens include skin, parenteral routes, and the mucous membranes of the respiratory, gastrointestinal, and genitourinary tracts.

- Pathogens use virulence factors such as adhesins to establish themselves in the host.

- To avoid being killed by the host's defenses, pathogens use an array of virulence factors, including capsules, M proteins, mycolic acid, leukocidins, hemolysin, coagulase, kinases, hyaluronidase, and collagenase.

- Some bacterial pathogens cause damage to host cells by releasing exotoxins; these include types such as cytotoxins, neurotoxins, and enterotoxins.

- Endotoxins on the outer membrane of the cell wall of Gram-negative bacteria cause damage to the host and are released and disseminated when the bacterial cell dies or during division.

Overall, the basis for our discussion here was the ability of a pathogen to invade a host, remain in that host, and protect itself against host defenses so as to cause disease. As you organize and reflect on the concepts in this chapter, keep in mind what you have learned in the previous chapters. We explore the structures of the bacterial cell in Chapter 9, and there you will be able to reaffirm the connection with the disease processes you have learned here and those structural components.

(Q) SELF EVALUATION AND CHAPTER CONFIDENCE

Multiple Choice

Answers are given in the back of the book and help can be found in the student resources at:

www.garlandscience.com/micro2

1. A man nicks himself while shaving and the area becomes infected. Which of the following portals of entry did the pathogen probably use?

   A. Gastrointestinal tract
   B. Skin (parenteral)
   C. Respiratory tract
   D. Genitourinary tract
   E. Exotoxin tract

2. The adherence of *E. coli* involves

   A. Cell surface adhesion molecules binding to receptors on cells lining the respiratory tract
   B. Adhesion molecules attaching to carbohydrates on respiratory cells

   C. Bacterial proteins attaching to cells in the reproductive tract
   D. Adhesin molecules on bacterial flagella attaching to gastrointestinal cells
   E. Adhesin molecules on bacterial pili attaching to gastrointestinal cells

3. A pathogen has entered the body. All of the following will have a role in its establishment except

   A. Releasing endotoxin that will cause clotting
   B. Boring through tissues
   C. Using adhesins to attach to tissues
   D. Creating a biofilm on a body surface
   E. Using fimbriae to attach to cell receptors

4. The LD_{50} of a pathogen is the number of organisms required to

   A. Benefit 50% of the hosts
   B. Infect 50% of the hosts
   C. Kill 50% of the hosts
   D. Produce lytic deaminase toxin
   E. None of the above

5. Imagine you are working in a lab and read reports about two different bacteria. Organism A has an ID_{50} of 20 cells, whereas organism B has an ID_{50} of 100 cells. Which of the following conclusions would you make?

   A. Organism A could be considered more virulent than organism B
   B. Organism A must have endotoxin
   C. Organism B could be considered more virulent than organism A
   D. Organism B has a portal of entry but organism A does not

6. A bacterial toxin that causes damage to the plasma membrane of host red blood cells and results in lysis is

   A. Coagulase
   B. Leukocidin
   C. Hemolysin
   D. Collagenase
   E. Hyaluronidase

7. A bacterial enzyme that breaks down connective tissue is

   A. Coagulase
   B. Hyaluronidase
   C. Hemolysin
   D. All of the above
   E. None of the above

8. An invasin would be used by a microbe to

   A. Change the nuclear structure of host cells
   B. Inhibit the functions of host ribosomes
   C. Increase the rate of bacterial division
   D. Change the structure of actin filaments in host cells
   E. Change the shape of microtubules in host cells

9. A culture of cells is producing exotoxins. On the basis of this information, the organisms are probably

   A. Gram-negative bacteria
   B. Viruses
   C. Gram-positive bacteria
   D. All dead
   E. None of the above

10. Three types of exotoxin are

    A. Lipopolysaccharide, enterotoxins, and neurotoxins
    B. Neurotoxins, cytotoxins, and lipid A
    C. Neurotoxins, cytotoxins, and enterotoxins
    D. Cytotoxins, disseminating clotting toxin, and neurotoxins
    E. None of the above combinations

11. Botulism toxin is a(n)

    A. Neurotoxin
    B. Muscular toxin
    C. Febrile toxin
    D. Enterotoxin
    E. None of the above

12. Many people refer to tetanus infection with the pathogen *Clostridium tetani* as 'lockjaw.' Which of the following best explains why? These bacteria

    A. Produce a capsule that hardens the surface of the jaw
    B. Produce an enzyme that breaks apart jaw muscle cells
    C. Accumulate in the jaw and prevent proper jaw movement
    D. Produce a toxin that causes jaw muscles to remain contracted

13. Which of the following best describe endotoxins?

    A. They are toxins that are produced by dead Gram-negative and Gram-positive cells
    B. They are toxins that are released by Gram-positive cells
    C. They are toxins found on Gram-negative cell walls
    D. They are toxins found on Gram-positive cell walls
    E. They are identical in structure to exotoxins

14. A new patient in the intensive care unit has disseminated intravascular clotting (DIC), which is probably caused by

    A. Exotoxins
    B. Enterotoxins
    C. Endotoxins
    D. Vascular toxins
    E. Neurotoxins

Q DEPTH OF UNDERSTANDING

Questions listed here require you to bring together the concepts you have learned in this chapter into a discussion format. This helps you to increase your depth of understanding of the material you have learned. Help can be found in the student resources at: www.garlandscience.com/micro2

1. Compare the portals of entry used by bacteria and give examples of organisms that use each portal.

2. Discuss the mechanisms that bacteria use to avoid or defeat the host defenses.

3. Describe the properties of exotoxins and compare them with endotoxins.

Q CLINICAL CORNER

Help can be found in the student resources at: www.garlandscience.com/micro2

1. *Neisseria meningitidis*, commonly known as meningococcus, is a Gram-negative bacterium that not only can cause meningitis but also can enter the bloodstream and cause a deadly infection of the blood. If the bacteria do invade the bloodstream, the death rate can be very high. Luckily, if it is detected early, doctors can administer antibiotics that kill the bacteria. However, once antibiotics have been administered and the bacteria have been killed, the patient can often get much sicker before recovering.

   A. Can you identify the virulence factor that might be responsible for this 'turn for the worse' after antibiotic killing of these Gram-negative bacteria?
   B. Imagine you must explain to the patient's concerned family what is happening inside the patient's body after antibiotic treatment. What would you say, in simple terms?

2. Imagine that a new drug has been designed that targets and destroys the M protein of a deadly strain of *Streptococcus*. It is experimental and your very ill patient might be able to benefit from this new treatment. The doctor has presented the idea to the patient's family but needs a signed consent form. The family is confused, however, about how this treatment will actually work. They've never heard of something called M protein. They turn to you to see whether you know how this might help. How would you explain this to them in simple terms that they can understand? (Be sure to explain to them what you know about M protein, what it does, and what the benefits of destroying it would be.)

Transmission of Infection, the Compromised Host, Epidemiology, and Diagnosing Infections   Chapter 6

Why Is This Important?

Our understanding of the ways in which infectious diseases are transmitted, the role of a compromised host in the process, and rapid and correct diagnosis of infections is vital for developing methods to prevent and monitor the spread of disease.

Mary is on her way to the hospital to visit her boyfriend Peter, who is cared for in the burns unit after an accident at work. She is looking forward to seeing him, but she is also very anxious about his recovery. So she chose a large beautiful bunch of cut flowers, that should brighten Peter's days and remind him of Mary when she is not available to visit. She misses him and does not want him to worry. When she enters his room, he welcomes her with a smile. After embracing Peter as much as she can, given his injuries, and catching up on his progress, she is trying to find a vase. There is nothing in the room and she leaves to ask a nurse. To Mary's surprise the nurse tells her in no uncertain terms, that she must not take the flowers into Peter's room. Mary is very upset and finds the nurse's instructions harsh and totally unnecessary. The nurse is sympathetic and explains that this requirement is in the best interest of the patients and therefore of their families and friends. Mary learns that water in the vases may contain *Pseudomonas aeruginosa*, which originate from the flowers. The soil where flowers grow is a reservoir of large numbers of such bacteria. *P. aeruginosa* prefers damp places and is resistant to many antibiotics and disinfectants. This makes it an even worse idea to introduce the opportunistic pathogen into the ward, where burn patients with large open wounds are particularly at risk. Mary is glad she was stopped in good time so Peter will soon be healthy and back home. Then they can enjoy eating a picnic and looking at all the flowers in the park.

OVERVIEW

In the previous chapter we discussed the principles involved in the infectious disease process. In this chapter we focus on the transmission of diseases by looking at where the pathogens responsible for infection are found and at the mechanisms by which those pathogens are transmitted from the environment to a host, and from one host to another. We also consider the compromised host, which is a very important and integral part of the infection process. The more compromised the host, the greater is the risk of successful infection. We review some of the basic principles of epidemiology, the discipline that helps us to understand how infectious diseases are spread and thus to develop methods for prohibiting their spread. As this requires detecting infections in the first place, we will look into traditional and modern diagnostics, and how genetic engineering and some areas of biotechnology provide us with the information and tools to diagnose sooner and more reliably.

We will divide our discussions into the following topics:

Transmission of Infection, the Compromised Host, Epidemiology, and Diagnosing Infections

| TRANSMISSION OF INFECTION | THE COMPROMISED HOST | EPIDEMIOLOGY | DIAGNOSING INFECTIONS |

TRANSMISSION OF INFECTION

The transmission, or spread, of an infection is the final requirement necessary for a successful infection. We can best study this process by looking at two factors: reservoirs of infectious organisms (which are places where pathogens can grow and accumulate) and mechanisms of transmission (which are the various ways in which organisms move from place to place). We begin with reservoirs.

Reservoirs of Pathogens

There are three potential reservoirs where pathogens can accumulate: humans, other animals, and nonliving reservoirs. With human reservoirs, it is obvious that a sick person is a reservoir for the pathogens causing the infection and will continue to be a reservoir for as long as the infection continues. However, determining that a given individual is acting as a reservoir can be difficult because some infectious diseases are transmitted before the symptoms are manifested. Take measles as an example. In a classroom full of second-grade students, by the time one student shows the symptoms—aches, fever, red spots—there is a good possibility that the rest of the class and the teacher have already been exposed.

Human reservoirs are also found in infected people who are asymptomatic carriers. These people can carry such pathogens as HIV, *Corynebacterium diphtheriae*, *Staphylococcus aureus* and *Salmonella typhi* but never show any signs of disease. However, they can readily infect others.

Our second type of reservoir of pathogens involves animals other than humans. Diseases that are transferred from animals to humans are called **zoonotic diseases**. Two well-known zoonotic diseases are rabies, in which the virus is transferred from a rabid animal to a human, and Lyme disease. Others are shown in Table 6.1. Zoonotic diseases usually occur after direct contact with the animal, but there are also ways in which indirect contact can cause infection. For instance, the waste material of a litter box, fur, feathers, infected meats, or vectors can indirectly transmit pathogens from an animal to a person.

Table 6.1 Examples of zoonotic diseases.

Disease	Causative agent	Mode of transmission
Anthrax	*Bacillus anthracis*	Direct contact with infected animals; inhalation
Bubonic plague	*Yersinia pestis*	Flea bite
Lyme disease	*Borrelia burgdorferi*	Tick bite
Endemic typhus	*Rickettsia prowazekii*	Louse bite
Rocky Mountain spotted fever	*Rickettsia rickettsii*	Tick bite
Relapsing fever	*Borrelia* species	Tick bite
Malaria	*Plasmodium* species	*Anopheles* mosquito bite
African sleeping sickness	*Trypanosoma brucei gambiense*	Tsetse fly bite
American sleeping sickness (Chagas' disease)	*Trypanosoma cruzi*	*Triatoma* species (kissing bug) bites and feces
Rabies	Rabies virus	Bite of infected animal
Hantavirus pulmonary syndrome	Hantavirus	Inhalation of virus in dried feces and urine
Yellow fever	Flavivirus	*Aedes* mosquito bite

Our third reservoir, nonliving, includes water, food, and soil. Of these three, water may be the most dangerous. In many developing countries there is poor sanitation, and opportunities for personal hygiene can be limited, which result in the fecal contamination of water. Because water is something that we cannot live without, it becomes the key component in the fecal–oral route of contamination. Many diseases are spread from this reservoir—typhoid fever and cholera, for instance—and these diseases are endemic in many parts of the world. In the case of food acting as a pathogen reservoir, contamination is part of the spoilage that occurs naturally. (Note that refrigeration only slows pathogen growth; it does not stop it.)

As far as soil is concerned, it is a normal habitat for many organisms that are potentially pathogenic, but these organisms must find a way to pass the physical barriers that protect the body if they are to cause disease. A case in point would be *Clostridium tetani*, a Gram-positive rod-shaped bacterium found in soil that can be transferred to humans through breaks in the skin, as, for example, by stepping on a rusty nail. In fact, the rust has nothing to do with it. It is the contamination of the nail with *C. tetani* that causes the infection. Recall from our discussions in Chapter 5 that this organism produces a neurotoxin that can be lethal.

Fast Fact Next time you are cleaning the litter box for Fluffy the cat, remember that the waste material you are dealing with can indirectly transmit pathogens such as *Toxoplasma gondii* from Fluffy to you! By the way, for those of you who carry the 'pooper scooper,' Fido the dog's fecal material can also transmit pathogens indirectly.

Keep in Mind

- Transmission is the final requirement for a successful infection.
- Reservoirs are places where pathogens grow and accumulate.
- There are three potential reservoirs of infection: humans, other animals, and nonliving reservoirs.
- Human reservoirs can be infected people who are asymptomatic carriers.
- Animal-to-human infections are referred to as zoonotic disease.
- Nonliving reservoirs include water, soil, and food.

Mechanisms of Transmission

There are three mechanisms that can be used to transfer infectious organisms: contact transmission, vehicle transmission, and vector transmission.

Contact Transmission

In the mechanism known as **contact transmission**, a healthy person is exposed to pathogens by either touching or being close to an infected person or object. We can subdivide the contact transmission mechanism into three types: direct, indirect, and droplet. In direct contact transmission, there is no intermediary between the infected person and the uninfected person (**Figure 6.1**). This mechanism encompasses such things as touching, kissing, and sexual interactions. The diseases transmitted through direct contact include hepatitis A, smallpox, staphylococcal infections, infectious mononucleosis, and of course sexually transmitted diseases.

Figure 6.1 shows that the direct-contact mechanism can also involve zoonotic disease. In that case, the infected organism is not a person but an animal. Indirect contact transmission occurs through intermediates that are usually nonliving articles, such as tissues, handkerchiefs, towels, bedding, and contaminated needles, the latter easily transferring HIV and hepatitis B virus.

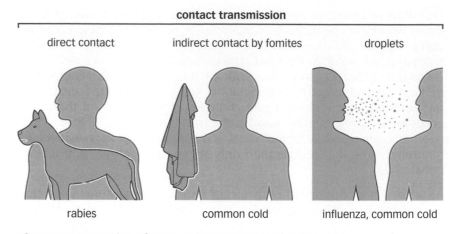

contact transmission

direct contact   indirect contact by fomites   droplets

rabies   common cold   influenza, common cold

Figure 6.1 Examples of contact transmission of pathogens.

Fast Fact Contaminated needles have become a major factor in the spread of HIV among intravenous drug users, and many cities have instituted needle-exchange programs, where authorities swap clean needles for used ones to try to slow this type of transmission.

We refer to the nonliving intermediates that act as the agents of transmission by indirect contact as **fomites**.

Droplet contact transmission is seen in the transfer of respiratory diseases such as influenza and whooping cough. Droplet transmission can occur through sneezing (**Figure 6.2**), coughing, and even laughing. Although it is confined to short distances, the size of the droplet is important. Large droplets will fall to the ground quickly, but smaller droplets can stay airborne for long periods. The smaller the droplet, the more dangerous it is as an agent of disease.

Vehicle Transmission

As the name implies, **vehicle transmission** involves pathogens riding along on supposedly clean components (**Figure 6.3**). For example, food can carry pathogenic organisms and is therefore considered a vehicle of disease transmission. Other possible vehicles are water, air, blood, body fluids, drugs, and intravenous fluids given to a hospitalized patient. The transmissibility is obvious for each of these vehicles, but let's look more closely at water, food, and air.

Food can transmit microorganisms that cause many infectious diseases that usually manifest themselves as food poisoning. Food contamination is usually the result of poor preparation or poor refrigeration, both of which are easily corrected. We have already mentioned water as a pathogen reservoir, and now we see it can also act as a vehicle. Obviously, proper sanitation can limit water's role as a vehicle of infection.

Perhaps the most difficult transmission vehicle to deal with is the air, because dust (which can contain huge numbers of microbes), microbial spores, and fungal spores can all use the air to travel from host to host. In fact, biological warfare strategies include transforming pathogens into an aerosol form so as to achieve maximal transmission effectiveness.

Vector Transmission

The third mechanism for transmitting infectious disease is **vector transmission**. In this mechanism, pathogens are transmitted to a healthy person by a carrier known to be associated with some disease. Vectors that can be used for disease transmission are usually arthropods, such as fleas, ticks, flies, lice, and mosquitoes.

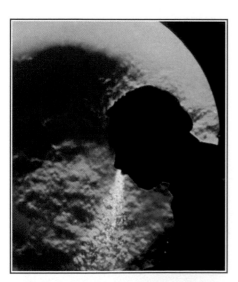

Figure 6.2 The sneeze, an example of droplet transmission of pathogens. Droplets can travel quite far and surround the sneezing person (and everything and everyone around) in a cloud.

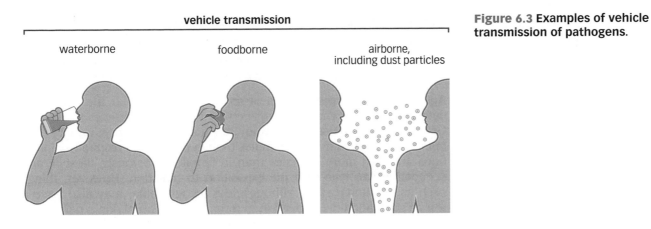

vehicle transmission

waterborne   foodborne   airborne,
                                       including dust particles

Figure 6.3 Examples of vehicle transmission of pathogens.

There are two ways in which this type of transmission occurs: mechanical and biological (Figure 6.4). In **mechanical vector transmission**, the vector's body parts are contaminated with the infecting microorganisms and they are passively brushed off the vector and onto the host, or they contaminate an object (such as food). This transmission mode is seen with houseflies, which are known to frequent fecal material and can get pathogens from this source onto their feet, wings, or other body parts. Biological vector transmission occurs through a bite, as seen with, for instance, the fleas that had a primary role in the transmission of *Yersinia pestis* (the plague of the Middle Ages); ticks, which can transmit *Borrelia burgdorferi* (the causative agent of Lyme disease); and mosquitoes, which carry such pathogens as *Plasmodium* (the malaria parasite) and West Nile virus.

Fast Fact   Some pathogens multiply in their vector and can arrive at the new host in larger numbers, which helps in establishing the infection.

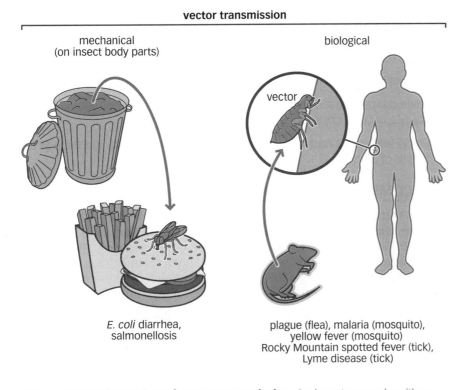

vector transmission

mechanical   biological
(on insect body parts)

vector

E. coli diarrhea,   plague (flea), malaria (mosquito),
salmonellosis   yellow fever (mosquito)
                                    Rocky Mountain spotted fever (tick),
                                    Lyme disease (tick)

Figure 6.4 An illustration of vector transmission. Such vectors can be either mechanical (involving pathogens found on the body of insects such as houseflies) or biological (through the bite of an insect such as the flea or mosquito).

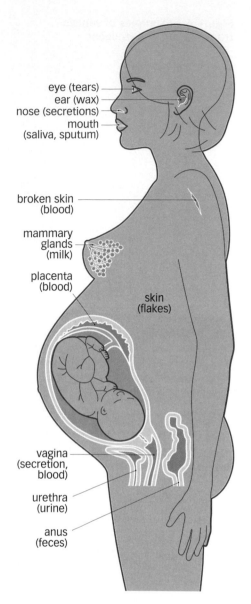

eye (tears)
ear (wax)
nose (secretions)
mouth
(saliva, sputum)

broken skin
(blood)

mammary
glands
(milk)

placenta
(blood)

skin
(flakes)

vagina
(secretion,
blood)

urethra
(urine)

anus
(feces)

Figure 6.5 Portals of exit (female) with examples of infectious or contaminated material.

Factors Affecting Disease Transmission

In addition to the overall health of the host, there are several other factors that can influence the disease process, including age, gender, lifestyle, occupation, emotional state, and climate.

Age affects the overall health of the host, in particular the decline in defensive capability, which translates to increases in disease as we age. Gender bias can be seen in certain infections; for instance, urinary tract infections are seen more in females than in males, and respiratory infections are seen more in males than in females. Obviously, lifestyle and occupation can lead to a predisposition to infection. However, things such as emotional state, although less recognizable, may also have a role in susceptibility to infection in terms of decision making, risk taking and immune status. Climate can have a role in infection in three ways: (1) there seems to be a greater incidence in respiratory infections associated with colder climates, (2) warmer climates allow longer periods when vectors are present, and (3) climate can effect changes in the variety of infectious organisms that may be present in a given location.

Portals of Exit

In addition to portals of entry (Chapter 5), the spread of disease also depends on portals of exit (**Figures 6.5** and **6.6**), which in many cases are identical to the portals of entry. Pathogens often exit from the body of an infected host in secreted materials, such as nasal secretions, saliva, sputum, and respiratory droplets. Pathogens can also exit in the blood, vaginal secretions, semen, urine, and feces.

Both portals of entry and portals of exit are important considerations for health care providers, especially in controlling the spread of disease within a patient population.

Keep in Mind

- Mechanisms of transmission are ways in which organisms move from place to place. There are three mechanisms: contact, vehicle, and vector transmission.

- Contact transmission can be direct, indirect, or by droplet transmission.

- There can be predisposing factors for disease such as the host's age, gender, lifestyle, occupation, and emotional status.

- Organisms leave a host by portals of exit, which are essentially the same as the portals of entry.

THE COMPROMISED HOST

In this section, we discuss one of the most important parts of infectious disease: the host. As we have repeatedly said, the spread of infectious diseases depends to no small extent on the hosts involved. The fact that we humans live in a cloud of potentially dangerous infectious organisms yet rarely get seriously ill illustrates that our defensive systems are very good at controlling the number and severity of infections we experience.

For most of this section, we consider the ability of a host to mount a defensive response against invading pathogens as a measure of that host's competence. As noted in Chapter 1, host defenses can be divided into innate and adaptive immune responses. These two powerful defensive systems can deal with most of the pathogens we come into contact with. As long as these mechanisms are in place and working adequately, infectious disease is in many cases only an annoyance. If these defenses are in some way compromised, however, the potential for damaging infectious disease increases. The relationship between host defense and infectious disease is dramatically illustrated by acquired immune deficiency syndrome (AIDS), in which the loss of host defense leads to opportunistic infections that were previously of no consequence but now cause the death of the host.

The loss of immunity in AIDS patients is obvious, but there are other situations in which the host's defense mechanisms can be compromised. There are specific kinds of compromise associated with particular risks for infectious disease. In addition to physiological compromise, meaning an actual impairment of the immune system, there are other risk factors for infection, including lifestyle, occupation, trauma, and travel. Aging is also a major factor in susceptibility to infection. For example, if you were looking at a 75 year-old previously healthy man who had suffered a stroke, you normally would not have to think about infection in the central nervous system. However, if the patient were a 30-year-old with a depressed immune system and was manifesting the symptoms of a stroke, it might be important to consider an infectious disease as the potential cause.

We can list several groups of people who can be considered as having a compromised immune system and therefore being at increased risk of infection:

- People with AIDS
- People with genetic immunodeficiency diseases
- People undergoing chemotherapy
- Patients taking broad-spectrum antibiotics
- Transplant patients
- Burn patients
- Premature infants
- Newborn infants
- The elderly
- Surgical patients
- Patients on artificial ventilators, with catheters or cannulas

We discuss immunodeficiency in detail in Chapter 17, covering the first two groups on this list. Here, we look at some of the other situations listed, keeping in mind that the defensive competence of the host has a major role in the extent and severity of many infectious diseases.

Neutropenia

The condition known as **neutropenia** is defined as a lower-than-normal number of neutrophils (less than 1500 cells per microliter of blood in adults) in the blood. As we will see, this type of white blood cell is a very important component of a host's innate immune response. Common causes of profound neutropenia are the administration of **cytotoxic chemotherapy**, cancer or other diseases that damage bone marrow, congenital disorders characterized by poor bone marrow function, viral infections that disrupt bone marrow function, autoimmune disorders that destroy neutrophils or bone marrow cells, and overwhelming infections that use up neutrophils faster than they can be produced. The drugs used in this type of chemotherapy destroy rapidly growing cells such as those associated with the tumors, but as a side effect they render the patient

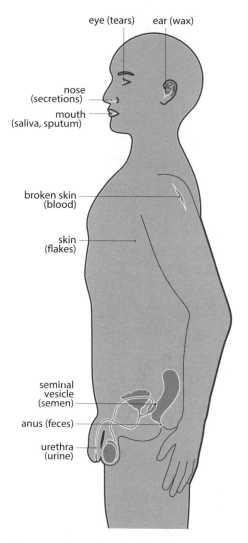

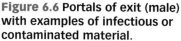

Figure 6.6 Portals of exit (male) with examples of infectious or contaminated material.

temporarily immunoincompetent. Consequently, the chance of infection during the administration of these drugs is very high.

The infections associated with neutropenia are primarily bacterial and fungal. Bacterial infections can begin as soon as the neutrophil level decreases, but fungal infections occur only in people who have been neutropenic for long periods. Although neutropenia is a transitory compromise of host defense and disappears when chemotherapy ends, patients must be monitored carefully for the onset of infection. Neutropenic patients also risk infection with any invasive surgical procedure and whenever an indwelling catheter is necessary.

Gram-negative bacteria from either the patient's colon or the environment can cause life-threatening septicemias in neutropenic patients, and some of these infections can be especially difficult to deal with because of resistance to antibiotics. Gram-positive pathogens, such as staphylococcal and streptococcal species (which are normally found on the skin and in the nasal cavity), may cause infections originating in either the bloodstream or central-venous catheters.

Infections caused by the fungus *Candida albicans*, which is commonly present on the mucosa of the human gastrointestinal tract and in the vagina, can cause mucocutaneous candidiasis in patients with neutropenia. *Aspergillus fumigatus*, a fungus found in many places that can disperse its spores through the air, can cause fungal pneumonia in neutropenic patients.

Organ Transplantation

Unless a patient needing an organ transplant has an identical twin, there will always be immunological differences between the patient and a transplanted organ. These differences are immediately recognized by the recipient's immune system, and, depending on how closely the donor and recipient were matched, significant differences cause a reaction intended to destroy the transplanted organ. The organ is being rejected. Consequently, transplant patients are placed on drug regimens (such as cyclosporine) designed to lessen the immune response against the organ, to prevent rejection. These drugs reduce the chances of rejection by diminishing a patient's overall immune capability. However, this causes the patient to be more susceptible to infection.

Infections in organ-transplant patients require the administration of broad-spectrum antibiotics. As we will see in Part IV, the overuse of these antibiotics leads to **superinfections** and, more importantly, to increased bacterial resistance to antibiotics. This cycle of overuse of antibiotics and increased resistance is seen in many clinical situations and remains one of the most serious concerns of modern health care.

Burn Patients

These patients are at risk because of the loss of large areas of the primary physical barrier to infection, the skin. When skin is lost, there is a greater chance of **septicemia**. Infections with *Pseudomonas aeruginosa* are a particular problem in burn patients because this organism is very resistant to methods used to control bacterial growth, such as antiseptics and disinfectants, and is becoming increasingly resistant to antibiotic treatment. Burn patients are also very susceptible to fungal infections.

The categories just mentioned each refer to one group of patients, namely chemotherapy patients, organ-transplant patients, and burn victims. Now let us look at two types of infection—opportunistic and nosocomial—that can attack multiple groups of susceptible people.

Opportunistic Infections

Many infections are caused by microorganisms that are opportunists, and any infection acquired from such microorganisms is referred to as an **opportunistic infection**. To understand this phenomenon, we must first consider the normal bacterial flora that is part of us. As previously mentioned, there are many bacteria that are residents of the body and provide essential services to the host such as protection from pathogens. For example, the large intestine is filled with organisms that are supposed to be there. These normal bacterial floras found in different areas of the body also prevent other organisms from inhabiting those areas. In some cases, the presence of bacteria in various regions of the body is associated with the production of bacteriocins, which are chemicals (protein toxins) similar to antibiotics that inhibit the growth of bacteria similar or closely related to the bacteria that produced them. Loss of resident bacteria allows the entry and growth of pathogens, which then cause opportunistic infections.

It is interesting to note that many of the resident microorganisms in the body also have the potential to be pathogenic given the right circumstances. This switch in character occurs when these bacteria move to a place in the body where they are normally not found. The best example of this is urinary tract infections. One of the primary causes of this type of infection is *E. coli* traveling from the large intestine, where it is part of the normal flora, to the urethra, where it is pathogenic.

The ability of pathogens to become established, or of resident bacteria to move to unfamiliar places and infect opportunistically, can be compounded by the improper use of antibiotics because such use destroys normal populations of resident bacteria, opening the way for both superinfections and opportunistic infections. In addition, improper use of antibiotics by individuals fosters the development of resistant strains of bacteria. Similarly, improper use of drug therapy by professionals can result in the emergence of resistant bacteria in the hospital environment, paving the way for nosocomial infections.

Nosocomial Infections

As noted in Chapter 1, a **nosocomial infection** is any infection acquired in a hospital or other medical facility. Because they occur during medical treatment, nosocomial infections can affect not only patients but also medical workers. In the United States, the Centers for Disease Control have estimated that as many as 2 million infections a year are acquired in hospitals, resulting in 90,000 deaths. This is more than twice the number of people killed in traffic accidents. The costs associated with these infections are estimated to be over $5 billion per year.

Any patient with a break in the skin (even a **bed sore**) is susceptible to a nosocomial infection, although these infections are usually associated with intravenous applications, urinary and other catheters, invasive tests, and surgical procedures. In fact, anything that allows the entry of organisms into the body is a potential source of nosocomial infections.

The same factors used when considering any other type of infection apply to hospital-borne infections. Therefore, hospitals must consider:

- The source of the infection
- The mode of transmission of the pathogen
- The susceptibility of the patient to infection
- Prevention and control

Fast Fact In some cases, even soaps and cleaning solutions can become a problem if pathogens such as *Pseudomonas* have become resistant to the antibacterial compounds contained in the soap or other cleaning products.

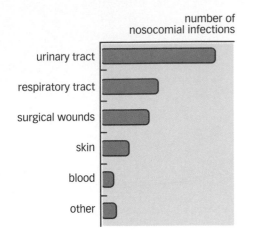

Figure 6.7 Relative frequencies of sites of nosocomial infections.

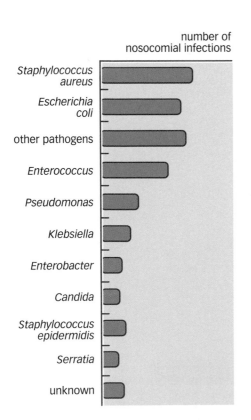

Figure 6.8 Common causative agents of nosocomial infections.

The most common sites of nosocomial infection are shown in **Figure 6.7**. Sources of the infection are usually the environment, other patients, hospital staff, or visitors. In addition, insects such as ants, cockroaches, and flies can cause the problem, and fomites such as waste and unsanitary toilet facilities can also be involved. Last, water supplies, respiratory equipment, plastic supplies, and catheters can all act as sources of nosocomial infections.

Opportunistic organisms can have a role in nosocomial infections, and once again we must consider the patient's degree of immune competence. Any disease or medical procedure requiring convalescence in a hospital will have an effect on the strength and defensive competence of a patient. This combination of a compromised host and an opportunistic pathogen increases the possibility of a nosocomial infection, and the potential for serious infection increases.

Most nosocomial infections are caused by *Escherichia coli*, *Enterococcus* species, *Staphylococcus aureus*, *Clostridium difficile*, or *Pseudomonas aeruginosa* (**Figure 6.8**). These organisms are ubiquitous in hospitals and can easily be moved from place to place and patient to patient by staff, visitors, and other patients. More importantly, many of these organisms are now resistant to antibiotics. For example, **MRSA** (methicillin-resistant *S. aureus*) and **VRSA** (vancomycin-resistant *S. aureus*) have become major medical problems because there are few alternative antibiotic treatments for these infections, which can be deadly in compromised hosts.

Of the equipment and procedures that contribute to nosocomial infections, catheters and respiratory equipment are the most likely to be involved. For example, nebulizers that administer oxygen gas and expand the passageways in the lungs can also distribute pathogens deep into the lung, where infection is much more serious. In addition, pathogens can grow in humidifiers and be transmitted to patients in aerosol form. All respiratory equipment must therefore be either disinfected or sterilized daily when in use and also between uses.

Infections can also easily result from contaminated catheters and tubing used for medical procedures, from loose connections in catheters or tubing, and from inadequate preparation of injection or surgical sites. Organisms that are part of the normal flora (for example *S. epidermidis*) can establish themselves and grow inside catheters and then enter the body. The equipment-contaminating organisms often form biofilms (remember Chapter 4), which makes removal and treatment more difficult. A major concern is patients undergoing dialysis, because there are many potential places for the initiation of a nosocomial infection. This procedure is essential for many people with renal disease, and dialysis clinics routinely reuse equipment for multiple patients daily.

Universal Precautions

In 1988, the Centers for Disease Control (CDC), concerned about the spread of HIV in hospitals, published a set of universal procedures requiring all medical facilities in the United States to conform to specific guidelines for patient care (**Table 6.2**). These procedures apply to all patients, and they include protocols that deal with blood, semen, vaginal secretions, and tissue specimens, and cerebrospinal, synovial, plural, peritoneal, pericardial, and amniotic fluids. They do not apply to feces, nasal secretions, sputum, sweat, tears, urine or vomit, as long as these do not contain blood.

1.	Wear gloves and gown if hands, skin, or clothing is likely to be soiled with patient's blood or other body fluids
2.	Wear nose/mouth mask and protective eyewear or chin-length plastic face shield whenever splashing or splattering of blood or body fluids is likely. A nose/mouth mask alone is not sufficient
3.	Wash hands before and after patient contact and after removal of gloves. Change gloves between patients
4.	Use disposable mouthpiece/airway for cardiopulmonary resuscitation
5.	Discard contaminated needles and other sharp items immediately in puncture-proof container. Needles must not be bent, clipped, or re-capped
6.	Clean spills of blood or contaminated fluids by (1) putting on gloves and any other barriers needed, (2) wiping up with disposable towels, (3) washing with soap and water, and (4) disinfecting with 1:10 solution of household bleach and water. Bleach solution should not have been prepared more than 24 hours beforehand

Table 6.2 CDC recommendations for preventing nosocomial infections.

Preventing and Controlling Nosocomial Infections

Hospitals are very aware of the problems associated with nosocomial infections. In the United States every hospital must have control programs in place if it is to be accredited by the American Hospital Association. These programs involve:

- Surveillance of nosocomial infections in patients and staff

- On-site microbiology laboratory plus standardized isolation procedures

- Standardized procedures for the use of catheters and hospital equipment

- Proper decontamination and sanitary procedures

- Mandatory nosocomial-disease education programs

- In some cases, infection-control specialist on staff

Keep in Mind

- One of the most important factors of the infectious process is the host defense.

- Compromise of the host defense can be due to a variety of problems such as infection with HIV resulting in AIDS, congenital immunodeficiency, transplantation, chemotherapy, or other conditions that debilitate the patient.

- Some infections are caused by opportunistic pathogens, which are normally harmless but can be pathogenic if the right conditions exist.

- Infections that occur in hospitals are referred to as nosocomial infections.

- Universal precautions are specific guidelines for patient care, which provide proper procedures for dealing with blood, semen, vaginal secretions, and other samples of tissue and body fluids.

EPIDEMIOLOGY

Epidemiology is the study of the factors and mechanisms involved in the frequency and spread of diseases or other health-related problems, and is an important part of our understanding of disease. These health-related problems can vary from infection to cigarette smoking or lead poisoning, but we will limit our discussion to infectious diseases, which means those caused by infectious pathogens invading a formerly healthy host. Epidemiology can be used not only as a tool to study disease but also as a way to design methods for the control and prevention of diseases.

As we have seen in all our discussions so far, there is tremendous variability associated with disease. When you combine things such as contagiousness, latency, and virulence with the large population on our planet, it is easy to see the need for up-to-date information on and strategies for combating infectious disease.

Incidence and Prevalence

The **incidence of a disease** is defined as the number of new cases contracted within a set population in a specific period of time. This kind of information can provide us with a reliable indication of the spread of a particular disease. **Prevalence** is defined as the total number of people infected within a population at any given time. This means new cases and already existing cases. Prevalence data can be used to measure how seriously and for how long a particular disease affects the population (**Figure 6.9**). The number itself does not indicate how severe that disease is, just the number of people affected.

Morbidity and Mortality Rates

The **morbidity rate** of a disease is the number of individuals affected by the disease during a set period divided by the total population. The **mortality rate** is the number of deaths due to a specific disease during a specific period divided by the total population.

When epidemiological studies are used to examine parameters such as particular geographic areas and the degree of harm caused by a disease, they classify diseases as sporadic, endemic, epidemic, and pandemic. **Sporadic diseases** occur in a random and unpredictable manner and pose no threat to public health. **Endemic** refers to diseases that are constantly in the population but in numbers too low to be a public health problem. As an example, the common cold is an endemic disease because there is always a percentage of any given population that has this infection. However, this percentage of the total population is low.

An **epidemic** occurs when the incidence of a disease suddenly becomes higher than the normally expected number of individuals affected by the disease. Over the past hundred years, for example, there have been several epidemics of influenza, such as the one that occurred in 1918 and killed millions. An epidemic causes morbidity and mortality rates to increase, and these increases signal a public health problem. It is important to note that there are several factors that contribute to the potential for the development of an epidemic. For example, the large populations that inhabit major cities allow the rapid spread of disease within a population. Increased aging of a population leads to a higher percentage of people who are immunologically or otherwise compromised and are therefore more predisposed to infection. Lack of **herd immunity** as a result of incomplete or inadequate vaccination is another factor in epidemics. Perhaps the most important factor of all in our twenty-first-century lives

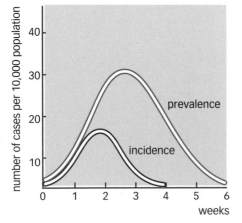

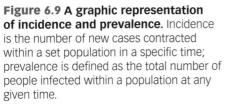

Figure 6.9 A graphic representation of incidence and prevalence. Incidence is the number of new cases contracted within a set population in a specific time; prevalence is defined as the total number of people infected within a population at any given time.

is the ability to travel easily and rapidly and thereby transmit diseases around the world.

When a disease occurs in epidemic proportions throughout the world, it is called a **pandemic**.

Epidemics are affected by the type of pathogen causing the disease and also by the mode of transmission. There are two basic types of epidemic, graphically represented in **Figure 6.10**: common-source outbreaks and propagated epidemics.

- A **common-source outbreak** is an epidemic that arises from contact with contaminated substances and most commonly occurs either when a water supply is contaminated with fecal material or when food is improperly prepared or contaminated in processing. Although common-source outbreaks affect large numbers of people, they quickly subside when the source of contamination has been identified and resolved.

- A **propagated epidemic** is one that results from amplification of the number of infected individuals as person-to-person contacts occur. Propagated epidemics stay in the population for long periods and are more difficult to deal with than common-source outbreaks.

Herd Immunity

One of the most important preventative aspects of infection and the spread of infection is herd immunity, illustrated in **Figure 6.11** This immunity can be conferred through vaccination or it can be developed after infection. Let's look at smallpox as a way of understanding this concept. In years past smallpox was responsible for millions of deaths. When a vaccination against the disease was developed and given widely, the immunity of the population as a whole to this infection increased. (Many older people carry the vaccination mark on an arm.) It does not mean that every individual in the entire population is immune, but individual protection is conferred by limiting the number of possible hosts. When the population, the herd, had become immune, the disease essentially disappeared because there were insufficient susceptible hosts to maintain transmission of the pathogen. When smallpox was eradicated, vaccinations stopped. As a result, there are now lots of potential (non-vaccinated) targets and the infection could spread once again, if the pathogen were to re-emerge.

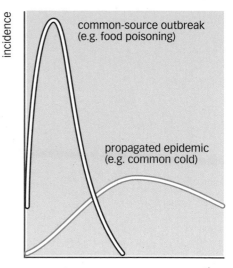

Figure 6.10 A graphic representation of the two types of epidemic. These are the common-source outbreak arising from contact with contaminated substances such as water supplies, and the propagated epidemic resulting from amplification of the number of infected individuals as person-to-person contact occurs.

Figure 6.11 Herd immunity. Notice that the lower the percentage of immune individuals in the group, the greater is the number of potential susceptible targets and therefore the greater is the number of cases of disease.

○ infected ● susceptible ● immune

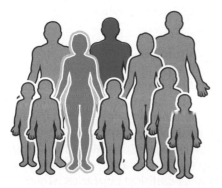

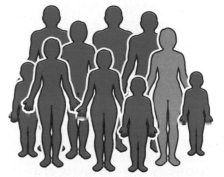

10% of the population is immune   50% of the population is immune   90% of the population is immune

Herd immunity is a major and fundamental parameter for infectious diseases. However, it also presents a socioeconomic problem. Should we vaccinate against diseases that are no longer present in the population? What about poliomyelitis, a disease that killed and crippled thousands in the past but is today essentially nonexistent in the United States and many other countries worldwide, yet not globally eradicated (as was the case with smallpox)? What is the herd immunity like for this disease? What would happen if we stopped vaccinating against it? Polio could be contracted in other countries in the world, where it is still prevalent, and therefore could resurface somewhere it is not currently found at any time, causing human suffering and financial loss.

Types of Epidemiological Study

There are two types of epidemiological study: descriptive and analytical. **Descriptive epidemiological studies** are concerned with the physical aspects of patients and the spread of the disease. These studies include data on the number of cases and on which segment of a population was affected. They also include the location and time of the infections, as well as the age, gender, race, marital status, and occupation of those affected. Careful examination of this type of information can allow epidemiologists to trace the outbreak of the disease and possibly identify the **index case**, which is the first person to have been infected with the disease. Identifying an index case is not mandatory for a successful descriptive epidemiological study. In some cases, it may be impossible to identify the index case because that person has recovered, moved away, or died.

In **analytical epidemiological studies**, the focus is on establishing a cause-and-effect relationship. These studies are always done in conjunction with a control group, which is a population known to be free of the disease being studied. A **retrospective analytical study** is one in which the records of patients who have already contracted the disease are studied. Using this type of already-available patient information, workers running retrospective studies can take into account a wide variety of factors that preceded an epidemic so that they can narrow down the potential causes. In contrast, **prospective analytical studies** do not allow the benefit of hindsight; in this type of study, data are analyzed as they are collected. In other words, prospective studies consider factors that occur as the epidemic proceeds. Knowledge-based mathematical models are invaluable tools. Retrospective studies allow epidemiologists to develop and refine the models to get more accurate predictions for the future, which then allow taking preventative or limiting measures for the benefit of everyone. You might want to compare this with a weather forecast. The more we learn from the past occurrence of rain, the better we can predict whether it is going to be sunny tomorrow or whether we should take an umbrella.

Health departments at the local and state levels require that doctors and hospitals report certain diseases. This type of information has been able to show how the effects of infectious diseases have changed over the years (**Figure 6.12**). In addition, some diseases are classified as **nationally notifiable** (**Table 6.3**), the ones that are particularly important to control and contain, which means they must also be reported to national centers such as the Centers for Disease Control in the United States, the clearing house for epidemiological research. Without question, the United States is one of the world's healthiest countries, which is due in no small measure to epidemiology.

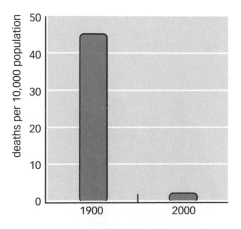

Figure 6.12 Deaths caused by infectious diseases changed dramatically over the course of a century in the United States.

Anthrax	Gonorrhea	Mumps	Streptococcal toxic shock syndrome
Babesiosis	*Haemophilus influenzae*, invasive	Novel influenza A infections	*Streptococcus pneumoniae*, invasive
Botulism	Hansen's disease (leprosy)	Pertussis	Syphilis, congenital
Brucellosis	Hantavirus pulmonary syndrome	Plague	Syphilis, contracted
Chancroid	Hemolytic uremic syndrome	Poliomyelitis	Tetanus
Chlamydia trachomatis	Hepatitis A	Psittacosis	Toxic shock syndrome (other than streptococcal)
Cholera	Hepatitis B	Q fever	Trichinellosis
Coccidiomycosis	Hepatitis C	Rabies, animal	Tuberculosis
Cyclosporiasis	HIV infection	Rabies, human	Tularemia
Cryptosporidiosis	Influenza-associated pediatric mortality	Rocky Mountain spotted fever	Typhoid fever
Dengue virus infections	Legionellosis	Rubella	Vancomycin-intermediate *Staphylococcus aureus* (VISA)
Diphtheria	Listeriosis	Rubella syndrome, congenital	Vancomycin-resistant *Staphylococcus aureus* (VRSA)
Ehrlichiosis	Lyme disease	Salmonellosis	Varicella (fatal)
Arboreal encephalitis/meningitis	Malaria	SARS	Vibrosis
Enterohemorrhagic *E. coli*	Measles	Shigellosis	Viral hemorrhagic fever
Giardiasis	Meningococcal disease	Smallpox	Yellow fever

Table 6.3 Infectious diseases that are nationally notifiable in the United States as of 2013.

Keep in Mind

- Epidemiology is the study of factors and mechanisms involved in the frequency or spread of diseases or health-related problems.

- Incidence describes the number of new cases contracted within a set population in a specific period of time.

- Prevalence is the total number of people infected within a population at any given time.

- The morbidity rate of a disease is the percentage of individuals affected by the disease during a set period.

- The mortality rate is the percentage of deaths due to a specific disease during a specific period.

- Diseases can be sporadic (occurring only occasionally), endemic (constantly in the population), epidemic (a higher than normal incidence of a disease), or pandemic (a worldwide epidemic).

- Herd immunity can limit the spread of infection.

- There are two types of epidemiological study: descriptive and analytical.

- Analytical epidemiological studies always contain a control group and can be retrospective or prospective.

DIAGNOSING INFECTIONS

You have now learned that infectious disease control, prevention, and therapy are closely linked to limiting any vectors and their reservoirs, immunization, ideally pathogen eradication, and also monitoring spread and transmission patterns. All these actions are highly dependent on a quick and correct diagnosis of the species or even strain of pathogen involved.

Since Koch established his postulates, which allow us to identify the causative agent of an infectious disease, we have come a long way in developing procedures and technologies for diagnostics. There is a common framework of steps to take, but what is done has changed over time and depends on the pathogen and the circumstances. Is the patient near a well-equipped laboratory? Is time of the essence? Are we working with a small budget? Sometimes we have to compromise, but the more we know about diagnostics and the factors affecting diagnosis the better we are at making the right choice in the interest of the patient and society.

The two key criteria of any diagnostic method are **specificity** and **sensitivity**. We want our test to be specific, which means it allows us to identify a certain pathogen reliably and correctly. Additionally, we need it to be sensitive. A very sensitive test permits the detection of very small amounts of the pathogen or antibodies against it. This makes a big difference, because we can analyze small samples (sometimes it is impossible to take large amounts of sample from a patient) or low-concentration samples (for example, at an early stage an infection).

Samples should be taken at or near the site of infection or where we expect the highest pathogen load. It is important to circumvent the patient's normal flora, because otherwise it can make it difficult or impossible to isolate and identify the pathogen. Common samples are blood, urine, feces, sputum, swabs, and biopsy samples. Then appropriate transport media and conditions have to be chosen to make sure that the pathogens in the sample do not die or grow further. Obviously everything needs to be properly and correctly labeled. We cannot risk lives by swapping or losing samples.

Growth-Based Diagnostics

Growth-based diagnostics depend on the fact that microorganisms grow differently on different media. When the sample arrives at the laboratory, growth media are **inoculated** from the sample. Culturable microorganisms such as bacteria can be grown on solid or in liquid media, which contain all nutrients and other necessary ingredients. We basically have to mimic the conditions in the host, so we do not have extreme growth conditions as are necessary for environmental samples. Some pathogens are very challenging to grow (for example *Legionella pneumophila*). There are media that support the growth of many bacteria, but of more use in diagnostics are **selective media** (which suppress the growth of some organisms), **elective media** (which support the growth of only the organism of interest), and **differential media** (which show differences between organisms in mixed samples). Some of these media are described in more detail in Chapter 10. Microorganisms can then be grown under aerobic conditions (with oxygen) or anaerobic conditions (without oxygen).

After the mixture of organisms from the sample has been grown in the chosen media, the suspected pathogen needs to be isolated in pure culture. The stakes are high when health is concerned, so work is usually done according to standard operating procedures: results need to be consistent and comparable. For **cultures** containing only the pathogen,

Figure 6.13 Example of a commercially available bacterial identification strip. Growth and color change are based on the physiology of the bacterium.

colony morphology can be assessed. The cell morphology is analyzed using a microscope (usually a light microscope; in rare cases an electron microscope) after the cells have been stained (see Chapter 4).

Growth-based diagnostic methods can also include analyzing the pathogen's physiology. We know the metabolic pattern of bacteria and can use this knowledge to differentiate them further via their biochemical reactions. A range of media are used so that a diagnostic pattern emerges. It is almost like taking a metabolic or physiological fingerprint of a pathogen (do not confuse this with the actual term "metabolic fingerprinting," which is used in large-scale metabolic analyses). In Chapter 3 you learned about microbial fermentations in which bacteria can grow using various compounds (such as certain sugars) and produce a variety of fermentation products (such as lactic acid). In diagnostic tests, a range of nutrient media is prepared, which are the same except for one component such as the **carbon source**. By determining where the pathogen can or cannot grow, we can narrow down what species it could be, or even identify it. Often there are components added to the test medium that result in a change of color during fermentation (**Figure 6.13**). This allows quicker or even automated reading of the results. Over time these tests have been made smaller, and most are not done in test tubes any longer but in cupules on strips or discs, which are manufactured by companies and sold to the diagnostic laboratories. Again, this helps with consistency and we only need very small sample volumes.

Serology-Based Diagnostics

A well-established alternative to growth-based methods is identifying the pathogen based on certain of its **antigens** (molecules on the pathogen that cause an immune response in the host, including the production of **antibodies** that specifically bind to this antigen). We can either use isolated or manufactured antibodies to detect pathogen antigens in the sample or to detect host antibodies in the sample. Such **serological** analysis usually uses blood as the sample and it can be tested directly (there is no need to grow cultures of the pathogen).

Blood samples can be used directly for serological analysis. Antibodies are used in diagnostics as the target for identification or as a tool for the diagnostic method. We need large amounts of specific, well-characterized antibodies to use them in reliable routine diagnostics. This demand has been met for some time by advances in **biotechnology** and the pharmaceutical industry. To help you understand the basic concepts and principles that are involved in developing and producing the modern tools of diagnostics, we will now take a little detour via **genetic engineering** and other relevant aspects of biotechnology before we discuss the actual serology-based diagnostics. These concepts and techniques are also used for research into infectious diseases and their prevention, and in the development of treatment and therapeutics.

Biotechnology and Health

Biotechnology has made routine what used to be fantastic, and is now ubiquitous in our day-to-day existence. It affects every aspect of our

1953
Watson and Crick define the structure of DNA
1966
Definition of the human genetic code
1970s
Recombinant DNA experimentation
1980s
Successful recombination of genetic material
First production of human drug by bacteria—insulin
First recombinant vaccine for humans—hepatitis B
1990s
Human Genome Project started
Animals successfully cloned
Human embryonic stem cells produced
2000s
Human genome sequenced
Gene therapy used in humans

lives, from food to reproduction to politics. There is no area that cannot somehow be related to some aspect of new biological technology. The tools of biotechnology are used in everything from manufacturing soaps to genetically engineering crops to potentially turning bacteria into weapons. Because these advances will impact the health care community, health care professionals should have a basic understanding of how they are being made.

Over time, scientists have developed the ability to use the basic cellular processes in microbes to create a large number of useful biotechnological tools. Despite the diversity of cell types in nature, most have the same basic properties: (1) they all need energy to function, (2) that energy is supplied in chemical form, and (3) they all synthesize proteins from a template. These shared basic characteristics provide opportunities for external manipulation by scientists. To see the limitless applications of biotechnology, you need a basic understanding of the fundamental molecules and techniques used in developing biotech products. Basic techniques are genetic engineering, the use of recombinant DNA, and the various specialties that fall under the umbrella name molecular biology (**Figure 6.14**).

In the outline to the left, we can see an overview of the beginnings of what we now call biotechnology.

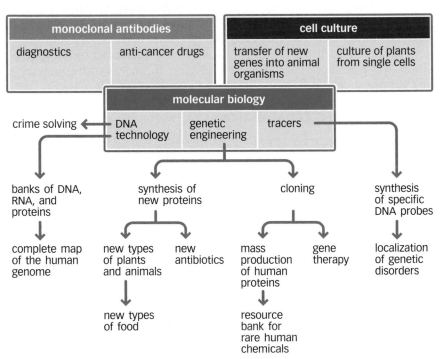

Figure 6.14 An overview of some health-related areas in which biotechnology is used. Companies involved in the development of products and processes hire specifically from the biological sciences for their employees. Students of microbiology, molecular biology, biochemistry, and pharmacology, to name just a few disciplines, are being actively recruited in ever-greater numbers. Even if you have no interest in the field of human health care you can work in *product discovery and development*. An advanced knowledge of biotechnological systems is essential for researchers trying to create both new diagnostic tests for diseases and protocols for gene-based therapies. To be able to deconstruct a disease and find ways to treat it, you must first understand the pathways and processes of the disease. This understanding requires a more laboratory-research-oriented profession.

Progress has been astonishing and the biotechnology industry has grown explosively in the past 30 years. It is an international enterprise with almost $100 billion dollars in annual revenues. More than 200 diseases such as Alzheimer's, AIDS, cancer, multiple sclerosis, diabetes, and arthritis are currently being targeted as possibly being amenable to treatment by drugs produced by biotechnology. In the United States, the biotech industry is regulated by the Food and Drug Administration (FDA), the Environmental Protection Agency (EPA), and the Department of Agriculture (USDA). The level of governmental regulation allows a certain amount of consumer confidence. Worldwide regulatory bodies do not exist as such.

Biotechnology has enormous potential to improve the health and well-being of people worldwide. However, it also has the potential to create repercussions of unforeseen magnitude, both in our own society and globally. **Recombinant DNA** technology has become accepted, and ethical considerations have moved on to issues such as cloning and stem cell research. The attempted regulation of access to embryonic stem cell lines has reached international proportions.

Science is going to have to apply itself to the rigorous standards of public scrutiny. Whether we, as professionals, pass that test will depend on our willingness and ability to understand the issues, defend our positions, and hold on to our view of society as a whole. Without knowledge, we fall prey to ignorance and misunderstanding and fear. Bearing all this in mind, we will first discuss recombinant DNA and genetic engineering in more detail.

Recombinant DNA Technology

Recall from Chapter 2 that DNA is a double-stranded molecule and is the genetic blueprint for every living organism. The two-strand structure of DNA has an essential role in the ability of scientists to take apart a DNA molecule and to recombine it either in different configurations or with DNA from different organisms. Scientists are able to cut out a piece of DNA (a gene), insert it into DNA from another organism (extrachromosomal entities known as a vector) to make a piece of recombinant DNA. The most commonly used vector is the plasmid: circular DNA that exists in the cytoplasm, can replicate in the host cell, and is easily transferable from one cell to another. For recombinant DNA amplification, these are ideal characteristics. Bacteriophages (viruses that attach to bacteria) can also be used.

The **recombinant plasmid** can then be introduced into a host cell to be amplified, and sometimes the protein coded for by the original piece of DNA can be made (**Figure 6.15**). This is the fundamental principle of recombinant DNA technology.

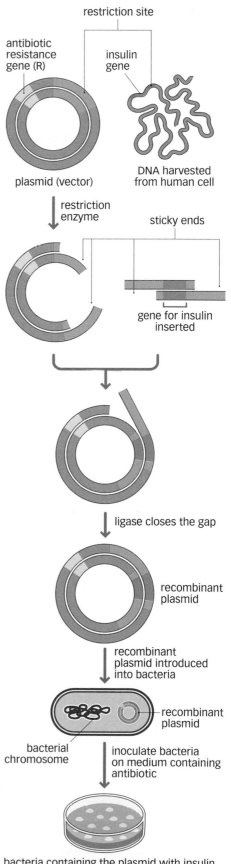

Figure 6.15 The construction of a recombinant plasmid vector. The insulin gene is used as the DNA sequence of interest. Restriction enzymes are used to open the plasmid at a specific location and also cut the DNA sequence of interest ready for insertion. Some restriction enzymes produce ends where one strand is longer than the complementary one. We call these 'sticky ends.' The DNA to be replicated is inserted and the strands are reconnected with an enzyme called ligase. The recombinant plasmid is then introduced into a host cell (for example a bacterium) that will amplify and potentially produce protein. The antibiotic resistance gene on the plasmid permits the selection of cells containing the plasmid.

Figure 6.16 A few of the many uses of recombinant DNA technology.

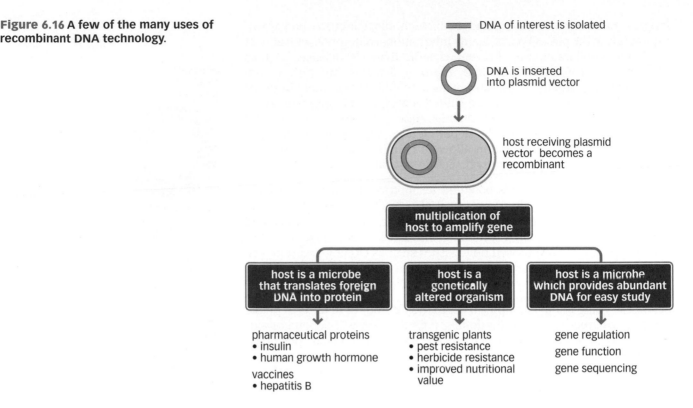

DNA of interest is isolated

DNA is inserted into plasmid vector

host receiving plasmid vector becomes a recombinant

multiplication of host to amplify gene

host is a microbe that translates foreign DNA into protein

host is a genetically altered organism

host is a microbe which provides abundant DNA for easy study

pharmaceutical proteins
• insulin
• human growth hormone
vaccines
• hepatitis B

transgenic plants
• pest resistance
• herbicide resistance
• improved nutritional value

gene regulation

gene function

gene sequencing

Antibiotic resistance genes on the vectors are commonly used to help select cells with the recombinant plasmid. Yes, they could easily be transferred further and thus add to the problem of antibiotic resistances that we face when treating infections. Legislation regarding the creation and use of genetically modified organisms ensures containment, unless the law is broken or best practice in the laboratory is not followed. Industrial use and containment are often regulated even more strictly. Given the scale at which antibiotic-resistant organisms evolve every day outside the laboratory as a result of incorrect treatment or the use of antibiotics, this approach does not add greatly to the problem. But why do we take the risk at all and create such a double-edged sword? To answer that question, think about this: what if your child had juvenile diabetes and there was no insulin to be found? Now think about the alternative: picking up the recombinant insulin from the pharmacy. There are many applications of recombinant DNA technology (**Figure 6.16**). It is the foundation for virtually all biotechnological research and development related to pharmaceutical manufacturing, one of which is the production of antibodies for research, therapy, and diagnostics.

Nowadays the standard method for amplifying desired stretches of DNA is the **polymerase chain reaction** (PCR), in which a large amount of DNA can be replicated in only a few hours. It has proved a revolutionary technique, and its inventor, Kary B. Mullis, was awarded the Nobel prize in 1993. We will revisit this technology for diagnostic purposes later in this chapter. Basically, the segment of DNA to be amplified (the target) is mixed with special short stretches of DNA (primers), the four nucleotides, and an enzyme called DNA polymerase (**Figure 6.17**) to produce large numbers of copies of the target DNA.

Monoclonal Antibodies

When a host's immune system is exposed to an antigen, specific host cells respond by producing antibodies. Many decades ago, antibodies

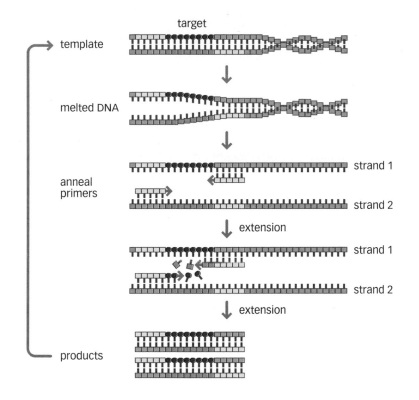

target

template

melted DNA

anneal
primers

strand 1

strand 2

extension

strand 1

strand 2

extension

products

Figure 6.17 The polymerase chain reaction (PCR), which uses heat, complementary base pairing, and polymerase to amplify a target DNA strand. The first step is to separate (melt) the target segment into its two single strands by heating the mixture to about 96°C. One primer is complementary to one end of strand 1, and another primer is complementary to one end of strand 2. In the next step, the temperature is lowered and the primers attach (anneal) to a complementary stretch. The extension step takes place at the best operating temperature for DNA polymerase, which extends each primer, using up nucleotides and ATP in the mixture, giving two double-stranded DNA segments, each identical to the original target segment. These three steps are repeated in many cycles until all the nucleotides and ATP are used up. The products of a completed cycle are the targets of the next cycle, making it a chain reaction that synthesizes exponentially increasing products.

were produced by challenging an animal with an antigen, and then harvesting the antibodies from the ascites, the abdominal fluid of the animal. Aside from the use of animals, the additional problem was that the harvested product was not always pure and so might contain molecules dangerous to humans. There were also insufficient quantities produced and the antibodies were a mixture of various antibodies to start with. To solve these problems, the goal of biotechnology scientists was a fast, clean, safe, humane procedure that would produce large amounts of good-quality antibodies and do so repeatedly for long periods. In 1975, Köhler and Milstein developed an advanced type of cell culture for creating antibodies from a single cell such that all populations of those antibodies are exactly the same in their specificity and binding abilities, and no longer a mixture. The single cell from which the antibodies are produced is called a clone cell, so the antibodies are called monoclonal antibodies. They have exceptional purity and specificity, are capable of binding to very specific antigens, and are also capable of recognizing other nonself molecules, such as drugs and viral or bacterial products. Meanwhile, technology has even advanced to recombinant antibody production, which we will also discuss briefly.

Antibody-producing cells have a relatively short life span. Malignant tumor cell lines, in contrast, are virtually immortal but are incapable of producing antibodies. Fusing them creates a hybridoma (a hybrid of nuclei from the two cell lines; **Figure 6.18**). Successful hybridomas became the tool for producing monoclonal antibodies.

Monoclonal antibodies can be used for everything from diagnostic tests to cancer therapy. Scientists are even able to attach, or conjugate, toxins and drugs onto these antibodies and then use the conjugated molecules as delivery vehicles. Monoclonal antibodies can be used to attack specific parts of the immune system of an organ-transplant patient to prevent rejection while leaving the rest of the immune system free to fight off the common cold! There is a very high commercial demand for monoclonal antibodies because their uses today include diagnosing infectious diseases, detecting harmful microorganisms in food, distinguishing cancer cells from normal cells, delivering chemotherapeutic drugs to cancer cells while avoiding the normal healthy cells, and locating environmental pollutants.

The simplest method of monoclonal antibody production via hybridomas still requires the use of mice, and it is prohibited in many countries now. Alternative FDA-approved methods use bulk tissue cultures or indeed recombinant vectors in cells such as *E. coli*. Recombinant technology permits the large-scale production of antibody fragments and their combinations with other biologically active molecules; even full-length antibodies can be produced and then customized depending on the application. The recombinant production capacity has been based largely on microbial cultures and mammalian cells, but meanwhile alternative systems were developed for plants. These green bioreactors recombinantly produce antibodies (called plantibodies), and **clinical trials** are being undertaken for approval.

We can produce completely functional antigen-binding fragments in bacterial systems, but there is even further use of recombinant *E. coli* in this context. The genetic information for the antibody component is inserted into the phage genome joined to DNA that codes for a protein on the outside of the phage, so the protein produced is the antibody fragment joined to the viral protein. New phages display the antibody fragment to the environment (**Figure 6.19**), and this technology is called **phage display**.

Usually in phage display we do not start with just the genetic information for one antibody fragment, but with a library of genetic information. This helps to identify the fragment that binds best to the antigen. The binding abilities of the displayed antibody fragment can be improved by repeating selection rounds, which is another advantage to using hybridomas. After the selection, the genetic information of the antibody fragments is taken out of the phage vector and inserted into a vector that allows the production of the antigen fragment only.

It does not stop there: the genetic information of the antibody fragments can be genetically fused to marker proteins such as **fluorescent proteins** or **alkaline phosphatase** and produced together. These markers are the basis of detecting results in serology-based diagnostics, as you will see in the next section. Traditionally, markers were added *in vitro* by means of a chemical reaction with the antibodies used in diagnostics, requiring additional effort, cost, and time. Phage particles still fused to antibodies can also be used for diagnostics, and by adapting the antibody fragment displayed or adding a further marker, it can be optimally tailored for the desired diagnostic assay.

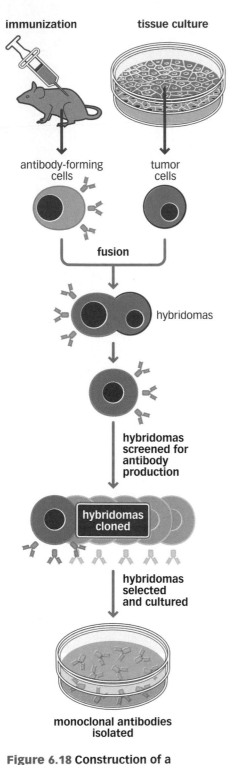

immunization   tissue culture

antibody-forming   tumor
cells   cells

fusion

hybridomas

**hybridomas
screened for
antibody
production**

hybridomas
cloned

**hybridomas
selected
and cultured**

**monoclonal antibodies
isolated**

Figure 6.18 Construction of a hybridoma that secretes a specific monoclonal antibody. After the antibody-producing cell has been fused with the tumor cell, the resulting hybridomas are screened and then used to produce large quantities of monoclonal antibodies.

Serology-Based Diagnostic Methods

You have already learned that the serology-based diagnosis of infectious diseases is based on an *in vitro* antigen–antibody reaction. This reaction is optimal if both antigen and antibody are present at the same concentration. Samples are usually diluted across a range of concentrations (**serial dilution**) when analyzed, to ensure successful detection. The lowest concentration of patient sample at which a reaction can be detected is called its **titer**. We have to know that a reaction took place and therefore need to make it visible somehow. Traditional serological methods use large volumes, and the reaction can be seen with the naked eye. When protein precipitates (becomes insoluble) we can see it turning white, in the same way as when you boil an egg. When antibody and antigen bind in large amounts, they become insoluble and therefore visible. We can analyze samples in either direction. An unknown antigen (for example a viral protein) in the sample can be identified or detected by using a known antibody, which is called the **Gruber reaction**. The opposite is the **Widal reaction**, in which a known antigen is used to spot an unknown antibody in the sample. If we use much smaller amounts we have to develop a method that produces another type of signal (such as fluorescence), which we can use to measure binding.

Depending on the amount of sample available, a suitably sensitive method is used. The techniques we are going to discuss now cover quite a range. **Precipitation** detects 0.1–1 mg of antigen, and **immunofluorescence** 50 μg, making them the least sensitive methods. **Active agglutination** detects about 2.5 μg of antigen, whereas **passive agglutination** is up to five times more sensitive. Both **radioimmunoassay** (RIA) and **enzyme immunoassay** (EIA) allow the detection of 0.1–1 ng of antigen.

Precipitation

Precipitation is performed in test tubes. A successful reaction can be seen as a white ring, which is why it is called the ring test (**Figure 6.20**). If we trap antibody in a gel matrix and put the antigen sample in a well within that gel, we can use much smaller amounts (**Figure 6.21**). To speed up the movement of the proteins in the gel (after all, we need to know soon what the patient is suffering from and how to help), we can apply an electric current, a method known as electroimmunodiffusion.

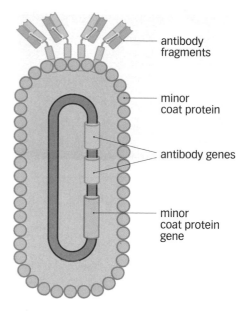

Figure 6.19 A phage displaying antibody fragments on its surface.

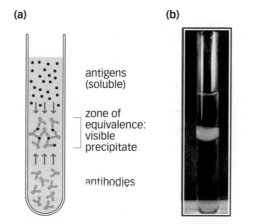

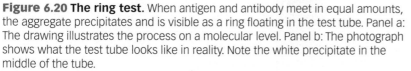

Figure 6.20 The ring test. When antigen and antibody meet in equal amounts, the aggregate precipitates and is visible as a ring floating in the test tube. Panel a: The drawing illustrates the process on a molecular level. Panel b: The photograph shows what the test tube looks like in reality. Note the white precipitate in the middle of the tube.

Figure 6.21 Photograph of the result of double-diffusion precipitation. Antibody is placed in the well in the center to diffuse into the gel and react with different antigens diffusing in from the peripheral wells. At equal amounts a line becomes visible.

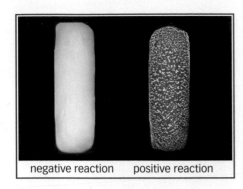

Figure 6.22 Photograph of the result of latex agglutination. The negative result appears as a milky solution, whereas we can see agglutination in the positive result.

Agglutination

Agglutination works on a smaller scale and allows faster results, by coupling one of the reaction partners (for example the antibody) to a bulky component such as erythrocytes or latex beads. Without a reaction the solution appears milky, but when a reaction takes place we can observe that the latex beads agglutinate (**Figure 6.22**).

Immunoblotting

The immunoblot permits the analysis of a mixture of antigens or antibodies. Proteins are separated by using an electric current (gel electrophoresis) and then transferred to a membrane. The membrane is then incubated in a solution containing antibody. This entire procedure is called a **Western blot**, and that is why we call the method immunoblotting (**Figure 6.23**).

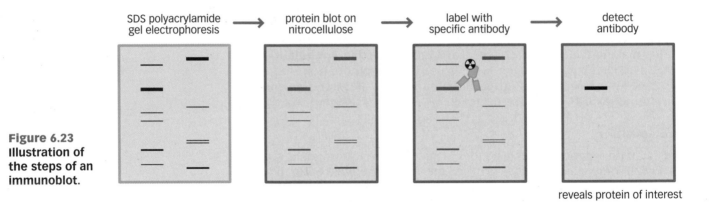

SDS polyacrylamide gel electrophoresis → protein blot on nitrocellulose → label with specific antibody → detect antibody

reveals protein of interest

Figure 6.23 Illustration of the steps of an immunoblot.

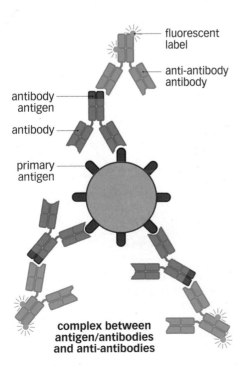

fluorescent label

anti-antibody antibody

antibody antigen

antibody

primary antigen

complex between antigen/antibodies and anti-antibodies

The antibodies bind to the antigen on the membrane. The binding is detected then via a signal that can be measured. The antibody can be conjugated to a **fluorescent molecule**, a **radioisotope**, or an enzyme that catalyzes a chemical reaction producing a color change or light. Alternatively, this antibody is not labeled itself, but a second labeled antibody is used that binds to the first antibody. We know that the first antibody is from a human (the patient); the second antibody would be sourced from an animal (such as rabbit or sheep) whose immune system had been challenged with human antibodies. The use of this secondary antibody amplifies the signal and therefore increases the sensitivity of the detection. The use of fluorescent antibodies in serology-based diagnostic is generally called direct immunofluorescence. Indirect immunofluorescence uses a secondary fluorescent antibody (**Figure 6.24**). We can analyze antigens in a sample taken from the patient. They would be on the membrane, and we use known antibodies for detection. Yet more commonly, antibodies in a patient's sample are analyzed by using known

Figure 6.24 Illustration of the principles of indirect immunofluorescence. The human antibody against the primary antigen (purple) has antigens itself (red circles). Anti-human antibody from another species is labeled with a fluorescent molecule (green). It binds the antibody that bound to the antigen.

antigens on the membrane. An example is HIV diagnosis. Anti-HIV antibodies in the patient's serum bind to HIV proteins on the membrane. Such a diagnostic result would then be confirmed with another, more sensitive diagnostic test, the enzyme immunoassay.

ELISA

In the enzyme immunoassay (or enzyme-linked immunosorbent assay; ELISA) the solid matrix for the antibody–antigen binding reaction is not a membrane but the bottom of a well. The well, together with many other wells, is part of a microtiter plate (Figure 6.25). You can imagine that this allows even smaller sample amounts and automated reading of the signal. And of course the diagnosis can again go either way: for a direct diagnosis the assay detects the antigen in the sample (wells are coated with known antibody), whereas antigen-coated wells are used in the indirect diagnosis, where the assay (Figure 6.26) detects antibodies in the sample (such as those used in the diagnosis of HIV). The analytical antibody can be is conjugated to an enzyme, just as discussed above, and used directly for detection. If the analytical antibody is unlabeled, a further labeled antibody needs to be involved, again as discussed above, for this indirect detection. If a radioisotope-labeled antibody is used, the method is called radioimmunoassay. There are also other variations of ELISA.

Even though this method can deal with a high throughput of analyses, as in a large hospital, it takes some time to complete all the incubation and wash steps. A very rapid EIA is the dipstick, which is used in health care to get a very quick answer to a diagnostic question on the spot (for example at the bedside or in point-of-care analysis). We know about dipsticks for use at home: this rapid EIA principle applies, for instance, to home pregnancy tests, which detect the hormone human chorionic gonadotropin in the urine of a pregnant woman instead of detecting a pathogen's antigen, which is the focus here.

Diagnostic Methods Based on Nucleic Acids

As a further alternative, we will now discuss methods based on nucleic acids. Here, identifying the pathogen is based on a nucleic acid sequence, specific to a certain pathogen, in a patient's sample. You can find an overview of the main approaches and principles of diagnostics based on serology and nucleic acids in Table 6.4.

PCR

As we learned above, PCR requires specific primers to amplify a stretch of DNA. If we use primers complementary to a specific pathogen gene, the PCR will produce a product only if nucleic acid from the pathogen was in

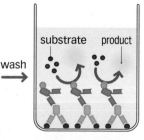

Figure 6.25 Example of a microtiter plate and multichannel pipetter.

Figure 6.26 Illustration of the steps of an indirect ELISA. The specific antibody is the one that we need to identify in the patient's sample. The enzyme-linked analytical antibody is against the human antibody.

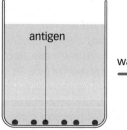
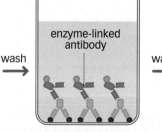

antigen	antibody	enzyme-linked antibody	substrate product
→ wash	→ wash	→ wash	
antigen-coated well	specific antibody binds to antigen	enzyme-linked antibody binds to specific antibody	substrate is added and converted by enzyme into colored product

Basis	Approach	Principle	Method
Serology Direct assays (Gruber reaction) use antibody to detect antigen in a sample. Indirect assays (Widal reaction) use antigen to detect antibody in a sample	Precipitation	Components are in solution	Ring test
		Some components are embedded	Diffusion precipitation
		Electric current accelerates the movement of components	Electroimmunodiffusion
	Agglutination	One component is coupled to a bulky entity for better visualization	Active agglutination Passive agglutination
	Immunoblotting	Components within a mixture are separated before detecting binding on a membrane	Immunofluorescence (fluorescence) Radioimmunoassay (radiation)
	Immunoassay	The solid matrix for binding reaction is not a membrane but the bottom of wells in microtiter plates, permitting small sample amounts and automated reading of the signal	Enzyme immunoassay/enzyme-linked immunosorbent assay (light production, fluorescence) Rapid enzyme immunoassay/dipstick (color change) The signal output depends on the type of labeling of the second analytical antibody
Nucleic acids Nucleic acid sequence (such as a gene) specific to a certain pathogen is detected in a sample	Pairing of complementary bases	The successful binding of diagnostic primers allows amplification	PCR Reverse transcriptase PCR Real-time PCR Product is visualized after gel electrophoresis, or measured via a fluorescence signal using a labeled probe in the reaction
		The binding of the probe (hybridization) takes place on a solid matrix	Southern blotting (for DNA) Northern blotting (for RNA) Microarray analysis (high-throughput analysis) Various rapid and miniaturized assays The detected signal output depends on the type of labeling of the probe (radiation, fluorescence, chemiluminescence, color)

Table 6.4 Overview of main approaches and principles of diagnostics based on serology and nucleic acids.

the sample. To visualize the result, the PCR reaction mixtures are loaded on a gel, and gel electrophoresis is carried out. If a segment of DNA of the correct size is seen, as a band in the gel, the patient had been exposed to the pathogen. Not all pathogens contain DNA: some viruses use RNA as their genetic material and in this case a **reverse transcriptase** PCR (RT-PCR) is done, which starts with producing the DNA target from the RNA. Diagnostic tests should ideally be completed quickly and reliably with little need for manual work, and **real-time PCR** has the advantage of not needing gel electrophoresis. DNA amplification is measured via a fluorescence signal as the chain reaction goes on. Another short piece of DNA, an **oligonucleotide probe**, complementary to part of the pathogen target DNA, is added to the PCR reaction mixture. During amplification, the probe (labeled with a fluorescent molecule) binds proportionally to the amount of product.

Hybridization

This binding of the probe is called **hybridization**. In a way it is comparable to the antigen–antibody binding principle we discussed above with all its applications. Similarly, hybridization of complementary nucleic acids is used in diagnostic methods. The hybridization can take place on a solid matrix. Traditional methods are **Southern blotting** (for DNA) and **Northern blotting** (for RNA). Detection of hybridization depends on the type of labeling of the probe: radiation, fluorescence, chemiluminescence, or color can be used. We also have rapid dipstick methods available.

Techniques based on hybridization to allow the high-throughput analysis of small DNA samples potentially containing a very large number of genes have been developed. This **microarray analysis** is well established. An array (also called a chip) is basically the solid support (glass, silicone, or nylon) on which thousands of specified DNA segments have been attached at fixed locations. A labeled nucleic acid molecule is added and binds to any complementary DNA sequence on the array. The complementary nucleic acids on these two molecules combine to form a hybrid molecule. The hybrid molecules are detected, and the data are analyzed by computational methods.

Microarray technology is useful in identifying whether a certain DNA sequence is included in a sample mixture. We know what DNA was bound at any given location on the array, so that when we detect binding at that spot, we know what was in the mixture. This is the basis of diagnostic microarrays. Arrays are also used for comparing healthy and diseased tissue samples to find a difference. The steps for creating and using such a microarray are shown in **Figure 6.27**.

Microarrays permit the analysis of hundreds of thousands of genes at once and therefore the parallel detection and identification of low-abundance pathogens within a complex microbial community such as those found in patient samples. Identifying pathogens in patient samples is necessary to enable us treat the patient. Even better is preventing anyone from catching a pathogen. Microarrays also help with this by allowing the detection of the most relevant bacterial foodborne and waterborne pathogens for food safety checks.

The technology is used in laboratory diagnostics, even as point-of-care analysis, for **genotyping** and **straintyping**, and the development is continuing toward further miniaturization and automation (**Figures 6.28** and **6.29**). Tests can nowadays achieve an absolute detection sensitivity of as few as 10,000 bacteria. High-density microarrays (such as the Virochip panviral microarray) and **deep sequencing** can test for thousands of potential pathogens simultaneously, and they are invaluable for broad surveillance for novel infectious agents.

Sequence Identification

The DNA that codes for a part of bacterial ribosomes (known as 16S rRNA) is routinely used for bacterial identification. Known sequences are stored in databases, and we can search those databases for a match with a 16S rRNA gene sequence we have produced from a clinical isolate. If an excellent match is found, we have reliably identified the pathogen we

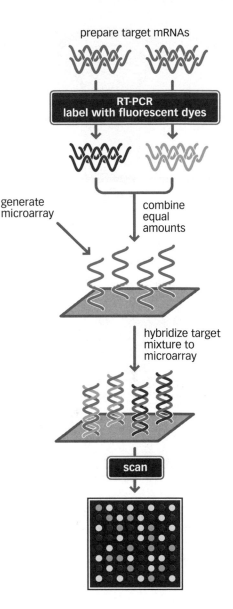

Figure 6.27 The microarray procedure permits rapid and economical analysis of a DNA mixture.

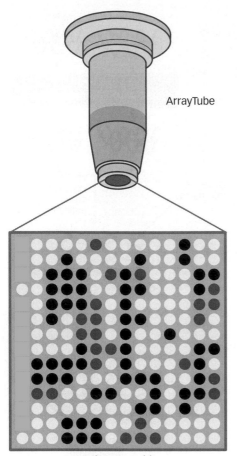

ArrayTube

probe array chip

Figure 6.28 The ArrayTube is an example of a micro probe array (biochip) integrated into a micro reaction vial.

isolated. Now let us make the link back to the microarrays. If we have thousands of 16S rRNA gene fragments attached on the array, and we know which location stands for which species, we can incubate it with a DNA fragment sample of our isolate. When detecting hybridization we can easily read the corresponding species, and we have therefore identified the pathogen very quickly without having to do all the sequencing and database searching.

Successful identification of the pathogen that infected the patient does not necessarily stop at the species level. We often need to identify the bacterial strain (the subspecies), in particular for epidemiological studies and disease monitoring. Whereas one strain of *E. coli* can be a harmless commensal, another strain can be a highly virulent pathogen. All the diagnostic methods we have got to know here can be adapted and used for straintyping. In addition, antimicrobial susceptibility patterns can be very useful, as can **plasmid fingerprinting** and restriction endonuclease analysis of genomic DNA. **Phagetyping** is only used for straintyping. Bacteriophages are very specific to their host and can only infect a few strains of a certain bacterial species. This is used for diagnostics by incubating bacteria from clinical isolates with known phages and testing the susceptibility of the isolate to lysis by a particular phage.

The Future Is Here: New Approaches and Opportunities

On our journey through all the diagnostic methods, you will have noticed that the developments were driven by the need for more sensitive technology (thus requiring a smaller sample), the results being available faster and with as little manual effort as possible while allowing high specificity, consistency, and reliability; altogether we produce ever more data.

Bioinformatics deals with the information overload. It is a combination of biology, computer science, and information technology. Some of its main goals are to create searchable databases of information (such as sequences of amino acids in proteins and nucleotides in DNA segments) that continuously receive curated data and allow analyzing global interactions in cells, tissues, and organisms. From databases, pieces of DNA can be cross-referenced between species, and similar base sequences can be searched for in different organisms. Once any similar base sequences are found, the protein products from the organisms being researched can be assigned to various sequences and cross-referenced between various species as well. Remember the use of database searching we discussed above in the context of comparing 16S rRNA gene sequences.

It is not enough merely to concentrate on the large-scale analysis of DNA (**genomics**) and RNA (**transcriptomics**): the presence of functional protein in the cell is what matters in the end. And this cannot be deduced just from genomics and transcriptomics. **Proteomics** analyzes all of the proteins expressed and modified in a cell and their relationships to one another and to the host organism. High-resolution two-dimensional electrophoresis and protein characterization are widely used to study global protein synthesis. The new knowledge we gain from research using proteomics is invaluable in our search for diagnostic markers and in studying virulence factors, identifying suitable antigens for vaccine development, developing and validating targets of new anti-infective compounds, investigating the epidemiology and taxonomy of human pathogens, and identifying new novel pathogenic mechanisms and mechanisms of drug resistance.

The use of two-dimensional sodium dodecyl sulfate polyacrylamide-gel electrophoresis (2D SDS-PAGE) resolves thousands of proteins

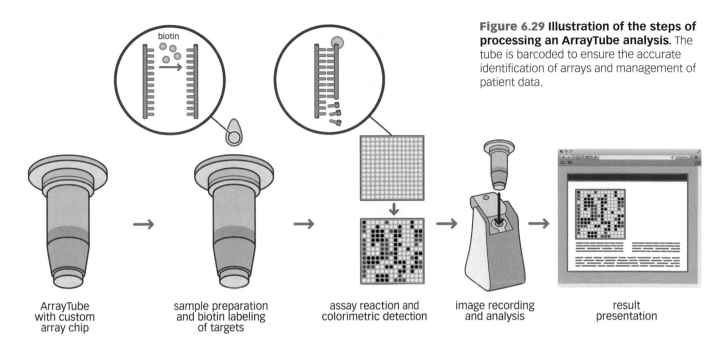

Figure 6.29 Illustration of the steps of processing an ArrayTube analysis. The tube is barcoded to ensure the accurate identification of arrays and management of patient data.

ArrayTube with custom array chip

sample preparation and biotin labeling of targets

assay reaction and colorimetric detection

image recording and analysis

result presentation

in a sample, resulting in a global view of the state of the proteome. Proteins are separated in the first dimension on the basis of their charge (**isoelectric focusing**; IEF) and then by mass in the second dimension (SDS-PAGE). Proteins of interest can be cut out of the gel and identified and quantified by **mass spectrometry** and database searches. Post-translational modifications can also be mapped. If we combine several methods of protein analysis, we can come to promising conclusions: for instance, antigens from *Helicobacter pylori* and *Chlamydia trachomatis* were discovered when 2D SDS-PAGE was combined with Western blotting. High-throughput analysis is possible using **protein microarrays**.

Biosensors and **nanotechnology** open exciting opportunities for further advancement in the detection and quantification of biological molecules, either remotely or in minute quantities. Electronic noses (certain sensors) can detect volatile organic compounds directly from clinical materials. In biosensors we see the fastest-growing use of a technology in pathogen detection, including optical, electric, and electrochemical detection formats. Meanwhile, we can manufacture tools and materials as small as 1–100 nm. Magnetic immunoassays and bio-barcode assays can detect both DNA and protein. The detection limit (sensitivity) is 1000 times better than the best ELISA. Standard desktop scanners are able to read the barcode.

Nanocantilever arrays can be used in biosensors to measure extremely small quantities of biomarkers in biological fluids. A cantilever is basically a mass sensor. A beam is immobilized at one end, and the other end is free. Molecules that allow the binding of others (for example by hybridization) are localized on the beam. If binding happens, the mass on the tiny beam increases, and this can easily be detected. There are nanocantilevers that can detect a single *E. coli* bacterium. Remember that the already highly sensitive tube array discussed above needed 10,000 cells to trigger a measurable response.

Nanomaterials enable us to detect previously unattainably minute changes in concentration, and provide new opportunities for producing cost-efficient and more accurate point-of-care home tests. This illustrates the trend of moving modern molecular diagnostics from hospital laboratories to the patient.

| Keep in Mind |

- The control, prevention, and therapy of infectious diseases depend on quick and correct diagnosis of the species or even the strain of the pathogen.

- After a sample has been taken, an enrichment step often follows for growth-based methods, then isolation and identification.

- Key criteria of any diagnostic method are specificity and sensitivity.

- Blood samples can be used directly for serological analysis.

- Serology-based diagnostics are based on antigen–antibody reaction.

- Antibodies used in diagnostics can be produced by using biotechnology.

- Basic biotech techniques are genetic engineering and recombinant DNA technology.

- Diagnostics based on nucleic acids are based on the hybridization of complementary nucleic acids.

- Drivers in diagnostics are the most sensitive technology, getting results faster with little manual effort.

- Nanomaterials enable us to detect previously unattainably minute changes in concentration.

- The obvious trend is moving modern molecular diagnostics from hospital laboratories to the patient.

SUMMARY

- Transmission is the fifth requirement for a successful infection.

- There are three major reservoirs for infections: humans, animals, and nonliving.

- Transmission of infection can occur through contact, vehicular, and vector transmission.

- Contact transmission can be direct, indirect, or by droplets.

- Several factors, including age, life style, overall health, and emotional state, can influence the disease process.

- The portals of exit involved in the transmission of disease are the same as the portals of entry.

- One of the most important factors in the infection process is the immunocompetence of the host.

- Opportunistic infections are more likely in people with lowered immunity.

- Infections occurring in the hospital are referred to as nosocomial infections, and many of these are associated with antibiotic-resistant pathogens.

- There are universal precautions that are required for preventing nosocomial infections.

- Herd immunity can limit the spread of infection.

- Epidemiology is used to study the factors and mechanisms involved in the spread of disease. It can provide an important tool for developing strategies to combat infectious diseases.

- Quick and correct diagnosis are crucial for the control, prevention, and therapy of infectious diseases.

- Nucleic acids and antibodies can be used to identify pathogens.
- Modern molecular techniques are moving diagnostics from hospital laboratories to the patient.

On the basis of our studies up to now, we can say that we live in a fog of potentially harmful if not lethal microorganisms that, given the opportunity, can cause us great harm. You might be asking yourself how it is that any of us are still left alive. The answer lies in later chapters, where we will learn about the elegant, and in some cases mind-boggling, mechanisms that protect us from diseases every minute of every day. It is important to note that new infectious diseases are constantly emerging and that some diseases that we thought had been eradicated are re-emerging. It is therefore important to examine some of these new threats, our main focus in Chapter 8.

 SELF EVALUATION AND CHAPTER CONFIDENCE

Multiple Choice

Answers are given in the back of the book and help can be found in the student resources at:
www.garlandscience.com/micro2

For questions **1–5** indicate in which of the reservoirs **A–C** the organisms that cause the diseases can be found. Answers may be used more than once.

1.	Lyme disease	A.	Nonliving
2.	Cholera	B.	Other human
3.	Typhoid fever	C.	Animal
4.	Measles		
5.	Anthrax		

6. Nonliving reservoirs include all of the following except

   A. Food
   B. Water
   C. Ticks
   D. Soil
   E. Air

7. Direct contact transmissions occurs because of

   A. The shared use of dirty towels
   B. Inhaling air containing droplets from a sick person's cough
   C. An infected person kissing a non-infected person
   D. A mosquito bite
   E. A tick bite

8. HIV transmitted by a contaminated needle is an example of

   A. Direct contact transmission
   B. Droplet contact transmission
   C. Indirect contact transmission
   D. Vector transmission

9. The most common cause of profound neutropenia is

   A. The administration of cytotoxic chemotherapy
   B. Cancer or other diseases that damage bone marrow
   C. A congenital disorder characterized by poor bone marrow function
   D. Viral infections
   E. Autoimmune disorders that destroy neutrophils or bone marrow cells

10. Examples of vehicles for disease transmission include

    A. Water
    B. Air
    C. Blood
    D. Intravenous fluids
    E. All of the above

11. To prevent the transmission of a disease from human to human, health workers immunized 90% of the population. However, 99% of the population was free of the disease. The reason for this was

    A. Transmission was by air droplets
    B. Transmission was by the fecal–oral route
    C. The portal of entry was the skin
    D. To enable herd immunity to function
    E. The vector population was large

12. Nosocomial infections are almost always seen

    A. In the very young
    B. In the elderly
    C. In debilitated patients
    D. In healthcare settings

13. The most common site of a nosocomial infection is
    A. The brain
    B. The digestive system
    C. The eyes
    D. The urinary tract

14. The most likely fomite for a nosocomial infection is a
    A. Catheter
    B. Nebulizer
    C. Bed sheet
    D. Humidifier
    E. Contaminated needle

15. Universal procedures are required
    A. Only when an infection occurs
    B. Only in major hospitals
    C. In all medical facilities in the United States
    D. Only in nursing homes

16. The number of new cases of a disease contracted per a set number of people per unit of time is its
    A. Prevalence
    B. Mortality rate
    C. Morbidity rate
    D. Incidence rate
    E. None of the above

17. Diseases that occur in a random and unpredictable manner are referred to as
    A. Pandemic
    B. Sporadic
    C. Epidemic
    D. Endemic

18. The total number of people infected within a population at any given time is its
    A. Mortality rate
    B. Prevalence
    C. Incidence
    D. Epidemic number

19. The total number of cases of polio in the world was 1300 in 2007, 1550 in 2008, 1400 in 2009, and 1350 in 2010. From this information we describe polio as _____ in the world population.
    A. Epidemic
    B. Pandemic
    C. Sporadic
    D. Endemic
    E. None of the above

For questions **20–24**, if the reason is correct, answer **A**; if the reason is incorrect, answer **B**.

20. The sensitivity of a diagnostic test is important because it discriminates between the true pathogen and other similar pathogens.

21. It is important to circumvent the patient's normal flora, because otherwise it can make the pathogen isolation and identification difficult or impossible.

22. Diagnostic work is usually done according to standard operating procedures, because every effort is made to prevent legal complications.

23. Because biotechnological advances will impact the health care community, health care professionals should have a basic understanding of how they are being made.

24. Monoclonal antibodies are commonly used today because they are powerful tools for diagnosing infectious disease and cancer and for detecting harmful microorganisms in food.

For questions **25–29** compare the hybridoma technique with the phage display techniques **A–D**.

25. The host animal is the source of antibody C	A. Phage display only
26. Used to enlarge antibody library A	B. Hybridoma only
27. Selects genes with the desired binding capacity A	C. Both
28. Uses only mouse immunization B	D. Neither
29. Used to produce polyclonal antibody	

Q DEPTH OF UNDERSTANDING

Questions listed here require you to bring together the concepts you have learned in this chapter into a discussion format. This helps you to increase your depth of understanding of the material you have learned. Help can be found in the student resources at: www.garlandscience.com/micro2

1. Compare and contrast the three major types of reservoir associated with infectious disease.

2. Discuss the predisposing factors that can contribute to the infectious disease process.

3. Evaluate the potential for infectious disease in burn patients, patients undergoing chemotherapy, and transplant patients.

Q CLINICAL CORNER

Help can be found in the student resources at: www.garlandscience.com/micro2

1. Eight children, all under the age of 9 years, are brought into the emergency room complaining of various degrees of diarrhea and vomiting. After initial examination, it is learned that all of the children are from the same elementary school class and that the class was taken on a field trip to a nearby water park on the previous day.

   A. What is the likely reason for their condition?
   B. How would you confirm your suspicions?
   C. What are the epidemiological possibilities?

2. Shady Grove retirement village has experienced an outbreak of bacterial pneumonia. This is a common problem seen in nursing home facilities. Eighteen of the 50 residents of the facility have severe symptoms and must be sent to the hospital.

   A. What are the possible reasons that only 18 of the 50 residents became ill?
   B. How could the disease have spread?
   C. Explain whether this is a common-source outbreak or a propagated epidemic.
   D. What are the main concerns you would have for the residents sent to the hospital?

Principles of Disease

Chapter 7

Why Is This Important?

This chapter introduces you to the principles of disease, and in particular how diseases are caused (the etiology), how they can be characterized, and the concepts of sepsis and shock. These concepts are important for developing an in-depth understanding of infections.

Jean is getting really excited. Soon she will embark on her university exchange term to the UK. How she wishes she had done all the boring packing already! Of course she will miss her friends, but thanks to social media and online video calls they can be frequently in touch. The time difference might be a bit inconvenient, but anyway she did not plan to sleep away her adventures. What will the food be like? Jean is not a fussy eater, but her cat Mike is not that easy to please. Maybe her mother has some good suggestions. Jean joins her in the kitchen to ask about the cat food. Her mother's response is more a shock than a help.

Apparently she cannot take Mike to the UK at all! He would need to spend six months in quarantine when entering. Inconsolably she returns to her room to share the news with her friends; but before that she wants to read more about rabies—what a thought that Mike might have been exposed to it! Rabies is an inflammation of the brain and generally affects the central nervous system. It is caused by the rabies virus (a rhabdovirus). Then Jean finds an amendment to the information and cannot believe her luck. From 2012 a cat entering the UK from the United States still needs to be microchipped and vaccinated against rabies,

but no longer needs a blood test and crucially will only have to wait 21 days before they travel. Improved rabies vaccines allowed that change to the UK's quarantine laws, which had been introduced in the nineteenth century. The last indigenous animal rabies case was reported in the 1920s, and 24 humans had died in the UK from imported rabies since 1902. Estimations using the new rules do not expect anyone to die from rabies contracted from a pet in the next several thousands of years. That is excellent news indeed. And just in good time. Now Jean only has to worry about finding Mike's favorite cat food.

OVERVIEW

This is an important chapter because here is where we begin to look at the principles of infectious disease. Armed with the things we learned in the previous chapters, we are well prepared for these discussions, and the things we learn here will become part of the foundation for all of the other topics we examine. Our discussions will be broken down into three major parts. The first will examine how infectious disease can get started, which is referred to as its etiology. The second will examine how disease can develop, including communicability and contagiousness. The final discussion will be about infections that move into the systemic circulation.

We will divide our discussions into the following topics:

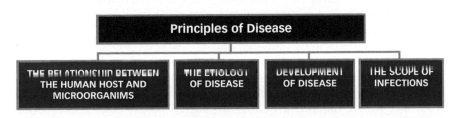

Now that we have looked at the primary requirements for a successful infection (Chapters 5 and 6), we can examine some of the principles associated with disease. Table 7.1 lists several of the terms used when discussing disease. We can start off by defining **disease** as any negative change in a person's health or, from the pathogen's perspective, any damage to the host. This change has a cause, and we refer to the cause of the disease as its **etiology**. Thus, for infectious disease, we can say:

- That the decline in health—in other words, the disease—is caused by microorganisms invading a host's body

- That these microorganisms are the etiological agents of the disease

Table 7.1 Some common terms used when describing infections.

Term	Defining characteristic
Acute disease	Symptoms develop quickly but infection lasts only a short time
Chronic disease	Symptoms develop slowly and disease can last a long time
Subacute disease	Symptoms are between those of acute disease and those of chronic disease
Latent disease	Symptoms can continue to reappear long after initial infection
Local infection	Confined to a small area of the body
Focal infection	Initial site of a spreading infection
Systemic infection	Pathogens use blood or lymph to move around body
Bacteremia	Presence of bacteria in the blood
Septicemia	Organisms multiplying in the blood
Viremia	Presence but not multiplication of viruses in the blood
Toxemia	Presence of toxins in the blood
Primary infection	Initial infection
Secondary infection	Immediately follows primary infection; can be more dangerous than initial infection

THE RELATIONSHIP BETWEEN THE HUMAN HOST AND MICROORGANISMS

One of the concepts we need to re-examine is the presence of normal microbial flora, which we mentioned in Chapter 6. The normal flora is also called the **microbiota** of humans. There is no denying that we exist in an environment of microorganisms, that our bodies contain a variety of them, and that they have a useful relationship with us. These microorganisms form complex communities. It has been suggested that in a healthy adult there are roughly ten times more microbial cells than human cells. Even though the microbial cells are too small to be seen with the naked eye, they can make up as much as 5 pounds (about 2.3 kg) of adult body weight. We humans are truly two-legged ecosystems, also called a **superorganism**. The microbes (bacteria, viruses, and single-cell eukaryotes) inhabiting the human host, their genetic information (called the **metagenome**), and the environment in which they interact are all together called the human microbiome. The term microbiome was introduced in 2001 by Joshua Lederberg, who discovered bacterial recombination.

There are many microorganisms located in many places in the body that make up the normal microbial flora (Table 7.2). For example, the large intestine has the largest population of bacteria (in fact, fecal material is mostly bacteria). These organisms are supposed to be there because they provide necessary functions that are helpful to us. The human microbiome affects our development, physiology, immunity, and nutrition.

Region of the body	Representative microorganisms
Skin	*Staphylococcus epidermidis, Staphylococcus aureus, Propionibacterium acnes, Candida* species, and *Corynebacterium* species
Conjunctiva	*Staphylococcus aureus, Staphylococcus epidermidis,* and *Corynebacterium* species
Nose	*Staphylococcus aureus* and *Staphylococcus epidermidis*
Throat	*Staphylococcus aureus, Streptococcus pneumoniae, Haemophilus influenzae, Corynebacterium,* and *Neisseria* species
Mouth	*Streptococcus* species, *Lactobacillus,* and *Corynebacterium*
Large intestine	*Lactobacillus, Enterococcus, Escherichia coli, Enterobacter, Proteus, Klebsiella,* and *Corynebacterium* species
Genitourinary tract	*Staphylococcus epidermidis, Enterococcus, Lactobacillus, Pseudomonas, Klebsiella,* and *Proteus* in the urethra; *Lactobacillus, Streptococcus,* and *Staphylococcus* in the vagina

Table 7.2 Representative normal microflora.

To use another example, the skin is covered with bacteria of many kinds, including normal resident bacteria, transient organisms, and even pathogens. As we saw in Chapter 6, however, unless these organisms can find a way through the skin (and some do), they are essentially harmless because the intact skin is such an impenetrable barrier. The mouth, nose, and throat are also places where large numbers of microorganisms are found, and once again many of these are harmless. Some are pathogenic, though, so it is important to remember that this area of the body is the most common portal of entry for bacteria. The respiratory system is readily accessible to potential pathogens from the nasopharynx and oropharynx.

There are three types of relationship between a host and bacteria living in that host (Table 7.3). The first is **commensalism**, in which one partner is benefited but the other is unaffected. The relationship between humans and the microbial flora named in Table 7.2 is one form of commensalism. The second type of host–bacteria relationship, **mutualism**, is one in which the host offers benefits to the bacteria and the bacteria offer benefits to the host. Some bacteria provide us with vitamins such as K and B, for instance, and we provide them with nutrients and a place to stay.

Table 7.3 Host–microorganism relationships.

Type of relationship	Microorganism	Host	Example
Commensalism	Benefits	Neither benefits nor is harmed	Saprophytic bacteria that live off sloughed-off cells in the ear and external genitalia
Mutualism	Benefits	Benefits	Bacteria in colon
Parasitism	Benefits	Is harmed	*Mycobacterium tuberculosis* in lungs

The third type of relationship is called **parasitism**, and here one of the partners benefits at the expense of the other. This is the relationship we see with pathogens, which benefit at the expense of our health.

There are two points that are important to remember about host–bacteria relationships. First, microbial flora can protect us from disease. This is accomplished by **microbial antagonism**. Like a dog that marks its territory, bacteria that reside in certain areas of the body will fight to prevent outsiders (such as pathogens) from taking up residence. This prevention is easy for resident bacteria to accomplish because they are well established in the host. For instance, our microbial flora competitively inhibits the growth of pathogenic interlopers by more efficiently using all the available nutrients and all the available oxygen, leaving none for the pathogens. In addition, the flora (for example *Lactobacillus* species) can acidify the region and make it less hospitable to any infringing pathogen. In fact, many bacteria, such as *Streptococcus* species and *Escherichia coli*, can produce **bacteriocins**, which are essentially localized bacterial antibiotics. These bacteriocins can kill invading organisms but do not affect the bacteria that produce them. (Bacteriocins are specific for the bacteria that produce them, and for that reason they are useful as diagnostic tools for identifying bacteria.)

The second important point about host–bacteria relationships is that many normally harmless resident bacteria are in fact opportunistic pathogens. Recall from Chapter 6 that this type of pathogen is defined as one that is harmless in its normal location but can cause disease if it moves to an area where it is not normally found, or if the host's immune defenses are weakened. An example is *Escherichia coli*, which moves from the large intestine, where it is part of the normal microbial flora, to the urethra, where it is a pathogen. Urinary tract infections can occur repeatedly and become increasingly dangerous if the infection moves into the bladder or the kidneys. (We discuss this in more detail in Chapter 23.)

As you saw in Table 7.2, we have identified species that usually inhabit certain areas of our body, but the community overall and the important interactions have largely remained a black box until recently. The Human Microbiome Project has started to shed light on what is in this black box and what is going on there. Sampling, identification, and characterization projects are undertaken on a massive scale to define the healthy human microbiome, the microbes involved, and their contributions and roles in this community. Up to 18 different sites representing main body areas (nose, skin, mouth, stool, vagina) of nearly 300 totally healthy people from 18 to 40 years old were sampled up to three times over the course of a year. Interestingly, there is no such thing as the normal microbiome! In fact we have an individual microbial fingerprint and there is

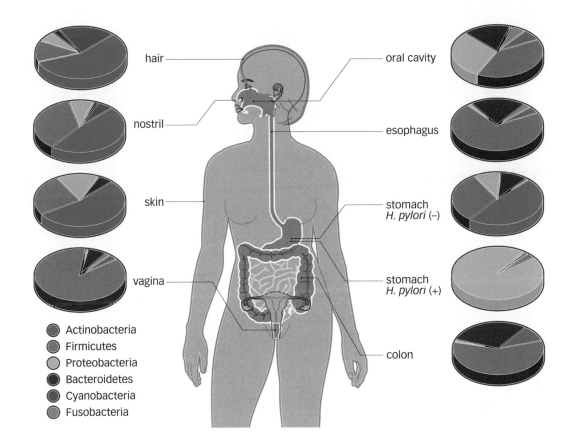

hair

nostril

skin

vagina

oral cavity

esophagus

stomach
H. pylori (–)

stomach
H. pylori (+)

colon

- Actinobacteria
- Firmicutes
- Proteobacteria
- Bacteroidetes
- Cyanobacteria
- Fusobacteria

much variation in the microbiome composition from person to person. Individual microbiomes also change over time. Despite this variance, the microbial communities have similar metabolic tasks. The microbiome composition naturally depends on the site of the body: we find distinct communities (with some signature bacteria, also called indicator organisms), which do not mix (**Figure 7.1**).

The microbiome fluctuates due to dietary change (for example the ratio of animal protein to carbohydrates), seasonal change (for example affecting the nasopharynx flora of young children), hormonal change (for example menstrual cycle and vaginal flora), and exposure to antibiotics. The mother's microbiome may affect child health, and the child's microbiota composition is also related to the delivery (Cesarean section or vaginal delivery) (**Figure 7.2**).

You have learned that humans co-evolved with microorganisms, which help us to survive and function (for example in digestion and in vitamin production), and that a healthy microbiome may prevent infections. A pattern is now emerging that the microbiome composition possibly imparts susceptibility to certain infectious diseases. There is also evidence that microbiome composition could contribute to obesity, diabetes, atherosclerosis, autism, allergies and asthma, chronic gastrointestinal diseases such as Crohn's disease, irritable bowel syndrome, and celiac disease, and that it can affect our response to drug treatment. The microbiome even affects our choice of partner! Mating preference depends on—among other things—body odor, and the microorganisms inhabiting us contribute to this odor.

A healthy microbiome is ecologically stable and functional. This means there is very little change to its composition under stress, or it quickly returns to the composition it had before the stress started.

Figure 7.1 The microbiome of various anatomical sites. Shown are the relative ratios of some major bacterial taxa.

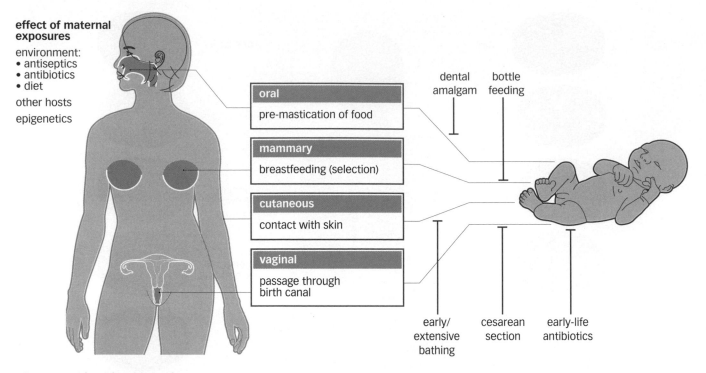

effect of maternal
exposures

environment:
• antiseptics
• antibiotics
• diet

other hosts

epigenetics

oral
pre-mastication of food

mammary
breastfeeding (selection)

cutaneous
contact with skin

vaginal
passage through birth canal

dental amalgam   bottle feeding

early/ extensive bathing   cesarean section   early-life antibiotics

Figure 7.2 The microbiome is introduced to the newborn from the mother, and then further acquired in early life from the environment and affected by various factors.

Any impaired communities are defined as dysbiotic. If we know the connection between microbiome composition, health, and disease, then microbiome analysis can permit early diagnoses, and altering the microbiota (for example by a change of diet or by drug treatment) could maintain or improve our health.

THE ETIOLOGY OF DISEASE

To understand the etiology of disease, it is important to mention one of the fundamental paradigms of microbiology, **Koch's postulates**. Robert Koch was a physician in the nineteenth century who studied anthrax in cattle. His studies were the first to close the loop between cause and effect for infectious disease and became a prerequisite for establishing the etiology of bacterial diseases based on evidence obtained from rigorously planned and conducted experiments. His postulates can be listed as follows:

- The same pathogen must be present in every case of the disease, but not in healthy individuals.

- The pathogen must be isolated from the sick host and grown in a pure culture.

- The pure pathogen must cause the same disease when given to uninfected hosts.

- The pathogen must be re-isolated from these newly infected hosts, and shown to be the same organisms as isolated initially.

Some experiments involved in testing the four postulates are illustrated in **Figure 7.3**.

These postulates allow us to place the responsibility for infection clearly and squarely on the organism that caused it, and they are still required today for determining the etiology of infections. Establishing the postulates was a milestone in medical microbiology and started the golden age of microbiology. If we know the causative agent of an infectious disease, we can research the features of the pathogen. This knowledge eventually allows us to make a correct diagnosis, to develop treatment and use it efficiently, and ideally to prevent infection by developing a vaccine and establish disease control measures. However, there are several exceptions, in particular the requirement for isolation and purification of the potential pathogen, which means that we need to adapt the tests when using the postulates in our modern times. There are many organisms that will not grow on artificial media and therefore cannot be purified. For example, the spirochete *Treponema pallidum* (which causes syphilis) and the Gram-positive acid-fast rod *Mycobacterium leprae* (which causes **leprosy**) cannot be researched traditionally with Koch's postulates; neither can *Rickettsia* (for example *Rickettsia prowazekii*, which causes epidemic typhus) or viruses. Viruses require cells in which to be produced. Alternatively, detecting the presence of the pathogen's genomic information in the sample can substitute for cultivation. Sometimes healthy individuals carry the pathogen but do not develop symptoms. In addition, some organisms can cause a variety of diseases, making the use of Koch's postulates more difficult. Consequently, there are some diseases in which we can identify the etiological agent directly by using Koch's postulates, such as tuberculosis (etiological agent *Mycobacterium tuberculosis*) and Lyme disease (etiological agent *Borrelia burgdorferi*), and some in which we have to use twenty-first-century technology to close the loop and fulfill the general principles of the postulates.

Keep in Mind

- The cause of a disease is referred to as its etiology.

- The body contains normal microbial flora made up of bacteria that are beneficial to the host.

- There are three types of relationship between bacteria and their hosts: commensalism, mutualism, and parasitism.

- Koch's postulates are an important way of evaluating the etiology of a disease.

DEVELOPMENT OF DISEASE

The course of a disease can be broken down into five specific periods (**Figure 7.4**). The first is called the **incubation period** and covers the time between the initial infection and the first **symptoms of the disease**. The length of the incubation period depends on the virulence of the pathogen: the lower the virulence, the longer this period will last (**Figure 7.5**). The second period is the **prodromal period**, during which the first mild unspecific symptoms appear. Once major symptoms (specific to the disease) are noted, the disease has moved into the **period of illness**. It is here that the immune response is highest, and depending on the severity of the illness the patient may die during this period. The fourth period is called the **period of decline**, during which the symptoms subside. Although this period is an indicator of the end of the illness, it is also a time during which the patient can acquire a secondary infection. This is particularly true of opportunistic infections and of nosocomial infections. In many cases, because of the debilitated condition of the patient, these secondary infections can be more dangerous than the initial problem. The final period is called the **period of convalescence**, during which the patient regains strength and proceeds to a full recovery.

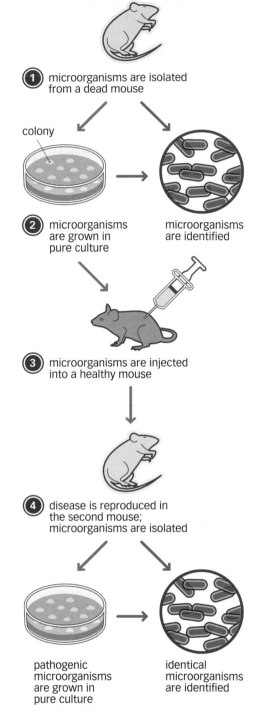

1 microorganisms are isolated from a dead mouse

colony

2 microorganisms are grown in pure culture

microorganisms are identified

3 microorganisms are injected into a healthy mouse

4 disease is reproduced in the second mouse; microorganisms are isolated

pathogenic microorganisms are grown in pure culture

identical microorganisms are identified

Figure 7.3 Koch's postulates test system. This test is part of a series of experiments used to determine the etiology of infectious diseases.

Figure 7.4 A graphical representation of the five periods of infection. The duration of the incubation period varies with the virulence of the pathogen, and secondary infections can occur during the period of decline. Because the patient is already debilitated, secondary infections can be more dangerous than the primary infection.

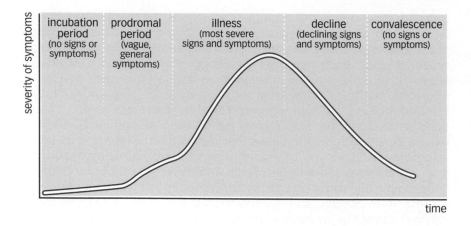

Communicable and Contagious Disease

Some diseases are **communicable**, which means that they can spread from one person to another. Tuberculosis is one example. Others are non-communicable, which means that they cannot spread from one person to another, with tetanus being a common example. In addition, communicable diseases that spread very easily, such as chickenpox or measles, are classified as **contagious**, meaning that they are communicable on contact with an infected individual. It depends on the pathogen and the mode of transmission how long a patient is infectious during the course of a disease.

There are three methods available for full or partial control (remember also Chapter 6) of communicable and contagious diseases: isolation, quarantine, and vector control.

Isolation means preventing an infected person from having contact with the general population. There are seven categories of isolation, which is usually done in a hospital. These are: strict isolation, respiratory isolation, protective isolation, enteric precautions, wound and skin precautions, discharge precautions, and blood precautions. However, patients with diseases such as tuberculosis can be difficult to isolate

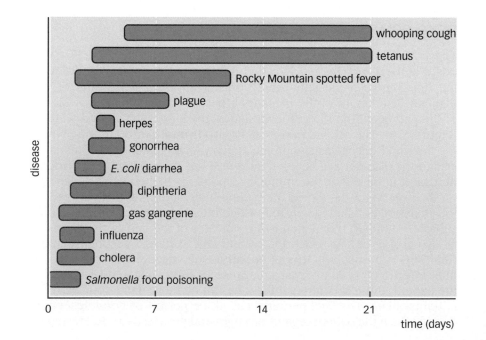

Figure 7.5 Ranges of the periods of incubation for selected diseases.

before they have been diagnosed. As we saw in Chapter 1, in a busy emergency room, preference is given to individuals who are bleeding or who have chest pain. Someone with tuberculosis might be asked to be seated in a crowded waiting room for hours before being examined, diagnosed, and isolated. In a situation like this, the term contagious becomes easily understandable.

Quarantine involves separating from the general population healthy individuals who may have been exposed to a communicable disease. Quarantine usually lasts as long as the expected incubation times for the suspected disease and is lifted if symptoms do not present during that period. Although quarantine is the oldest method of dealing with communicable diseases, it is now generally used only for very severe diseases, such as cholera and yellow fever. In fact, quarantine is rarely used today because of the difficulty in enforcing it.

The third method for controlling communicable diseases is vector control, where **vector** is defined as any organism that carries disease pathogens from one host to another. The principal means of vector control are the destruction of vector habitats and the inhibition of vector breeding habits and feeding behaviors. In addition, measures such as window screens, mosquito netting, and insect repellents can be used to protect against transmission. For instance, malaria control in the United States is accomplished through control of the mosquito population.

Keep in Mind

- The development of a disease can be broken down into five periods: incubation, prodromal, illness, decline, and convalescence.
- Communicable diseases can be spread from one person to another.
- Contagious diseases are communicable on contact with an infected individual.
- Methods for the control of communicable and contagious diseases include isolation, quarantine, and vector control.

Duration of Disease

The duration of a disease can vary depending on several factors, in particular the overall health of the host. When we look at duration, there are four basic categories we can use. **Acute diseases** such as chickenpox and measles develop quickly but last for only a short time, whereas **chronic diseases**, such as **hepatitis**, mononucleosis, and tuberculosis, develop slowly but remain for long periods. Diseases such as **subacute sclerosing panencephalitis (SSPE)** fall in the category of **subacute**. This disease has an insidious onset that can take 6–12 months or even longer to develop after initial viral infection and can be fatal. Latent diseases are those in which the pathogen remains in the host after the initial signs and symptoms disappear and can be become reactivated after long periods. Latency is seen in several viral diseases, the most common example being **chickenpox**. The patient initially shows the classical signs of this pox infection, and after a while these disappear. However, the virus has taken up residence in the patient's neurons. At some point it can be reactivated and the patient will present with the rash identified as **shingles**. The reasons for reactivation of the virus involve stress of the host.

Persistent Bacterial Infections

Some pathogenic bacteria are capable of maintaining infections in hosts, even in the presence of inflammatory and specific antimicrobial

Pathogen	Disease	Site of persistence
Mycobacterium tuberculosis	Tuberculosis	Granulomas and in other sites in macrophages
Salmonella enterica serovar Typhi	Typhoid fever	Macrophages in the bone marrow and possibly the gallbladder
Chlamydia trachomatis	Trachoma; genital tract infections and *Lymphogranuloma venereum*	Epithelial and endothelial cells
Helicobacter pylori	Gastritis, ulcers, and gastric cancer	Extracellular and possibly intracellular in the stomach
Neisseria gonorrhoeae	Genital tract infection, which can lead to pelvic inflammatory disease and infertility	Extracellular; intracellular at mucosal sites

Table 7.4 Selected persistent infections in human.

mechanisms as well as a perfectly good immune response (Table 7.4). Even though they are infected, some people with persistent infections show no clinical signs of the disease. *M. tuberculosis*, *Salmonella enterica* serovar Typhi, and *Helicobacter pylori* are good examples of bacteria that can cause persistent infections, and there are many questions as to how a persistent infection can occur in spite of the innate and adaptive immune responses of the host. In an effort to understand this, let's look at two examples of persistent infections: tuberculosis and typhoid fever.

Tuberculosis is one of the oldest known diseases (it used to be called consumption) and infects one-third of the world's population. The infection starts at a site in a lung and can move throughout the lung. It is interesting to note that there is some evidence that this movement is performed by dendritic cells that are part of the host defense. Most people resolve infections with *M. tuberculosis* after the onset of the adaptive immune response, but in some individuals the organisms are never completely cleared. Persistently infected hosts can harbor the pathogens for life, and in some cases the tuberculosis is reactivated later in life. This reactivation usually occurs in patients who have become immunocompromised, for instance through the aging process. It is for this reason that tuberculosis is seen in so many nursing homes.

In persistent tuberculosis infections, pathogens survive inside **granulomas**, which are bodies made up of host defensive cells, such as macrophages, T cells, B cells, dendritic cells, neutrophils, fibroblasts, and matrix components (**Figure 7.6**). Granulomas form as activated macrophages aggregate into gigantic cells similar to the syncytia seen in viral infections. The question is: How do the pathogens survive initial contact with cells that are programmed to phagocytose and kill them? *M. tuberculosis* gets taken into a phagosome in the normal fashion but then remodels this structure and prevents the development of an acidic, hydrolytic environment. In addition, the pathogens inhibit the formation of phagolysosomes by preventing the fusion of the phagosome with cellular lysosomes. So the phagocytes remain unactivated. Several genes have been identified in *M. tuberculosis* that seem to be involved in this process.

Typhoid fever is caused by *S. enterica* serovar Typhi and can cause a variety of problems in the intestinal tract, such as localized gastritis. However, if this pathogen becomes systemic, it can cause typhoid fever. This disease starts with a localized infection and inflammatory response, usually in connection with **Peyer's patches** in the intestine. The *Salmonella* organisms infect the phagocytic cells in the lamina propria

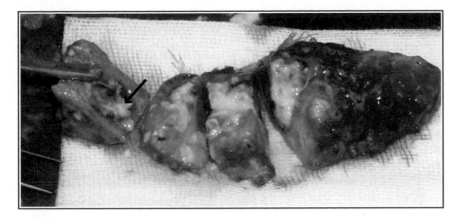

Figure 7.6 Gross pathology of human lung tissue granuloma removed during resection surgery. The arrow indicates a granuloma.

of the intestine and gain access to the blood and lymph. Once in those liquids, the bacteria can spread to the liver and spleen and can become persistent in the gallbladder and bone marrow (**Figure 7.7**).

Infection with *Salmonella* is a major public health problem and is endemic in regions of the world in which there are unclean water supplies and a general lack of sanitation. It is also difficult to treat, because the level of antibiotic resistance is rising. Between 1% and 6% of people infected with *S. enterica* serovar Typhi will become carriers who will

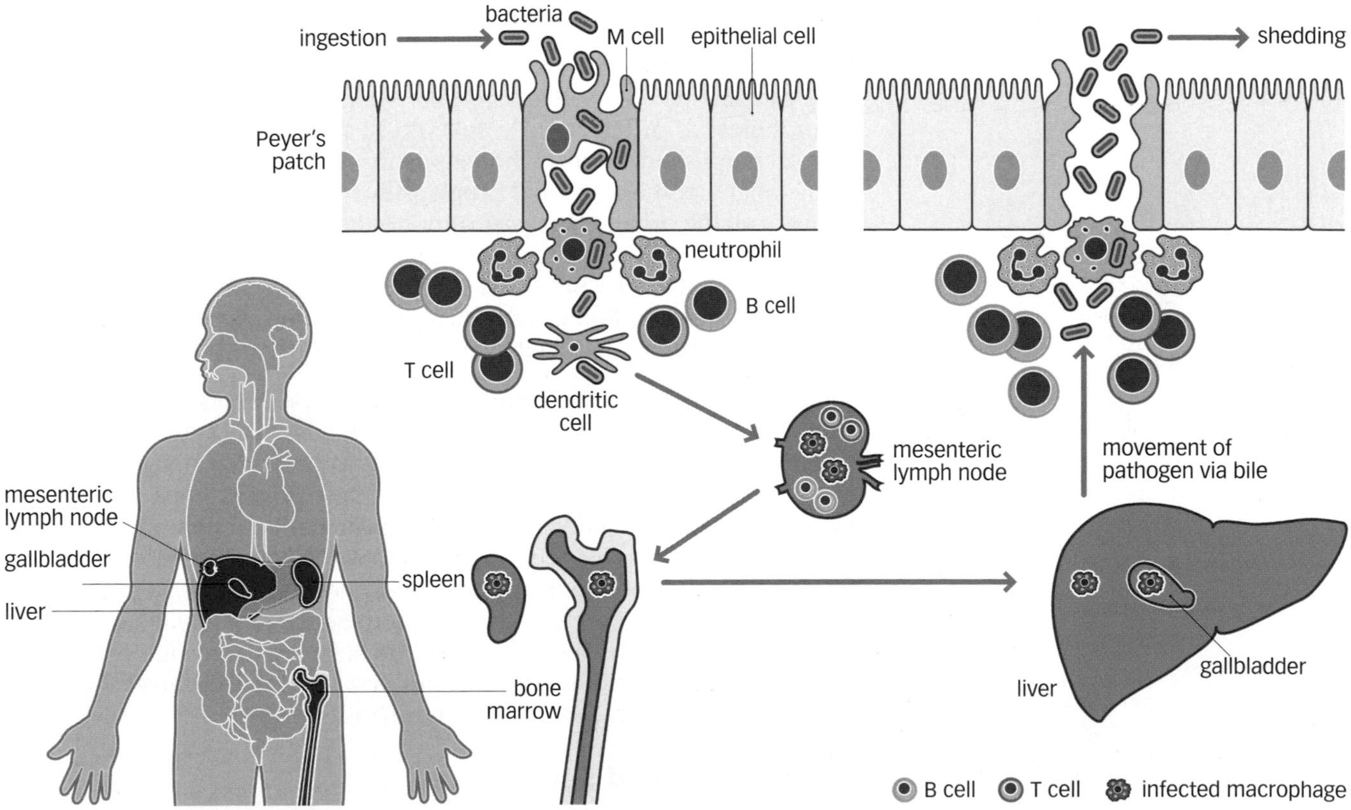

Figure 7.7 A schematic representation of persistent *Salmonella* infection in humans caused by *Salmonella enterica* serovar Typhi. Bacteria enter the intestinal tract by invading M cells (microfold cells, localized in the digestive tract). This invasion is followed by inflammation, phagocytosis of bacteria by neutrophils and macrophages, and recruitment of immune T and B cells. In typhoid fever, *Salmonella* may use dendritic cells to disseminate to lymph nodes and deep tissues, which can lead to transport of the pathogens to the spleen, bone marrow, liver, and gallbladder. Re-infection and transmission occur by way of pathogens that are released from the gallbladder and re-enter the digestive system with bile.

shed large numbers of pathogens in their stool and urine for the rest of their lives but will never show symptoms of the infection (asymptomatic carriers). The process by which *Salmonella* survives the host defensive response seems to be similar to the process for *Mycobacterium*: the invaders are phagocytosed, but the phagosomes never join with lysosomes to form phagolysosomes.

It is important to remember that several other mechanisms are used by pathogenic bacteria to survive a host's defensive responses and cause persistent infection. Some pathogens evade the innate response by enzymatically destroying the antimicrobial toxin nitric oxide as it is produced. Others, such as *Helicobacter pylori*, form megasomes inside a host macrophage by fusing together many phagosomes. Megasomes seem to protect the pathogens even though they are inside a phagocytic cell. Some persistent pathogens subvert the adaptive immune response by blocking the activation of T cells, whereas others take advantage of genetic diversity (similar to that seen in influenza virus) to confuse the adaptive immune response. In fact, some pathogens change the level and type of lipopolysaccharide on their cells so that it will not bind as well to receptors on defensive cells, thereby lessening the host response against them.

THE SCOPE OF INFECTIONS

Infections can be local or systemic. In many cases the host defenses wall off an infection and keep it a **local infection**, as occurs with boils or abscesses. From a host defense perspective, these infections are the easiest to deal with. However, when pathogens find a way to move from their original location in a host, a point called the **focus of infection**, they can become far more dangerous. An infection that has moved away from the focus of infection is called a **systemic infection** and is usually accomplished when pathogens gain access to the blood or lymph. Remember that the blood and lymph go everywhere in the body and interact with all the organs. The occurrence of bacteria in the blood is called a **bacteremia**, and if the organisms are growing in the blood, the condition is referred to as a **septicemia**. When bacterial toxins move through the blood, the condition is referred to as **toxemia**, whereas viruses being present in the blood is called **viremia**.

Infections can also be primary, subclinical, or secondary. A **primary infection** is the one causing the initial acute onset of symptoms, whereas a **subclinical infection** is one in which symptoms do not appear even though the infection is ongoing (for example poliomyelitis and hepatitis A). Even though individuals with subclinical infections do not show signs of the disease or feel any symptoms, they are carriers who can infect others. A **secondary infection** is one that establishes itself in a host weakened by some primary infection. Secondary infections, such as *Pneumocystis* pneumonia, which is often seen in connection with AIDS, can be more dangerous than the primary infection because they take advantage of the weakened state of the host.

Toxic Shock and Sepsis

It is important that we take a minute here to discuss sepsis and toxic shock (remember Chapter 5). These are two distinctly different clinical situations that arise as a result of infection. In **toxic shock**, there is massive leakage of plasma from the circulation that causes the blood pressure to plummet. This condition is fatal for between 30% and 70% of patients in which it occurs. It has been shown that toxic shock is caused by the activation of neutrophils, which occurs when the neutrophils come into contact with such bacterial surface proteins as the M proteins found on streptococcal species. In fact, when *Streptococcus pyogenes* begins to reproduce in the blood, it can cause **streptococcal toxic shock syndrome (STSS)**, a condition that can rapidly cause death.

The mechanism by which STSS occurs is interesting and may provide researchers and clinicians with possible methods to prevent a negative outcome. The M proteins of *S. pyogenes* are shed while the pathogens are growing in the blood. These proteins then bind to the plasma protein fibrinogen, which normally promotes clotting. The M protein–fibrinogen complexes then bind to and activate neutrophils that begin to release potent inflammatory molecules called heparin-binding proteins. It is these proteins that induce dynamic changes in vascular permeability in the host, changes that cause a rapid loss of fluids from the body, a catastrophic decrease in blood pressure, and difficulty in ventilating the lungs. In fact, strains of *S. pyogenes* that are rich in M proteins have been shown to be far more virulent than those containing only small amounts of these proteins.

The dynamic changes in the host just described result in a state of shock for the host. This type of shock is not the same as that seen in infections in which Gram-negative endotoxins are released, however. Although a similar rapid decrease in blood pressure is seen with some Gram-negative endotoxins, the cause of the decrease does not seem to be related to activated neutrophils.

Sepsis syndrome is a general term indicating the presence of either pathogenic organisms or their toxins in the blood. There are two categories of sepsis: severe sepsis and acute septic shock. **Severe sepsis** is defined by signs of systemic inflammation and organ dysfunction accompanied by abnormal temperature, heart rate, respiratory rate, and leukocyte count as well as elevated levels of liver enzymes and altered cerebral function. Severe sepsis kills slowly over several weeks, and autopsy shows minimal tissue inflammation or necrosis. In contrast, **acute septic shock** occurs suddenly, and patients can die within 24–48 hours. At autopsy after acute septic shock, there is widespread indication of tissue inflammation and cell damage.

| **Fast Fact** | Sepsis is the most common cause of death in hospital intensive care units, and there are about 225,000 deaths from sepsis each year in the United States alone.

Keep in Mind

- Disease can be acute, chronic, or subacute.

- In latent disease, pathogens remain in the host after signs and symptoms have disappeared but can be reactivated after long periods.

- Pathogens such as *Mycobacterium tuberculosis* can cause persistent disease in which infections continue even though the host has a working immune defense.

- Infection can be localized or systemic and can be classified as primary (with acute initial symptoms), subclinical (without symptoms), or secondary (occurring after a primary infection).

- Infection can result in toxic shock or sepsis.

SUMMARY

- Normal microbial flora helps to protect against opportunistic pathogens.

- Etiology is defined as the cause of a disease.

- Koch's postulates can be used to evaluate and identify the etiology of a disease.

- Disease can be acute, chronic, subacute, or latent.

- Infection can be local or systemic.

- Relationships between bacteria and hosts can be described as commensal, mutual, or parasitic.

- There are five periods in the development of an infection: incubation period, prodromal period, period of illness, period of decline, and period of convalescence.

- Communicable diseases can be spread from one person to another.

- Contagious diseases are spread on contact with an infected person.

- Infection can cause septic shock or sepsis.

This chapter has taught us about how different infections get started, how diseases can develop and be characterized, and how infections can spread throughout the body. In addition, we learned about the importance of herd immunity and also about communicable and contagious diseases. Coupled with Chapter 5 on the requirements for infection and with Chapter 6 on the transmission of infection, the material presented here completes our discussion of the infection process.

Ⓠ SELF EVALUATION AND CHAPTER CONFIDENCE

Multiple Choice

Answers are given in the back of the book and help can be found in the student resources at:

www.garlandscience.com/micro2

1. A student is ill. Her friend asks her how she feels. Which of the following would be the most likely response if she were in the prodromal period?

   A. I'm fine
   B. I have a little nausea and I'm tired
   C. I have diarrhea and am running a temperature
   D. I was sick last night, but I'm feeling much better now

2. A young chicken farmer, working alone in an isolated farm, develops a contagious influenza infection, caused by a new strain that has a vastly different genetic composition from the old strains. She visits her local community regularly. Her community all received an influenza vaccination that did not include a vaccine against this strain. What outcome would you predict?

   A. Her disease will not spread
   B. Only her close relatives and contacts will get the disease
   C. The disease will spread to most people in the community but not beyond it
   D. Her disease might spread over the entire human population

3. Suppose that the farmer in the previous question was isolated immediately upon detection of her disease. Which outcome would you predict?

   A. Her disease will not spread
   B. Only her close relatives and contacts will get the disease
   C. The disease will spread to most people in the community but not beyond it
   D. Her disease might spread over the entire human population

4. Which of the following is a mismatch?

   A. Acute disease—symptoms develop quickly but infection lasts only a short time
   B. Chronic disease—symptoms develop slowly and disease can last a long time
   C. Subacute disease—secondary infection
   D. Latent disease—symptoms can continue to reappear long after initial infection
   E. Local infection—confined to a small area of the body

5. Etiology refers to the study of

    A. Viral infection
    B. The cause of disease
    C. The portal of exit
    D. The results seen after a disease occurs
    E. None of the above

6. Microbial antagonism refers to

    A. An increase in bacterial metabolism
    B. A decrease in bacterial metabolism
    C. An increase in infection symptoms
    D. Enhanced disease due to resident bacteria
    E. Protection by normal resident bacteria

7. A person develops influenza as a result of infection with the influenza virus. The relationship between the virus and the person is

    A. Mutualism
    B. Parasitism
    C. Commensalism
    D. Discordance
    E. None of the above

8. Koch's postulates were first used to determine the

    A. Portals of entry for pathogens into the body
    B. Portals of exit for pathogens leaving the body
    C. Existence of exotoxins
    D. Etiology of viral infections
    E. Etiology of bacterial infections

9. The time between the initial infection and the onset of symptoms is called the

    A. Disease period
    B. Illness period
    C. Prodromal period
    D. Decline period
    E. Incubation period

10. Which method cannot be used to fully or partly control a communicable disease?

    A. Isolation
    B. Antiseptics
    C. Quarantine
    D. Vector control
    E. Vaccination

11. The period during which major disease occurs is called the

    A. Period of illness
    B. Period of decline
    C. Prodromal period
    D. Period of convalescence
    E. Major period

12. Chronic diseases

    A. Develop quickly and subside quickly
    B. Develop quickly but subside slowly
    C. Develop slowly and subside quickly
    D. Develop slowly and remain for a long time
    E. Develop quickly and remain for a long time

13. Latent diseases

    A. Develop quickly and subside quickly
    B. Remain in the host after symptoms have gone but are able to be reactivated
    C. Develop quickly and subside slowly
    D. Remain in the host only when symptoms are present
    E. Remain in the host after symptoms have gone but are never reactivated

14. Septicemia refers to pathogens growing in the

    A. Focus of infection
    B. Tissues
    C. Urinary tract
    D. Blood
    E. None of the above

15. Which of the following is true about human microbiomes?

    A. Microbiomes can protect against infections
    B. Each individual has a unique microbiome
    C. Healthy individuals differ in the microbes that occupy the gut, skin, and vagina
    D. Microbiomes change with diet, climate, and age
    E. All the above are true

DEPTH OF UNDERSTANDING

Questions listed here require you to bring together the concepts you have learned in this chapter into a discussion format. This helps you to increase your depth of understanding of the material you have learned. Help can be found in the student resources at: www.garlandscience.com/micro2

1. There are three methods used to control communicable diseases. Which of them is most effective, and why?

2. Tuberculosis has been referred to as both a persistent infection and a re-emerging infection. Justify this duality of classification.

 CLINICAL CORNER

Help can be found in the student resources at: www.garlandscience.com/micro2

1. Your patient is a 78-year-old male who is in the hospital because of a serious bladder infection. He has been on antibiotic therapy for 4 days and seems to be getting better. Although he says he feels better, he is tired and not eating well. On the fifth day, you are surprised to find that he has been transferred to the intensive care unit. The doctor informs you that he has developed a severe upper respiratory infection and is in a grave condition. How will you explain to his wife and family what has happened?

2. Your neighbor has had frequent bouts of severe gastritis. She has been to the doctor and was diagnosed with a salmonella infection each time. After she was given antibiotics, the infection subsided but always seemed to recur. Finally the doctor explained that she needed to have her gallbladder removed. She is confused by all this and frightened by the possibility of surgery.

   A. How would you explain her condition to her?
   B. Why does the doctor want to remove her gallbladder?
   C. Why did the infection keep recurring even after antibiotic therapy?

3. Montrose elementary school is located in southern Florida, a region that sees a seasonal influx of migrant workers who help during the fruit-picking season. Two weeks into the orange-picking season, there was an outbreak of measles in the second grade. Of the 26 students, only 4 came down with the infection and one of those was the daughter of a migrant worker.

   A. Explain why only four students got the measles.
   B. Why did the daughter of the migrant workers get infected?
   C. Why did the three other children who were not from migrant worker families get sick?

Emerging and Re-Emerging Infectious Diseases

Chapter 8

Why Is This Important?

The world is facing the challenge of both new diseases and re-emerging ones. As a health care professional it is important that you know about them and about the potential for changes in health care they might cause.

Sam has recently learnt how to swim and cannot wait to impress her grandparents. She is visiting them in their house near the lake. What could possibly take her grandmother that long to pack everything? Not even putting on her new swimming suit is distraction enough to pass the time. Finally, they leave the house and pick up her grandfather en route to the lake. The sun is shining and Sam is happy. They spread the blanket on the grass and get ready to swim. Sam is watching her grandparents and notices that they both have scars on their left upper arms. That is too much of a coincidence. So Sam asks what happened. Her grandfather explains that the scars originate from smallpox vaccinations, which Sam has not had. Smallpox is caused by a DNA virus. A rash develops and spreads all over the body, and then pustules form. The subsequent multiorgan failure can be fatal. The vaccination is based on Edward Jenner's work in the late eighteenth century. The WHO officially declared smallpox eradicated in 1980 after a global vaccination program from 1967. Vaccinations of the public were stopped in the United States in 1972 after eradication. The vaccine is given using a bifurcated needle pricking the skin repeatedly to cause breaks and blood droplets to form. If the vaccination is successful, an itchy bump develops within 3–4 days and then becomes a blister. Once the blister begins to heal, a scab forms and then falls off to reveal a small scar. Sam is so intrigued by this story that she even forgets all about the swimming.

OVERVIEW

In this chapter, we look at the general concept of emerging infectious diseases, and important examples of human pathogens that have appeared over the past 30 years or so. These diseases are important because they will have a great impact on health care in the future, and have socio-economic relevance locally and globally. Traveling and food have a role here, as well as climate change and bioterrorism. Emerging infections such as SARS (severe acute respiratory syndrome) not only make headlines but are also a wake-up call for the health care community. They remind us that there are always new and potentially dangerous diseases cropping up in various populations. We will also discuss re-emerging infectious diseases by taking a close look at influenza and tuberculosis.

We will divide our discussions into the following topics:

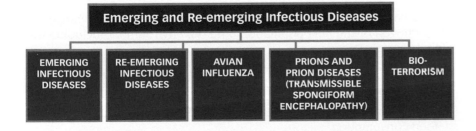

EMERGING INFECTIOUS DISEASES

"There will come yet other new and unusual ailments in the course of time. And this disease will pass away, but it later will be born again and be seen by our descendants."

Girolamo Frascatoro (c.1478–1553) (speaking about syphilis)

Even though the quoted passage was written more than 450 years ago, it is just as valid today as we take our first look at emerging and re-emerging infectious diseases. When we look at infectious disease from a historical perspective, we can see that it has had a prominent role in shaping world events and society. *Yersinia pestis*, the causative agent of the Black Death, had wiped out about half of the population in plague epidemics in Europe by the fourteenth century. The spread followed initial Roman invasions and caravan routes. This large-scale loss of human lives led to the stalling of agriculture, to food shortages, and eventually to the redistribution of wealth and an end of the feudal system. The measles virus contributed to the decline of the entire Aztec civilization, and many of the indigenous peoples of North and South America were decimated by smallpox brought unknowingly by explorers from Europe, who in turn brought new diseases back home. Syphilis is believed to be an example of the latter. Smallpox is also documented to have been used with intent as the first biological weapon when blankets from smallpox victims were given to Native Americans by the British in pre-Revolutionary War days in the seventeenth century.

More recently, the ability of disease to affect history has continued. Almost 90 pathogen species were discovered between 1980 and 2010, causing diseases such as legionnaires' disease, AIDS, hepatitis C, Creutzfeldt–Jakob disease, Nipah virus infection, several types of hemorrhagic fever, SARS, and avian influenza. In addition, we have had a re-emergence of diseases such as tuberculosis and cholera that can be attributed to changes in ecology, migration, mobility, and resistance to antibiotics.

Emerging infectious diseases are those whose incidence in humans have either increased in the past 30 years or threaten to increase in the near future. These may be diseases that had previously been unseen in humans or had been seen only rarely. The category also encompasses diseases that had previously been recognized but whose cause was unknown until recently (such as hepatitis C and the association between *Helicobacter pylori* and ulcers). Diseases well known to us were emerging diseases hundreds or even thousands of years ago. Some diseases also disappear over time.

In addition to emerging infectious diseases, we are faced with the problems presented by **re-emerging infectious diseases**. This category comprises diseases that were previously controlled (for example through treatment with antibiotics, or prevention through vaccine programs) but now have returned (such as when a pathogen has become resistant to treatment). Examples of emerging and re-emerging infectious diseases are given in Tables 8.1 and 8.2.

As you can see, the (re-)emergence of a pathogen is due to changes in the pathogen or the host population, and the environment including the global climate. Drivers are:

- Biological factors such as the immune status of the host and the virulence of the pathogen (for example resistance to treatment)

- Social factors such as trade, travel, migration, sexual practices, and food consumption

Disease	Infectious agent	Year recognized	Contributing factors
Lassa fever	Lassa virus (belongs to *Arenaviridae*)	1969	Urbanization and consequent increased rodent population; increased nosocomial transmission
Ebola hemorrhagic fever	Ebola virus (belongs to *Filoviridae*)	1976	Wild animals; nosocomial transmission
Legionnaires' disease	*Legionella pneumophila* (belongs to Proteobacteria)	1976	Cooling and plumbing systems
Lyme disease	*Borrelia burgdorferi* (belongs to spirochetes)	1975	Environments that favor tick and deer populations
AIDS	Human immunodeficiency virus (belongs to *Retroviridae*)	1981	Global travel, intravenous drug abuse, multiple sexual partners
Cholera	*Vibrio cholerae* 0139 (new strain of the known species belonging to Proteobacteria)	1992	Increased virulence
Hantavirus pulmonary syndrome	Hantavirus (belongs to *Bunyaviridae*)	1993	Encroachment into rodent territories

- Political factors such as public health access and resources, prophylaxis and treatment, education, and technology and communication infrastructure

- Economic factors such as research and development budget, disease control, land use, domestication of animals, and size of settlements

We will explain some of those aspects in more detail later in this chapter.

Along with the rise in new types of infectious disease, there has naturally been a concurrent rise in interest in them. In 1972 the Australian immunologist Frank Macfarlane Burnet summed up the future of infectious diseases in this remarkably hopeful way: "The most likely forecast about the future of infectious disease is that it will be very dull." This quote must be taken in the context of what medical research had accomplished in the

Table 8.1 Examples of emerging infectious diseases.

Table 8.2 Re-emerging infectious diseases.

Disease	Infectious agent	Contributing factors
Cryptosporidiosis	*Cryptosporidium parvum*, a protozoan parasite	International travel; contaminated water supplies
Diphtheria	*Corynebacterium diphtheriae*	Interruption of immunization program due to political changes
Influenza	Influenza virus, an orthomyxovirus	Genetic re-assortment
Malaria	*Plasmodium* species	Drug resistance; inadequate mosquito control
Pertussis	*Bordetella pertussis*	Refusal to vaccinate; decreased vaccine efficiency; waning immunity in adults
Rabies	Rabies virus, a rhabdovirus	Breakdown in public health measures; travel; changes in land use
Tuberculosis	*Mycobacterium tuberculosis*	Antibiotic resistance; increased immunocompromised populations
Yellow fever	Yellow fever virus, a flavivirus	Urbanization; insecticide resistance

Figure 8.1 Proportions of total deaths from major causes in Chile. Panel a: 1909. Panel b: 1999. This country is a good illustration of the full transition from a developing nation to a developed status in the twentieth century. Notice the dramatic decrease in infectious diseases, which can be attributed to the development of antibiotics.

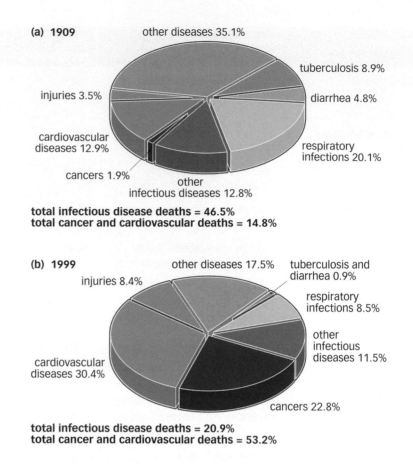

(a) 1909

other diseases 35.1%

tuberculosis 8.9%

diarrhea 4.8%

injuries 3.5%

respiratory infections 20.1%

cardiovascular diseases 12.9%

cancers 1.9%

other infectious diseases 12.8%

total infectious disease deaths = 46.5%
total cancer and cardiovascular deaths = 14.8%

(b) 1999

other diseases 17.5%　tuberculosis and diarrhea 0.9%

injuries 8.4%

respiratory infections 8.5%

other infectious diseases 11.5%

cardiovascular diseases 30.4%

cancers 22.8%

total infectious disease deaths = 20.9%
total cancer and cardiovascular deaths = 53.2%

years before 1972. During that period, antibiotics had become commonplace, and we had conquered so many diseases that deaths from infectious disease had fallen dramatically (**Figure 8.1**), which actually led to an increase in life expectancy (**Figure 8.2**). Smallpox was near eradication (declared in 1980). In fact, this trend continues today except where HIV infection is widespread.

In the past 15 years, however, falling living standards and the decline of infrastructure due to political change have aided the re-emergence of some infectious diseases such as poliomyelitis.

Twenty-five to thirty percent of the approximately 60 million deaths that occur worldwide each year are caused by infectious disease, although many of the infections are rarely seen for instance in the United States. Infectious diseases, both established ones and emerging and re-emerging ones, will remain major global health threats that will affect society for the foreseeable future.

It is estimated that 61% of the 1415 species of organisms known to infect humans are transmitted by animals. In these cases, the human is the terminal, or dead-end, host. However, occasionally a zoonotic infection will adapt to human-to-human transmission and in doing so will diversify away from its animal origin. Examples of this sequence are measles, smallpox, and HIV.

Throughout history, emerging diseases have followed a pattern of four transitions. The first was called a crowd transition and occurred when people began living close to one another in crowded places, allowing the easy transmission of diseases. These crowded conditions also allowed the movement of zoonotic diseases from animals to humans. A good example is measles, which was confined to cattle 7000 years ago but

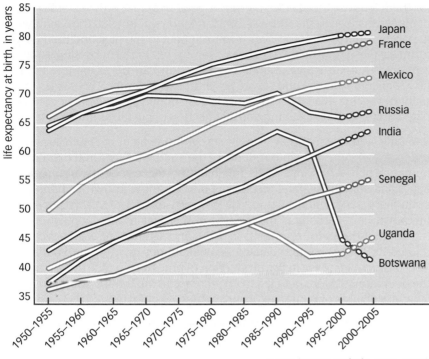

Figure 8.2 Changes in life expectancy at birth for both sexes in eight representative countries during the period 1950–2005. The trend toward longer life expectancies is a direct result of our success in controlling infectious disease and improved socio-economic conditions. Decreases are usually related to limited access to health care and education, involvement in wars, and emerging pathogens such as HIV.

then through a series of mutations was able to diverge into a pathogen that infects only humans. Although this first transition involves ancient history, it is still relevant today, as we will see below when we look at the development of SARS in China.

The second transition occurred in classical times as European civilization came into contact with other societies through either war or trade. This contact allowed various groups to exchange pools of infections and vectors. For example, in 430 BC during the Peloponnesian War, Thucydides wrote the first reports describing typhus.

The third transition coincided with worldwide exploration and colonization from the late fifteenth century. There are many examples of emerging diseases as humans came into contact with pathogens that they had never seen before. Cortez writes of smallpox and measles helping to defeat the Aztec empire. Captain Cook and his crew destroyed many Pacific island populations by bringing syphilis, measles, and tuberculosis along with them. Pacific islanders who had never encountered these pathogens had no defense against them.

The fourth transition is currently under way and can be classified as global urbanization. The increase in densely populated settlements coupled with poverty, social upheaval, air travel, long-distance trade, technological development, land clearance, and climate changes are contributing to the emergence and re-emergence of dangerous diseases that have the potential to spread more quickly than ever before.

Environment and Infectious Disease

As humans encroach on uncultivated environments, there are new contacts between humans, domestic animals, and wild fauna (including disease vectors). This increases the risk of contact with previously unknown pathogens for both humans and other animals. Even if these infectious organisms develop in animals first (zoonotic diseases), there is the potential

Figure 8.3 The female western blacklegged tick *Ixodes pacificus*, a vector for Lyme disease caused by *Borrelia burgdorferi*.

for them to cross the species barrier. A good example is the establishment of piggeries close to the tropical forest in Malaysia. In 1988 the Nipah virus (a paramyxovirus) crossed over from fruit bats to the pigs and then began to infect the pig farmers. This was seen in 1999. The next step is direct human-to-human transmission. The infection can be asymptomatic but also fatal (encephalitis and severe respiratory problems). There is no treatment or vaccine available. The virus is transmitted by droplets or contaminated fruits. Routine cleaning and disinfection of pig farms limit the spread.

Similarly, a hantavirus that was a pathogen prevalent in South American rodents worked its way into rodents of the southwestern United States. As humans encroached into areas inhabited by these rodents, outbreaks of acute and sometimes fatal respiratory disease (now known as hantavirus pulmonary syndrome) occurred, first observed in 1993. This previously unknown hantavirus was maintained in the deer mouse population and transmitted through excrement. In 1991 and 1992 there was a weather-related El Niño event that caused heavy summer rains in the Southwest. This led to an increase in the production of pine nuts, which are a staple of the deer mouse. More nuts led to an increase in the mouse population and a greater chance of human exposure, which resulted in the 1993 outbreak.

This connected series of events is also seen with the dengue virus, which is spreading because its mosquito vector is now breeding rapidly in congested urban environments in developing countries. Lack of proper mosquito control and poor living standards set the stage for outbreaks of this disease. For Lyme disease, seen today in many regions of the United States, the vector is the deer tick (**Figure 8.3**). As the number of deer increases, the number of ticks increases and so, too, does the potential for increased outbreaks of Lyme disease. Lyme disease also occurs in Europe, and the vectors are other tick species.

Climate change also affects pathogen vectors. Some will be able to spread into new territories as a result of increased temperatures. For example, we will soon see an increased malaria risk in countries near the present distribution, and small increases in current malaria regions.

Foodborne Infections

Enterohemorrhagic *Escherichia coli* (EHEC) causes severe foodborne diseases (described in detail in Chapter 22). Common sources of outbreaks are raw or undercooked ground meat products, raw milk, and fecal contamination of vegetables. EHEC can grow in temperatures ranging from 7°C to 50°C and in acidic foods. Spread of the pathogen can be limited by adhering to basic food hygiene and particularly by cooking everything thoroughly. Waterborne transmissions are seen as a result of contaminated drinking water and recreational water use. Person-to-person transmissions occur through the oral–fecal route. After an incubation period of 3–8 days the illness, with symptoms of abdominal cramps, diarrhea, fever, and vomiting, is usually self-limiting, but in up to 10% of patients (especially young children and the elderly) it can be life-threatening and can cause serious conditions such as hemolytic uremic syndrome (HUS) with acute renal failure, hemolytic anemia, and thrombocytopenia. Fatality rates range from 3% to 5%. EHEC produces exotoxins (verotoxins or Shiga-like toxins; see Chapter 5). *E. coli* O157:H7 is the most important EHEC serotype in relation to public health. The first outbreak in the United States was seen in 1982.

Limited outbreaks of enteroaggregative *E. coli* (EAEC) O104:H4 have been reported since 1996 in Japan, Sweden, and the United States. In 2011 the pathogen had spread to 15 countries from northern Germany—almost

3500 cases were recorded, including 50 deaths in just 3 months, and 25% of these involved HUS. *E. coli* O104:H4 produces Shiga toxin. This virulence factor was acquired via a bacteriophage that carried the Shiga toxin genes into *E. coli*. Production of the toxin is linked to the bacteria's response (known as the SOS response) to antibiotic and actually increases during the early phase of treatment with antibiotic (ciprofloxacin) if the concentration of antibiotic is too low to kill the bacteria. *Salmonella* can also cause fatal diseases in humans and is transmitted from food sources.

As the population grows worldwide, there is increased pressure to produce more meat. This has led to the emergence of infections that are transmitted from farm animals to humans. One well-known example is the 1986 outbreak of mad cow disease in Britain, although this disease is caused by a prion, which is not an organism (but is discussed later in this chapter). The problem of infections transmitted from animals to humans is compounded by the increasing demand for exotic and wild animals. SARS, for instance, is associated with such exotic animals as the palm civet and the raccoon dog.

Globalization and Transmission

Changing patterns in human behavior and a changing ecology on our planet contribute to the emergence of infectious disease in two ways:

- Increased opportunity for animal-to-human infection because of greater exposure

- Increased opportunity for the transmission from one human to another once a person is infected

Genetic changes in pathogens can occur through a process known as **re-assortment**. This is a type of gene shuffling in which an organism's genes rearrange themselves and cause changes in the characteristics of the organism. Re-assortment can also take place between different pathogens. Genetic re-assortment between pathogens could give rise to new, rapidly spreading strains of pathogens. For instance, as we will see later in this chapter, there is the potential for deadly avian strains of influenza to recombine genetically with strains that commonly infect humans, resulting in an influenza outbreak that could be comparable to or even worse than the pandemic of 1918 (Spanish Influenza), which killed 30–50 million people worldwide. Influenza pandemics have been observed every 10–50 years since the sixteenth century. There were three in the twentieth century.

Fast Fact Novel infectious disease can emerge in any part of the world at any time. For example, HIV and Ebola came out of Africa, avian influenza and SARS from Hong Kong and China, Nipah from Malaysia, and hantavirus from South America.

Some of the most sinister infections are latent ones that have long incubation periods, such as with HIV. By the time these diseases are recognized, the infection is already established in humans, potentially out of control, and the pathogen has been spread farther, exacerbated by the marvel of modern air travel, which can disperse infections worldwide in a matter of days. The West Nile virus, for instance, which was unknown in the United States before 1999, arrived in the northeastern part of the country. It had circulated in Israel and Tunisia before it was imported, either in an infected traveler or in a mosquito that hitched a ride on an airplane. Whichever is the case, it took less than 4 years for this virus to move from isolated counties in New York State to the Pacific coast of the United States.

If we combine all the factors mentioned above, especially travel, the potential for deadly infections to appear quickly worldwide is not small. Keep in mind that the immunodeficiency associated with HIV infection as well as increasing numbers of compromised hosts can amplify the entire process by presenting increasing numbers of potential targets.

Hurdles to Interspecies Transfer

It is important to remember that crossing the species barrier is not a simple task. A pathogen must overcome two major hurdles to replicate successfully in a human host:

- It must adapt in such a way that it can replicate in human cells. This can be a complex problem for the pathogen.

- It must be able to configure itself so that it can be easily transmitted from one human to another.

Many pathogens, such as the hantavirus and avian influenza, have overcome the first hurdle and have jumped from animals to humans, but they have not yet been able to overcome the second. Other pathogens such as HIV or swine influenza (in 2009) successfully cleared both hurdles.

For a pathogen to clear these two hurdles it must undergo extensive genetic changes. Viruses are prone to mutation because of the lack of fidelity in replication (especially RNA viruses such as the influenza virus), making it easier for them to acquire genetic changes.

Emerging Viral Diseases

The great majority of today's emerging infectious diseases are caused by viruses rather than by bacteria or other pathogens. Because viral replication requires the presence of particular specific factors in the host cell, it may be either the absence or the foreignness of these host factors that prevents virus replication in foreign species. If adjusting to this absence or foreignness requires only minor genetic change in the virus, there is an increased risk that the jump to a new host species will be made. All in all, given the extent of the obstacles facing the virus, it is amazing that a jump ever occurs. In fact, the number of viral emerging infectious diseases is quite small when you consider the number of viruses that exist in animal reservoirs.

The jump does occur, however, and RNA viruses are the best at it because they are best adapted to the three mechanisms that can overcome the problem: mutation, re-assortment, and recombination (Figure 8.4). We know that both the Asian influenza strain (H2N2) of 1957 and the 1968 Hong Kong influenza (H3N2) pandemic arose through the re-assortment of human and avian strains of the virus. Now there is increasing global concern that re-assortment between genes from the avian strain known as H5N1 and the human influenza virus will result in a repeat of the 1918 influenza (H1N1) pandemic.

Let's look at a few of the emerging viral diseases.

SARS (Severe Acute Respiratory Syndrome)

SARS was a previously unrecognized animal coronavirus that exploited the wet markets of southern China, places where live animals such as palm civets and raccoon dogs are kept in close proximity to humans until sold. In these hospitable locations, the virus adapted and became readily transmissible to humans during the 1990s. The course and spread of the original infection are a great example of how infectious disease spreads in the modern world. The first human case of SARS was seen in Guangdong Province, China, and was transmitted unknowingly from the

Fast Fact Within weeks, SARS had spread to more than 8000 people in 30 countries on five continents by July 2003. It killed 744 people.

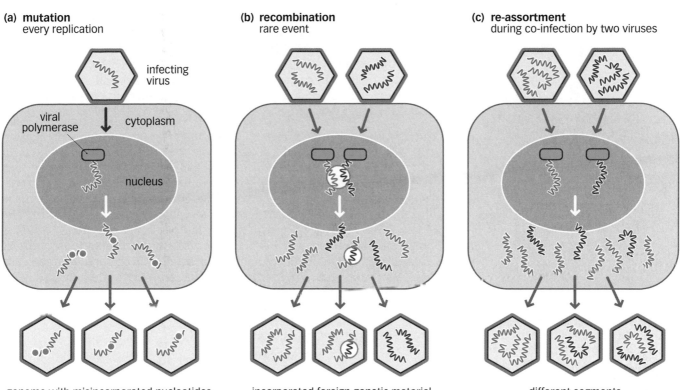

(a) mutation
every replication

infecting virus

viral polymerase

cytoplasm

nucleus

genome with misincorporated nucleotides packaged into progeny viruses

(b) recombination
rare event

incorporated foreign genetic material is packaged into progeny viruses

(c) re-assortment
during co-infection by two viruses

different segments packaged into progeny viruses

Figure 8.4 Three molecular mechanisms for generating viral diversity. Panel a: Single-point mutations incorporated into one or more positions as a result of the lack of proofreading by the viral polymerase. Panel b: During recombination, foreign genetic material is incorporated into the viral genome. Panel c: Re-assortment, in which whole gene segments can be swapped. All three mechanisms, which are not mutually exclusive, may result in viruses with new biological properties such as host ranges and pathogenic potentials.

patient to a local physician, who then traveled to Hong Kong for a wedding in early 2003. He spent one night in a hotel in Hong Kong, during which time he transmitted the virus to 16 other guests, most of whom were foreigners. They in turn spread the virus, causing outbreaks in Hong Kong, Singapore, Vietnam, and as far away as Toronto.

It is thought that the first humans to be infected were workers involved with wild game such as palm civets and raccoon dogs. It is likely that the virus stumbled many times in overcoming the hurdles of crossing to a new species before it was finally successful in adapting to replication in humans. The second and perhaps more difficult hurdle—being transmissible from human to human—took longer but was also finally cleared. SARS can now be transmitted by droplet, aerosol, and fomite and is deposited on the respiratory mucosal epithelium, where it initiates infection.

Pathogenesis of SARS

After an incubation time of 2–4 days, SARS causes infection of the lower respiratory tract accompanied by fever, **malaise**, and **lymphopenia** involving T cells. Coagulation times are lengthened, and hepatic enzyme levels are increased. X-rays show infiltrates and sub-pleural consolidation consistent with atypical viral pneumonia (**Figure 8.5**). Twenty to thirty percent of patients infected with SARS require intensive care, and approximately 10% will succumb to the disease.

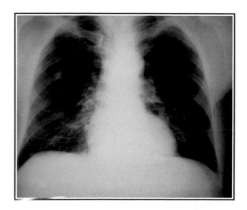

Figure 8.5 Chest X-ray of a patient with SARS (severe acute respiratory syndrome). The white shadow seen in the lungs is caused by the infection.

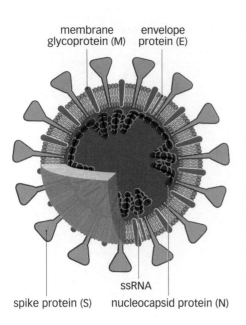

Figure 8.6 A schematic representation of the SARS coronavirus structure. Viral surface proteins (labeled M, E, and S) are embedded in a lipid bilayer envelope derived from the host cell. The single-stranded viral RNA is associated with the nucleocapsid protein (labeled N). The spike proteins are used to bind to receptors on host cells.

The pathogenesis of SARS is due to a high viral load in the lower respiratory tract. This viral load is low in the first 4 or 5 days of the infection and peaks at about day 10. There is also a direct correlation between viral load and prognosis, higher loads being associated with poor prognosis. On about the tenth day of illness, the virus can be found in nasopharyngeal aspirates, feces, and serum. Poor clinical outcome is associated with continued uncontrolled viral replication.

The SARS virus enters host cells by attaching spike proteins present on the virion (**Figure 8.6**) to receptors on the host plasma membrane. In the first SARS viruses to be isolated, the spike proteins on the virions were genetically very diverse, but in later viral isolates the spike proteins became more homogeneous. This is probably because the virus is becoming more adapted to human hosts after selective pressure. The main SARS-CoV receptor is angiotensin-converting enzyme 2, which is expressed by pneumocytes and enterocytes. In addition to those primary target cells, renal tubule epithelium, cerebral neurons, and immune cells can also be infected. The virus causes direct injury to the infected cells, whereas indirect injury is mainly due to immune responses, circulatory dysfunction, and hypoxia.

Host Response and Treatment

Although the levels of several inflammatory **cytokines** and **chemokines** are elevated in patients with SARS, there is a prolonged immunological impairment during the disease. Treatments vary; antiviral chemotherapeutics (such as ribavirin) are effective if administered during the first few days of the infection. Attempts to develop a vaccine for SARS are under way, and it has also been shown that interaction of the receptors and the spikes can be inhibited, thereby preventing the virus from attaching to host cells and achieving successful infection. So far, however, there is no vaccine for this disease.

West Nile Fever

The coronaviruses exemplified by SARS are not the only instance of emerging viral diseases today. We also have arboviruses, which are those carried by arthropods (**ar**thropod-**bo**rne viruses; hence **arbo**viruses). The best examples of this group of emerging viruses are the West Nile and dengue viruses. The arthropod vectors that carry these pathogens must get blood if they are to complete their life cycle, and it is during this feeding event that the virus is transmitted. There are more than 100 viruses in this group. They can be mosquito-borne, tick-borne, sandfly-borne, or borne by an as yet unknown vector. Arboviruses include several taxonomic groups of RNA viruses that are thought to have emerged in the human population as a result of continued deforestation in several African nations. These viruses most probably dispersed from the original sites by means of wind-blown mosquitoes, migrating birds, or infected travelers. When viruses circulate only between vectors and nonhuman animals we call it the sylvatic cycle, but if circulation is between vectors and humans we call it an urban cycle.

West Nile virus is a flavivirus (family *Flaviviridae*) and was first isolated in 1937 from the blood of a woman in the West Nile district of Uganda. Today, this virus is widespread throughout Africa, the Middle East, and parts of Europe, North America, west Asia, and Australia. The largest outbreaks have been in Greece, Israel, Romania, the Russian Federation and the United States, but it was not detected in the United States until 1999, when there was an outbreak in New York City. Epidemiologists believe that the virus arrived by means of an infected traveler from the Middle

East. The rapid spread of this virus throughout the United States, Mexico, and the Caribbean in only 5 years is truly remarkable (**Figure 8.7**).

Birds are the primary host of the West Nile virus, and it is spread from bird to bird by mosquitoes. Humans and animals such as horses are incidental hosts; they can be infected by mosquitoes carrying the virus, particularly near major bird migratory routes. The virus has been identified in more than 50 species of mosquito, but it is still unclear whether there is a specific mosquito species that transmits the virus to humans. Nearly all infections have resulted from mosquito bites, but transmission through contact with other infected animals, their blood, or other tissues, or by blood transfusion, transplanted organs, and breast milk has occurred. There has been one reported case of transplacental (mother-to-fetus) transmission, but direct human-to-human transmissions have not been seen.

Pathogenesis of West Nile Virus

The pathogenesis of West Nile infection is not fully understood, but it is clear that the virus is transmitted to humans in the saliva of the mosquito. Most infected people are asymptomatic unless the infection causes an invasive neurological disease called West Nile fever, which occurs in 20–30% of those infected. There is a 2–14-day incubation period. Symptoms include fever, headache, back pain, **myalgia**, nausea, vomiting, and anorexia (weight loss). There may also be profound fatigue and skin rash. Severe infection can cause **myocarditis**, **pancreatitis**, and hepatitis, and 1 in 150 patients develop either **encephalitis** or **meningitis**. Approximately 10% of these latter patients will succumb to the disease.

Advanced age (over 50 years old) or being immunocompromised (for example transplant patients) is a predictor of poor prognosis, and survivors of severe symptoms can have long-term neurological impairment. West Nile virus can also cause acute flaccid paralysis syndrome, from which complete recovery is uncommon. In the neurological conditions caused by West Nile virus, viral antigens are most commonly associated with neurons and neuronal processes. There is both a cellular and a humoral host defense response against this viral infection. Vaccines are available for use in horses but are not yet available for humans. Treatment involves supportive care, and prevention focuses on vector control.

Dengue Fever

This severe, flu-like illness, which occasionally leads to potentially fatal complications, has been known since the 1950s, but the incidence has increased thirtyfold over the past decades. Up to 100 million infections are estimated per year in more than 100 countries, with more than one-third of the world's population at risk in tropical and subtropical climates. Annual outbreaks are now found in the United States, mainly in southern Florida and Texas, and it is also prevalent in Puerto Rico and many popular tourist destinations in Latin America and Southeast Asia.

Four serotypes of the dengue virus (a flavivirus) are known. Cross-immunity to other serotypes after recovering from dengue fever is only partial and temporary, so subsequent infections with other serotypes can take place, and may increase the risk of developing a more severe form of illness. The usual vector is the infected female mosquito *Aedes aegypti*. The vector is found in urban habitats and bites early in the morning and in the evening. *Aedes albopictus* is a secondary vector, which has permitted spreading of the disease to cooler regions of Europe. There is no vaccine and no specific treatment. Vector control measures are used for prevention.

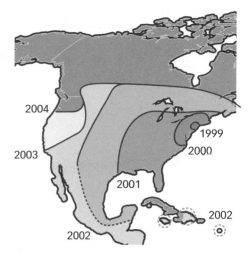

Figure 8.7 The approximate geographical distribution of West Nile virus in the Americas from 1999 to 2004. This was determined by virologic surveillance of dead birds reported to the Centers for Disease Control in the United States and to the Department of Prevention and Health in Canada. The dashed lines indicate estimated range limits.

Fast Fact A genetic mutation has been identified that increases a person's susceptibility to infection with West Nile virus. Ironically, it is the same mutation that seems to protect people from infection with HIV.

Pathogenesis of Dengue Fever

After an incubation period of 4–10 days, a severe flu-like illness is seen with high fever, severe headache, pain behind the eyes, muscle and joint pain (dengue is also known as breakbone fever), and a rash. Sometimes the disease progresses to severe dengue (dengue shock syndrome) and dengue hemorrhagic fever in which plasma leaks from blood vessels leading to a decrease in blood pressure, fluid accumulation in the chest and abdominal cavity, respiratory distress, and organ impairment or failure. There may also be severe bleeding from the gastrointestinal tract.

Viral Hemorrhagic Fever (VHF)

The emerging infectious diseases classified as viral hemorrhagic fevers include the conditions caused by the Ebola, Marburg, and yellow fever viruses. Some of these diseases, in particular those caused by Ebola and Marburg viruses, are frequently fatal. VHF infections are some of the most exotic emerging diseases and are characterized by fever, bleeding, and **circulatory shock**. Ebola virus infections occur in Central and West Africa near tropical rainforests. It first appeared in 1976 in two simultaneous outbreaks. Since 1994, Ebola outbreaks have been found in chimpanzees and gorillas; and since 2008 the Ebola Reston virus was found during outbreaks of fatal pig diseases. Marburg hemorrhagic fever was first seen in 1967 during epidemics in Marburg and Frankfurt (Germany) and in the former Yugoslavia as a result of the importation of infected monkeys from Uganda. All of the viruses are enveloped single-stranded RNA types, but they vary in morphology (**Figure 8.8**).

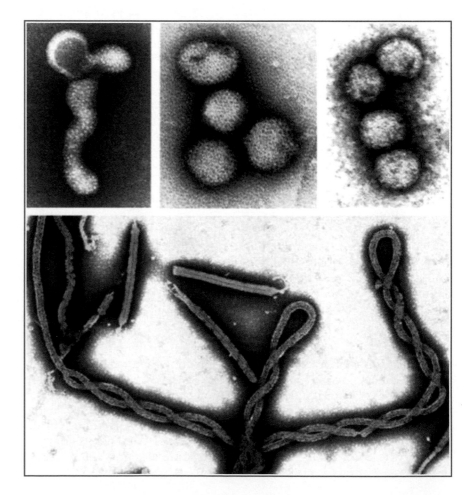

Figure 8.8 Transmission electron micrographs showing the variety of morphologies seen in the viruses responsible for viral hemorrhagic fever (VHF). Top left, Junin virus; top center, Rift Valley virus; top right, yellow fever virus; bottom, Ebola virus.

These viruses are transmitted in diverse ways from wild animals to humans. Arthropod and rodent vectors can also play a role. Fruit bats are the natural reservoir for Ebola virus. All of the hemorrhagic viruses have developed the ability to be transmitted from human to human, usually through direct contact with infected blood or bodily fluids. Working with these viruses or with infected patients as well as handling ill or dead infected wild animals requires the highest levels of protection possible.

Pathogenesis of Viral Hemorrhagic Fever

The hemorrhagic viruses have incubation periods of between 2 and 21 days, and the severity of the infection depends on the virulence of the virus, the route of exposure, the viral load, and the competence of the host defense. The symptoms are fever, myalgia with prostration, **flushing**, **petechial hemorrhaging**, vomiting, diarrhea, rash, impaired kidney and liver function, and in some cases, both internal and external bleeding. Fatality rates average between 5% and 20% for all VHFs. However, for Ebola infection, the death rate is 50–90%. Fatality rates related to Marburg virus vary considerably, from 25% in the initial laboratory-associated outbreak to more than 80% from 1998 to 2000, and to even higher in the outbreak in Angola in 2004.

VHF viruses target monocytes, dendritic cells, endothelial cells, hepatocytes, and adrenal cortical cells. They use endocytosis as the mechanism of penetration and cause varying degrees of cell destruction. All VHF viruses target and impair host antiviral responses, a strategy that permits high levels of viremia and also immunosuppression (**Figure 8.9**). Sudden, severe shock-like symptoms are seen in all VHF fatalities.

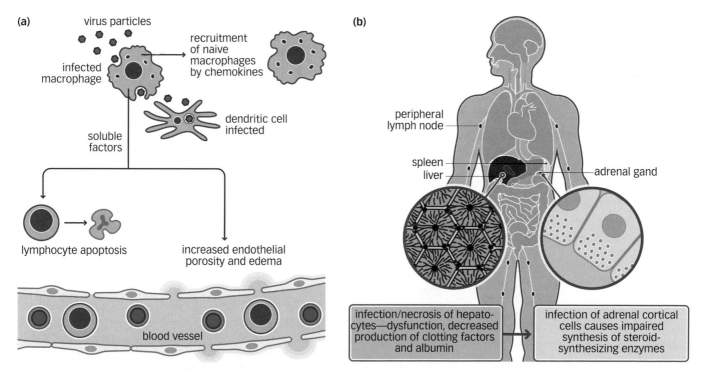

Figure 8.9 A model for VHF pathogenesis. Panel a: Virus spreads to regional lymph nodes, liver, and spleen. At these sites, virus infects macrophages (including Kupffer cells) and dendritic cells. Soluble factors are released and act both locally and systemically. This recruits more macrophages to the site, and these also become infected. This process continues and the host defenses begin to malfunction with a loss of lymphocytes. Panel b: Hemodynamic and coagulation disorders common to all VHF infections are exacerbated by the infection of hepatocytes and adrenal cortical cells, which results in a reduced synthesis of albumin by hepatocytes and leads to decreased plasma osmotic pressure. The result is edema and eventual hypotension.

The immunosuppression seen in VHF infections is caused by varying degrees of lymphocyte depletion and necrosis of the spleen and lymph nodes. Although lymphopenia occurs in all VHF infections, the cause remains unclear. However, in Ebola and Marburg infections there is evidence of increased apoptosis of lymphoid cells. Rapid loss of lymphocytes results in a failure to control viral replication, leading to enormous systemic viral burdens.

VHF infections in humans trigger the expression of several inflammatory mediators, including interferon and interleukins, but these agents can contribute to the disease process rather than inhibiting it. However, it seems that the earlier and more robust a host defensive response, the greater the chance of survival. Although the mechanisms involved are yet to be fully worked out, it seems that there is a balance between the protective and deleterious cytokines produced during these infections. The result of tipping this balance can be either recovery or death.

The production of nitric oxide (NO), a powerful antimicrobial agent, also seems to have an important role in the pathogenesis of these infections. Increased blood levels of NO are associated with mortality because they cause increased apoptosis, tissue damage, and loss of vascular integrity, all of which may contribute to shock. NO is also an important mediator of hypotension, which is a prominent problem seen in VHF infections.

Abnormalities in blood coagulation and fibrinolysis (the destruction of fibrin) during VHF infection result in **petechiae**, **ecchymoses**, and mucosal hemorrhage. However, massive loss of blood is atypical and is restricted largely to the intestinal tract; it is not enough to cause death but poses an infection risk for anyone caring for the patient and coming into contact with bodily fluids and excrements. Cutaneous flushing, rash, **thrombocytopenia**, and **DIC (disseminated intravascular coagulation)** characterize most VHF infections. DIC is characterized by an increased rate of blood coagulation, an increase so great that the large number of clots in the blood vessels can lead to organ failure. In humans, a large proportion of hemorrhagic manifestations are associated with fatal outcome.

Outbreaks of VHF infections are very infrequent, small in size, and usually confined to remote areas. However, an outbreak of Ebola in West Africa in February 2014 spread to four countries (Guinea, Sierra Leone, Liberia, and Nigeria) by February 2015, causing 23,000 known cases and 9000 deaths.

Several health workers were infected and airports throughout the world were put on high alert for anyone arriving with a fever. There have been scares in hospitals in the United States and UK, and in August the WHO declared it an international health emergency. Increased international travel and a heightened interest in bioterrorism (we discuss this in more detail later in this chapter) has changed the perspective on these diseases, however, and consequently there is a renewed interest in vaccine and drug development. Since August 2014 vaccine candidate evaluation has been prioritized. By January 2015 no vaccines were licensed. Some of the infected health workers have been treated with experimental (not yet approved) drugs. The cases are too few to assess safety and efficacy. It is impossible to scale up use of these therapies quickly. Convalescent whole blood and plasma has been transfused to patients in some Ebola treatment centers.

Reliable point-of-care diagnostic tests are required and should be rapid, sensitive, safe, and easy to use. Fewer than three steps should be performed in the test and results should be available in 30 minutes, while workers should only need to wear personal protective equipment to

Fast Fact The date of the end of the outbreak in a country will be declared by the WHO once twice the maximum incubation period has passed without detection of new cases. This 42-day period is counted from the last day of anyone having been in contact with a confirmed or probable Ebola case. At least 22 African countries are at risk of future outbreaks based on ecological conditions and social behaviors.

Fast Fact Candidates ChAd3-ZEBOV and rVSV-ZEBOV were safe and successful in animals, and Phase I clinical trials are taking place. A full set of results is expected for March 2015. Phase II clinical trials for ChAd3-ZEBOV are expected from February. These include larger numbers and broader populations including elderly, children, and HIV carriers. Phase III clinical trials are planned to start also in countries most affected by Ebola to further assess the level of protection and safety. A 2-dose vaccination approach is considered (using Ad26-EBOV and MVA-EBOV) in a Phase I trial. There are also two recombinant protein Ebola vaccine candidates with one to start Phase I trials after Summer 2015.

perform the test. All should be easy to store and handle, with staff training taking less than half a day. The equipment should be portable and no power supply or maintenance should be necessary. Two rapid tests (see Chapter 6) are expected to be available in early 2015, based on either integrated nucleic acid PCR or antigen detection. The latter is easier to use but potentially less reliable.

Therapy for VHF Infection

Currently, there are no useful therapeutic strategies to deal with VHF infections. Early detection seems to give the greatest chance of success if the viral load can be kept low. However, it has been shown that even decreasing the viral load by 50% may not be sufficient to stop the disease. Remember, Ebola infections have a fatality rate of 50–90%. If left unchecked, VHF infection will move to hepatocytes and to adrenal cortical cells. This leads to the coagulation abnormalities seen with these infections and results in multiple organ failures similar to those seen in septic shock. Therefore, general supportive therapy is required.

> **Keep in Mind**
>
> - Emerging infectious diseases are those whose incidence in humans has increased in the past 30 years or threatens to increase in the near future.
>
> - The environment can contribute to emergence of disease when humans come into contact with previously unknown pathogens.
>
> - Genetic mutation, re-assortment, and recombination can give rise to more dangerous pathogens.
>
> - Pathogens that can affect humans must first be able to adapt to humans and then be easily transmitted from one human to another.
>
> - Most emerging infectious diseases are caused by viruses.
>
> - SARS, West Nile virus, and viral hemorrhagic fevers are examples of emerging infectious diseases.

RE-EMERGING INFECTIOUS DISEASES

As Table 8.2 indicates, several diseases we once thought were no longer a threat to humans have bounced back in recent years. All of them, from cryptosporidiosis to yellow fever, present important challenges to health care workers today. Coverage of all the re-emerging diseases listed in Table 8.2 would be beyond the scope of this chapter. Instead, let us consider two of them: tuberculosis and influenza. One is caused by a bacterium and one by a virus, and they illustrate two different crucial reasons for re-emergence.

Tuberculosis

It is estimated that about 2 billion people (one-third of the world's population) are infected with *Mycobacterium tuberculosis*, and about 10% will develop tuberculosis (TB). According to World Health Organization (WHO) estimates, each year 8–9 million people worldwide are infected with TB, and 1.4–2 million die. In 2011, the Centers for Disease Control and Prevention (CDC) reported 10,528 cases of active TB in the United States. Even though there has been a slight decrease, *Mycobacterium tuberculosis* is still a leading killer of young adults as well as women of childbearing age worldwide. It is second only to HIV/AIDS as the greatest killer due to a single infectious agent. Of people living with HIV, a TB infection causes death in one-quarter. In high-income countries, minorities are affected disproportionately by TB, which occurs nine times more frequently among foreign-born individuals living in the United States than in people born in the United States. More than 95% of TB deaths occur in low-income and middle-income countries. Although TB is curable, and infection with *M. tuberculosis* is preventable, about 10 million children were orphaned by TB in 2010.

Symptoms of active lung TB are cough (producing sputum and blood at times), chest pains, weakness, weight loss, fever, and night sweats, but may be mild for many months. Untreated illness is fatal in two out of three cases. Someone with active TB can infect up to 15 others over the course of a year by airborne transmission.

The development of effective antibiotics in the 1950s slowed the spread of TB for some years, and the death rate had dropped significantly, but by the year 2000 the incidence of TB had begun to rise again. Several factors (not exclusive), are behind the resurgence of TB:

Immune Status and Fitness of the Host

People with HIV are particularly vulnerable to infection with *M. tuberculosis* and to developing the active form of TB rather than the more common latent form. Elderly people, especially in long-term care facilities, are also at risk and their increased number contributes to an increase in TB.

Increased poverty and homelessness helps TB to spread rapidly in crowded shelters and prisons, where people are weakened by poor nutrition, drug addiction, tobacco use, and alcoholism while being constantly exposed to *M. tuberculosis.*

Virulence/Fitness of the Pathogen

Drug-sensitive TB disease is usually treated for 6 months with a combination of four antimicrobial drugs under supervision to ensure that the medication is taken properly. However, patients often fail to take all prescribed antibiotics, because they actually feel better and do not see why they should put up with the considerable side effects of the treatment any longer. Patients with TB who do not complete the required drug treatment can stay infectious for longer periods and spread the infection to more people. In addition, failure to follow the treatment protocols can result in the evolution of strains of *M. tuberculosis* that are resistant to the standard antibiotic treatments. In fact, the primary cause of multi-drug resistant TB (MDR-TB) is inappropriate treatment, and MDR-TB is now seen in all countries surveyed. In 2011, of all notified pulmonary TB cases in the world, 310,000 cases were related to MDR-TB. Almost 60% of these cases were reported in India, China, and the Russian Federation.

MDR-TB does not respond to first-line drugs (such as isoniazid or rifampicin) but is still curable by using second-line drugs. As you can imagine, it does not stop there. Further drug resistance can develop in the same way, and first cases of extensively drug-resistant TB (XDR-TB) were reported in 2006 in Italy. These do not respond to second-line anti-TB drugs either. About 9% of MDR-TB cases are XDR-TB, and a global spread is ongoing (**Figure 8.10**). And again in parallel we observe further increases in resistance. In 2009, clinical isolates resistant to all anti-TB drugs were found in Iran. In India in 2011 and 2012, extremely drug resistant (XXDR-TB) and totally drug resistant tuberculosis (TDR-TB) cases were reported, but those two terms are not yet recognized by the WHO.

Early detection is critical to prevent the spread of TB and to increase the chances of successful treatment. Because the initial signs are similar to those seen in other respiratory infections, it is important to look for signs

Fast Fact In sub-Saharan Africa it is estimated that 50% of HIV-infected patients also have TB.

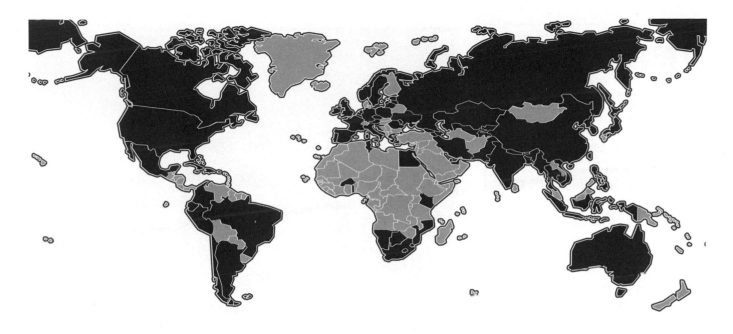

such as fever, fatigue, weight loss, chest pain, and shortness of breath in conjunction with coughing as indications of potential TB. We will cover the pathogenesis and treatment of TB in Chapter 21, which deals with diseases of the respiratory system.

Figure 8.10 Global distribution of countries (shown in red) with at least one reported XDR-TB case by 2011.

Influenza

Although seasonal influenza does not usually result in high death rates, it remains a major public health problem that causes losses in work and school time. The outcome of the infection depends on both the virus and the health status of the host. A host's previous exposure can lead to partial protection, but in a naive host (one that has never been exposed to the virus) it is the virulence of the influenza subtype that determines the outcome of the infection. The virus has what are called **gene constellations**, clusters of genes that determine its virulence, and single mutations in these constellations can markedly affect the level of virulence of the strain. Unlike measles or smallpox, which are fairly stable genetically, influenza is caused by an RNA virus that is continuously undergoing mutations that change the characteristics of the virus an infected host must fight off. When we couple this high mutation rate with the fact that the virus has a stable animal reservoir (aquatic birds), it becomes apparent that new epidemics and pandemics are likely to occur, and the eradication of the disease would be very difficult to achieve.

The influenza virus contains eight segments of RNA and has two major glycoproteins on its surface—hemagglutinin and neuraminidase—which are required for successful infection. There are at least 17 hemagglutinin subtypes and 10 neuraminidase subtypes. Human infection has been linked to hemagglutinins H1, H2, and H3 and to neuraminidases N1 and N2. All the other subtypes of the virus are found primarily in birds. As a result of changes in the surface glycoproteins, there have been devastating epidemics and pandemics in humans, and major epizootics (animal epidemics) in poultry, pigs, horses, seals, and even camels.

As we mentioned previously, there were three human influenza pandemics in the past century. The 1918 pandemic of **Spanish influenza** was the most highly contagious and deadly of the three and had a major global

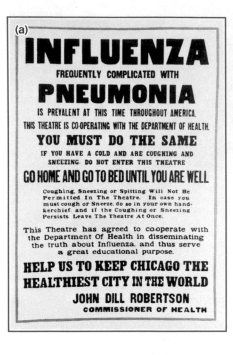

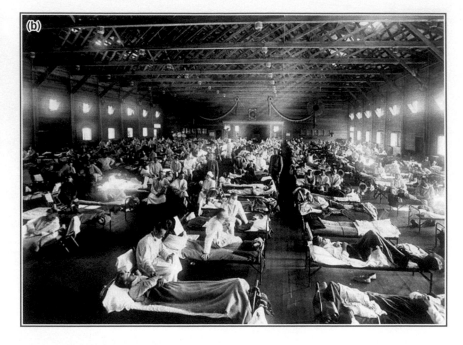

Figure 8.11 The 1918 influenza pandemic. Panel a: A public health poster from the Spanish influenza pandemic of 1918. Panel b: Hospital wards filled with patients during the 1918 influenza pandemic. Note the absence of any isolation facilities.

impact affecting mainly young adults (**Figure 8.11**). It was caused by a virus containing H1 and N1 surface glycoproteins and was considered the most deadly in recorded history, killing an estimated 30–50 million people in less than one year. The outbreak in 1957, known as **Asian influenza**, was characterized by H2 and N2 glycoproteins. Because these proteins were different from those found in the Spanish virus, there was no herd immunity to the Asian variety and mainly children were infected. Recall from Chapter 6 that herd immunity results from having individuals in the population who are immune to the infection, thereby limiting the number of potential targets for infection. It is estimated that the Asian virus of 1957 caused 70,000 deaths in the United States, and up to 4 million worldwide. In 1968, **Hong Kong influenza** killed an estimated 34,000 people in the United States. The virus contained H3 and N2 glycoproteins. All age groups were affected this time.

An outbreak of the influenza known as **swine flu** emerged in Mexico in March 2009, reached the United States by the end of April 2009 and then rapidly spread around the world. Cases were first seen in Europe and within days had been recorded on five continents. In June 2009, there had been over 27,000 reported cases worldwide, including over 140 deaths, and the World Health Organization declared the outbreak a pandemic, making it the first influenza pandemic of the twenty-first century. Person-to-person transmission was confirmed in Europe.

Swine flu is subtype H1N1, the same as the Spanish flu of 1918, but not a human virus and therefore would not normally infect humans. Initially humans were exposed to infected pigs; in hosts infected with influenza viruses from different species, these viruses mixed and (through re-assortment) created a new virus that could spread from person to person. Such a major change in the influenza viruses is known as **antigenic shift** and can cause pandemics. In the 2009 pandemic an H1N1 virus with swine, avian, and human genes had emerged. Fortunately, it caused only a moderate illness and most experienced mild symptoms. Influenza vaccines against human influenza viruses usually do not protect from swine influenza viruses. A vaccine was subsequently developed and was then included in seasonal flu vaccines. We will cover the pathogenesis and treatment of influenza in Chapter 21.

Keep in Mind

- Some infectious diseases are re-emerging. They were previously thought to have been controlled but now have returned.
- Tuberculosis is one of these re-emerging infectious diseases because of several contributing factors such as HIV infection, poverty, and failure to follow recommended treatment protocols.
- Tuberculosis is becoming more resistant to antibiotic therapy.
- Influenza is another good example of a re-emerging infectious disease because of the variety of strains that continue to cause infection.
- Influenza virus contains gene constellations that help to determine its virulence.

AVIAN INFLUENZA

One of the most potentially devastating emerging infections is **avian influenza**—bird flu. We have therefore chosen to separate this from the discussion of other emerging infections. Recall that viruses must be able to overcome two hurdles: they must be able to jump from some non-human species into humans, and they must be transmissible from human to human. The avian influenza virus has already overcome the first hurdle. But the virus has not yet cleared the second hurdle and is as yet not easily transmitted from person to person. So the serious infections seen in people are a result of direct or indirect contact with infected live or dead poultry.

The avian virus has the subtype designation H5N1 and uses birds as a reservoir. The H5N1 virus mutates at an extremely high rate and can acquire genes from other viruses. This is what makes it so dangerous. H5N1 has pandemic potential.

In 1996 and 1997, the first cases of avian influenza were observed in humans in Hong Kong. This led to the killing of more than 1 million birds, a measure that initially squelched the outbreak. Six years later, however, the H5N1 virus reappeared and was much more virulent, causing deaths in both birds and people. Although the human fatalities were small in number, they were scattered all across Asia, and then the infection was spreading to Europe and Africa. Today H5N1 is spreading at an alarming rate because migratory birds are carrying the virus in ever-expanding patterns. The only good news concerning H5N1 is that nearly all confirmed cases so far have resulted from direct contact with infected birds.

The H5N1 virus is more deadly than any of the viruses seen in previous influenza pandemics and in seasonal influenza. It is estimated that H5N1 could have a 50% mortality rate. A 20% fatality rate would translate to half a million dead and 2 million hospitalized in the United States alone. Thus the effects of a pandemic having a 50% mortality rate are truly frightening. Hospitals in New York City alone could see 300–400 cases of influenza a day, many of which will be patients in a serious condition. Numbers of this magnitude will quickly overwhelm the health care system. Furthermore, the cost of a pandemic to the US economy could be several hundred billion dollars, a figure that does not include the health care costs, which could exceed 100 billion dollars.

Adding to the problem is the fact that the H5N1 strain is resistant to amantadine and rimantadine, two of the most effective antiviral drugs. This resistance is attributed to efforts in Asia to control the epidemic by treating birds with massive doses of these drugs. This approach did little to solve the problem, however, and in fact worsened it by

Fast Fact Since we know that H5N1 is the most dangerous subtype, concerted efforts to develop a vaccine for the subtype have been under way for several years. Candidate vaccine viruses are proposed and may be considered for pilot vaccine production.

fostering resistance. The drug Tamiflu® has some effect on the H5N1 strain but is in extremely short supply. Since we know that H5N1 is the most dangerous subtype, concerted efforts to develop a vaccine for this subtype have been under way for several years. Unfortunately, so far these efforts have been unsuccessful.

Both the H5N1 avian virus and the human influenza virus can be transmitted to pigs. This presents the potential for an increased rate of gene re-assortment using the pig as an incubator (**Figure 8.12**). If the genes for transmissibility from human to human (which come from the human influenza virus) combine with avian influenza genes, the resulting virus would be extremely virulent and highly transmissible in humans. And it does not end there. Other avian influenza virus subtypes reported to have infected people include avian H7 and H9. The governments of the world are becoming more and more aware of the danger, and are working together to deal with this potentially devastating problem and be prepared.

Keep in Mind

- Avian influenza (bird flu) has the potential to become a dangerous problem.
- The virus can jump from birds to humans but has not yet become transmissible from human to human.
- If this infection becomes epidemic in the United States, it could cripple the health care system.

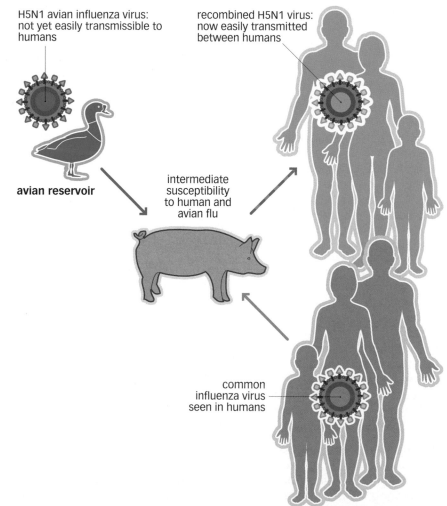

Figure 8.12 An illustration of how the avian influenza virus (H5N1) could become easily transmissible between human hosts. Both the avian virus and the influenza virus that is easily transmitted by humans can infect the pig. Re-assortment of viral genes within the pig could lead to an H5N1 virus that could easily move from human to human.

PRIONS AND PRION DISEASES (TRANSMISSIBLE SPONGIFORM ENCEPHALOPATHY)

We have separated **prion** diseases from the other emerging infectious diseases because they are not caused by a nucleic acid-containing agent at all. In 1982, Stanley Prusiner proposed the existence of diseases caused by infectious proteins called prions, and this work won him the Nobel Prize in Physiology or Medicine in 1997.

The Prion Hypothesis

The prion hypothesis is centered on the premise that all normal nerve cells contain prions, proteins that are known as **PrP^C** (prion protein cellular) and found predominantly on the outer membrane of neurons. The function of PrP^C prions is not fully understood, but it has been proposed that they may help in some way to maintain the body's population of stem cells.

Infectious prions, designated **PrP^{SC}**, seem to be prions that become folded improperly and then cause normal PrP^C prions to fold improperly, which means they cannot function normally in the body. You can compare it to a domino effect. The abbreviation PrP^{SC} stands for **prion protein scrapie**. (Scrapie is a neurological disease known for many years to affect sheep.) PrP^{SC} prions aggregate into fibrous structures that disrupt the cell membrane and cause cell death.

Abnormal (PrP^{SC}) prions have a predominance of flat, pleated-sheet structure, whereas normal PrP^C prions contain mostly helices (**Figure 8.13**). This makes PrP^{SC} prions quite hydrophobic. They are therefore less likely to interact with their hydrophilic environment, and instead they adhere to each other. The abnormal prions are found inside the nerve cells rather than on the membrane and, as noted above, can catalyze PrP^C prions to convert into the abnormal form.

Hydrophobic PrP^{SC} prions are practically indestructible. Technically they are already destroyed as far as our cells are concerned, because being intact means being functional within the human body. PrP^{SC} prions are not destroyed by normal cooking, and even the standard autoclave is not enough to destroy them. Abnormal prions survive disinfectants and antiseptics and are resistant to strong alkali solutions for over an hour. They can also survive in soil for more than 3 years and still be infective. To effectively disinfect abnormal prions requires treatment for 1 hour with bleach containing 2% chlorine or 1 hour of autoclaving in an alkali solution.

The Biology of Prion Diseases

Infective (PrP^{SC}) prions can be ingested with prion-containing material. These prions can then pass through the intestinal wall rapidly and enter lymph nodes, where they incubate. Researchers believe the next step in infection to be that the abnormal prions are picked up by peripheral nerves and transported to the spinal cord and brain. Infectious prions can be transmitted between species, but when they do cross species the incubation time becomes significantly longer.

Prion diseases are referred to as **transmissible spongiform encephalopathy** (TSE). The characteristics of TSE are long incubation periods, spongiform effects in brain tissue, and protein deposits (called plaque) in the brain. There are no antibodies produced in response to TSE, because the proteins are not recognized as foreign, and there is no

(a) PrP^C

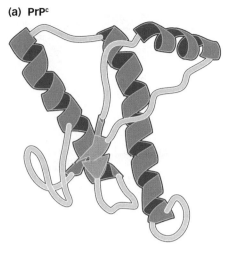

(b) PrP^{SC}

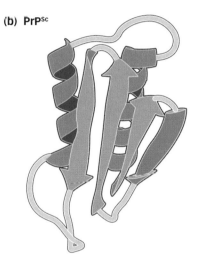

Figure 8.13 An illustration of the differences in structure between the normal and abnormal forms of prion protein. Panel a: The normal form, PrP^C. Panel b: The abnormal form, PrP^{SC}. Note the increase in β-sheets in the secondary structure.

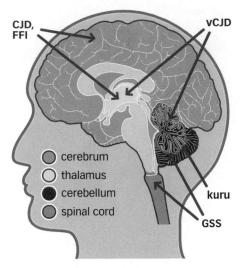

CJD: Creutzfeldt–Jakob disease
vCJD: variant Creutzfeldt–Jakob disease
FFI: fatal familial insomnia
GSS: Gerstmann–Sträussler–Scheinker

Figure 8.14 Different forms of prion disease affect different sites in the brain.

Fast Fact In 1976, the Nobel Prize in Physiology or Medicine was awarded to D. Carlton Gajdusek for his work on kuru.

inflammatory response observed during the infection. Five forms of TSE are seen in humans: **kuru**, Creutzfeldt–Jakob disease (CJD), variant CJD (vCJD), **Gerstmann–Sträussler–Scheinker syndrome (GSS)**, and **fatal familial insomnia** (FFI). These different forms affect different areas of the brain, as illustrated in **Figure 8.14**. They are neurodegenerative diseases for which there is no treatment or cure. TSE can affect cattle and sheep. It is known as BSE (or mad cow disease) in bovines, and as scrapie in sheep.

Prion disease was first observed in humans in the Fore people of New Guinea in the 1920s. It was the custom of these people to honor their dead by cannibalizing them. Women, infants, and children ate the brains and internal organs of the deceased, while men consumed the muscles. Women also spread the tissues of the dead over their skin. This material dried and remained on the skin for weeks. In the 1920s a mysterious and horrifying disease spread among the Fore people, a disease that incapacitated and killed many of them. The men were rarely affected, and the disease was primarily found in women and children. The symptoms of the disease included lack of coordination, staggering, slurred speech, dramatic mood swings, paralysis, and death within one year. The Fore people called the disease kuru (to tremble with fear). Autopsy showed that the brains of the afflicted were full of abnormally large deposits of protein, called plaques. Study of the Fore people showed that kuru had a long incubation period (40 years or more) and that the infectious agent could survive cooking. When the practice of cannibalism was stopped, kuru disappeared among the Fore people.

BSE was first seen in the UK in 1986 and was made a notifiable disease in 1988. The incubation time for cattle seems to be shorter than that observed with kuru and is estimated to be 5 years. The infection in cattle has been attributed to feed supplemented with sheep brains, a product used in an effort to increase the weight of the cattle. The belief is that the feed supplement was made from sheep infected with scrapie. A mass slaughter of cattle followed, and many countries banned British beef.

In 1996, the British government announced that BSE had infected a human. Ten young patients showed the symptoms, and all of them degenerated clinically and died quickly. So far, more than 175 people have died of TSE in the UK and in other countries between 1996 and 2011. Cases have declined since 2000, most probably due to successful containment of the BSE epidemic in cattle. Consumption of food contaminated with the agent of BSE is strongly linked to the observed TSE in humans, which is now believed to be variant CJD. Not all tissues are equally infectious. Brain and spinal cord are considered high-infectivity tissues. Lower-infectivity tissues are peripheral nerves, lymph organs and associated tissues, some blood vessels, and muscular tissue. Reproductive organs and bodily fluids have no detectable infectivity.

Creutzfeldt–Jakob Disease

This form of TSE was first identified in the 1920s and is the most common form in humans. The rate of infection with CJD is one person per million per year throughout the world; the infection is predominantly seen in people aged 55–75 years. Signs are rapid progressive dementia, visual problems, speech abnormalities, muscle tremors, agitation, and depression. Death usually occurs within 12 months after signs first appear. The brain of those infected with CJD shows widespread spongiform changes and moderate protein deposits. The incubation time for CJD is estimated to be from 3 to more than 20 years.

There are three forms of CJD: sporadic, familial, and iatrogenic. The sporadic form encompasses 85% of the cases seen, seems to occur spontaneously and has no identifiable risk factors. Familial CJD accounts for 5–15% of the cases and is genetically linked to mutations in the genes coding for normal prions (Prnp). The iatrogenic form of CJD is transmitted through medical treatments such as blood transfusions, cornea transplants, and dura mater grafts. It can also occur as a result of contaminated instruments or growth hormones obtained from the pituitary glands of infected individuals. It accounts for less than 5% of CJD cases.

Variant CJD (vCJD)

This form of TSE was first reported in 1996 with the link to ingesting the BSE agents, and has unique properties that distinguish it from CJD. With vCJD, patients are usually young (16–39 years old) and the illness takes longer. The first symptoms of vCJD are behavioral and include aggressiveness, anxiety, apathy, depression, and delusions. The first neurological symptoms, such as shaking, incontinence, and immobility, are not seen for the first 6 months. On autopsy, brain tissue from infected patients resembles that seen in kuru. vCJD is also different from CJD in that the plaque deposits found in the brain tissue of patients with vCJD are larger than those found in the brain tissue of patients with CJD. There are no reliable diagnostic tests before the onset of clinical symptoms. However, magnetic resonance scans and tonsillar biopsy can be useful.

In the United States and other countries, new regulations have been enacted to make sure that the blood supply is safe from infectious prions. These laws include banning donors who resided in the UK for 3 months or more, or who have spent a total time there that adds up to 6 months or more between 1980 and 1996. Blood donors who have resided in continental Europe for 5 years or more or received blood transfusions in the UK between 1980 and the present are also banned from donating blood. This has caused blood shortages and has also had an impact on organ donation. Further regulations have also been introduced in several areas: stringent labeling of surgical equipment, using disposable surgical instruments where possible, using recombinant growth hormones (rather than those of animal origin), limiting use of animal products in the pharmaceutical industry, and banning ruminant tissues in ruminant feed.

Keep in Mind

- Prions are proteins that cause transmissible spongiform encephalopathy in humans and other animals.

- Kuru was one of the first human prion diseases to be described in detail.

- Transmissible spongiform encephalopathy comes in five forms in humans, but all have the same general characteristics: long incubation periods, spongiform defects in brain tissue, and plaque deposits in the brain.

Very clearly, emerging and re-emerging infectious diseases pose a threat to human health directly, but also in general to global socio-economics. Our knowledge and experience help us to be protected or at least prepared (by emergency management and disaster preparedness plans). We have disease surveillance, trade regulation, legislation, and administration in place; we work on protective technologies, focus on prevention, public health education, and vaccination programs, and use weather forecast and warning systems.

Sadly, this does not yet factor in malicious intent.

BIOTERRORISM

Biological weapons were around centuries before the discovery of bacteria in the nineteenth century and antibiotics in the twentieth century. The Greeks probably initiated the practice of intentional contamination of drinking water, when they threw animal carcasses into wells in about 300 BC. During the Middle Ages, Tartar forces would catapult plague victims into besieged villages to infect and decimate the enemy population. European conquerors entering the New World did not need to rely on firepower alone to fight the indigenous peoples already there. Smallpox devastated various indigenous tribes in both North and South America. Europeans gave Indians blankets infested with smallpox during the French and Indian Wars. The wars of the twentieth century also saw biological weapons, but now the strategies were informed and more directed. Biological weapons have also been used outside wars accidentally, or even worse as acts of terror like in the 2001 anthrax attacks through the US postal system.

It is essential for health care providers to be familiar with the symptoms and presentations of bioterrorism agents because hospitals will probably be the first to respond to a bioterrorist attack. In this section we also discuss clues and indications that a bioterrorist attack has occurred and how to respond to such an event (**Figure 8.15**).

The intent of a bioterrorist attack is to cause public panic and social disruption. With this goal in mind, we can ask: What does it take to be an effective biological weapon (commonly referred to as a **bioweapon**)? First, it must be an agent that can either be easily disseminated over a large population or is highly contagious so that it will spread quickly even when only a few individuals are infected by the initial attack. Second, high mortality rates are preferred, but some survivors are necessary so as to transmit the disease from person to person. Last, an efficient biological weapon will cause disease that will have a significant impact on the resources of the health care system. Biological agents used as weapons can be spread in a variety of ways, including from person to person, as occurs with smallpox and plague.

The CDC (Centers for Disease Control) in Atlanta defines **bioterrorism** as "the intentional release of bacteria, viruses, or toxins for the purpose of harming or killing civilians." Although these agents can be found in nature, they may be altered to make them more effective in causing disease, spreading disease, or resisting treatment. This is why we consider bioterrorism in the context of emerging and re-emerging diseases. Biological agents may appeal to terrorists because they are difficult to

Figure 8.15 Workers in protective garments.

detect and may not cause disease for hours or days after initial exposure, making it challenging to trace the illness back to the source. Yet, large numbers of deaths are not necessary to cause widespread panic. Just the suggestion of a bioterrorist attack can quickly disrupt society. During the 2001 anthrax attack, 33,000 individuals were given prophylactic antibiotics for possible exposure. Think of the implications of providing care for thousands of individuals who were exposed to a biological agent and may or may not develop the disease. Think of the difficulty of treating those who are not killed by the disease but remain critically ill for weeks or months.

The CDC separates bioterrorism agents into three categories based on how easily they can be spread and on the severity of illness they cause (Table 8.3). Category A agents are given the highest priority because they pose the greatest risk to national security. They are easily disseminated, have high mortality rates, cause public panic, and require special preparations by public health authorities. Category B agents are moderately easy to disseminate, have low mortality rates, and require increased disease surveillance. Category C agents are emerging infectious diseases that could be engineered in the future as bioweapons because of their availability, ease of dissemination, and high mortality rates.

Category	Risk	Agent
A	Greatest	Anthrax (*Bacillus anthracis*)
		Botulism (*Clostridium botulinum* toxin)
		Plague (*Yersinia pestis*)
		Smallpox (variola major)
		Tularemia (*Francisella tularensis*)
		Viral hemorrhagic fevers (filovirus and arenaviruses)
B	Lower than with category A agents	Brucellosis (*Brucella* species)
		Epsilon toxin of *Clostridium perfringens*
		Food safety threats (*Salmonella* species, *Escherichia coli* O157:H7, *Shigella*)
		Glanders (*Burkholderia mallei*)
		Melioidosis (*Burkholderia pseudomallei*)
		Psittacosis (*Chlamydophila psittaci*)
		Q fever (*Coxiella burnetii*)
		Ricin toxin from *Ricinus communis* (castor beans)
		Staphylococcal enterotoxin B
		Typhus (*Rickettsia prowazekii*)
		Viral encephalitis (alphaviruses)
		Water safety threats (*Vibrio cholerae*, *Cryptosporidium parvum*)
C	Lower than with category B agents	Emerging infectious diseases

Table 8.3 CDC categories of biological weapons.

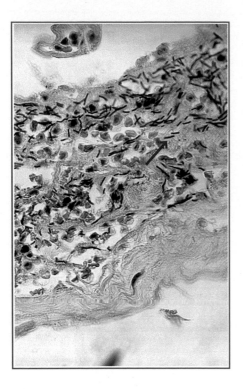

Figure 8.16 A photomicrograph of the meninges, showing the presence of *Bacillus anthracis* (black rods) in a case of fatal inhalation anthrax.

We will now outline some of the diseases caused by potential biological weapons, giving you an overview of their features to help you understand the danger, why they have potential as weapons and how they are (re-) emerging. Biology, pathogenesis, and treatment are described in more detail in Part VI. Remember what you learnt about toxins in Chapter 5.

Anthrax

Bacillus anthracis is an encapsulated, aerobic, Gram-positive, rod-shaped bacterium (**Figure 8.16**). It produces spores that are highly resistant and can survive for decades. It was one of the first bacterial pathogens recognized when the German scientist Robert Koch discovered the life cycle of the organism in the 1870s. From his studies of *B. anthracis* came Koch's postulates, which, as we saw in Chapter 7, were the basis on which was established the theory that a specific pathogen causes a specific disease.

There are three routes by which anthrax can infect the body: through the skin (cutaneous anthrax), through the digestive tract (gastrointestinal anthrax), and through inhalation (inhalation or respiratory anthrax). Cutaneous anthrax is the most common form of the disease and occurs when the skin comes into direct contact with the *Bacillus* spore. The infected individual develops localized itching; a papular lesion forms and turns into a black eschar. It is not considered lethal and can easily be treated with antibiotics. Gastrointestinal anthrax occurs after eating undercooked meat or dairy products from infected animals. Symptoms are nausea, vomiting, anorexia, fever, bloody diarrhea, bloody vomit, and abdominal pain. Shock and death occur 2–5 days after the onset of symptoms. It would be difficult to infect a large number of individuals through the cutaneous or digestive routes of infection, and consequently these routes would most probably not be used by terrorists. Instead, anthrax would most probably be released as an aerosol in a terrorist attack. This would result in inhalation anthrax, which is the rarest and most deadly form of the disease. The incubation period is usually 1 week but can be as long as 2 months. Infected individuals first develop nonspecific signs and symptoms such as fever, nonproductive cough, malaise, fatigue, muscle aches, and chest discomfort. Because these signs and symptoms mimic those of much milder infections, a diagnosis of inhalation anthrax could easily be delayed. There is frequently a short respite during which the person seems to be getting better, much as one would recover from a cold or the flu. However, in a couple of days the individual begins to feel much worse, developing a high fever and respiratory distress. Without antibiotic or supportive treatment, shock and death ensue in 24–36 hours. The mortality rate can be up to 90%.

All forms of anthrax require treatment with intravenous antibiotics, but the amount of supportive care required is different for the three types. Inhalation anthrax requires the skills of a critical care team. The patient may be admitted to an intensive care unit that can offer ventilator assistance. If the patient survives the initial complications, several weeks of care may be needed for a full recovery. For this reason, a mass casualty involving large numbers of cases with inhalation anthrax will cause an enormous strain on the health care system, especially in areas where intensive care beds are in short supply.

A vaccine for all forms of anthrax has been developed and given to US military personnel. It is a series of six inoculations given over a period of 18 months. For any significant increase in survivability rates, however, even vaccinated patients require antibiotic therapy after exposure. The civilian use of the anthrax vaccine is limited to those at a high risk of exposure. New treatments and vaccines are being studied.

Botulism

Clostridium botulinum is a Gram-positive, spore-forming, obligate anaerobic rod-shaped bacterium found mainly in soil and in marine and agricultural products. The spores are heat-resistant, surviving at 100°C for several hours, but the exotoxin is heat labile and easily denatured at 80°C. Because the spore is dormant, the disease cannot be passed from person to person. There are five ways in which botulism can be acquired. Food-borne botulism occurs through the ingestion of toxin-contaminated food. Infant botulism and adult infectious botulism develop when spores are ingested. In wound botulism, the wound is infected with *C. botulinum*. Inhalational botulism occurs only if the toxin has been aerosolized and released.

It is the toxin that *C. botulinum* produces that causes serious physical effects in individuals unlucky enough to come into contact with it (remember Chapter 5). Absorption of the toxin can be through a mucosal surface, such as the gut or lung, or through a wound, but the toxin is unable to penetrate the skin. Botulinum toxin blocks the release of acetylcholine at cholinergic receptors, leading to descending paralysis that begins with the face and progresses to the upper and lower extremities and eventually the respiratory muscles. Neurological symptoms begin with cranial neuropathies and descending symmetric weakness, but the central nervous system and sensory nerves are not affected.

The incubation period for botulism is most commonly between 12 and 36 hours. The initial symptoms—dry mouth, double vision, difficulty in speaking, difficulty in swallowing, and facial weakness—commonly proceed to shortness of breath, loss of muscle tone, and eventually complete paralysis (**Figure 8.17**). Intubation and mechanical ventilation are often required to combat respiratory difficulties. Other symptoms include decreased salivation, intestinal obstruction, and urine retention. Clinical manifestations can vary widely from one person to the next, with some patients mildly affected and others, at the opposite extreme, dying within 24 hours of developing symptoms. Because of these variables, botulism is frequently misdiagnosed as Guillain–Barré syndrome or some other neurological disorder.

The mortality rate for botulism has decreased from 25% to 5–10%, most probably because of modern medical care. However, paralysis can last for weeks or months and require extensive medical support. The only treatment for botulism is supportive care and administration of antitoxin after exposure. Antitoxin can reduce the severity of the disease but will not necessarily decrease the duration of the paralysis.

Inhalational botulism, the most likely form to be used in a bioterrorist attack, would not easily be detected initially. The first patients who were maximally affected would not present to a medical facility until a few days after exposure. This delay in diagnosis and the limited supply of antitoxin would result in a significant number of patients requiring ventilator support, which would immobilize intensive care units for months. Despite the low mortality rate, botulinum toxin makes a good biological agent because of its significant morbidity and the toll it will take on the health care system.

Plague

Yersinia pestis is a nonmotile, Gram-negative, rod-shaped bacterium causing plague, which is rapidly fatal, highly contagious, and not easily contained. These properties make *Y. pestis* an excellent bioweapon candidate. Reservoirs for this organism are usually rodents, and the vector for

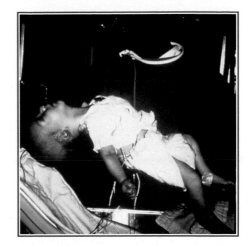

Figure 8.17 A limp infant showing the poor muscle tone that is an early symptom of botulism.

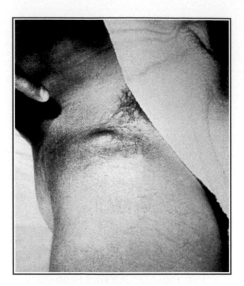

Figure 8.18 This patient with a Yersinia pestis infection shows symptoms that include a swollen inguinal lymph node referred to as a bubo (bubonic plague).

transmission is the oriental rat flea. There are three main types of clinical syndrome associated with plague: bubonic, septicemic, and pneumonic. Bubonic plague, the most common type, occurs when a person is bitten by a *Y. pestis*-infected flea or when the skin comes into direct contact with an infected animal or its tissues. In septicemic plague, bacteremia develops usually after the bacterium enters through a break in the skin. Primary pneumonic plague is the result of inhaling the microorganism in an aerosolized form (for example from respiratory secretions from infected animals). Pneumonic plague can also develop in patients with bubonic plague.

The bacteria migrate to regional lymph nodes and are phagocytosed but not destroyed. They quickly cause destruction and necrosis of the lymph nodes, followed by bacteremia and septicemia that can lead to shock, disseminated intravascular coagulation, and coma. The incubation period for bubonic plague is 2–8 days. Patients present with a sudden onset of fever, chills, weakness, and headache. Tender lymphadenopathy usually occurs in the groin, axilla, and neck. It can take hours to days for a bubo to develop (**Figure 8.18**), which is characterized by surrounding edema (swelling) and extreme pain limiting the range of motion of the affected limb. The buboes are 1–10 cm in diameter, firm, and tender. In the 2–4 days after buboes form, the patient deteriorates rapidly, with a high fever, tachycardia, malaise, headache, vomiting, chills, and altered mental status as bacteremia develops. In some patients, *Y. pestis* are contained in the lymphatic system and bacteremia never develops, but without treatment the estimated mortality rate is 50%. Treatment consists of antibiotics and supportive treatment. The cyanosis (dark blue color of the skin as a result of lack of oxygen in the blood) and gangrene of the peripheral tissues that are characteristic of bubonic plague led to the disease being called the Black Death during the historic European epidemics

Septicemic plague can be difficult to diagnose because there may be no detectable involvement of the lymph nodes and therefore no buboes to alert medical personnel. Patients have a high concentration of bacteria in their blood, and apparently the bacteremia overwhelms patients before lymphadenitis occurs. In septicemic plague, infected individuals are febrile without localized signs or symptoms. Without treatment, hypotension and multiorgan failure lead rapidly to death. Standard treatment includes antibiotics and supportive treatments that will often require the resources of an intensive care unit.

Pneumonic plague has an incubation period of 1–3 days and is rapidly fatal. Patients present with fever, chills, headache, body pains, weakness, and chest discomfort. As the disease progresses, sputum production, chest pain, hemoptysis (bloody sputum), and respiratory failure develop. Buboes do not present. This form of plague is 100% fatal without antibiotic treatment, with death occurring 18–24 hours after the onset of symptoms. Patients are highly contagious from the onset of disease. Therefore strict respiratory isolation is crucial for those patients in whom pneumonic plague is suspected. Antibiotics are very effective but should be given in the first 24 hours after symptoms appear, because delay in treatment greatly increases the mortality rate. In a bioterrorist attack, anyone with a temperature greater than 38.5°C or a new cough should be treated with prophylactic antibiotics. Precautions should be taken to isolate respiratory droplets, which usually means having patients wear facemasks.

The World Health Organization studied the potential consequences of using plague as a biological weapon. Plague would most probably be disseminated in an aerosolized form in the event of a bioterrorist attack, which would probably result in an outbreak of pneumonic plague. Symptoms would initially resemble those of other respiratory illnesses. Therefore a patient waiting in a crowded emergency department could potentially infect dozens of individuals. Providers would not initially think of pneumonic plague because its natural occurrence is so rare. The patient would most probably be treated for a simple respiratory infection and would continue to infect others until properly treated. Clues that a bioterrorism attack had taken place would include the occurrence of disease in nonendemic locations, disease in persons without risk factors, and the absence of previous rodent deaths.

Smallpox

As a result of the WHO global vaccination program, the last documented case of naturally occurring smallpox was in Somalia in 1977, and in 1980 smallpox was declared eradicated and vaccination was stopped. Reference stocks of the virus are kept in laboratories in Moscow and at the CDC in Atlanta.

Smallpox is a large contagious DNA virus that infects the respiratory tract and migrates to nearby lymph nodes. An asymptomatic viremia develops as virions move to the spleen, bone marrow, and lymph nodes. As the virions replicate in these organs, a secondary viremia develops, and it is at this time that signs and symptoms appear. The incubation period is typically 12–14 days with initial signs or symptoms of fever, malaise, and headache. A maculopapular rash begins with lesions on the tongue and palate, and small red spots on the face. The rash spreads to the proximal extremities, trunk, and distal extremities. After a couple of days, the rash becomes vesicular and pustular (**Figure 8.19**). The pustules are round, tense, and deeply implanted in the dermis. Approximately 8–9 days later, crusts form over the pustules, followed by scabs and scarring. As the oral lesions form vesicles and break down, a large number of virions are released. This is when infectivity is at its greatest, which occurs in the days when the maculopapular rash first appears. Death is secondary to hypotension and multiorgan failure, and usually occurs during the second week.

The only treatment for smallpox is supportive care. Because of the threat of a bioterrorist attack, new treatments are being studied. Before 1972, every child in the United States received a smallpox vaccination, but antibodies seem to decline over a 5–10-year period, which means that the vaccine does not confer lifelong immunity. Therefore, adults who were vaccinated as children are probably no longer protected. Military

Figure 8.19 Patients with smallpox lesions.

personnel serving in combat environments are being vaccinated. The vaccine can also be administered within 4 days of exposure, and may prevent or decrease the effects of disease.

If used as a bioterrorist agent, smallpox virus would most probably be aerosolized. The long incubation period means that many humans could be infected before detection of the virus. Because very few individuals would still have immunity against smallpox, a bioterrorism attack would require administering the limited amount of vaccine we have on hand to thousands of exposed individuals, a task that would prove difficult after an attack. This is why the initial diagnosis of smallpox is so critical.

Tularemia

Tularemia is a zoonotic disease with a large range of hosts covering many different environments in several countries. The causative agent, *Francisella tularensis*, is a Gram-negative, facultatively anaerobic, intra-cellular rod-shaped bacterium. It is highly resistant and can survive for long periods in soil, water, and animal carcasses. The most significant sources of human infection are rodents, hares, and rabbits, especially the cotton-tailed rabbit. Humans can be infected by the bite of an infected animal or through vectors such as ticks and deerflies. Often, outbreaks of human infection parallel outbreaks in other animals. Humans can also be infected by ingestion or inhalation of the bacterium.

In 60–80% of cases, ulceroglandular syndrome develops with an abrupt onset of fever, chills, malaise, sore throat, and headache after an incubation period of 2–5 days. If the route of infection is through the skin or mucous membranes, a primary ulcer will develop at that site. The ulcer begins as a solitary papule that changes to a pustule with surrounding inflammation. Sometimes the lesion will go unnoticed and will heal in a week. If antibiotics are not given within 7–10 days of the original infection, in 30–40% of cases lymph node enlargement can occur with suppuration (formation of pus). If the route of infection is by inhalation, the signs and symptoms of respiratory tularemia are variable and can resemble those of several less threatening infections. Patients present with a mild respiratory infection or high fever, chills, malaise, and cough, with the severity of symptoms being dependent on the subspecies of the infecting *F. tularensis*.

Isolation is not necessary for patients with tularemia because human-to-human transmission does not occur. Tularemia can be treated effectively with antibiotics. A previously used live vaccine is no longer approved, and options for a viable tularemia vaccine are being researched.

Francisella tularensis subspecies *tularensis*, is the most likely to be used as a bioweapon through aerosolization. It takes just 10–50 bacteria to cause disease in humans, which makes *Francisella* much more infectious than *Bacillus anthracis*, and is 30% fatal if untreated. Even if the bacterium were not altered and was sensitive to antibiotics, the spread of tularemia would quickly overwhelm health care communities.

Viral Hemorrhagic Fevers

As we already discussed in this chapter, a principal sign of hemorrhagic fever is hemorrhaging of the capillaries of the patient's circulatory system. There are at least 18 viruses that cause human hemorrhagic fevers and several are listed as category A agents (see Table 8.3), including the filoviruses Ebola and Marburg and the arenaviruses Lassa and Machupo.

Treatment is mainly supportive, and patients should be handled with great care to prevent bleeding and transmission. The preexisting antivirals, favipiravir and brincidofovir, will be entering clinical trials in early 2015. Since August 2014, Ebola vaccine candidate evaluation has been ongoing. By January 2015 no vaccines were licensed yet. The only vaccine for viral hemorrhagic fevers is indicated for yellow fever, and all travelers to areas where yellow fever occurs must receive this vaccine. Infected individuals should be isolated because the virus is transmissible through close contact. Respiratory isolation is ideal but may not be possible in a large outbreak. Contact-barrier precautions are essential in these situations to prevent spread of the disease.

The United States and the Russian Federation have experimented with weaponizing these viruses and have found that aerosolizing them produces a high infectivity rate in nonhuman primates. In the 1990s, the Aum Shinrikyo cult in Japan attempted to weaponize the Ebola virus. There is no known incident in which these viruses have been used in a biological attack, but the interest exists.

The Danger of Bioweapons

Biological attacks are likely to continue in the future. Bioweapons are cheaper than chemical weapons and can cause mass destruction. Several countries have established bioweapons programs for experimentation, and countries that sponsor international terrorism are suspected of possessing harmful biological agents. Additionally, individuals who possess knowledge of genetic engineering could alter simple biological agents to make them more virulent and resistant to antibiotics.

The biggest consequence of a bioterrorist attack may be not the physical casualties but the psychological impact. Even the suspicion of a biological weapon being released could instigate mass panic and disruption of communities, health care systems, and governments. Global disorder is a serious threat because people travel routinely and quickly around the world, potentially spreading infectious diseases far from where they originated.

Perhaps more importantly, the initial symptoms may not lead health care providers to suspect bioterrorism. As a result, proper precautions may not be used at first, potentially increasing the number of people exposed and infected. The majority of any population is susceptible to infection with a category A agent. A few initial exposures could quickly turn into mass casualties, especially when the pathogen can be transferred through human contact. The economic impact of a bioterrorism attack will most probably be devastating.

Warning Signs

In any location hit by a bioterrorism act, the public health system will probably be the first to detect the attack and respond. It may not be realistic to wait for confirmation of a diagnosis, because the potential for spread of the disease increases. An emergency response may need to be activated on the basis of patterns, timing, and patient presentation. There are important clues that can help alert hospitals to a bioterrorist attack. This information can be passed on to the appropriate authorities, who can coordinate health care facilities, local and state departments, and emergency response teams. Some clues are listed in **Table 8.4**, but every health care professional should be suspicious of any unusual activity.

Bioterrorism continues to be a real threat, and most experts believe it is simply a matter of time until the next attack occurs. Scientists, epidemiologists, doctors, and nurses will all need to work together to battle the impact of a biological disaster. Familiarity with organisms that are likely

Table 8.4 Clues to a bioterrorist attack.

Clues
Increased incidence of a specific disease in an otherwise healthy population
Presence in a given geographical area of a disease not typically found in that area
Disease outbreak at an atypical time of year
Large numbers of people presenting to medical facilities with similar symptoms
Increase in number of rapidly fatal cases
Infectious diseases not responsive to antibiotics that were formerly successful in treating the disease
Single case of a relatively uncommon disease (smallpox, pneumonic plague, tularemia)
Increased numbers of emergency calls
Discovery of a package that has been tampered with

to be used as bioweapons will prepare you to act efficiently and effectively during an emergency. It is important for you to review your medical facility's emergency response plan before an actual event. Preparation is the key to controlling the effects of a bioterrorist attack. Law enforcement will be particularly important when it comes to reporting disease and controlling public reaction. Bioterrorism is a matter of national and international security that will require the coordination of local, state, and federal agencies.

SUMMARY

- Emerging and re-emerging diseases will provide new and challenging problems for health care professionals.

- Genetic mutation, environmental changes, and travel can contribute to these types of infection.

- Most emerging diseases are caused by viruses.

- Re-emerging diseases are those that were thought to be under control but have returned. In many cases this is due to an increase in resistance to antibiotics.

- Avian influenza has the potential to become a very dangerous problem that could cripple the health care systems.

- Prions are proteins that are not folded correctly and cause transmissible spongiform encephalopathy in humans and BSE in cattle.

- The use of bioweapons can lead to the emergence or re-emergence of an infectious disease.

- Global collaboration, disease monitoring and control, and preparedness plans help us in managing the risks.

Emerging and re-emerging infectious diseases will have an increasing impact on health care, and some of these diseases could have devastating effects on the health care systems around the world. The resulting problems will directly affect you as a health care professional. It is therefore important for us to include these diseases in our study of microbiology.

Ⓠ SELF EVALUATION AND CHAPTER CONFIDENCE

Multiple Choice

Answers are given in the back of the book and help can be found in the student resources at:

www.garlandscience.com/micro2

1. All of the following are emerging infectious diseases except
   A. Creutzfeldt–Jakob disease
   B. Legionnaires' disease
   C. Mumps
   D. SARS
   E. Hepatitis C

2. Which of the following properties do AIDS, legionnaires' disease, and Creutzfeldt–Jakob disease have in common?
   A. They have all emerged in the past 40 years
   B. They are all bacterial
   C. They are all viral diseases
   D. They are all caused by microorganisms
   E. They all have long latency periods

3. The Nipah virus originated in
   A. Pigs
   B. Chickens
   C. Fruit bats
   D. Monkeys
   E. Rats

4–8. Match the disease with its place of origin.

Disease	Origin
4. SARS	A. Hong Kong
5. Hantavirus disease	B. South America
6. Spongiform encephalopathy	C. Africa
7. HIV	D. Middle East
8. MERS	E. New Guinea

9. Which of the following has not achieved the ability to be transmissible from one human to another?
   A. Influenza
   B. HIV
   C. SARS
   D. Hantavirus
   E. Both **B** and **C** above

10. SARS causes infection of the
    A. Digestive system
    D. Nervous system
    C. Respiratory system
    D. Skin
    E. Reproductive system

11. West Nile virus causes infection in the
    A. Respiratory system
    B. Skin
    C. Digestive system
    D. Nervous system
    E. Urinary tract

12. West Nile virus is initially spread from mosquitoes to
    A. Humans
    B. Birds
    C. Pigs
    D. Rodents
    E. None of the above

13. The virulence of viral hemorrhagic fever infections depends on
    A. The virulence of the virus
    B. The route of exposure
    C. The viral load
    D. The competence of the host defence
    E. All of the above

14. Which of the following is a characteristic seen in viral hemorrhagic fever infection?
    A. Rash
    B. Cutaneous flushing
    C. Fatigue
    D. Muscle aches
    E. All of the above

15. The influenza genome is unique in that it is made of
    A. DNA only
    B. RNA only
    C. Eight segments of DNA
    D. Eight segments of RNA
    E. Both DNA and RNA

16. We are currently in a period of emerging disease transition called global urbanization. Which of the following describes this transition?
    A. Increase in densely populated settlements
    B. Social upheaval
    C. Air travel
    D. Long-distance trade
    E. All of the above

17. In 1993, hantavirus infections began to appear in the United States. Which of the following was not involved?

    A. Heavy summer rains in the Southwest
    B. More deer mice
    C. More pine nuts
    D. Mutation of the hantavirus

18. Which of the following is not a way for a virus to acquire the ability to infect humans?

    A. They undergo mutation
    B. They undergo genetic rearrangement
    C. They become latent
    D. They undergo genetic re-assortment
    E. Both **A** and **B** above

19–23. For each of the emerging diseases listed, find its matching mode of transmission. Modes can be used more than once or not at all.

Emerging disease	Mode of transmission	
19. SARS	**A.**	Mosquito
20. West Nile fever	**B.**	Body fluids, bush meat
21. Viral hemorrhagic fever	**C.**	Droplets
22. Dengue fever	**D.**	Food contamination
23. Mad cow disease	**E.**	Flea

24. The virulence of influenza is determined by

    A. The health of the host
    B. Segments of its DNA chromosome
    C. Its ability to integrate in the host genome
    D. DNA constellations
    E. None of the above

25. How is H1N1 influenza different from H5N1?

    A. They have different neuraminidases
    B. One can infect humans, the other cannot
    C. They have different hemagglutinins
    D. One is a current health concern, the other is not
    E. None of the above

26. Prion diseases are caused by

    A. Bacteria
    B. Viruses
    C. Fungi
    D. Abnormal proteins

27. Which of the following is a characteristic of prion disease?

    A. Fever
    B. Long incubation period
    C. Defects in brain tissue
    D. Plaque deposits
    E. **B**, **C**, and **D** above.

28. Which of the following is not likely to be an indication of a bioweapons attack?

    A. Outbreak of influenza in local schools in December
    B. Increased incidence of smallpox in Seattle, Washington
    C. Many patients arriving at a hospital with terminal encephalitis
    D. Increased number of emergency calls about infectious diseases
    E. Outbreak of systemic anthrax

DEPTH OF UNDERSTANDING

Questions listed here require you to bring together the concepts you have learned in this chapter into a discussion format. This helps you to increase your depth of understanding of the material you have learned. Help can be found in the student resources at: www.garlandscience.com/micro2

1. Before people wishing to donate blood can do so, they must answer several questions. One of these questions asks whether they have lived in the UK for longer than 3 months between 1980 and 1996 or more than 5 years in Europe between 1980 and the present. What is the purpose of these questions?

2. What is the relationship between socio-economic conditions and emerging infectious diseases?

3. Three influenza pandemics have occurred in the United States since 1918. Were they connected to one another in any way, and if so, how?

Q CLINICAL CORNER

Help can be found in the student resources at: www.garlandscience.com/micro2

1. Mr. Johnson, a 75-year-old male, was admitted to the hospital complaining of flu-like symptoms, shortness of breath, and chest pain. He was producing large amounts of mucus that contained blood. On examination, he was diagnosed with tuberculosis. Mr. Johnson's daughter tells you that her father had TB more than 20 years ago. He was treated for it and has been healthy ever since.

   A. What would you tell Mr. Johnson's daughter about his current condition?

   B. Should Mr. Johnson be put on the antibiotic therapy routinely prescribed for TB?

   C. Can you think of any tests that could help in the treatment of Mr. Johnson?

2. Your patient is a 50-year-old white male who has been admitted to the psychiatric ward because of dramatic mood swings and bouts of uncoordinated movements. His speech is also slurred. He has been prescribed medication to stabilize his mood swings. During his admission interview he mentions that he has lived in England for most of his life and has never had any medical problems except for complications associated with a corneal transplant when he was 35 years old.

   A. Is there another possible explanation for his condition?

   B. How would you test for that other possibility?

PART III

Characteristics of Disease-Causing Microorganisms

Now that we have studied the disease process, our journey now takes us to **Part III** and a detailed discussion of the organisms that can infect us. We divide our discussions into bacterial pathogens, viruses, and parasitic organisms, and include chapters on growth requirements and bacterial genetics.

In **Chapter 9**, we thoroughly investigate the anatomy of the bacterial cell. In this chapter you will learn about the structures that make up bacteria and how they can be used in the infection process. You will see that many of these structures satisfy several of the five requirements for satisfactory infection—getting in, staying in, and defeating the host defenses—that we learned about in Parts I and II.

Chapter 10 teaches you about the requirements for growth of bacteria, and here you will see that many pathogenic bacteria have very strict requirements for their division and growth.

Chapter 11 gives you a brief description of bacterial genetics. The chapter has four major themes: replication, gene expression, mutation, and transfer of genetic information. All of these subjects are intimately involved in the infection process of pathogens, but many students do not realize how important this information is.

In **Chapter 12**, we look at the structures of viruses that infect us and find that there is a distinct cycle involved in infection. In **Chapter 13**, we look at the pathogenesis of viral infection and the development of vaccines.

Chapter 14 gives you an overview of protozoan parasites, worms, and fungi, each of which can infect humans.

When you have finished **Part III** you will:

- Understand how the structures of the bacterial cell can provide assistance in the infection process and also help to defeat the defenses of the infected individual.

- See how bacterial structures are also targets for antibiotic treatment.

- Have an understanding of how important genetic mechanisms are to the infection process, and how mutations and the transfer of genetic information can increase the severity and duration of an infection.

- See how variable the structure of viruses are and how these structures have an important role in the process of viral infection.

- Realize how widespread and diverse protozoan and parasitic infections can be, and also how fungal infections are on the rise as a result of compromised host defenses.

The Clinical Significance of Bacterial Anatomy

Chapter 9

Why Is This Important?

This chapter details the structures of microorganisms that are involved with infectious disease. These structures will have a significant role in the five steps (entry, colonization, defeating host defenses, damaging the host, and transmission) required for infection.

When Jourdan was a sophomore student at the university, she developed a sore throat and had difficulty swallowing. The on-campus health care center diagnosed her with strep throat and gave her a prescription of penicillin. After two days, she began to feel better, but after two more days, her throat became sore again. She went to her family doctor, who prescribed a different antibiotic called cephalexin. Her doctor explained to Jourdan that her infection was probably caused by a penicillin-resistant strain of *Streptococcus*. Within a week her throat was without pain, and after the full course of cephalexin was complete, the pain and infection were resolved.

The problem of antibiotic-resistant strains of bacteria is of enormous consequence to health care professionals and their patients. Bacteria are constantly changing their genetic makeup in response to environmental stresses (including exposure to antibiotics), and as a result, they produce new enzymes via altered genes. In Jourdan's case, the bacteria produced a penicillinase (penicillin-destroying) enzyme called β-lactamase. Penicillin normally inhibits the synthesis of peptidoglycan in the bacterial cell wall, resulting in bacterial cell bursting (lysis). When the penicillin is destroyed by penicillinase, however, the bacteria survive and continue to cause infection. In this chapter, we will learn about bacterial cell walls, peptidoglycan, and other bacterial structures that are targets for antibiotics or otherwise have significance in health care.

OVERVIEW

In this chapter, we examine the anatomy of the bacterial cell. We need to understand the structures associated with bacterial cells because so many of them have an important role in the generation of disease. As you read this chapter, keep in mind what you learned in previous chapters, especially structure–function relationships. As we saw in Part I, at the molecular level the three-dimensional structures of proteins are directly responsible for their functions. The same can be said for the cell wall, capsule, fimbriae, pili, and flagella, which are all structures associated with the bacterial cell. The structure–function relationships of the bacterial cell show us how bacterial structures influence the development of disease. These structures will become core objectives for our studies and will be mentioned repeatedly throughout subsequent chapters.

We will divide our discussions into the following topics:

The Clinical Signifcance of Bacterial Anatomy

| THE BACTERIAL CELL WALL | STRUCTURES OUTSIDE THE BACTERIAL CELL WALL | STRUCTURES INSIDE THE BACTERIAL CELL WALL |

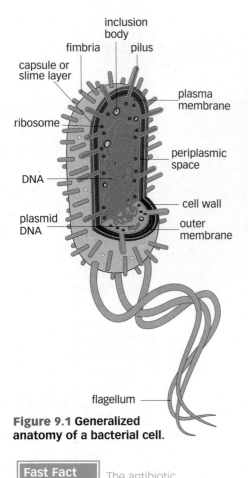

Figure 9.1 Generalized anatomy of a bacterial cell.

Fast Fact The antibiotic **fosfomycin** targets *murA* gene activity and thereby prevents peptidoglycan subunits from being produced.

"In addition to its possible use in the treatment of bacterial infections, penicillin is certainly useful … for its power of inhibiting unwanted microbes in bacterial cultures so that penicillin insensitive bacteria can readily be isolated."

Sir Alexander Fleming (1881–1955)

In the quotation above, Fleming (the discoverer of penicillin) acknowledges the resistance of some bacteria to the antibiotic he helped develop. As we have learned, pathogenic bacteria use a variety of structures (**Figure 9.1**) and substances to overwhelm host defenses, while the host uses nonspecific and specific defense mechanisms to stop the infection. Our discussions in this chapter include not only a description of how the structures that pathogens use to overwhelm a host are relevant to the infectious process but also a description of how some of these structures can be targets for antibiotics. It is important to remember that the bacterial structures identified in this chapter have evolved over time, and they continue to evolve, as seen in antibiotic resistance. This evolution occurs through genetic mutations such that the host-against-pathogen struggle is continuously unfolding.

THE BACTERIAL CELL WALL

Bacteria must develop mechanisms to handle changes in osmotic pressure, which can be a lethal event for the organism. To accomplish this, bacteria construct a complex cell wall. This wall is composed of several parts and is different in Gram-positive and Gram-negative bacteria. The primary component of the wall is **peptidoglycan** (also called **murein**), a structure composed of repeating molecules of the sugars **N-acetylglucosamine** (**NAG**) and **N-acetylmuramic acid** (**NAM**), which meshes together with small peptide chains to form the basis of the cell wall. **Figure 9.2** shows the structure of NAG and NAM and how they bind together in repeating units. The binding is covalent and is therefore very strong. This strong bonding is a major part of the mechanical strength of the cell wall and provides a barrier to osmotic pressure.

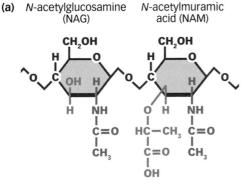

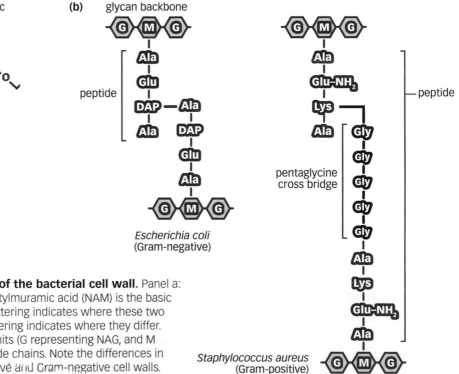

Figure 9.2 The structural components of the bacterial cell wall. Panel a: *N*-acetylglucosamine (NAG) bound to *N*-acetylmuramic acid (NAM) is the basic building block of peptidoglycan. The blue lettering indicates where these two sugar molecules are the same; the gold lettering indicates where they differ. Panel b: The binding of peptidoglycan subunits (G representing NAG, and M representing NAM) with attached polypeptide chains. Note the differences in the amino acid compositions of Gram-positive and Gram-negative cell walls.

Building the Cell Wall

Peptidoglycan is an important component of the bacterial cell wall and has implications for antibiotic therapy and antibiotic resistance. As with most cellular construction, NAG and NAM are linked together by enzymes, which place the subunits in the proper orientation for promoting the growth of the NAG/NAM chain. There are three phases of peptidoglycan assembly: a cytoplasmic phase, a membrane-associated phase, and an extracytoplasmic phase. The events associated with each phase are dynamic and complex, but understanding the basic events associated with each phase will give us an opportunity to understand the strength and durability of the cell wall and also how it can be a prime target for antibiotic therapy.

The Cytoplasmic Phase

During this phase, the NAG/NAM building blocks for peptidoglycan are fashioned in the cytoplasm of the cell. The cell wall must be constantly refurbished, especially when the cell is ready to divide. This requires that peptidoglycan layers be laid down both transversely (along the long axis of the cell) and vertically as the intracellular partition (septum) that divides the cell into the two daughter cells develops (**Figure 9.3**).

The NAG and NAM components of peptidoglycan are formed by enzymes that are encoded by the *mur*A–F genes. The enzymes coded for by these genes attach amino acids to each NAM molecule, hence the name peptido (peptide) glycan (sugar). Because peptidoglycan subunits must be assembled if a bacterial cell is to build new cell wall, this cytoplasmic phase is an important potential target for antibiotic therapy.

The Membrane-Associated Phase

In the membrane-associated phase of cell wall building, specific enzymes link the NAG and NAM subunits with a lipid portion of the bacterial cell plasma membrane. Recall from Chapter 4 that the plasma membrane is the outer layer of eukaryotic animal cells. However, in bacterial cells it is the cell wall that is the outer layer, with the plasma membrane lying inside. The insertion of NAG and NAM into the plasma membrane is performed via the **lipid carrier cycle**. The first step in the cycle is the formation of a bond between the peptidoglycan and the side of the plasma membrane that faces the cytoplasm. This is efficient because the subunits were put together in the cytoplasm adjacent to this layer.

The Extracytoplasmic Phase

Because the cell wall grows outward from the cell, the new subunits must be moved from the inner side of the plasma membrane to the outer, but the mechanism is not yet understood. We do know that on arriving at the outer lipid layer of the plasma membrane, the peptidoglycan subunits react with a series of membrane-bound enzymes, and it is these reactions that allow the orderly incorporation of the new subunits into the growing cell wall. These enzymes are also very efficient targets for antibiotic therapy, as we will see in Chapter 19.

The last step in the process is the joining together of peptidoglycan layers to form a meshwork that is the foundation of the cell wall (see **Figure 9.4**). These connections can give the cell wall many layers and consequently increased strength. The amount of peptidoglycan in cell walls differs between Gram-positive bacteria (many layers) and Gram-negative bacteria (few layers).

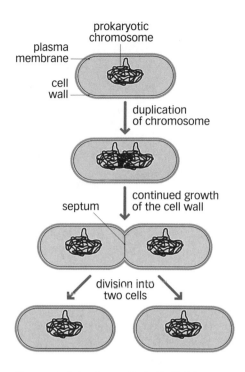

Figure 9.3 Prokaryotic cell division shows the development of new cell wall as one cell becomes two through the process of binary fission.

Fast Fact The lipid carrier cycle is also a target for antibiotic therapy with drugs such as bacitracin and the antibiotic peptide mersacidin.

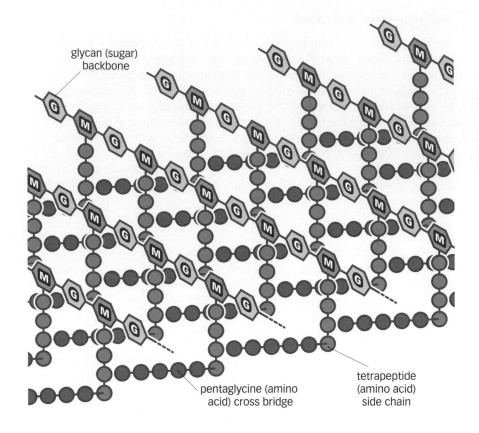

glycan (sugar) backbone

pentaglycine (amino acid) cross bridge

tetrapeptide (amino acid) side chain

Figure 9.4 An illustration of the cross-linking by which layers of peptidoglycan are joined together during cell wall formation in Gram-positive bacteria. The cross-linking by pentaglycine bridges and tetrapeptides between layers, and the covalent bonds contained, make peptidoglycan a strong cell wall component.

Keep in Mind

- Many of the structures associated with the bacterial cell wall are involved in the infection process.

- The cell wall is the protective outer structure of the bacterial cell.

- The cell wall is made of peptidoglycan, which is a series of repeating disaccharides and polypeptide pieces.

- The wall is thicker in Gram-positive bacteria than in Gram-negative bacteria.

- Peptidoglycan is constructed inside the cell and transported to the outside of the cell.

- Peptidoglycan forms a meshwork foundation for the cell wall and is a target for antibiotics.

Additional Cell Wall Components

Now that we have seen how peptidoglycan, the basic building block of the cell wall, is constructed, let's turn our attention to the other elements of the cell wall. In Chapter 4, we learned that most bacteria can be divided into two major groups on the basis of the differential Gram stain. This staining effect is based on the components of the cell walls.

The Gram-positive cell wall is rich in peptidoglycan, with multiple layers of meshwork, whereas the Gram-negative cell wall contains very little peptidoglycan.

Wall Components in Gram-Positive Bacteria

In addition to many layers of peptidoglycan, the cell wall of Gram-positive bacterial cells also contains **teichoic acid** molecules. These are repeating subunits of sugar-phosphate molecules (either ribitol or glycerol) to which various other sugars and amino acids are attached. Teichoic acids come in two types (**Figure 9.5a**): those that go partway through the peptidoglycan layers (**wall teichoic acids**) and those that go completely though the peptidoglycan layer and link to the plasma membrane (**lipoteichoic acids**). Both types of teichoic acid protrude above the peptidoglycan layer.

Wall Components in Gram-Negative Bacteria

The cell walls of Gram-negative bacteria are much more complex than those of Gram-positive bacteria and contain only a thin layer of peptidoglycan (**Figure 9.5b**). Consequently, the kind of environmental protection seen with Gram-positive bacteria is not available to Gram-negative organisms. However, the latter acquire additional protection through the presence of an outer membrane. The outer membrane includes a **lipopolysaccharide layer** (LPS layer) composed of lipids, proteins, and polysaccharides. Lipoprotein molecules are used to fasten the outer membrane of the cell to the peptidoglycan layer of the cell wall.

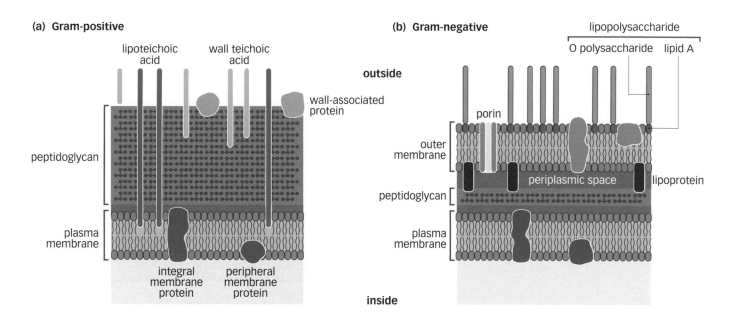

Figure 9.5 A comparison of bacterial cell walls. Panel a: The cell wall of a Gram-positive bacterium. Gram-positive cell walls are easy to identify because of the numerous layers of peptidoglycan. In addition to wall-associated proteins, the Gram-positive wall contains wall teichoic acid and lipoteichoic acid. The latter type spans the entire layer of peptidoglycan and fastens the cell wall to the underlying plasma membrane. The structure of this wall has a direct bearing on how certain antibiotics function and also on how the Gram stain works. Panel b. The Gram-negative cell wall contains very little peptidoglycan and no teichoic acids. One of the most prominent features of the Gram-negative wall is the outer membrane, which is composed of lipopolysaccharides, lipoproteins, and phospholipids. The Gram-negative cell wall contains endotoxin components that can cause toxic shock, and the O polysaccharides in the wall can be used diagnostically.

The structure of the outer membrane is unusual in that the lipid portion, although similar to the phospholipid bilayer seen in many cell types, has a unique outer layer. This layer is composed of lipopolysaccharides instead of the standard phospholipid molecules. It serves as a major barrier to the outside world and contains specialized proteins called **porins**. These proteins contain channels that vary in size and specificity, and they are responsible for the passage of molecules and ions into and out of the Gram-negative cell. The size of the channel in the porin has a role in determining which substances can be moved into or out of the cell.

The outer membrane also contains a variety of **translocation protein systems** that move substances out of the cell. Many of these translocation systems are found in the **periplasmic space**, which is the space between the plasma membrane and the outer membrane. This space is filled with a gel-like material and contains a variety of proteins secreted by the cell. Many of these proteins have important roles in the breakdown of nutrients and transport, whereas others are found only transiently in this space on their way to the outer membrane.

The difference between the Gram-positive cell wall and the Gram-negative cell wall is significant. Table 9.1 compares the two types of cell wall. It is important to understand the components that make up these walls and how the two types of wall differ from each other, because these walls can have a role in the disease process and can also be targets for antibiotics.

It should also be noted that some bacteria, such as *Mycoplasma* (which can cause a mild pneumonia called walking pneumonia), have a cell membrane but no cell wall. As a result, these bacteria have no distinct shape, are difficult to culture, and do not respond to antibiotics such as penicillin and cephalosporin, which use the cell wall as a target.

Keep in Mind

- Gram-positive cells have many layers of peptidoglycan and also contain teichoic acid.

- Teichoic acid is a glycoprotein that can be found in two forms: wall teichoic acid and lipoteichoic acid.

- Gram-negative organisms have no teichoic acid and very little peptidoglycan but have an outer membrane layer made up of phospholipids and lipopolysaccharides.

Characteristic	Gram-positive bacteria	Gram-negative bacteria
Peptidoglycan	Thick layer	Thin layer
Teichoic acid	Present	Absent
Lipids	Very little	Lipopolysaccharide layer
Outer membrane	No	Yes
Toxins produced	Exotoxins	Endotoxins
Sensitivity to antibiotics	Very sensitive	Moderately sensitive

Table 9.1 Characteristics of Gram-positive and Gram-negative bacteria.

Clinical Significance of the Bacterial Cell Wall

Gram-Positive Bacteria

As we discussed in Chapter 6, some of the most troubling types of respiratory infection are nosocomial ones caused by *Staphylococcus aureus*. These infections begin with colonization of the nasal passages by *S. aureus*. It has been determined that the wall teichoic acid of these Gram-positive organisms mediates the colonization of nasal epithelial cells and is an essential element for the infection to be successful. Sadly, these infections often lead to bacteremia, where bacteria are found in the blood, and septicemia, where the organisms are actively growing in the blood (see Chapter 7). These conditions can result in high rates of mortality.

Although colonization of the nasal passages with *S. aureus* can be inhibited by the antibiotic **mupirocin**, the appearance of mupirocin-resistant Gram-positive organisms is on the rise. The gene that codes for the synthesis of wall teichoic acid has been identified and is referred to as the *tag* O gene. Strategies are in hand to develop new antimicrobials that target this gene and its products. In addition, it may be possible to use teichoic acid as an immunizing agent for the development of a vaccine that can be used in hospitals to protect against these Gram-positive infections.

The lipoteichoic acid component of the Gram-positive cell wall can also elicit an inflammatory response in the host, a response that is a by-product of Gram-positive cell death and fragmentation of the cell walls.

Streptococcus pyogenes, which causes strep throat, necrotizing fasciitis, and other diseases, incorporates virulence factors into its cell wall. One of these factors is the **M protein**, which protrudes from the cell wall. This protein is required for infection, as illustrated by the fact that antibodies against it can inhibit infection. M protein is anti-phagocytic and is a major virulence factor produced by certain species of *Streptococcus*. Unfortunately, this protein is highly susceptible to mutations, and there are currently more than 80 variants. Because antibodies against one variant do not bind sufficiently to others, the organism can use M-protein variants to defeat the adaptive immune responses of infected individuals.

Mycobacterium, species of which cause tuberculosis (*M. tuberculosis*) and leprosy (*M. leprae*), synthesizes a waxy lipid called **mycolic acid** that is incorporated into the cell wall. In fact, as much as 60% of the cell wall of these organisms is mycolic acid, which makes the organisms extremely resistant to environmental stress. More importantly, the incorporation of mycolic acid into cell walls provides a barrier that prevents the actions of many antibiotics and many host defense mechanisms.

Gram-Negative Bacteria

The complexity of the Gram-negative cell wall has a clinical role in a variety of ways because the outer membrane protects bacteria from antiseptics and disinfectants. More importantly, this membrane also provides a barrier against the uptake of antibiotics.

Fast Fact The multiple drug resistance of *M. tuberculosis* has become a major health care concern.

Therefore, infections with Gram-negative bacteria are on the whole less sensitive to antibiotic treatment. The porin channels that run through the outer membrane exclude large antibiotic molecules but not smaller ones. As antibiotics are administered, bacterial genes encoding porin molecules can mutate and specify smaller openings, thereby inhibiting the exposure of the cell to even smaller antibiotic molecules.

Perhaps the most important clinical aspect of Gram-negative bacteria is the LPS layer of the outer membrane, which is an endotoxin. Many of the clinical effects associated with Gram-negative infections are due to the release of endotoxin molecules (see Chapter 5). Two parts of the outer membrane have clinical relevance: lipid A, and O polysaccharides.

Lipid A anchors the LPS portion of the outer membrane to the phospholipid bilayer portion (see Figure 9.5b). When lipid A is released as a by-product of bacterial cell death, it is referred to as an endotoxin. The body's innate immune defenses overreact to this endotoxin, and if enough of it is present, these defensive reactions can eventually have an adverse effect on host cells and cause shock.

O polysaccharides are carbohydrate chains that are part of the outer membrane located on the side of the membrane that faces the extracellular fluid. These molecules vary from one bacterial species to another and can be recognized by the immune system. In addition, they can be used as a diagnostic tool for the identification of certain bacteria. For example, in the name *Escherichia coli* O157:H7, which causes a potentially lethal infection, the O indicates the O polysaccharide (designated 157) found on the outer membrane of this pathogenic strain of *E. coli*.

Our examination of the bacterial cell wall configuration has given us a good idea of how this structure provides the environmental protection required for bacteria. We have also seen that the wall has important implications for infections and therapy. Because the major focus of our discussions is on clinical considerations, we now return our attention to the five requirements discussed throughout previous chapters: entry, colonization, defeating host defenses, damaging the host, and transmission. There are structures associated with the bacterial cell that can help fulfill some of these requirements.

Keep in Mind

- The bacterial cell wall has clinical significance.
- Gram-positive teichoic acid is involved with the production of inflammation.
- Some Gram-positive bacteria incorporate virulence factors known as M proteins in their cell walls.
- *Mycobacterium* species are Gram-positive bacteria that incorporate mycolic acid (a waxy substance) in their cell walls, which makes them resistant to antibiotics.
- In Gram-negative organisms the outer membrane layer protects against disinfectants and antibiotics.
- The outer membrane of Gram-negative organisms functions as an endotoxin.

STRUCTURES OUTSIDE THE BACTERIAL CELL WALL

There are five structures that can be found outside the cell wall of a bacterial cell. Not all bacteria have all of these structures. Three of them—the glycocalyx, fimbriae, and pili—are primarily involved with adherence (the requirement of entry and staying in); the flagella, axial filament, and also pili are involved with motility (which can be used to defeat host defenses and damage the host, as well as aiding in transmission).

The Glycocalyx

Colonies of some pathogens grown on agar plates look wet and shiny. This is because each cell in the colony is surrounded by the **glycocalyx**, a sticky substance composed of polysaccharides and/or polypeptides. The polysaccharides and polypeptides are produced in the cytoplasm and secreted to the outside of the cell, where they become fastened to the cell wall in two distinct ways. If the molecules are loosely attached to the cell wall, the glycocalyx they collectively form is referred to as a **slime layer**. If the molecules produce a highly organized structure that adheres tightly to the cell wall, the glycocalyx is referred to as a **capsule**.

Both the capsule and the slime layer give the bacterial cell the ability to adhere to surfaces. For example, certain bacteria can grow on rocks in the middle of river rapids. The ability to adhere to surfaces in the presence of fast-moving water is due to the presence of a glycocalyx. In addition, the glycocalyx protects the organism from desiccation and in desperate times can be used as a source of nutrition. This structure also has an important role in pathogenesis.

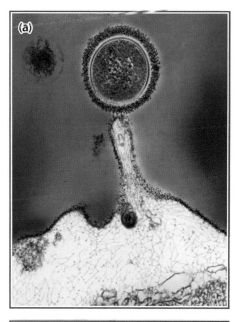

Clinical Significance of the Glycocalyx, Slime Layer, and Capsule

The glycocalyx is a prime factor in any successful invasion of a host by respiratory and urinary tract pathogens. After gaining entry to these anatomical systems, the invading bacteria face the possibility of being swept away by the natural defensive mechanisms of the host. This is where the adherence properties of the glycocalyx have a significant role in the infectious process.

The slime layer associated with bacteria is seen in those forms of dental decay that depend on the formation of a biofilm (Chapter 4) on the tooth surface. The slime layer permits the initial organisms to adhere to the tooth surface, and this population of initial invaders can be rapidly followed by a progression of organisms that adhere to one another, forming a biofilm. The final result is the breakdown of the dental surface.

The capsule form of the glycocalyx is clearly required for infection by some bacteria. For example, *Streptococcus pneumoniae* is not infectious unless it is encapsulated. The same holds true for *Bacillus anthracis* and *Klebsiella pneumoniae*. Examples of encapsulated bacterial cells are shown in Figure 9.6. As we will see in Chapter 11, transformation, a type of genetic transfer, was demonstrated with encapsulated and nonencapsulated *S. pneumoniae* and clearly showed that the development of a capsule conferred infectivity on this organism.

One of the primary reasons for the infectivity associated with the presence of a capsule is the inhibition of phagocytosis.

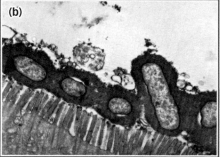

Figure 9.6 Bacterial capsules. Panel a: A colored transmission electron micrograph of a section through a *Streptococcus* bacterium attached to a human tonsil cell. The capsule form of the glycocalyx (thick red outer layer) can be seen tightly attached to the outside of the cell wall (thin yellow layer). Panel b: An electron micrograph showing how capsules facilitate the attachment of bacteria to cells of the intestine.

As we will see in Chapter 15, this defense is nonspecific and requires a sequence of events to destroy invading organisms. The first step in this sequence is the adherence of the phagocytic cell to the bacterium. The presence of a capsule can prevent this initial step, allowing the bacterium to continue with the infection process.

Fimbriae and Pili

Fimbriae (singular fimbria) and **pili** (singular pilus) are two more cell wall components involved in adherence. These sticky projections, which are shorter than flagella, are found on Gram-negative organisms, with the number of projections varying from species to species. Fimbriae (**Figure 9.7a**) and pili (**Figure 9.7b**) are composed of the same **pilin protein** subunits. Both structures are involved in adherence to host cells and to other bacterial cells. Although adherence seems to be the only function of the fimbriae, pili are also involved in the transfer of genetic material from a donor bacterium to a recipient bacterium through the process of conjugation (Chapter 11).

Fast Fact Pili have been shown to have a specificity for host cell receptors that seems to be associated with specialized protein molecules found only at the tip of the pili.

Figure 9.7 Structures exterior to the bacterial cell wall. Panel a: An electron micrograph showing a dividing *Proteus vulgaris* organism, which has both flagella and fimbriae. Notice the difference in length between these two structures. Fimbriae are made of pilin protein and are used for attachment; flagella are made of flagellin protein and are used for movement. Panel b: A transmission electron micrograph of a bacterial pilus during conjugation.

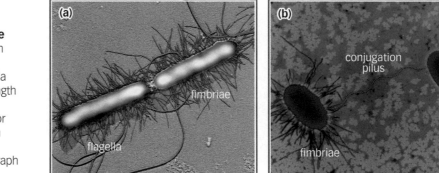

Clinical Significance of Fimbriae and Pili

Because of the adherence characteristic of the fimbriae and pili, they are important in fulfilling the requirement to stay in. For example, *Neisseria gonorrhoeae* is infectious only if fimbriae are present. Adherence factors are obviously important for organisms that infect the urinary tract, because without this ability they could easily be flushed out of the urethra. Similarly, successful *E. coli* infections of the small intestine require the presence of fimbriae because the rapid movement of the intestinal contents would eliminate the bacteria from the intestines.

In bacteria that use pili for adherence, the binding ability of the pili provides a possible target for vaccine development, because antibodies that bind to pili could block the binding to host cell surface receptors. Early studies on the binding of *Vibrio cholerae* have confirmed this possibility.

Pili also give bacteria motility, which is important for moving the organisms to different areas of the host. In Gram-negative organisms, this movement is a specialized twitching or gliding that occurs by retraction of the pili. A bacterium binds to nonspecific host cell surface determinants (such as proteins) and then retracts its pili, a motion that allows the bacterium to pull itself across the surface. The mechanism responsible for the motion is the detachment of pilin subunits, an action that pulls the cell forward, followed by pilin subunit reformation and elongation (the result is similar to the movement of a caterpillar). This ability to move also allows organisms to form biofilms (see Chapter 4), which help colonies become resistant to immune responses, antibiotics, and other therapies, leading to persistent infections.

Pili can also be used for **immune escape**, which is the ability to evade a host immune response. For example, *Neisseria* species use **phase variation**, a tactic in which the number of pili decreases after initial infection, taking away the target for antibodies. They can also use **antigenic variation** or **post-translational modification**, in which they change or mask the structure of the pili so that antibodies no longer recognize the *Neisseria* invaders. Lastly, in what is perhaps their most amazing tactic, they secrete fragments of pili, referred to as **S pili**, that bind to antibody molecules and essentially inactivate them. These strategies for the use of pili as both a pathogenic factor and a protective mechanism show the ingenuity that bacteria have evolved to function more efficiently and also to protect themselves.

Pili are also clinically important because they facilitate the transfer of genetic material from one bacterial cell to another (the process called conjugation, described in Chapter 11). This mechanism provides Gram-negative organisms with a method for genetic modification and self-defense. These genetic modifications can involve the production of dangerous toxins and the transfer of genes responsible for antibiotic resistance.

Axial Filaments

Axial filaments wrap around the bacterial cell and give it motility (**Figure 9.8**). These filaments, often referred to as endoflagella, are found on spirochetes and are one of two structures exterior to the cell wall that are responsible for bacterial motility. They are confined to the space between the plasma membrane and the outer membrane. Axial filaments can produce a rotational movement, and the rotation of all the axial filaments causes the entire organism to rotate in a corkscrew manner.

Figure 9.8 The structure of the axial filament. Panel a: A scanning electron micrograph of two *Leptospira interrogans* spirochetes. The filament runs inside the cell wall of this spirochete. Panel b: A diagrammatic representation of the axial filament wrapping around a spirochete. Panel c: A colorized transmission electron micrograph of a cross section of a spirochete. Note the numerous axial filaments (red circles).

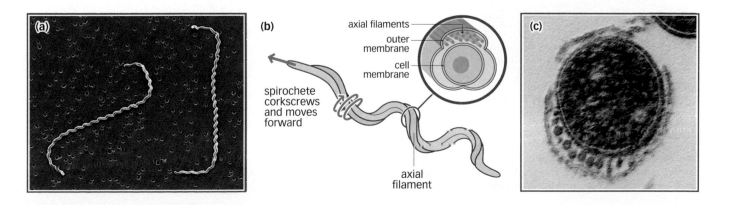

Clinical Significance of Axial Filaments

The corkscrew rotational motion created by axial filaments gives the bacterium the ability to bore through tissue. Evidence for this boring capability can be seen in the infectious process associated with *Treponema pallidum* (the causative agent of syphilis; Figure 9.9a) and *Borrelia burgdorferi* (the causative agent of Lyme disease; Figure 9.9b). In both cases, the infection involves movement of the organism from an external location through tissue, into the blood, and back into tissue. The design and structure of the axial filament facilitate this function.

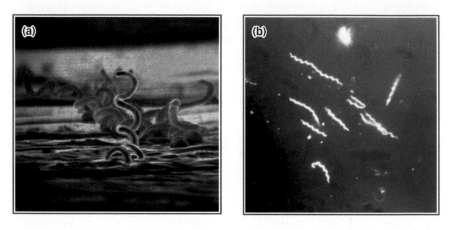

Figure 9.9 Bacterial axial filaments. Panel a: Electron micrograph of the organism *Treponema pallidum* (the causative agent of syphilis) invading rabbit epithelial tissue. The organism uses its axial filament to produce a corkscrew motion, enabling it to enter the tissue. Panel b: A dark-field micrograph showing the corkscrew-shaped bacterium *Borrelia burgdorferi*, which is the causative agent of Lyme disease.

Flagella

A **flagellum** is a long structure that extends far beyond the cell wall and even beyond the glycocalyx (Figure 9.10). Bacterial cells being propelled by flagella can be seen in **Movie 9.1**. This structure is used only for motility and is a classic example of the relationship between structure and function. Remember that in nature, structures are usually configured to maximize function. This reduces energy wastage, and, as we saw in Chapter 3, energy is a very valuable and much-in-demand commodity.

Structure of the Bacterial Flagellum

The bacterial flagellum contains three parts: the filament, the hook, and the basal body. The filament is made up of molecules of the protein **flagellin** (Figure 9.11). These molecules attach to one another to form a chain that is twisted into a helical structure, giving the flagellum a hollow core. The hook is the structure that links the flexible filament to the basal body (described in a moment), which anchors the structure to the bacterial cell interior.

Before we discuss the basal body, it is important to consider how the flagellum moves. This occurs through rotational movement that can be as fast as 100,000 rotations per minute and can move the organism up to 20 cell lengths in seconds. Although this movement rate may not seem like much, it equates to a 6-foot-tall human running at 80 or more miles per hour! With this in mind, it is easy to see that there is considerable stress and torque on the connection between the filament and the cell. It is at this junction that we find the basal body, which fastens the flagellum to the cell. The structure of the basal body and how it connects in Gram-positive and Gram-negative cells constitute another excellent example of structure matching function.

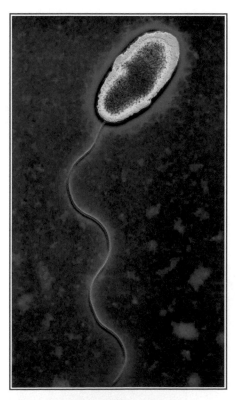

Figure 9.10 Colorized scanning electron micrograph of a rod-shaped bacterium with a prominent flagellum.

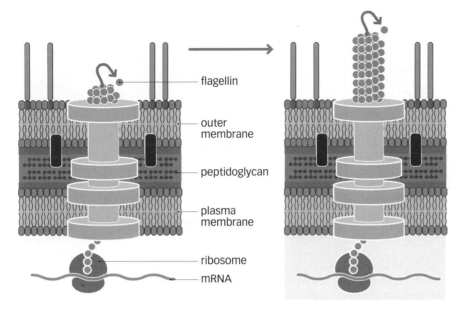

Figure 9.11 Deposition of flagellin during the construction of a bacterial flagellum. The protein is produced in the cytoplasm and moved through the developing flagellum, to be deposited at the top of the elongating filament.

The basal body is a rod that has rings strategically fastened to it. Depending on which type of cell (Gram-positive or Gram-negative) the flagellum is attached to, there are either one or two pairs of rings. Remember that the cell wall of the Gram-positive cell has multiple layers of peptidoglycan, which confers strength and stability. Consequently there is only one pair of rings (an inner ring and an outer ring) on the basal body of a flagellum attached to this type of cell. This pair of rings is fastened to the plasma membrane, with the flagellum passing through the entire layer of peptidoglycan (**Figure 9.12a**).

Because the Gram-negative cell wall contains very little peptidoglycan, the point at which the flagellum connects to the cell needs reinforcement. This is accomplished by two pairs of rings on the basal body.

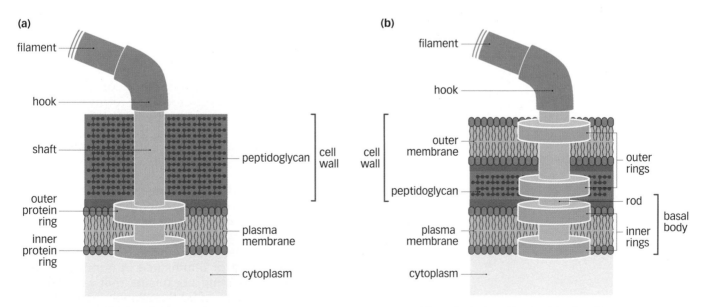

Figure 9.12 Bacterial flagella. Panel a: Anchoring of the flagellum to a Gram-positive cell. The basal body of the flagellum on a Gram-positive organism has only one pair of rings, and these rings are both fastened to the plasma membrane. The many layers of peptidoglycan found in the Gram-positive cell allow this one pair to be sufficient for stability. Panel b: The anchoring of the flagellum to a Gram-negative cell requires two pairs of rings, with one pair binding to the plasma membrane and the other pair to the cell wall and outer membrane. The Gram-negative cell wall contains very little peptidoglycan and therefore requires additional stabilization.

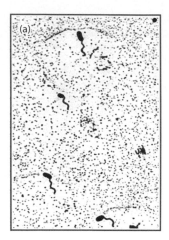
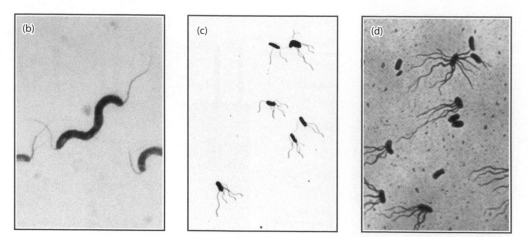

Figure 9.13 Photomicrographs of flagellar arrangements. Panel a: Monotrichous. Panel b: Amphitrichous. Panel c: Lophotrichous. Panel d: Peritrichous.

| Fast Fact | If we think about the energy required to propel a cell through a medium such as water or blood, we can understand the need for the propelling structure to be as energy-efficient as possible. To get a better picture of this process, remember that a bacterium moving through water is like a human swimming through molasses. Obviously, moving through a viscous substance such as blood is even more difficult for a bacterial cell.

Figure 9.12b and **Movie 9.1** show that one pair is attached to the plasma membrane and the other is attached to the peptidoglycan/LPS layer of the Gram-negative cell.

Flagellar Configurations

Bacteria have four distinct patterns of flagella arrangement (**Figure 9.13**):

- **Monotrichous**—the most common form, in which the bacterium has one flagellum located at one end of the cell.

- **Amphitrichous**—in which the bacterium has two flagella, one at each end of the cell.

- **Lophotrichous**—in which the bacterium has two or more flagella located at the same end or both ends of the cell.

- **Peritrichous**—another common form, in which the entire cell is surrounded by flagella.

Some bacteria lack flagella and are nonmotile. These bacteria are called **atrichous**.

Clinical Significance of Flagella

A good example of the clinical importance of flagella is *Helicobacter pylori*, the causative agent for some forms of gastric ulcers in humans. This organism lives in the folds of the stomach and uses powerful flagellar movements to propel itself into the heavy layer of mucus that protects the epithelial cells of the stomach from stomach acids.

In general, as we discussed previously, the movement of bacteria to places where they do not ordinarily reside results in opportunistic infection. In healthy individuals, when invading organisms enter regions already occupied by normal flora, the resident bacteria inhibit the growth of the newcomers. However, if resident populations are destroyed or diminished in number, as in the overuse of antibiotics, invading organisms that are motile because of the presence of one or more flagella can gain a foothold and initiate an opportunistic infection. Similarly, any organism that can move from tissue to the lymph or blood systems has the opportunity to travel throughout the body, generating a systemic infection.

Intracellular Movement

Any infection in which bacteria gain access to the inside of a host cell is good for the bacteria because then they are shielded from defenses such as phagocytosis and host immune reactions. Some pathogens—*Salmonella* is one example—use **cytoskeletal structures** of the host cell, such as actin fibers, not only to gain access to the host cell but also for intracellular movement within the host cell and to move from one host cell to another. This type of movement also occurs with *Shigella*, a bacterium associated with a type of **dysentery** (bloody diarrhea). Although *Shigella* is classified as nonmotile, it can use cytoskeletal structures of the host to move from cell to cell.

Keep in Mind

- Structures found outside the cell wall are all involved in the infection process.

- The glycocalyx can be found in either the capsule or slime-layer form and can satisfy the requirement of staying in by facilitating adherence to host cells.

- In many cases, an organism is not pathogenic unless it has a capsule.

- Fimbriae also satisfy the requirement to stay in by adhering to host cells.

- Pili can be used to adhere to host cells but also allow the transfer of genetic information between bacteria in the process known as conjugation.

- Axial filaments allow organisms to penetrate tissues by means of rotational movement.

- Flagella permit movement, which helps organisms stay in.

- There are four forms of flagella seen in bacteria: monotrichous, amphitrichous, lophotrichous, and peritrichous.

- The host cell's cytoskeleton can be used for intracellular movement during the infection.

STRUCTURES INSIDE THE BACTERIAL CELL WALL

The bacterial structures just described deal with external environmental stress and, in many cases, pathogenic processes. In this section we discuss the six major structures inside the cell wall, all involved with intrinsic cellular functions: the plasma membrane, the nuclear region, plasmids, ribosomes, inclusion bodies, and endospores.

The Plasma Membrane

Just like the plasma membrane of eukaryotic cells (see Chapter 4), the plasma membrane of a bacterium is a delicate flexible structure that holds in the internal cellular matrix of cytosol. The most important feature of any plasma membrane—bacterial or eukaryotic—is **selective permeability**. The membrane is constructed in such a way that only certain small molecules, such as gases and hydrophilic molecules, can pass through easily. To understand the structure of this membrane fully, we must continue to think about structure–function relationships. The bacterial plasma membrane, like the cell wall that lies outside it, provides a barrier between the inside and the outside of the cell. However, one important difference between the two is that, unlike the cell wall, the plasma membrane is involved in the physiological functions of the bacterial cell, such as transport, secretion, DNA replication, and the generation of energy. It is the unique design of the membrane that allows it to participate in these functions.

Figure 9.14 A diagrammatic representation of the phospholipid bilayer of the plasma membrane. The arrangement of the hydrophobic lipid tails provides a barrier between the outside and the inside of the cell.

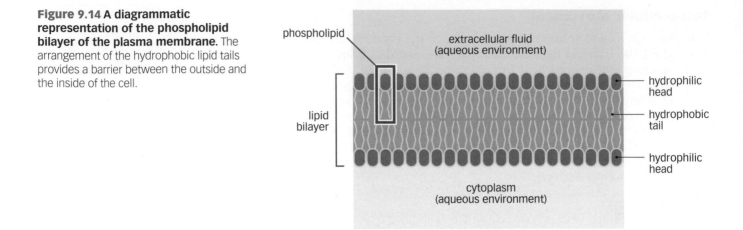

Structure of the Plasma Membrane

The plasma membrane is a phospholipid bilayer (**Figure 9.14**). Recall that phospholipids contain a hydrophilic head and a hydrophobic tail. In a phospholipid bilayer, the two layers of phospholipids are arranged such that the two tail regions are adjacent and the heads are pointing outward in opposite directions. Remember, the environment both inside and outside the cell is water-based, and there are components on both sides that need to be moved across the membrane. Because the two outward-facing sides of the bilayer are hydrophilic, they interface with water-soluble molecules on both sides of the membrane. The unique part of the membrane structure is that because the hydrophobic tails of both layers face each other, a barrier to water is produced.

The many functions required of the bacterial plasma membrane are performed by a variety of proteins that float in the phospholipid bilayer. There are two basic types of membrane protein (**Figure 9.15**). Peripheral proteins are found on either the inside or outside layer of the bilayer, whereas **integral proteins** penetrate the membrane completely and in some cases contain a pore that connects the interior of the cell to the external environment. Although all of the functions of these membrane proteins are important for cell function, we will concentrate our discussions on energy and transport.

Fast Fact There are more than 200 different proteins in the plasma membrane of *E. coli*.

Energy Production

Recall from Chapter 3 that cellular energy, in the form of ATP, is required for all metabolic functions. To understand the magnitude of this requirement,

Figure 9.15 The bacterial plasma membrane, showing membrane proteins and other components. Some of the proteins extend through the entire membrane (integral proteins), whereas others are restricted to either the inner or outer layer (peripheral proteins).

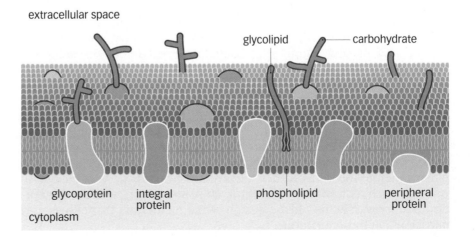

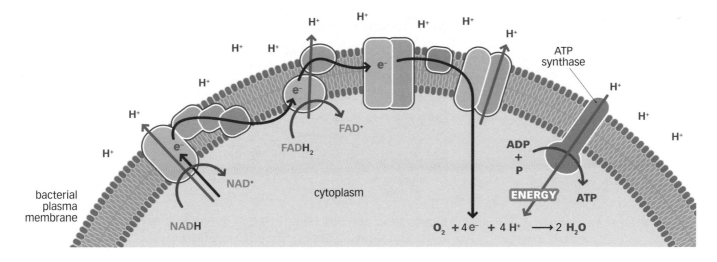

Figure 9.16 Membrane electron transport. This important membrane function is performed in bacteria by a series of compounds located in the plasma membrane. These compounds eject hydrogen ions (H+) from the cell, thereby creating a concentration gradient that results in a proton motive force. As the concentration of hydrogen ions outside the cell membrane increases, the ions begin to flow back into the cell interior and energy is released.

think back to our discussion of the flagellum and how many rotations it produces. Each rotation requires energy. Unlike eukaryotic cells, in which mitochondria produce the energy required for cellular functions, bacteria use the plasma membrane to house the electron transport chain, the series of membrane proteins that generate ATP through the sequential transfer of electrons and hydrogen ions from protein to protein. As we saw in Chapter 3, this forms a concentration gradient across the membrane and results in the proton motive force, a form of energy that can be either used immediately or stored in ATP by reactions involving the enzyme ATP synthase (**Figure 9.16**). (Because a hydrogen ion (H+) is nothing more than a single proton, the terms 'proton' and 'hydrogen ion' mean the same thing, so we can call it a proton motive force but talk about hydrogen ions being moved.)

Membrane Transport

Transport across the plasma membrane should be viewed in the context of overall bacterial cell structure. The complex bacterial cell wall helps bacteria cope with environmental stresses. The plasma membrane serves a similar function. It is a barrier between the cytoplasm and the exterior environment. Keep in mind that the cell wall is a meshwork structure that lets things percolate through, but it is the plasma membrane that regulates what gets into the cell and what does not.

Through everyday functions, cells generate waste that must be removed from the cell, and they also require access to nutrients and other materials located in the extracellular fluid. The mechanisms of transport and the membrane proteins involved are designed to accomplish these functions. There are three types of membrane transport: osmosis, passive transport, and active transport.

The plasma membrane is selectively permeable, and small molecules such as water can move across it. For the purposes of our discussion, we can define **osmosis** as the movement of water across a selectively permeable membrane from low to high solute concentration. If the concentration of solutes inside the cell is lower than that outside, water will

Fast Fact A high concentration of solutes, as occurs in salty and sugary foods, results in the plasmolysis of bacterial cells. This explains why these foods have long shelf lives and why salt was used in food preservation for centuries before refrigeration and preservatives were developed.

Figure 9.17 Plasmolysis. Plasmolysis occurs when cells are placed in a hypertonic environment, where the solute concentration is higher outside the cell than inside. Water moves out of the cell, causing the plasma membrane to pull away from the cell wall and the cell volume to shrink.

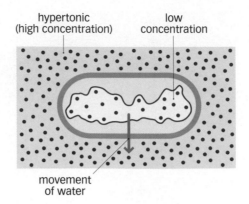

move out of the cell (**Figure 9.17**). When this happens, the cell loses water and shrinks. This is referred to as **plasmolysis**. If the concentration of solutes is greater inside the cell than outside, water will move into the cell (**Figure 9.18**). This influx of water causes the cell to expand, become turgid, and possibly lyse.

When a cell is placed in a solution in which the solute concentration outside the cell is higher than the solute concentration inside the cell, the solution is referred to as **hypertonic**. When a cell is placed in a solution in which the solute concentration is lower than the solute concentration inside of the cell, the solution is called **hypotonic**. A hypotonic solution results in osmotic lysis, and a hypertonic solution causes the cell to shrink, or undergo plasmolysis.

It is important to note that cells are normally found in an **isotonic** environment, in which the solute concentrations inside and outside the cell are essentially equal (for example, approximately 0.9% NaCl, as seen in standard intravenous saline solutions). In this situation, water does not move overall in either direction; the water is moving equally into and out of the cell.

Passive transport

There are two types of **passive transport**: simple diffusion and facilitated diffusion. The most important thing to remember about both types of passive transport is that they do not require energy.

Simple diffusion depends on the development of concentration gradients. You can think of a concentration gradient as a truck going downhill with no brakes. The steeper the hill, the faster the truck goes. The higher a concentration gradient between two regions, the faster the solute moves from the region of higher concentration to the region of lower concentration. The movement of a solute by simple diffusion slows down as the concentration gradient becomes smaller but continues until the concentration of solute is the same on both sides of the membrane and there is no longer a gradient.

Simple diffusion occurs only with solute molecules that are either soluble in lipids or small enough to pass through the cell wall and the plasma membrane.

The second type of passive transport, **facilitated diffusion**, is similar to simple diffusion in that it requires no ATP. However, in facilitated diffusion, molecules are brought across the plasma membrane through molecules called **permeases**. Permeases are specialized transport proteins found in the plasma membrane whose function is based on their three-dimensional shape.

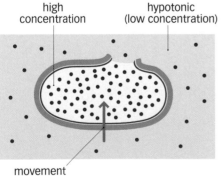

Figure 9.18 Osmotic lysis. Osmotic lysis results when cells are placed in a hypotonic environment, where the solute concentration is higher inside the cell than outside. Water moves into the cell and causes it to expand. This expansion can cause the cell to lyse (burst).

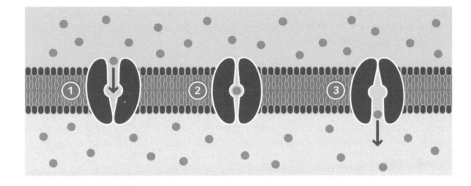

Figure 9.19 Facilitated diffusion.
A permease assists in the transport of molecules across the plasma membrane. This is based on the three-dimensional structure of the permease (carrier) proteins. (1) In this simplistic illustration with no solute attached, the three-dimensional structure causes the end of the protein facing the outside of the cell to be open and the end facing the inside of the cell to be closed. (2) When the solute to be transported binds to the permease, the structure of the permease changes, allowing the end facing the cell exterior to close and the end facing the interior to open. (3) The solute is released from the permease and enters the cytoplasm, causing the three-dimensional structure of the permease to revert to its original shape.

The binding of a solute to a permease changes the three-dimensional shape of the protein (**Figure 9.19**). To see how this works, let's look at a situation in which the concentration of a particular solute is higher outside the cell than inside. A permease in the cell's plasma membrane has a specific shape when not bound to any solute. When a solute attaches to a binding site located on the side of the permease facing the extracellular fluid, the shape of the permease changes such that the opening facing the extracellular fluid closes and the side of the permease facing the cell interior opens to allow the solute to move into the cytoplasm. On release of the solute, the three-dimensional structure of the permease regains its original shape and the binding site on the extracellular side is again available.

If the solute molecule is too large to fit into the binding site of a permease, the cell secretes enzymes into the extracellular fluid, and these enzymes break the solute into smaller pieces that can then be accommodated by the permease.

Active transport

The carrying of solute either into or out of a cell against the concentration gradient is known as **active transport**. It is the most common form of membrane transport but requires the expenditure of energy, either from ATP or from the proton motive force. Active transport uses specific carrier proteins found in the plasma membrane (**Figure 9.20**).

There are three types of active transport: efflux pumping, ABC transport systems, and group translocation.

Efflux pumping involves proteins that are part of what is called the super-family of transporters and is seen in both prokaryotic and eukaryotic membrane transport. Efflux pumping employs a revolving door mechanism in which membrane pumps bring in certain molecules and expel

Fast Fact Efflux pumping is routinely used by bacteria to expel antibiotic molecules (as we shall see in Chapter 20).

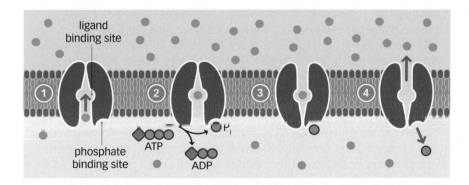

Figure 9.20 Active transport. This transport mechanism uses the energy in ATP to move a solute against its concentration gradient. In this drawing, the movement of solutes is from inside the cell to outside, even though the solute concentration is higher outside. Note that the carrier protein has one binding site for the solute to be transported and a separate binding site for an organic phosphate group (P_i) from ATP. As in facilitated diffusion, the three-dimensional changes that occur allow the transport of solute across the plasma membrane while ATP is broken down into ADP + P_i.

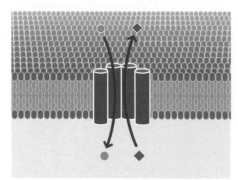

Figure 9.21 Efflux pumping, a type of membrane transport. The membrane protein pump brings one type of molecule into the cell as another type of molecule is simultaneously transported out of the cell. Bacteria regularly use this mechanism as a protection against antibiotics.

others at the same time (**Figure 9.21**). The energy source for these membrane pumps is not ATP but rather the proton motive force that comes directly from electron transport.

The second type of active transport, **ABC transport systems**, is a complex system that involves several proteins. (ABC stands for **A**TP-**b**inding **c**assette.) During ABC transport, the molecule to be transported forms a complex with a binding protein on the outside of the plasma membrane and is then handed off to a complex of proteins located in the plasma membrane. Together these proteins then transport the molecule into the cytoplasm.

The third type of active transport, **group translocation**, is unique to bacteria and involves a clever way of making sure that molecules transported into the cell remain inside. Let's use glucose as an example. Glucose is the preferred sugar for many microorganisms because it is the easiest to break down (Chapter 3) and most efficient source of energy for them. With this in mind, it is easy to see that once a glucose molecule is inside a bacterial cell it must be kept there. As the concentration of glucose inside the cell increases, a concentration gradient forms. This gradient allows glucose molecules inside the cell to move by simple diffusion back across the cell's plasma membrane and out of the cell. To prevent this migration out of the cell, bacteria use a phosphotransferase enzyme to attach a phosphate group to the glucose molecule once it has entered the cell (**Figure 9.22**). This modification of the glucose molecule removes it from the diffusion equilibrium reaction so that it stays inside the cell and is available as an energy source.

As with all other forms of active transport, energy is required to bind the phosphate to the glucose. In group translocation, the energy source is the high-energy molecule **phosphoenolpyruvate** (PEP) instead of ATP.

Secretion

Up to now, we have talked about the methods by which bacteria bring in molecules from the outside. The plasma membrane is also involved with secretion, which involves moving substances from the inside of the cell to the extracellular fluid. Secretion involves several plasma membrane proteins that act in a specific sequence. The exact interactions are not yet understood, although it has been shown that protein molecules destined to be moved out are recognized through a signal sequence of about 20 amino acids. It is thought that this sequence helps the protein molecule to interact with the plasma membrane and is clipped off as the molecule crosses the plasma membrane and exits from the cell.

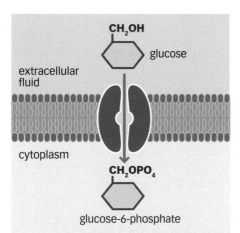

Figure 9.22 Group translocation, a type of transport seen only in bacteria. The molecule being transported into the bacterial cell is chemically altered as it crosses the plasma membrane. Alteration of the molecule avoids the buildup of a concentration gradient of the original molecule, with the result that the molecule cannot leave the cell by simple diffusion. This mechanism is often used for molecules such as glucose, which are in high demand by the cell.

Clinical Significance of the Plasma Membrane

Although the plasma membrane does not contribute any pathogenic qualities to a bacterium, it is a primary target for therapeutic strategies. Obviously, any damage to this membrane inhibits the transport of essential components into the cell, not to mention the ability to produce energy and replicate DNA. In a broader sense, loss of integrity of the bacterial plasma membrane causes the outside to come in or the inside to come out, and either is lethal for the microorganism.

In Chapter 19, we will see that the bacterial plasma membrane is one of the five primary targets for antibiotics. However, because of the similarity to plasma membranes in eukaryotic cells, this target does not demonstrate selective toxicity, and antibiotics directed against it can have major side effects.

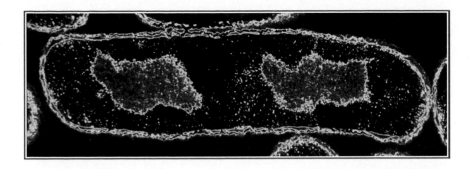

Figure 9.23 A colorized transmission electron micrograph showing an _E. coli_ organism in the process of cell division. The bright red area inside the cell is the nucleoid region where DNA is localized.

Keep in Mind

- The plasma membrane surrounds the cytoplasm in bacterial cells.

- This membrane is composed of a phospholipid bilayer, which is selectively permeable.

- The membrane contains many different types of protein that are either localized on the inner or outer side of the membrane (peripheral proteins) or span the entire membrane (integral proteins).

- The membrane in bacteria is involved with energy production (ATP is formed here) and also in the replication of the bacterial chromosome.

- Membrane transport can involve osmosis and passive transport (which require no energy) or active transport (which requires ATP).

The Nuclear Region

As prokaryotes, bacteria have no nucleus and we refer to the region of the cell containing DNA as the **nuclear region** (or sometimes the nucleoid region). It is this region that is the most discernible area inside the cell wall (**Figure 9.23**). Bacteria usually have only one circular chromosome, which contains all of the genetic information required by the organism. It is supercoiled and associated with specific positively charged proteins that stabilize it. We discuss DNA in detail in Chapter 11.

Plasmids

Plasmids are extrachromosomal segments of DNA that can be seen separated from the main DNA structure (**Figure 9.24**). The size of a plasmid ranges from 0.1% to 10% of the size of the chromosome, and some bacteria carry more than one plasmid. Plasmids often carry genes for toxins and resistance to antibiotics, and they can be transferred from one cell to another through pili during the process of conjugation.

Clinical Significance of DNA and Plasmids

Because DNA is the genetic blueprint for the organism, any disruption or damage to it can be a lethal event. It is therefore one of the five targets for antibiotic therapy. Like the plasma membrane, DNA is not a selectively toxic target because the DNA of pathogenic microbes is similar to the DNA of the hosts they infect. However, several therapeutic strategies are routinely used for viral infections. These employ chemicals that are reactive with viral nucleic acids (DNA or RNA) but not toxic to the host.

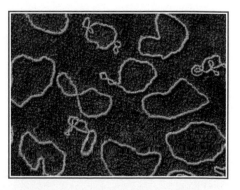

Figure 9.24 A color-enhanced electron micrograph showing bacterial plasmids.

The clinical significance of plasmids is not only as a potential therapeutic target but, more importantly, as a direct pathogenic factor in antibiotic resistance. Plasmids carry genes for toxins that are in many cases responsible for the clinical symptoms of an infection. Equally important is the fact that many of the genes for resistance to antibiotics are carried on plasmids, and the ease of movement of plasmids from one bacterial cell to another is a major cause of the spread of antibiotic resistance among bacteria.

Ribosomes

Recall from Chapter 4 that ribosomes are non-membrane-enclosed structures involved in protein synthesis. More active bacteria contain more ribosomes. For example, an *E. coli* cell can have anywhere between 7000 and 25,000 ribosomes.

In bacteria, each ribosome is composed of two subunits that remain apart until a messenger RNA molecule approaches. At that time, the two subunits enfold the mRNA in a very specific way, and the process of protein synthesis, referred to as **translation**, takes place (see Chapter 11). The two subunits of bacterial ribosomes are different in size from those of eukaryotic ribosomes and can be distinguished.

Clinical Significance of Ribosomes

Because protein synthesis is required for any cell to live, any disturbance or inhibition of this process can be lethal. Bacterial ribosomes are therefore another of the five major targets for antibiotic therapy. As we shall see in Chapter 19, several antibiotics target different parts of the ribosome so as to disrupt or damage protein synthesis. Pathogens have developed several efficient ways of resisting these antibiotics and preventing the loss of protein synthesis (see Chapter 20).

Inclusion Bodies

Other structures found in the interior of a bacterial cell are **inclusion bodies**, which are used to store materials that the cell may require. There are several types of inclusion body, including those for glycogen, which is a stored form of the energy-rich glucose molecule (see Chapter 2). Some organisms, for example, *Corynebacterium* (which causes diphtheria), store phosphates in inclusion bodies called **metachromatic granules**. These inclusions, also called volutin, are called metachromatic because they induce a color change in the stained sample.

Endospores

Fast Fact Endospores have been found that are 25 million years old.

An amazing characteristic of some bacteria is the ability to form **endospores** through a process called **sporulation**. An endospore forms whenever a spore-forming bacterium is exposed to too great an environmental stress. The process is restricted mainly to Gram-positive rods, including pathogenic species of *Bacillus* and *Clostridium*. Keep in mind, however, that the highly pathogenic Gram-negative bacterium *Coxiella burnetii* can also form endospores.

Once an endospore has formed, it is easily identifiable in the cytoplasm of the cell (**Figure 9.25**) and confers on the cell a kind of dormancy. Organisms can return from the endospore state to the vegetative (growing) state through the process of **germination**.

The endospore is extremely resistant to heat, desiccation, toxic chemicals, antibiotics, and ultraviolet irradiation. It can also survive long periods of boiling.

Sporogenesis involves several steps (see Figure 9.25). The first step is replication of the bacterial chromosome, followed by sequestering of the copy along with some cytosol. This volume becomes cut off from the rest of the cell as a wall referred to as a **septum** forms. The larger part of the cell then wraps itself around this newly formed smaller volume, forming what is called a **forespore**. The forespore at this point is composed of two membranes, one on top of the other. As sporogenesis continues, large amounts of peptidoglycan begin to be deposited between the two membranes of the forespore. Finally, the rest of the original cell deteriorates and degrades, leaving the bacterial genetic information protected inside the endospore.

Germination of the endospore into a vegetative cell occurs when the environmental stress has subsided, or in response to chemical cues. At this time, the endospore accepts water molecules, swells, and cracks. The water also activates metabolic reactions involving dipicolinic acid and calcium, two of the few chemicals packaged in the endospore.

Clinical Significance of Sporogenesis

Endospores represent a significant clinical challenge. Because they are so resistant, organisms that can produce them can withstand treatments that would kill other bacteria, and continue to cause serious problems. For example, *Clostridium botulinum* is a problem for the food industry because the endospores it produces can survive many food processing procedures, and when the endospores germinate, the resulting cells produce a botulinum toxin that causes significant damage to the host. Because of the resistance of endospores to heat, simple cooking cannot deal with the problem. For this reason, when items are sterilized, heat is combined with high pressure. This combination effectively deals with both vegetative *C. botulinum* cells and *C. botulinum* endospores.

Sadly, endospores have also become a tool for terrorists, as demonstrated in the anthrax attacks that occurred through the mail in 2001. *Bacillus anthracis* is commonly found in fields where cattle graze. This organism forms endospores that can be manipulated in specific ways (referred to as machining, or weaponizing) to produce a highly dispersible endospore powder. It was this type of material that was found on the letters involved in the 2001 anthrax incidents that killed 5 people and injured 17. The machined anthrax endospores were extremely small (3.5 μm or less in diameter) and therefore remained airborne for a long time. This small size also allowed them to penetrate deeper into the lungs, making the infection much more serious. The endospore form of these potential weapons is resistant to antibiotics, making it difficult to prevent environmental contamination that can lead to infection.

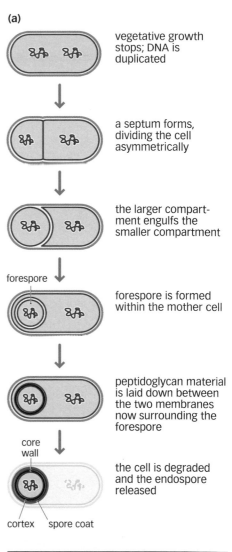

(a)

vegetative growth stops; DNA is duplicated

a septum forms, dividing the cell asymmetrically

the larger compartment engulfs the smaller compartment

forespore

forespore is formed within the mother cell

peptidoglycan material is laid down between the two membranes now surrounding the forespore

core wall

the cell is degraded and the endospore released

cortex   spore coat

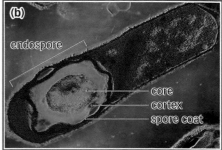

(b)
endospore
core
cortex
spore coat

Figure 9.25 Endospore formation. Panel a. The sequence of events in sporulation, the process by which a bacterial endospore is formed. Panel b: Colorized electron micrograph of a bacterium undergoing sporulation.

Table 9.2 Clinical relevance of bacterial cell structures.

Structure	Direct clinical relevance	Target for antibiotics
Cell wall	Yes	Yes
Glycocalyx	Yes	No
Fimbriae	Yes	No
Pili	Yes	No
Axial filaments	Yes	No
Flagella	Yes	No
Plasma membrane	No	Yes
DNA	No	Yes
Ribosomes	No	Yes
Inclusion bodies	No	No
Endospores	Yes	No

Table 9.2 presents an overview of the clinical significance of all the bacterial cell structures described in this chapter.

Keep in Mind

- There is no nucleus in bacteria: the DNA is found freely floating in the cytoplasm in an area referred to as the nuclear region.

- Bacteria can contain extrachromosomal DNA segments known as plasmids.

- Plasmids are very important in clinical disease because they can carry genes for antibiotic resistance as well as genes for the production of toxins.

- Ribosomes of bacteria are different from those found in eukaryotic cells and are consequently a selective target for antibiotic therapy.

- Endospore formation can protect bacteria from environmental pressures and are also very resistant to antibiotics and disinfectants.

- After environmental pressures subside, endospores can undergo germination into the original bacteria.

- Endospore formation makes pathogens extremely difficult to deal with.

SUMMARY

- Many of the structures associated with the bacterial cell wall are involved in the infection process.

- The cell wall is a meshwork made up of layers of peptidoglycan; there are many layers in Gram-positive organisms and few in Gram-negative organisms.

- The cell wall is a primary target for attack by antibiotics.

- Some Gram-positive bacteria incorporate M protein or mycolic acids in the cell wall, which are virulence factors.

- Gram-negative cells contain an outer layer made up of lipoproteins, lipopolysaccharides, and phospholipids that protect against antibiotics and function as an endotoxin.

- The five structures found outside the cell wall—capsules, fimbriae, flagella, axial filaments, and pili—are involved in the infection process.

- The plasma membrane surrounds the cytoplasm and is the place where the replication of DNA and the production of ATP take place. It is also a target of antibiotics.

- Bacteria have no nucleus, and the DNA floats freely in the cytoplasm.

- Plasmids are extrachromosomal structures made of DNA that contain the genes for toxins and antibiotic resistance. These plasmids can be transferred from one bacterial cell to another.

- Endospore formation protects bacteria from environmental pressure and also from antibiotics and disinfectants. This structure can have a major role in clinical settings.

In this chapter, we have looked at the anatomy of bacterial cells. It is important to keep in mind what we have learned as we continue our discussions, in particular how these structures are involved in the infectious process. When we look at bacterial genetics, antibiotics, and resistance, these anatomical structures will be revisited.

 SELF EVALUATION AND CHAPTER CONFIDENCE

Multiple Choice

Answers are given in the back of the book and help can be found in the student resources at:

www.garlandscience.com/micro2

1. Which of the following is an attribute of the bacterial cell wall?

    A. It stops growing during bacterial elongation
    B. It protects the cell from the environment
    C. It is used for DNA replication
    D. It is used for mitochondrial attachment
    E. It is the same in Gram-positive and Gram-negative bacteria

2. The primary substance making up the bacterial cell wall is

    A. Phospholipid
    B. Carbohydrate
    C. Glycerides
    D. Protein
    E. Peptidoglycan

3. List the phases of peptidoglycan biosynthesis in the order in which they occur.

    A. Membrane-associated phase, cytoplasmic phase, flagellar phase
    B. Membrane-associated phase, extracytoplasmic phase, lipid phase
    C. Membrane-associated phase, peptidoglycan phase, lipid phase
    D. Membrane-associated phase, cytoplasmic phase, extracytoplasmic phase
    E. None of the above

4. Peptidoglycan is made up of

    A. *N*-acetylglucosamine only
    B. Repeating *N*-acetylglucosamine molecules
    C. *N*-acetylmuramic acid only
    D. Repeating *N*-acetylmuramic acid molecules
    E. Repeating *N*-acetylglucosamine and *N*-acetylmuramic acid molecules

For questions **5–10**, match the property to Gram-negative or Gram-positive bacteria. If the property describes Gram-positive bacteria, mark it as A; if it describes Gram-negative organisms, mark it as B.

5.	Thick peptidoglycan layer
6.	Outer membrane
7.	Contains teichoic acid
8.	Source of endotoxins
9.	Contains lipopolysaccharides
10.	Produces exotoxin

11. M protein

    A. Has no clinical significance
    B. Is involved in sexually transmitted diseases
    C. Is a virulence factor in streptococcal infections
    D. Is produced by Gram-negative organisms
    E. Is the primary cause of bacterially induced headaches

12. *Mycoplasma* are organisms that have

    A. A unique cell wall
    B. A cell wall similar to that found in Gram-negative bacteria
    C. No cell wall
    D. Three layers of cell wall
    E. A cell wall during part of its life cycle as it grows

13. Which of the following is not true of mycolic acid?

    A. It can be found in the cell wall of acid-fast bacteria
    B. It is a waxy substance
    C. It is a minor cell wall component
    D. It protects bacteria from antibiotics
    E. It protects bacteria from the host immune system

14. The properties of porin proteins found in Gram-negative (and some Gram-positive) cells include all of the following except that they

    A. Become larger during attack with antibiotics
    B. Control the kinds of molecules that may pass onto the cell
    C. Create channels that vary in size and specificity
    D. Make bacteria resistant to some antibiotics
    E. Are molecules whose mutation can lead to antibiotic resistance

15. Lipid A is

    A. Part of the Gram-positive cell wall
    B. Part of the exotoxins of Gram-positive bacteria
    C. Part of the endotoxins of Gram-negative bacteria
    D. Part of the plasma membrane of all bacteria
    E. None of the above

16. O polysaccharides are

    A. Found on all bacteria
    B. Not part of the LPS found in Gram-positive cells
    C. Used to identify certain bacteria
    D. Found only on Gram-positive cells
    E. Found in the plasma membrane of all bacteria

For questions **17–20** match the cell part with the disease for which it is required.

Cell part	Disease
17. Capsules	A. Streptococcal infection
18. Fimbriae	B. Syphilis
19. Flagella	C. Gonorrhea
20. Axial filaments	D. *Helicobacter* infection in stomach

21. Which of the following is a target for attack by antibiotics?

    A. Fimbriae
    B. Pili

C. Axial filaments
D. Flagella
E. Plasma membrane

22. Which of the following **is not** known to have clinical relevance?

    A. DNA
    B. RNA
    C. Axial filaments
    D. Bacterial inclusion bodies
    E. Ribosomes

23. Flagella located around an entire bacterial cell are called

    A. Circumferential flagella
    B. Amphitrichous flagella
    C. Peritrichous flagella
    D. Monotrichous flagella
    E. Lophotrichous flagella

24. The plasma membrane is composed of

    A. A single phospholipid layer
    B. A phospholipid bilayer
    C. Only lipids
    D. Lipid A
    E. None of the above

25. Integral proteins

    A. Contain no lipid-soluble regions
    B. Go all the way through the membrane
    C. Connect the plasma membrane to the cell wall
    D. Are found only in the nuclear membrane of bacteria
    E. None of the above

26. Membrane transport can occur by all of the following processes except

    A. Passive transport
    B. Facilitated diffusion
    C. Active transport
    D. Plasmolysis
    E. Osmosis

27. Simple diffusion involves all the following except

    A. Hypotonicity
    B. Hypertonicity
    C. A concentration gradient
    D. Active transport
    E. A partly permeable membrane

28. Group translocation involves

    A. ATP
    B. PEP
    C. ADP
    D. None of the above

29. Plasmids are

   A. Specialized parts of the chromosome
   B. Parts of bacterial mitochondria
   C. Found in the cell wall
   D. Extrachromosomal pieces of RNA
   E. Extrachromosomal pieces of DNA

30. Bacterial ribosomes

   A. Are identical to those found in eukaryotes
   B. Are involved in protein synthesis
   C. Are enclosed by their own individual membrane
   D. Made up of three subunits
   E. Made up of four subunits

31. Bacterial ribosomes are clinically important because they

   A. Are directly involved in causing infection
   B. Produce plasmids
   C. Are targets for antibiotic therapy
   D. Are resistant to antibiotics
   E. Are part of endotoxins

32. Bacterial endospores are

   A. Resistant to heat
   B. Sensitive to heat
   C. Part of the division process seen during bacterial growth
   D. Targets of antibiotics
   E. None of the above

 ## DEPTH OF UNDERSTANDING

Questions listed here require you to bring together the concepts you have learned in this chapter into a discussion format. This helps you to increase your depth of understanding of the material you have learned. Help can be found in the student resources at: www.garlandscience.com/micro2

1. Discuss the clinical relevance of the flagella and the glycocalyx.

2. Compare and contrast the Gram-positive and Gram-negative cell wall.

3. Describe the problems associated with endospore formation.

 ## CLINICAL CORNER

Help can be found in the student resources at: www.garlandscience.com/micro2

1. A nosocomial infection has broken out in the hospital. The organism responsible is a Gram-positive spore-forming bacillus. You are asked to devise a strategy to deal with this problem.

   A. What is the most important consideration?
   B. Should you recommend changes in hospital disinfection procedures?
   C. Should you consider any recommendations about patient therapy protocols?

2. Your patient is suffering from an intestinal infection. After several days on broad-spectrum antibiotics without improvement, the patient was admitted to the hospital. A routine lab workup reveals that the pathogen responsible for the infection is a Gram-negative bacillus.

   A. What other characteristics of the organism would be useful for you to know about?
   B. What other information could be useful for developing a plan for the treatment of the patient?

Bacterial Growth

Chapter 10

Why Is This Important?

One of the requirements for a successful bacterial infection is an increase in the number of bacteria in the infected host. Also, to study bacteria it is often necessary to maintain them in the laboratory, and that requires knowledge of the factors that influence the growth of bacteria.

Paul was admitted to the hospital after suffering from severe diarrhea, vomiting, abdominal pain, and dehydration for almost two days at home. Before that, he had spent the weekend camping in the mountains and drank unfiltered water from a pond. His symptoms suggested a severe gastrointestinal infection, but because there was no blood in his stool sample and he was vomiting, the doctors ruled out dysentery and suspected that a fecal coliform might be responsible. Fecal coliforms are Gram-negative, rod-shaped, facultatively anaerobic bacteria (described in this chapter) associated with fecal contamination. These bacteria, including *Escherichia*

coli, ferment lactose and do not have endospores. To identify the pathogen, a medical technologist at the hospital obtained a fecal sample from Paul, used it to inoculate a Petri plate of MacConkey agar, and incubated it overnight. The next morning, bright pink colonies were observed on the agar and the confirmation of fecal coliform contamination was made.

As we will see in this chapter, MacConkey agar is a selective and differential culture medium that can help quickly discern coliform bacteria from other bacteria. Coliforms ferment the lactose in the medium and release acid; the acid causes a pH indicator in the

medium to turn red. Only Gram-negative bacteria can grow on MacConkey agar because Gram-positive-inhibiting bile salts are added to it. The pink colonies were thus clear evidence of fecal coliform bacteria. Paul was given oral rehydration therapy, in which he drank lots of fluids containing electrolytes. Antidiarrheal drugs were not prescribed because it would have delayed Paul's recovery (his diarrhea was actually helping eliminate the bacteria). By the next day he had improved dramatically and was released from the hospital. He learned a tough lesson: bring your own water or use a water filter when camping!

OVERVIEW

In this chapter, we examine the characteristics of bacterial growth. This information is important because infectious organisms have specific growth requirements, and understanding the requirements for growth will help us to understand the infectious process better. Recall that one of the requirements for a successful infection is being able to defeat host defenses. One way to accomplish this is to overwhelm it numerically. As we examine bacterial growth, you will also learn about specific techniques used by clinical microbiologists that have a key role in identifying and diagnosing bacterial diseases.

We will divide our discussions into the following topics:

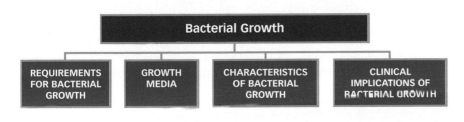

"... it can be shown that in a single day, one cell of E. coli *could produce a super-colony equal in size and weight to the entire planet Earth."*

Michael Crichton (1942–2008)

For decades, mathematicians have applied various models to test Crichton's quote (above) from his best-selling book *The Andromeda Strain*. If there were unlimited nutrients and enough room to grow, and if toxic waste products did not accumulate, bacterial growth would indeed be staggering. Under more realistic conditions, however, their growth is significantly reduced. In Chapter 4 we saw that bacteria have specific sizes, shapes, and staining reactions, and we have learned that most bacteria use the process of binary fission for growth. In Chapter 9 we examined the structures of bacterial cells and the functions associated with these structures. In addition, the structure–function relationships seen in membrane transport proteins showed how bacteria can import the nutrients and other components needed for growth. We will build on these concepts as we explore bacterial growth in this chapter.

REQUIREMENTS FOR BACTERIAL GROWTH

How well bacteria grow depends on the environment in which the organisms live, and we can divide growth requirements into two major categories: physical and chemical.

Physical Requirements for Growth

The physical requirements for growth include temperature, pH, and osmotic pressure. Each of these factors can markedly, and in some cases lethally, affect bacteria.

Temperature

One way of classifying bacteria is to separate them according to the temperature ranges at which they grow best (**Figure 10.1**).

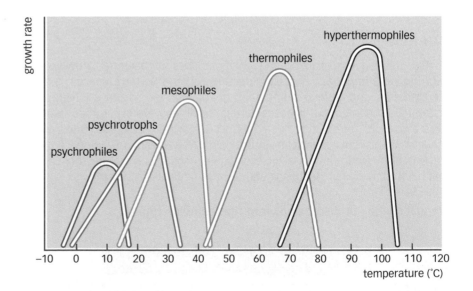

Figure 10.1 Bacteria can be classified according to temperature ranges at which they grow best. Psychrotrophs are considered a subset of the major group psychrophiles, and hyperthermophiles are considered a subset of the major group thermophiles. Note that there is a range of temperatures for each group and that in some cases neighboring ranges overlap. Most of the bacteria that are harmful to humans are in the mesophilic group.

Bacteria known as **psychrophiles** grow at cold temperatures (the prefix psychro- comes from the Greek word meaning cold). There are two subsets of psychrophiles. One subset, which has no special name, is made up of psychrophiles that grow at temperatures from 0 to 15°C and are found in the deep oceans and in arctic environments. The other subset, referred to as **psychrotrophs**, is made up of psychrophiles that grow best between 20 and 30°C.

Mesophiles are bacteria that grow best at moderate temperatures, in the range 25–40°C. Most common types of bacteria, including human pathogens, are mesophiles.

Thermophiles are bacteria that grow only at temperatures above 45°C. As with psychrophilic bacteria, there are two subsets of thermophiles: those that grow between 45 and 60°C and those that grow above 80°C. Members of the latter group are called either **extreme thermophiles** or hyperthermophiles. No thermophilic bacteria are pathogenic to humans; to those bacteria, the temperature of the human body is like a refrigerator!

The **minimum growth temperature** is the lowest temperature at which an organism grows. For example, the minimum growth temperature for a thermophile is 45°C. If you place a thermophilic bacterium in an environment below this temperature, there will be no growth. The **maximum growth temperature** is the highest temperature at which growth occurs, and the **optimal growth temperature** is the temperature at which the highest rate of growth occurs.

Table 10.1 lists the optimal growth temperatures for a variety of bacteria. Note that some pathogens such as *Mycobacterium tuberculosis* have an exceptionally long **generation time** (the time interval between cell divisions). Their slow growth makes organisms of this type difficult to culture. In contrast, pathogens such as *Staphylococcus aureus* as well as opportunistic pathogens such as *Escherichia coli* have short generation times. Keep in mind that growth rates are important because the number of pathogens can have a direct effect on the disease process.

Fast Fact Psychrophiles are not pathogenic to humans. However, many of these organisms can cause spoilage of refrigerated food, which can result in illness.

Fast Fact For most human pathogens, the optimal growth temperature is approximately 37°C (normal body temperature).

Microorganism	Disease	Optimal growth temperature (°C)	Generation time (hours)
Escherichia coli	Diarrheal diseases, opportunistic infections, and urinary tract infections	40	0.35
Staphylococcus aureus	Skin, respiratory, and other infections	37	0.47
Pseudomonas aeruginosa	Nosocomial infections	37	0.58
Clostridium botulinum	Botulism	37	0.58
Mycobacterium tuberculosis	Tuberculosis	37	12.0

Table 10.1 Comparison of optimal growth temperatures and generation times for several pathogenic bacteria.

Figure 10.2 Indicators of syphilis. The appearance of chancres on the penis (panel a) or on the external female genitalia (panel b) are indicators of syphilis, the infection caused by the bacterium *Treponema pallidum*. Although this microorganism is mesophilic, it initially grows best on the external genitalia, where the temperature is slightly cooler than within the body.

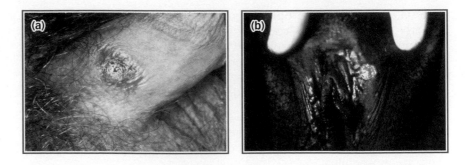

To see how temperature can affect growth, we have to think about protein structure (discussed in Chapter 2). Remember that proteins are integral parts of the bacterial cell and can function as structural components, as membrane transport mechanisms, and, perhaps most importantly, as enzymes. Because the three-dimensional structure of a protein is the key to its function, any change in this structure can have drastic effects on function. Remember also that the three-dimensionality of the protein is due to hydrogen bonds that form during the development of secondary and tertiary structure. Because they are weak, these hydrogen bonds are easily broken as the temperature rises. Consequently, exposure to high temperatures can destroy the three-dimensional structure of a protein, thereby destroying its ability to function in the bacterial cell.

An interesting observation is that variable temperature requirements are seen in certain diseases. For example, *Treponema pallidum*, the organism that causes syphilis, prefers temperatures that are on the low side of normal body temperature. Therefore, lesions of this disease are initially seen on the lips, tongue, and genitalia, where the temperature of the body is slightly cooler than 37°C (**Figure 10.2**). This is also true for leprosy, which is caused by the organism *Mycobacterium leprae* and is initially manifested on the extremities of the body, in particular the face, ears, hands, feet, and fingers (**Figure 10.3**).

pH

Bacteria grow in a wide range of pH values. In Chapter 2 we saw that pH represents a measurement of the acidity or alkalinity of an environment, and it ranges from 1.0 (very acidic) to 14.0 (very alkaline). Although most bacteria prefer the neutral pH of 7.0, some have adapted so that they are able to grow in environments in which the pH is far from 7.0. For instance, **acidophiles** are bacteria that grow at extremely low pH values.

Helicobacter pylori is found in the stomach, where the pH can be as low as 1.0. As we saw in Chapter 5, *H. pylori* is known to be a cause of stomach ulcers and survives for long periods in this low-pH environment by moving into the heavy mucous layer that coats and protects the digestive epithelium of the stomach (**Figure 10.4**). *H. pylori*, like many bacteria, also produces chemical buffers to protect against pH changes in the environment.

The mechanisms by which changes in pH affect bacteria are similar to those we discussed for temperature. Just as hydrogen bonds are disrupted by elevated temperatures, increased quantities of free hydrogen ions (H^+) in low-pH environments or of free hydroxyl ions (OH^-) in high-pH environments also disrupt the hydrogen bonding that gives proteins their three-dimensional structure. As we have emphasized, changes in

Fast Fact Although many bacteria can deal with acidic conditions fairly well, alkaline environments inhibit the growth of most of them. For this reason, disinfectants usually contain either ammonia or ammonia-like compounds because these substances are alkaline.

protein structure lead to changes in protein function, and these changes can be lethal for bacteria.

Osmotic Pressure

Osmotic pressure is the pressure exerted on bacteria by the water inside and outside the cells, and this can affect bacterial growth. Not only do bacteria get nutrients in aqueous solution, but the bacterial cytoplasm is also 80% water.

Osmotic pressure can be used to inhibit bacterial growth. For example, food can be preserved with high salt concentrations, which create a hypertonic environment, causing organisms that might spoil the food to undergo plasmolysis. However, even this method of food preservation can be overcome by bacteria classified as **halophiles**. (Solutes are frequently ions of salts, halo- is a prefix meaning salt, and so halophiles are bacteria that prefer to be in a high-salt, or high-solute, environment.) Halophilic organisms can be divided into three categories: obligate, facultative, and extreme. **Obligate halophiles** require high salt concentrations for growth, **facultative halophiles** can live with or without high salt concentrations, and **extreme halophiles** can grow in the presence of very high levels of salt.

In summary, the main physical factors that affect bacterial growth—temperature, pH, and osmotic pressure—are relatively clear-cut. We can also consider these physical parameters in the context of pathogenic bacteria and infection. The normal human body temperature falls within the range required by pathogenic bacteria that are mesophilic. Although most bacteria grow best in a pH range from 6.5 to 7.5, acidophiles can survive even in the highly acidic environment of the stomach (which can have a pH of less than 2) and some bacteria have developed mechanisms to cope with such an extreme environment. As far as osmotic pressure is concerned, remember that the same requirements for osmotic pressure exist for the cells of our bodies. Consequently, pathogens rarely have to deal with hypertonic or hypotonic environments during the infectious process. When they exit from the body and enter a different environment, however, pathogens must adapt rapidly to changes in osmotic pressure.

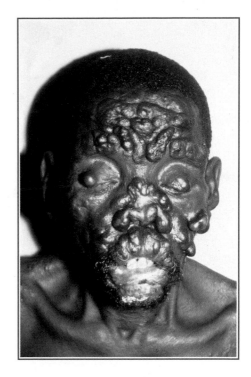

Figure 10.3 Leprosy. *Mycobacterium leprae*, the etiological agent of leprosy, is a mesophilic organism that prefers to grow at the lower end of this temperature range. Consequently, it is seen on the ears, lips, and nose.

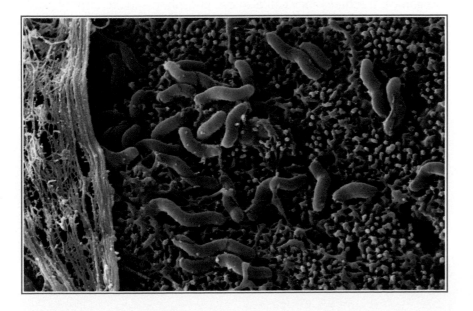

Figure 10.4 A color-enhanced scanning electron micrograph of *Helicobacter pylori*, the cause of gastric ulcers, growing on the surface of the human intestine. This organism is not an acidophile but can withstand very low pH. In the stomach, *H. pylori* survives by burrowing into the mucus layer that protects the cells of the stomach.

Chemical Requirements for Growth

When we look at the chemical requirements for bacterial growth, we see that they are almost as variable as bacterial species themselves. Table 10.2 summarizes several of the core chemicals required for bacterial growth. Let's look at some of them in detail.

Carbon

The biosphere of the planet Earth is carbon-based, which means that biological molecules all contain carbon. Therefore all living things require carbon atoms for the construction of biological molecules (carbohydrates, lipids, proteins, and nucleic acids; Chapter 3). In fact, except for water, carbon is the most important chemical requirement for bacterial growth.

Nitrogen

Many of the cellular tasks carried out by bacteria, as well as some of the molecules that make up bacterial cells, require nitrogen. This element is an integral part of the structure of amino acids and so is involved in protein synthesis. Nitrogen is also part of the structure of DNA and RNA molecules.

Bacteria obtain nitrogen in a variety of ways, including the decomposition of existing proteins and from the ammonium ions (NH_4^+) found in organic materials. Nitrogen can also be obtained from atmospheric nitrogen through nitrogen fixation.

Sulfur

Bacteria must have sulfur to make certain amino acids and vitamins. This element is acquired primarily by decomposing existing proteins in which one or more of the component amino acids contain sulfur. Sulfur can also be procured in the sulfate ion form (SO_4^{2-}) and from H_2S, which is a by-product of many biochemical reactions carried out by bacteria.

Phosphorus

The element phosphorus is essential for the synthesis of nucleic acids, AMP, ADP, and ATP molecules (it's the P in these compounds). It is also a major chemical requirement for the development of the plasma membrane (remember the phospholipid bilayer).

Bacteria obtain the phosphorus they need by cleaving ATP to ADP plus inorganic phosphate. They can also use phosphorus in the form of phosphate ions (PO_4^{3-}).

Chemical	Function
Carbon, oxygen, hydrogen	Major components of all organic molecules (proteins, carbohydrates, lipids, nucleic acids)
Nitrogen	Required for making bacterial amino acids and nucleic acids
Sulfur	Required for making some bacterial amino acids
Phosphorus	Required for making bacterial nucleic acids, membrane phospholipid bilayer, and ATP
Potassium, magnesium, calcium	Required for maintaining cell osmolarity and the functioning of certain bacterial enzymes
Iron	Required for bacterial metabolism

Table 10.2 Chemical requirements for bacterial growth.

Growth factor	Function
Amino acids	Components of bacterial proteins
Heme	Functions in electron transport system in bacteria
NADH	Electron carrier
Niacin (nicotinic acid)	Precursor of bacterial NAD^+ and $NADP^+$
Pantothenic acid	Component of bacterial coenzyme A
para-Aminobenzoic acid	Precursor of folic acid, which is involved in the bacterial metabolism of carbon compounds and in nucleic acid synthesis
Purines, pyrimidines	Components of bacterial nucleic acids
Pyridoxine (vitamin B_6)	Used in the synthesis of bacterial amino acids
Riboflavin (vitamin B_2)	Precursor of bacterial FAD
Thiamine (vitamin B_1)	Used in some bacterial decarboxylation reactions

Table 10.3 Growth factors and their functions.

Organic Growth Factors

In addition to the major chemical requirements we have discussed, bacteria also use organic growth factors, such as certain vitamins. Some bacteria cannot synthesize these growth factors, so they must be obtained from the environment.

In addition, bacteria also require access to the elements potassium, magnesium, and calcium, which function as enzyme cofactors, and such trace elements as iron, copper, molybdenum, and zinc.

The functions of several growth factors are listed in Table 10.3.

Oxygen

Oxygen would seem to be one of the most important chemical requirements for bacterial growth. However, many bacteria not only do not require oxygen for growth, they actually die in the presence of oxygen. Although we know from Chapter 3 that using oxygen as a final electron acceptor in aerobic metabolism maximizes the yield of ATP, oxygen also has a dark side because during normal bacterial respiration, a small number of oxygen molecules can assume a superoxide free-radical form (O_2^-). These free radicals are very unstable and steal electrons from other molecules, a process that causes the electron-deficient molecules to steal electrons from other molecules. This cascading free radicalization of molecules eventually leads to the death of the bacterial cell.

Two types of bacteria can grow in the presence of oxygen: **aerobes**, which are bacteria that require oxygen for growth, and **facultative anaerobes**, which can grow in the presence or absence of oxygen (see Chapter 3). Both aerobic bacteria and facultative anaerobic bacteria produce an enzyme called **superoxide dismutase (SOD)** that can convert free-radical oxygen to molecular oxygen and peroxide:

$$O_2^- + O_2^- + 2H^+ \xrightarrow{\text{SOD}} H_2O_2 + O_2$$

Unfortunately, the product H_2O_2 (hydrogen peroxide) contains the peroxide anion O_2^{2-}, which is just as toxic to the bacteria as free-radical

Figure 10.5 Gas gangrene in the left foot. *Clostridium perfringens*, an obligate anaerobe, causes this highly destructive tissue infection. Exposure of this bacterium to oxygen is a lethal event for the bacterium.

Fast Fact Treatment options for patients with obligate anaerobe infections, such as *Clostridium perfringens*, include subjecting the patient to a high oxygen atmosphere, for example inside a hyperbaric chamber.

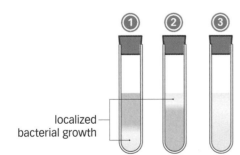

localized bacterial growth

Figure 10.6 Sodium thioglycolate tubes are used for the incubation of bacteria with specific oxygen requirements. Tube 1 shows the growth of an obligate anaerobe; note the absence of growth in the top portion of the broth where oxygen is present. Tube 2 shows the growth of an obligate aerobe; note that the growth is only in the top portion of the tube where oxygen is present. Tube 3 shows the growth of a facultative anaerobe; note the uneven distribution of growth from top to bottom (more growth at the top).

oxygen. (This toxic peroxide anion is the active component in antimicrobial agents such as hydrogen peroxide and benzoyl peroxide.) To avoid this toxic anion, bacteria have developed two enzymes to neutralize it. The enzyme **catalase** converts hydrogen peroxide to water and oxygen:

$$2H_2O_2 \xrightarrow{\text{catalase}} 2H_2O + O_2$$

The enzyme **peroxidase** converts hydrogen peroxide not to water plus molecular oxygen, but rather to just water:

$$H_2O_2 + 2H^+ \xrightarrow{\text{peroxidase}} 2H_2O$$

We divide bacteria into three major groups based on oxygen use: **obligate aerobes**, which require oxygen for growth, **obligate anaerobes**, which cannot survive in the presence of oxygen, and the facultative anaerobes mentioned earlier, which grow either with or without oxygen. Remember that because ATP production is greater under aerobic conditions, facultative anaerobes grow better in the presence of oxygen. Many of the bacteria that, like *E. coli*, make up our normal flora as well as many pathogenic bacteria are facultative anaerobes.

There are two smaller groups of bacteria that are classified according to oxygen use. **Aerotolerant bacteria** can grow in the presence of oxygen but do not use it in metabolism, and **microaerophiles** are aerobic bacteria but require only low levels of oxygen for growth.

Although many bacteria are facultative anaerobes, several are obligate anaerobes. For example, *Clostridium perfringens* is an obligate anaerobe that causes **gas gangrene**, a condition in which there is significant tissue destruction (**Figure 10.5**). Because this organism is an obligate anaerobe, one of the most effective methods of treatment is the removal of infected and dead tissue (debridement), exposing any remaining *C. perfringens* to oxygen in the air, which is lethal to the bacteria.

Growth of Anaerobic Organisms

Special growth media and incubation are required for anaerobic bacteria to grow. Considering these requirements is especially important when working with anaerobic bacteria that could easily be misidentified if not cultured properly.

Two methods are used for culturing anaerobic bacteria. The first uses a medium called **sodium thioglycolate**, which binds to oxygen. This medium forms an oxygen gradient such that the farther into the medium we go, the less oxygen there is. When bacteria with different oxygen requirements are compared using this medium (**Figure 10.6**), obligate anaerobes will always grow in the area of the tube where oxygen is absent, whereas obligate aerobes grow only where the oxygen concentration is highest. Facultative anaerobes, which can grow either with or without oxygen, grow throughout the sodium thioglycolate medium even though the oxygen concentration decreases steadily from top to bottom of the tube. In these bacteria, more growth occurs at the top of the culture medium, where there is more oxygen present.

The second method for culturing anaerobic organisms uses what is called a **GasPak™ jar**. This incubation container provides an oxygen-free environment that is developed through a series of chemical reactions, as shown in **Figure 10.7**. Because the jar is totally devoid of oxygen, only obligate and facultative anaerobic organisms can grow in it.

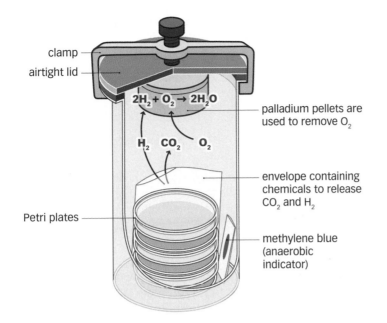

clamp
airtight lid

$$2H_2 + O_2 \rightarrow 2H_2O$$

palladium pellets are used to remove O_2

H_2 CO_2 O_2

Petri plates

envelope containing chemicals to release CO_2 and H_2

methylene blue (anaerobic indicator)

Figure 10.7 The GasPak™ jar is used for incubating anaerobic organisms. The oxygen-free environment results from the use of palladium pellets that catalyze the removal of oxygen through the production of water (as shown in the equation written just under the jar lid). An indicator of the presence of oxygen is included to ensure that the environment is oxygen-free during incubation.

Fast Fact   Before the invention of GasPak™ jars, microbiologists used candle jars, in which a lit candle was placed. The flame consumed most of the oxygen inside the jar before it went out, leaving a low-oxygen environment inside the airtight jar.

Keep in mind that organisms that do not use oxygen make less ATP, which causes them to grow more slowly. Incubation times for anaerobes must therefore be longer than those for aerobes. Abbreviated incubation times could cause anaerobic bacteria to be missed, leading to mistakes in diagnosis and treatment.

Table 10.4 summarizes the temperature, pH, and oxygen requirements of bacteria.

Table 10.4 Bacterial requirements for temperature, pH, and oxygen.

Requirement and bacterial type	Description
Temperature	
Psychrophile	Grows well at 0°C and optimally at 15°C or lower
Psychrotroph	Can grow at 0–7°C but optimally between 20 and 30°C; maximum 35°C
Mesophile	Optimal growth between 25 and 45°C
Thermophile	High temperatures between 45 and 65°C
Hyperthermophile	Optimal growth between 80 and 130°C
pH	
Acidophile	Can grow between pH 0 and 5.5
Neutrophile	Growth between pH 5.5 and 8.0
Alkalophile	Optimal growth at pH 8.5 and 11.5
Oxygen	
Obligate aerobe	Dependent on the presence of oxygen
Facultative anaerobe	Does not require oxygen but grows better in its presence
Obligate anaerobe	Dies in the presence of oxygen
Microaerophile	Requires oxygen levels below 2–10%
Aerotolerant anaerobe	Grows the same with or without oxygen

| Keep in Mind |

- Bacterial growth has an effect on disease.

- Successful bacterial growth depends on the physical and chemical environment.

- Physical requirements include temperature, pH, and osmotic pressure.

- Chemical requirements include carbon, nitrogen, sulfur, phosphorus, organic growth factors, and trace elements.

- Bacteria have different requirements for oxygen: obligate aerobes require oxygen to grow, whereas obligate anaerobes are killed in the presence of oxygen and facultative anaerobes can grow with or without oxygen.

GROWTH MEDIA

As noted above, bacteria need a variety of growth factors. Any bacterium that requires not one or two but rather a large number of these growth factors is said to be a **fastidious bacterium**. Fastidious bacteria are usually slow growing, and to grow them in the laboratory, the growth medium must provide all of the essential growth factors. Also, laboratory conditions never exactly reproduce the conditions found in nature, and so you must keep in mind that any laboratory results may not completely mimic what is seen in clinical disease.

There are two general types of medium used to grow bacteria: complex and chemically defined.

Complex Media

Complex media contain numerous ingredients of known chemical composition as well as digested proteins and extracts derived from plants or meat. Such media are referred to as complex because the exact chemical composition of these digests and extracts is not known. Table 10.5 lists a recipe for a complex medium. Don't be puzzled by the fact that the table shows an exact amount for each ingredient. Even though the manufacturer knows the amount of each extract and digest added to the medium, the chemical composition of each is not known. (Contrast this with the chemically defined medium of Table 10.5, where each ingredient is pure and thus its chemical composition is known.)

In complex media, the energy, carbon, sulfur, and nitrogen required for bacterial growth are all provided by proteins. These proteins are digested into fragments referred to as peptones. These fragments are more easily used than the intact protein molecules, which are relatively insoluble in water. In addition, the meat or plant extracts in complex media provide many of the vitamins required for optimal growth. A complex medium is usually referred to as a **nutrient medium** and is available both as a liquid (**nutrient broth**) and as a solid with agar added (**nutrient agar**). **Agar** is a polysaccharide found in marine algae. It is used as a thickener or solidifier in the manufacture of jams, jellies, and—of most interest to us—laboratory growth media.

Chemically Defined Media

A **chemically defined growth medium** is one for which the chemical composition is precisely known. This is critical when doing research, where the consistent duplication of culture conditions is necessary. An example is shown in Table 10.6. Because all the ingredients in chemically

Table 10.5 Recipe for a complex medium.

Ingredient	Quantity
Water	1 liter
Peptone	5 g
Beef extract	3 g
Sodium chloride	8 g
Agar	15 g

defined media are precisely known, these media can be used for the laboratory analysis of compounds produced by specific bacteria.

Ingredient	Quantity
Carbon and energy sources	
Glucose	9.1 g
Starch	9.1 g
Sodium acetate	1.8 g
Sodium citrate	1.4 g
Oxaloacetate	0.3 g
Salts	
Potassium phosphate dibasic	12.7 g
Sodium chloride	6.4 g
Potassium phosphate monobasic	5.5 g
Sodium bicarbonate	1.2 g
Potassium sulfate	1.1 g
Sodium sulfate	0.9 g
Magnesium chloride	0.5 g
Ammonium chloride	0.4 g
Potassium chloride	0.4 g
Calcium chloride	0.4 g
Ferric nitrate	0.006 g
Amino acids	
Cysteine	1.5 g
Arginine and proline (each)	0.3 g
Glutamic acid and methionine (each)	0.2 g
Asparagine, isoleucine, and serine (each)	0.2 g
Cystine	0.06 g
Organic growth factors	
Calcium pantothenate	0.02 g
Thiamine	0.02 g
Nicotinamide adenine dinucleotide	0.01 g
Uracil	0.006 g
Biotin	0.005 g
Hypoxanthine	0.003 g
Reducing agent	
Sodium thioglycolate	0.00003 g
Water	1.0 liter

Table 10.6 An example of a chemically defined medium.

Media as a Tool for Identifying Pathogens

In an ill patient, pathogens must be identified before a medical diagnosis can be made and effective treatment options can be explored. As we saw in Chapter 6, there are several ways in which media can be used to help establish the presence of pathogenic bacteria and then identify them. These methods all use selective and differential media. A **selective medium** is one that contains ingredients that prohibit the growth of some organisms while fostering the growth of others. A **differential medium** is one that contains ingredients that can allow microbiologists to discern visually between types of bacterial colonies (Table 10.7).

It is informative to look at several types of medium that have a role in identifying pathogenic bacteria. The selective medium bismuth sulfate agar selects for the growth of the bacterium *Salmonella enterica* serovar Typhi, which causes typhoid fever. This disease is a serious epidemic infection associated with poor sanitation and the fecal–oral route of infection. Although normally rare in the United States, this pathogen has been found in contaminated shellfish harvested off the coast of the United States. The bismuth sulfate in this nutrient agar selects for the growth of *S. enterica* serovar Typhi by inhibiting the growth of all other bacteria. Consequently, fecal samples from patients who present with symptoms consistent with typhoid fever can be inoculated onto this agar; if growth occurs, a preliminary diagnosis can be formulated.

The selective medium brilliant green uses a dye to inhibit the growth of most microorganisms and selects for the genus *Salmonella*. Another selective medium, Sabouraud's agar, takes advantage of a low pH (5.6) to select for the growth of fungi while inhibiting the growth of bacteria. This type of medium has become increasingly important because the number of fungal infections has risen as a result of AIDS.

Table 10.7 Examples of differential media.

Medium	Use	Interpretation of results
MacConkey agar	Isolation, culture, and differentiation of Gram-negative bacteria	Lactose fermenters form red to pink colonies; nonfermenters form colorless or transparent colonies
Eosin methylene blue (EMB) agar	Isolation, culture, and differentiation of Gram-negative bacteria	Lactose fermenting bacteria form green metallic sheen; nonfermenting bacteria form colorless or light purple colonies
Triple sugar iron agar	Differentiation of Gram-negative bacteria on the basis of their fermentation of glucose, sucrose, and lactose and on their production of H_2S gas	Red slant/red butt, no fermentation; yellow slant/red butt, glucose fermentation; yellow slant/yellow butt, glucose and lactose fermentation; butt turns black, H_2S produced
Blood agar	Culture of fastidious bacteria and differentiation of hemolytic bacteria	Partial digestion of blood, alpha hemolysis; complete digestion of blood, beta hemolysis; no digestion of blood, gamma hemolysis

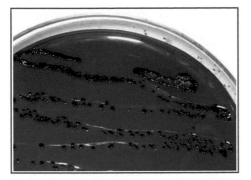

Figure 10.8 (*left*) Differential medium. An eosin methylene blue (EMB) plate inoculated with *E. coli*, showing the characteristic green sheen that results from the vigorous fermentation of either lactose or sucrose.

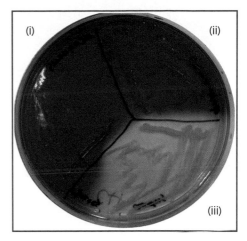

Figure 10.9 An example of selective and differential media. Mannitol salt agar culture plate inoculated with (i) *Micrococcus* spp and (ii) *Staphyloccocus epidermidis* showing growth without mannitol fermentation (no yellow zone), and (iii) *Staphyloccocus aureus* showing growth with fermentation of mannitol (yellow zone). This mediums high salt concentration (7.5%) makes it selective for Gram-positive *Staphyloccoci* and *Micrococcaceae*, and the mannitol sugar makes it differential between species of *Staphylococcus*, with mannitol-fermenting species producing a yellow zone around the growth.

Many selective media are also differential. These media use specific ingredients to differentiate between organisms. As an example, *E. coli*, which can be an opportunistic pathogen, is an enteric Gram-negative rod that is difficult to distinguish from other enteric Gram-negative rods. However, when grown on **eosin methylene blue** (**EMB**) agar, the colonies of *E. coli* exhibit a bright and easily identifiable green sheen created as the *E. coli* ferment lactose (**Figure 10.8**). Because other enteric bacteria do not ferment this sugar in the same way, the growth of *E. coli* is easily distinguishable from that of other enteric organisms that could grow on this medium. From a clinical perspective, because many urinary tract infections are caused by *E. coli*, EMB can be used to preliminarily identify *E. coli* as a source of the infection. (EMB is also a selective medium because the eosin and methylene blue dyes inhibit the growth of Gram-positive organisms.)

Mannitol salt agar (MSA) is another selective/differential medium that is useful for identifying Gram-positive organisms. The medium uses a high salt concentration (7.5%) to select for *Staphylococcus* species while inhibiting the growth of other bacteria. In addition to this selection, the mannitol sugar in MSA permits differentiation between species of *Staphylococcus*. As an example, *S. aureus* can be differentiated from other staphylococcal species because it readily ferments mannitol. When grown on MSA, *S. aureus* presents a bright yellow zone around the growth (**Figure 10.9**). Preliminary identification of this pathogen in clinical isolates is easily accomplished with MSA.

Another example of a clinically useful differential medium is **blood agar**. This medium is not selective, because blood is a nutrient for most microorganisms. The medium is used to differentiate between organisms on the basis of the hemolysis of red blood cells. Many pathogenic organisms produce hemolysins as part of the infectious process. These hemolysins can be separated into three groups (**Figure 10.10**) on the basis of how completely they destroy red blood cells, and they can be used as preliminary diagnostic tools:

- **Gamma hemolysis** shows no destruction of red blood cells around the bacterial growth.

- **Alpha hemolysis** shows incomplete hemolysis of red blood cells, resulting in a greenish halo around the bacterial growth. This type of hemolysis is characteristic of *Streptococcus pneumoniae* (one microbe that causes pneumonia),

- **Beta hemolysis** causes complete destruction of red blood cells, resulting in a clear area around the bacterial growth. This type of hemolysis is characteristic of *Streptococcus pyogenes* (the cause of

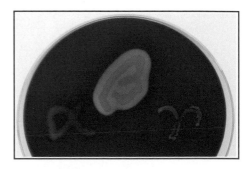

Figure 10.10 Hemolysis This blood agar plate has been inoculated with three types of streptococcal organisms: group D, which shows gamma hemolysis; group B, which shows alpha hemolysis; and group A, which shows beta hemolysis.

strep throat and necrotizing fasciitis, or flesh-eating disease). It is produced either by streptolysin O, an enzyme that is antigenic (it can elicit an immune response) or by the enzyme streptolysin S, which is nonantigenic.

Keep in Mind

- Identification of organisms is important for the study of infectious disease and requires that organisms be grown in a laboratory setting.

- A culture medium is used to grow bacteria.

- There are two types of medium: chemically defined and complex.

- Selective medium contains ingredients that prohibit the growth of some organisms while fostering the growth of others.

- Differential medium contains ingredients that can differentiate between organisms.

CHARACTERISTICS OF BACTERIAL GROWTH

As we have seen in Chapters 5 and 9, bacteria divide primarily by binary fission, in which a parent cell divides into two daughter cells (**Figure 10.11**).

Generation times vary between bacterial species and are heavily influenced by environmental pH, oxygen level, the availability of nutrients, and temperature. With proper levels of nutrients present, the number of bacteria in a colony increases exponentially from generation to generation.

The human body is a very suitable incubator for bacteria because of its constant temperature, excellent environmental conditions, and large supply of available nutrients in tissues and blood. Knowing this, it is easy to understand how quickly an invading bacterial population can multiply. If we magnify the problem by looking at toxin production or antibiotic resistance, the difficulty in dealing with infectious organisms becomes even more apparent.

Fast Fact For a microorganism with a generation time of 20 minutes, incubation for 24 hours yields 10^{21} daughter cells.

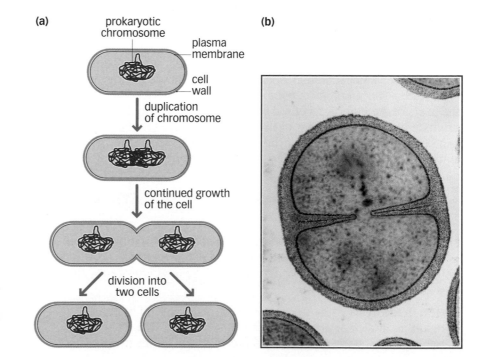

Figure 10.11 Bacterial cell division. Panel a: A graphic illustration of binary fission, the most common type of bacterial growth. Panel b: A colorized electron micrograph of a dividing cell.

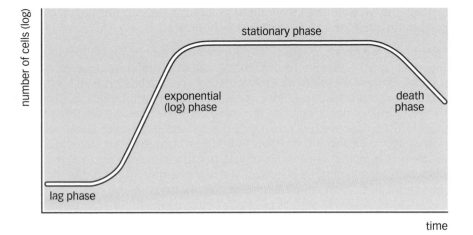

Figure 10.12 The bacterial growth curve. Specific functions and numbers of cells differ at each phase. During the lag phase, there is essentially no cell division. In the log (exponential) phase, the growth is exponential, and in the stationary phase the number of dying organisms equals the number produced during binary fission. The death phase, which inversely correlates with the log phase, is also referred to as the logarithmic decline phase.

The Bacterial Growth Curve

As shown in **Figure 10.12**, there are four phases of bacterial growth. We can examine each of these by using bacterial numbers and functions. During the **lag phase** of bacterial growth, the bacteria are adjusting to their environment and may have to synthesize enzymes to utilize the nutrients available in the environment. In this phase, little if any binary fission occurs, indicated by the fact that the growth curve is horizontal. How long this phase lasts varies between bacterial species and also depends on culture conditions.

During the **log phase** (log as in logarithmic; also known as the exponential phase) the number of bacteria doubles and increases exponentially (1 becomes 2, 2 become 4 and so on) and will have reached the constant minimum generation time. This level of growth can be sustained only while environmental conditions remain favorable and, more importantly, only if an adequate supply of nutrients remains available.

In some cases, especially with bacteria that have been genetically engineered to produce specific products, the conditions required for log-phase growth can be artificially maintained. This is accomplished using a **chemostat**, a device that permits the replacement of depleted medium with fresh, nutrient-rich medium (**Figure 10.13**). Under these conditions,

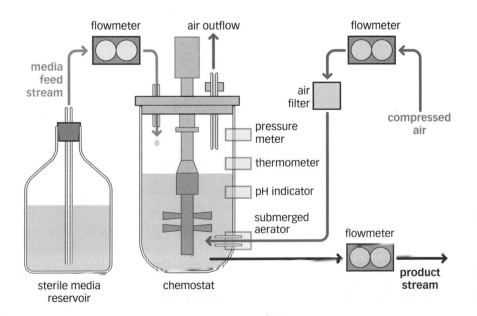

Figure 10.13 A chemostat. This instrument can maintain organisms in the log phase of growth for extended periods. This is accomplished by the removal of wastes and the constant replacement of nutrients. This technology is widely used by the pharmaceutical industry.

organisms can be kept in the log phase for long periods during which the product in question can be continually harvested.

During the log phase, bacteria are the most metabolically active and are therefore most sensitive to antibiotics, radiation, and adverse environmental changes. Recall from Chapter 9 that when a bacterial cell is dividing, cell wall production is at its peak. Therefore, antibiotics that target the synthesis of peptidoglycan are extremely effective during this phase of growth. So are antibiotics and/or treatments that target DNA replication or the transcription of RNA. In addition, because the translation of proteins is maximal during this phase, treatments targeting the bacterial ribosome are also highly effective. We will examine these antibiotic target interactions in more detail in Chapters 19 and 20.

At the point where nutrient levels begin to fall and waste levels (resulting from high rates of cell division) begin to rise, a **stationary phase of growth** begins. This phase can be characterized as the phase in which the number of cells dying is essentially equal to the number being produced through cell division. This phase is relatively short because it is dependent on the availability of nutrients, which continue to disappear as the growth curve shifts to the last phase.

The final phase of the bacterial growth curve—the **death phase**—represents a continuous decline in the number of dividing cells. This decline is caused by exhaustion of the nutrient supply as well as collapse of the environment due to the buildup of toxic waste materials. Notice in Figure 10.12 how the death phase inversely parallels the log phase. Because of this relationship, the death phase is sometimes called the **logarithmic decline phase**.

Measurement of Bacterial Growth

Because the severity of bacterial infections can be correlated with the number of bacteria, it is important to be able to determine the number of organisms. This can be accomplished by either **direct** or **indirect methods**.

One direct method (in which cells or colonies are actually observed) for determining the bacterial population of an infecting medium is filtration, which can be used to examine water or air samples for bacterial contamination. As we have previously mentioned, a common route of infection for many pathogens is the fecal–oral route. These types of infection are routinely seen in developing countries where sanitation is poor and as a result human waste may be commingled with available water supplies. Filtration is routinely used to examine water supplies for contamination with potentially pathogenic bacteria. This testing generates the **fecal coliform count**, which is used as an indicator of the level of fecal contamination. The method is shown in **Figure 10.14**.

A water sample is passed through a filter in which the holes are small enough to exclude bacteria. The filter is then placed on a selective medium. After incubation, potentially pathogenic bacteria filtered out of the water form visible colonies on the filter. With this method, contaminated water sources can be easily and inexpensively identified and bacterial concentrations measured.

The most common indirect method (in which the number of cells are estimated, not observed) for measuring bacterial populations is turbidity, or spectrophotometry. As illustrated in **Figure 10.15**, a spectrophotometer is an instrument that measures light at specific wavelengths. A liquid containing bacteria becomes turbid (cloudy) as the number of bacteria

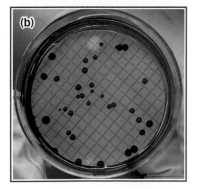

Figure 10.14 Filtration testing. Panel a: A water filtration system used to detect and quantify the level of fecal contamination in a water supply. Panel b: After water has been passed through it, the filter is placed on a nutrient agar plate and incubated. The number of colonies that grow indicates the level of contamination of that water sample.

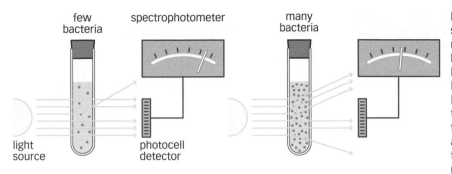

few bacteria   spectrophotometer   many bacteria

light source   photocell detector

Figure 10.15 An illustration of spectrophotometry, an indirect method of measuring the number of bacteria in a sample. As the number of bacteria increases, more and more of the light is deflected from the pathway to the light detector. How much light is lost from the pathway as the beam passes through the turbid sample is registered on the meter and is compared with standard curves to arrive at the number of bacteria per milliliter of sample.

increases, and when a turbid liquid is placed in the path of light entering a spectrophotometer, some of the light is deflected from the path and is therefore not measured by the light detector. How much of the light is deflected is a measure of how turbid the liquid is. The turbidity is then used in conjunction with a standard curve (a prediction equation determined in a separate experiment) as a method for determining the number of bacteria.

Direct and indirect methods for measuring bacteria are described in Table 10.8.

CLINICAL IMPLICATIONS OF BACTERIAL GROWTH

Many factors affect the rate at which bacterial populations grow. For example, the generation time for a given bacterial species is not one unchanging number. Instead, this number can fluctuate depending on the nutrients available to the bacteria, and this variation influences the slope of the log phase of the bacterial growth curve. Because this slope is directly related to growth, a change in slope means a change in growth.

In addition, many pathogenic bacteria are fastidious in laboratory settings. For example, *Mycobacterium tuberculosis*, which grows well *in vivo*, requires a much longer time to grow in a laboratory, and *Mycobacterium leprae* (the causative agent of leprosy) cannot be grown in a microbial culture at all and must instead be grown *in vivo*.

Some pathogenic bacteria have stringent medium requirements. A good example is *Haemophilus influenzae*, which inhabits the mucous membranes of the upper respiratory tract, mouth, vagina, and intestinal tract. This bacterium is particularly fastidious and requires medium supplements, such as the X factor of blood hemoglobin and the cofactor nicotinamide adenine dinucleotide. These substances are liberated from blood by heating, and the resulting brown agar is referred to as **chocolate agar** (Figure 10.16).

Specimen Collection

Because many bacteria have specific requirements for growth, it is easy to understand how these requirements can affect the diagnosis and treatment of diseases. If clinical specimens are not handled and cultured properly, the presence of some pathogenic bacteria can be missed, leading to errant diagnoses and potential exacerbation of the disease. The problem can be compounded by improper isolation of clinical specimens. It should therefore be noted that samples taken from different parts of the body require different methods of isolation. For example, sputum samples are intrinsically important for the diagnosis of respiratory diseases and can be collected by coughing or catheterization. However, needle

Fast Fact *Mycobacterium leprae* bacteria grow well on armadillos, because these mammals have a suitably low body temperature (33 to 36°C). Decades ago, leprous armadillos escaped from a research facility in Louisiana! Whether their infection came from the research setting or was naturally occurring, some armadillos living today in the Southeastern United States carry these pathogenic bacteria and can pass the disease to humans who handle them.

Figure 10.16 Chocolate agar is used for growing fastidious bacteria such as *Haemophilus influenzae*. To grow and divide this organism requires certain substances which are released by heating blood giving the medium a brown color.

Type	Method and process	Characteristics and limitations
Direct		
	Direct microscopic count	Bacteria on a slide are counted using a microscope. Rapid, but at least 10^7 cells/ml must be present to be counted effectively. Counts include living and dead cells
	Cell-counting instruments	Coulter counters and flow cytometers count total cells in dilute solutions. Flow cytometers can also be used to count organisms to which fluorescent dyes or tags have been attached
	Viable cell counts	Used to determine the number of viable bacteria in a sample, but includes only those that can grow in given conditions. This requires an incubation period of about 24 hours or longer. Selective and differential media can be used to enumerate specific species of bacteria
	Plate count	A time-consuming but technically simple method that does not require sophisticated equipment. Sample is grown on a Petri plate and the bacterial colonies are counted. Generally used only if the sample has at least 100 cells/ml
	Membrane filtration	Concentrates bacteria by filtration before they are plated; thus can be used to count cells in dilute environments
	Most probable number	Statistical estimation of likely cell number; it is not a precise measurement. This method can be used to estimate numbers of bacteria in relatively dilute solutions
Indirect		
	Measuring biomass	Biomass can be correlated with cell number
	Turbidity	Very rapid method; used routinely. A one-time correlation with plate counts is required in order to use turbidity for determining cell number
	Total weight	Tedious and time-consuming; however, it is one of the best methods for measuring the growth of filamentous microorganisms
	Chemical constituents	Uses chemical means to determine the amount of a given element, usually nitrogen. Not routinely used
	Measuring cell products	Methods are rapid but must be correlated with cell number. Frequently used to detect growth, but not routinely used for quantification
	Acid	Titration can be used to quantify acid production. A pH indicator is often used to detect growth
	Gases	Carbon dioxide can be detected by using a molecule that fluoresces when the medium becomes slightly more acidic. Gases can be trapped in an inverted Durham tube in a tube of broth

Table 10.8 Methods used to measure bacterial growth.

aspirations are used for blood and cerebrospinal fluids. Table 10.9 lists several methods for collecting clinical samples.

In the end, it is interesting to speculate on how many diseases may go undiagnosed even today because of the unknown growth requirements of their causative agents.

Approximately 10% of the infections seen in humans are caused by bacteria, but more than 80% result from viral infections. Because viruses are obligate intracellular parasites, they have growth requirements that are very different from those for bacteria. We will examine the growth of viruses in detail in Chapter 12.

Table 10.9 **Methods used to collect clinical specimens.**

Type or location of specimen	Collection method
Skin, accessible mucous membrane (including eye, outer ear, nose, throat, vagina, cervix, urethra), or open wound	Sterile swab brushed across the surface; care should be taken not to contact neighboring tissue
Blood	Needle aspiration from vein; anticoagulants are included in the specimen tube
Cerebrospinal fluid	Needle aspiration from subarachnoid space of spinal column
Stomach	Intubation, which involves inserting a tube into the stomach, often via a nostril
Urine	In aseptic collection, a catheter is inserted into the bladder through the urethra; in the 'clean catch' method, initial urination washes the urethra, and the specimen is midstream urine
Lungs	Collection of sputum either dislodged by coughing or acquired via a catheter
Diseased tissue	Surgical removal (biopsy)

Keep in Mind

- Bacteria grow by binary fission in which one cell divides into two.

- Different bacteria have different generation times.

- The bacterial growth curve has four phases: lag, log, stationary, and death.

- The number of organisms can be determined directly through methods such as direct microscopic counting, filtration, and automated cell counting.

- Indirect measurement of bacterial numbers involves methods such as spectrophotometry, total weight, and measurement of cell products.

- Samples taken from different parts of the body are collected with different methods.

SUMMARY

- Bacterial growth is an important part of the infection process.

- Growth of bacteria requires proper physical and chemical environments.

- Different bacteria have different requirements for oxygen. Obligate anaerobic organisms cannot grow in the presence of oxygen, aerobic organisms cannot grow without it, and facultative anaerobes can grow with or without oxygen.

- Identification of organisms is important for the study of bacteria and the treatment of disease.

- Media, either complex or chemically defined, can be used to grow bacteria in a laboratory setting.

- Selective medium prohibits the growth of some organisms while fostering the growth of others.

- Differential medium contains ingredients that can differentiate between organisms.
- Most bacteria grow by the process of binary fission.
- Different bacteria have different generation times.
- The bacterial growth curve has four phases: lag, log, stationary, and death.
- Different methods are used to take samples from different parts of the body.

In this chapter we have examined the characteristics of and requirements for bacterial growth. Remember that there are both physical and chemical requirements and that bacterial growth can be subdivided into four basic phases: lag, log, stationary, and death. Bacteria carry out different functions in each phase, and the log phase (where the bacteria are undergoing cell division at the highest rate) can be the most important for targeting by inhibitors of bacterial growth, such as antibiotics. We saw that bacteria have generation times that can vary on the basis of factors such as nutrient and incubation requirements. We also learned that selective and differential media, as well as specific methods of measurement, can be important in the identification and isolation of pathogenic bacteria

(Q) SELF EVALUATION AND CHAPTER CONFIDENCE

Multiple Choice

Answers are given in the back of the book and help can be found in the student resources at:

www.garlandscience.com/micro2

1. Psychrophiles are organisms that grow

   A. Between 0 and 15°C
   B. Between 45 and 60°C
   C. At room temperature
   D. Between 25 and 40°C
   E. Above 80°C

2. Most human pathogens are

   A. Psychrophiles
   B. Psychrotrophs
   C. Thermophiles
   D. Extreme thermophiles
   E. None of the above

3. Minimum growth temperature is

   A. The highest temperature at which an organism will grow
   B. The lowest temperature at which an organism will grow
   C. The temperature at which there will be minimum growth
   D. The temperature at which there will be the greatest amount of growth
   E. The temperature at which cells will grow the best

4. Optimal growth temperature is

   A. The highest temperature at which an organism grows
   B. The lowest temperature at which an organism grows
   C. The temperature at which the highest rate of growth occurs
   D. The temperature at which bacteria grow poorly
   E. None of the above

5. Acidophilic bacteria grow best at

   A. pH 7.0
   B. pH 2.0
   C. pH 9.0
   D. pH 14
   E. None of the above

6–9. The growth of *E. coli* in enriched medium was measured and the data in the table at the top of the next page were obtained. Examine the data and answer questions 6–9 below.

6. At which letter does the death phase begin?

7. At which letter does the stationary phase begin?

Time (min)	0	10	20	30	40	50	60	70	80	90	100
Cell number	100	100	200	400	800	1600	3200	3200	3200	2800	1400
Choice		A	B			C				D	

8. At which letter does exponential cell growth begin?

9. The doubling time for the culture is

   A. 10 minutes
   B. 20 minutes
   C. 15 minutes
   D. 30 minutes

10. The term facultative anaerobe refers to an organism that

    A. Is killed by oxygen
    B. Doesn't use oxygen but tolerates it
    C. Uses oxygen but can grow without it
    D. Requires less oxygen than is present in air
    E. Prefers to grow without oxygen

11. Which type of organism produces catalase and superoxide dismutase?

    A. Obligate anaerobe
    B. Aerobe
    C. Aerotolerant anaerobe
    D. Anaerobic halophile
    E. Anaerobic extreme halophile

Medium A	Medium B	Medium C
Na_2HPO_4	Laundry detergent	Glucose
KH_2PO_4	Na_2HPO_4	Nutrient broth
$MgSO_4$	KH_2PO_4	$(NH_4)_2SO_4$
$CaCl_2$	$MgSO_4$	KH_2PO_4
$NaHCO_4$	$(NH_4)_2SO_4$	Na_2HPO_4

12. Which medium or media in the table above is/are chemically defined?

    A. A
    B. B
    C. A and B
    D. A and C
    E. None of the above

13. Which enzyme catalyzes the reaction
    $H_2O_2 + H_2O \rightarrow 2H_2O + O_2$?

    A. Peroxidase
    B. Superoxide dismutase
    C. Catalase
    D. Oxidase
    E. None of the above

14. An organism that cannot grow in the presence of oxygen is called

    A. An obligate anaerobe
    B. An obligate aerobe
    C. A dual-purpose organism
    D. A facultative aerobe
    E. A facultative anaerobe

15. Mannitol salt agar (MSA) is a medium that is

    A. Selective only
    B. Differential only
    C. Both selective and differential
    D. Neither selective nor differential
    E. Used for culturing organisms from the ocean

16. Eosin methylene blue (EMB) medium

    A. Differentiates organisms by glucose fermentation
    B. Is selective because of the lactose it contains
    C. Differentiates organisms by lactose fermentation
    D. Is not selective
    E. Is not differential

17. Gamma hemolysis on a blood agar medium shows

    A. Complete destruction of red blood cells
    B. Partial destruction of red blood cells
    C. Complete destruction of platelets
    D. No destruction of red blood cells
    E. Partial destruction of platelets

18. Logarithmic decline is another way of describing which phase of bacterial growth?

    A. Lag
    B. Log
    C. Stationary
    D. Death

In questions **19–24**, mark the items as either direct (A) or indirect (B) ways of measuring bacterial growth:

19. Biomass

20. Viable count

21. Microscopic count

22. Turbidity

23. Plate count

24. Acid production

DEPTH OF UNDERSTANDING

Questions listed here require you to bring together the concepts you have learned in this chapter into a discussion format. This helps you to increase your depth of understanding of the material you have learned. Help can be found in the student resources at: www.garlandscience.com/micro2

1. When you arrive at the pool, there is a sign indicating it is unsafe to swim because of coliform contamination. Using what we have learned in this chapter, discuss how this was determined.

2. Discuss the relationship between bacterial growth and pathogenicity, and the relationship between bacterial growth and treatment for infections.

3. Compare and contrast organisms that are obligate anaerobes and those that are facultative anaerobes.

Q **CLINICAL CORNER**

Help can be found in the student resources at: www.garlandscience.com/micro2

1. Mr. Rodriguez is in the hospital because of an intestinal infection. He has been having bouts of diarrhea and vomiting for 2 days and has become seriously dehydrated.

   A. What samples would you take from this patient and how would you do so?
   B. What tests would you do?

2. Amy Smith has been coughing for a week and is producing a greenish mucus when she coughs. Examination of her breathing indicates she may have a lower respiratory tract infection that could be pneumonia.

   A. What samples would you collect from Amy, and how would you collect them?
   B. What tests should you do on the samples you collect from her?

Microbial Genetics and Infectious Disease

Chapter 11

Why Is This Important?

Understanding genetic mechanisms lets us study how microorganisms can mutate and change in ways that allow them to defeat host defenses. These changes are one of the most important topics in health care today.

Joe is walking along the benches in the teaching laboratory, keeping a close eye on his students. He wants to make sure they learn as much as they can while being safe. Every so often he needs to remind them of good aseptic technique. He stops at one bench and observes how a group of his students drop some 3% KOH solution onto a microscope slide. Using a wooden toothpick they then carefully rub a small amount of bacteria taken from a colony into the alkaline solution. Slowly one student pulls the toothpick upward and to everyone's amazement draws out a short stringy transparent thread. What the students are performing is a Gram test, and their bacterial sample is Gram negative. The alkaline solution breaks open the cells of Gram-negative bacteria. DNA is released and aggregates to form a very viscous solution, which can be seen with the naked eye as this thread. If the students had performed an actual Gram stain, these Gram-negative cells would have appeared pink under a microscope. The cell walls of Gram-positive cells do not break with KOH; DNA is not released and the suspension remains granular. Joe hears the 'ohs' and 'ahs' everywhere and looks around. He has never forgotten how thrilled he was, when he literally held DNA in his hands for the first time just like his students now. He smiles and thinks that fortunately some things never change: curiosity is in the genes.

OVERVIEW

One of the most difficult problems in medicine today is antibiotic resistance. Pathogenic microorganisms can become resistant to antibiotics through mutation, a process that causes changes in the genetic information of the organisms. These changes can be transferred from one organism to another, which means that microbes have the capacity to take up genetic information, exchange it, and develop it further. The genetic changes resulting from these exchanges can make harmless organisms dangerous (pathogenic) and pathogenic organisms lethal (more virulent). To understand pathogenesis and virulence, we must be familiar with microbial genetics. In this chapter, we begin to acquire this familiarity by looking at the structure of the nucleic acids DNA and RNA. We then examine the processes of replication, gene expression, mutation, and gene transfer and end the chapter by taking a look at the genetics of pathogenesis and virulence.

We will divide our discussions into the following topics:

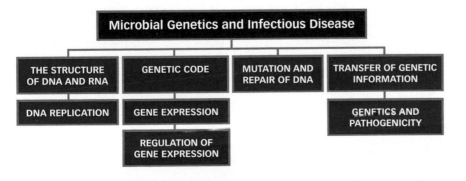

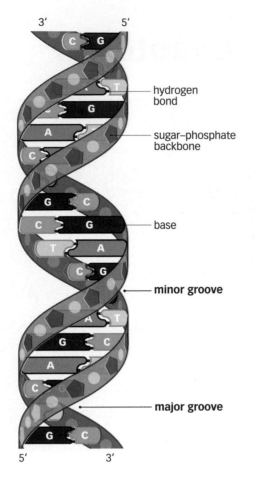

Figure 11.1 The double-helical structure of DNA. The two strands are complementary and wind round each other to form a double helix with two grooves between the twists of the helix. The sugar–phosphate backbone is on the outside, and the bases are on the inside. The bases of the two strands are joined by hydrogen bonds.

THE STRUCTURE OF DNA AND RNA

If we are to understand the mechanisms discussed in this chapter, we must first review some of the fundamental information about nucleic acids. In the following sections we examine the structure of DNA and RNA.

Deoxyribonucleic Acid (DNA)

As we begin our discussions of DNA structure, remember that, in nature, structure and function are always tightly connected to each other. The function of DNA is a blueprint for all of the components of the cell that can be faithfully passed on to subsequent generations of cells. It is the structure of DNA that allows this to be accomplished. This structure–function relationship maximizes the ease of replication and makes **gene expression**, defined as the process of transcription (DNA to RNA) and translation (RNA to protein), efficient and almost error-free. To understand the structure of DNA, we must first examine its components and how they orient themselves together.

As we saw in Chapter 2, deoxyribonucleic acid (DNA) is a double-stranded helical molecule with a sugar–phosphate backbone and inwardly projecting bases joined together by hydrogen bonds (**Figure 11.1** and **Movie 11.1**). Although the bases project inward, they are still accessible because the helical nature of the molecules naturally produces two grooves (a major groove and a minor groove) between the twists of the helix.

The building blocks of DNA are nucleotides, which are combinations of a phosphate, a sugar (which in DNA is **deoxyribose**), and a nucleotide base (**Figure 11.2**). It is important to note that these components bind together in a very specific way chemically, and that binding permits the correct and precise orientation of the nucleotide. It is this orientation that permits the building of the strands of DNA. Nucleotides are joined together when a phosphodiester bond forms between the 3′ hydroxyl group of the sugar of one nucleotide and the 5′ phosphate group of another nucleotide. This imparts inherent polarity and structural orientation to the growing chain, which is required for the proper addition of nucleotides.

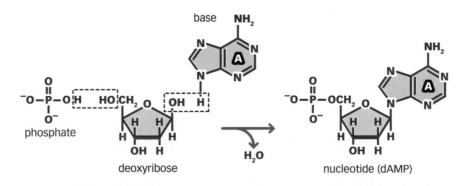

Figure 11.2 Formation of a deoxynucleotide. The nucleotide is synthesized from phosphate, deoxyribose sugar, and a nucleotide base (adenine shown here) through dehydration synthesis (removal of water).

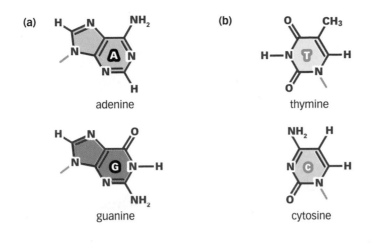

Figure 11.3 The structures of purines and pyrimidines. Panel a: The structure of the purines adenine and guanine. Panel b: The structure of the pyrimidines thymine and cytosine. For each base, the orange line shows the bond attaching it to the deoxyribose sugar.

The bases in DNA are of two types: the **purines** (adenine and guanine) and the **pyrimidines** (thymine and cytosine). Purines are large double-ring structures, whereas the pyrimidines have single rings (**Figure 11.3**). DNA has a helical geometry governed by how the bases pair up. In DNA, adenine always pairs with thymine, via two hydrogen bonds, and cytosine always pairs with guanine, via three hydrogen bonds (**Figure 11.4**). The size, structure, and bonding of these bases are important because they affect this helical geometry. The bonding between the bases also requires that one of the strands is oriented upside down relative to the other. This arrangement is referred to as anti-parallel (**Figure 11.5**). Proper base pairing also keeps the sugars and phosphates in alignment so that there is no distortion of the helical DNA structure. This causes the bases to stack on top of each other. All of these elements help to ensure the thermodynamic stability of DNA. Consequently, any mismatched base pairing becomes chemically unstable. Remarkably, this chemical instability is one of the ways in which errors are identified and corrected.

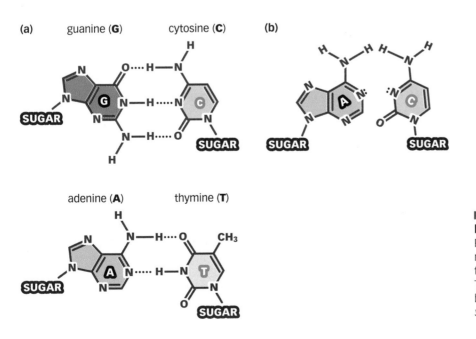

Figure 11.4 Specific pairing of the bases of DNA via hydrogen bonding. Panel a: The chemical nature of the nucleotide bases allows the formation of two hydrogen bonds between the A and T bases and three between the C and G bases. Panel b: Mismatching of bases is structurally inhibited.

Figure 11.5 Detailed structure of DNA.
The phosphodiester bonding between the sugars and phosphates of the backbone gives the two chains their anti-parallel orientation and allows the bases to pair properly.

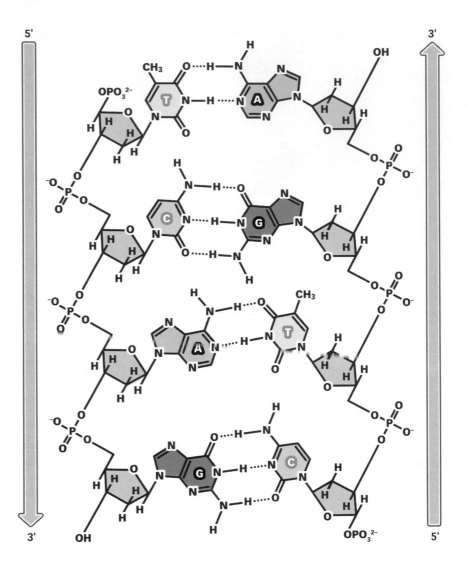

Ribonucleic Acid (RNA)

The second nucleic acid we are interested in when studying genetic mechanisms is ribonucleic acid (RNA), which differs from DNA in the following ways:

- RNA contains the sugar ribose rather than deoxyribose (see Figure 2.17).

- The bases in RNA are adenine, cytosine, guanine, and uracil (see Figure 2.18), and the base pairings are adenine with uracil, cytosine with guanine.

- RNA is usually found in a single-stranded form. However, RNA can fold on itself and form double-stranded areas.

RNA functions in three ways (Table 11.1). **Messenger RNA** (mRNA) is the form containing information derived from DNA and is used in the construction of proteins. In eukaryotes, an mRNA molecule is an RNA copy of the information from one gene (the instructions for making one protein). In **polycistronic** mRNAs found in prokaryotes, an mRNA molecule encodes more than one protein. **Transfer RNA** (tRNA) carries amino acids to the ribosome where protein is being constructed. Finally, **ribosomal RNA** (rRNA) helps in maintaining the proper shape of the ribosome (composed of ribosomal RNA and proteins) and the orientation of the protein under construction.

Type of RNA	Properties and function
Messenger	Carries information used to construct proteins; uses a three-base combination (called a codon) that codes for a specific amino acid; complexes with the ribosome for translation
Transfer	Found in cytoplasm and used to pick and transfer amino acids to the ribosome; has a distinct three-dimensional shape (cloverleaf-like) with an attachment arm for the amino acid and an anticodon region that guarantees the proper codon match and orientation at the ribosome
Ribosomal	Combines with specific proteins to form the proper three-dimensional configuration of the ribosome

Table 11.1 Types of RNA.

As we examine each of these roles of RNA, keep in mind the complementary nature of base pairing—adenine with uracil, cytosine with guanine—because this pairing ensures the construction of exactly the right proteins needed by a cell.

If complementary bases are close to each other on an RNA strand, the bases can pair with each other by means of hydrogen bonds, forming a loop in the strand. In addition, because of the flexibility and rotation of this single-stranded molecule, RNA can also form a three-dimensional structure by folding on top of itself. This tertiary structure is seen in both tRNA and rRNA.

Keep in Mind

- DNA stands for deoxyribonucleic acid.
- DNA is the informational molecule of the cell and is a double-stranded molecule made up of nucleotides.
- A DNA nucleotide is composed of a phosphate, a deoxyribose sugar, and one of the four nucleotide bases (adenine, thymine, guanine, cytosine).
- RNA stands for ribonucleic acid.
- In RNA, thymine is replaced by uracil and the sugar is ribose instead of deoxyribose.
- There are three types of RNA: messenger, transfer, and ribosomal.

DNA REPLICATION

DNA replication is the process by which the DNA is copied, and is a carefully controlled and regulated operation involving several specific components and mechanisms. It is a critical cellular procedure accomplished with remarkable accuracy, at astounding speed.

DNA Separation and Supercoiling

One of the characteristics of a double-helical molecule is the potential for **supercoiling**, which occurs when the helix twists around itself (like a coiled phone cord or extension lead that becomes coiled over again and again). It allows a large amount of information to be accommodated within very small cells. Although the chemical and physical details of supercoiling are beyond the scope of this discussion, it is important to note that before strands can be unwound and separated for replication or

Figure 11.6 The general structure of a primer:template junction. The primer region, which is complementarily base paired to the DNA template strand, provides a free 3′ hydroxyl group for addition of the appropriate nucleotide. The primer is made of RNA rather than DNA (and thus contains the pyrimidine base uracil instead of thymine) and will eventually be excised and replaced with DNA.

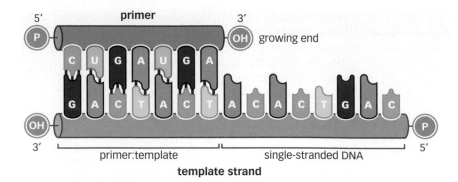

transcription (a process we discuss below), the DNA supercoiling must be relaxed. This is accomplished by **topoisomerase**, a remarkable enzyme that unwinds the supercoils by first breaking the DNA chain so that the supercoil relaxes, and then precisely resealing the break. Once the supercoiling has been relaxed, the enzyme **helicase** can unwind and separate the chains.

There are two basic requirements for replication once the helix has separated:

• An ample supply of each of the four nucleotides

• A **primer:template junction**. Once the double-stranded DNA has been unwound, each unwound single DNA strand is called a template. A portion of this template is then paired with a short segment of RNA called a primer. The point at which the primer and template join is called a primer:template junction (**Figure 11.6**).

DNA Polymerase

DNA polymerase is the enzyme responsible for creating the new strand of DNA, but it can only add nucleotides to an existing strand. The RNA primer at the primer:template junction gives the DNA polymerase a place to attach the first nucleotide. DNA polymerase creates the new DNA strand in one direction only: it attaches the 3′ end of one sugar (on the growing strand) to the 5′ end of the next. This one-way addition of nucleotides is a universal feature of both DNA and RNA synthesis.

The template strand dictates which nucleotide will be added to the elongating strand, because the new base has to pair correctly with the base on the template strand, ensuring that a faithful copy of the template strand is made (**Figure 11.7**). Forming the new phosphodiester bond releases a two-phosphate molecule (pyrophosphate), which is immediately cleaved into two individual phosphate molecules, releasing energy. The released energy is used to elongate the growing strand of DNA, and this coupling process makes DNA synthesis essentially irreversible.

Fast Fact An important and interesting characteristic of DNA polymerase is that it will never pick up ribonucleotide (the building block for RNA), even though there may be more than 1000 times as many available.

DNA polymerase takes any of the four nucleotides and attempts a fit via the complementary bases. This allows the enzyme to work incredibly fast, adding one suitable nucleotide after the other, because it does not have to wait for the right one to appear at the active site. It takes about 1 second for DNA polymerase to find and bind to the primer:template junction. Once it has bound, the actual addition of a nucleotide is in the millisecond range. Furthermore, the longer the enzyme is associated with the DNA being elongated, the faster the addition of nucleotides occurs. There are actually several types of DNA polymerase that perform specific

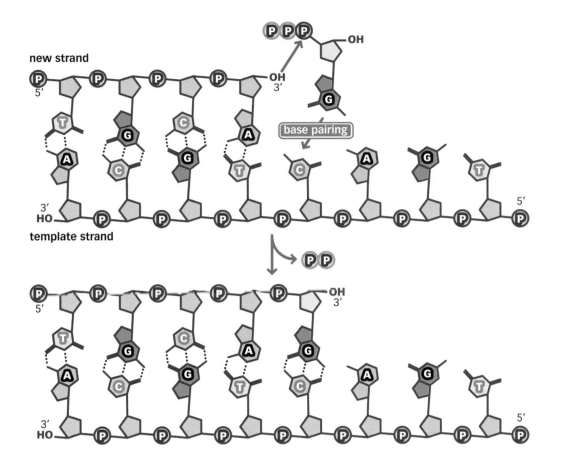

Figure 11.7 An illustration of DNA synthesis. The phosphate of the new nucleotide is attached to the free 3′ hydroxyl group of the previous nucleotide on the growing strand. The base of the new nucleotide pairs with the base on the template strand. Two phosphates are cleaved from the incoming nucleotide, which is then attached to the free 3′ hydroxyl group of the previous nucleotide.

functions and have different levels of processivity (how many nucleotides they can add), ranging from a few to more than 50,000 per binding event.

Proofreading By DNA Polymerase

Because the same genetic information is passed down from generation to generation of cells, it is vital to the survival of the organism that this information remains correct. Fortunately, replication of DNA is extraordinarily accurate. However, occasionally DNA polymerase does incorporate wrong nucleotides into the new DNA, and these genetic changes (mutations) have a pivotal role in the evolution of the organism.

As we consider what we have learned so far, we can see that the system of DNA replication, which is based on the structure and the proper pairing of bases, can be fraught with potential problems if nucleotides are added incorrectly. Experiments have shown that during DNA replication, it is possible for incorrect base pairing to occur at a frequency of about one in every 10^5 pairings. This frequency is significantly higher than the actual observed error for replication, which is approximately once in every 10^{10} pairings. The difference between the number of errors possible and the actual number is due to the proofreading capability of the DNA polymerase.

This proofreading of bases added takes place at the newly synthesized strand and is possible because the DNA polymerase contains an exonuclease component. The DNA proofreading exonuclease is located in the DNA polymerase and can remove the growing end (3′) of the DNA strand being synthesized. Exonuclease activity is strongly attracted to nucleotides that are improperly incorporated into to the growing DNA strand and preferentially removes them.

(a)

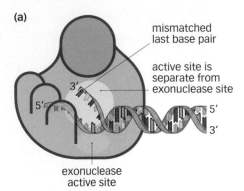

mismatched last base pair

active site is separate from exonuclease site

exonuclease active site

(b) **removal of mismatched nucleotide**

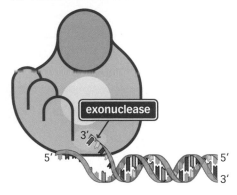

exonuclease

(c) **resumed DNA synthesis**

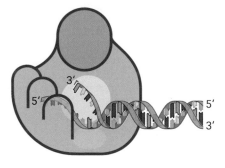

Figure 11.8 A representation of the proofreading capability of the DNA polymerase. Mismatching of base pairs at the active site of the enzyme causes replication to slow down (panel a), and causes the growing strand to move from the active site to the exonuclease part of the enzyme (panel b). Here, the mismatched base is removed and the strand moves back into the active site (panel c).

DNA polymerase activity slows down if a wrong nucleotide tries to attach to the growing DNA strand. An incorrect base changes the configuration of the growing DNA strand in the active site of the polymerase, causing it to move to the exonuclease active site (**Figure 11.8**). The distorted configuration also causes an increase in exonuclease activity, which speeds up the removal of the wrong nucleotide when it arrives at this site. Once the wrong nucleotide has been removed, the growing DNA strand changes shape again (back to normal), leaves the exonuclease active site and moves back into the polymerase active site.

One of the most important aspects of this proofreading mechanism is that the corrections are made without having to disengage the DNA polymerase. If removal were required, the process could be very time-consuming because the polymerase would have to disengage and then re-engage once the correction had been made.

The Replication Fork

The **replication fork** is the location along a DNA double helix where replication is going on. It is easily recognizable because at the fork, the double helix is unwound and the strands are separated from each other. It is important to note that the fork is moving along the DNA as the strands are unwound. When we examine the elements involved at this location, we can get a good idea of the overall process of DNA replication. However, before we examine these events, we need to review three facts:

- Both strands of DNA are replicated at the same time.

- The two strands are antiparallel.

- Each template strand is replicated by the addition of nucleotides to the 3′ end of the growing strand. This presents a problem because the antiparallel alignment of the strands makes it impossible to replicate both of them simultaneously in the same direction.

In DNA replication, one strand is in a perfect position for the addition of nucleotides to its 3′ end. This is called the **leading strand**. Looking at the leading strand (**Figure 11.9**), it is easy to understand that the polymerase can add nucleotides and move toward the advancing replication fork (in effect, the polymerase on this strand is chasing the fork). In contrast, because of the antiparallel alignment of the two strands of DNA, the DNA polymerase replicating the other strand—called the **lagging strand**—is moving away from the fork. This movement of the enzyme away from the replication fork necessitates replicating the lagging strand in pieces, which are called **Okazaki fragments** (after the scientist who discovered them).

Remember that for nucleotides to be added to a growing DNA strand, there has to be a free 3′ end that the polymerase can use. This is no problem for the leading strand because the 3′ end of the growing strand is all that is required for replication of this strand. In contrast, for the lagging strand, each new Okazaki fragment needs to have its own RNA primer and the DNA polymerase has to wait for the replicating fork to move

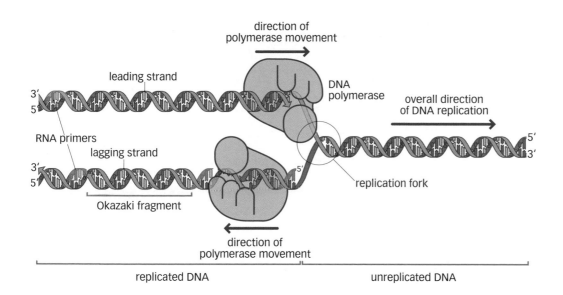

far enough to expose a substantial segment of the template DNA. The primer is made by a type of RNA polymerase called **primase**, which can synthesize RNA without a 3′ end being present (this synthesis can occur at any place along the DNA sequence). As the replication fork moves, the lagging strand of DNA elongates, and a primase molecule attaches to the strand and synthesizes the RNA primer, using the lagging strand as a template.

Once the short RNA primer segments have been synthesized, the primer is ready for the DNA polymerase to begin adding nucleotides, which it does until it reaches the end of the previously synthesized Okazaki fragment. (Okazaki fragments are usually between 1000 and 2000 nucleotides long.) When the synthesis of an Okazaki fragment has finished, an enzyme called **RNAase H** removes the primer RNA. The gap that results from the missing primer is then easily filled in by DNA polymerase by adding to the 3′ end of the previous Okazaki fragment. Once the gap has been filled, the ends of the DNA pieces are linked together by the enzyme **DNA ligase**. The whole process of replication can be seen in **Movie 11.2** and **11.3**.

Initiation and Termination of Replication

Initiation of DNA replication in bacteria begins at a specific spot on the chromosome called the **origin of replication**. It is thought that there are two components that control this initiation. The first is the **replicator sequence**, a specific set of DNA sequences that includes a long string of A-T base pairs. Remember that there are fewer hydrogen bonds between A and T than between G and C, so this stretch is more easily opened. The second component in initiation is the **initiator protein**, which specifically recognizes the replicator sequence of DNA. The series of events involved in initiation includes the initiator protein binding to the replicator sequence and unwinding the DNA strands adjacent to that site.

Replication will terminate when the entire DNA chromosome has been copied; it also requires a specific set of events in which the components involved in replication dissociate from the DNA. At this point, the replicated circular chromosomes will still be linked together (**Figure 11.10**). If the daughter cells that result from cell division are to acquire one of these chromosomes, they must be separated from each other. This separation is accomplished by topoisomerase, the same enzyme involved in relaxing the supercoils in DNA before replication.

Fast Fact Primase is different from the RNA polymerase that normally transcribes DNA into RNA, in that its *only* function is to synthesize short primer segments on the lagging strand.

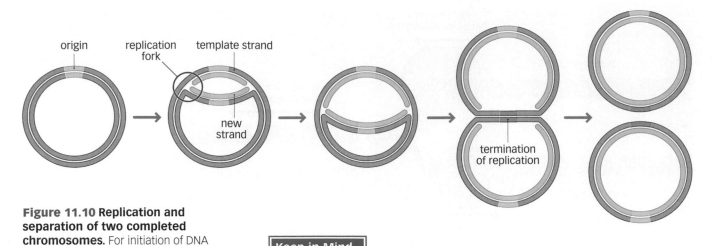

Figure 11.10 Replication and separation of two completed chromosomes. For initiation of DNA replication, the initiator protein has to bind to the replicator sequence at the origin of replication. After replication is complete, the two identical chromosomes are separated by the topoisomerase enzyme in a reaction similar to that seen in the relaxation of supercoiled DNA.

Keep in Mind

- DNA is faithfully replicated so that the same genetic information is passed on from generation to generation.

- The enzymes topoisomerase and helicase unwind and separate the strands of DNA to be replicated.

- DNA polymerase copies both DNA strands so that each daughter cell will contain a chromosome made up of an original strand and a daughter strand.

- DNA polymerase has a proofreading capability to prevent mistakes during replication, and it replicates in only one direction (from the 3′ end of the new strand).

- The replication fork is the site at which replication is occurring.

- At the replication fork there is a leading strand, which is replicated continuously, and a lagging strand, which is replicated in pieces known as Okazaki fragments.

- Replication is initiated at a site on the DNA called the origin of replication and proceeds until the entire chromosome has been copied.

THE GENETIC CODE

Before we can examine gene expression, we have to have an understanding of the **genetic code**. The information contained in DNA that is used for gene expression (transcription and translation) is based on a four-letter alphabet—A, T, G, and C (the four bases)—which uses combinations (called **codons**) of three letters. These codons code for specific amino acids. If you list all the possible three-letter combinations using the four available letters A, T, G, and C, you come up with 64 possible codons. However, only 20 amino acids are used to make all the proteins found in all organisms. We use the term **degenerate** to describe this redundant feature of the genetic code.

It is thought that the genetic code evolved in this degenerate fashion as a way to minimize the potentially devastating effects of mutations. Consequently, if you look at the combinations shown in **Figure 11.11**, you see that changes in the third letter of the codon do not necessarily change the designated amino acid. The insignificance of a last letter change in the codon is part of the **wobble hypothesis** that was originally put forward by Francis Crick (one of the discoverers of the structure of DNA). The DNA code is converted into its corresponding mRNA code, where the letter T is replaced by the letter U, before it is used to make protein. In addition to specifying amino acids there are three **stop codons** and a start (**initiation**) codon, which indicate where protein synthesis starts and stops.

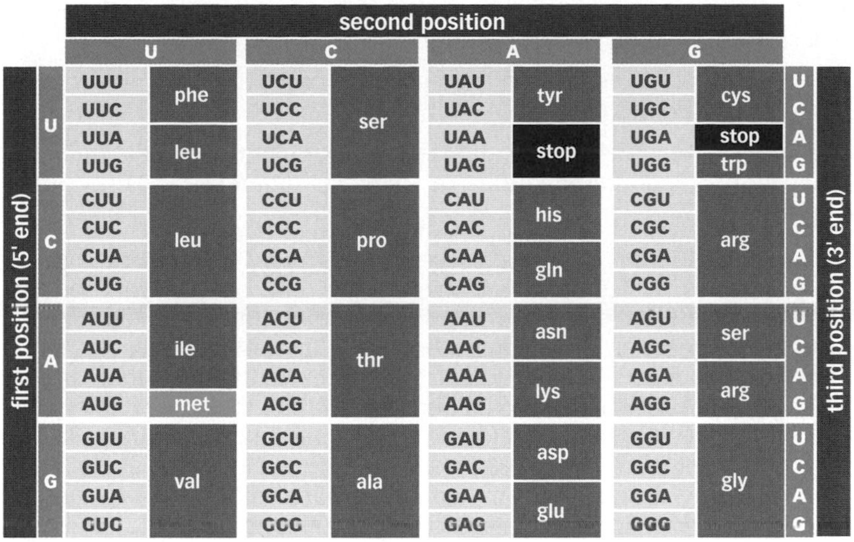

second position									
	U		**C**		**A**		**G**		third position (3' end)
U — UUU / UUC	phe	UCU / UCC / UCA / UCG	ser	UAU / UAC	tyr	UGU / UGC	cys	U / C	
UUA / UUG	leu			UAA / UAG	stop	UGA	stop	A	
						UGG	trp	G	
C — CUU / CUC / CUA / CUG	leu	CCU / CCC / CCA / CCG	pro	CAU / CAC	his	CGU / CGC / CGA / CGG	arg	U / C / A / G	
				CAA / CAG	gln				
A — AUU / AUC / AUA	ile	ACU / ACC / ACA / ACG	thr	AAU / AAC	asn	AGU / AGC	ser	U / C	
AUG	met			AAA / AAG	lys	AGA / AGG	arg	A / G	
G — GUU / GUC / GUA / GUG	val	GCU / GCC / GCA / GCG	ala	GAU / GAC	asp	GGU / GGC / GGA / GGG	gly	U / C / A / G	
				GAA / GAG	glu				

Figure 11.11 The genetic code. Each codon specifies a particular amino acid, but a particular amino acid is specified by more than one codon; in other words, the code is degenerate. Changes in the third letter of the codon (the 3' end) are less important in designating the amino acid. The code also contains start (green) and stop (red) codons.

There are three rules that govern the arrangement and use of codons in mRNA. First, codons are always read in only one direction: from 5' to 3'. Second, the message is translated in a fixed **reading frame**, which is set by the initiation codon. Third, there is never any overlap or gap in the code. If this were to occur, it would throw off the reading frame and cause aberrant changes in protein structure.

Keep in mind that any deviations in the DNA sequence, such as a mutation resulting in a change of amino acid, would ultimately affect the structure (and therefore the function) of the protein that is coded for.

GENE EXPRESSION

Now that we have examined DNA replication and the genetic code, we can turn our attention to the expression of genes. A gene is a segment of DNA that codes for a functional product (often a protein such as an enzyme), and the production of that product based on the information contained in the DNA is called gene expression.

As we have seen, replication ensures that the DNA of one cell is available for the next generation. Any changes (mutations) that occur as a DNA strand is replicated are therefore carried in subsequent cell generations. This is important because changes in the genetic information can help organisms become and remain pathogenic and more or less virulent. As we examine the expression of genes into functional proteins, we will see that the mechanisms of expression (1) involve specific interactions between proteins, RNA, and DNA and (2) are highly regulated.

There are two parts to gene expression: transcription (construction of RNA from a DNA template) and translation (construction of protein by using RNA instructions).

Transcription

Transcription is the process whereby RNA is made from DNA. This process is similar to DNA replication in that DNA is the template for the production of RNA. However, there are several important differences:

- Only one strand of DNA serves as the template.

- RNA synthesis does not require a primer.

- RNA does not remain base-paired to the DNA template once transcription is complete.

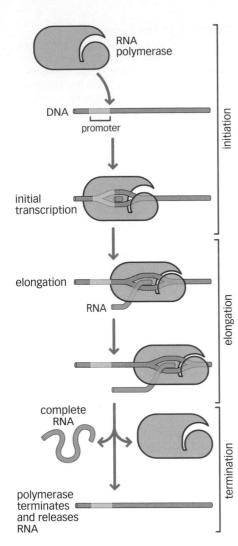

Figure 11.12 The three phases of transcription. Initiation, elongation, and termination result in the synthesis of a complete RNA strand. Notice the claw shape of the RNA polymerase, which binds to the promoter region of the DNA strand to begin transcription.

- RNA polymerase is a poor proofreader. This makes RNA synthesis less accurate than DNA replication.

- Transcription copies only certain portions of the DNA strand, whereas replication copies all of it.

The process of transcription proceeds through three steps—initiation, elongation, and termination—in which different events occur (**Figure 11.12**).

During initiation, RNA polymerase first binds to a DNA sequence called the **promoter**. This complex is converted to a different configuration that allows the initiation phase to continue; the DNA strands are separated in front of the complex, producing a bubble in the DNA (**Figure 11.13**) in which RNA is made. It is important to remember that the bubble is only as large as needed for the production of RNA, and the separated strands of the DNA are immediately rejoined right behind the polymerase. Just as we saw in the replication of DNA, nucleotides are added only at the 3′ end of the growing strand, and complementary base pairing ensures that the correct nucleotide is added. It is the base pairing between the DNA template and growing RNA strand that ensures that the information carried in the DNA strand is transferred correctly to mRNA.

After about 10 ribonucleotides have been added, the polymerase transitions from the initiation phase to the elongation phase. During elongation, the RNA polymerase multitasks in that it unwinds the DNA ahead of it, adds ribonucleotides to the growing end of the RNA strand and re-anneals (closes) the DNA strands behind. During this process, the polymerase does limited proofreading as well. The initiation and elongation phases of transcription can be viewed in **Movies 11.4** and **11.5**.

Of all the steps in the synthesis of RNA, termination is the least well understood. For our purposes, it is enough to say that when the polymerase stumbles over a segment of the DNA that signals the end of the required RNA segment it detaches from the DNA.

Translation

Now that we have seen how RNA is assembled, we need to examine **translation**, the process by which a cell makes proteins by reading the information contained in its mRNA. This process is directly affected by any errors in either the DNA template or the mRNA itself.

Translation uses the sequence of nucleotides in mRNA to generate a specific sequence of amino acids and thereby manufacture a protein needed by the cell. Like transcription, translation is a very highly conserved function seen in all cells. Translation also requires a high energy investment. In a rapidly dividing cell, up to 80% of all the energy is invested in the making of new proteins by translation.

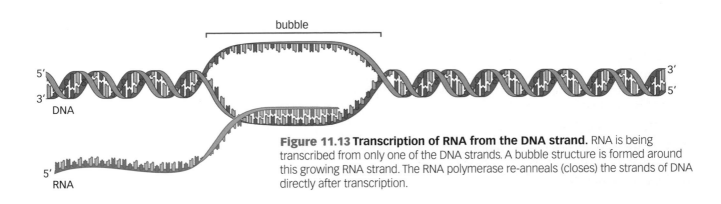

Figure 11.13 Transcription of RNA from the DNA strand. RNA is being transcribed from only one of the DNA strands. A bubble structure is formed around this growing RNA strand. The RNA polymerase re-anneals (closes) the strands of DNA directly after transcription.

Translation requires all the three types of RNA—messenger, transfer, and ribosomal—each of which is transcribed from DNA during transcription.

Messenger RNA in Translation

As we mentioned in our discussion of the genetic code, the mRNA message is translated in a fixed reading frame (known as an **open reading frame** or ORF). The initiation (start) codon at the beginning of each ORF indicates the start of a specific protein's amino acid sequence. Translation starts here and moves along the mRNA strand in the 5′ → 3′ direction until a stop codon is reached and the process stops.

Recall from Chapter 4 that ribosomes are non-membrane-enclosed cytoplasmic structures made up of protein and ribosomal RNA and are the site of protein production in the cell. Prokaryotic mRNA molecules contain a sequence of nucleotides that bind to the ribosomal components of the ribosome through complementary base pairing. This allows the mRNA to be properly oriented along the ribosome.

Transfer RNA in Translation

Transfer RNA (tRNA) molecules form a physical link between mRNA attached to a ribosome and the amino acids being added to the growing protein chain. They can achieve this because each tRNA carries a specific amino acid that corresponds to a specific codon in the codon table. Each tRNA molecule is about 75–95 nucleotides long, and ends with the sequence 3′-CCA-5′. This codon is important because it represents the site at which the amino acid is coupled to the tRNA by the enzyme aminoacyl-tRNA synthetase. It is also essential that each tRNA is specific for only one amino acid. This specificity is essential because if the tRNA could carry just any amino acid, the genetic code would not work.

As mentioned earlier, a tRNA molecule can fold back on itself and form loops. The folded tRNA molecule can have a cloverleaf structure, as shown in **Figure 11.14**. The stem part of the cloverleaf, called the acceptor arm, is where the amino acid specific for that particular tRNA binds.

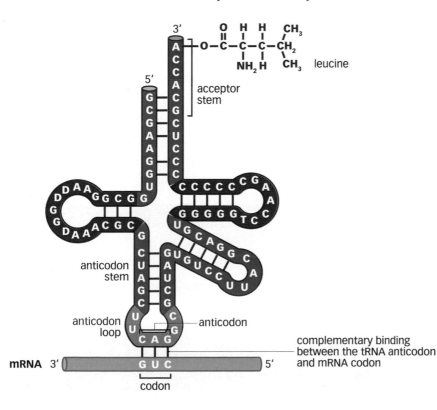

Figure 11.14 A diagram of the transfer RNA (tRNA) molecule. The acceptor arm sequence of 3′-ACC that is present on all tRNAs signifies the amino acid-binding section of this carrier molecule. Also note the anticodon region, which uses complementary base pairing with the messenger RNA (CUG in the figure; remember that the mRNA entering the ribosome is read from 5′ to 3′) to ensure that the correct amino acid (leucine) is placed in the growing polypeptide chain.

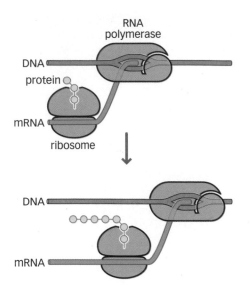

Figure 11.15 The prokaryotic ribosome and the RNA polymerase can work in unison. As the mRNA is transcribed, ribosomes begin to translate protein.

The loop farthest from the acceptor arm is called the **anticodon loop** and is the most important loop because it is where codon recognition occurs. The anticodon loop contains an anticodon, which uses complementary base pairing to attach to the codon of mRNA. This ensures that the correct amino acid is placed in the protein sequence. So the translation from nucleotides to amino acids essentially occurs on the basis of the interaction between the codon and the anticodon.

How does tRNA pick up the amino acid? This may be the most important reaction in protein synthesis and is a two-step process. Step one is the coupling of the amino acid to AMP. The second step is the transfer of the amino acid to the 3′ end of the tRNA. These reactions require the presence of enzymes called **aminoacyl-tRNA synthetases**, with each amino acid having its own synthetase. Because there are 20 amino acids, most organisms have 20 different synthetase enzymes.

The process of specifically coupling amino acids has a high rate of fidelity and is associated with an interaction between the acceptor arm and the anticodon loop of the tRNA molecule. In fact, the incorrect amino acid is attached on less than 1 in 1000 occasions. This specificity is very important because the ribosome is unable to discriminate right from wrong amino acids. It blindly accepts any tRNA if there is proper codon–anticodon recognition. The enforcement of the specificity of translation is therefore due to aminoacyl-tRNA synthetases.

The bond joining the amino acid to the tRNA is a high-energy bond. The energy released in breaking this bond is used to form the peptide bonds between amino acids as a protein is created.

The Ribosome in Translation

As already noted, the ribosome contains the machinery that directs protein synthesis. It is composed of three molecules of rRNA and more than 50 proteins and is a highly efficient machine that can add 2–20 amino acids per second. Because prokaryotes have no nucleus, the ribosome attaches to mRNA as the latter is being made. Recall that RNA polymerase moves along the DNA template strand and transcribes the information encoded in the DNA sequence into a strand of mRNA. As this mRNA leaves the polymerase molecule, it can be immediately bound to a ribosome (**Figure 11.15**). In fact, as one ribosome moves along the mRNA strand, another binds onto the same strand right behind the first. In an elegant piece of cellular engineering, transcription and translation can be linked because transcription occurs in the 5′ to 3′ direction, adding nucleotides to the 3′ end of mRNA, while translation starts back at the 5′ end. Efficient linking and high speed of the processes is necessary to permit the sometimes extremely fast growth of bacteria.

The ribosome has two subunits of different sizes (**Table 11.2**). The large subunit is responsible for peptide bond formation; the small subunit

Fast Fact The shape of the ribosome is a primary targeting strategy for antibiotics.

Table 11.2 The structure of prokaryotic ribosomes.

Property	Value
Overall size	70S
Small subunit	30S
Proteins in small subunit	Approximately 21
Size of RNA in small subunit	16S (1540 bases)
Large subunit	50S
Proteins in large subunit	Approximately 34
Size of RNA in large subunit	23S (2900 bases) and 5S (120 bases)

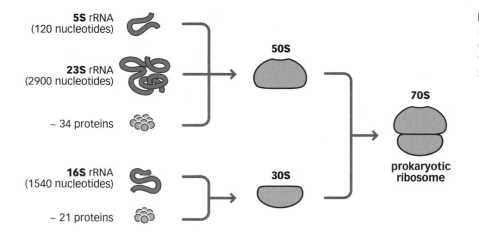

Figure 11.16 Composition of the prokaryotic ribosome. Notice the difference in weights (S units) between the intact ribosome (70S) and each of the subunits (30S and 50S).

contains the decoding center, where tRNA binds to the mRNA (it is here that codon–anticodon recognition occurs). We designate the large subunit as the 50S subunit and the small one as the 30S subunit. The S stands for Svedberg units, a measurement of the mass of the subunit. Interestingly, the mass of the whole ribosome is only 70S rather than the 50S + 30S = 80S you might expect, because shape and volume also affect the measurement.

The 50S subunit is made up of a 5S rRNA and a 23S rRNA, which are coupled with many different proteins. The 30S subunit also contains a variety of proteins but contains only one 16S rRNA molecule, which is very specific and can be used for bacterial species identification (**Figure 11.16**).It is important to once again recall that the combination of rRNA molecules and protein molecules confers on these subunits a three-dimensional shape.

Translation occurs in cycles. During each cycle, the large and small ribosome subunits associate with the mRNA, translate the information contained in the mRNA sequence, and then dissociate from it (**Figure 11.17**).

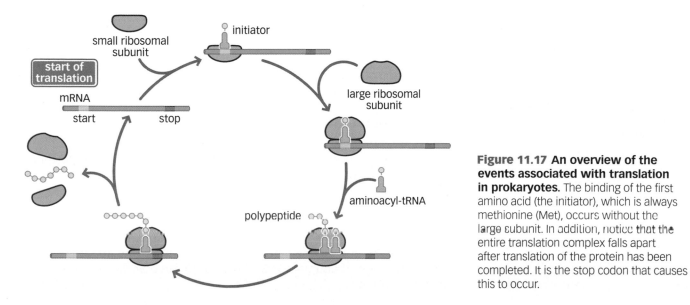

Figure 11.17 An overview of the events associated with translation in prokaryotes. The binding of the first amino acid (the initiator), which is always methionine (Met), occurs without the large subunit. In addition, notice that the entire translation complex falls apart after translation of the protein has been completed. It is the stop codon that causes this to occur.

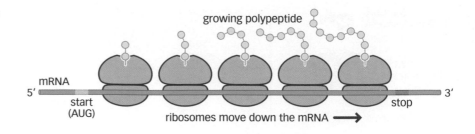

Figure 11.18 A polyribosome. More than one ribosome can complex with messenger RNA as it is being transcribed. Therefore, many ribosomes moving down the same mRNA can be translating. This is important if large amounts of a specific protein are required by the organism.

As mentioned above and illustrated in **Figure 11.18** and **Movie 11.6**, more than one ribosome can attach to the same mRNA molecule. When this occurs, the complex is referred to as a **polyribosome** (or polysome).

Formation of Peptide Bonds in Translation

Recall that proteins are composed of sequences of amino acids linked to one another by peptide bonds. Although the ribosome catalyzes the formation of these bonds, they are actually formed on the tRNA molecule attached to the ribosome. The reaction that bonds each amino acid being added to the growing peptide chain is called the **peptidyl transferase reaction**. Remember that the energy needed to create these peptide bonds comes from the energy released when the amino acid is released from its tRNA.

The ribosome has three binding sites for tRNA molecules (**Figure 11.19**). The **A site** is for tRNA carrying a single amino acid. The **P site** is for peptidyl-tRNA, which is holding on to the growing peptide chain, and the **E site** is for tRNA that is in the process of exiting from the ribosome.

The ribosome is honeycombed with tunnels that facilitate the movement of the components involved in protein synthesis. Both the decoding center on the small subunit and the site of peptide bond formation on the large subunit are buried deep in the interior of the ribosome. The mRNA strand enters through one tunnel and exits through another, both located in the small subunit. The entry tunnel is wide enough for a single mRNA strand only, so that there can never be problems with bunching up of the mRNA molecules. Between the mRNA entry and exit tunnels is an area of the ribosome accessible to incoming tRNA molecules. This area is the location of the A and P sites. Finally, there is a separate tunnel through the large subunit that is the exit path for the newly synthesized polypeptide chain, and the width of this tunnel restricts premature folding of the polypeptide.

Just as with transcription, there are three stages of translation: initiation, elongation, and termination.

Translation Initiation

Initiation of translation requires three things: (1) recruitment of the ribosome to the mRNA to form what is referred to as the **translational apparatus**, (2) placement of a methionine-carrying tRNA on the P site (note that this first tRNA does not go to the A site first), and (3) precise positioning of the ribosome over the mRNA start codon (Figure 11.19). This last step is the most crucial because it establishes the open reading frame. If this positioning step is off by even one position, the synthesized protein will be completely inaccurate. The result of this initiation stage is the formation of an intact 70S ribosome situated at the start codon of the mRNA strand.

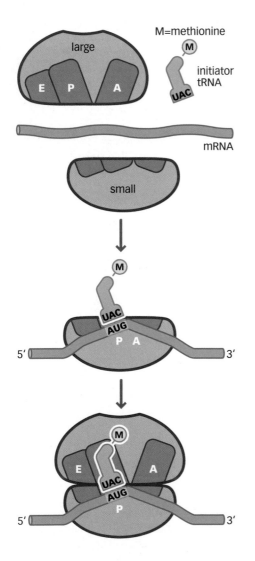

Figure 11.19 Initiation of translation showing the three binding sites on the ribosome. Each of the sites (A, P, and E) spans both the large and small subunits. Note that the polypeptide always begins with the tRNA for methionine binding at the P site.

Figure 11.20 A summary of the steps involved in translation. The ribosome moves along the message, causing the tRNA molecules to move from site to site. Note that the peptide bond is formed between the newly arrived amino acid in the A site and the amino acid in the P site. Recall that the energy needed to form this bond is carried by the charged tRNA entering the A site.

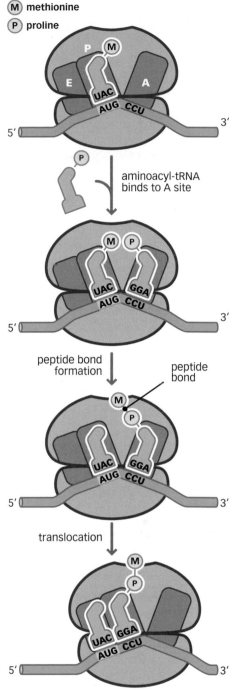

Translation Elongation

After initiation, three steps must be taken in order for the amino acids to be added to the methionine recruited to the mRNA during initiation (**Figure 11.20**). First, a tRNA carrying the next amino acid coded for in the mRNA sequence is loaded into the A site (remember that this placement requires matching at the anticodon region). Second, a peptide bond is formed between this amino acid and the amino acid (methionine) in the P site. Third, each tRNA moves—the one at the A site to the P site, and the one at the P site to the E site. The process of elongation can be seen in **Movies 11.7** and **11.8**.

Ribosomes have several checks to ensure that the correct tRNA binds, which in turn helps prevent the wrong amino acid from being added to a sequence. Additional hydrogen bonds form at the site where the mRNA codon and the tRNA anticodon interact to help orient and hold the codon and anticodon in place. If there is disagreement in base pairing between the mRNA codon and the tRNA anticodon, these hydrogen bonds cannot form, and the interaction is unstable (**Figure 11.21**). The process of accommodation uses the conformational strain placed on the tRNA anticodon when it is aligning with the mRNA strand. Only the correct tRNA–amino acid complex can withstand the strain of interacting with the mRNA codon. The wrong tRNA is too strained to remain in place and so is automatically dissociated from the ribosome. Lastly, the proteins that regulate elongation help prevent the incorrect binding of amino acids.

Translation Termination

Translation continues until a stop codon enters the A site. These codons are recognized by specialized proteins that cause the translation complex (ribosome plus newly synthesized peptide chain) to break down. At this time, the peptide chain is released from the ribosome and begins

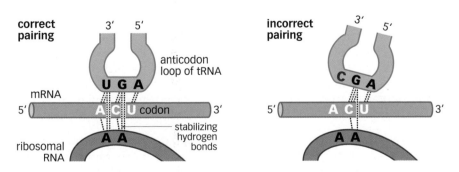

Figure 11.21 The formation of extra hydrogen bonds ensures the correct pairing of the messenger RNA codon and the anticodon region of the transfer RNA. Note that the stabilizing hydrogen bonds can only form if there is complementary bonding between the mRNA codon and the anticodon region of the tRNA. If there is a mismatch, these bonds do not form and the strain involved causes the mismatched tRNA to break away from the ribosome.

to form its secondary and tertiary structure. In addition, the large and small subunits of the ribosome dissociate from the mRNA and begin to recirculate in the cytoplasm of the cell in search of a new mRNA strand.

Keep in Mind

- The genetic code is based on combinations of three letters called codons.
- Each codon codes for a specific amino acid except for the three stop codons.
- There are three stop codons and one start codon in the code.
- Gene expression is the process of making a functional product based on the genetic information contained in the DNA and consists of transcription and translation.
- Transcription proceeds through three steps: initiation, elongation, and termination.
- Translation uses messenger RNA, transfer RNA, and ribosomal RNA and occurs at the ribosome of the cell.
- The ribosome is made up of two subunits that contain RNA and protein.
- Amino acids are brought to the ribosome by transfer RNA molecules, which are specific for certain amino acids.
- While at the ribosome, peptide bonds form between adjacent amino acids.
- Translation stops when a stop codon enters the ribosome.

REGULATION OF GENE EXPRESSION

Now that we have examined DNA structure, the genetic code, transcription, and translation, it is important to put the expression of genes into context. As we have seen, genes code for functional proteins, and the synthesis of these proteins is energetically expensive. Because bacteria are required to be energetically frugal, not all genes in a bacterium are expressed all the time, and gene expression is therefore highly regulated. For example, in *Escherichia coli*, about 80% of genes are classified as **constitutive**, meaning that they are always on (being expressed). The other 20% are genes that are expressed only when required.

Gene expression is regulated by mechanisms that either increase or decrease the rate at which a gene is expressed. There are various stages where this regulation can occur, but the most common is during transcription. There are two types of gene expression at this level, both controlled by regulatory proteins. Positive regulation (induction) involves activators, which increase the rate at which the DNA in a gene is transcribed to mRNA. Negative regulation (repression) uses repressors to either decrease the transcription rate for a given gene or prevent any transcription of that gene. Another form of positive regulation is de-repression, which involves molecules that interfere with the repressor.

These activators and repressors are DNA-binding proteins that recognize two sites on the DNA double helix at or near the genes they control. The **promoter site** is where RNA polymerase binds to the DNA to begin transcribing specific genes. This site is adjacent to a region of the DNA strand referred to as the **operator site**, which is where the regulatory proteins bind. These sites overlap to a degree such that when the operator site is occupied by a regulatory protein, the RNA polymerase cannot bind properly to the promoter site, thereby inhibiting transcription and gene expression.

Induction

Induction turns on genes that are usually turned off. The best example of positive regulation of gene expression (that is, induction) is the *lac* operon (**Figure 11.22**). An **operon** is a set of genes that are co-regulated. The *lac* genes code for enzymes that enable bacteria to utilize the sugar lactose. It is a system that employs both repressor and inducer mechanisms.

As we mentioned above, making any protein requires energy: if that protein is not used, its synthesis is a waste of energy. Glucose is the preferred energy source for most bacteria because it is easily broken down to yield high levels of energy (as we saw in Chapter 3, if broken down using oxygen, one molecule of glucose can provide 38 ATPs), so the genes coding for enzymes that break down glucose are constitutive (always on), even if glucose is not present. You might think that this is wasteful on the part of the bacteria, but glucose breakdown is important enough for this to be acceptable.

So what does the cell do if no glucose is present? It uses whatever it can, for example lactose. However, because lactose is used only if glucose is not present, the genes that code for enzymes that break down lactose are normally off. The genes for lactose are said to be inducible (normally off but can be turned on). There are three structural genes used for the breakdown of lactose, namely *lacZ*, *lacY*, and *lacA*. These lie adjacent to one another on the chromosome of *E. coli*. The promoter for these genes is located at the 5′ end of *lacZ* (see Figure 11.23). The mRNA from these genes is polycistronic, meaning that it contains the information for all three proteins on one message. *LacZ* codes for β-galactosidase, an enzyme that cleaves lactose into galactose and glucose. *LacY* codes for permease, which is a membrane transport protein for lactose, and *lacA* codes for transacetylase, which is thought to have a role in ridding the cell of toxic by-products of galactose.

It is important to remember that *lac* genes are expressed only when there is no glucose present. When glucose is present, they are repressed (turned off). There are two regulatory proteins involved in this repression: the *lac* repressor protein and the *lac* activator called CAP (for catabolite activator protein; also known as CRP, for cAMP receptor protein). The *lac* repressor protein is encoded by *lacI* and is expressed constitutively, meaning that it is always being produced. This makes sense because the cell does not want to use lactose as long as glucose is present, and having the genes to utilize lactose turned on would be a waste of energy.

The *lac* repressor protein binds to the DNA in such a way as to prevent the transcription of the structural genes unless lactose is present and glucose is absent. When bound to the operator site, the *lac* repressor overlaps part of the promoter region and physically interferes with the binding of the RNA polymerase, resulting in no mRNA being made (**Figure 11.23a**).

For the genes of the *lac* operon to be turned on (induced), the repressor must first be inhibited. This occurs through a mechanism called allosteric control. Recall that the three-dimensional structure of proteins can be

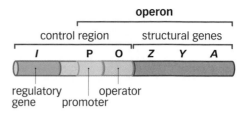

Figure 11.22 The *lac* operon. The *lac* operon contains the regulatory elements O and P and the structural genes Z, Y, and A. The regulatory gene, *I*, is located nearby. Expression of the structural genes is controlled by *I*, P, and O.

(a)

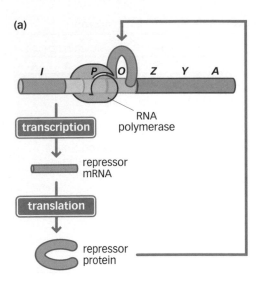

(b)

Figure 11.23 The operation of the lactose operon. Panel a: The lactose operon in the off (repressed) position. The repressor protein is able to bind to the operator segment of the operon and prohibit transcription of the structural genes. Panel b: The lactose operon is in the on position when allolactose (a fragment of lactose) binds to the repressor protein and prevents it from binding to the operator gene. RNA polymerase in combination with CAP protein can now bind to the promoter region, and transcription of the structural genes can proceed.

affected in a variety of ways. For example, binding of a molecule at one site can change the capacity to bind another molecule at a separate site on the same protein. This is the mechanism used for induction of the lactose operon.

Although there is continuous repression of the lactose genes when glucose is present, a few transcripts are usually made, which permits the entry of small amounts of lactose into the cell when it is present in the environment. The lactose is then converted to **allolactose**, which binds to the lactose repressor protein and changes its shape such that it can no longer bind to the operator site on the DNA (**Figure 11.23b**). It is important to note that the binding site for allolactose on the repressor protein is away from the active site that binds to the DNA molecule. When lactose, and thus allolactose, is no longer present, the repressor protein can return to its original shape and bind again to the operator region of the DNA, effectively reestablishing repression of the operon.

The CAP protein acts in a similar fashion. When glucose is present, levels of cyclic AMP (cAMP) are low. cAMP is a product that accumulates as available energy decreases. It serves as a hunger signal to the cell that alternative methods for energy production are required. As glucose disappears, the level of cAMP rises and it binds to the CAP protein. This causes a change in the three-dimensional shape such that the CAP–cAMP complex now binds to the DNA. This in turn recruits the RNA polymerase to the promoter site and transcription of the *lac* operon begins. The binding of cAMP to the CAP protein is also allosteric. The protein configuration will therefore return to normal when cAMP levels decline and CAP falls off the DNA. Without the CAP protein, RNA polymerase cannot bind efficiently to the promoter region of the DNA. This CAP protein has been shown to be involved with more than 100 genes in *E. coli* and combines with a wide variety of protein partners for gene expression regulation.

Repression

In addition to the induction (turning on) of genes that we have discussed above, there are also cellular mechanisms that repress (turn off) genes. This is also a very important factor for the energy-conscious

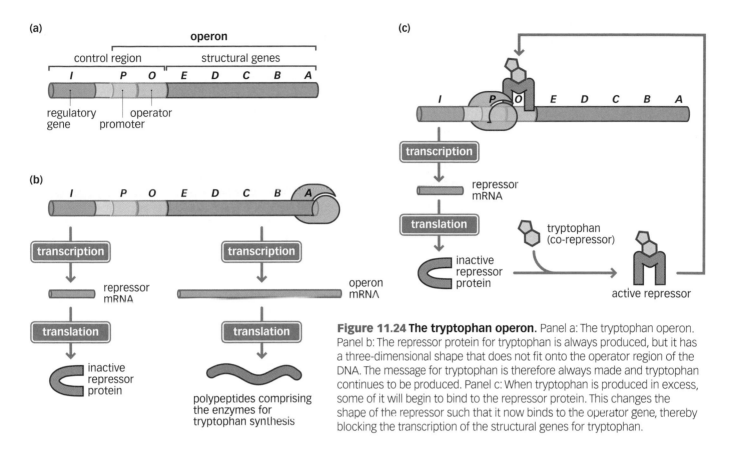

Figure 11.24 The tryptophan operon. Panel a: The tryptophan operon. Panel b: The repressor protein for tryptophan is always produced, but it has a three-dimensional shape that does not fit onto the operator region of the DNA. The message for tryptophan is therefore always made and tryptophan continues to be produced. Panel c: When tryptophan is produced in excess, some of it will begin to bind to the repressor protein. This changes the shape of the repressor such that it now binds to the operator gene, thereby blocking the transcription of the structural genes for tryptophan.

cell, because protein synthesis costs energy, and the synthesis of excess amounts of protein is wasteful. Consequently the cell can regulate the expression of genes in a negative fashion. This is similar to feedback inhibition and occurs post-transcriptionally. For an example of this repression we can look at the synthesis of the amino acid tryptophan. In this situation, tryptophan is made until it begins to accumulate in excessive amounts. At this point the tryptophan actually becomes a co-repressor of its own synthesis.

This occurs because the cell produces a repressor protein specific for the production of tryptophan, and like the repressor for lactose it is always produced. However, this protein is produced in a configuration that cannot bind to DNA to repress the production of messenger RNA. Excess tryptophan will bind to this repressor protein and change the shape such that the repressor will now bind to the DNA; eventually the production of tryptophan ceases (**Figure 11.24**). This is an elegant mechanism because when the amount of tryptophan decreases, there is none left to bind to the repressor. This causes the configuration of this protein to change back to its original shape such that it can no longer bind to the DNA, and tryptophan is again produced.

Keep in Mind

- The expression of a gene is carefully regulated.
- Genes can be constitutive (always on), inducible (off and can be turned on), or repressible (on but can be turned off).
- An operon is a set of structural genes that share a common promoter and operator and are regulated together by a control gene.

- Regulatory proteins control induction and repression through binding on the DNA at the site known as the operator site.
- Gene expression is regulated at the level of mRNA production. When induced, message is made; when repressed, message is not made.

MUTATION AND REPAIR OF DNA

One of the most important topics in bacterial genetics is the process of mutation, which can cause changes in the DNA sequence. As we have seen throughout our discussions, any change in the DNA sequence can result in changes in the proteins coded for by that gene. In many cases, these changes can have adverse effects on the cell and could have lethal consequences (Table 11.3). However, mutations can also be responsible for increased virulence as well as antibiotic resistance. Remember also that although the level of mutations has to be kept to a minimum, some mutations must occur if the organism is to evolve. In fact, adaptation of a species, which is a cornerstone of evolution, is directly related to mutations. The bacterial cell therefore depends on a balance between mutation and repair. Mutations can occur because of errors in the accuracy and fidelity of DNA replication; although the proofreading capability of DNA polymerase can cope with most of these errors, some escape correction.

Mutations can also result from chemical damage to DNA, and more importantly from insertions that are generated by **transposons** (mobile genetic elements). For these reasons, there are multiple overlapping safety systems that enable the bacterial cell to cope with changes in DNA. In the following discussion, we examine some of these systems.

Replication Errors

The simplest mutations are switches of one base for another. These are called **point mutations** because they affect only one nucleotide. There are two kinds of point mutation: transitions, in which one pyrimidine is switched for the other, or one purine is switched with the other, and transversion, in which a pyrimidine is switched for a purine or vice versa.

Table 11.3 Types of mutation.

Type of mutation	Effect
Point mutation	
Change at one nucleotide in DNA sequence with no change in amino acid	No effect on the protein. Called a silent mutation
Change at one nucleotide in the DNA sequence that results in a change in the amino acid sequence	Causes a change in the protein that can cause significant alteration of protein function. Called a missense mutation
Change at one nucleotide in DNA sequence that creates a stop codon	Produces truncated and nonfunctional protein. Called a nonsense mutation
Frameshift mutation	
Deletion or insertion of one or more nucleotides in the DNA sequence	Changes the entire sequence of codons and greatly alters the amino acid sequence. Transposition is a form of insertion that can cause a frameshift and change the genetic makeup of a bacterium

More drastic mutations are those that involve insertions or deletions of nucleotides in the DNA sequence. These may be the result of transposition of genes from one place to another on the chromosome, or they may result from erroneous recombination. When one or more than one nucleotide is involved, as in insertion or deletion, the mutation is referred to as a **frameshift**.

The rate of **spontaneous mutations** ranges from 1 in 10^{-6} to 1 in 10^{-11} per round of replication. However, certain sections of the chromosome, called hot spots, seem to be more susceptible to mutation and therefore have a higher rate of spontaneous mutations. Unless corrected, any mismatch that occurs during replication can become a permanent change and will be propagated through subsequent generations.

It is important to note that there are also **suppressor mutations** that can reverse the primary mutation and reestablish the construction of a functional protein (a case of two wrongs making a right). These suppressor mutations can be intragenic (occurring in the same gene) or intergenic (occurring in separate genes).

How DNA Damage Occurs

DNA can be damaged by hydrolysis, by deamination (the loss of an amino group, $-NH_2$), by alkylation, by oxidation by chemicals called **mutagens**, and by radiation. For example, water can cause the deamination of cytosine, which causes it to become uracil, but this is not a normal event and water is not a mutagen. Although uracil is not normally found in DNA, it can bind to adenine and cause a mutation that is then carried from generation to generation.

In alkylation, a G-C base pair can be changed to an A-T pair after replication. An example of damage by oxidation is the oxidation of guanine to a derivative that can bind to adenine as well as to cytosine. After replication, the G-C pair is replaced by A-T.

Gamma radiation and ionizing radiation cause double-strand breaks in the DNA backbone, which are very difficult to repair. In most cases, these types of radiation are lethal to bacteria. Ultraviolet radiation can also cause DNA damage through the formation of thymine dimers, which are two thymine bases on the same strand of DNA bound to each other. When these bases link together on the same strand, they cause DNA polymerase to stop at that site, and replication stops.

In addition, mutations can be caused by planar compounds taken up from the environment that can slip between two bases on the DNA strand and cause errors in replication. Examples of these compounds are acridine and ethidium, which can cause shifts in the reading frame and negatively affect transcription and translation. There are also base analogs that are so similar to the DNA bases that they are mistakenly placed in the growing DNA strand during replication. **Figure 11.25** shows two examples.

Repair of DNA Damage

The most frequently used method for repairing damaged bases in DNA is the removal and repair of altered bases. There are two principal mechanisms. In the first, base excision, a damaged base is removed (in other words, excised) from a DNA double helix. After the base has been removed, DNA polymerase fills in the gap, and DNA ligase repairs the break in the strand.

The enzymes involved in base excision recognize different types of damage and diffuse along the DNA double helix, scanning for a specific type of damage. X-ray crystallographic studies have shown that when a damaged base is detected by an enzyme, the base is flipped away from the

Figure 11.25 Analogs of nucleotide bases can become erroneously inserted into growing DNA strands. Notice the structural similarity between the chemical 2-aminopurine and the base adenine (panel a), as well as that between 5-bromouracil and thymine (panel b).

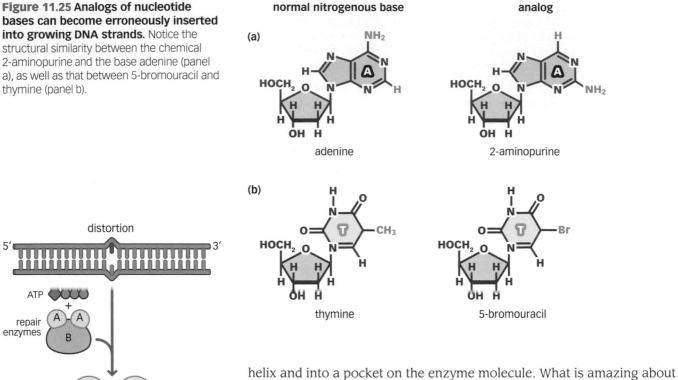

normal nitrogenous base

analog

(a)

adenine

2-aminopurine

(b)

thymine

5-bromouracil

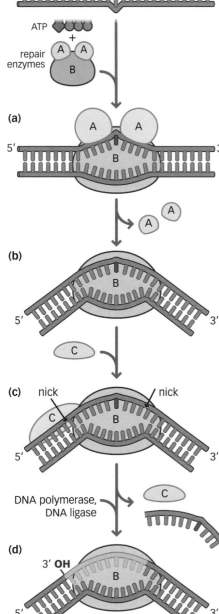

distortion

ATP

+

repair enzymes

(a)

(b)

(c) nick   nick

DNA polymerase, DNA ligase

(d)

3' OH

helix and into a pocket on the enzyme molecule. What is amazing about all this is that this flipping out of the base does not cause major distortion of the DNA helix, reaffirming the flexibility of DNA.

The second repair mechanism is called **nucleotide excision** (**Figure 11.26**). Unlike the enzymes involved in base excision, those involved in nucleotide excision do not recognize any particular type of damage. Instead, the repair enzymes look for distortions in the double helix. These distortions trigger a chain of events leading to the removal of short single-strand sections of DNA surrounding the distortion. As with base excision, nucleotide excision is followed by the actions of DNA polymerase and DNA ligase to fill in the gap.

Another DNA repair mechanism is called **photoreactivation**. This mechanism unlinks thymine dimers formed when DNA is exposed to ultraviolet radiation. Photoreactivation is accomplished by an enzyme called photolyase. In the dark, this enzyme binds to the dimer. When then exposed to light, the photolyase becomes activated and breaks the thymine–thymine bond apart (**Figure 11.27**).

Keep in Mind

- Mutations have an important role in the infection process because pathogens can become resistant to antibiotics or generally more virulent through mutation.
- Bacteria depend on a balance between mutation and repair.
- Mutations can result from transposition of genes in the chromosome, point mutations, or frameshift mutations.

Figure 11.26 Nucleotide excision repair. Panel a: This repair process uses proteins (A) to scan DNA, looking for chemical distortions (remember the stereochemistry of the DNA molecule). Panel b: An enzyme (B) unzips the area around the distortion. This is followed by another enzyme (C) nicking the strands on each side of the distortion (panel c). Lastly, DNA polymerase fills the gap and DNA ligase reseals the area (panel d).

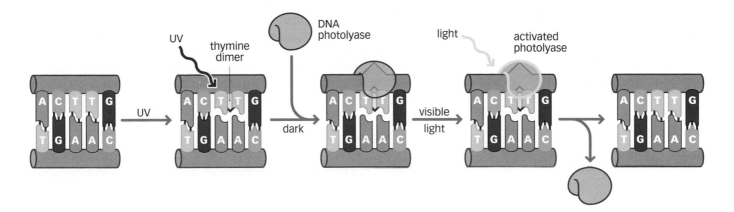

Figure 11.27 Photoreactivation. This is a mechanism that can repair thymine dimers. There are two reactions that take place in this repair mechanism, the binding of the enzyme DNA photolyase, which occurs in the dark, and the second reaction, which requires visible light to activate the enzyme to break the thymine-thymine bond apart.

- Suppressor mutations can reverse the primary mutation.
- DNA can be damaged by chemicals called mutagens, by chemical reactions, or by radiation.
- Repair of damaged DNA can be accomplished by excision repair or nucleotide excision systems in the cell.

TRANSFER OF GENETIC INFORMATION

In the last part of this chapter, we see that it is very important to examine how bacteria can transfer genetic information. This discussion is based on the fact that bacteria can shuffle their genes in a process referred to as genetic **recombination**. In previous chapters, as well as in those that follow, we find that there are many factors that can influence pathogenicity and virulence of bacteria. Recall from Chapter 9 that the presence of fimbriae or a capsule can make organisms more virulent, and that the capacity to form endospores can protect organisms. In addition, many disease symptoms result from the bacterial production of toxins. Each of these traits is genetically controlled, and the transfer of these genetic determinants can make harmless bacteria dangerous.

There are four ways in which genetic recombination can occur. Three of the four mechanisms—transformation, transduction, and conjugation—are transfers of genetic material from one bacterium to another. The fourth, transposition, is a way in which the genes in the DNA of a bacterium are shuffled into a different order.

Transposition

In the recombination process called **transposition**, genetic elements called **transposons** move from one place on a bacterial chromosome to another. (An analogy would be taking a card from one place in the deck and putting it back in another place in the deck.) This rearrangement is accomplished in a random fashion, so that the result can be either detrimental (if genes are disrupted) to the bacterium or beneficial (if functional genes are created). If changes to DNA result in negative consequences for the cell, the end result is usually the death of that cell. Therefore negative consequences are rarely, if ever, seen. In contrast, any genetic change that enhances a bacterium's ability to survive and grow is selected for and maintained throughout subsequent generations.

Transpositions are the most common cause of new mutations in many organisms. They may be important causes of mutations that lead to genetic diseases in humans. In bacteria, transposons can carry genes encoding proteins that promote resistance to one or more antibiotics.

For a transposed gene to function in its new location, the gene or operon structure needs to be maintained. Transposons move in and out of DNA strands by using cleavage and rejoining mechanisms akin to cutting and pasting. These mechanisms involve excision of the transposon from its initial location in the DNA, followed by integration into a new DNA site. Even viruses seem to use methods similar to transposition to integrate into host chromosomes.

The rest of our gene-transfer discussion involves the transfer of genetic information from one cell to another. There are three ways in which this transfer can happen: transformation, transduction, and conjugation. Successful recombination can result from a combination of transfer events.

Transformation

The major point to remember about **transformation** is that it involves naked DNA, which can be taken up from the environment, for example when released from cells that have died. This DNA is taken up by a bacterial cell and recombines with that cell's DNA. Therefore traits that are encoded on these DNA pieces can become part of the recipient cell's genetic repertoire.

For this to happen, the recipient cell must be competent, which means that the wall of this cell can accommodate the uptake of large molecules such as pieces of DNA. Some bacteria are naturally competent, whereas others can become competent after chemical treatment and this is routinely used in research laboratories.

The mechanism that confers competence on a bacterial cell is not well understood but may involve Ca^{2+} ions, which seem to affect the conformation of cell wall proteins and also potentially shield the electrical charge on the naked DNA pieces, allowing them to move through the membrane.

Transformation is an inefficient process because only a small percentage of competent cells take up naked DNA from their surroundings, and those few take up only a small percentage of the total amount available to the cell.

Transformation was discovered by the English scientist Fredrick Griffith in 1928 (**Figure 11.28**). He used two strains of *Streptococcus pneumoniae*, one that was encapsulated and caused pneumonia and death, and one that was nonencapsulated and harmless. When encapsulated organisms that had been killed were injected there was no disease, indicating that the dead organisms were harmless. However, if dead encapsulated organisms were mixed with live nonencapsulated organisms and injected into mice, the mice died. Furthermore, when *S. pneumoniae* cells were isolated from these dead mice, they were encapsulated. Clearly, something (naked DNA) from the dead bacteria had been taken up by the live bacteria. Keep in mind that this observation was made decades before the structure and function of DNA were completely elaborated.

Transduction

Transduction is a common event in both Gram-positive and Gram-negative organisms; it involves the transfer of genetic information via a bacterial virus (called a **bacteriophage**). Transduction permits the movement of genetic information from one infected host (the bacterium) to

Fast Fact Transformation is routinely used to genetically engineer bacteria.

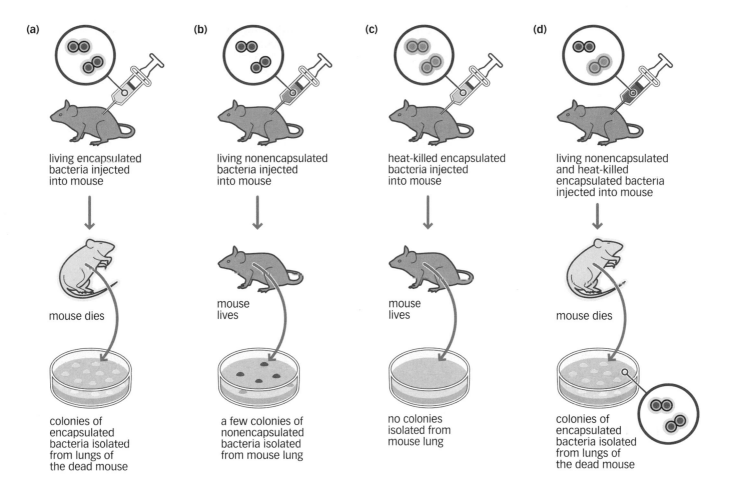

(a) living encapsulated bacteria injected into mouse → mouse dies → colonies of encapsulated bacteria isolated from lungs of the dead mouse

(b) living nonencapsulated bacteria injected into mouse → mouse lives → a few colonies of nonencapsulated bacteria isolated from mouse lung

(c) heat-killed encapsulated bacteria injected into mouse → mouse lives → no colonies isolated from mouse lung

(d) living nonencapsulated and heat-killed encapsulated bacteria injected into mouse → mouse dies → colonies of encapsulated bacteria isolated from lungs of the dead mouse

another followed by recombination of the newly arrived DNA into the host chromosome. This integration into the new host DNA actually uses the same cut-and-paste mechanisms as transposition.

Transduction can be random, in which case it is referred to as **generalized transduction**, or specific, in which case it is referred to as **specialized transduction**. In generalized transduction (**Figure 11.29**), after a phage infects a bacterium, a phage-encoded enzyme called DNase is activated. This enzyme cleaves DNA contained in the phage into the proper size so that it can be enclosed in new viral particles as the infection proceeds. The DNase also chops the bacterium's chromosome into small pieces, and some of these pieces get enclosed in viral particles. Although these latter viral particles contain bacterial DNA instead of viral DNA, they are still capable of infecting other bacterial cells. When one of the virus particles containing bacterial DNA infects a new host cell, the bacterial DNA it carries recombines with the newly infected host cell DNA in the recipient's chromosome.

In specialized transduction, the phage DNA initially becomes incorporated into an infected cell's chromosome and is called a **prophage**. At some later time, the prophage can excise itself from the chromosome and become enclosed in viral particles as they are produced in the infected cell. The excision of the prophage is usually in a precise location on the chromosome and involves only the prophage DNA. However, excision occasionally happens in a slightly different position, so that the excised segment contains both prophage DNA and bacterial DNA. This bacterial DNA is incorporated into viral particles, just as in generalized

Figure 11.28 An illustration of the Griffith experiment that demonstrated transformation.

Fast Fact The genes that code for production of the diphtheria toxin (eventually potentially causing the death of the infected patient) are transferred by transduction. The non-phage-infected bacterium (*Corynebacterium diphtheriae*) causes only minor illness.

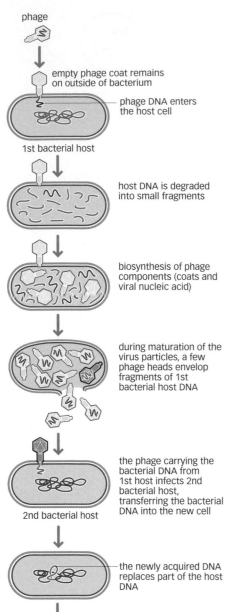

phage

empty phage coat remains
on outside of bacterium

phage DNA enters
the host cell

1st bacterial host

host DNA is degraded
into small fragments

biosynthesis of phage
components (coats and
viral nucleic acid)

during maturation of the
virus particles, a few
phage heads envelop
fragments of 1st
bacterial host DNA

the phage carrying the
bacterial DNA from
1st host infects 2nd
bacterial host,
transferring the bacterial
DNA into the new cell

2nd bacterial host

the newly acquired DNA
replaces part of the host
DNA

this new genetic
material will be carried
through further
generations

Figure 11.30 Bacterial pilus during conjugation. Although the connection shown here is relatively long, it is thought that in addition to permitting the passage of DNA between cells, the pilus may also act as a grappling hook, pulling the recipient cell closer to the donor cell.

Figure 11.29 Generalized transduction. Follow the color designations of the DNA through the process. The outcome is recombination (gene shuffling) between the sequences of DNA from host 1 (blue) with the chromosome of host 2 (black). Phage DNA is colored red.

transduction. The infected bacterium then lyses and releases the viral particles—some containing exclusively viral DNA, and some containing a small amount of donor bacterial DNA along with the viral DNA. When one of the latter infects a recipient bacterium, the donor DNA recombines with the recipient chromosome.

Transduction-related development of pathogenicity or increase in virulence is called phage conversion.

Conjugation

Our final mechanism for gene transfer is **conjugation**, which requires direct contact between donor (F⁺, in the case of the F plasmid) and recipient (F⁻) cells. This important mechanism occurs in both Gram-positive and Gram-negative bacteria. In Gram-positive bacteria, direct contact involves the cell walls of two bacteria sticking together. In Gram-negative bacteria, the process is more elaborate, so that is where we focus our discussion.

The major element involved in any type of conjugation is the plasmid. Recall that a plasmid is a genetic element that is separate from the bacterial chromosome and can replicate independently.

In Gram-negative bacteria, conjugation is facilitated through the formation of a bacterial sex pilus on the donor cell (**Figure 11.30**). Although the pilus is at times relatively long, as shown in Figure 11.31, it does not remain that way during conjugation. It is thought that although the pilus acts as a bridge, it also functions like a grappling hook that pulls donor and recipient close together.

Conjugation in *E. coli* has four steps, as described in **Figure 11.31** and **Movie 11.9.**

- The sex pilus of the donor cell recognizes specific receptor sites on the cell wall of the recipient.
- An enzyme in the donor cell causes the plasmid DNA double helix to unwind, beginning at a specific site called the origin of transfer.

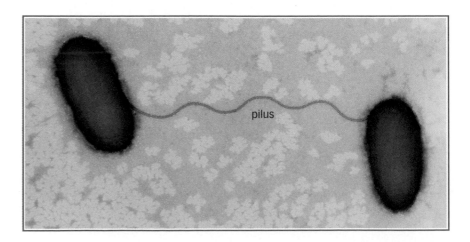

pilus

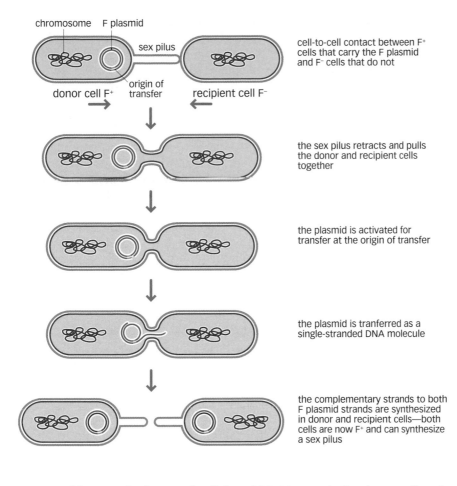

chromosome   F plasmid

sex pilus

origin of transfer

donor cell F⁺   recipient cell F⁻

cell-to-cell contact between F⁺ cells that carry the F plasmid and F⁻ cells that do not

the sex pilus retracts and pulls the donor and recipient cells together

the plasmid is activated for transfer at the origin of transfer

the plasmid is tranferred as a single-stranded DNA molecule

the complementary strands to both F plasmid strands are synthesized in donor and recipient cells—both cells are now F⁺ and can synthesize a sex pilus

Figure 11.31 The steps involved in conjugation. Remember that only one strand of the plasmid is passed from the F⁺ donor cell to the F⁻ recipient cell. DNA replication results in complete plasmids in both donor and recipient cells, which causes the F⁻ cell to become F⁺.

- One of the two single strands of plasmid DNA stays in the donor cell and the other moves from donor to recipient.

- The single strand of plasmid DNA remaining in the donor and the single strand now in the recipient are both replicated. Because of the complementary nature of DNA, after this replication is complete, the donor and the recipient contain identical plasmids.

At the completion of this transfer, the recipient cell becomes a donor and can conjugate with another recipient cell.

It should be noted that cells can also spontaneously lose their plasmid, and when this happens the cell is referred to as cured. Some of the traits that can be transferred on plasmids are listed in Table 11.4.

Table 11.4 Examples of traits that can be encoded on plasmids.

Trait	Examples of organisms with the trait
Antibiotic resistance	*E. coli, Salmonella* species, *Neisseria* species, *Staphylococcus* species, *Shigella* species
Synthesis of pilus	*E. coli, Pseudomonas* species
Utilization of unusual nutrients	*Pseudomonas* species
Virulence	*Yersinia enterocolitica*
Toxin production	*Corynebacterium diphtheriae*
Antibiotic synthesis	*Streptomyces* species

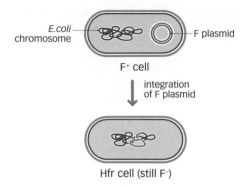

Figure 11.32 Plasmid from the donor cell is incorporated into the host cell's chromosome. Conjugation can result in the formation of an Hfr recipient cell if the plasmid DNA that is received from the F+ donor cell becomes incorporated into the chromosome of the F− cell. Note that in this case the recipient cell remains F−.

As we saw above, the plasmid produced after a transferred single strand has been replicated in the recipient can remain in the cytoplasm as a plasmid, capable of being transferred to another recipient. Alternatively, the replicated plasmid in the recipient can become incorporated into the cell's chromosome (**Figure 11.32**). This occurs because of the presence of recombination sites on the chromosome, which are similar to the insertion sites seen in transposition. As a result of the incorporated plasmid, the recipient's chromosome becomes larger than it was before conjugation. A bacterial cell that harbors the F plasmid integrated into its chromosome is referred to as an **Hfr cell** (the letters stand for **high frequency of recombination**). The way in which genetic material is transferred from this type of cell is illustrated in **Figure 11.33**. DNA from an Hfr cell integrates into the recipient cell's chromosome not by adding to the chromosome but rather by replacing an existing DNA segment of the recipient, which is then enzymatically destroyed. In other words, some genes in the recipient chromosome are replaced by genes from the donor. Most cells that receive DNA transfer from Hfr cells cannot then become donor cells and therefore cannot pass the information they receive to other cells.

Table 11.5 summarizes the types of horizontal genetic transfer we have discussed.

Keep in Mind

- Genetic recombination occurs in bacteria through transposition, transformation, transduction, or conjugation.

- Transposition is a specific form of recombination in which genetic elements called transposons move from one place in the chromosome to another in the same cell.

- Transformation involves the uptake of naked DNA by a cell.

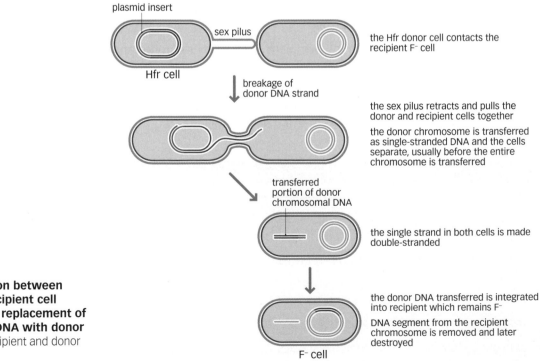

Figure 11.33 Conjugation between an Hfr cell and an F− recipient cell results in the complete replacement of recipient segments of DNA with donor DNA. Interestingly, the recipient and donor cells remain F−.

Transfer mechanism	Effect
Transformation	Uptake of naked DNA. Requires the recipient cell to be competent. Occurrence is very rare
Transduction	Transfer of genetic material by a bacterial virus (bacteriophage). In specialized transduction, only genes that are near the integrated prophage are transferred. In generalized transduction, the host chromosome is fragmented and random genes can be incorporated into new phage particles for transfer
Conjugation	Transfer of genetic material contained in plasmid. When recipient genetic material remains in plasmid form, recipient can become donor. When recipient material is integrated into host chromosome, cell becomes Hfr. Recipient conjugating with Hfr cannot become a donor

- Transduction is caused by a virus transferring pieces of DNA from one cell to another.

- Conjugation occurs when DNA is moved from a donor cell (designated F⁺) to a recipient cell (designated F⁻).

- Each of the transfer mechanisms causes genetic recombination in the recipient cell and therefore can be important in making a pathogen more dangerous.

Table 11.5 Horizontal genetic transfer mechanisms.

GENETICS AND PATHOGENICITY

Although the discussions in this chapter are complex, they are important for understanding disease processes because some of the mutations that occur during DNA replication and the various processes of gene expression can lead to a harmless bacterium becoming pathogenic or to increased virulence in a pathogen, or to antibiotic resistance. The genes coding for toxins, which in many cases are the main causes of successful infection, disease development, and disease progression, are found on plasmids. As we have seen, this information is easily transported from one bacterial cell to another through conjugation. Genes for antibiotic resistance are also found on plasmids and so can be easily transferred between bacteria. Furthermore, certain plasmids known as **dissimilation plasmids** can contain genetic information that allows organisms to become more resistant to disinfectants and more able to adapt to otherwise destructive environments.

SUMMARY

- DNA is the informational molecule of the cell.

- DNA is a double-stranded molecule made up of nucleotides (which consist of a phosphate, a deoxyribose sugar and one of the four bases adenine, thymine, guanine, and cytosine).

- Nucleotides are bound together through complementary base pairing.

- RNA is a single-stranded molecule and contains uracil instead of thymine. It can be found in the form of messenger RNA, transfer RNA, and ribosomal RNA.

- DNA is faithfully replicated by the enzyme DNA polymerase.

- DNA polymerase has a proofreading capability that prevents errors in replication

- Replication occurs at the replication fork (a separation of the DNA strands) and is continuous on one strand and discontinuous (made in pieces) on the other.

- DNA is transcribed into RNA by the enzyme RNA polymerase.

- The genetic code is based on codons, which are combinations of three nucleotides.

- Gene expression is the highly regulated process of making a functional product by transcription and translation.

- Mutations are changes in the DNA and are important in infectious disease because they can lead to antibiotic resistance.

- Genetic recombination can occur in bacteria through transposition, transformation, transduction, or conjugation and can result in pathogenicity or increased virulence.

In this chapter, we have looked at four major topics: DNA replication, gene expression, mutations, and the transfer of genetic material. We have spent so much time on these topics because pathogenic bacteria use these methods to infect their hosts successfully. One of the most obvious examples is the development of antibiotic resistance. Mutations can bring about this resistance, and the genes for antibiotic resistance are located on these easily transferable genetic elements. We shall see in later chapters that antibiotic resistance may be the most important problem facing health care today. As we move through subsequent topics, keep in mind the things we have discussed in this chapter and how they relate to the issues that follow.

(Q) SELF EVALUATION AND CHAPTER CONFIDENCE

Multiple Choice

Answers are given in the back of the book and help can be found in the student resources at:
www.garlandscience.com/micro2

1. A nucleotide is composed of

   A. A phosphate and a nucleotide base
   B. A nucleotide base
   C. A sugar and a nucleotide
   D. A phosphate and a sugar
   E. A phosphate, a sugar, and a nucleotide base

2. In complementary base pairing

   A. A preferentially binds to C
   B. T preferentially binds to G
   C. A preferentially binds to G
   D. T preferentially binds to C
   E. A preferentially binds to T

3. The purines are

   A. Adenine and thymine
   B. Guanine and cytosine
   C. Adenine and cytosine
   D. Adenine and guanine
   E. None of the above pairs is correct

4. The enzyme that catalyzes the removal of the supercoiling of DNA is

   A. DNA polymerase
   B. DNA ligase
   C. RNA polymerase
   D. Topoisomerase
   E. Reverse transcriptase

5. Transcription is the process in which DNA is copied into all of the following except

   A. Messenger RNA
   B. Ribosomal RNA
   C. A new DNA molecule
   D. Transfer RNA

For questions **6–9**, use the following choices: **A.** Found in DNA; **B.** Found in RNA; **C.** Found in both DNA and RNA; **D.** Found in neither nucleic acid. (You can use each choice more than once or not at all.)

6. ATA

7. Codons with two bases

8. Codons with three bases

9. AUG (the initiation codon)

For questions **10–14**, use the following choices, organizing the events in the order in which they occur in bacterial DNA replication: **A.** The initiator protein binds to the replicator sequence; **B.** The entire DNA chromosome is copied; **C.** DNA partly uncoils; **D.** Daughter chromosomes are separated from each other; **E.** The components involved in replication dissociate from the DNA.

10. First event

11. Second event

12. Third event

13. Fourth event

14. Last event

15. A gene is best defined as

   A. A segment of DNA that contains inherited information that defines the structure of a protein or RNA
   B. Three nucleotides that code for an amino acid
   C. A transcribed unit of DNA
   D. A sequence of nucleotides in RNA that codes for a functional product

16. The base that is added to the growing DNA strand is determined by

   A. The base itself
   B. The free end of the growing DNA strand
   C. The template strand
   D. The DNA polymerase
   E. The RNA polymerase

17. The processivity of DNA polymerase has to do with

   A. Reading the genetic code
   B. The fidelity of RNA transcription
   C. The speed of DNA replication
   D. The speed of translation
   E. The speed of transcription

18. Proofreading

   A. Minimizes errors in DNA
   B. Maximizes errors in RNA
   C. Is done by DNA helicase
   D. Is done by the ribosome
   E. None of the above

19. The replication fork is the place where

   A. RNA polymerase attaches to the DNA
   B. DNA replication is interrupted
   C. DNA replication is ongoing
   D. DNA replication ends
   E. Ribosomes attach to the DNA

20. The leading strand of DNA is

   A. Made first
   B. Made last
   C. Made in pieces
   D. Made continuously
   E. Made in the 3′ → 5′ direction

21. Okazaki fragments are found

   A. On the leading strand
   B. On the lagging strand
   C. In some places on both DNA strands
   D. Only where the DNA has been damaged
   E. Where transcription ends

22. Okazaki fragments are joined together by the enzyme

   A. DNA polymerase
   B. Helicase
   C. Topoisomerase
   D. RNA polymerase
   E. DNA ligase

23. Which of the following is not a product of transcription?

   A. Messenger RNA
   B. A new strand of DNA
   C. Ribosomal RNA
   D. Transfer RNA

24. The steps involved in transcription are

   A. Initiation
   B. Elongation
   C. Termination
   D. A and C only
   E. A, B, and C

25. A transfer RNA molecule carries

    A. New bases for DNA replication
    B. Any amino acids
    C. Links to plasmids
    D. Only one specific amino acid
    E. Only two amino acids at a time

26. The ribosome is the place at which

    A. Transcription occurs
    B. Replication occurs
    C. Translation occurs
    D. All of the above
    E. None of the above

27. Ribosomes are composed of

    A. DNA and protein
    B. DNA, RNA, and protein
    C. Ribosomal RNA and protein
    D. Ribosomal RNA, transfer RNA, and protein
    E. Ribosomal RNA, messenger RNA, and protein

28. The ribosome

    A. Has one site used for translation
    B. Has two sites for translation
    C. Has three sites for translation
    D. Has four sites for translation
    E. Is not involved in translation

29. Genes that are always expressed are referred to as

    A. Active
    B. Inducible
    C. Repressible
    D. Constitutive
    E. None of the above

30. Genes that are expressed only under certain conditions are referred to as

    A. Constitutive
    B. Unregulated
    C. Inducible
    D. Capable
    E. Armed

31. For the structural *lac* enzymes to be transcribed, the

    A. Repressor must not be synthesized
    B. End product must not be in excess
    C. Repressor must not bind to the operator
    D. Inducer must bind to RNA polymerase

32. Point mutations are errors in

    A. The reading frame of the RNA
    B. The reading frame of the DNA
    C. A single base in the RNA
    D. A single base in the DNA
    E. None of the above

Codon on mRNA	Corresponding amino acid
AUG	Methionine
UUA, CUA	Leucine
GCA	Alanine
AAG	Lysine
GUU	Valine
UUU, UUC	Phenylalanine
UAA	Stop
AAU	Asparagine
UGC	Cysteine
UCG, UCU	Serine

33–35. Use the table above in your answers. The following sequence is found in normal mRNA:

...AUGUUUUCUACU...

Analysis of mutations in this open reading frame found the following sequences:

33. AUGUUCUCUACU...

34. AUGUAAUCUACU...

35. AUGUUUUUACU...

Which of the above sequences contains (A) a nonsense mutation; (B) a silent mutation; (C) a frameshift mutation?

36. Mutagens are chemicals that

    A. Cause mutations
    B. Repair mutations
    C. Destroy the DNA polymerase
    D. Improve proofreading
    E. Decrease the number of mutations

37. Thymine dimers are caused by

    A. X-rays
    B. Ionic radiation
    C. Ultraviolet radiation
    D. Abnormal thymine molecules
    E. Replication errors

38. Conjugation differs from reproduction because conjugation

    A. Transcribes DNA to RNA
    B. Transfers DNA vertically, to a new generation of cells
    C. Replicates DNA
    D. Transfers DNA horizontally, to cells in the same generation
    E. None of the above

39. Transposition is

   A. A form of conjugation
   B. A type of bacterial 'sex'
   C. Seen only in Gram-positive bacteria
   D. Seen only in Gram-negative bacteria
   E. A specific form of recombination

40. Transformation is

   A. A type of genetic transfer
   B. Carried out by competent cells
   C. Done with uptake of naked DNA
   D. Different from conjugation
   E. All of the above

41. Transduction

   A. Is carried out by a virus
   B. Is facilitated by a pilus
   C. Involves naked DNA
   D. Is not seen in Gram-negative cells
   E. Is not involved in genetic recombination

42. In order for conjugation to occur, one of the cells must have

   A. DNA fragments
   B. A latent virus
   C. A flagellum
   D. An F factor plasmid
   E. A large enough chromosome

43. Information that makes organisms more resistant to disinfectants and antiseptics is found on

   A. F factor plasmids
   B. Transposons
   C. Transducing phage
   D. Dissimilation plasmids
   E. R factor plasmids

(Q) DEPTH OF UNDERSTANDING

Questions listed here require you to bring together the concepts you have learned in this chapter into a discussion format. This helps you to increase your depth of understanding of the material you have learned. Help can be found in the student resources at: www.garlandscience.com/micro2

1. Discuss the arrangement and orientation of DNA and how this affects replication.

2. Explain how an inducible operon works, including the genes and molecules that are involved.

3. Explain how point and frameshift mutations affect the protein structure and function.

(Q) CLINICAL CORNER

Help can be found in the student resources at: www.garlandscience.com/micro2

1. A recent outbreak of diphtheria has occurred at the local junior college. When students were examined, it was found that both those infected and those not infected were harboring the *Corynebacterium diphtheriae* organisms.

   A. Why were some students sick and others not?
   B. What would you recommend be done to help with this outbreak and to prevent future outbreaks?

2. You are working in Dr. Richard's bacteriology laboratory. The work he is doing involves potentially pathogenic microorganisms that could cause respiratory infections. Consequently, all work with these organisms is done under a sterile hood. You have been instructed to thoroughly wipe down the surfaces of the work area under the hood and turn on the UV light when you are finished. Although these instructions are part of the standard operating procedures for all lab personnel, you know that the last technician to work at this hood became sick.

   A. Why do you have to turn on a UV light when you finish working?
   B. Will the UV light alone be effective at preventing contamination?
   C. Why did the other technician become sick?

The Structure and Infection Cycle of Viruses   Chapter 12

Why Is This Important?

More than 80% of infectious diseases are caused by viruses. As a health professional, most infectious diseases that you see will be caused by viruses. Therefore, it is important that you have an understanding of viral structure and the infection cycle used by viruses.

A few days before turning 19, Vanessa started having redness, swelling, and itching around her upper lip. Then she had a small open sore that she thought was acne. But after several more open sores erupted, she checked online and found that her symptoms matched those of cold sores, or oral herpes, a condition caused by a virus. She made an appointment with her doctor and had to wait an anxious 9 days to get the test results. When she finally got a call from her doctor, the results were positive. She was upset by the news and every time she thought about it, it made her want to cry.

Vanessa is now 21 and has started taking an antiviral medication called valaciclovir that decreases the frequency of outbreaks and reduces the possibility of passing it on to partners. Even though it will be with her forever, she realizes it's not the end of the world. She has learned that 90% of adults have contracted oral herpes, although the period of latency (inactivity) is highly variable among individuals and some people never have outbreaks. Her outbreaks have become much less frequent and she doesn't get upset about living with the virus, even though it will be with her forever.

OVERVIEW

Recall that in Part III (Chapters 9–14) we are looking at the characteristics of disease-causing microorganisms. In Chapter 9, we surveyed bacterial structures and related them to potential for infection, and in Chapter 10 we looked into what bacteria require for growth and infectivity. In Chapter 11, we examined the genetics of bacteria in an effort to understand how they can undergo changes that can make them more infectiously formidable. However, only about 10–15% of all infectious diseases are caused by bacteria, with the vast majority of the remaining 85–90% being caused by viruses. As you begin Chapter 12, it is important to remember that viruses succeed by commanding a host cell to make as many copies of themselves as possible so that they can continue infecting host cells. In this chapter, we examine the structure of viruses and the events associated with the viral infection cycle. Then, in Chapter 13, we concentrate on the pathogenesis of viral diseases.

We will divide our discussions into the following topics:

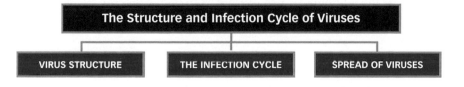

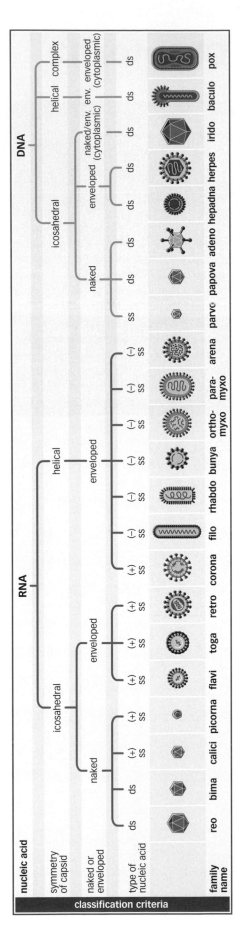

"After all, just one virus on a computer is one too many."

Glenn Turner

Just as computer viruses contain code sequences that need a 'host' computer to replicate themselves, send them out, and cause damage to other computers, biological viruses contain genetic code sequences that require host cells to replicate, disseminate, and cause damage. Viruses are defined as **obligate intracellular parasites**. This means that they cannot live outside a cellular host. Keep in mind that viruses are not actually alive, and the term live in this context refers to being able to multiply and produce new viral particles.

VIRUS STRUCTURE

To understand the infection process associated with viruses, we have to examine the structure of viruses. Viruses are composed of proteins surrounding genetic material. There are two types of virus, DNA viruses and RNA viruses, a division based on the type of nucleic acid that they contain. This genetic material is enclosed in a protein coat called a **capsid**, and some viruses will also wrap themselves in an envelope structure made up of lipids obtained from internal membranes or the plasma membrane of the infected host cell.

Viruses come in many sizes and shapes (**Figure 12.1**), with both size and shape determined by the molecules from which they are constructed.

The Virion

A single, mature, infective viral particle is referred to as a **virion**. The virion must be strong enough to withstand the environment in which it must survive until it successfully infects a host cell. This will depend on the strength of the capsid. The virion must be able to shed this same protective protein coat easily on entry into a host cell. The overall design of the virion is based on the ability to meet these two challenges.

Close examination of the virion shows that the protein coat is what protects the viral genome from chemical agents and physical conditions such as pH, drying out, or temperature changes. The presence of a lipid envelope surrounding the virion can also protect viruses. The assembly of virions includes very stable molecular interactions that help protect the genome, but these interactions are also reversible so that the virus can disassemble the structure after entering the host cell.

Viral Nomenclature

Some of the terms used to describe virus structure are listed in **Table 12.1**. A widely used method for studying viruses is electron microscopy, but even with this very powerful tool, detailed study of the viral structures is very difficult. Therefore, viruses have been studied by computerized image analysis and three-dimensional reconstruction methods. X-ray crystallography has also proved to be excellent for observing the three-dimensional structure of virions.

Figure 12.1 Classification of animal viruses. This illustration shows a summary of the major characteristics of the 22 families of animal viruses that infect humans. Note the shapes of these viruses. ds stands for double-stranded nucleic acid; ss stands for single-stranded nucleic acid.

Table 12.1 Viral structures.

Term	Synonym	Definition
Capsomere		Protein molecule forming capsid
Capsid	Protein coat	Protein shell surrounding nucleic acid
Nucleocapsid		Nucleic acid plus capsid
Envelope	Viral membrane	Phospholipid bilayer with embedded glycoproteins surrounding capsid in enveloped virus
Virion	Viral particle	Complete infectious viral structure: nucleic acid plus capsid for non-enveloped virus; nucleic acid plus capsid plus envelope for enveloped virus

The Capsid

Capsids are built from identical protein subunits called **capsomeres**, which are arranged to provide maximal contact and bonding among the subunits. The bonding of these subunits occurs in such a way that there is structural symmetry. There are three categories of virus shape: **helical**, **icosahedral**, and complex.

Helical Viruses

Helical viruses have either a rod shape (relatively rigid; straight) or a filamentous shape (relatively flexible; curved or coiled). A classic example of a helical RNA virus is the influenza virus, which also has a lipid envelope. The RNA in this virus is segmented, with each segment enclosed in its own helical capsid held in place by multiple proteins called **nucleoproteins**. This virus also has a second layer of protein called **matrix protein** just inside the envelope (**Figure 12.2**).

As we saw in Chapter 8, embedded in the influenza virus envelope are the glycoprotein **hemagglutinin** (which serves in binding to and entering the host cell) and the enzyme **neuraminidase** (which serves in release from the host cell). Influenza viruses, such as the ones that cause swine flu and bird flu, are categorized by virologists according to these proteins (H1N1 and H5N1, respectively).

Fast Fact   Vaccines that protect against seasonal influenza are based on strains of influenza viruses (such as H1N1, H3N2, and H1N2) that are most likely to proliferate in the upcoming flu season. Influenza kills more than 30,000 people in the United States each year.

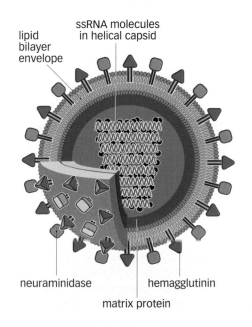

labels: lipid bilayer envelope; ssRNA molecules in helical capsid; neuraminidase; hemagglutinin; matrix protein

Figure 12.2 A diagrammatic representation of the influenza virus in cross section. The RNA is in eight segments, each enclosed in a helical capsid and surrounded by a further layer of protein (matrix protein) and a lipid bilayer envelope. Note the neuraminidase and hemagglutinin molecules embedded in the envelope.

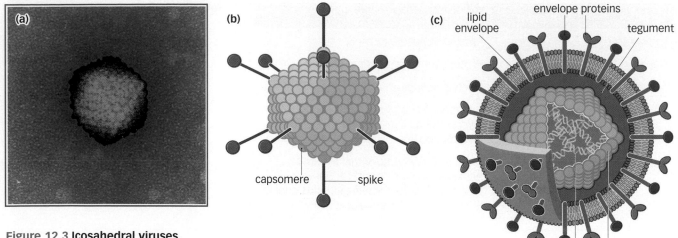

Figure 12.3 Icosahedral viruses.
Panel a: Colorized transmission electron micrograph of adenovirus, a non-enveloped simple icosahedral virus. Panel b: A diagrammatic representation of the non-enveloped complex icosahedral virus. The spikes on the vertices of the capsid are glycoproteins used to attach the virion to receptors on a host cell. Panel c: A diagrammatic representation of the herpes virus, which has an enveloped complex icosahedral structure. Notice that this virus has three distinct groups of proteins: the capsid, the tegument, and the envelope.

Icosahedral Viruses

The shape of icosahedral viruses is derived from 20 triangular faces that make up the capsid (**Figure 12.3a** and **b**). There are two types of icosahedral virus: simple and complex. In the simple icosahedral virus, the capsid is made up of repetitive polypeptide capsomeres. A good example is the poliovirus, which is a picornavirus. This virus is one of the smallest and simplest of the viruses that infect humans and is composed of only 60 copies of four proteins.

Poliovirus uses the digestive tract in some of its infection strategies, and therefore it must be able to withstand the harsh environment that it encounters there. Survival is possible because of the very stable protein–protein interactions that take place in the capsid. Keep in mind the functional dichotomy of the capsid: even though poliovirus must be strong enough to withstand harsh environmental conditions, the protein coat is still designed to disassemble when contact is made with a receptor on a host cell.

In complex icosahedral viruses, the complexity comes from additional proteins and lipids surrounding the capsid. The best example of this type of virus is the enveloped herpesvirus (**Figure 12.3c**), in which the virion contains three groups of proteins: the proteins of the capsid; those forming a structure called a **tegument**, which is a protein layer located between the capsid and the envelope; and glycoproteins that form spikes on the surface of the envelope.

Complex Viruses

Viruses that do not have either helical or icosahedral symmetry are categorized as **complex viruses**. The best example is the variola (smallpox) virus, which has a smooth rectangular shape encased in an envelope (**Figure 12.4**).

Viral Envelopes

Many viruses that infect humans and other animals are **enveloped**. Envelopes are formed when viral glycoproteins and oligosaccharides associate with membranes of the host cell. As we will see later, movement of a virion out of an infected host can occur in such a way that the virion gets wrapped in host membranes. Envelopes vary in size, morphology, complexity, and composition, but the foundation of all envelopes is

Figure 12.4 The structure of the variola virus. Panel a: Colorized electron micrograph of the complex virus variola, the agent of smallpox. Panel b: Diagrammatic cross section of the variola virus, showing significant structural features.

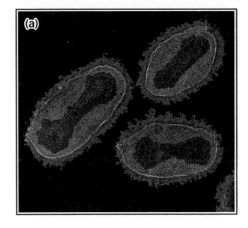

the phospholipid bilayer. The precise amount of lipid in a viral envelope varies depending on the virus and on the mechanism by which it captures lipid from the host membrane.

Envelope Glycoproteins

The glycoproteins in viral envelopes are integral membrane proteins that are firmly embedded in the lipid bilayer of the envelope. Part of the protein faces inward toward the virus capsid, and part faces outward from the envelope. On the exterior of the membrane, the glycoproteins can form spikes or other surface structures that the virion can use to attach to a host cell.

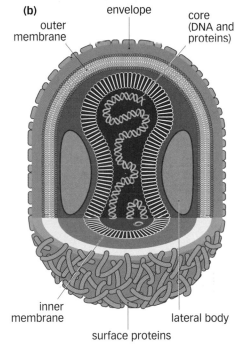

Keep in Mind

- Viruses are obligate intracellular parasites because they cannot live outside a host cell.

- Viruses come in a variety of sizes and shapes.

- Viruses contain either DNA or RNA but never both.

- The nucleic acid in a virion is surrounded by a protein coat called a capsid.

- The capsid is made up of repeating protein subunits known as capsomeres.

- Some viruses are surrounded by an envelope composed of viral glycoproteins and oligosaccharides complexed with host cell membranes.

THE INFECTION CYCLE

The stages of the infection cycle of viruses have been worked out, and a great deal of information has been accumulated for each step. In this chapter, our primary focus is on animal viruses, by which we mean those that infect eukaryotic (animal) cells.

During the infection cycle of animal viruses, the host cell is used to produce more virions; when the host cell is completely filled with new virions, the host cell releases the viruses through exocytosis (as we will see later in this chapter, a process called budding), or it simply bursts. The latter is called **lysis** and is the hallmark of the **lytic infection cycle** shown in **Figure 12.5a** and **Movie 12.1**.

However, infection by animal viruses does not always result in the production of new virions; the virus can be in a dormant state, called **latency**, for long periods (**Figure 12.5b** and **Movie 12.1**). For example, herpesviruses can lie dormant in the body for decades! During latency, the host cell DNA and viral genome combine to make a **provirus**. The host cell reproduces itself as usual, without making any new virions, but it continues to incorporate the viral genome into new host cells. Latency can last through many host cell generations, or it can be interrupted suddenly and new virions can be produced and released in large quantities.

As we saw in Chapter 11, there are also viruses that infect bacteria, called either bacteriophages or simply phages. They also have two alternative infection cycles. The lytic infection cycle destroys bacteria, so their

Fast Fact Some pathogenic bacteria, including *Corynebacterium diphtheriae* (which causes diphtheria), *Vibrio cholerae* (cholera), *Shigella dysenteriae* (shigellosis), and *Clostridium botulinum* (botulism), only produce toxins when they are infected by bacteriophages. The bacteriophages carry toxin-encoding genes that are then expressed by the bacteria cells.

Figure 12.5 The lytic infection cycle and latency. In the lytic cycle, the virus-infected host cell produces new virions and releases them as it bursts open (lysis). During latency, the viral genome is incorporated into the host cell genome (becoming a provirus), but no new virions are produced. Latency can be interrupted, at which point the virus enters the lytic cycle.

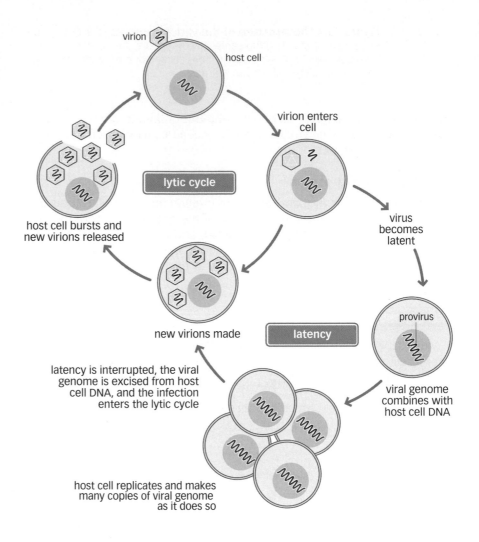

virion

host cell

virion enters cell

lytic cycle

host cell bursts and new virions released

virus becomes latent

new virions made

latency

provirus

latency is interrupted, the viral genome is excised from host cell DNA, and the infection enters the lytic cycle

viral genome combines with host cell DNA

host cell replicates and makes many copies of viral genome as it does so

potential application in treating bacterial diseases, called **phage therapy**, is currently being studied. Some phages can also integrate their genome into that of the host cell, producing a dormant state known as lysogeny, in which viral DNA is incorporated into bacterial DNA to form a prophage.

For animal virus infections in which new virions are being produced, the infection cycle contains the following steps: attachment, penetration, uncoating, biosynthesis, maturation, and release.

Attachment

Attachment occurs as a virion binds to specific receptors on a host cell. For some viruses, only one receptor interaction is required, whereas for others there needs to be a co-receptor involved. Receptor–virus interactions are important for understanding the pathogenesis of viruses and also for examining potential blocking mechanisms that might be useful for therapy.

Virus entry into the host cell is not a passive process but instead relies on the virus taking over normal cell functions such as vesicular transport. Remember that the capsid is made up of capsomeres, which are three-dimensional proteins. In some cases the interaction between the virion and the host cell's receptor initiates a conformational change in the capsomeres that prepares the virion for the uncoating step. In other viruses, the host cell's receptor is more like a tether that holds the virion in place next to the host cell until the virion has gained entry.

When Virus Meets Host

Viral interactions with host cells occur through random collisions and are essentially governed by chance. Consequently, the concentration of viral particles is important in determining whether an infection occurs. The production of large numbers of viral progeny in each infected host cell increases the chance that the number of random encounters with new host cells will be high enough to continue the infection. It is interesting to note that viruses can be either promiscuous (not specific) or highly specific when it comes to attachment to a host cell. The presence of a receptor makes identification of the host cell easy, but binding to the receptor may not be enough to cause infection. In addition to having the proper receptor, the host cell must also have the cellular components necessary to produce new virions.

Viral infections occur most often at the apical surface (top) of epithelial cells, and such infections are usually localized. In contrast, viruses that are transported to the basolateral (bottom) surface of the host cell and released into the underlying tissues usually spread to other sites.

Many viruses attach only to specific areas of the host plasma membranes, called **lipid rafts**. These areas of the membrane are rich in cholesterol, fatty acids, and other lipids, and consequently they are more densely packed and less watery, making them more reliable for stable attachment. These areas of the membrane are also the release site for many viruses, such as Ebola and HIV type 1.

Many different plasma membrane molecules on the host cell can serve as receptors for virus attachment. Some viruses can attach to more than one type of receptor molecule, and some receptors can be shared by different types of virus. In fact, for some viruses, the receptor determines the host range.

Initial virus–receptor interactions are probably electrostatic. However, these initial reactions are followed by high-affinity binding, which occurs because of the conformational interactions between virion and receptor.

In some cases, the host's response to the attachment of virus can amplify the attachment step. For example, rhinovirus, which causes upper respiratory infections (including 50% of common colds), uses a glycoprotein host cell receptor called **ICAM-1**. It turns out that this is a common adhesion molecule normally involved in a variety of host cell inflammatory responses. Ironically, inflammation seen with upper respiratory infections caused by rhinovirus increases the number of ICAM-1 molecules, which facilitates more binding to host cells by the virus. Consequently, the initial defensive response of the host to rhinovirus infection leads to more infection by the virus.

Some viral infections require the involvement of co-receptors. These molecules are host components that interact with the virus and the primary receptor to allow continuation of the infection cycle. Co-receptors have been shown to be involved in HIV type 1, picornavirus, adenovirus, and herpesvirus infections. Viruses that require co-receptors cannot successfully complete the infection without these molecules.

Fast Fact It has been determined that, in many cases, a single rhinovirus–receptor binding event is all that is required to mediate the entry of the virus into the host cell.

Receptor Binding

Examination of non-enveloped viruses gives us a good idea of receptor binding mechanisms. For these viruses, attachment is between the host

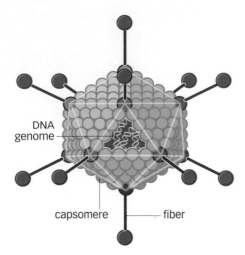

Figure 12.6 An illustration of adenovirus showing the complex icosahedral capsid. Notice the fibers that extend from each of the pentamers. These fibers, which are similar to the spikes seen on many viruses, are used for attachment of the adenovirus to receptors on host cells.

cell's receptor and structures on the viral capsid. For example, adenoviruses have fibers that protrude from each point of the capsid that attach the virus to the host cell's receptor (**Figure 12.6**).

For enveloped viruses, remember that the lipid envelope surrounding the virus originates from host cells previously infected by the virus. During the assembly of these viruses, specific viral proteins are inserted into these envelopes. It is the attachment site on one or more of these viral envelope proteins that binds to specific host cell receptors. Two of the most studied enveloped viruses are influenza and HIV type 1. In the influenza virus, hemagglutinin binds to the host cell's receptor (sialic acid). For HIV type 1, attachment is accomplished by similar spikelike structures that are located on the outer surface of the envelope. These spikes bind to the CD4 receptor found on the helper T cell population of lymphocytes. With HIV, there is also a requirement for co-receptor binding.

Penetration and Uncoating

Once attached to the host cell, the virion must gain access to the interior of the cell in the penetration step. However, the viral genetic material is still enclosed in either a capsid or a capsid and envelope, and so the next step in the infection cycle must be uncoating.

Penetration and Uncoating by Non-enveloped Viruses

In some cases, uncoating is a simple process that occurs at the plasma membrane (**Figure 12.7a**). If uncoating takes place within the host

Figure 12.7 Three virus uncoating strategies. Panel a: Uncoating at the plasma membrane. Panel b: Uncoating in endosomes. Panel c: Uncoating at the nuclear membrane. In the first two, the viral genome is released into the cytoplasm. In the last example, uncoating occurs at the nucleus, which causes the release of the viral genome directly into the nucleus.

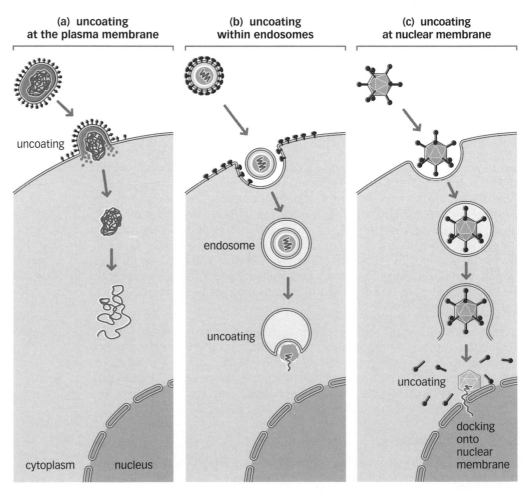

(a) **uncoating at the plasma membrane**

uncoating

(b) **uncoating within endosomes**

endosome

uncoating

(c) **uncoating at nuclear membrane**

uncoating

docking onto nuclear membrane

cytoplasm   nucleus

cell, help is required because even the smallest viral particle is too big to move across the host plasma membrane without help. Non-enveloped viruses take advantage of normal cellular mechanisms, such as receptor-mediated endocytosis, to gain entry into the host cell. Using receptor-mediated endocytosis, a virion binds to a receptor on the host cell's plasma membrane, causing a pit to form in the membrane. The membrane then encloses the virion to form a vesicle called an **endosome**. These vesicles travel through the host cytoplasm and contain an acidic interior environment that begins uncoating the virion (**Figure 12.7b**). The endosome then fuses with lysosomes in the cytoplasm, and uncoating is completed. Keep in mind that the phospholipid bilayer structure of the plasma membrane facilitates the fusion of the endosomes with one another and with lysosomes.

Some non-enveloped viruses form a pore in the host plasma membrane to gain entrance to the cytoplasm. Binding of the virus to the host cell's receptor causes conformational changes to the capsomeres such that they become hydrophobic and are attracted to the plasma membrane. The hydrophobic part of the virus inserts itself into the membrane, forming a pore that allows the viral genetic material to enter the cytoplasm of the host cell.

Uncoating can also be a very complex mechanism that includes endocytosis through the plasma membrane and a docking event at the nucleus (**Figure 12.7c**). In still other cases, the nucleus of the host cell is involved in uncoating the virus.

Penetration and Uncoating by Enveloped Viruses

With enveloped viruses, penetration of the virus into the host cell is relatively straightforward because the envelope is essentially a plasma membrane and can fuse with the host plasma membrane. This fusion is mediated by specialized fusion proteins, and the fusion is believed to result in the formation of a large opening called a fusion pore in the host cell. This opening permits the virion to move across the plasma membrane and into the host cell.

In some cases, virion–receptor binding causes conformational changes in the host cell that expose the fusion proteins. In other instances, fusion requires the presence of co-receptor molecules. Perhaps the best example of the need for co-receptor molecules is seen in HIV type 1 infections. In these infections, not only does HIV require specific binding at the CD4 receptor of the helper T cell, but other host cell proteins are also required to complete the fusion event. In fact, all of the co-receptors for HIV are normal cell surface receptors for small molecules called chemotactic cytokines. Several of these co-receptors have been identified and it seems that all of them are required for the fusion to occur.

Cytoplasmic Transport of Viral Components

We have chosen to mention cytoplasmic transport as part of the penetration and uncoating process because viral infections involve compartmentalization. In this process, DNA or RNA synthesis of viral genomes as well as synthesis of the new capsids and other viral proteins occurs in specific locations of the host cell. After synthesis, these components are finally moved to other specialized sites for final assembly of the intact virions. Therefore, there need to be mechanisms that allow the movement of viral components through the host cell's cytoplasm. After the virus has uncoated, there are two ways in which this is done. The first uses membrane-enclosed vesicles in conjunction with host cell cytoskeletal structures such as microtubules. The second method employs the direct association of viral components with cellular transport mechanisms.

In some cases there are actually specialized lipids and proteins called chaperones that will facilitate the movement of virus through the host cell's cytoplasm. The actual methods employed by these chaperones are not yet understood, but they seem to be associated with maintaining stabilization of the three-dimensional structure of viral components during the journey through the cytoplasm.

Transport of the Viral Genome into the Nucleus

Some viruses must ultimately enter the host cell's nucleus for replication of the viral genome. Recall from Chapter 4 that the structure of the nucleus in eukaryotic cells involves a double phospholipid bilayer membrane. Viral DNA must penetrate this double membrane for a successful infection to occur. This transport is accomplished using pathways that are routinely used by host cell proteins, and it involves pores that are part of the nuclear membrane.

Schemes for entry into the nucleus depend on particular viruses. For example, DNA viruses are too large to get into the nucleus intact, so the capsid of these viruses moves through the cytoplasm and docks on the outside of the nucleus and then delivers the viral genome into the nucleus.

There are RNA viruses such as the **retroviruses** (including HIV) that also need to replicate their genomes in the nucleus. These viruses carry the enzyme **reverse transcriptase**, which can convert RNA into DNA. The DNA copy of the viral RNA then moves into the nucleus and integrates with the host chromosome (**Figure 12.8**). It turns out that the movement of this DNA actually correlates with the breakdown of the nuclear membrane during mitosis. The timing of this host cell mitotic event is neatly connected to two important viral requirements: first, the entry into the nucleus, which is made easy because the nuclear membrane is breaking down during mitosis, and second, integration into the host chromosome, which is in the process of replicating.

Fast Fact Reverse transcriptase was first discovered independently by David Baltimore and Howard Temin in 1970, and both men were subsequently awarded the Nobel Prize in Physiology or Medicine for this work.

Keep in Mind

- Viruses can go through a lytic cycle, in which new virions are produced and then released as the host cell bursts.

- Viruses can also exhibit latency, in which they infect host cells but new virions are not produced.

- Host cell receptors used by viruses represent a small fraction of the cell membrane proteins.

- Many different viruses can share the same receptor.

- Virus–receptor interactions facilitate the infection process by enhancing virus entry into host cells.

- For enveloped viruses, penetration occurs through a fusion event between the viral envelope and the host cell's plasma membrane.

- Once fusion has occurred, the virus is released into the cytoplasm of the host cell, where various mechanisms allow the uncoating of the virus.

- Retroviruses are RNA viruses that contain the enzyme reverse transcriptase.

- Reverse transcriptase converts RNA into DNA, which is then integrated into the host cell's DNA.

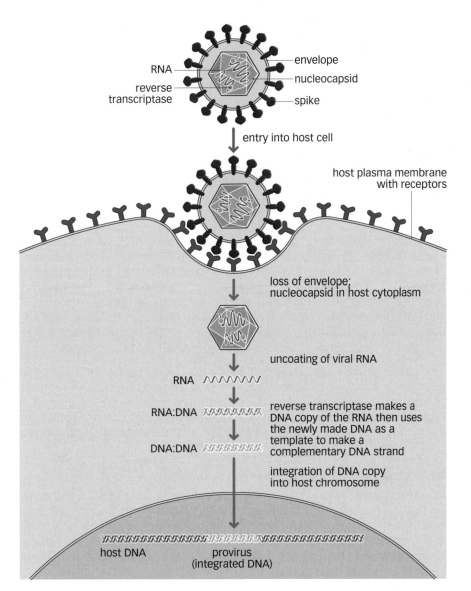

Figure 12.8 Illustration of the integration of proviral DNA into the host cell chromosome after infection with a retrovirus. Notice the action of reverse transcriptase as it copies viral RNA into DNA.

Biosynthesis

Now that we have seen how viruses attach, penetrate, and uncoat, we need to consider the way in which new virions are made. Remember, viruses contain either DNA or RNA, and the genomes of both types of virus can be either single-stranded or double-stranded. The production of new virions is complex and involves a variety of biochemical pathways, depending on the type of virus. Simply put, the viral DNA (or DNA copy from RNA) is transcribed at very high rates inside the host cell, producing large quantities of proteins to make as many new virions as possible within an infection cycle. The host cell produces all of the viral components, such as capsomeres and viral genomes, as well as the enzymes needed to orchestrate the reactions.

DNA Viruses—Replication

DNA viruses have DNA genomes, so the same mechanisms that are used for replication of the host cell's DNA can also be used for replication of the viral genome. However, replication of viral genomes requires the synthesis of at least one viral protein and the expression of several viral genes. Consequently, viral DNA synthesis cannot begin immediately after

Fast Fact Some antiviral drugs function by interfering with viral biosynthesis. Nucleoside analogs, such as acyclovir (used in the treatment of herpetic diseases), get incorporated into the viral DNA or RNA and deactivate the enzymes that are needed for the production of new viruses.

infection but must wait until these viral components have been made in sufficient quantity.

Once viral DNA synthesis begins, there are many cycles of replication during which large numbers of viral DNA molecules begin to accumulate. Because the replication of viral DNA is performed by host cell machinery, this avoids the need for the virus to devote any of its limited genetic capacity to coding for unnecessary enzymes.

Remarkably, DNA synthesis in the host cell is inhibited by the virus, so that all polymerases and proteins that are involved in replication can concentrate on viral DNA synthesis. In fact, adenovirus and herpesvirus actually induce the arrest of the entire cell cycle of the host cell so that viral DNA replication can be maximized.

In DNA virus infections, specialized sites form in the host cell nucleus. These distinct replication compartments are essentially viral factories. They contain both the DNA templates and also the viral replication machinery. This localization of the components needed for viral DNA synthesis at a limited number of cellular locations facilitates exponential viral DNA replication. This arrangement also increases the local concentration of the proteins involved in replication and allows each newly synthesized DNA to serve as an immediate template for the next molecule.

The same type of concentration phenomenon is useful for viral gene expression because large numbers of DNA templates and the proteins required for transcription are all concentrated in the same location. It is important to note that viral replication compartments do not form randomly but are actually formed as a result of viral colonization of specialized areas of the host cell.

DNA Viruses—Transcription

In transcription, a viral DNA molecule acts as a template for the creation of viral mRNA. Transcription of viral DNA is performed by the host cell's RNA polymerase (an exception being the poxviruses, which bring their own RNA polymerase into the host cell), but double-stranded and single-stranded viruses have different processes because of the different conformations of their genomes.

Double-stranded DNA (dsDNA) viruses

Because the viral DNA is in the same conformation as the host cell's DNA, transcription begins as soon as viral DNA reaches the host nucleus. The mRNA required to make viral proteins is produced by transcribing one strand of the viral DNA. This is accomplished by using the host cell's machinery, although some viruses bring in their own RNA polymerase molecules. With some double-stranded DNA viruses, such as adenovirus, herpesvirus, and papillomavirus, transcription continues after the replication of new viral DNA. The transcription and replication strategies of dsDNA viruses are shown in **Movie 12.2**.

Single-stranded DNA (ssDNA) viruses

Transcription of single-stranded DNA viruses is more complicated than for dsDNA viruses because the single DNA strands must first be converted to double strands before they can be used for transcription. This is accomplished by using the host cell's DNA polymerase. The new DNA strand is then used as the template for transcription and for new viral genomes, which will be incorporated into new virions. The transcription and replication strategies of ssDNA viruses are shown in **Movie 12.3**.

Expression of the genes in both types of DNA virus occurs in a sequential and reproducible order. One of the hallmarks of DNA virus transcription

Fast Fact Antiviral drugs that inhibit viral DNA synthesis also inhibit the production of proteins and therefore the construction of new virions.

is the coordination of transcription with the synthesis of viral DNA. Viral proteins needed for replication are made first, but proteins that will be needed to build the new virion are made only after viral DNA synthesis has begun. It is as if the virus does not make the proteins needed to build the new virions until there is something to put inside them (that is, the newly synthesized viral DNA). This is different from gene expression in RNA viruses, in which all the genes are expressed continuously so that enzymes, regulatory proteins, structural proteins, and new viral RNA are all being made at the same time.

One last point about the transcription of viral DNA is important. On infection with a DNA virus, host cell transcription is shut down, and part of the transcriptional machinery is subverted for use by the virus. This offers the virus several advantages. For instance, inhibition of host cell function can allow host resources that were required for these functions to be devoted exclusively to the needs of the virus. In addition, there is no competition between the transcripts of the host and those of the virus when it comes to translation. These advantages amplify the primary goal of the virus, which is a productive infection.

Viral genes are transcribed at very high rates. This is necessary because large quantities of proteins must be produced so as to make as many new virions as possible within an infection cycle. Although the rate of transcription is high, there is considerable regulation of the process. In most cases this regulation is performed by regulatory proteins of the host cell. However, some viruses do bring along their own regulatory proteins. The carefully regulated process of DNA virus transcription ensures the coordinated production of the structural proteins required to make new virions. This regulation also serves another important function in that it prevents the overproduction of components that are not needed, making the infection process very efficient.

Keep in Mind

- Replication of DNA viral genomes uses the same mechanism as the host cell but also requires several viral components.

- Viral DNA replication occurs at specialized sites, which contributes to the efficiency and productivity of the viral infection.

- Transcription of dsDNA genomes can begin immediately that viral DNA enters the nucleus; however, for ssDNA genomes, the single strands must be converted to double strands before transcription can begin.

- Transcription of viral DNA is performed by the host cell's RNA polymerase (except for poxviruses, which bring their own RNA polymerase).

- Viral genomes are transcribed at a very high rate, to make as many new virions as possible.

RNA Viruses—Replication and Transcription

The RNA strands in viruses are classified using a (+) or (–) depending on whether the strand can be used directly as mRNA. Before we look further at RNA biosynthesis, it will be helpful to revisit the concept of complementary base pairing. If one strand (the original) is used as a template to synthesize a new strand, the new strand (new strand 1) will have a complementary sequence: C in one will be G in the other, and A in one will be U (we are using RNA) in the other, as shown below. If that strand is used as a template to synthesize a further strand, then the sequence of the next new strand (new strand 2) will be complementary to its template (new strand 1) but it will be the same as the original strand.

Original strand ⟶ AUGACCAGUACC

New strand 1 ⟶ UACUGGUCAUGG

New strand 2 ⟶ AUGACCAGUACC

The first new strand (new strand 1 above) can therefore be used as a template to make more copies of the original strand. This template mechanism is used by RNA viruses.

Double-stranded RNA (dsRNA) viruses

Double-stranded RNA virus genomes contain both a (+) and a (–) strand. During infection, the (–) strand is copied into mRNA by a viral RNA polymerase; this mRNA is used to produce viral proteins. The newly synthesized strand is also used as a template to make a double-stranded genome, which will be placed into new virions. See **Movie 12.4**.

(+) Single-stranded RNA (+ssRNA) viruses

The (+) strand is essentially already mRNA and as such it can be directly translated into viral proteins by the host cell's ribosomes. Genome replication takes two steps: first, the (+) strand is copied into a (–) strand; then this (–) strand is used as a template to produce more (+) strand genomes to place into new virions. See **Movie 12.5**.

(–) Single-stranded RNA (– ssRNA) viruses

Things are more complicated for viruses with a (–) strand RNA genome, as the (–) strand cannot be directly used as mRNA. It must first be copied into a (+) strand by viral RNA polymerase, which is brought in with the virus. This (+) strand copy is mRNA and can be used for the synthesis of viral proteins. The (+) strand is also used as the template for new (–) strand genomes that are packaged into new virions. See **Movie 12.6**.

Retroviral Transcription and Integration

Retroviruses are RNA viruses that contain the enzyme reverse transcriptase. This enzyme is an RNA-dependent DNA polymerase and it transcribes RNA into DNA. It is called reverse transcriptase because normal transcription, the sort we learned about in Chapter 11, transcribes DNA to RNA.

There are between 50 and 100 copies of the reverse transcriptase enzyme in one retrovirus virion. Because it is carried in with the virus during infection, it can begin to function after the virus has uncoated and nucleotides have become available. Reverse transcription occurs in the cytoplasm of the host cell, and the newly copied DNA molecules are transported from the cytoplasm into the nucleus.

About 4–8 hours after the initial infection of the host cell, newly synthesized viral DNA transcripts have begun entering the nucleus, and integration of this DNA into the host cell's chromosome commences immediately. After the viral RNA has been copied, it is automatically degraded because there is no longer any need for the viral RNA template.

Viral Control of Translation

Viruses are completely reliant on the host cell's translational machinery for translating mRNA templates into proteins. In fact, viral infection often results in modification of this machinery so that viral RNA is translated preferentially.

Alterations in the translation apparatus of an infected host cell can signal to host defenses that a viral infection is under way. Two host genes code for enzymes that prevent any RNA, whether viral or host, from associating with ribosomes. This effectively stops all translation. Obviously the cell will die without protein synthesis, but, as we will see in the next chapter, cell death may be preferable to viral infection. Viruses have developed ways to combat these host defenses, and many viruses have genes that code for proteins that neutralize the host defense and restore translational capability. Unfortunately, these mechanisms are not yet well understood.

| Keep in Mind |

- RNA virus replication is more complicated than DNA virus replication.
- RNA viruses have either single-stranded or double-stranded RNA.
- Single strands can be (–) or (+).
- In all cases, RNA viruses use a template strand of RNA to make new viral genomes.
- Retroviruses are RNA viruses that contain the enzyme reverse transcriptase.
- Reverse transcriptase can convert RNA into DNA, which can then be integrated into the host cell's chromosomes.

Maturation

The next step in the infection cycle is maturation. Up to now we have seen how the viruses attach and penetrate host cells and how new components (capsids and genomes) are synthesized. In this section, we will see that the maturation step of the infection cycle involves the movement of newly made viral components to specific sites in the host cell (referred to as intracellular trafficking) where the assembly of the new virions takes place.

Intracellular Trafficking

Some viral components are synthesized in the host cytoplasm, and some are synthesized in the host nucleus. They then have all to be brought to one site for assembly into new virions. Some viral components must therefore travel great distances, relative to their size, to get to assembly sites within the cell. Simple diffusion mechanisms are not able to cover these distances. Instead, newly synthesized viral components are transported through the cell by means of microtubules, a process that requires a considerable expenditure of energy. However, this is of no concern to the virus because the energy is all supplied by the infected host cell!

Assembly sites in the cell are determined by several factors, such as whether the virions will have an envelope, the type of genome (DNA or RNA), and the mechanism of genome replication. Because viral envelopes are derived from the host plasma membrane, many enveloped viruses assemble at sites adjacent to the plasma membrane. Other viral envelopes are derived not from the host's plasma membrane but rather from organelle membranes, and in these cases the virion's assembly site is near the organelle.

Some non-enveloped viruses assemble in the host cytoplasm, and some assemble in the nucleus. All of the structural proteins for the latter type must be transported from the cytoplasm, where they are constructed, to the nucleus. Viral structural proteins seem to enter the host nucleus through cellular pathways that are normally used to import host nuclear proteins.

Figure 12.9 A diagrammatic illustration of the localization of viral proteins to the plasma membrane. The viral glycoproteins (in red) are translocated from ribosomes into the lumen of the endoplasmic reticulum (ER). They then travel to the plasma membrane by vesicular transport and are united with other viral proteins and new viral genomes.

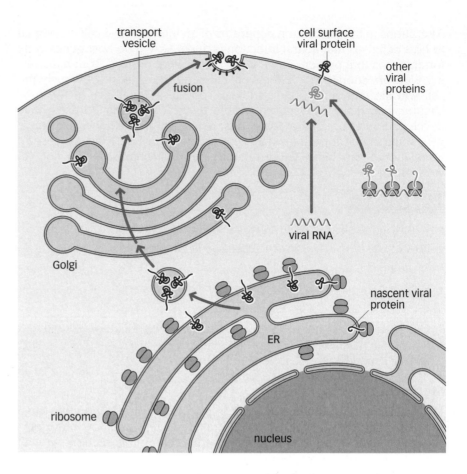

Fast Fact Vesicular movement from compartment to compartment means that viral proteins are never again exposed to the cytoplasm of the host cell, thereby avoiding cytoplasmic enzymes that might degrade them.

Viral proteins travel from their site of assembly to the cell surface through a series of membrane-enclosed compartments and vesicles. The first step is the transport of correctly folded viral proteins from the endoplasmic reticulum to the Golgi apparatus (**Figure 12.9**). In the Golgi, the viral proteins are sorted according to 'delivery addresses' in their protein sequences, using transport vesicles that bud from one compartment and move to the next. Because of the fusion that occurs between phospholipid bilayer membranes, it is easy for viral components to move in vesicles and use the fusion with other membranes to release these components into the lumen.

Just as protein components are synthesized and transported to assembly sites, the newly synthesized viral genomes must also be moved to the site of final assembly. For enveloped viruses, this site is near the host plasma membrane; for non-enveloped viruses, it is at the host nucleus. During this movement, the newly constructed genome becomes dormant because it must not be allowed to start transcription or replication mechanisms. Doing so would either slow down or inhibit the assembly of new virions.

Keep in Mind

- Virus replication is a complicated process in which the host cell manufactures new viral components from the template provided by viral DNA or RNA.
- Viruses are completely reliant on host cell machinery for translation.
- Intracellular trafficking is crucial for viral reproduction.
- Intracellular trafficking requirements can be quite complex, with transport of viral macromolecules over long distances in the cell.
- The assembly of different viral components occurs at different sites and requires that all viral proteins be sorted in the Golgi.

Assembly

All virions must complete a common set of assembly reactions to ensure reproductive success. For non-enveloped viruses these reactions are:
- Formation of structural subunits for the capsid
- Assembly of the capsid
- Association of viral genome within the capsid

For enveloped viruses, the sequence is:
- Formation of structural subunits for the capsid
- Assembly of the capsid
- Association of viral genome within the capsid
- Assembly of viral envelope glycoproteins

Assembly is a remarkable process that requires exquisite specificity as well as the coordination of multiple reactions. The structure of the virion determines how it is assembled and also affects the mechanism of entry into the next host cell.

Assembly of capsid subunits

Recall that the viral capsid is made up of protein molecules called capsomeres. These capsomeres are assembled first, and there can be differences in how this process works between RNA viruses and DNA viruses. There are several mechanisms for forming these structural subunits, and in some cases other proteins may be involved in the process. Some viruses use common cellular techniques to produce capsid protein subunits that then fold into the proper three-dimensional structure.

Interestingly, the number of subunits produced is always far in excess of the number required. This is because the subunits must find one another in the host cell's cytoplasm, which is filled with irrelevant host proteins. Remember, if the capsids cannot be assembled, the infection will fail. The chances of random interactions between viral subunits are increased by locating the production of capsid subunits at distinct assembly sites and producing more of the subunits than are necessary. Together these increase the chances for successful virion assembly.

In some viruses, the assembly of capsids can be assisted by host chaperone proteins. These proteins facilitate viral protein folding by preventing the improper association of subunits. In some cases, these chaperones participate in the formation of the capsid subunits.

Assembly of the viral genome

Perhaps the most important part of assembling a new virion is the placement of the viral genome inside the capsid. If this step is either incomplete or inaccurate, the potential for continued infection is compromised. During assembly, either the virion is assembled while the viral genome is being synthesized, or the viral genome is inserted into already-formed capsids. When you consider the potential for errors in capsid development or difficulties in pushing or pulling the genome into the capsid, it is easy to see that some capsids wind up empty. It turns out that these empty capsids have an important role as a viral defense against the host immune response (see below).

Release

New virions are released from the host cell in one of two ways: by lysis or by budding off from the host cell (**Figure 12.10**). When release is by lysis, the host cell dies immediately. When release is by budding, the host cell remains alive for a while and intermittently releases newly formed virions.

Figure 12.10 A colorized transmission electron micrograph of influenza virus budding from the host cell.

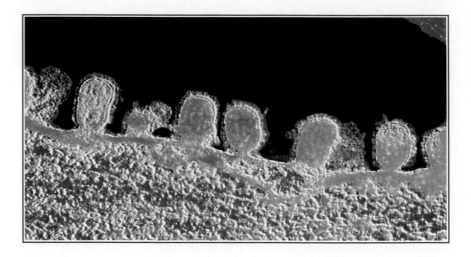

Budding from the Plasma Membrane

The mechanisms by which enveloped virions bud off from a host plasma membrane are not completely understood. However, it is known that this process has four steps: bud formation, bud growth, fusion of the bud membrane, and separation from the host. In the formation and growth steps, the viral bud forms and begins to push the host plasma membrane out. Many viral proteins are required to make this happen, but the exact mechanisms are not understood.

Membrane fusion reactions pinch off the viral bud from the host plasma membrane. This is also a complex mechanism that is not well understood, but it has been shown that host proteins are involved.

Last, in the least understood of these processes, the viral bud uses what seem to be exocytosis mechanisms to free itself from the infected host cell. Suffice it to say that this simple process of exiting from the infected host cell without killing it involves a complex series of reactions that include both viral and host cell components.

It is important to point out that in some viral infections, virions are completely assembled but remain non-infectious. These virions are referred to as immature, and proteolytic enzyme processing must be employed to convert them to mature infectious virions. These conversion reactions are performed by virally encoded enzymes and take place either very late in the assembly process or after the release of the virions from the host cell.

Fast Fact Proteolytic proteins responsible for the conversion of immature virions into mature infectious virions are very good potential targets for antiviral therapy because without their function, viral particles will remain immature and non-infectious.

Keep in Mind

- All virions complete a set of assembly reactions.

- Capsomere proteins are assembled first.

- The number of capsomeres produced is always far in excess of the number required for the number of virions to be assembled.

- Some viruses use host cell proteins called chaperones to assemble viral capsids.

- New virions are released through lysis, which kills the host cell, or by budding from the host cell, which allows the host cell to survive for a period.

- Some viruses are released in an immature non-infectious state and must be activated enzymatically before they can infect host cells.

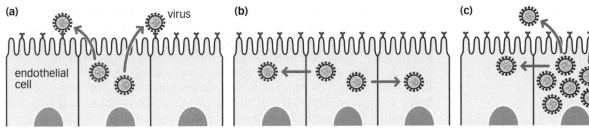

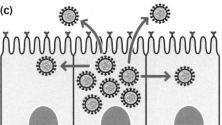

SPREAD OF VIRUSES

Before we leave the infection cycle, it is important to discuss how viruses spread from host cell to host cell. Remember, viruses have only one goal—to continue the infection. For a newly made virion, the next host cell target can be right next door (a neighboring cell) or far away in the body. Once they leave the host cell and enter the extracellular environment, they are susceptible to host defense mechanisms. Exposure to host defenses does not occur if the next host cell is a neighbor and the virus can move directly between cells (**Figure 12.11**).

Viruses can also spread through the formation of **syncytia**, which are masses formed by the fusion of many infected cells into one gigantic cell. A syncytium allows virions to move to places far from where they were synthesized without leaving the confines of the infected syncytium and being exposed to host defenses. The presence of a syncytium is easily visible by light microscopy and can be used as a diagnostic tool for the identification of virus-infected cells.

In another very clever trick, some viruses produce decoy virions, which are either empty capsids or non-infectious virions, and release large numbers of them from infected cells. The decoys confuse and distract the host defenses so that the real virus particles can achieve infection of new host cells. In the same way, some viruses keep certain host cell proteins, such as major histocompatibility proteins, in the virion as a type of camouflage.

Figure 12.11 The spread of viral particles. Panel a: The movement of virions through the apical surface of the host causes infection of neighboring cells. Panel b: The spread of viral particles from the basolateral surface of the host cell can cause infection of neighboring cells and can also keep the virus from being identified by the host's immune system. Panel c: Some viruses use both extracellular and cell-to-cell contact.

SUMMARY

- Viral structures include nucleic acids (either DNA or RNA), and a capsid protein coat made of capsomere subunits.

- Some viruses are surrounded by envelopes composed of viral glycoproteins, oligosaccharides, and host cell membrane lipids.

- Viruses may lyse the host cells by using a lytic infection cycle or they may be dormant in a latent infection.

- A lytic infection cycle involves the steps of attachment, penetration, uncoating, biosynthesis, maturation, and release.

- Host cell receptors are used to facilitate virus attachment and entry into host cells.

- The synthesis of new viral components involves complex replication mechanisms of the viral DNA or RNA.

- Intracellular trafficking and assembly of viral components in the host cell occur during the maturation stage of the viral infection cycle.

- Newly assembled viruses may be released through lysis of the host cell or by budding from the surface of the host cell.

In this chapter, we have focused on viral structure, the mechanisms by which viruses are constructed, and the events associated with the infection cycle. These events and mechanisms are important because they show us how viral particles are made and how many of the structural components of viruses are required for successful infection. We have also referred repeatedly to the notion that productive infection is the ultimate aim of any virus. It is important that you organize and reflect on the key concepts in this chapter because this information provides the foundation for the subject of viral infection, which is covered in the next chapter.

Ⓠ SELF EVALUATION AND CHAPTER CONFIDENCE

Multiple Choice

Answers are given in the back of the book and help can be found in the student resources at:

www.garlandscience.com/micro2

1. List in the correct order the steps of the lytic infection cycle.

    A. Attachment, uncoating, penetration, maturation, biosynthesis, and release
    B. Release, attachment, penetration, biosynthesis, maturation, and uncoating
    C. Attachment, uncoating, penetration, biosynthesis, maturation, and release
    D. Attachment, penetration, uncoating, biosynthesis, maturation, and release
    E. None of the above

2. Viruses contain

    A. Either RNA or DNA
    B. Both RNA and DNA
    C. No nucleic acids
    D. The RNA of host cells they infect
    E. Only proteins

3. Viral components can include all of the following except

    A. Proteins
    B. A capsule
    C. Nucleic acids
    D. An envelope
    E. A capsid

4. Viral envelopes are formed

    A. When the virus synthesizes them
    B. When the virus is first replicated
    C. When the virus leaves the host cell
    D. When the virus enters the host cell

5. Bacteriophages are viruses that

    A. Infect other viruses
    B. Infect human cells
    C. Infect plant cells
    D. Infect bacteria
    E. Are no longer able to infect

6. A lytic virus has infected a patient. Which of the following would best describe what is happening inside the patient?

    A. The virus is causing the infected cells in the patient to burst and release many virus particles
    B. The virus is incorporating its nucleic acid with that of the patient's cells
    C. The virus is infecting cells and then releasing only small amounts of virus
    D. The virus is causing the infected cells in the patient to commit suicide

7. Imagine you are a virologist studying how a type of virus attaches to its host cells. You have found a virus attachment area on the host cell membrane that appears to be dense and full of fatty acid molecules. This area on the host cell is most probably a

    A. Lipid shaft
    B. Envelope
    C. Lipid raft
    D. Spike
    E. Capsomere

8. The common cold virus (rhinovirus) uses which of the following host-cell features for attachment?

    A. ICAM-1 molecules
    B. Hemagglutinin
    C. Capsids
    D. Capsomeres
    E. Host cell nucleic acids

9. The influenza virus uses which of the following viral features for attachment?

    A. Capsids
    B. Protein coats
    C. ICAM-1 molecules
    D. Hemagglutinin
    E. Host cell nucleic acids

10. Non-enveloped animal viruses penetrate host cells by

    A. The injection of nucleic acid into the host cell
    B. Wrapping themselves in host cell lipids
    C. Endocytosis
    D. Exocytosis
    E. None of the above

11. Enveloped viruses penetrate their host cells by

    A. Endocytosis
    B. Exocytosis
    C. The injection of nucleic acid into the host cell
    D. Fusion of their envelope to the host cell membrane
    E. None of the above

12. The enzyme reverse transcriptase catalyzes the synthesis of

    A. DNA using an RNA template
    B. RNA using a DNA template
    C. The reverse sequence of DNA
    D. Messenger RNA
    E. None of the above

13–17. For each organelle indicate the event in viral replication that occurs there.

13. Golgi	**A.** Virion assembled here
14. Plasma membrane	**B.** Viral proteins synthesized here
15. Endoplasmic reticulum	**C.** Viral protein transit vesicle
16. Endosome	**D.** Viral proteins packaged for transit to plasma membrane
17. Nucleus	**E.** Synthesizes viral nucleic acid

18. Construction of viral proteins is carried out by

    A. The host cell ribosomes
    B. Proteins brought into the host cell by the virus
    C. A combination of host cell and viral proteins
    D. The Golgi apparatus of the host cell

19. The assembly of a virus requires all of the following except

    A. Formation of capsomeres
    B. Formation of capsids
    C. Formation of viral ribosomes
    D. Association of viral genomes with the capsid
    E. Coordination of multiple assembly reaction

20. Newly formed virus can be released from the cell by which of the following?

    A. Budding
    B. Cell division
    C. Cell lysis
    D. A and C
    E. All the above

21. Viruses can spread through which of the following methods?

    A. Host cell fusion with a neighbor cell
    B. The formation of syncytia
    C. The rupture of the host cell
    D. A and C
    E. A, B, and C

22–27. Indicate which of the following statements are true (**A**) or false (**B**).

22. Non-enveloped viruses can use endocytosis to enter cells.

23. Viral genomes are transcribed slowly.

24. Viruses are completely reliant on host cell machinery for translation.

25. Final assembly of viral components occurs at the plasma membrane.

26. Virus–receptor interactions inhibit the infection process.

27. In some cases the interaction between the virion and the host cell's receptor initiates a conformational change in the host cell.

DEPTH OF UNDERSTANDING

Questions listed here require you to bring together the concepts you have learned in this chapter into a discussion format. This helps you to increase your depth of understanding of the material you have learned. Help can be found in the student resources at: www.garlandscience.com/micro2

1. Discuss the structures associated with an enveloped virus and how each plays a role in the infection process.

2. One of the mechanisms used to design vaccines against viral infections is to introduce free host cell receptor molecules. How would this influence the viral infection cycle.

3. Discuss which parts of cellular machinery are used by viruses in the course of an infection.

Q **CLINICAL CORNER**

Help can be found in the student resources at: www.garlandscience.com/micro2

1. Imagine you are administering an HIV patient's drug cocktail treatment for the day. The patient has become curious about how the drugs he's taking actually help him. He points to one of the drugs and says he read on the Internet that it is a reverse transcriptase inhibitor but doesn't know what that means.

   A. Explain reverse transcriptase to him.
   B. Then explain what the reverse transcriptase inhibitor does.
   C. Explain what overall effect this will have on the virus and patient.

2. Your friend is feeling the onset of a cold and runs to the store to buy some zinc nasal spray that she saw in an advertisement. She hopes that it will work. She calls you to ask about what she is reading on the box, because you are the only person she knows in the medical field. It says that the zinc in the spray has been clinically tested, and "binds to ICAM-1 receptors on cells" to block the cold virus and "shorten the duration" of the infection.

   A. On the basis of this brief description, which stage in the virus infection cycle is this treatment targeting?
   B. Explain why this treatment may actually shorten the time that your friend is sick.

Viral Infection

Chapter 13

Why Is This Important?

Because viral diseases are so prevalent in humans, health care professionals must understand the patterns and characteristics of viral infection.

In 2012, a 60-year-old patient in Saudi Arabia was hospitalized with acute pneumonia and renal failure. Before he died, doctors retrieved a sample from his lungs. Analysis of this sample revealed the presence of a new virus, called Middle East Respiratory Syndrome Coronavirus (MERS-CoV). As its name implies, the virus has been isolated from patients living or traveling in the Middle East; all cases have occurred in or near the Arabian Peninsula. To date, it has claimed hundreds of lives.

Patients with MERS-CoV infection exhibit fever, cough, and shortness of breath. The virus has an estimated incubation period of 12 days. It rapidly leads to kidney failure, and about half of infected patients have died. The sudden appearance of this new virus has caused serious concern because of the high mortality rate of infected individuals, its mysterious origin (it has been discovered in bats and camels, but the source of infection has not been confirmed), and the possibility that it might spread to other geographic regions. MERS-CoV can spread from person to person, and in multiple cases, the virus was acquired in health care settings.

The Centers for Disease Control, the World Health Organization, and the regional council of health in Qatar are still investigating this new virus and have advised residents and travelers in the region to avoid close contact with animals (such as camels) and with people suffering from respiratory illnesses. Good hygiene, such as frequent handwashing, is also recommended for those residents and travelers.

OVERVIEW

Recall from Chapter 12 that most infections are caused by viruses. In that chapter, we learned about the mechanisms responsible for the viral infection cycle. With this information in mind, we can now turn our attention to viral infection patterns. It is important to keep in mind the events associated with the infection cycle from the previous chapter as we look at viral infection and pathogenesis.

We will divide our discussions into the following topics:

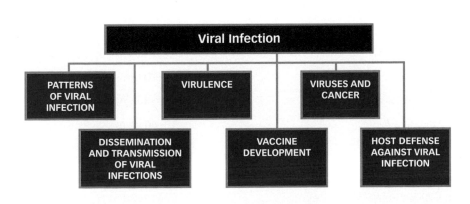

"Our virus is a lot smarter than the ones you see in zombie movies. It doesn't make its victims stagger around slobbering and moaning so anyone in their right minds would run the other way. It gets you cozying up to people so you cough and sneeze it right into their faces."

Megan Crewe

The above quotation comes from the book *The Way We Fall*, in which a fictitious virus sweeps through a community, resulting in huge death rates and the destruction of an entire community. As far-fetched as this may sound, viruses have indeed resulted in epidemics throughout history and have had a profound impact on human civilization. And it can happen again, at any time. Viruses are expert dispersers, traveling by means of sneezes, coughs, food and drink, casual contact, and sexual activity. As we see in this chapter, the interactions between human hosts and viruses are complex and fascinating, and viral infection is an unfortunate reality that health care professionals need to be prepared to deal with.

PATTERNS OF VIRAL INFECTION

Viral infections can be acute (rapid and self-limiting) or persistent (long-term). There are various types of persistent infection.

The **incubation period** is the time between exposure to a pathogen and the onset of symptoms and signs. There is considerable variation in incubation periods among viruses: some periods are as short as days, whereas others, such as HIV, are as long as years (Table 13.1). During the incubation period, the virus is replicating and the host is beginning to respond.

Table 13.1 Incubation periods of some common viral diseases.

Viral disease	Incubation period
Influenza	1–2 days
Common cold	1–3 days
Acute respiratory disease (adenoviruses)	5–7 days
Herpes	5–8 days
Enterovirus disease	6–12 days
Poliomyelitis	5–20 days
Measles	9–12 days
Smallpox	12–14 days
Chickenpox	13–17 days
Mumps	17–20 days
Mononucleosis	1–2 months
Hepatitis A	15–40 days
Hepatitis B and C	2–5 months
Rabies	1–3 months
Papilloma (warts)	2–5 months
AIDS	1–10 years

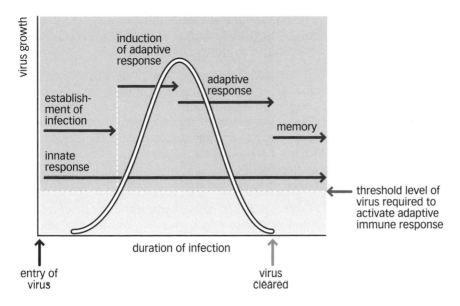

Figure 13.1 A graphic representation of the course of a typical acute viral infection. In this graph, viral growth is shown as a function of how long the virus is in the body. Notice that during the early stages of the infection, only the innate defenses are at work. When the level of virus reaches a certain point referred to as the threshold, the adaptive immune response is mobilized, and immunological memory remains after the virus has been cleared.

Acute Infections

These infections are the best-understood infections and involve the rapid production of virions followed by rapid resolution and elimination of the infection by host defenses, or host death (**Figure 13.1**). Two excellent examples of this type of infection are flu (caused by influenza virus) and the common cold (caused by rhinovirus).

Some acute infections are asymptomatic because the innate immune response limits and contains most acute viral infections. For example, 90% of people infected with poliovirus are asymptomatic.

Acute viral infections are severe public health problems and are usually associated with epidemics such as those caused by the measles, influenza, and polio viruses. The main problem with these viral infections is the incubation period. In many cases, by the time people feel sick and begin to show symptoms, they have already transmitted the infection. In fact, by the time that symptoms appear in a patient, the infection is essentially over for that patient. This delay in the appearance of identifiable symptoms makes it very difficult to control the transmission of these diseases in crowded populations such as those found in schools, military bases, prisons, and nursing homes.

Antigenic Variation

In most cases, hosts that survive acute infections are immune to re-infection for life. This is the reason that we vaccinate against many viral diseases. We will see in later chapters that the adaptive immune response comes with memory, and it is this memory that confers long-term protection from viral re-infection. However, some diseases escape this restriction. For example, with rhinovirus (which causes about 50% of all common colds) and influenza virus, re-infection can occur because of structural changes in the virions. Much of the host defense against viral infection revolves around the recognition and destruction of virally infected cells and the elimination of free virions. The adaptive response has, for any given virus, a specific memory based on the structure of the virus. If that structure changes, adaptive memory is no longer effective.

Changes in virion structure are referred to as **antigenic variation**. We will see in Chapter 16 that any protein that can induce an immune response is classified as an antigen. When the immune system is triggered by a

Fast Fact Asymptomatic hosts can still transmit the virus to others.

particular antigen, the response against that antigen is very specific, but when antigenic variation occurs, the response to that antigen is changed and becomes either less effective or nonexistent.

Antigenic variation can occur in two forms. **Antigenic drift** involves a slight change in the virion structure, resulting from mutations, and occurs after the infection has begun. **Antigenic shift** involves major changes in the structure of the virion as a result of the acquisition of new genes either from co-infection or through recombination events. These mechanisms are shown in **Movies 13.1** and **13.2**.

Persistent Infections

Persistent infections (**Table 13.2**) are caused when the defenses of the host are either modulated or completely bypassed, and the viruses are never cleared from infected hosts. There are three types of persistent infection: **chronic infections**, in which there is continuous and long-term production of virions; **latent infections**, in which there is no virion production between periodic outbreaks; and **slow infections**, in which there is a prolonged incubation period followed by a progressive disease state and virion production.

> **Fast Fact** Substantial recombination between genes of the influenza virus leads to antigenic shift, which is the reason why we need new flu shots every year.

Table 13.2 Some persistent viral infections of humans.

Virus	Site of Persistence	Consequence
Adenovirus	Adenoids, tonsils, lymphocytes	None known
Epstein–Barr virus	B cells, nasopharyngeal epithelia	Lymphoma, carcinoma
Hepatitis B virus	Liver, lymphocytes	Cirrhosis, hepatocellular carcinoma
Hepatitis C virus	Liver	Cirrhosis, hepatocellular carcinoma
Human immunodeficiency virus	CD4 T cells, macrophages, microglia	AIDS
Herpes simplex virus types 1 and 2	Sensory and autonomic ganglia	Cold sore, genital herpes
Papillomavirus	Skin, epithelial cells	Papillomas, carcinomas
Polyoma virus BK	Kidney	Hemorrhagic cystitis
Polyoma virus JC	Kidney, central nervous system	Progressive multifocal leukoencephalopathy
Measles virus	Central nervous system	Subacute sclerosing panencephalitis, measles inclusion body encephalitis
Rubella virus	Central nervous system	Progressive rubella panencephalitis
Varicella-zoster virus	Sensory ganglia	Zoster (shingles), postherpetic neuralgia

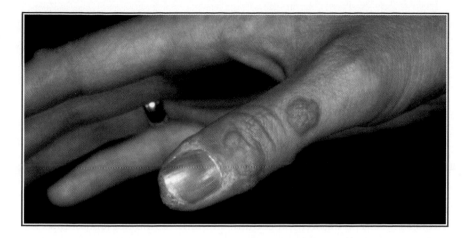

Figure 13.2 Seed warts caused by infection with papillomavirus.

Changes in the immune response to a viral infection can induce a persistent infection. For this reason, many individuals who are receiving immunosuppression therapy after they have undergone an organ transplant come down with persistent infections.

It has been shown that a host's immune system responds only to certain viral peptides present on infected cells. Because this response is so narrow, it is easily bypassed, and any change in these immunodominant peptides causes the infected host cells to become essentially invisible to the immune response. Viruses that can mutate their immunodominant peptides are called cytotoxic T lymphocyte (CTL) escape mutants. These mutants are very important in HIV pathogenesis and arise because of error-prone viral replication and the constant selective pressure that comes from exposure to the immune response. If CTL escape mutations occur early, a persistent infection occurs; if the mutations do not occur early, there is no persistent infection.

To avoid a CTL response by a host, some viruses infect only tissues that have either limited or reduced immunosurveillance ability. The best example of this is papillomavirus, which causes skin warts (**Figure 13.2**). These infections occur in terminally differentiated outer skin layers, where there is no local immune response. This same strategy is seen in the central nervous system and the vitreous humor of the eye, both of which are areas that are normally sequestered from the immune system.

Chronic Infections

A chronic virus infection is a long-term illness that occurs when the virus remains in a host's body indefinitely. For example, hepatitis C infection can be chronic and last a lifetime. Although serious liver problems can result, including cirrhosis (scarring of the liver) and hepatocellular carcinoma (liver cancer), most people do not know they are infected because they don't look or feel sick. An estimated 3.2 million persons in the United States have chronic hepatitis C virus infection.

Latent Infections

Latent infections have three general characteristics:

- Absence of a productive infection; in other words, no large-scale production of virions
- Reduced or absent host immune response
- Persistence of an intact viral genome so that productive infections can occur later

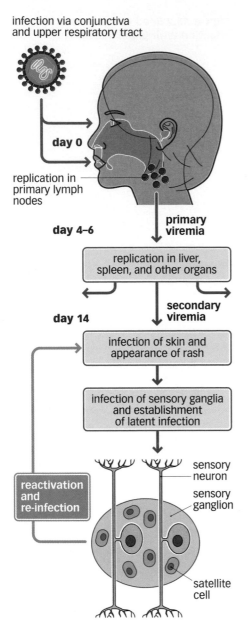

infection via conjunctiva and upper respiratory tract

day 0

replication in primary lymph nodes

day 4–6 — **primary viremia**

replication in liver, spleen, and other organs

day 14 — **secondary viremia**

infection of skin and appearance of rash

infection of sensory ganglia and establishment of latent infection

reactivation and re-infection

sensory neuron

sensory ganglion

satellite cell

Figure 13.3 A model of viral infection with varicella-zoster (chickenpox) virus. The infection is initiated either in the conjunctiva of the eye or in the mucosa of the upper respiratory tract. It then moves quickly to regional lymph nodes, where it can infect T cells. About four to six days later, infected T cells move into the blood, causing a primary viremia. These cells move to the liver, spleen, and other organs, causing a second round of infection and then back into the blood, causing a secondary viremia. It is during this period that the skin lesions appear. The virus is not finished once the lesions disappear, however. It now moves into sensory ganglia in the peripheral nervous system, where it becomes latent. If reactivated at a later date, the virus moves back to the skin, re-infects it and the patient will exhibit skin lesions called shingles.

Several viruses produce latent infections, but perhaps the best examples are herpes simplex and varicella-zoster (Figure 13.3). Both use host neurons as a vehicle for their latency, and the genomes of both of these viruses remain as extrachromosomal elements in the nucleus of infected host cells. Latent viruses have the ability to be reactivated years after they enter a host. As an example, varicella-zoster can cause chickenpox in a child, and shingles (Figure 13.4) in that same individual many years later. The reasons for reactivation are not completely understood but may have to do with trauma, stress, or any condition that indicates to the virus that the host cell is no longer a suitable place to stay. Reactivation can also be a way of establishing new or improved latency by moving the virus to non-infected neurons.

Slow Infections

Slow infections are usually associated with fatal brain infections and show signs such as **ataxia** (loss of motor control) and/or dementia. Because they are variants of persistent infections, signs may not be seen until years after the primary infection. Once signs appear, death usually follows very quickly. It should be noted that viral diseases such as measles, certain polyoma infections, and HIV can establish slow infections with severe nervous system pathology in the end stages of the disease.

Keep in Mind

- Viral infections can be acute or persistent.
- Viruses differ in their incubation periods.
- Most acute viral infections result in lifelong immunity.
- The process by which virions change structure is called antigenic variation.
- Persistent infections last for long periods and can be chronic, latent, or slow.

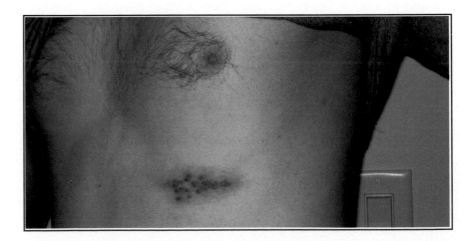

Figure 13.4 Shingles, a reaction to reactivation of latent varicella-zoster virus.

DISSEMINATION AND TRANSMISSION OF VIRAL INFECTION

In this section we look at how viral infections are **disseminated** within the infected host and **transmitted** from one host to another. To establish an infection, a virus must have access to susceptible and permissive host cells—cells that are capable of being infected. When we speak of susceptibility, we mean that the potential host cell must have the appropriate receptor for viral attachment (see Chapter 12). Permissiveness requires that the host cell contain the gene products used by the virus for a successful infection. For a successful infection, there must also be sufficient virus present in the host's body. How many virions in a host are enough? In theory, it takes only one virion to get the infection initiated, but in practice the required number is considerably higher and depends on the virus, the site of infection, and the age and health of the host.

Portals of Entry

We can use the entry points to the host body that we discussed in Chapter 5 as starting points for exploring dissemination. Some of the common sites of viral entry are shown in **Figure 13.5**.

Respiratory Tract

This is the most common of all the sites of entry. The human respiratory tract is exposed to large numbers of foreign particles and aerosol droplets, all of which are pulled into the respiratory tract with almost every breath we take. Among this aerosol stew are many viruses, including the rhinoviruses responsible for more than 50% of the common colds

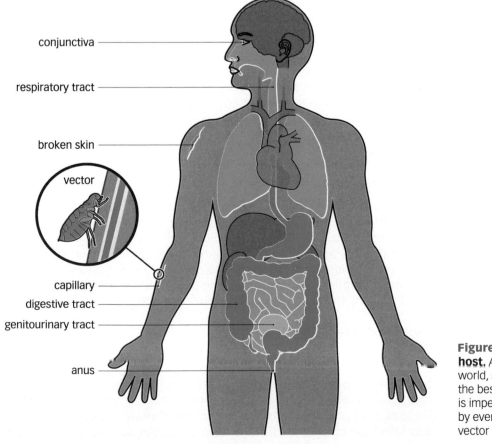

conjunctiva

respiratory tract

broken skin

vector

capillary

digestive tract

genitourinary tract

anus

Figure 13.5 Portals of entry into a host. Areas that have access to the outside world, such as the respiratory system, make the best portals of entry. Note that the skin is impermeable to virions unless it is broken by events such as a scratch, needle stick, vector bite, or other injury.

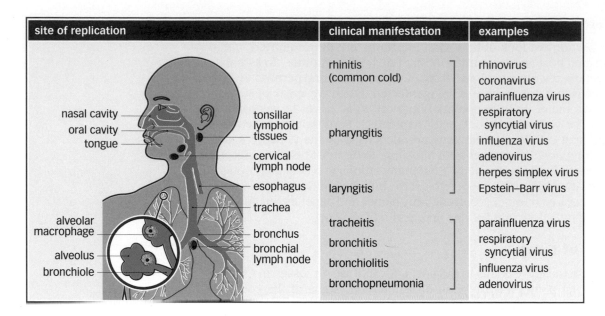

site of replication	clinical manifestation	examples
nasal cavity, oral cavity, tongue, tonsillar lymphoid tissues, cervical lymph node, esophagus, trachea	rhinitis (common cold)	rhinovirus, coronavirus, parainfluenza virus
	pharyngitis	respiratory syncytial virus, influenza virus, adenovirus, herpes simplex virus
	laryngitis	Epstein–Barr virus
alveolar macrophage, alveolus, bronchiole, bronchus, bronchial lymph node	tracheitis, bronchitis, bronchiolitis, bronchopneumonia	parainfluenza virus, respiratory syncytial virus, influenza virus, adenovirus

Figure 13.6 Viral entry into the respiratory tract. Notice that viruses can replicate in different parts of the system, producing different symptoms as indicated in the clinical manifestation column.

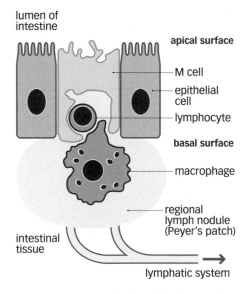

Figure 13.7 Intestinal epithelium and M cells. Notice that M cells have a different shape from normal intestinal epithelial cells and are in close contact with lymphocytes and macrophages. Viruses enter through the top (apical surface) of the cell and exit via the bottom (basal) surface.

we get. Viruses can infect all parts of the respiratory tract (**Figure 13.6**), with infections in the lower tract being more dangerous because they can cause pneumonia. Once in the respiratory tract, viruses can be disseminated from there to other parts of the body.

Gastrointestinal Tract

This is an obvious portal of entry for viruses, and one used by many viruses. Although the digestive tract is a common route of infection, it is extremely hard on the virus because of the hostility of the environment, which contains stomach acid, bile, strong digestive enzymes, immunoglobulins, and phagocytic cells. Consequently, viruses that infect the digestive tract must be resistant and resilient. Examples of such durable viruses are the picornaviruses. It is interesting to note, however, that for some viruses the harsh environment of the digestive tract is required for infection. For example, reoviruses enter the digestive tract as noninfectious viral particles and use intestinal protease enzymes for their conversion to infectious particles.

Viruses are collected by specialized cells (M cells) found in the intestinal epithelium that transport the viruses to lymphocytes and macrophages for destruction, either by these cells or by a regional lymph nodule (Peyer's patch) (**Figure 13.7**). However, some viruses can evade destruction and enter the lymphatic system or the bloodstream to become systemic. Some viruses stay in the M cells and replicate, eventually destroying the M cells and causing inflammation and diarrhea.

Genitourinary Tract

This is the entry point for sexually transmitted viruses. The genitourinary tract has some primary defenses in the mucous membranes that line it, but these defenses can be compromised. For instance, in women, sex can result in minute tears and abrasions in the vaginal epithelium, and these injuries can allow viruses to enter. Some viruses remain in the tract and infect the epithelium, producing local lesions such as vaginal warts

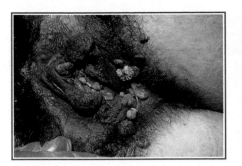

Figure 13.8 Genital warts on the labia of an infected woman. These eruptions can also occur on anal areas of both sexes as well as on the lining of the vagina.

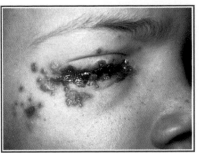

Figure 13.9 Ophthalmic herpes is caused by reactivation of latent herpesvirus on the trigeminal ganglion.

(Figure 13.8). In both women and men, viruses can use the genitourinary tract to gain access to underlying tissues. HIV virions, for instance, infect lymphocytes in the tissue underlying the genitourinary tract, whereas herpes simplex infects sensory neurons in that tissue. These types of virus disseminate from this location to the rest of the body and establish lifelong persistent infections.

Eyes

The sclera and the inner surfaces of the eyelids can be portals of entry for viruses. However, there is little opportunity for infection here because of the secretions and constant flushing action of tears across the eye. Tears contain IgA antibody, a substance that makes it difficult for virions to gain a foothold. Herpes viruses can cause an eye infection if the cornea is scratched, and this infection can lead to total destruction of the cornea and to blindness. Herpes infections that enter the body via the eye can also set up a latent infection of the optic neurons (Figure 13.9).

Skin

Although a host's skin is the most effective barrier to invasion by microorganisms, it is also an effective entry point if the barrier is broken in some way. This type of break can easily occur when the body is exposed to biting insects, such as mosquitoes, which transmit dengue and yellow fever viruses. Certain poxviruses use insects as a way into the body. The insect thus acts as a **vector** for the virus. By using a vector, the virus avoids exposure to the environment and the potential hardships that this can entail. If the virus remains in the epidermis, the infection is local. However, if the virus moves to the dermis, which is rich in blood and lymph, the infection can become systemic and be disseminated to other parts of the body. Needle sticks can also cause infection in the skin. In a similar fashion, the bite of a dog can transmit the rabies virus (Figure 13.10) through this port of entry.

Dissemination Pathways

Virions can remain localized at their portal of entry or they can disseminate throughout the host body to other tissues. For example, the rhinovirus remains a localized respiratory infection, whereas the measles virus moves from the respiratory system to other tissues. When a virus infects other organs it is said to be a systemic virus.

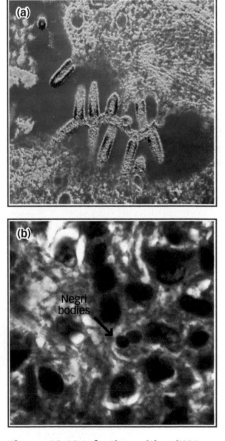

Figure 13.10 Infection with rabies virus. Panel a. A transmission electron micrograph of the rabies virus. Note the bullet shape of the virion. Panel b: Rabies-infected cells develop easily identifiable Negri bodies.

Directional movement is an important factor in viral dissemination. If the virion is released from the apical surface of the infected host epithelial cell, it will establish a localized limited infection. However, if the virion is released from the basal surface into underlying tissue, it can spread systemically. Viruses use the host's systems for dissemination.

Bloodstream

The bloodstream is the best route by which systemic viruses can disseminate; this is referred to as **hematogenous dissemination** (**Figure 13.11**). Recall from Chapter 7 that viremia is defined as a condition in which virions are present in the blood. If the virions replicate in the blood, the condition is called **active viremia**. Although most systemic viruses spend very little time in the blood (about 1–60 minutes), some, such as hepatitis B and C, can spend years traveling through the blood. This makes sense because the target for these viruses is the hepatocytes of the liver, which is the major blood-filtering organ in the body.

Figure 13.11 Dissemination of virus using the blood. Notice the target organs for certain viruses.

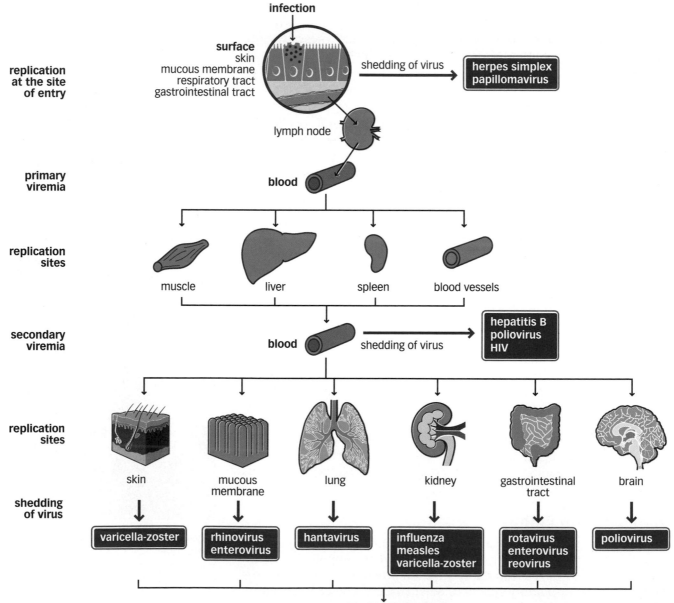

Pathway	Viruses
From peripheral neurons to neurons of central nervous system	Poliovirus, yellow fever virus, Venezuelan encephalitis virus, rabies virus, reovirus (type 3 only; type 1 disseminated by viremia), herpes simplex virus types 1 and 2, pseudorabies virus
From olfactory neurons to neurons of central nervous system	Herpes simplex virus, coronavirus
From blood to neurons of central nervous system (hematogenous route)	Poliovirus, coxsackievirus, arenavirus, mumps virus, measles virus, herpes simplex virus, cytomegalovirus

Table 13.3 Viral dissemination using the central nervous system.

Nervous System

Because neurons are found throughout the body, many viruses use the neuronal network to disseminate through a host's body (Table 13.3). For certain viruses, the neuron is the target. In rabies, for instance, both mature virions and nucleocapsids have been identified in the axons and dendrites of neurons. For other viruses, such as poliovirus, neurons are not the primary target (although polio virions are sometimes found on a patient's neurons). Virions move through neurons via microtubule structures. Obviously, virions disseminated from neurons connected to the spinal cord and brain can have devastating effects on the host, causing viral meningitis or encephalitis.

Viruses can also move from the central nervous system into the periphery of the body. When they arrive in the periphery, they cause local infections, such as the **cold sore** seen with herpes (Figure 13.12). Fortunately for humans, the direction of viral dissemination is most frequently from the central nervous system to the peripheral system and not the other way around.

Internal Organs

The liver, spleen, and bone marrow are all good candidates for systemic viral infection. In the liver, virions enter the organ via the blood that is being filtered there. The presence of the infected blood leads to viral infection of the liver's Kupffer cells followed by translocation through these cells to the hepatocytes. This process engenders a potent inflammatory response that destroys liver tissue, resulting in hepatitis.

In some parts of the brain, the capillary epithelium is fenestrated with a sparse basement membrane. Viruses such as mumps virus can pass through this area and move into the choroid plexus, where cerebrospinal fluid is being made. From here, the viruses can be disseminated throughout the central nervous system.

Another dissemination tactic used by viruses is to adhere to blood vessel walls. In this way, the virions can invade such organs as the pancreas, renal glomerulus, and colon with relative ease. In fact, some viruses—the herpes simplex, yellow fever, and measles viruses are examples—move out of the blood vessels and into the tissues during the inflammatory response.

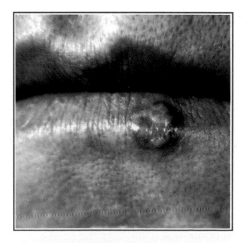

Figure 13.12 The 'cold sores' that are a common sign of oral herpes. About 90% of cold sores are caused by herpes simplex type 1 infections.

Viral Transmission

There are two general patterns of virus transmission. The first is perpetuation of the infection in only one species, as is seen with measles and hepatitis A, both of which occur only in humans. The second is perpetuation of the infection by transmission from other animals to humans. Rabies and influenza are two examples of this interspecies pattern. Recall that diseases caused in this manner are referred to as zoonotic diseases.

Viruses that cause acute infections must be transmitted efficiently. Efficient transmission means that enough virus must be produced to permit a productive infection. This requirement is easily satisfied in acute infections because of the huge numbers of virions released. In contrast, viruses that cause persistent infections do not require efficiency in transmission because the virions are produced continuously over many years. Many acute infections also have a striking seasonal variation.

Transmission via the Respiratory Tract

Respiratory infections can be spread from individual to individual by coughing or sneezing (**Figure 13.13**), as well as by contact with saliva. Large droplets of fluid are found in the nose, and smaller droplets in the lungs. When these droplets are expelled from an infected individual, the larger ones fall to the ground more quickly than the smaller ones; some small droplets stay airborne for very long periods. Furthermore, inhaling the smaller droplets can increase the risk of a more severe infection because they can more easily find their way into the alveoli of the lungs.

For respiratory-tract infection, the sneeze is probably the best transmission method; however, transmission also occurs to a smaller degree in coughing, in laughing, and even in normal exhalation. It has been estimated that the volume of air expelled in a single sneeze can contain more than 20,000 droplets, whereas the volume expelled in a typical cough contains only a few hundred droplets. Obviously, infectivity in this situation requires that the exposed individuals be in close proximity to the aerosol.

Transmission can also be geographically and seasonally influenced: respiratory viral infections are seen more often in the winter than in the summer months. The dry, cold air in winter reduces the moisture content of virus-containing airborne droplets, keeping them suspended in the air for longer periods. Furthermore, the low humidity of winter dries out the nasal passages, making minute cracks in the nasal mucosa where viruses can more easily enter the host.

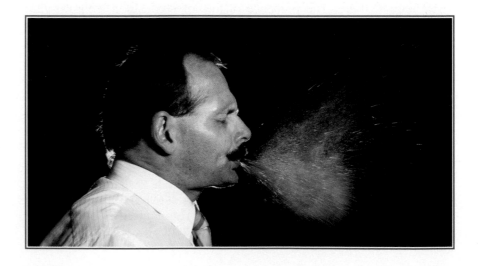

Figure 13.13 The sneeze is the main mechanism by which respiratory tract virus is spread from one person to another. Notice how large the volume occupied by droplets can be. Small droplets can circulate in the air for prolonged periods, increasing the potential for viral transmission.

Virus	Disease	Rash type
Coxsackievirus A16	Hand-foot-and-mouth disease	Maculopapular
Measles virus	Measles	Maculopapular
Parvovirus	Erythema infectiosum	Maculopapular
Rubella virus	German measles	Maculopapular
Varicella-zoster virus	Chickenpox, zoster	Vesicular

Table 13.4 Viruses that cause skin rashes in humans.

Transmission via the Epidermis

Many systemic viruses leave a telltale reminder in the form of a skin rash when they leave blood vessels. Maculopapular rashes are characterized by red splotches or bumps on the skin, and vesicular rashes are shown as fluid-filled or pus-filled blisters (Table 13.4). All of the skin lesions seen in systemic viral infections are the result of the viral destruction of host cells.

Some lesions form in the mucosal tissue of the mouth and throat. For example, measles also forms vesicles in the mouth known as Koplik's spots (Figure 13.14), which begin to ulcerate before the familiar red-spot skin lesions appear. In fact, by the time spots appear on the skin, the infection is on the wane. This means that by the time a patient develops the spots, they have already transmitted the virus to anyone who has been in close contact and has no immunity to it.

Viral transmission can also occur from the skin through direct contact, as occurs with poxvirus, herpesvirus, papillomavirus, varicella-zoster virus, and Ebola virus.

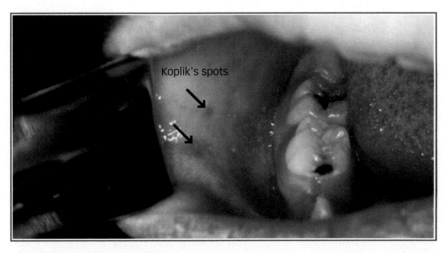

Figure 13.14 Koplik's spots on the mucosa lining the mouth interior are the earliest signs of infection by the measles virus.

Koplik's spots

Transmission via Bodily Fluids

Viruses such as HIV, herpes, and hepatitis B are found in semen and vaginal secretions, and are sexually transmitted. **Genital herpes** virus (Figure 13.15) and papillomavirus infect the genital mucosa, causing lesions that can be transmitted through genital secretions.

HIV can also be transmitted via blood, as can other viruses, such as hepatitis C virus. Transmission from the patient to the health care worker can be facilitated by poor techniques, for example, via accidental needle sticks.

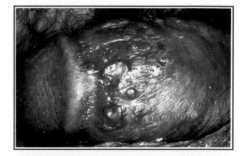

Figure 13.15 Genital herpes lesions, which are primarily the result of herpes simplex type 2 infections.

Transmission via the Fecal–Oral Route

The fecal–oral route of contamination seen in so many bacterial infections is also a common route for the spread of viral infections, and viral transmission via urine can contaminate food and water supplies. In fact, several human viruses that replicate in the kidneys are shed into the urine, although this is not the most desirable transmission vehicle. Viruses transmitted by the fecal–oral route must be able to withstand hard environments. Such viruses can also be transmitted by fomites, on hands, and in saliva.

Fetal Infection

Viremia in pregnant women can expose the fetus to dangerous virions that can cause developmental defects in the fetus. With rubella, for instance, the fetal infection rate during the first trimester is more than 80%. Babies can be infected by HIV *in utero* from infected mothers, during natural childbirth, and during breast feeding.

Fast Fact Recent studies have shown that aggressive treatment of infected mothers can dramatically reduce the incidence of babies born with the HIV virus.

Keep in Mind

- Viruses are disseminated between tissues, organs, and organ systems within a host.

- Viruses can be transmitted between hosts.

- Portals of entry, including the respiratory, gastrointestinal, and genitourinary tracts, and the eyes and skin, allow viruses access to the host's body.

- For an infection to be established, there must be adequate numbers of virions, permissive host cells, and an ineffective host defense response.

- Dissemination occurs through the blood stream, nervous system, and internal organs.

- Viral transmission occurs via respiratory tracts, epidermis, bodily fluids, the fecal–oral route, and fetal infection.

VIRULENCE

Recall from Chapter 4 that virulence refers to how fit a pathogen is when it comes to fighting the host, and thus how harmful a given pathogen is to a host. There are three ways to measure the virulence of viruses; two of these are LD_{50} (lethal dose 50%), a measure of how much virus is required to cause the death of 50% of infected individuals, and ID_{50} (infectious dose 50%), a measure of how much virus it takes to infect 50% of a population. A third way, called **PD_{50}** (paralytic dose 50%), indicates how much virus is needed to paralyze 50% of infected individuals. Although all these measurements may seem to be somewhat arbitrary, they are important indicators of the potential virulence of viruses.

Virulence varies from one virus to another and can be directly affected by the route of infection and by the age, the health, and in some cases the gender of the host. In addition, alterations in the ability of the virus to replicate can affect virulence. For example, a genetic mutation that decreases the number of virions produced in an infected host cell can lower the virulence of the virus. In addition, viral cytopathology genes can be mutated and turned off. This limits the destructiveness of the virus and thereby lowers its virulence. Last, any change in a host cell's function required by the virus can affect virulence.

Virulence and Host Susceptibility

As we have mentioned previously, virulence is affected by the health of the intended host. There are two types of host: susceptible and immune. A susceptible host is one who can be infected and can also transmit the disease. This will depend on how efficient the virus infection is and how prevalent the disease is in the prospective population. In groups in which most individuals have been immunized against a particular viral disease, that disease cannot take hold. As we saw in Chapter 6, this is referred to as herd immunity. If the potential populations all have healthy immune response capabilities, any infection will be resolved quickly and the transmission of the infection will be limited. In contrast, if the potential population is immunocompromised, transmission of the infection can be rapid. This is amply demonstrated in nursing homes, where the population has limited immunity. In this case, viral infections that would be easily cleared by younger people spread with devastating quickness.

Interestingly, some viral infections are milder in the elderly. This may be due to physical as well as physiological changes that take place with age. For example, the alveoli of elderly individuals are smaller than those of young people, allowing less area for infection. In addition, the muscles of elderly people are no longer strong enough to propel viral particles long distances during coughing and sneezing.

Gender may also have a role because males seem to be slightly more susceptible to viral infections, and pregnant women seem to be more susceptible to hepatitis A, B, and E, for reasons that are not clear. Last, the physical condition of the host seems to have a significant role in viral disease; malnourished individuals, for example, are much more susceptible to infection.

VACCINE DEVELOPMENT

As we will see in Chapter 19, there are very few drugs that are effective against viral infections, and none provide a definitive cure. Consequently, the most effective strategy in dealing with viruses is prevention through vaccination. With many viral diseases, the memory property of the adaptive immune response translates into lifelong immunity. In addition, immunization increases herd immunity. Together these help to control serious diseases that could easily reach epidemic proportions.

Let us look at smallpox as an example of how vaccination works. This is a disease that was perhaps the most devastating the world has ever seen. It caused the death of more than 300 million people in the twentieth century alone. Yet it was the first disease to be eradicated through vaccination. Edward Jenner, a rural physician, noticed that milkmaids (women who milked and maintained the family's cows) did not seem to contract smallpox. He noticed that these women had poxlike lesions on their hands. We now know that these lesions came from cowpox, a harmless virus related to the smallpox virus. In 1796, Jenner took pus from the hand lesions of milkmaid Sarah Nelmes and injected the pus under the skin of a young boy named James Phipps. Then, in what would today be considered an unethical, immoral, and definitely illegal decision, he infected young James with smallpox. James did not get sick, however. It was the beginning of the end for smallpox and the beginning of the era of vaccinations for humans.

Since Jenner's time, many vaccines have been developed and are routinely administered (Table 13.5). Vaccines can be remarkably effective in limiting viral diseases, as demonstrated most convincingly by the polio

Fast Fact In developing countries where many individuals are malnourished and have poor immune function, measles can be 300 times as lethal as it is in the United States. In these cases, the Koplik's spots become massive, and mortality can be as high as 50%.

Fast Fact Mary Wortley Montagu, after seeing smallpox inoculations in Turkey where her husband was ambassador, brought the idea of vaccination for smallpox back to England and had her children inoculated before Jenner's work took place.

Disease or virus	Type of vaccine	Population vaccinated	Schedule
Adenovirus	Live, attenuated, oral	Military recruits	One dose
Hepatitis A	Inactivated whole virus	Universal vaccination of infants, also travelers, other high-risk groups	0, 1, and 6 months
Hepatitis B	Recombinant virus-like particles (proteins)	Universal in children, exposure to blood, sexual promiscuity	0, 1, 6, and 12 months
Human papilloma	Recombinant virus-like particles (proteins)	Recommended in boys and girls before they become sexually active	Three shots over a 6-month period
Influenza	Inactivated viral subunits	Recommended annual vaccination, especially in elderly and other high-risk groups	Two-dose primary series, then one seasonal dose
Measles	Live attenuated	Universal vaccination of infants	12 months; 2nd dose, 4–6 years
Mumps	Live attenuated	Universal vaccination of infants	Same as measles, given as MMR[a]
Polio (inactivated)	Inactivated whole viruses of types 1, 2, and 3	Universal vaccination in the US; also commonly used for immunosuppressed for whom live vaccine cannot be used	2, 4, and 12–18 months; then 4–6 years
Polio (live)	Live, attenuated, oral mixture of types 1, 2, and 3	Universal vaccination; no longer used in United States	2, 4, and 6–18 months
Rabies	Inactivated whole virus	Exposure to rabies, actual or prospective	0, 3, 7, 14, and 28 days after exposure
Rotavirus	Live attenuated	Universal vaccination of infants	2, 4, and 6 months
Rubella	Live attenuated	Universal vaccination of infants	Same as measles, given as MMR
Smallpox	Live vaccinia virus	Certain laboratory workers; military personnel	One dose
Varicella	Live attenuated	Universal vaccination of infants	12–18 months; second dose, 4–6 years of age
Yellow fever	Live attenuated	Travel in areas where infection is common	One dose every 10 years
[a]MMR: measles, mumps, and rubella.			

Table 13.5 Viral vaccines licensed in the United States.

and measles vaccines, which have essentially eradicated these diseases in the United States (**Figure 13.16**). Unfortunately, vaccines can have unpredictable side effects, but for the most part, they are minor and resolve in a few days. The medical benefits and safety of vaccines are substantial and well-established.

Vaccines can be broadly classified into three groups:

- A **live attenuated vaccine** is made up of intact viral particles that have been mutated and selected for their poor growth in humans. Examples are the MMR (measles, mumps, and rubella) vaccine routinely given to children, and the oral polio vaccine. Because these vaccines are composed of infectious virions, there is the potential danger of the vaccine causing symptoms and signs of the disease, particularly in immunocompromised individuals.

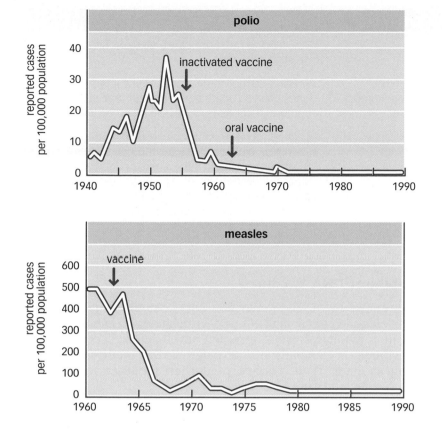

- An **inactivated vaccine**, also called a **killed vaccine**, is composed of virus that is either 'dead' or made non-infectious by chemical or physical treatment. This type is potentially safer than a live attenuated vaccine because there are no virions present, or there are virions present but they have been rendered non-infectious.

- A **subunit vaccine** is composed of immunogenic parts of the virus and is usually derived through the use of genetic engineering and recombinant DNA techniques. This is the safest type of vaccine because there are no intact virions present; however, this type can also be less effective than live attenuated vaccines or inactivated vaccines, depending on the viral proteins that have been engineered.

Vaccination can be active or passive. In **active immunization**, the antigen representing the infectious agent is administered and causes the onset of an immune response. In **passive immunization**, an already-formed antiviral product, such as antibody, is administered. Of these, only active immunization provides long-lasting immunological immunity.

Table 13.6 shows the basic and fundamental requirements for an effective vaccine. It is estimated that the cost of developing an approved vaccine is in the hundreds of millions of dollars. Consequently, not many companies are willing to risk the cost of development. This problem is compounded by the fact that accidental infection during vaccination can bring about considerable litigation and expense. Therefore, in general, there is little effort to make vaccines for many viral diseases. Another factor is the economics of producing vaccines for diseases that do not affect industrialized nations.

Fast Fact Before the 1970s, everyone in the United States was vaccinated against smallpox, and the resulting herd immunity essentially eliminated it from this country. However, relatively few people have been vaccinated for smallpox since then, resulting in a vulnerable population if the disease were to make a comeback through natural or bioterrorist means.

Table 13.6 Requirements of an effective vaccine.

Requirement	Comments
Safety	The vaccine must not cause disease. Side effects must be minimal
Induction of protective immune response	Vaccinated individual must be protected from illness due to pathogen. Proper innate, cellular, and humoral responses must be evoked by vaccine
Practical issues	Cost per dose must not be prohibitive. The vaccine should be biologically stable (no genetic reversion to virulence; able to survive use and storage in different surroundings). Vaccine should be easy to administer (oral delivery preferred to needles). The public must see more benefit than risk

Recombinant DNA Vaccines

Recent advances in vaccine production have focused on recombinant DNA technology. Recombinant DNA (rDNA) technology combines genetic material from multiple sources, inserting novel genetic sequences into cells or viruses. In comparison with older, slower methods of mass-producing vaccines, rDNA vaccine production promises to be much quicker, cheaper, and safer.

H5N1 typically infects birds, but it has also infected and killed hundreds of people in the past decade. Symptoms of H5N1 infection start within a few days of exposure and include fever, cough, shortness of breath, diarrhea, vomiting, and abdominal pain. Mortality rates exceed 60%, and many die from respiratory failure. If it were to acquire the ability to pass from person to person, H5N1 could result in a pandemic capable of killing millions of people worldwide. In the case of such an outbreak, effective vaccines would need to be developed and produced in large supplies and in a short period; modern methods of vaccine production would be too slow to meet the demand. To understand rDNA vaccine production better, let's consider how a new vaccine against bird flu (H5N1) has been produced.

To make the new H5N1 vaccine, researchers obtained the genetic sequence data that codes for the hemagglutinin gene (the H surface protein on all influenza viruses, including H5N1; see Chapter 12). Using rDNA technology, they spliced this gene into a human adenovirus, which was then used as a vaccine in mice and chickens. Intranasally, the rDNA vaccine provided partial immunity to H5N1, but when injected it provided 100% protection. It also resulted in a **dual immunological response**, in which virus-neutralizing antibodies and T cells were produced inside the vaccinated animals' bodies. The researchers developed the vaccine in less than a month and are moving ahead with further trials. This would have taken many months using traditional methods.

Viral Culture

Many viral vaccines are made from viruses or viral components grown in embryonated chicken eggs. However, growing cells outside living organisms has become a necessary and routine technique for the manufacture of vaccines. For example, vaccines for measles, mumps, rubella, chickenpox, and polio are made using cell cultures. Cell culturing is used to propagate viruses and intracellular bacteria, to harvest useful products made by cells, and to study how various chemicals and drugs affect certain cells. Some types of cell are relatively easy to grow, such as certain plant cells and fibroblasts (skin cells). However, some specialized cells are still in experimental stages for growth *in vitro*. In all types of cultured **cell lines**, enzymes are used to isolate some cells from the tissue of an organism, and the cells are then cultured in a medium that supplies them with the nutrients and growth factors needed for their growth. These cultures may one day remove the need for testing new pharmaceuticals on live animals.

We can categorize cell lines as **primary cell lines** (which have to be grown from scratch every time from the organ or tissue, for example monkey kidney), **semi-continuous cell lines** (which can be grown for some time until cells are too old or differentiated, for example human embryo kidney), and **continuous cell lines** (which can be grown for a very long time, for example HeLa). The polio vaccine, for example, was one of the first products to be mass produced using monkey kidney cell cultures.

VIRUSES AND CANCER

Our discussion would not be complete without mentioning how viruses are implicated in the development of cancer. Cancer is the leading cause of death in the developed world. In the United States alone, there are more than 500,000 deaths a year from this disorder. It is estimated that viruses are involved in about 20% of human cancers, and there is a clear cause-and-effect relationship between viruses and cancer of the liver and the cervix. However, because the induction of malignancy is not a requirement for the propagation of viruses, malignancy can be viewed as a side effect of the infection.

Oncogenic Viruses

Specific members of several virus families have been implicated both in cancers in humans and in experimentally induced cancers in laboratory animals. Retroviruses, which were initially called RNA tumor viruses, have been shown to be able to inactivate genes that suppress tumor formation.

Several human cancers are associated with infection by one of five oncogenic viruses, or **oncoviruses**: Epstein–Barr, hepatitis B, hepatitis C, human lymphotropic virus, and human papillomavirus (HPV). In many of these infections, it seems that viral proteins or transcriptional controls override the mechanisms that normally ensure that cells divide only when necessary. Such overriding leads to the uncontrolled cell growth that we know by the generic term cancer.

Keep in Mind

- Virulence refers to the capacity of a virus to cause disease.

- Virulence varies from one virus to another and can be affected by the route of infection, by the age and health of the host, and in some cases by the sex of the host.

- Susceptible hosts can be infected and transmit the disease, but immune hosts cannot be infected.

- Vaccines have been very effective in limiting viral diseases.

- Vaccines can be composed of live attenuated virus, inactivated virus, or subunits of the virion (parts of the virion that can elicit an immune response).

- There are strict requirements for vaccines, including minimal side effects coupled with maximum protection from infection.

- Viral vaccine development and production rely on embryonated chicken eggs; on primary, semi-continuous, or continuous cell lines; on adult or embryonic stem cells; or on recombinant DNA technology.

- Oncogenic viruses (oncoviruses) can induce cancer in humans.

HOST DEFENSE AGAINST VIRAL INFECTION

One of the cornerstones of human health is the presence of a fully functional immune system. It keeps us safe from myriad infectious organisms and diseases. This concept is well illustrated in the battle against viral infections. In point of fact, we live most of our lives in a cloud of infectious organisms, and it is the combination of innate and adaptive immunity that keeps us safe. The battle between virus and host is a fascinating one in which we humans have developed weapons to defeat viruses, and viruses have developed ways to defeat our weapons. In fact, the genomes of successful pathogenic viruses code for many products that modify or block almost every step we take to defend ourselves. Put another way, for every host defense, there is a viral offense. We examine our defense mechanisms against viral infection in Chapters 15 and 16.

SUMMARY

- Viral infections can be acute or persistent.

- Persistent infections last longer and can be chronic, latent, or slow.

- Latent viral infections do not produce large numbers of virions, but these infections can be reactivated later and release virions.

- Viruses can be disseminated (move to other parts of the infected host's body) through the respiratory, digestive, and genitourinary tracts as well as the nervous system and internal organs.

- Transmission of virus from one host to another can be through the fecal–oral route of infection, fomite transmission, and iatrogenic mechanisms (by health care workers).

- Virulence varies from one virus to another and can be affected by the age of the host, the health status of the host, and the route of infection.

- Vaccination has been very effective in limiting viral diseases.

- Vaccines can contain live attenuated virus, inactivated virus, or viral subunits.

- Some oncogenic viruses (oncoviruses) have been implicated in the development of cancer.

In this chapter, we have looked at the pathogenesis of viral disease and explored the diverse and complex ways in which viruses can disseminate through the body or be transmitted from one host to another. Keep in mind that pathogenic viruses have only one objective: a productive infection. Even latent and persistent viral infections begin in this way, but they produce virions for the life of the host cell rather than all at once at the expense of the host cell. Protection against viral infection is through vaccination, and there are many safe and effective vaccines that are routinely administered as a way of preventing viral infection.

Multiple Choice

Answers are given in the back of the book and help can be found in the student resources at:
www.garlandscience.com/micro2

1. Which of the following terms is not used to describe a viral infection?

   A. Slow
   B. Latent
   C. Acute
   D. Temporary
   E. Persistent

2. Acute infections are represented by

   A. Slow production of virus but rapid resolution of the infection by host defense
   B. Rapid production of virus and slow resolution of infection by host defense
   C. Rapid production of virus and rapid resolution of infection by host defense
   D. Slow production of virus and slow resolution of infection by host defense
   E. None of the above

3–6. Arrange the following viruses in order of increasing incubation period, beginning with the one that has the shortest incubation period: **A.** Papilloma; **B.** Mumps **C.** Influenza; **D.** HIV.

3. Shortest incubation period (days)

4. Second shortest incubation period (weeks)

5. Third shortest incubation period (months)

6. Longest incubation period (years)

7. Major changes in the structure of the virus are referred to as

   A. Viral shifts
   B. Antigenic drift
   C. Antigenic shift
   D. Antigenic shuffling
   E. Viral camouflage

8. Latent infections have all of the following characteristics except

   A. The absence of an early productive infection
   B. A reduced or absent immune response
   C. The viral genome remains intact
   D. The presence of an early productive infection
   E. No large-scale production of virus

9. Latent viruses

   A. Can never be reactivated
   B. Occur only in adults
   C. Can integrate into the host chromosome
   D. Destroy the host chromosome
   E. None of the above

10. Which of the following help to determine virulence?

   A. The ability of the virus to replicate
   B. The route of the infection
   C. The function of the host cell
   D. The gender of the infected person
   E. All of the above

11. Which of the following is a danger for viruses that infect the digestive tract?

   A. Stomach acid
   B. Digestive enzymes
   C. Bile
   D. Immunoglobulins
   E. All of the above

12. When a virus is released from the basal surface of a host cell it can

   A. Cause a localized infection
   B. Not infect other cells
   C. Cause a systemic infection
   D. Be deactivated
   E. None of the above

13. Iatrogenic transmission of virus is caused by

   A. Ingestion of virally contaminated food
   B. Mosquitoes
   C. Health care workers
   D. Family members
   E. Co-workers

14. The target site for rabies virus is

   A. The digestive epithelium
   B. The conjunctiva of the eye
   C. The muscle cells of the heart
   D. The neuron
   E. Both **A** and **C**

15. Hepatitis C virus

    A. Causes a persistent infection of the kidneys
    B. Can spend years traveling round the blood
    C. Can be transmitted by mosquitoes
    D. **B** and **C** only
    E. All of the above

16. Viral vaccines are usually composed of

    A. Virulent virus
    B. Non-virulent virus
    C. Attenuated virus
    D. **B** and **C**
    E. All of the above

17. PD_{50} defines the

    A. Lethal dose of a particular virus in 50% of infected individuals
    B. Paralytic dose of a virus in 50% of infected individuals
    C. The pandemic dose of a virus in 50% of infected individuals
    D. The prior dose of a virus in 50% of infected individuals
    E. None of the above

18. Koplik's spots are seen in

    A. Herpes infections
    B. Hepatitis infections
    C. Polio
    D. Measles

19. The first successfully tested viral vaccine was for

    A. Polio
    B. Smallpox
    C. Rubella
    D. Measles
    E. Mumps

20. The vaccine that is most potentially dangerous

    A. Consists of live attenuated virus
    B. Contains killed virus
    C. Contains inactivated virus
    D. Is a subunit vaccine
    E. None of the above

21. All of the following are requirements for an effective vaccine except

    A. The vaccine must be safe
    B. The vaccine must be made from killed pathogens
    C. The vaccine must induce a protective response
    D. The vaccine must be stable
    E. The vaccine should be inexpensive if possible

22. Oncogenic viruses

    A. Cause acute infections
    B. Are genetically unstable
    C. Are associated with cancers
    D. Are lytic viruses that kill the host cells
    E. Can only have a dsDNA genome

23. Which type of cell in the gastrointestinal tract is used by some viruses as an initial point of entry in the host?

    A. Peyer's cells
    B. M cells
    C. Lymphocyte cells
    D. All of the above

24. A cold sore is a symptom of which of the following?

    A. Herpes simplex virus
    B. HIV
    C. Poliovirus
    D. Influenza

25. Which of the following is an example of a live attenuated vaccine?

    A. Influenza
    B. Hepatitis A
    C. Hepatitis B
    D. MMR

26–29. Match the following with their counterpart:
    A. A vaccine with either dead or non-infectious particles; B. A vaccine of viral immunogenic parts; C. A vaccine with intact viral particles; D. The use of harmless virus as vaccine.

26. Live attenuated vaccine

27. Inactivated vaccine

28. Subunit vaccine

29. Cowpox vaccination

DEPTH OF UNDERSTANDING

Questions listed here require you to bring together the concepts you have learned in this chapter into a discussion format. This helps you to increase your depth of understanding of the material you have learned. Help can be found in the student resources at: www.garlandscience.com/micro2

1. Compare and contrast the latent and acute infections, with regard to the maximum amount of new virions produced.

2. Discuss the effects of herd immunity on viral infection and how this affects the transmission of viral disease.

3. Using what you have learned, design the best vaccine against viral infection.

 CLINICAL CORNER

Help can be found in the student resources at: www.garlandscience.com/micro2

1. Viruses such as smallpox have become of interest to terrorists as a bioweapon. As a result, military personnel are routinely vaccinated against smallpox.

   A. Why would smallpox be a good choice for a bioweapon?
   B. What precautions could be taken to prevent a terrorist attack with these weapons, and would these precautions be effective?

2. Millicent's grandmother has lived at the Shady Grove retirement home for more than 6 years, and Millicent visits her as often as possible. At her last visit she found out that two of her grandmother's friends had come down with viral pneumonia and one had died. Fortunately, her grandmother seemed to be fine.

   A. Should she be worried about her grandmother? If so, why?
   B. Should she be worried about her own health?

Parasitic and Fungal Infections

Chapter 14

Why Is This Important?

Over the previous five chapters, we have looked at the pathogenicity of bacteria and viruses. In this chapter, we look at the last two groups of infectious organisms, the parasites and the fungi. Although fungal infections are usually opportunistic, parasitic infections affect billions of people in the world.

In August 2013, a few days after playing in a water-filled ditch with friends, a 12-year-old boy in Florida started having headaches, fever, nausea, and vomiting. He was brought to the hospital, and when the disease rapidly progressed, he was placed in the intensive care unit. Doctors diagnosed him with primary amebic meningoencephalitis (PAM), which is caused by the 'brain-eating ameba' *Naegleria fowleri*. Multiple antibiotics were administered and the ameba was eradicated from his system. However, he had suffered severe brain damage and died later that year.

Infection by *Naegleria* is not common (only a few dozen cases have been reported in the United States in the past decade), but it is almost invariably fatal. Only a few victims in North America have survived the deadly infection. Victims acquire the ameba while swimming or playing in freshwater ponds, lakes, and rivers during the warm summer months. The protozoan enters the nose and migrates to the brain, where it causes the symptoms and signs of PAM. Several cases have occurred from people using household water in neti pots to flush their sinuses. Unfortunately, almost everyone who gets this parasite will die from the infection, but like many infectious diseases, early diagnosis and treatment increase the overall chances of survival.

OVERVIEW

In this chapter, we begin by looking at some general features of parasites and then take a closer look at some infections caused by protozoan parasitic pathogens, which account for millions of infections in the world. Next, we look at some detail about the infections caused by helminths (worms), and in the last part of the chapter, we look at diseases caused by fungi. We have chosen to look at fungal infection last because they are for the most part opportunistic and rarely bother us if we have intact and functioning defenses. However, fungal infections can be very dangerous for individuals who are immunocompromised.

We will divide our discussions into the following topics:

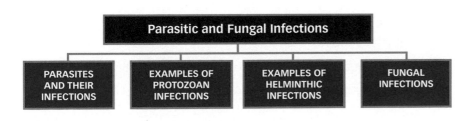

'This parasite [the malaria-causing protozoan] is so lethal that it is estimated to have killed at least half of the human beings that have ever existed on this planet..."

Dr. Robert Buckman

In his book *Human Wildlife: The Life That Lives On Us*, Dr. Buckman reminds us that parasites (such as the malaria parasite mentioned in the above quote) have had a tremendous impact on humankind and continue to kill millions and millions of people each year. Parasites have coevolved with humans and have even altered our genetics and immune systems. In this chapter we examine malaria and other protozoan diseases, parasitic worm infestations, and several types of fungal infection.

PARASITES AND THEIR INFECTIONS

The parasites described in this chapter are divided into two main groups, protozoans and **helminths** (a third group, parasitic lice, which are insects that attach to the skin, will be discussed briefly in Chapter 26). Protozoans are microscopic, single-celled eukaryotes. Helminths are macroscopic, multicellular worms that possess differentiated tissues and complex organ systems. Most protozoans and helminths are free-living (in other words, not parasitic) in the environment, have an important role in ecology, and are incapable of infecting humans. The disease-causing parasites live in association with, and depend on, their infected host for survival.

Significance of Parasitic Infections

In the industrialized world, we hardly ever think about parasitic infections. However, they remain among the major causes of human misery and death in the world (Table 14.1). For example, more than 500 million people are infected with malaria, a disease caused by the protozoan parasite *Plasmodium*, and more than 2 million people, mostly children, die of this disease each year. Perhaps more incredibly, 2.5 billion people live in areas of the world where malaria is endemic. All of the *Plasmodium*

Fast Fact It has been estimated that as many as 25–50% of Americans harbor parasitic worms.

Table 14.1 The prevalence of parasitic infections.

Disease	Estimated population affected
Ascaris infection	More than 2 billion
Hookworm and whipworm infection	1.3 billion
Amebiasis	600 million
Malaria	500 million
Enterobius vermicularis infection	400 million
Giardiasis	200 million
Schistosomiasis (blood fluke)	200 million
Cestodiasis (tapeworm)	65 million
American trypanosomiasis	24 million
Clonorchis (liver fluke) infection	13.5 million
Leishmaniasis	12 million
Paragonimiasis (lung fluke)	2.1 million
African trypanosomiasis	100,000

species are becoming resistant to the drugs used to prevent and treat malaria. Other species of pathogenic protozoans also develop resistance to drugs, and vaccine development is hindered by the ability of these parasites to quickly mutate and become resistant.

Two other important causative agents of protozoan parasitic diseases are *Entamoeba* and *Trypanosoma*. Members of the genus *Entamoeba*, which causes amebiasis, are intestinal parasites that infect 10% of the world's population; and 2–3% of Americans are infected with this organism. In Latin America, the parasite *Trypanosoma cruzi*, which causes Chagas' disease (American trypansomiasis), infects an estimated 16 million people every year. This infection can cause heart and gastrointestinal lesions that can be very serious. A related species, *T. brucei*, causes sleeping sickness (**African trypanosomiasis**), one of the most lethal diseases in humans.

Whereas certain species of protozoans infect millions of humans, some parasitic helminths infect billions. For example, approximately one-sixth of the human population suffers from ascariasis (infection by species of *Ascaris*). These worms have eggs that are highly resistant to harsh environmental conditions and antimicrobial chemicals, and thus are among the most difficult pathogens to kill. The eggs survive for years and, as will see later in this chapter, gastrointestinal infection occurs when they are consumed in fecally contaminated food or water. The human whipworm (*Trichuris trichiura*) is another parasitic helminth that infects more than a billion people via consumption of its eggs, and it will be discussed more in Chapter 22.

Hookworm infection (caused by *Ancylostoma duodenale* and *Necator americanus*) affects more than half a billion people and is most common in tropical, developing countries. In contrast to the infective eggs of *Ascaris* and *Trichuris*, it is the larval stage of hookworms that is infective. Infection can occur from people ingesting the larvae, but more often the larvae penetrate the skin of human feet as they walk barefoot over fecally contaminated soil. Such poor sanitation is responsible for the persistence of hookworm infections in impoverished regions. As described more in Chapter 22, hookworm eradication campaigns focus on building outhouses, as well as maintaining clean water, encouraging the wearing of shoes, and deworming.

Protozoan Morphology and Pathogenesis

Protozoans vary in size, ranging from 2 to 100 μm in diameter, and contain membrane-bound nuclei and cytoplasm. Most infectious protozoans are facultative anaerobes and heterotrophs that engulf food into digestive vacuoles either through pinocytosis or through phagocytosis. Many protozoans possess a **trophozoite** (active) stage, in which the organism is metabolically operational, growing, and reproducing, as well as a **cyst** (dormant) stage that protects them from environmental challenges and serves as a mechanism to move from host to host. Reproduction of many protozoan organisms is by binary fission, but some go through a cycle of simple fission (called **schizogony**) followed by a sexual reproductive phase called **gametogony**.

The four groups of protozoa that infect humans are classified by their means of locomotion:

- **Ameboids** move by bulges of cytoplasm called **pseudopods**.
- **Ciliates** move through the use of short, hair-like **cilia**.

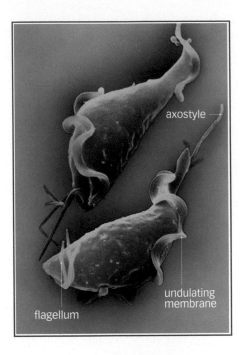

Figure 14.1 Scanning electron micrograph of *Trichomonas vaginalis*, a protozoan parasite associated with genitourinary tract infections. Notice the axostyle, which may be used for tissue attachment and destruction.

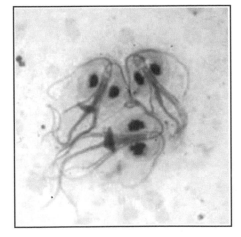

Figure 14.2 Light micrograph of the trophozoite form of *Giardia intestinalis* (formerly known as *lamblia*), which causes giardiasis. This protozoan has eight flagella and two very prominent nuclei. *Giardia* can also be found in a cystic form for protection from the environment.

- **Flagellates** move via whip-like appendages called **flagella**.

- **Sporozoans** only exhibit **motile structures** (used for movement) in the gamete stage.

The pathogenesis of protozoan parasitic diseases is quite variable. For example, *Plasmodium* destroys red blood cells, whereas *Leishmania* invades the bloodstream and causes skin lesions as well as chronic illness. *Trichomonas vaginalis* (**Figure 14.1**) inhabits the human genitourinary tract and causes vaginal epithelial cell damage and inflammation, and *Giardia intestinalis* (**Figure 14.2**) inhabits the gastrointestinal tract and interferes with the absorption of nutrients from the intestine.

Helminth Morphology and Pathogenesis

Helminths are worms, characterized by elongated, cylindrical or flat bodies. They can vary in length from less than 1 mm to 10 m or more. The body of the worm is covered by a tough cuticle, which may be smooth or may possess ridges, spines, or nodules called tubercles. At the anterior end of some helminths there can be suckers, hooks, or plates, which are used for attachment. All helminths have differentiated organs, including primitive nervous and excretory systems and highly developed reproductive tracts. None have circulatory systems.

The three classes of helminth that infect humans are nematodes, cestodes, and trematodes:

- **Nematodes** (roundworms) have a cylindrical body and an alimentary canal that goes from the anterior mouth to the posterior anus. There are two types of parasitic nematode: those that dwell in the gastrointestinal tract and use only one host to complete their life cycle, and those that infect blood and tissues and use multiple hosts to complete their life cycle.

- **Cestodes** (tapeworms) have a flat, ribbon-shaped body. At the anterior end of the body is a head, which contains suckers and frequently has hooks for attachment (**Figure 14.3**). In these worms the neck region generates reproductive segments called **proglottids**, each of which contains both male and female gonads. These worms have no digestive tract, and nutrients are absorbed across their cuticle. Some of these helminths use one host to complete their life cycle, and others use two.

- **Trematodes** (flukes) are leaf-shaped and have a blind branched alimentary tract (**Figure 14.4**). (A blind tract is one that has an opening at one end only.) They have two suckers, an oral sucker through which nutrients are taken in and waste material is regurgitated, and a distal sucker responsible for attachment.

Because most worms cannot increase their numbers while in a host, the severity of a helminthic infection is related to the number of worms acquired by repeated infection over time—the smaller the number, the greater the chance of asymptomatic infection. However, many worms are long-lived, and repeated infections with these worms will drive up the number in the host. The resulting large numbers of worms cause the host to become increasingly incapacitated.

Helminthic parasites are nourished by the ingestion or absorption of host bodily fluids, lysed tissue, or intestinal contents. Hookworms can cause a loss of iron by their feeding mechanisms, and schistosomes can compromise organ function by obstruction, by secondary infection, and by causing cancer.

It is important to note that in addition to the damage caused by the invading organisms, the immunological defense of a host against these parasites can also cause extensive tissue damage and clinical symptoms. For example, allergic and anaphylactic cutaneous reactions are seen in response to hookworms, whipworms, and *Ascaris*, and fever and swollen lymph nodes are associated with the response to *Schistosoma* larvae.

Their tough cuticle and the enzymes they secrete protect helminths from host defense responses. In fact, *Schistosoma* will incorporate some of the host antigens into its cuticle as a way of protecting itself from host immune responses. Although the life span of most of these worms is usually short (weeks or months), some species, such as hookworms and flukes, can survive in the host for decades.

Life Cycles and Transmission Pathways of Protozoans and Helminths

The life cycles and transmission mechanisms of parasitic protozoans and helminths are highly variable (**Table 14.2**), including a difference in the number of hosts that must be infected to complete the parasite's life cycle. In some cases, parasitic organisms use a single host; in other cases multiple hosts are required.

Parasites That Use a Single Host

Many parasites require only a single host to complete their life cycle, and transmission of these parasites from one host to another depends on the ability of the parasites to survive in the external environment. For example, the protozoan *Trichomonas vaginalis* can survive outside a host for only a few hours and requires direct genital contact through sexual intercourse to be transmitted. In contrast, the protozoan *Entamoeba histolytica* lives in the human gut and produces strong cysts that are passed in the stool and transmitted through the fecal–oral route of contamination. These cysts can survive in the environment for long periods and may eventually contaminate food or drinking water.

The parasitic helminth *Ascaris lumbricoides* produces highly resistant eggs that are passed in the stool of the infected host. These eggs are not immediately infectious and must mature in the soil for a period. Consequently, this worm cannot be directly transmitted from host to host.

Parasites That Use Multiple Hosts

A few protozoans and many helminths need more than one host to complete their life cycle, including the **definitive host**, in which sexual reproduction occurs, and the **intermediate host**, in which asexual development occurs. In some cases, such as the beef tapeworm *Taenia saginata*, both hosts are vertebrates (with humans being the definitive host and cattle the intermediate host). However, in parasites that live in

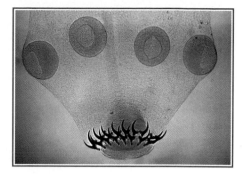

Figure 14.3 The rostellum of a helminth. These hooks, which can come in one or two rows, are found on the scolex of the worm and are used to attach and stay fixed to the tissue. Notice the four suckers above the rostellum.

Fast Fact Most of the energy required by parasitic helminths is used for reproduction, and the number of eggs that some of these worms produce is 250,000 per day under ideal conditions.

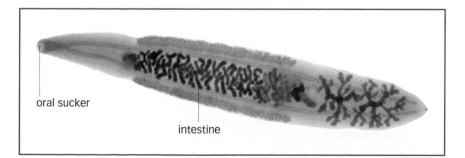

Figure 14.4 Light micrograph of the Asian liver fluke *Clonorchis sinensis*, which like all flukes has a leaf-like shape and an oral sucker. This fluke has an incomplete digestive system.

Table 14.2 Transmission, hosts, and distribution of common parasites.

Organism	Mechanism of transmission	Number of hosts	Distribution
Trichomonas vaginalis (protozoan)	Sexual contact	1	Worldwide
Entamoeba histolytica (protozoan)	Fecal–oral route	1	Worldwide
Ascaris lumbricoides (helminth)	Fecal–oral route	1	Worldwide
Trichinella spiralis (helminth)	Eating infected, undercooked pork	1 or 2	Worldwide
Plasmodium falciparum (protozoan)	Vector—Anopheles mosquito	2	Tropical and subtropical regions
Trypanosoma (protozoan)	Vector—tsetse fly (in Africa) or kissing bug (in the Americas)	2	Tropical and subtropical regions of Africa and the Americas
Clonorchis sinensis (helminth)	Eating infected, undercooked fish	3	Mainly China, Japan, Southeast Asia
Diphyllobothrium (helminth)	Eating infected, undercooked fish	4	Mainly Europe and Asia

the blood and tissue of humans, it is more common to find blood-feeding arthropods serving as hosts and transmitting vectors.

The most important example of a parasite that uses multiple hosts is the protozoan *Plasmodium*, the causative agent of malaria, which is transmitted by the *Anopheles* mosquito. In this case, the mosquito is the organism in which sexual reproduction occurs and *Anopheles* is therefore the definitive host, making the human an intermediate host. The areas where malaria is endemic are restricted by the availability of mosquito vectors, an availability that depends on a warm climate. Consequently, in tropical and subtropical climates, the transmission of malaria is constant and intense.

Keep in Mind

- Parasites can be protozoans or helminths (although not all protozoans and not all helminths are parasites).
- Parasitic infections affect hundreds of millions of people throughout the world and cause millions of deaths each year.
- There are three classes of helminth that infect humans: nematodes (roundworms), cestodes (tapeworms), and trematodes (flukes).
- Some parasites have a life cycle that involves a single host, whereas others use more than one host.
- Pathogenic mechanisms for both protozoan and helminthic infections vary and depend on the specific parasite.

EXAMPLES OF PROTOZOAN INFECTIONS

In this section, we look at diseases caused by several protozoan parasites. In each case we discuss the life cycle of the parasite, and then the pathology and treatment of the disease.

Malaria (*Plasmodium* species)

Malaria is a febrile illness caused by a parasitic infection of human red blood cells. It results from infection by the sporozoan protozoan *Plasmodium* and is transmitted through the bite of the *Anopheles* mosquito. This disease is found throughout the world, and in particular in areas with warm climates. There are four species of *Plasmodium* that can cause malaria symptoms to different degrees, with *P. falciparum* being the most pathogenic and the dominant species found in the tropics. *Plasmodium vivax* is the most widespread species. Spread of the disease depends on the density and feeding habits of the mosquito vectors, and mortality is largely restricted to infants and immunocompromised adults. In some areas, the transmission of malaria can be seasonal, and in these areas the infection can be seen in people of all ages. In the United States, about 100 cases are reported each year, and these occur mostly in immigrants coming from endemic areas and people who travel to those areas. Clinical symptoms are usually seen within six months of returning from those areas.

Life cycle of *Plasmodium*

The sexual cycle of *Plasmodium* begins when a female mosquito (the definitive host) ingests circulating male and female *Plasmodium* **gametocytes** from the blood of an infected person (the intermediate host). In the gut of the mosquito, the male gametocyte fertilizes the female gametocyte, and the resulting zygote penetrates out of the gut wall of the mosquito and forms an **oocyst**. Inside this oocyst, thousands of **sporozoites** are formed, and the cyst eventually ruptures (**Figure 14.5**), releasing the sporozoites into the body cavity of the mosquito. Some

Fast Fact It is estimated that 2 billion people live in areas endemic for malaria, and 25–50% of this group are believed to carry the parasite, resulting in more than 2 million deaths a year from malaria.

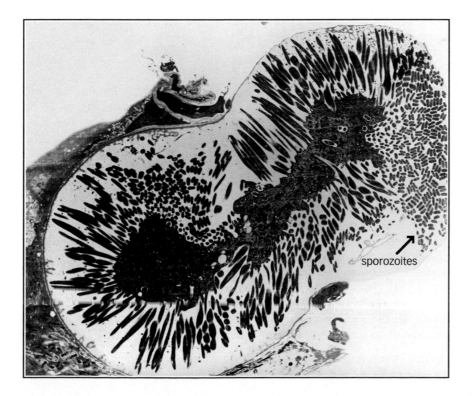

Figure 14.5 Malaria parasites. Colored transmission electron micrograph (TEM) of malaria sporozoites in a mosquito (*Anopheles* species) gut. The sporozoites (*Plasmodium* species, purple) are seen bursting from an oocyst (white/yellow) at upper right.

sporozoites

Figure 14.6 A photomicrograph of *Plasmodium falciparum* (yellow) emerging from red blood cells.

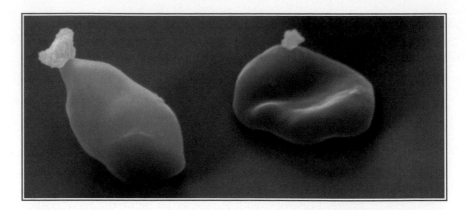

of these sporozoites will penetrate the salivary glands of the mosquito, and this is the form responsible for infection of humans. One life cycle takes about one to three weeks to complete, depending on temperature and humidity. The life cycle of *Plasmodium* in the mosquito is shown in **Movie 14.1**.

The asexual reproductive cycle begins in the human host when sporozoites from the saliva of the biting mosquito are injected into the human bloodstream and begin to circulate. Within 30 minutes, they find their way to the liver and attach to hepatocytes. Once inside a hepatocyte, each sporozoite can produce 2000–40,000 daughter cells called **merozoites**. The infected hepatocytes begin to rupture within two weeks and release the merozoites, which then attach to specific receptors on red blood cells and are carried into the cells by endocytosis. This stage of the infection is referred to as the **ring stage** because of the appearance of the merozoites in the red blood cells (they form a ringlike structure). Within 72 hours, the infected red blood cells begin to rupture, releasing more merozoites (**Figure 14.6**). Some of the released merozoites continue to invade and destroy red blood cells, whereas others transform into the gametocyte form. This gametocyte form is not capable of destroying red blood cells but circulates in the peripheral blood until it is ingested by a mosquito, and the infectious cycle begins again (**Figure 14.7** and **Movie 14.2**).

Pathogenesis of Malaria

Several symptoms occur during this disease, including fever, anemia, and circulatory changes. Fever is the hallmark of malaria and seems to be initiated by the rupture of red blood cells, which in turn liberates new parasites to continue the infection. However, the actual mediators of the fever have yet to be identified. As it turns out, temperatures above 40°C destroy mature parasites, and at these temperatures the sporulation of merozoites eventually becomes synchronized, with the result that fever occurs about every 48 hours.

The anemia seen in malaria is caused by the destruction of red blood cells. When the destruction of red blood cells is severe, the patient gets hemoglobinuria, which causes the urine to become dark. This is why malaria is sometimes referred to as black water fever. One common circulatory change seen during the infection is hypotension (low blood pressure), which occurs as the high fever causes the blood vessels to dilate. Thrombocytopenia (low levels of platelets) is common in malaria and seems to be due to the shortened life span of platelets.

During infections with *P. falciparum*, red blood cells stick to the walls of capillaries, especially in the brain (**Figure 14.8**). The involvement of the

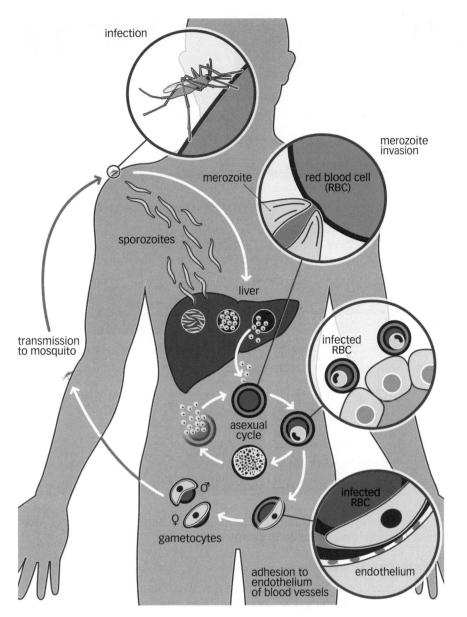

infection

merozoite
invasion

merozoite

red blood cell
(RBC)

sporozoites

liver

transmission
to mosquito

infected
RBC

asexual
cycle

infected
RBC

♂

♀

gametocytes

adhesion to
endothelium
of blood vessels

endothelium

Figure 14.7 The life cycle of malaria.
Keep in mind that the mosquito is the
definitive host and that sexual reproduction
occurs in this vector. In contrast, humans
are the intermediate host and the parasite
undergoes asexual reproduction here.

central nervous system results in the most deadly form of malaria, called
cerebral malaria. It is marked by delusions, convulsions, paralysis, coma,
and rapid death. Acute **jaundice** (yellowing of the skin and eyes, often
associated with liver failure), renal failure, and coma are also common.
The mortality for cerebral malaria is 80%, with most patients succumbing
in as little as three days.

The incubation period between the bite of the mosquito and the onset of
disease is about two weeks, and the clinical manifestations vary depending on the species of *Plasmodium* involved in the infection. The cycle
of the disease (referred to as **malarial paroxysm**) begins with a cold
stage, which lasts for 20–60 minutes. During this period the patient experiences continuous shaking and chills. The body temperature increases for
3–8 hours, and the hot stage begins as the shaking and chills subside.
During this period, the temperature can reach 40–41.7°C and causes profuse sweating. This sequence of recurring hot and cold stages leaves the
patient exhausted, but well, until the next cycle begins. By the second or
third week of infection, this cyclical sequence becomes repetitively synchronized. Eventually, the paroxysms diminish and finally disappear as the

Fast Fact As a malaria infection
continues, the host defense response to
the infection can cause decreased renal
function.

Figure 14.8 The pattern of sequestration and microvascular pathology seen in cerebral malaria. Notice the multiple petechial hemorrhages in the cortical white matter, and also the classical ring shape of hemorrhage surrounding a parasitized vessel (bottom left and right).

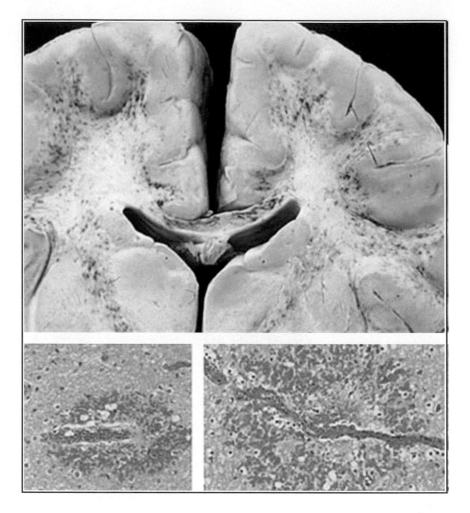

parasites in the blood disappear. The stages and beneficial aspects of fever are discussed in more detail in Chapter 15.

Treatment of Malaria

The treatment of this infection rests on two factors: which species of *Plasmodium* is causing the infection, and the immune status of the patient. If the infecting organism is *P. falciparum*, the most dangerous of the species, treatment must be started as soon as possible. Regardless of which species is involved, successful treatment requires the destruction of all forms of the parasite. This breaks the transmission cycle. At present, there is no single drug that affects all three forms.

Toxoplasmosis (*Toxoplasma gondii*)

This infection is caused by the parasitic protozoan *Toxoplasma gondii*, which is an obligate intracellular sporozoan. The definitive host for this parasite is the domestic housecat, and transmission can occur through the accidental ingestion of microscopic oocysts present in the feces of infected cats (for example, by cleaning the litter pan).

Life Cycle of *Toxoplasma*

The life cycle of *Toxoplasma* begins in the intestine of the cat. Here the trophozoite form of the parasite enters the epithelial cells of the ileum, with entry into the cells being aided by specialized enzymes secreted by the parasite for this purpose. The trophozoites remain in cellular vacuoles and undergo schizogony into merozoites. This change causes the epithelial cell to rupture and release the parasites. The merozoites then differentiate into female and male gametocytes and begin gametogony. Millions of cysts containing zygotes (called oocysts) are released each day for two to three weeks, and these oocysts mature in the external environment and are stable in the soil for months.

In the intermediate host (humans), sporozoites are released from disrupted oocysts and enter macrophages present in the host's blood, using these cells to travel to all the organs of the body. Eventually the macrophages rupture and release new parasites, which invade adjacent host cells and begin another turn of the asexual portion of the cycle. In the brain, heart, and skeletal muscles, trophozoites produce a membrane that surrounds and protects the tissue cyst. These cysts can eventually hold more than 1000 organisms and persist for the life of the host.

Pathogenesis of Toxoplasmosis

In the primary infection, the proliferation of trophozoites leads to the death of host cells, which initiates an immune response. Normally this response controls the infection, but in patients that are immunodeficient there can be continuous tissue death. During this defensive response, extracellular parasites are killed and intracellular multiplication is inhibited. Serious disease can lead to the inhibition of cell-mediated immunity, and intense host defense can also exacerbate the pathology associated with this infection. Infants infected while in the womb may have no symptoms at birth, but can show signs of the disease later in life. Remarkably there is some evidence (mostly circumstantial) that infection with *Toxoplasma* can result in behavioral changes in humans.

Treatment of Toxoplasmosis

No treatment is usually required unless the symptoms are severe and persistent, or unless vital organs are involved. In the United States, the most common treatment is a regimen of combinations of pyrimethamine and sulfonamides.

Amebiasis (*Entamoeba histolytica*)

Most amebas are free-living and do not infect humans. However, there are several species, including *Entamoeba*, that are obligate intracellular parasites of the human alimentary canal and cause the disease called amebiasis. These amoeboid parasites are passed from host to host as cysts via the fecal–oral route of contamination. Only *E. histolytica* produces disease in humans, and it has been subdivided into two genetic forms, *E. histolytica*, which is always pathogenic, and *E. dispar*, which is a harmless commensal organism.

The infectious dose for this parasite is usually 1000 or more organisms. However, the ingestion of a single cyst can sometimes cause an infection. Infection rates are higher in warmer climates, and *E. histolytica* is thought to produce more deaths worldwide than any other parasitic disease except malaria and **schistosomiasis**. In the United States, it is estimated that 1–5% of the population harbor *Entamoeba*, but the majority form is *E. dispar* rather than *E. histolytica*. Amebiasis has been on the rise in the United States, especially in institutionalized individuals, on Indian reservations, and among AIDS patients and migrant workers. It is also found among individuals who have traveled to parts of the world where this parasite is prevalent.

Life Cycle of *Entamoeba histolytica*

E. histolytica can be found in either the trophozoite form or the cyst form. The trophozoites dwell either in the lumen or the wall of the colon, where

Fast Fact Humans are the principal host and reservoir for *Entamoeba histolytica*, and infected humans can pass a staggering 45 million cysts a day in their feces!

they feed on the bacteria found there as well as on the tissue cells. They can multiply rapidly in the environment of the gut, and when diarrhea occurs, the trophozoites are passed in the watery stool. Electron micrographs show that the trophozoites contain microfilaments and an external glycocalyx, along with cytoplasmic projections that are thought to be important for attachment to host cells. Trophozoites usually encyst themselves before they leave the gut. These cysts can survive temperatures of 55°C as well as the chlorine concentrations usually found in municipal water supplies. Cysts are also able to survive the acidic conditions found in the stomach of infected individuals.

Pathogenesis of Amebiasis

The initial amebiasis infection is the result of direct person-to-person contact through the fecal–oral route (for example, by drinking or swimming in contaminated water), whereas systemic amebiasis usually occurs only after the parasite has colonized the colon. During the infection, the parasite produces several virulence factors and extracellular proteinases. The trophozoite form adheres to the target cell and then releases a protein that causes membrane lesions in the host cell. In most cases, tissue damage is minimal, and the host remains essentially asymptomatic. However, *E. histolytica* can create a portal of entry in the intestinal mucosa through which bacteria and viruses can readily enter and spread to other parts of the body.

After passage through the stomach, the cyst reaches the small intestine. Here the wall of the cyst disintegrates, releasing a quadri-nucleated (containing four nuclei) parasite that divides to form four trophozoites that move to the colon, where they attack the epithelial cells lining the colon and produce small mucosal ulcerations. There are rare instances in which the lesions occur in the brain, liver, lung, or spleen; in these areas, abscesses can form.

Symptoms are usually diarrhea, flatulence, cramping, and abdominal pain. The diarrhea is intermittent and can be accompanied by bouts of constipation that can last for months to years. During the episodes of diarrhea, it is common to find blood in the stool. The most virulent amebiasis has sudden onset, with a high fever, severe cramping, and profuse diarrhea.

Treatment of Amebiasis

Treatment involves treating the symptoms through blood and fluid replacement. The drug of choice for eradication of the parasite is metronidazole, which is effective against all forms of amebiasis. In addition, efforts toward the sanitary disposal of feces and good personal hygiene can help to prevent this infection.

Trichomoniasis (*Trichomonas vaginalis*)

Trichomonas infection is a sexually transmitted disease that in females produces the condition known as vaginitis, the symptoms of which are pain, discharge, and **dysuria** (painful or difficult urination). The infection can last for weeks to months and may cause prostatitis or urethritis in men. It is estimated that 3 million women in the United States and 180 million worldwide acquire this infection each year. In fact, 25% of sexually active women will be infected at some time in their lives, and this rate is 70% for prostitutes. The peak period of prevalence in women is between 16 and 35 years of age, but the infection can also be transmitted to newborn infants during passage down the birth canal.

The flagellated *Trichomonas vaginalis* trophozoite has a rounded anterior end and a pointed posterior end, and measures about 7 μm across (see Figure 14.1). It has five flagella and contains an **axostyle**, which is a microtubule believed to be used for attachment and which may also cause tissue damage in a host.

Life cycle of *Trichomonas*

Trichomonas does not form cysts, but the trophozoite form can survive outside a host for one to two hours. In urine, semen, and water, the trophozoite is viable for up to 24 hours.

Pathogenesis of Trichomoniasis

Direct contact of the parasite with the epithelial cells of the genitourinary tract results in the destruction of these cells and an inflammatory response accompanied by petechial hemorrhage. The exact mechanisms of pathogenicity are not understood, but *Trichomonas* is not invasive and does not produce a toxin. Interestingly, changes in the vaginal environment (such as pH, hormonal, and microbial) can affect the severity of the pathological changes that occur during *Trichomonas* infection.

The infection causes a persistent vaginitis with clinical symptoms that can last for months. These include discharge, itching or burning, dysuria, and in some cases disagreeable odor. In mild cases, there is little vaginal and mucosal tissue damage. However, in severe cases there can be hemorrhaging and extensive tissue erosion.

Fast Fact Men infected with *Trichomonas* are usually asymptomatic.

Treatment of Trichomoniasis

Oral metronidazole administered either in a single dose or over seven days cures 95% of these infections. However, this therapy should not be administered in the first trimester of pregnancy. Sexual partners should also be treated for the infection.

Trypanosomiasis (*Trypanosoma* species)

This infection is caused by the flagellated protozoan *Trypanosoma*, a parasite with a blunt posterior end and a sharp anterior end. It moves in a spiral fashion. Trypanosome parasites are transmitted to humans by an insect vector—either the tsetse fly (*Glossina* species; **Figure 14.9**) in Africa, or kissing bugs (family Reduviidae) in the Americas. There are several morphological changes in the parasite during the cycling between insect and human (similar to what happens in the malaria life cycle).

There are two forms of trypanosomiasis: the African form, which is called **sleeping sickness**, and the American form, which is referred to as **Chagas' disease**. We confine the discussion here to sleeping sickness, and discuss Chagas' disease in Chapter 25, which covers infections of the blood. Sleeping sickness is confined to central Africa, where 50 million people live and 10,000 to 20,000 of them get this disease each year. Although the tsetse fly is the vector, it is believed that the reservoir for this pathogen is humans.

Life Cycle of *Trypanosoma*

Trypanosoma reproduces by binary fission. There are three subspecies—*T. brucei gambiense*, *T. brucei rhodesiense*, and *T. brucei brucei*—and all of them undergo morphological changes as they cycle from insect to human host. In the mammalian host, they multiply extracellularly and eventually invade the blood. They also have the ability to change their antigens, expressing dozens to hundreds of variations, which makes it difficult for the host's immune system to respond effectively.

Figure 14.9 The tsetse fly is the vector for *Trypanosoma* (the etiological agent of African sleeping sickness). Notice that the fly has become engorged with blood.

Pathogenesis of Trypanosomiasis

Trypomastigotes (one of the morphological forms of the parasite) are deposited by the bite of the tsetse fly and begin to multiply, causing localized inflammation. This develops into a chancre from which organisms spread into the blood and lymph of the host, causing swollen lymph nodes and recurrent **parasitemia**. The host responds by producing antibodies that destroy the parasites, and the trypomastigotes disappear from the blood. Amazingly, they reappear three to eight days later with different antigenic markers. These reappearances become less frequent but can last for years.

During parasitemia, the *Trypanosoma* trypomastigotes localize in the small blood vessels of the heart and the central nervous system. In the brain, the infection can cause hemorrhage and inflammation and degradation. During recurrent bouts of parasitemia, the patient experiences fever, tenderness in the lymph nodes, skin rash, headache, and impaired mental status. Bouts of fever can last for years before gradual problems with the central nervous system appear. Eventually, alertness diminishes, attention wavers and the patient must be prodded to eat or talk. Speech becomes indistinct, tremors develop, and loss of sphincter control heralds the final stage, which includes coma and death.

Treatment of Trypanosomiasis

Because of the involvement of the central nervous system, agents that cross the blood–brain barrier, such as melarsoprol (an arsenic compound), must be used. If there is no central nervous system involvement, pentamidine or eflornithine is effective, and the cure rate is high with recovery being complete.

Keep in Mind

- Protozoan parasitic diseases include malaria, toxoplasmosis, amebiasis, trichomoniasis, and trypanosomiasis.
- Protozoan parasites may use humans as an intermediate host and another animal as their definitive host.
- Malaria (caused by *Plasmodium* species) is spread by the bite of *Anopheles* mosquitoes.
- Toxoplasmosis (caused by *Toxoplasma gondii*) is spread by housecat feces.
- Amebiasis (caused by *Entamoeba histolytica*) is acquired by the ingestion of fecally contaminated water.
- Trichomoniasis (caused by *Trichomonas vaginalis*) is a sexually transmitted disease.
- Trypanosomiasis (caused by *Trypanosoma* species) is spread by the bite of tsetse flies and kissing bugs.

EXAMPLES OF HELMINTHIC INFECTIONS

In this section, we look at diseases caused by the three types of parasitic helminth that cause disease in humans: nematodes, cestodes, and trematodes. In each case we discuss the life cycle of the parasite, and then the pathology and treatment of the disease.

Intestinal Nematodes

The group of roundworms known as nematodes has two subgroups: intestinal nematodes and tissue nematodes. Here we look at the intestinal variety first. These parasitic roundworms have a fusiform body consisting of a tough cuticle, a tubular alimentary canal, and a muscular layer. There are male and female forms, and the female can produce thousands of offspring. Eggs must incubate outside the host before they are infectious, and this maturation involves the development of a larval form.

There are several intestinal nematodes that routinely infect humans (Table 14.3), including pinworms (*Enterobius vermicularis*), whipworms (*Trichuris trichiura*), and large roundworms, such as *Ascaris lumbricoides* and *Strongyloides stercoralis*. These worm infections produce discomfort, malnutrition, anemia, and occasionally death. They also cause embarrassment because they can unexpectedly exit the body from the anus, nose, mouth, or ears (Figure 14.10).

It is interesting to note that, with intestinal nematodes, the severity of the disease depends on the level of adaptation to the host. The more adapted the worm becomes, the less severe the symptoms of the infection. Conversely, the less adapted the worm is to the host, the more serious the disease. This relationship can also be viewed from the perspective of worm load (the number of worms) in that infections resulting from smaller worm loads are normally asymptomatic, whereas those resulting from larger worm loads cause symptomatic disease. Overall, the immune response is slow to develop in parasitic infections caused by intestinal nematodes.

We will only discuss infections caused by two of these nematodes, *Enterobius* and *Ascaris*.

Enterobiasis

The pinworm *Enterobius vermicularis* is a ubiquitous parasite of humans and the cause of the condition known as enterobiasis. It is estimated that more than 200 million people (a large percentage of which are children) are infected with this worm every year. It is most frequently found in the temperate regions of Europe and North America and is relatively rare in the tropics. *Enterobius* is entirely restricted to humans, and infection can be readily transmitted in places where large numbers of children gather together (such as nurseries, child care facilities, and orphanages).

Pathogenesis of Enterobiasis

These pinworms lie attached to the mucosa of the cecum portion of the large intestine. The female migrates down the colon and through the anal canal (Figure 14.11), and deposits about 20,000 sticky eggs on the perianal skin, as well as on bedclothes and linens. These eggs are

Parasite	Human disease
Enterobius vermicularis (pinworm)	Enterobiasis
Ascaris lumbricoides (large roundworm)	Ascariasis
Necator americanus and *Ancylostoma duodenale* (hookworms)	Hookworm infections
Trichuris trichiura (whipworm)	Trichuriasis

Table 14.3 Some intestinal nematodes that cause human disease.

Fast Fact Nematode worms infect 25% of the entire human population.

Figure 14.10 *Ascaris lumbricoides* worms. These worms can exit from the anus, nose, mouth, or ears.

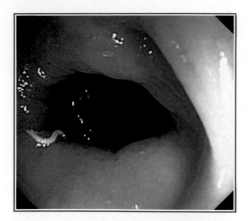

Figure 14.11 Pinworms (*Enterobius vermicularis*) pass through the intestines and leave the anus to lay eggs. These eggs are very sticky and can become affixed to the fingers during scratching of the irritation, as well as to bedclothes.

Fast Fact More than 1 billion people, including 4 million Americans, are infected with *Ascaris* annually, and it is estimated that a phenomenal 25,000 tons of *Ascaris* eggs are moved into the environment every year.

near maturity at the time they are deposited, and mature on exposure to oxygen. The main symptom of pinworm infection is itching in the anal region; scratching of the perianal area results in adherence of the eggs to fingers and eventual transfer to the oral cavity. In addition, the eggs can be shaken into the air (when making the bed) and inhaled or swallowed.

Eggs hatch in the upper intestine, and the larvae begin the migration down to the cecum. This cycle takes about two weeks and causes sleep deprivation and daytime irritability in children. In addition, the skin abrasion due to scratching can occasionally cause cutaneous bacterial infections. Female worms will occasionally migrate into the genitourinary tract of women.

Treatment of Enterobiasis

Several agents, including mebendazole and pyrantel pamoate, can effectively treat this infection. However, recurrence is common.

Ascariasis

Ascaris lumbricoides is the largest and most common of the intestinal nematodes, measuring 15–40 cm in length. The female can lay 250,000–500,000 eggs per day; these are very resistant to environmental stress and can be viable in temperate climates for up to six years. The medical condition is called ascariasis, and the infection is maintained by small children who defecate indiscriminately and pick up mature eggs from the soil while playing. These eggs can also contaminate food, and in dry, windy areas they can become airborne, then inhaled or swallowed.

Life cycle of Ascariasis

Adult *A. lumbricoides* parasites live in the human small intestine, and eggs are passed into the feces. The eggs require about three weeks in soil before the eggs can become infectious. Once the infectious eggs have been ingested, they proceed to the larval stage. The larvae of the worms penetrate the intestinal mucosa of the human host and invade the liver. Here the larvae are still small enough to exit through the hepatic vein, be carried to the right side of the heart and subsequently be pumped into the lung. By the time they reach the pulmonary capillaries, they are too large to pass through the capillaries to enter the heart, and they remain in the lung. Eventually, the larvae will rupture into the alveolar spaces and be coughed up and swallowed, regaining access to the intestine. See **Figure 14.12**.

Pathogenesis of Ascariasis

Clinical ascariasis may result either while the *A. lumbricoides* larvae migrate from host liver to host lung as just described or when the larvae establish themselves in the intestinal lumen. The symptoms include fever, coughing, wheezing, and shortness of breath. However, if the worm load is small, the patient will be asymptomatic. Worms can be vomited up or passed in the stool during episodes of fever and can be observed crawling out of the anus, nose, mouth, or ear. Heavy worm loads can cause malabsorption of fat, protein, carbohydrates, and vitamins. In addition, there can be abdominal pain and obstruction of the bile duct and pancreatic ducts. Worm loads of 50 are common in this infection, and as many as 2000 worms have been recovered from a single child.

Treatment of Ascariasis

Albendazole, mebendazole, and pyrantel pamoate are very effective in dealing with this infection. Sanitation is also important in preventing its spread.

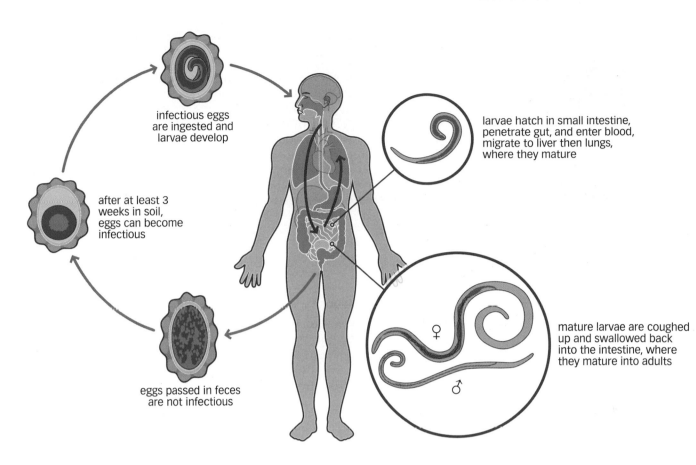

infectious eggs
are ingested and
larvae develop

after at least 3
weeks in soil,
eggs can become
infectious

eggs passed in feces
are not infectious

larvae hatch in small intestine,
penetrate gut, and enter blood,
migrate to liver then lungs,
where they mature

mature larvae are coughed
up and swallowed back
into the intestine, where
they mature into adults

Figure 14.12 The life cycle of *Ascaris*. The eggs of this helminth are highly resistant to harsh environmental conditions and must mature in the soil for a few weeks before becoming infective. The worms are not directly transmitted between human hosts.

Tissue Nematodes

These parasites induce disease through their presence in the tissue, blood, and lymph systems of the host's body. Some nematodes use humans as their definitive hosts, and the adults live for years in subcutaneous tissues and lymphatic vessels.

Trichinosis

The nematode *Trichinella spiralis* lives in the intestinal mucosae of flesh-eating animals, particularly swine and bears. In the intestinal mucosae, the tiny male worm couples with the larger female and from this one insemination the female can lay eggs for 4–16 weeks, generating up to 1500 larvae. The larvae enter the host's vascular system and are distributed throughout the body. Larvae that penetrate tissue other than skeletal muscle disintegrate and die, but those that find their way into skeletal muscle will continue to grow over a period of several weeks. Once embedded in the muscle, the larvae can remain viable for 5–10 years. The muscles most often invaded are the eye muscles, tongue, deltoid, pectoral, intercostal, diaphragm, and gastrocnemius.

The disease trichinosis is widespread in carnivores, with swine being the most often involved. Human infection results from eating undercooked meat, and it is estimated that there are more than 1.5 million Americans carrying either live *Trichinella* or dead encysted larvae in their musculature. Around a dozen cases are reported in the United States each year, but they are for the most part asymptomatic.

Pathogenesis of Trichinosis

Pathogenic lesions related to the presence of larvae can be found in striated muscle, muscle of the heart and also in the central nervous system.

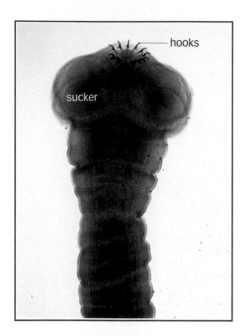

Figure 14.13 The scolex of a tapeworm showing the suckers used to obtain nutrients and also the rostellum bearing the hooks used for attachment.

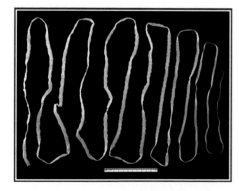

Figure 14.14 An illustration of the size of a tapeworm (recall that some can be 10 meters in length). Notice the segmented body of the worm, which broadens at the distal (left hand) end. These segments are the proglottids.

The invaded muscle cells undergo host defensive measures involving the immune system.

One or two days after a host has ingested tainted meat, the worms mature, and adult worms can perforate the intestinal mucosa, causing nausea, abdominal pain, and diarrhea. In serious infections these symptoms may persist for days. Larval invasion starts one week later and initiates the long period of disease, which can last for six weeks. Worm load is important here because 10 worms or fewer will cause no symptoms. However, a worm load of 100 or more can cause significant disease, and 1000–5000 can be lethal.

The most prominent symptoms are fever, muscle pain, tenderness, and weakness. In severe disease, there can also be pulmonary dysfunction. If the heart is involved, there can be **tachycardia** or congestive heart failure, and central nervous system manifestations involve encephalitis and meningitis.

Treatment of Trichinosis

In patients with severe edema, myocardial involvement, or disease of the central nervous system, the use of corticosteroids is required. Mebendazole and albendazole halt the production of new larvae, but in severe infection the destruction of larvae by the host defenses may cause the onset of dangerous hypersensitivity.

Cestodes

These long, ribbonlike helminths, commonly known as **tapeworms**, represent the largest of the intestinal parasites. The adult has three parts: a head, called the **scolex**; a regenerative neck region; and a long, segmented body.

The scolex of certain cestodes is equipped with four muscular sucking disks, whereas other species have only two. These disks serve to attach the worm to the intestinal mucosa of the host. In some species of cestode, the scolex can have a retractable **rostellum** armed with crowns of chitinous hooks (**Figure 14.13**).

The neck region of the worm is where the proglottid segments are generated, and each segment is a self-contained hermaphroditic reproductive unit containing male and female gonads. Sexual reproduction, fertilization, and maturation occur as a proglottid moves farther from the neck, and the proglottid eventually ruptures and releases eggs. The development of all except one type of cestode requires passage through more than one intermediate host.

Pathogenesis of Cestode Infection

The severity of these infections depends on whether the patient is the definitive or intermediate host. If a patient is the definitive host, the tapeworm stays in the lumen of the gut, causing only minor symptoms. If the patient is the intermediate host, the larval stages of the worm cause tissue invasion, with frequently serious disease. We will use the cestode *Taenia saginata* (beef tapeworm) as an example.

T. saginata inhabits the human jejunum and can live there for 25 years, growing to a length of 10 m (more than 30 feet; **Figure 14.14**). Mature *Taenia* worms can have six to nine terminal proglottids, each containing about 100,000 eggs, which break free and exit through the anal canal of the host. When these proglottids reach the soil, they disintegrate and release their eggs, which can survive in the soil for months. If the eggs are ingested by cattle, the resulting larvae will penetrate the intestinal wall and be carried by the blood to striated muscle of the tongue, diaphragm, or hindquarters, where they transform into a **cysticercus**, giving a mealy appearance to the meat.

Humans are infected when they eat improperly prepared meat and fish, but most patients are asymptomatic and become aware of their infection only by observing proglottids being passed in their stool or visible on their bedclothes. In some cases, there can be some gastric discomfort, nausea, diarrhea and weight loss.

Treatment of Cestode Infection

The drugs of choice for these infections are praziquantel and niclosamide, which either paralyze or kill the worm. Peristalsis then pushes the worm out of the host.

Trematodes

Adult trematodes, also known as **flukes**, live for decades in human tissue and blood vessels, producing progressive damage to vital organs. One of their suckers surrounds the oral cavity, and the other is located on the ventral surface of the worm.

There are two major categories of fluke, based on the reproductive systems: **hermaphrodites** and **schistosomes**. Eggs are excreted from the human host, and if they reach water, they hatch and release ciliated larvae called **miracidia**. These larvae find and penetrate snails. The snail is the intermediate host, and it is here that the larvae undergo reproduction into tail-bearing larvae called **cercariae**, which are continuously released from the snail for several weeks. What happens next depends on whether the species is a hermaphrodite or a schistosome. If it is a schistosome, the cercariae shed their tails and invade the skin of humans. If the species is a hermaphrodite, the cercariae encyst in or on an animal or plant, which is the second intermediate host. Here the larvae develop into the **metacercariae** form. Humans become infected when they eat the animal or plant contaminated with the metacercariae.

Although many flukes infect humans, we will focus our discussion on the lung flukes (*Paragonimus*, various species), the liver fluke (*Clonorchis sinensis*), which is hermaphroditic, and the blood flukes (*Schistosoma*, various species).

Paragonimiasis

Pathogenesis of Paragonimiasis

There are several species of *Paragonimus* that infect humans, and this lung fluke causes more than 5 million infections, mostly in the Far East. This infection is caused by eating infected crabs and is not a problem if the shellfish have been cooked properly.

Presence of the adult worm in a human host causes **eosinophilia**, inflammation, and eventually the formation of a fibrous capsule that surrounds one or more parasites. An infected individual may have as many as 25 of these capsules, which eventually swell and erode into the bronchioles of the lung, causing the coughing up of brownish eggs, blood, and inflammatory exudate. If the capsules rupture in the pleural cavity, chest pain can result. Eventually the capsules form cystic rings and calcify and can resemble tuberculosis lesions on X-ray. Adult flukes in the intestine can cause pain, bloody diarrhea, and occasional cutaneous masses, and in 1% of cases (mostly children) there can be brain invasion, causing epilepsy and paralysis.

Treatment of Paragonimiasis

The disease responds well to praziquantel or bithionol.

Clonorchiasis

Life Cycle of Clonorchiasis

The trematode *Clonorchis*, the liver fluke, forms cercariae that encyst on freshwater fish. When the infected fish is eaten by a human host, larvae are released into the duodenum and ascend to the common bile duct, where they mature over the course of 30 days. Eggs are passed in the feces and ingested by freshwater snails, where they eventually develop into cercariae that can swim and infect fish. *Clonorchis* therefore has three hosts (see **Figure 14.15**).

Pathogenesis of Clonorchiasis

Migration of the larvae from the duodenum may produce fever and chills, as well as mild jaundice, eosinophilia, and enlarged liver. Adult worms cause **epithelial hyperplasia**, inflammation, and fibrosis around the bile ducts, but with a low worm load the patient will be mostly asymptomatic. However, repeated infection can produce worm loads of up to 1000 and can lead to bile stones and bile-duct carcinoma. These flukes can occasionally move to the pancreas and cause obstruction of the pancreatic duct and **acute pancreatitis**.

Figure 14.15 The life cycle of *Clonorchis*. The liver fluke has a complex life cycle involving freshwater snails, fish, and mammals. Adult flukes develop in the bile duct of the definitive mammal host, in this case, a human.

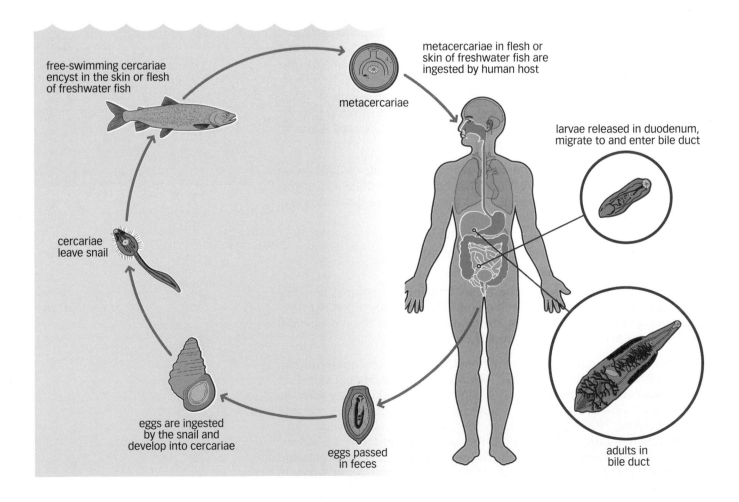

free-swimming cercariae encyst in the skin or flesh of freshwater fish

metacercariae in flesh or skin of freshwater fish are ingested by human host

metacercariae

larvae released in duodenum, migrate to and enter bile duct

cercariae leave snail

eggs are ingested by the snail and develop into cercariae

eggs passed in feces

adults in bile duct

The adult form of *Clonorchis*, the liver fluke, can survive for 50 years in a human host, feeding on mucosal secretions. Rats, cats, dogs, and pigs can also be definitive hosts in addition to humans.

Treatment of Clonorchiasis

Praziquantel and albendazole are effective for this infection. Prevention requires thorough cooking of fish as well as sanitary disposal of human feces.

Schistosomiasis

Pathogenesis of Schistosomiasis

The fluke has a cylindrical body (**Figure 14.16**); there are separate males and females, which copulate and stay conjoined for life. *Schistosoma* couples first mate in the portal vein of a host and then use their suckers to ascend the mesenteric vessels, traveling against the flow of blood, until they reach the ascending colon. Here they lay eggs in the submucosal venules (between 300 and 3000 eggs, depending on the species of schistosome) every day for the remainder of their life, which can be as long as 35 years.

The eggs deposited closest to the mucosal layer rupture into the lumen of the bowel or bladder and are excreted to the outside. If they reach fresh water, they hatch quickly into the miracidia form, which invades snails. Once in a snail, the miracidia transform into thousands of fork-tailed cercariae, which can penetrate human skin. The cercariae spend one to three days in the skin and then enter the small blood vessels. From there they move into the systemic circulation, on to the gut and through the intestinal capillaries to the portal vein, where they mature to the adult form.

This infection is so widespread worldwide that there is extensive morbidity. It continues to be a problem because of the ongoing practice of disposing of human excrement into fresh water. Although most of those infected will have low worm loads of less than 10 and will be asymptomatic, heavier worm loads result in serious clinical disease and death.

Some species of *Schistosoma* can cause bladder infections in their hosts, with progressive obstruction leading to renal failure and uremia. If the infection moves to the bowel, patients experience abdominal pain, diarrhea, and blood in the stool. If eggs reach the central nervous system, epilepsy or paraplegia can result.

Treatment of Schistosomiasis

There is no specific treatment for this infection, but treatment with corticosteroids may limit more severe infection. In late stages, therapy is directed at interrupting the deposition of eggs by killing or sterilizing adult worms.

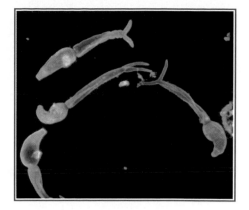

Figure 14.16 A micrograph of four *Schistosoma* blood flukes.

Fast Fact Schistosomiasis has become the single most important helminthic infection in the world, with 200 million people infected in 74 countries.

Keep in Mind

- Nematodes (roundworms) cause tissue, blood, and lymph infections and can be caused by intestinal nematodes, such as *Enterobius* and *Ascaris*, or by tissue nematodes, such as *Trichinella spiralis*.

- Cestodes (tapeworms) are the largest intestinal parasites and have a scolex, which incorporates both muscular sucking disks and in some cases attachment hooks called a rostellum.

- Trematodes (flukes) can infect the blood, liver, and lungs.

FUNGAL INFECTIONS

In this section, we look at the pathogenicity of fungi. It is important to remember that the fungi associated with the body are in most cases commensal organisms that live harmoniously with humans. Also keep in mind that these organisms have important roles in the environment and in the production of many foods and products used by humans. Consequently, when we look at fungi from a clinical perspective, they are for the most part viewed as opportunistic pathogens. Our discussion here deals with the structure of fungi and the pathogenic mechanisms used by these organisms, but we reserve discussion of the clinical diseases in which fungi are involved for the chapters on infections of specific body systems (Chapters 21–26).

Infections with fungi are usually either subacute or chronic with relapsing features. Acute fungal infections, such as those seen in bacterial or viral diseases, are uncommon.

Fungal Structure and Growth

The fungal cell possesses typical eukaryotic structures, such as a nucleus containing nucleoli and linear chromosomes. Fungal cytoplasm contains an actin cytoskeleton and organelles such as mitochondria, endoplasmic reticulum, and Golgi bodies.

The plasma membrane of fungi differs from that of bacteria in that it incorporates the sterol ergosterol, which helps make it stronger. Each fungal cell is surrounded by a cell wall that is different from the cell wall in bacteria. The fungal cell wall is composed of the polysaccharides mannan, glucan, and chitin. Mannan is found on the surface and in the structural matrix of the wall, where it is linked to proteins. It is these mannan–protein associations that make up the antigenic determinants of the fungal cell. Glucans are polymers of glucose, and some glucans form the fibrils that, in association with chitin, increase the strength of the cell wall. Chitin, which is composed of long, unbranched polymers of N-acetylglucosamine, is an inert, water-insoluble, rigid substance. It is the major component in cell walls of certain fungi and allows these fungi to form stable aerial hyphae.

Fungi are heterotrophs, requiring carbon for growth and obtaining nutrients from decaying organic matter. There is considerable metabolic diversity in these organisms. Most fungi are obligately aerobic, and although there are some facultative anaerobic forms, there are no obligately anaerobic fungi.

Fungi reproduce either asexually or sexually. The asexual reproductive elements are called **conidia**. Asexual reproduction involves mitotic division and is associated with the production of budding structures. For sexual reproduction, fungi produce spores. In sexual reproduction, haploid nuclei of the donor and recipient cells fuse to form a diploid nucleus, which may then divide by meiosis. It is during this fusion that genetic recombination occurs in fungi.

The sizes of fungi vary widely, and a single cell may vary from microscopic to a macroscopically visible structure. Growth can occur in colonies or in some of the most complex multicellular, colorful, and beautiful structures seen in nature.

Yeasts and Molds

Fungi can occur in two forms: as a **mold**, which is multicellular (Figure 14.17), or as **yeast**, which is unicellular (Figure 14.18). Yeast

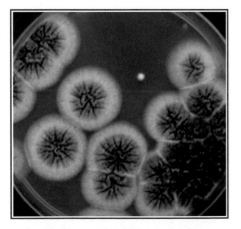

Figure 14.17 The mold form of fungi.

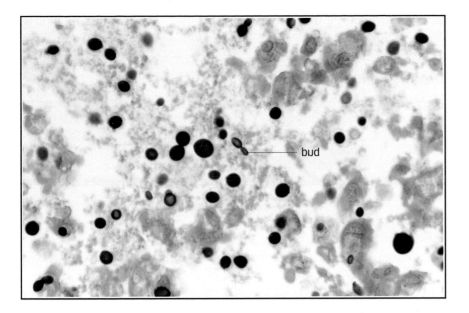

Figure 14.18 An example of the yeast form of fungi. Yeasts are single cells and they can divide by budding off new cells. Notice the buds that are forming on the individual yeast cells.

can grow by a simple form of growth called budding, in which a new cell projects from an existing cell. These buds are called **blastoconidia**.

Molds can form **hyphae**, which are tubelike extensions of the cytoplasm that have thick parallel cell walls. As the hyphae of the mold extend, they can intertwine to form a **mycelium**. Most molds form **septa**, or crosswalls, within the hyphae (Figure 14.19); however, some molds are **nonseptate** (lacking septa). Both septate and nonseptate hyphae contain multiple nuclei.

A portion of the mycelium will root itself in the nutrient medium (such as soil) and become an anchor, while the rest of the hyphae become aerial as they push upward. It is the aerial hyphae that contain the reproductive structures.

The reproductive conidia and spores of molds and the structures that bear them come in a variety of shapes and sizes and have a variety of relationships to the hyphae. Looking at the morphology and development of these structures is one way of identifying the medically important fungi. For example, conidia may arise either directly from the hyphae or on a specialized stalk known as the **arthroconidium**. Conidia that form on arthroconidia are delicately attached and break off and disseminate

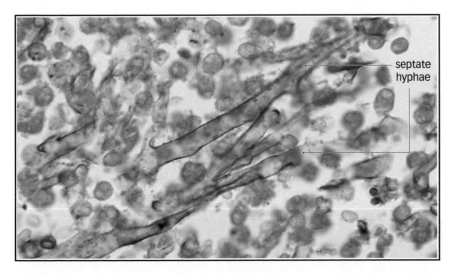

Figure 14.19 A photomicrograph of septate hyphae. Remember that not all forms of fungi will form these separations between cells.

when disturbed. Another way to identify medically important fungi is to use ribosomal RNA analysis (genetic testing) to distinguish one species from another.

Dimorphism

Some fungi can grow in either a yeast form or a mold form depending on the environment. This is referred to as **dimorphism**. The yeast phase of dimorphic growth requires conditions similar to those of the *in vivo* environment (that is, 35–37°C) and an enriched source of nutrients. In contrast, the mold form requires only ambient temperatures and minimal nutrients. Conidia produced in the mold form may be infectious and can disseminate the fungi.

Keep in Mind

- Fungi are mostly harmless free-living or commensal organisms that cause no problems for humans.

- Fungi are eukaryotes that have the sterol ergosterol incorporated in their plasma membrane and the polysaccharides mannan, glucan, and chitin in their cell walls.

- Fungi are heterotrophic, metabolically diverse, and either aerobic or facultatively anaerobic.

- Fungi reproduce either sexually or asexually.

- Fungal growth can be in a mold or yeast form, but some fungi are dimorphic and can grow in either form depending on the environmental conditions.

- Medically important fungi can be distinguished by their morphology or ribosomal RNA typing.

Fungal Infections

Fungal infections (**mycoses**) can be categorized on the basis of the types of infection they cause: superficial mycoses, subcutaneous mycoses, mucocutaneous mycoses, or deep mycoses. We will look at each in turn.

Superficial Mycoses

Superficial mycoses are fungal infections that occur in the nonliving outer skin, hair, and nails. Many superficial mycoses are caused by species of *Trichophyton*.

Piedra is a colonization of the hair shaft characterized by nodules affixed to the hair. The nodules can be either white or black, and the infection is caused by various species of fungi.

Tinea nigra results in brown or black superficial skin lesions found mostly on the palms. This condition is usually seen in women less than 20 years of age and is caused by the fungus *Hortaea werneckii*.

Pityriasis is caused by yeasts of the genus *Malassezia*, which are commonly found on the skin. Overgrowth leads to dermatitis characterized by redness of the skin, itching, and sloughing off of skin cells. The symptoms are caused by hypersensitivity to the fungus. The infection can lead to localized lesions on the chest and back, and more severe forms of the infection can lead to **folliculitis** (seen mostly in immunocompromised patients).

Tinea capitis occurs on the scalp and eyebrows. Folliculitis, which is an infection of the hair follicle, is common in this condition.

Favus causes hair loss due to permanent destruction of the hair follicle. The result is bald spots associated with crusty scarred skin.

Cutaneous and Mucocutaneous Mycoses

These fungal infections are associated with the skin, eyes, sinuses, oropharynx, external ears, or vagina.

Ringworm

This presents as skin lesions characterized by red margins, numerous scales and reddish itching skin (Figure 14.20). The infection is restricted to the stratum corneum of the skin and causes the development of an inflammatory response. Classification of ringworm is based on the location of the lesions:

*   **Tinea pedis**, which is found on the feet (Figure 14.21) and spaces between the toes
*   **Tinea corporis**, which is found between the fingers, in wrinkles in the palm, and on scaly skin
*   **Tinea cruris**, which causes lesions on hairy skin around the genitalia

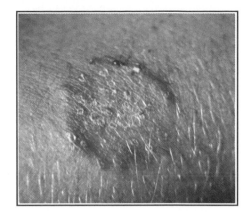

Figure 14.20 An example of ringworm, tinea corporis.

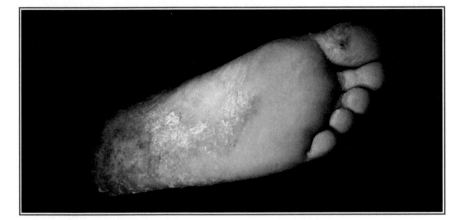

Figure 14.21 An example of athlete's foot, tinea pedis.

Onychomycosis

A chronic infection of the nailbed and nail (Figure 14.22), this is most commonly found in toenails. The infection expands from the periphery to the center of the nail and can cause the distal part of the nail to rise up and crumble.

Figure 14.22 Nail infection with *Trichophyton rubrum*.

Mucocutaneous candidiasis

This is caused by the yeast *Candida albicans*, which commonly colonizes the mucous membranes. The infection is associated with the loss of immunocompetence or a reduction of normal flora, as can occur when taking antibiotics. There are two clinical types of candidiasis. The first is thrush, in which there is fungal growth in the oral cavity. This is one of the first indications of acquired immune deficiency syndrome (AIDS). The second form of candidiasis is vulvovaginitis, which results in the accumulation of a dry, white, crumbling material in the vaginal canal and often causes irritation and erythema. This infection can in many cases be associated with changes in hormone balance.

Hyperkeratosis

This causes extended scaly areas on the hands and feet.

Keratitis

This involves the colonization or infiltration of corneal epithelium; it can occur after surgery, the use of corticosteroids, or the careless application of contact lenses. The affected eyes can become ulcerated and scarred.

Subcutaneous Mycoses

These mycoses present as localized infections of the subcutaneous tissues. Infection provokes an immune response that can cause the development of cysts and granulomas. Although not life-threatening, these growths can sometimes cause disfigurement.

- **Sporotrichosis** occurs after traumatic implantation of fungal organisms.

- **Paranasal conidiobolomycosis** is an infection of the submucosa of the paranasal sinuses, resulting in the formation of granulated fibrotic tissue filled with eosinophils. The infection progresses slowly but can be severe.

- **Zygomatic rhinitis** causes areas of the mucosa to become grayish black (similar to a blood clot). The fungus invades the tissue through the arteries and causes thrombosis (blood clotting). Severe cases of this infection can involve the nervous system, leading to meningitis and unilateral blindness.

Deep Mycoses

These infections are usually seen in immunosuppressed patients. There have been increases in opportunistic infections in immunocompromised patients who have AIDS, cancer, or diabetes and also in intravenous drug abusers. Deep mycoses can be acquired by the inhalation of fungi or fungal spores, as well as through ingestion. Infection can also be caused by contaminated surgical instruments, catheters, and hypodermic needles.

Deep mycoses can be either remain localized in the deep tissues and organs, or move from the initial site through the blood or lymph system. Secondary cutaneous mycoses are infections that move to the skin (Figure 14.23).

- **Coccidiomycosis** is caused by members of the genus *Coccidioides*. The primary infection enters the body through the respiratory system. The infection leads to fever, bronchial pneumonia, and erythema. Conjunctivitis may also occur. In most patients, coccidiomycosis resolves spontaneously as a result of the host's

Fast Fact Some eyedrops contain preservatives that fungi can use for growth, thereby increasing the potential for fungal infection of the eyes.

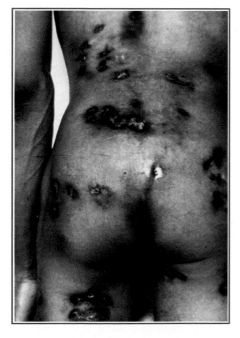

Figure 14.23 Surface lesions are characteristic of a disseminated fungal infection.

defensive systems. However, in a small percentage of cases there can be systemic spread to skin, bones, and visceral organs, which often results in a fatal outcome.

- **Histoplasmosis**, caused by the fungus *Histoplasma capsulatum*, is associated with the monocyte/macrophage components of the innate immune response and is often associated with immunodeficiency. Macrophages containing the fungal cells multiply in great numbers and give rise to granuloma formation. These granulomas necrotize and become calcified. If this infection becomes disseminated (Figure 14.24), the result is frequently fatal.

- **Aspergillosis** is also associated with reduced immune competence. There are several species of *Aspergillus* that can cause infection, with the worst form being invasive. In this case, the fungal mycelium grows between epithelial cells and disseminates through the blood to the lungs. This results in acute pneumonia with high fever. The mortality rate for this type of aspergillosis infection is very high, and death occurs in one to three weeks.

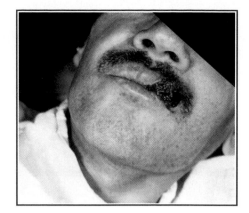

Figure 14.24 A skin lesion associated with a *Histoplasma capsulatum* disseminated infection.

Pathogenesis of Fungal Infections

We all have regular contact with fungi on a daily basis. In fact, thousands of fungal spores are inhaled or ingested every day. Most fungi are so well adapted to us that they are part of our normal microbial flora. Clinical fungal infections are very uncommon, but when they occur, progressive systemic fungal infections are some of the most difficult of all infectious diseases to diagnose and treat. This is especially true in immunocompromised individuals.

Fungal infections can be acquired from the environment through the inhalation or ingestion of infectious conidia from molds. Some of these molds are ubiquitous, but others are limited to specific geographical areas and can infect only individuals who travel through these areas. Keep in mind that pathogenic fungi constitute only a tiny fraction of the total number of fungi found in nature. Endogenous fungal infections are restricted to a few yeasts (primarily *Candida*) that have the ability to colonize by adherence to host cells; if given the opportunity to do so, these yeasts will invade more deeply.

The pathogenesis of fungal infections can be divided into three categories: adherence, invasion, and tissue injury.

Adherence

Several fungal species, particularly yeasts, are able to adhere to the mucosal surfaces of the gastrointestinal tract and the female genital tract. This adherence usually requires a surface adhesion molecule on the fungi and a receptor on the target cell. For example, in *Candida albicans* the mannan seems to be the adhesion molecule, which attaches to fibronectin receptors on target cells.

Invasion

This is a very important part of the infectious process, and some fungi are introduced to tissues through mechanical breaks in the skin. However, fungi that infect the lung must produce conidia that are small enough to be inhaled past the defenses of the upper respiratory system. For example, *Coccidioides immitis* conidia are between 2 and 6 μm in diameter. This small size allows the conidia to remain suspended in the air for long periods, a condition that allows them to reach the bronchioles of the lung directly and initiate pulmonary coccidiomycosis.

Some dimorphic fungi can be triggered to undergo a metabolic shift and change their morphology into a more invasive form by temperature change. For example, the yeast *Candida albicans* can convert and grow hyphae that invade tissues and spread the infection.

Tissue Injury

Fungi do not produce extracellular products that damage tissues in the way that bacteria produce exotoxins. Some fungi produce exotoxins in the environment but not inside the host. In fact, the primary tissue injury seen in fungal infections is associated with the host inflammatory and immune response stimulated by the prolonged presence of fungi during the infection.

Keep in Mind

- Medically important fungi can be divided into four categories of mycoses (diseases caused by fungi).

- Superficial mycoses do not involve tissue responses and include infection of the hair shafts and superficial skin.

- Mucocutaneous mycoses are associated with the skin, eyes, sinuses, oropharynx, external ears, or vagina.

- Subcutaneous mycoses are localized infections of the subcutaneous tissues.

- Deep mycoses can be localized or systemic, and are usually restricted to patients who are immunocompromised.

- The pathogenesis of fungal infections involves adherence, invasion, and tissue injury.

SUMMARY

- Parasitic infections in humans can be caused by protozoans and helminths.

- These infections affect millions or billions of people around the world, and millions die from these infections each year.

- Parasitic protozoans are classified into four groups: ameboids, ciliates, flagellates, and sporozoans.

- Parasitic helminths are classified into three groups: nematodes, cestodes, and trematodes.

- Medically important fungi can be divided on the basis of the types of infection they cause: superficial mycoses, subcutaneous mycoses, mucocutaneous mycoses, or deep mycoses.

- Deep mycoses are the most serious fungal infections and can be either localized or systemic.

In this chapter, we have looked at the pathogenesis of parasites and fungi. This completes our examination of the different types of infectious organism. Taken together with what we have learned in the previous chapters, we now have a good grounding in the disease process and in the different pathogens that are involved in the infectious disease process.

Ⓠ SELF EVALUATION AND CHAPTER CONFIDENCE

Multiple Choice

Answers are given in the back of the book and help can be found in the student resources at:

www.garlandscience.com/micro2

1. Protozoan parasites are

   A. Multicellular
   B. Prokaryotic
   C. Macroscopic
   D. Unicellular
   E. Worms

2. *Plasmodium falciparum*

   A. Causes sleeping sickness
   B. Is a helminth
   C. Causes the mildest form of malaria
   D. Is the most dangerous of the malaria-causing parasites
   E. Causes dysentery

3. Which is true of parasitic infections?

   A. Parasites can be protozoans or helminths
   B. Parasitic infections affect hundreds of millions of people
   C. There are three classes of helminth that infect humans
   D. Some parasites have a life cycle that involves a single host
   E. All the above are true

4. Sexual reproduction in protozoans involves all the following except

   A. Schizogony
   B. Binary fission
   C. Budding
   D. Gametogony

5. The body of a worm is covered by a

   A. Cell wall
   B. Membrane
   C. Cuticle
   D. Scolex
   E. None of the above

6. Nematodes are

   A. Tape worms
   B. Flukes
   C. Roundworms
   D. A type of fungus
   E. Single celled eukaryotes

7. Trematodes are

   A. Tapeworms
   B. Flukes
   C. Roundworms
   D. A type of fungus
   E. Cestodes

8. The severity of a helminth infection is directly related to

   A. Whether the patient Is the definitive or intermediate host
   B. A preexisting bacterial infection
   C. The size of the worm
   D. The number of worms that are present

9. The organism that causes malaria is transmitted by

   A. The bite of a fly
   B. The bite of a tick
   C. The bite of a mosquito
   D. The respiratory system
   E. The fecal–oral route of contamination

10. The intermediate host for *Plasmodium* species is

   A. A fly
   B. A mosquito
   C. A human
   D. A tick
   E. There is no intermediate host

11. Merozoites of *Plasmodium* are found in

   A. The saliva of a mosquito
   B. The human intestinal tract
   C. The intestinal tract of a mosquito
   D. The blood of an infected human
   E. None of the above

12. Sleeping sickness is characterized by all of the following events except

   A. Brain hemorrhage
   B. Fever
   C. A chancre
   D. Recurrent parasitemia
   E. Short-lived disease

13. *Enterobius vermicularis* is a

    A. Tapeworm
    B. Cestode
    C. Trematode
    D. Pinworm
    E. None of the above

14. Which is not true of *Ascaris lumbricoides* parasites?

    A. Their eggs are passed into the feces
    B. The eggs require about three weeks to form embryos
    C. Eggs must mature in the digestive tract before they can become infectious
    D. The larvae of the worms penetrate the intestinal mucosa of the human host
    E. The larvae can migrate from liver to lung

15. Cercariae, metacercaria, and miracidia, are stages in the life cycle of

    A. Fungi
    B. Nematodes
    C. Cestodes
    D. Trematodes
    E. All of the above

16. Which one of the following is not a protozoan?

    A. *Entamoeba*
    B. *Trypanosoma*
    C. *Cryptococcus*
    D. *Giardia*
    E. *Plasmodium*

17. What do tapeworms eat?

    A. Intestinal bacteria
    B. Intestinal contents
    C. Red blood cells
    D. Host tissues
    E. All of the above

18. Ringworm is caused by a

    A. Protozoan
    B. Nematode
    C. Trematode
    D. Cestode
    E. Fungus

19. Fungal plasma membranes contain

    A. Peptidoglycan
    B. Chains of *N*-acetylglucosamine
    C. Cholesterol
    D. Ergosterol
    E. Mannan

20. Fungal cell walls contain

    A. Mannan
    B. Ergosterol
    C. Glucan
    D. A and C
    E. B and C

21. The aerial structures seen in fungi are called

    A. Spores
    B. Conidia
    C. Hyphae
    D. Blastoconidia
    E. None of the above

22. Which is correct about superficial mycoses?

    A. They involve tissue destruction
    B. They are systemic infections
    C. They are caused by *Coccidioides*
    D. They are infections of the kidneys
    E. They are associated with hair follicles

23. All of the following are true about mucocutaneous candidiasis except

    A. It is caused by *Candida albicans*
    B. It is an infection of the mucous membranes
    C. It can be seen as thrush
    D. It is a deep mycosis
    E. It can be seen as vulvovaginitis

24. Tinea corporis is

    A. A subcutaneous mycosis
    B. A systemic infection
    C. A form of ringworm
    D. Found on the feet and between the toes

25. The most commonly seen yeast infections are caused by

    A. *Histoplasma*
    B. *Aspergillus*
    C. *Penicillium*
    D. *Saccharomyces cerevisiae*
    E. *Candida albicans*

For questions **26–33** answer **A** if the statement is true and **B** if the statement is false.

26. Parasites can be protozoans or helminths.

27. Parasitic infections are not a major health problem.

28. Malaria is an example of a protozoan infection.

29. Some helminths do not have differentiated organs.

30. Parasitic helminths (worms) are bilaterally symmetrical animals.

31. There are only two classes of helminth that infect humans: nematodes (roundworms) and cestodes.

32. Some parasites have a life cycle that involves a single host, whereas others use more than one host.

33. Pathogenic mechanisms for both protozoan and helminthic infections vary and depend on the specific parasite.

34. Which is not a characteristic or part of a nematode?

    A. A blind branched alimentary tract
    B. A cylindrical body
    C. It inhabits digestive tract
    D It can have two hosts
    E. It can produce anemia

35–38. Match the organism with its infective form. Choices may be used more than once.

Organism	Infective form
35. *Plasmodium*	**A.** Trophozoite
36. *Trichomonas*	**B.** Cyst
37. *Entamoeba*	**C.** Egg
38. *Ascaris*	**D.** Sporozoite

 DEPTH OF UNDERSTANDING

Questions listed here require you to bring together the concepts you have learned in this chapter into a discussion format. This helps you to increase your depth of understanding of the material you have learned. Help can be found in the student resources at: www.garlandscience.com/micro2

1. Describe the life cycle of *Plasmodium falciparum* and why there are not many cases in the United States.

2. Fungal infections are usually opportunistic. Explain how this relates to patients with (a) Acquired Immune Deficiency Syndrome (AIDS) and (b) patients who are undergoing prolonged treatment with antibiotics.

 CLINICAL CORNER

Help can be found in the student resources at: www.garlandscience.com/micro2

1. Ronald Johnson has just received a kidney transplant. With the exception of the kidney problem, he has been in relatively good health until now. After the transplant, he was put on drugs that suppressed his immune response. During his stay in the hospital, he has come down with *Pneumocystis* pneumonia.

    A. Explain how this could have happened.
    B. Could the administration of the immunosuppressive drugs have had a role in his infection?

2. You are working in a large county hospital emergency room. Two men have been brought in from a homeless shelter. Both are complaining about shortness of breath, but initial exam has ruled out heart attack. When you examine them, you notice that one of the men has oral thrush. Both men have a long history of alcoholism and have been homeless for several years.

    A. What does the oral thrush indicate?
    B. Why would one of the men have this condition but not both?
    C. What would you be concerned about in each of these patients?

3. You are working as a physician's assistant in a large family practice. Your patient is a 44-year-old female in apparent good health. She complains of a chronic cramping in the abdomen, excessive flatulence, and intermittent diarrhea. She has also lost weight for no apparent reason. She explains that she traveled extensively in Africa for several months more than two years ago.

    A. On the basis of her symptoms and what she has told you, what kind of problem do you think she may have?
    B. What would be the easiest way to find out what may be happening to her?

Up to now we have looked at the infection process in a variety of ways that focused on the organisms that infect us. Here in **Part IV** we look at how we defend ourselves from infection.

In **Chapter 15**, we take a detailed look at the innate immune response, which is the first response to any infection. This response is nonspecific and will work against any foreign material that comes into the body. We will see that this powerful protective response is often enough to overcome an infection before it is able to cause symptoms. Infection also initiates the adaptive immune response, which we discuss in **Chapter 16**. The adaptive response is very powerful and very specific. It involves the production of antibody and a potent cellular response to specific foreign antigens. The adaptive response also involves immunological memory, which in many cases will provide us with lifelong immunity to a particular infection.

Our next chapter, **Chapter 17**, shows you how things can go wrong with our protective systems and how the loss of protection can have catastrophic consequences. Failure of our immune response results in a greater possibility of infection. In this chapter, we take a detailed look at HIV infection and AIDS, which remain a significant problem for the health care community. We also take a brief look at autoimmune diseases, where the immune system incorrectly targets host tissues, and the four types of allergic response.

When you have finished **Part IV** you will:

- See how the host defends itself against infection by using the innate immune response, which is nonspecific and the first to fight when infection occurs.

- Know how the adaptive immune response not only gives us powerful protection against infection but also has memory. Therefore if an infection has been seen before, the response to it will be very fast and very powerful when the infection is seen a second time.

- Have an understanding of how the immune response can be ineffective, fail completely, incorrectly target host tissues, or provide an overreaction to a nonthreatening antigen in the form of allergies.

The Innate Immune Response

Chapter 15

Why Is This Important?

This chapter introduces you to the innate immune response, which includes all the nonspecific host defenses that are of paramount importance in fighting off infectious disease.

Sue is concerned. Her 3-year-old daughter, Amanda, is running a low grade fever. She takes Amanda to her pediatrician who explains to her that Amanda is probably suffering from a mild viral infection that is spreading through the local day-care centers. Sue is surprised that her pediatrician does not seem worried about the fever and does not recommend medications to bring the fever down. He explains that fever is actually beneficial in fighting an infection and is part of the innate immune response. According to current American Academy of Pediatrics recommendations, parents should administer medications such as acetaminophen or ibuprofen only to aid in the comfort of the child, not solely to treat a fever. He tells Sue to watch Amanda and make sure her behavior is normal and that she is drinking plenty of fluids because fever can contribute to dehydration. If all goes well, Amanda's immune system will defeat the infection and she should be back to normal in just a day or two. In this chapter we will learn about innate immune responses like Amanda's fever that are important in fighting infection.

OVERVIEW

In this chapter, we begin to examine the ways in which a host defends itself against any foreign material that enters the body. Materials that are seen by the host as foreign are called antigens. An antigen is any substance that can elicit an immune response in a host. During this discussion, it is important that you keep in mind what you learned in Chapters 5 and 6 about compromised hosts and about the mechanisms that pathogens use to undermine our defenses. The interaction between pathogen and host is one of the cornerstones of infectious disease. We will see in this chapter that there are several natural barriers to infection. There are also highly effective and lethal nonspecific defense mechanisms that hosts can employ as guards against infection. Along with the specific immune reactions that we will learn about in the next chapter, these nonspecific defense mechanisms help to protect us against a variety of potentially dangerous infections.

We will divide our discussions into the following topics:

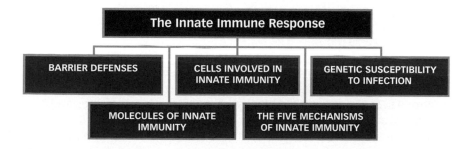

The body can defend itself against infection by using two types of immune response, the innate and the adaptive. The **innate immune response** is available to us when we are born and is immediately available. It is also nonspecific, where nonspecific means that this response can react against any infection or pathogen. In contrast, the **adaptive immune response** (Chapter 16) is specific, meaning that it responds against a particular pathogen. The adaptive immune response also has the benefit of memory, which allows it to remember any pathogen it reacted against in the past and to respond quickly and powerfully if that pathogen returns.

In this chapter, we look at the innate immune response. This response can be divided into two parts: (1) barriers that prevent infectious organisms from entering the body's environment, and (2) mechanisms that destroy any infectious organisms that manage to break through the barriers found in the body. Barriers can be looked on as the first line of nonspecific host defense, and the destruction of pathogens can be looked on as the second line of nonspecific defense.

The best way to look at host defense mechanisms is to consider what is happening in an infected patient. As we saw in previous chapters, bacterial and viral infections can result in cell and tissue damage. Host innate immune responses are triggered when this damage occurs, and a series of chemical messages are sent out from the damaged site signaling the injury. These messages can be chemotactic, meaning they give off a chemical signal. If we use the analogy of crumbs in the forest, defensive cells follow these chemical signals (the crumbs) back to the site of the injury. As these initial-responder cells arrive at the damaged site, they release further chemicals that amplify the response against the cause of the injury. This complex ballet of interactions works to maximize the defensive efforts such that the infection is brought under control as quickly and efficiently as possible.

BARRIER DEFENSES

Barrier defenses are our first line of defense against foreign invaders. There are several natural barriers to infection, most of which have other functions as well. Just their presence and normal function in the body inhibit entrance of and colonization by infectious organisms. These barriers can serve to mechanically block the entrance of microbes into the body or they can produce chemicals that kill or affect the microbes' ability to grow and divide. In fact, quite a few barriers do both. In addition to mechanical and chemical barriers, we also have our normal flora bacteria or human microbiota (discussed in Chapter 7) that help block the colonization of pathogens by simply being present or through the production of bacteriocins.

Skin

The most obvious barrier presented to microorganisms trying to get inside the body is the skin. The skin is a semi-watertight barrier made up of epidermis and dermis. The outer layer, the epidermis, is composed of many layers of tightly packed dead and dying cells that contain keratin, a tough protein that gives the cells strength and semi-waterproofs them (**Figure 15.1**). Aside from the places where glandular structures and hair follicles are located, the epidermis is essentially impenetrable to most microorganisms. Because there is no access to blood or lymph in the

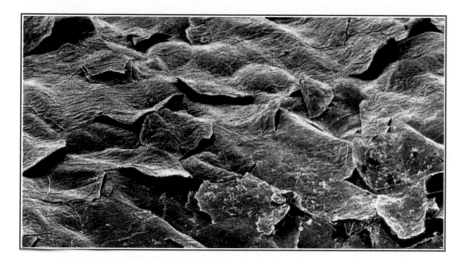

Figure 15.1 **A scanning electron micrograph of the surface of human skin.** Notice how tightly the epidermal cells are packed together. These cells are also keratinized, providing a very strong barrier to entry into the body.

epidermal layer, any intrusion of bacteria into this layer will be localized and not systemic.

Entry through the epidermal layer requires breaking the barrier—for example, by a cut, burn, insect bite, or even dry cracked skin. Of these intrusion routes, burns are the most deadly from the perspective of infection because large patches of the epidermis can be lost. The skin is constantly occupied by organisms such as *Staphylococcus* and *Streptococcus* (**Figure 15.2**). While these normal flora microbes usually provide a benefit by blocking the growth of other microbes, loss of large sections of this barrier allows these organisms to enter the dermis and underlying tissue. Because the dermis and tissue underlying it are vascularized, the chances of systemic infection are greatly increased when skin has been burned away.

Mucous Membranes

Recall from Chapter 5 that the major portals of entry for infectious organisms—the respiratory, gastrointestinal, and genitourinary tracts—all contain mucous membranes. Mucous membranes are made of epithelial tissue and an underlying connective tissue. Mucous membranes line body cavities that open to the outside environment. Although the primary function of the mucus produced by these membranes is to keep tissue moist, it also traps microorganisms. The respiratory tract is a good example of how this works. Although the lower respiratory tract is essentially sterile, the upper respiratory tract is constantly exposed to pathogenic bacteria and viruses that enter through the nasal and oral cavities.

Mucus helps defend against these intrusions through an elegant mechanism known as the **mucociliary escalator**. The lower respiratory tract is lined with ciliated cells (**Figure 15.3**) and goblet cells. The goblet cells produce mucus, which traps any microorganisms that have entered the tract. Then the ciliated cells rhythmically move this mucus up to the junction of the larynx and esophagus (hence the term escalator), where it is either swallowed or expectorated.

Mucus also has a role in the gastrointestinal tract. Produced in copious amounts in the stomach, it coats the stomach wall to protect it against the acidic fluids needed for digestion. Some microorganisms, such as *Helicobacter pylori* (as we saw in Chapter 5), use this mucus coating for their own purposes.

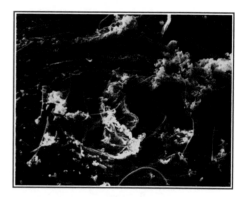

Figure 15.2 **Scanning electron microscopy shows microorganisms on the surface of the skin.** Notice the depression in the center, which is a pore. Any break in the skin will allow these organisms immediate access to the interior of the body.

Fast Fact The mucociliary escalator normally moves a small amount of mucus at a rate of about 1–3 cm per hour, but coughing or sneezing accelerates the movement.

Figure 15.3 Ciliated cells found in the lower respiratory tract are part of the mucociliary escalator. These cells, shown here by scanning electron microscopy, work with mucus-producing goblet cells to trap and remove organisms and also any other materials that enter the respiratory tract.

The Lacrimal Apparatus

Recall from Chapter 5 that the eye is a portal of entry for microbes. It is an immunologically protected site, meaning that the immune system does not protect this site like it does other parts of the body. Host defense is reduced, meaning that the eye should be a good region for microbial entry, but in fact few infections occur here. This is because of the **lacrimal apparatus**. Tears, which are produced in the **lacrimal glands**, constantly flush any foreign particles across the eye and into the lacrimal canal, which drains into the nasal cavity (**Figure 15.4**). The lacrimal apparatus increases the flow of tears whenever any irritant enters the eye. In addition to the flushing action, tears also contain three important components: immunoglobulin A (IgA), an antibody that prevents microbial attachment and neutralizes toxins; **lysozyme**, an enzyme that destroys bacterial cell walls by breaking the bonds between the NAG and NAM subunits in peptidoglycan; and **lipocalin**, which binds iron and has recently been shown to have antimicrobial activity by inhibiting pathogens as they scavenge for iron.

Saliva

Saliva is produced by the salivary glands of the mouth and is similar to tears. The main functions of saliva are to cleanse the teeth and mucous membranes of the mouth and prepare food for digestion. Saliva also washes microbes down the esophagus and into the stomach. However, this liquid can also inhibit microbial growth because, like tears, it contains both lysozyme and the antibody IgA. Saliva also contains **histatin**, a peptide that has antifungal activity and has a role in wound repair. It is important to remember that although saliva can reduce the extent of microbial colonization, the mouth still has a large amount of microbial flora.

Epiglottis

The epiglottis is a flap of tissue at the back of the throat and is a barrier that keeps food from entering the respiratory tract. In this way, organisms that come into the body through the mouth are kept away from the respiratory tract. Liquids or foods aspirated into the lungs almost always bring with them microorganisms that can result in aspiration pneumonia.

Figure 15.4 The lacrimal apparatus is the primary mechanical barrier protecting the eye. Tears, which contain the antimicrobial enzyme lysozyme, are constantly produced in the lacrimal gland and flush across the eye, taking organisms and debris with them. The tears are collected in the nasolacrimal duct. Any irritant that enters the eye increases the rate at which tears are produced.

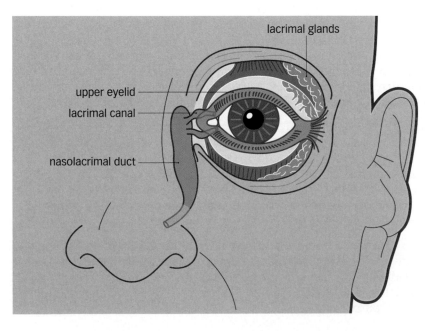

Sebum

The sebaceous glands are located in the skin and produce an oily secretion called **sebum** that prevents hair from drying out and forms a protective film over the skin to moisturize it and keep it from cracking. Remember that dry skin can provide a portal of entry for microbes. Sebum contains unsaturated fatty acids, which inhibit bacterial and fungal growth, and organic acids that make the environment of the skin naturally acidic (pH about 3–5) and inhibitory to bacterial growth. The combination of these chemical factors and dead bacterial cells on the skin can cause body odor and influence certain skin conditions, such as acne.

Perspiration

Perspiration is produced by sweat glands located in the skin. Perspiring is a continuous natural mechanism that regulates body temperature and eliminates wastes. The effect of perspiring is like the flushing action seen in the lacrimal apparatus of the eye. Because we are continuously perspiring, there is a flushing action by which the perspiration we produce clears many organisms from the surface of the skin. When we perspire profusely, this action is magnified. Just as tears do, perspiration contains the enzyme lysozyme, which breaks down microbial cell walls. Perspiration is also salty and acidic, both factors that can inhibit microbial growth.

Gastric Juice

Stomach acids and enzymes, collectively called gastric juice, break down the ingested materials that enter the stomach. This combination of substances produces a harsh chemical environment that is not conducive to microbial growth. However, bacteria hidden in food particles are protected from this environment and can therefore thrive in other regions of the digestive tract.

Urine and Vaginal Secretions

In Chapter 5, you learned that the genitourinary tract is a portal of entry for many infectious organisms and that women are more prone to infection in this tract than men. The acidity of urine inhibits most microbial growth in the urinary tract, and the flushing action of this body fluid keeps microbes from attaching to tissues. Vaginal secretions coupled with lactic acid produced by resident bacteria help make the vagina more acidic (lower pH), and therefore help prevent colonization by many potentially pathogenic bacteria. It is interesting to note that because low-pH conditions allow fungal organisms to grow, yeast infections are more common in women than in men.

Transferrins

In addition to the iron-binding substance contained in tears, the body produces another class of substances, called **transferrins**, that also bind iron as a defense mechanism. Many bacteria require iron for growth and division. Human blood contains transferrins, and bacterial growth is competitively inhibited when the body's transferrin molecules capture iron and prevent bacteria from using it.

Fast Fact Iron is required as a cofactor for many metabolic functions that take place in bacteria and is therefore required by bacteria for successful infection.

In this section we have seen that many of the body's anatomical and physiological mechanisms can prevent the entry and growth of infectious microorganisms. Table 15.1 summarizes the various barriers involved in the innate immune response's first line of defense. Keep in mind, however, that defense against infection is not the only function

Table 15.1 Barriers in the innate immune response.

Barrier	Defense function
Skin	The skin provides an impenetrable barrier to most microbes. Normal flora blocks the colonization of the skin by other microbes. The salty, acidic nature of the skin prevents the growth of most microbes
Mucous membranes	Movement of mucus can prevent many microbes from attaching. The mucociliary escalator works to keep microbes from colonizing the lower respiratory tract
Lacrimal apparatus	Tears contain lysozyme, lipocalin, and IgA; constant flushing across the eyeball keeps microbes from attaching
Saliva	Saliva washes microbes from teeth and gums; contains the antibacterial enzyme lysozyme, and also IgA and histatin
Epiglottis	The epiglottis prevents liquids, foods, and saliva from entering the lower respiratory tract
Sebum	Sebum contains oils that prevent dry skin, and low pH that inhibits growth of some bacteria
Perspiration	Perspiration flushes organisms from skin surface; contains lysozyme
Gastric juice	Stomach acids and enzymes digest microbes
Urine	Urine contains lysozyme; acidity inhibits most microbial growth; flushing action keeps microbes from attaching
Transferrins	These blood proteins bind iron and prevent microbes from using it in their growth and metabolism

of these mechanisms and that defensive results can be side effects of normal functions.

Keep in Mind

- There are two types of immune response that protect the host: the innate immune response and the adaptive immune response.

- The innate immune response is a nonspecific response that responds immediately to any type of infection.

- The first line of defense in the innate immune response consists of mechanical, chemical, and microbial barriers.

- Barrier defenses include the skin, mucous membranes, the lacrimal apparatus, saliva, the epiglottis, sebum, perspiration, gastric juice, urine, vaginal secretions, and transferrins.

Let's look now at the second line of defense in the innate immune response. This second line of defense against infection involves specific cells and molecules and can be broadly broken down into five distinct mechanisms that are among the most potent and lethal seen in nature.

MOLECULES OF INNATE IMMUNITY

We will first look at some of the cellular components and signaling molecules of the innate immune response.

Toll-like receptor	Ligand bound
TLR-1	Lipoproteins
TLR-2	Bacterial lipoproteins
TLR-3	Double-stranded RNA
TLR-4	Lipopolysaccharide, some viral proteins
TLR-5	Flagellar protein
TLR-6	Lipoteichoic acid
TLR-7	Single-stranded viral RNA
TLR-8	Single-stranded viral RNA
TLR-9	Bacterial DNA
TLR-10	Unknown
TLR-11 (mice only)	*Toxoplasma* profilin

Table 15.2 Some Toll-like receptors and the nonself components they bind.

Toll-Like Receptors

Before we go further in our discussion of the innate immune response, it is important to talk about how cells of the immune system can differentiate between self and nonself. Remember that an antigen is any substance that triggers an immune response in a host body, and the host defense provides protection against nonself antigens found on foreign invading organisms. How do our cellular defenses know what is nonself? One answer to this question is simple: **Toll-like receptors** (**TLRs**). These molecules are associated with defense cells and are a required part of the innate immune response. They bind to antigens called **pathogen-associated molecular patterns** (**PAMPs**) found on pathogens. The first TLR to be discovered was TLR-4, which recognizes and binds with lipopolysaccharide (LPS). The action of TLR-4 can be seen in **Movie 15.1**.

Thirteen TLRs have now been identified as part of the innate immune response, and humans express ten of them. These receptors are the way in which our defenses see the microbial world, and they are used to activate the innate response. **Table 15.2** shows that TLRs react to a variety of microbial structures.

A TLR is activated as soon as it binds to a target antigen. This activation triggers the host defense cell associated with the TLR to release a variety of chemical messengers. It can also stimulate dendritic cells, one of the cell types involved in innate immunity (see below), to remodel their actin cytoskeleton, which mediates a transient increase in antigen-dependent endocytosis. This increased importing of substances can enhance the ability of the dendritic cell to present antigens to other immune cells, which is required for an adaptive immune response to develop.

Fast Fact Toll-like receptors were initially discovered during studies on the growth of insects.

Cytokines

We will now look at some of the chemical messengers that serve to activate cells or allow communication between cells. These factors are produced both at the onset of an infection and throughout the course of the response to that infection.

Cytokines are low-molecular-weight proteins that are released by a variety of cell types in the body. There are two major families of cytokines:

the hematopoietin family, which includes growth hormones and **interleukins** (**ILs**), and the **tumor necrosis factor** (**TNF**) family. Both are involved in the innate and adaptive immune responses. When TLRs recognize a pathogen, a variety of cytokines are released in response, chiefly from the TNF family.

Cytokines can affect the cells that produce them, neighboring cells, or cells in other areas of the body. Cytokines act through specific receptors on their target cells and alter the activity of those cells. The action of a cytokine depends on the concentration of that cytokine. Individual cytokines have overlapping functions and act as a network to either induce or inhibit the effects of other cytokines.

Chemokines are cytokines that are released in the earliest part of the immune response and attract defense cells to the site of infection or injury. They also have a role in the destruction of pathogens. Chemokines are released by many types of immune-system cell. Some chemokines are also involved in building new blood vessels (**angiogenesis**) when the body is repairing damaged tissues.

Keep in Mind

- The second line of defense in the innate immune response involves specific cells and molecules that actively defend the body.

- Toll-like receptors (TLRs) identify molecules associated with antigens and enable immune cells to differentiate between self and nonself.

- Cytokines and chemokines are chemical mediators that are produced at the onset of and throughout the course of the response to an infection.

CELLS INVOLVED IN INNATE IMMUNITY

The innate immune response relies heavily on white blood cells (**leukocytes**), and the number of these cells can be directly correlated to the stage of disease. Clinical blood analysis usually includes a **complete blood count** (**CBC**), which includes the total number of white blood cells. A **differential blood analysis** gives the percentage of each type of white blood cell. Taken together, these two pieces of information can help in understanding the state of infection for a particular individual. For example, whether the CBC increases or decreases depends on the nature of the infection. It increases in cases of pneumonia, meningitis, appendicitis, and gonorrhea, but decreases in *Salmonella* infections, pertussis, and some viral infections. No matter whether the count increases or decreases, it should return to normal as the infection subsides. Consequently, in many cases, a count that is either higher or lower than normal tells the health care provider that an infection is under way, and a return to the normal level indicates that the infection is subsiding.

There are several types of white blood cell involved in the innate immune response, each derived from stem cells in the bone marrow (**Figure 15.5**), and each with a different role (**Table 15.3**).

Neutrophils

Neutrophils represent between 55% and 70% of the white blood cell population and guard the skin and mucous membranes. They protect against bacterial and fungal infection by sensing the site of the infection, migrating to it, and destroying the infectious organisms by phagocytosis, as can be seen in **Movies 15.2, 15.3,** and **15.4**.

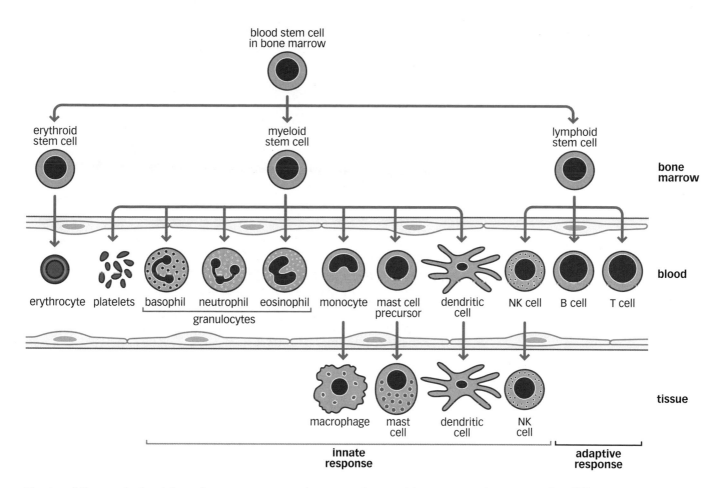

Figure 15.5 The differentiation of cell types in the blood. Stem cells in the bone marrow give rise to a variety of cell types. Erythrocytes (red blood cells) and platelets are shown on the right. The remaining cells are leukocytes (white blood cells). Basophils, neutrophils, eosinophils, monocytes, macrophages, dendritic cells, natural killer cells (NK cells), and mast cells are involved in the innate immune response. Lymphocytes (B and T cells) function in the adaptive immune response.

Neutrophils are derived from bone marrow and mature there. This process takes about two weeks, after which the mature neutrophils move from the bone marrow into the blood. At any given time, there are about 20 times as many neutrophils in the bone marrow as in the blood, kept in reserve in case of infection. In some infections, this reserve is exhausted, and in these cases immature neutrophils that cannot function at maximum level are moved out of the bone marrow. The number of these immature cells in the blood can indicate the severity of

Cell type	Function
Neutrophils	Destroy bacterial and fungal pathogens by phagocytosis
Basophils	Release histamine during inflammation
Eosinophils	Defend against parasites
Macrophages and monocytes	Destroy bacterial, fungal, and protozoan pathogens by phagocytosis; remove virus-infected cells (and tumor cells); present antigen to adaptive immune system
Dendritic cells	Present antigen to adaptive immune system; perform phagocytosis
Mast cells	Produce chemical mediators, recruit effector cells; influence adaptive response
Natural killer cells	Kill tumor cells and pathogen-infected host cells

Table 15.3 White blood cells involved in the innate immune response.

Figure 15.6 The movement of white blood cells through margination and diapedesis. The neutrophil shown in this figure rolls along the surface of a blood vessel until it binds tightly to the endothelium in a process called margination. The neutrophil then squeezes between the endothelial cells of the vessel in a process called diapedesis and migrates into the tissue.

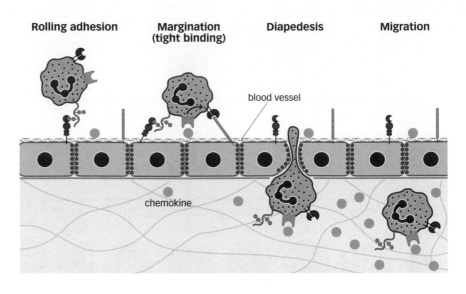

infection and the length of time that an infection has been in the body. Neutrophils use TLRs to detect lipoproteins, peptidoglycan, and lipoteichoic acid from Gram-positive pathogens, and endotoxins and LPS from Gram-negative organisms.

Neutrophils circulate in the blood for about 6–10 hours but can remain in tissues in an inactive state for 2–6 days. The passage of these cells from blood to tissue involves two processes called **margination** and **diapedesis** (**Figure 15.6**). Before they can leave the blood, the neutrophils have to settle down at the exit site, which is accomplished by margination. Endothelial cells in the area of the damaged tissue become activated and begin to express a substance called **selectin**. This substance causes a series of binding interactions with receptors on the surface of the neutrophils, causing them to slow down and begin to roll along the vessel wall. As the cells roll along, the binding events continue, eventually causing the neutrophils to stop. Once they have come to a stop, they begin to respond to chemotactic gradients that guide them through the vessel wall and into the tissue. Using diapedesis, the neutrophils squeeze through the gaps between the cells lining the capillary walls, without damaging the capillaries. The whole remarkable process can be seen in **Movie 15.5**.

Neutrophils have a wide range of toxic mechanisms for fighting invading organisms and are tightly controlled. They can produce toxic substances such as **perforins**, which produce pores in plasma membranes, and **granzymes**, which enter the pores and induce **apoptosis** (programmed cell death) in the target cell. These toxic substances can be released into extracellular fluids and damage host cells. Because of this, neutrophils are themselves programmed for apoptosis and have a relatively short life span. They undergo apoptosis within hours after they move into extravascular tissues. Dead neutrophils are eliminated from the tissue by macrophages, which are agranular white blood cells that we will discuss in more detail later.

Basophils

Like neutrophils, **basophils** are formed and mature in the bone marrow. Basophils circulate in small numbers in the blood and have a life span of only days. They are recruited into tissues as needed. They are cells with bilobed nuclei and granular cytoplasm, which is very visible when stained with a basic dye such as methylene blue (the cells stain a distinctive purple color). These granules contain histamine, a chemical

Figure 15.7 Eosinophils attacking the larval stage of a liver fluke. These white blood cells will discharge toxic enzymes that tear holes in the organism.

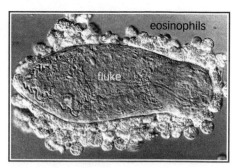

that is well known for its involvement in allergic reactions and inflammation but which can also increase and magnify innate immune responses. Basophils can be activated by bacteria, viruses, and parasites, all of which they detect by using TLRs.

Eosinophils

These cells are normally found in very small numbers in the blood and are easily identified by the red/orange color they display when stained with an acidic dye such as eosin. **Eosinophil** numbers increase greatly in cases of parasitic infection and in allergic reactions. They are the primary defense in parasitic infections, during which they produce powerful toxic enzymes. **Figure 15.7** illustrates this event dramatically. These cells also modulate the inflammatory response and may be involved in the detoxification of foreign substances.

Neutrophils, basophils, and eosinophils have a multilobed nucleus and visible granules in their cytoplasm. They are therefore known as **granulocytes**.

Macrophages and Monocytes

Monocytes are mononuclear white blood cells derived from stem cells in the bone marrow that circulate in the blood. Normally the number of monocytes circulating in the blood is quite small. During an infection, however, this number increases markedly When there is an infection, they are called to the site by chemotactic factors released from damaged tissue and from neutrophils already joined in battle at the site. Once monocytes reach the site, they begin to adhere to vessel walls and migrate out into the tissues, where they differentiate into **macrophages** (**Figure 15.8**). In this form, they are the most phagocytic of all white blood cells and can engorge themselves on bacteria as well as tissue debris.

Macrophages are responsible for recognizing, engulfing, and destroying bacteria, fungi, and protozoans. They are also involved in removing tumor cells, virus-infected cells, and normal cells that have undergone apoptosis. Equally importantly, macrophages function in wound healing, tissue repair, and bone remodeling. As we will see in Chapter 16, they also function as antigen-presenting cells in adaptive immune responses.

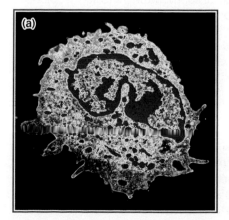
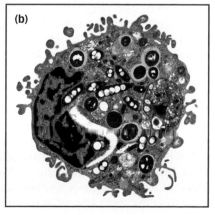

Figure 15.8 Colorized transmission electron micrographs showing the maturation (differentiation) of a monocyte into a macrophage. Panel a: Monocytes have no phagocytic capability, but after differentiation into macrophages (panel b) they are the most phagocytic of all white blood cells. Notice that the macrophage shown here has phagocytosed several *Legionella* bacteria (white areas surrounded by purple).

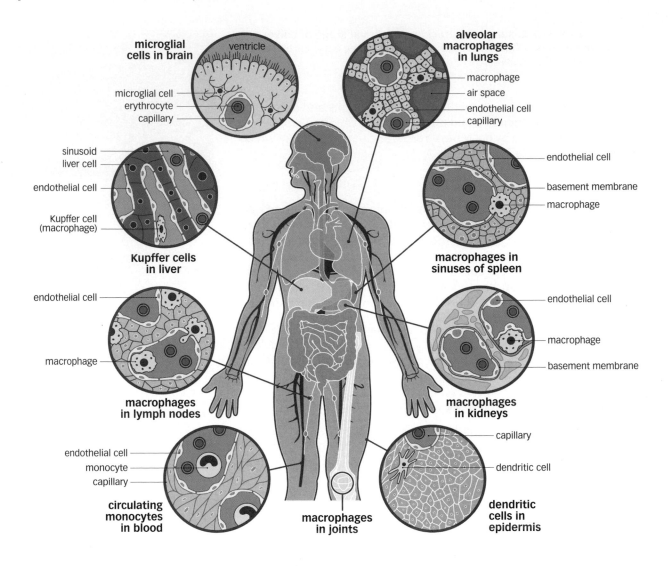

Figure 15.9 Resident macrophages are found in every one of the body's organ systems. They are sometimes given specific names, such as Kupffer cells (the liver's resident macrophages) and microglial cells (the brain's resident macrophages).

Almost all tissues and organs in the body contain macrophages (Figure 15.9). These **resident macrophages** maintain their position in the tissues through interactions with the tissue reticular fibers and form a network known as the **mononuclear phagocytic system** (formerly called the **reticuloendothelial system**).

Let's look now at some resident macrophages and their functions in the liver, lungs, and central nervous system. In the liver, resident macrophages are called **Kupffer cells** (Figure 15.10). They reside in the sinusoids of the liver, where they are anchored to the endothelium by long cytoplasmic projections. The liver contains the largest population of macrophages in the body. This makes sense because the liver is responsible for the filtration of the blood supply, which could be filled with microbes in the event of a systemic infection. The liver is also the site of routine clearance of defunct red blood cells, which are cleared and recycled by Kupffer cells.

In humans, there are two populations of resident macrophages in the lungs: alveolar and interstitial. **Alveolar macrophages** are the first-line defenders against inhaled microbes that enter the lung. These macrophages are extremely active and highly phagocytic cells that kill with great efficiency. **Interstitial macrophages** are found in the structural tissues of the lung and are smaller and less phagocytic than alveolar macrophages. It is important to keep in mind that the respiratory system is the most common portal of entry so there is a need to protect the alveolar spaces, where gas exchange occurs. We will see in Chapter 21 that there is a variety of infections that can affect the lung, many of them potentially fatal.

In the central nervous system, resident macrophages are called **microglial cells**. These cells are found in the brain and spinal cord, representing about 10% of brain tissue. There are two forms of microglia. The ameboid forms, which look and move like amebas and function in phagocytosis, are found traveling through developing brain tissue and also in diseased brains. The ramified forms are found in normal brain tissue and exhibit long branching processes and a small cell body. Microglial cells are the first line of defense for the central nervous system, which is a highly protected site in the body. These cells are easily activated, proliferate rapidly, and phagocytize aggressively.

Dendritic Cells

Dendritic cells participate in both the innate (by phagocytosis) and the adaptive (by presenting antigens) immune responses, depending on the local environment in which they are found, and form a vital link between the two. They are called dendritic cells because they have long membranous extensions that resemble the dendrites of nerve fibers (see Figure 16.4). The cells are continuously produced in the bone marrow and move from the marrow to all tissues. Dendritic cells are strategically located in mucosal tissues associated with routes of pathogen entry such as the oral, respiratory, and genital mucosae. They have different effects depending on the tissue in which they are located, the microbial environment, and the presence or absence of inflammation. Let's look next at some of the areas in which we find dendritic cells.

In the skin, dendritic cells are called **Langerhans cells**. They are located in the lower layers of the epidermis, where they connect to one another to form a network. Langerhans cells are maintained by a renewable population of progenitor cells located in the skin. This permits continuous defensive coverage of this primary barrier to infection. Langerhans cells are functionally mature when formed, but they must be activated to do their jobs. They are activated after capturing and processing signals from infectious organisms. Once activated, the Langerhans cells migrate from the site of activation in the epidermis to regional lymph nodes, where they trigger the adaptive immune response. Dendritic cells are also found in the body's mucous membranes. The function of these cells in the mucous membranes is essentially the same as in the skin. One difference is that, in contrast to the skin's dendritic cells, those in the mucous membranes are replaced by cells that come from the bone marrow.

In the intestines, dendritic cells are found in the Peyer's patches, where they are close to **M cells**, which are antigen-collecting cells in the Peyer's patches (see Figure 13.7). Intestinal dendritic cells are also found in the lamina propria, where they are distributed along the entire intestinal epithelium. The epithelial cells are connected with tight junctions. Remarkably, dendritic cells can open these tight junctions and extrude dendrites into the intestinal lumen to sample the antigens there. The tight junctions are resealed immediately after the dendritic process has been withdrawn.

The lungs are constantly bombarded with inhaled antigens. Most inhaled antigens and microbes are relatively harmless, but some are infectious. The lung dendritic cells must discriminate between the harmless antigens and infectious agents. During an inflammatory response, pulmonary dendritic cells are recruited into the airway epithelium to deal with pathogens entering the respiratory tract.

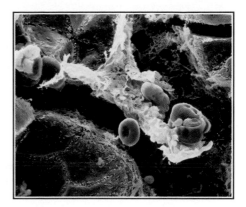

Figure 15.10 A colorized scanning electron micrograph of the resident macrophages known as Kupffer cells (yellow), which are found in the liver.

Dendritic cells in lymphoid tissue are functionally mature but less phagocytic. Lymphoid dendritic cells can produce inflammatory cytokines and chemokines quickly once an infection has taken place, usually within two or three hours. They use TLRs to identify nonself antigens, and their placement in lymphoid tissue is strategic for their function. They are located in areas of the lymph nodes where they can load up with antigens that are flowing in with lymph fluid and then move into the areas of the nodes that are filled with T cells. Here they prime antigen-specific T cells as part of the adaptive immune response.

Mast Cells

Sometimes referred to as sentinel cells, **mast cells** are responsible for allergic reactions and responses to parasitic infections. They are found throughout the body but most commonly in tissues that are exposed to the external environment, such as the mucous membranes (**Figure 15.11**). Mast cells leave the bone marrow in an immature form and differentiate into mature cells when they arrive at tissue sites. Interestingly, mature mast cells that are resident in the tissues can proliferate there. This replacement capability in the tissues makes possible a sustained level of effector cells during infection.

Mast cells have three distinct properties. First, they can rapidly and selectively produce mediators that work in host defense. Second, because they are positioned for long periods adjacent to blood vessels, they can enhance the recruitment of effector cells that respond to infection. Third, they can influence the adaptive immune response, as we will see in Chapter 16.

Activated mast cells produce a variety of mediators (**Table 15.4**). When directly activated by the presence of invading pathogens, the mediators

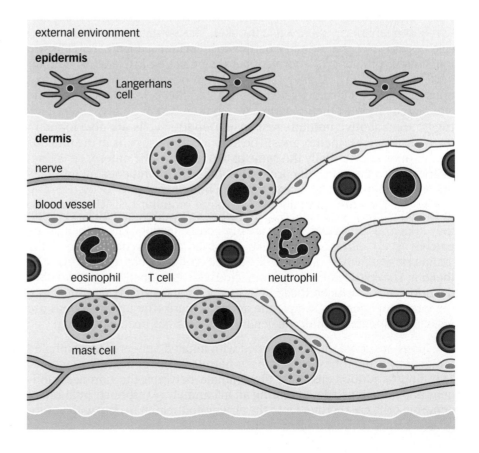

Figure 15.11 Mast cells (yellow) are found at sites in the body that are exposed to the external environment. These cells are located close to blood vessels and can regulate vascular permeability during an infection. They also recruit key cells required for the innate immune response.

Mediator	Function
Histamine, serotonin	Alter vascular permeability
Heparin, chondroitin sulfate	Enhance cytokine and chemokine function; enhance angiogenesis
Tryptase, carboxypeptidase, other proteases	Remodel tissue; recruit effector cells
Cytokines	Induce and regulate inflammation and angiogenesis; recruit effector cells; enhance angiogenesis
Chemokines	Recruit effector cells, including dendritic cells; regulate innate immune response

Table 15.4 Mediators released by mast cells activated directly or by complement.

produced cause alterations in vascular function and recruit effector cells. Mast cells can also be activated indirectly through interactions with the complement system. The combination of complement system and mast cells provides crucial initial signals for the recruitment of more cells to fight an infection. Mast cells can also recruit and activate dendritic cells. Using TLRs, mast cells can distinguish between different types of pathogen and generate a highly selective response to bacterial, viral, fungal, or parasitic infections. In addition, mast cells can reposition themselves during tissue repair and can initiate and maintain an effective adaptive immune response.

As with a variety of other defense mechanisms, the response by mast cells can be potentially damaging to the host. As a result of their close proximity to blood vessels and their ability to produce large and sustained amounts of inflammatory cytokines, mast cells have been implicated in allergic diseases, rheumatoid arthritis, vasculitis, and atherosclerosis.

Natural Killer Cells

Natural killer cells (**NK cells**) were initially thought to be restricted to the immune surveillance of tumor cells. Now it is known that they are also involved in the host response to microbial pathogens. NK cells fulfill the need for the early destruction of pathogen-infected host cells as well as the detection and destruction of tumor cells.

NK cells are derived from different stem cells from the cells we have discussed so far (see Figure 15.5). They are **lymphocyte** cells and form a unique population of lymphocytes found in peripheral tissues and the blood. They are normally seen as large granular cells that have no antigen receptors. NK cells use the same margination and diapedesis mechanisms used by other cells that migrate into tissues during the innate response. Once in the tissues, NK cells can kill tumor cells and pathogen-infected host cells.

NK cells are involved in the innate immune response in two ways: through target-cell killing and through the production of cytokines. Target-cell killing is mediated by the instigation of apoptosis in the target cell. This is accomplished by the triggered, directional release of enzymes called perforin and granzyme. The perforins polymerize in the target cell membrane and form a pore, which the granzyme then uses to enter the target cell and stimulate apoptosis.

When activated, NK cells also produce cytokines, such as TNF, **granulocyte–macrophage colony-stimulating factor** (**GM-CSF**), which recruits

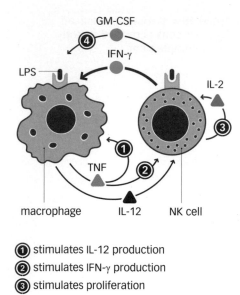

① stimulates IL-12 production
② stimulates IFN-γ production
③ stimulates proliferation
④ recruits more phagocytes

Figure 15.12 The interaction between NK cells and macrophages is a good example of the amplification seen in the innate immune response. The gamma interferon (IFN-γ) produced by NK cells amplifies the production of the cytokines tumor necrosis factor (TNF) and interleukin 12 (IL-12) by macrophages. The elevated levels of these cytokines then cause NK cells to produce more interleukin 2 (IL-2). The production of IL-2 increases the number of NK cells.

large numbers of additional phagocytic cells to fight the infection, and gamma interferon (IFN-γ). In addition, NK cells respond to cytokines released during the innate response to infection, such as interleukin 2 and interleukin 12. The complex amplification relationship is illustrated in **Figure 15.12.**

Unlike all other cells involved in the immune response, NK cells do not use TLRs. The question is, then, how do NK cells know friend (normal host cell) from foe (infected or tumor cell)? The recognition process for NK cells, which is very interesting and somewhat complex, takes advantage of two types of receptor on the surface of the NK cell. One of these is an inhibitory receptor, and the other is an activation receptor. The activation receptor activates the killing effect. The inhibitory receptor, which binds to self-antigens, inhibits killing by blocking activation. The fate of the target cell depends on its interaction with both of these receptors. Normal self-antigens will inhibit cell destruction. Abnormal or absent self-antigens will target the cells for destruction (**Figure 15.13**). It is

Fast Fact Antimicrobial toxins released by immune cells destroy pathogenic organisms and also break down damaged tissue, a process that results in the formation of the thick substance we know as pus.

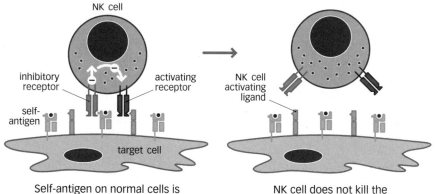

Self-antigen on normal cells is recognized by inhibitory receptors that inhibit signals from activating receptors

NK cell does not kill the normal cell

Figure 15.13 Activation of NK cells. Binding of self-antigens on the host cell to the inhibitory receptor on the NK cell prevents the actions of the activating receptor. When self-antigens are absent or altered, the NK cell will be activated and kill the target cell.

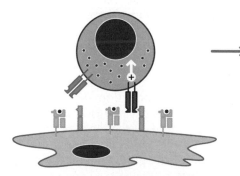

Altered or absent self-antigen cannot stimulate a negative signal. The NK cell is triggered by signals from activating receptors

Activated NK cell releases granule contents, inducing apoptosis in the target cell

becoming apparent that the NK response is a protective mechanism that takes place in the interval before the more powerful adaptive immune response is ready to begin.

| Keep in Mind |

- The cells involved in innate immunity include cells that originate in the bone marrow, and are found in the blood and in the tissues.

- Circulating white blood cells include neutrophils that destroy pathogens by phagocytosis, basophils that release histamine during inflammation, eosinophils that defend against parasitic infections, and monocytes that mature into macrophages.

- Tissue cells involved in innate immunity include: macrophages and dendritic cells that are phagocytic cells that present antigens to the adaptive immune system, mast cells that release chemical mediators, and natural killer cells that kill tumor cells and pathogen infected host cells.

- White blood cells are able to attach to the endothelial lining of the blood vessel in the process of margination, and are able to leave the blood vessel and move into the tissues in the process of diapedesis.

THE FIVE MECHANISMS OF INNATE IMMUNITY

As mentioned at the outset of this chapter, the second line of innate host defense can be broadly broken down into five distinct mechanisms. These are: **phagocytosis**, **inflammation**, **fever**, the **complement system**, and **interferon**. Each of these nonspecific mechanisms has characteristics that give it a unique role in host defense. Although we will examine them one by one, it is important to keep in mind that the five mechanisms are highly interrelated and work together to magnify the overall innate immune response.

Phagocytosis

Phagocytosis is the process by which cells ingest external objects (pathogens) by forming vesicles and is carried out primarily by neutrophils and macrophages. Any immune-system cell that conducts phagocytosis is called a **phagocyte**. When an infection begins, the traumatized tissue produces chemokines, and the presence of chemokines in the body signals phagocytes to move to the damaged site by **chemotaxis**. Neutrophils move immediately to the damaged site. Monocytes are the second cells to arrive at the scene; as they arrive, they differentiate into macrophages, which are more phagocytic than the neutrophils. Once some phagocytes reach the damaged site, they also release chemokines, thereby bringing even more phagocytes to the site and amplifying the effectiveness of the host defensive response. As phagocytes get closer to the site, the chemokine concentration becomes higher and higher, causing the phagocytes to move more quickly to the site.

Once phagocytes arrive at the site of infection, they can begin phagocytosis, which consists of four phases: adherence, ingestion, digestion, and excretion (**Figure 15.14** and **Movie 15.6**).

Adherence

As the plasma membrane of the phagocytic cell comes into contact with a pathogen cell, the cells adhere to each other. This involves a binding of

Figure 15.14 Four phases of phagocytosis: adherence, ingestion, digestion, and excretion. The fusion of phagosomes with lysosomes enables the destruction of the pathogen cells.

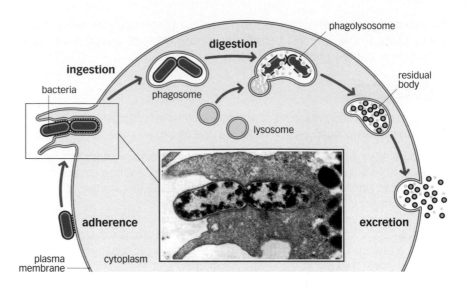

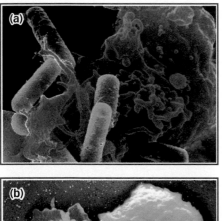

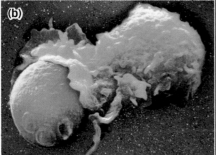

Figure 15.15 Phagocytosis. Panel a: Colorized scanning electron micrograph of a leukocyte attacking *Bacillus cereus* bacteria. The leukocyte (orange), part of the body's immune system, is attaching to and engulfing the *Bacillus* cells (blue, rod-shaped). Panel b: A colorized scanning electron micrograph showing phagocytosis of a yeast cell (green).

structures such as glycoproteins on the pathogen to surface receptors on the phagocyte. Recall from Chapter 5 that bacterial capsules can inhibit this attachment, making the pathogen cell resistant to phagocytosis.

Ingestion

During ingestion, the pathogen is taken inside the phagocytic cell. This process involves the formation of pseudopodia, which are projections that extend from the phagocyte and surround the pathogen. A pseudopod can engulf a pathogen only if receptors on the pseudopod surface bind to the pathogen surface. This binding also allows the pseudopod to fit tightly around the pathogen (**Figure 15.15**) and eventually forms a vesicle called a **phagosome**. This phagosome, like any other vesicle, can then move farther into the phagocytic cell and eventually fuses with lysosomes in the cell.

Digestion

Digestion of the pathogen begins when the phagosome fuses with a lysosome. Recall from Chapter 4 that the lysosome is a vesicle filled with such enzymes as lysozyme, lipase, protease, and ribonuclease in a low-pH (4.0) environment. Phagosomes fuse with lysosomes by virtue of their similar membrane structure to form **phagolysosomes**. This fusion allows the enzymes of the lysosome to come into contact with the bacterium and destroy it. This process can occur in as little as 30 minutes. Macrophages use oxygen derivatives called reactive oxygen species (ROS) such as hydrogen peroxide and hypochlorite ions for the digestion phase. (Hypochlorite ions are highly reactive ions also found in laundry bleach.) If a pathogen cell is too large to be ingested, toxic compounds from the lysosome are sent to the cell surface in vesicles and will be released onto the pathogen while it is outside the macrophage. While killing the pathogen, this can also damage the surrounding host tissues.

Excretion

This final phase of phagocytosis takes place after digestion of the pathogen. The phagolysosome, now filled with pathogen fragments and referred to as a **residual body**, moves to the surface of the phagocyte and discharges the debris via exocytosis. Before this discharge takes place, the phagocytic cell will scavenge any useful molecules to be recycled.

Avoiding Phagocytosis

Phagocytosis is a powerful tool in the host defense response. The infectious process is a tug-of-war between pathogen and host, however, and in the case of phagocytosis, bacteria have evolved ways to defeat this host defense mechanism. As we saw in Chapter 5, some bacteria produce enzymes called leukocidins that destroy white blood cells, thereby removing the phagocytic threat. Other bacteria have capsules that interfere with the adherence phase of phagocytosis.

Many bacteria can also resist the digestion phase. For example, *Yersinia pestis*, the organism that causes plague, has a tough capsule that resists enzymatic digestion, and as a result *Y. pestis* actually multiplies in the phagolysosome. Other pathogens that resist digestion are *Mycobacterium tuberculosis*, *Mycobacterium leprae*, and the parasite *Leishmania*. Some bacteria produce toxins that destroy the membrane of the phagolysosome, thereby releasing the toxins from the lysosome into the cytoplasm of the phagocytic cell, killing it.

There are several other ways in which phagocytosis can be compromised. For example, patients who are receiving chemotherapy and/or radiation for cancer have deficient phagocytic responses. Acquired immune deficiency syndrome (AIDS) also causes a loss of this response, and even drug treatments used after organ transplantation will decrease the ability of the innate immune system to respond. All have deficient responses due to a reduced number of phagocytic cells.

The function of phagocytosis is to destroy invading microorganisms. However, any microorganisms that can resist destruction while inside a phagocyte will remain hidden from the powerful and specific adaptive immune response.

Inflammation

Inflammation is a normal physiological response to infection and/or physical injury. It is beneficial because it helps to destroy infectious agents and also participates in the repair and replacement of damaged tissue. Irrespective of the cause, from bacterial infection to a burned hand or broken bone, an inflammatory response involves redness, pain, heat, swelling, and loss of function. Inflammation results from changes in the blood vessels in the area of injury or infection.

Vasodilation and Vascular Permeability

Vasodilation is characterized by an increase in the diameter of blood vessels and is a localized rather than systemic reaction. Because vasodilation causes an increased amount of blood to flow into the injured area, the area becomes redder and warmer. Vasodilation occurs in response to a variety of chemical signals sent from damaged tissues. Four types of factor associated with vasodilation are histamine, kinins, prostaglandins, and leukotrienes.

- **Histamine** is released by many cell types, including mast cells and basophils, and causes vasodilation of blood vessels.

- **Kinins** are chemotactic factors that are released from damaged tissue and draw phagocytes to the site of injury.

- **Prostaglandins** intensify the effects of both histamine and kinins, resulting in a magnification of the overall response. Prostaglandins also help in the migration of phagocytes through the walls of blood vessels.

- **Leukotrienes** are produced by mast cells. They increase vascular permeability and promote the adherence phase of phagocytosis.

Vasodilation also helps to deliver clotting elements to the site of injury. These elements prevent the spread of infection by walling off the site. This walling-off mechanism is part of the process that can lead to the formation of abscesses, such as those seen in boils.

The swelling and pain characteristic of inflammation are related to increased vascular permeability, which causes fluid to escape from the vessels into the tissues. This causes swelling, and the swelling puts pressure on nerve endings, signaling pain. The pain and swelling also cause a loss of normal function.

Phagocyte Migration

One of the most important aspects of the inflammatory response is the migration of phagocytes from the blood to the site of infection. The ability of phagocytes to leave the blood is very important in preventing further infection and repairing tissue damage. The increased blood flow that results from vasodilation leads to increased numbers of phagocytes moving to the site of injury. The localized increase in permeability of the vessel walls facilitates the exit of phagocytes from blood vessels.

The inflammatory process therefore permits margination (in which cells stick to the vessel walls at the site of the infection) and diapedesis (in which the phagocytes leave the blood and move into the tissues), as discussed in the section on neutrophils. Diapedesis occurs rapidly, with phagocytes leaving the bloodstream as early as 2 minutes after arrival at the site of injury. At this point, destruction of the invading organisms begins and does not stop until they are gone.

Acute-Phase Response

The acute-phase response to pathogen invasion is related to inflammation, trauma, and infection. It involves the production of proteins called **acute-phase proteins**, including cytokines, fibrinogen (used in clotting), and kinins (involved in vasodilation). The response begins when phagocytes somewhere in the body ingest pathogens, an event that stimulates the synthesis and secretion of several cytokines. One of these cytokines is **interleukin 6 (IL-6)**, which causes the liver to produce acute-phase proteins.

The presence in the body of these acute-phase proteins initiates a nonspecific host defense that is distinct from the classical inflammatory response (and in some ways similar to the specific defenses seen with the adaptive immune response). This acute-phase response involves recognition of foreign substances (as in the adaptive immune response) and acts as a type of early systemic inflammatory response that promotes the function of the innate immune system.

The best-understood acute-phase proteins are **C-reactive protein (CRP)**, which binds to phospholipids, and **mannose-binding protein (MBP)**, also known as mannose-binding lectin. MBP binds to the mannose sugars found in many bacterial and fungal membranes, and this makes the acute-phase response nonspecific. These two acute-phase proteins coat pathogens and by doing so make the pathogens more readily taken up by phagocytic cells. In addition, the proteins activate the complement system (we will discuss this shortly) and also stimulate the production of chemotactic factors that attract more phagocytes to an injury.

Fever

Fever is a systemic rise in body temperature that often accompanies and augments the effects of inflammation. The normal body temperature can vary to some degree but is usually around 37°C (98.6°F). The clinical definition of fever is an oral temperature above 37.8°C (100.5°F) or a rectal temperature above 38.4°C (101.5°F). A fever in which the body temperature rose to 43°C (109.4°F) would cause death.

Body temperature is controlled in the hypothalamus of the brain, and many pathogenic infections can reset body temperature, causing a fever. In some cases, fever accompanies certain immune responses, such as those seen after vaccination, but many types of active infection will elicit a fever reaction.

Fever is caused by chemicals called **pyrogens**. There are two forms: endogenous pyrogens, which are produced by the host; and exogenous pyrogens, which are produced by the invading pathogens but cause fever in the host. Bacterial toxins, and even just the cytoplasm from ruptured pathogen cells, act as exogenous pyrogens. **Interleukin 1 (IL-1)**, which is released by activated macrophages, is the best-known endogenous pyrogen. This pyrogen resets the hypothalamus temperature control within 20 minutes of release. It moves through the systemic circulation into the hypothalamus and triggers the release of prostaglandin, which resets the body temperature (Figure 15.16). The increased body temperature continues for as long as IL-1 is present. As the infection subsides, there are fewer phagocytes secreting IL-1 and the levels of prostaglandin drop, causing the thermostat in the hypothalamus to be reset to normal. The resulting drop in body temperature is referred to as the crisis phase of the fever.

Fever is a beneficial response because it raises the body temperature to levels that inhibit the growth of many bacteria. Fever also inactivates many bacterial toxins by changing their three-dimensional shape. The fever response can cause the release of **leukocyte endogenous mediator** (LEM), a factor that lowers plasma iron concentration; without iron, pathogens do not grow well. In addition, the elevated body temperature increases the speed at which host defenses work, because chemical reactions occur at a faster rate at higher temperatures. Finally, fever makes

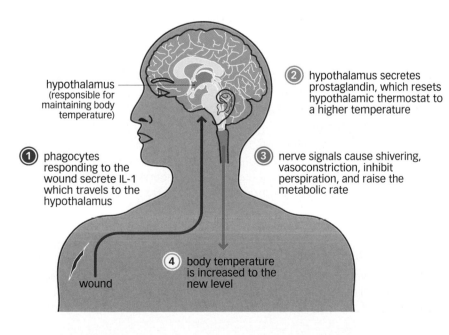

① phagocytes responding to the wound secrete IL-1 which travels to the hypothalamus

hypothalamus (responsible for maintaining body temperature)

② hypothalamus secretes prostaglandin, which resets hypothalamic thermostat to a higher temperature

③ nerve signals cause shivering, vasoconstriction, inhibit perspiration, and raise the metabolic rate

wound

④ body temperature is increased to the new level

Figure 15.16 The development of a fever. Interleukin 1 (IL-1) is secreted by phagocytic cells at the injury site (1). IL-1 travels to the brain and influences the hypothalamus to release prostaglandins (2), which reset the body temperature. The hypothalamus signals nerves to cause vasoconstriction and shivering and inhibit perspiring (3), all of which increase body temperature (4).

you feel ill, and this feeling takes away your energy and forces you to rest, allowing all available energy to be used for fighting off the infection.

Unfortunately, a fever that causes body temperature to go too high can be dangerous for the host. Furthermore, as long as body temperature is elevated, there is vasoconstriction. This inhibits the movement of phagocytes to the site of infection, thereby countering one of the most important parts of the innate immune response. Unchecked fever can also increase the rate of metabolism by 20%, making the heart work harder. Prolonged high temperature can cause denaturation of proteins, inhibition of nerve impulses, and electrolyte imbalance due to the loss of water. These changes can lead to hallucinations, convulsions, coma, and eventually death. Therefore, if fever goes beyond 40°C, antipyretic medications are given to the patient to counter the effects of pyrogens in the body.

The Complement System

Although the responses we have discussed so far are very effective in dealing with infection and tissue damage, they are not alone in the innate immune response arsenal. Another system—the complement system—also participates in our defense. This system not only has a lethal capacity but also enhances and magnifies other parts of the innate immune response. About 30 serum proteins are involved, and they are numbered in the order in which they were discovered, beginning with C1. Some of the proteins function in a cascade sequence that amplifies as it progresses because more product is formed with each reaction. The complement proteins are produced in the liver and then circulate in the blood in inactive forms. As the cascade proceeds, the proteins are activated. The proteins not directly involved in the cascade either activate or inhibit the response.

The complement system is activated as soon as an invading organism has been detected, and it functions in a variety of ways. Its proteins can trigger and modulate the inflammatory response, and they also act as **opsonins** (molecules that enhance the ability of phagocytes to attach and ingest) and chemotactic factors. However, the major function of this system is the lysis of bacterial cells and viral envelopes. This lysis is accomplished through the production of a **membrane attack complex** (**MAC**) that essentially punches a hole in the plasma membrane of the bacterium (**Figure 15.17**) or the envelope of a virus.

The complement response is a complex series of enzymatic interactions between complement proteins that can follow any one of three pathways: classical, alternative, or lectin-binding.

Activation of the Classical Pathway

The **classical pathway** is activated by antigen–antibody complexes. The presence of antibodies means that the host has previously seen the infecting organism and has already generated an adaptive immune response to it. Antigens associated with pathogens bind antibody molecules associated with complement protein C1, and a series of enzymatic reactions leads to the cleavage of complement protein C3 (**Figure 15.18** and **Movie 15.7**).

Activation of the Alternative Pathway

The **alternative pathway** works with pathogens that have never before infected the host. Because the infectious organism has not been seen before, it takes several days for the adaptive immune system to respond

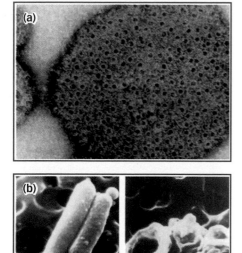

Figure 15.17 Complement-mediated bacterial lysis. Panel a: A transmission electron micrograph of a pathogen cell after attack by the complement membrane attack complex. Note the numerous holes produced in the membrane by the complex. Panel b: 'Before and after' scanning electron micrographs of a rod-shaped pathogen, clearly showing the cytolysis caused by the membrane attack complex.

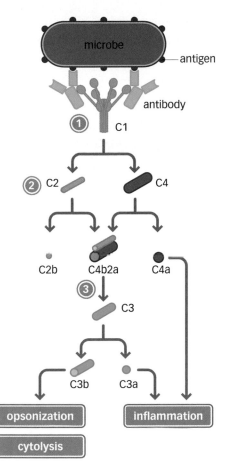

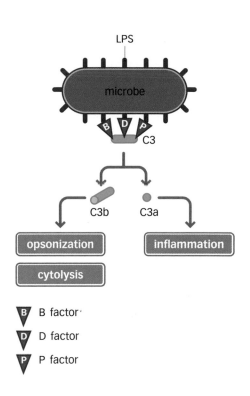

Figure 15.18 Activation of the complement system via the classical pathway. Step 1: binding of antibody with antigen causes the complement protein C1 to bind to the antibody–antigen complex. Step 2: this complex then enzymatically cleaves proteins C2 and C4 into two fragments each. Step 3: fragments C2a and C4b join and react with protein C3 to launch the rest of the pathway, culminating in the formation of the membrane attack complex.

Figure 15.19 Activation of the complement system via the alternative pathway, which requires no antibodies. Lipopolysaccharides (LPS) on the surface of the pathogen, and also endotoxins, interact with factors B, D, and P in the blood. These interactions attract complement protein C3 and launch the cascade.

maximally. During this time, if the only available pathway were the classical one, we would be without the benefit of the very powerful complement system.

The alternative pathway (also referred to as the **properdin pathway**) is activated by contact between LPS and endotoxins on the pathogen surface, and three factors found in the blood: **factor B**, **factor D**, and **factor P (properdin)**. Once all these substances have combined, complement protein C3 is attracted to the complex and cleaved (**Figure 15.19**).

Although less efficient than the classical pathway, the alternative pathway is still very useful in the early stages of infection.

Activation of the Lectin-Binding Pathway

The third and perhaps most important complement-activating pathway—the **lectin-binding pathway**—is stimulated by the carbohydrate mannose. The pathway involves **mannose-binding protein (MBP)**, a protein produced in the liver. MBP binds to mannose, which is found on pathogens such as *Candida*, HIV and influenza virus, on many bacterial pathogens, and also on parasites such as *Leishmania*. More important than where mannose is found, however, is the fact that it is not found on normal healthy host cells. MBP is activated when it binds to mannose, and it enzymatically cleaves complement protein C3, just as in the classical and alternative pathways.

C3 and Beyond

The C3 protein is the nexus of the complement system because once it has been cleaved into C3a and C3b (see Figures 15.18 and 15.19), the three complement-activating pathways all follow the same sequence as shown in **Figure 15.20** and **Movie 15.8**.

The interactions of the remaining proteins of the complement system involve conformational changes and enzymatic cleavages that result in the formation of the membrane attack complex. In addition, the cascade can influence other parts of the nonspecific host defense. The C3a fragment interacts with mast cells, causing them to release histamine, which amplifies the inflammatory response. The C3b fragment can act as an opsonin, coating pathogens and making them more susceptible to phagocytosis. C3b can also attract protein C5, which on binding to the complex is cleaved into two fragments. Here again we see the amplification of the host defense: the C5a fragment reinforces the production of histamine by mast cells, and the C5b fragment interacts with C6, causing the cascade to continue through to C9.

The combination of proteins C5 to C9 is the membrane attack complex, which creates holes in a pathogen's cell membrane, thereby destroying

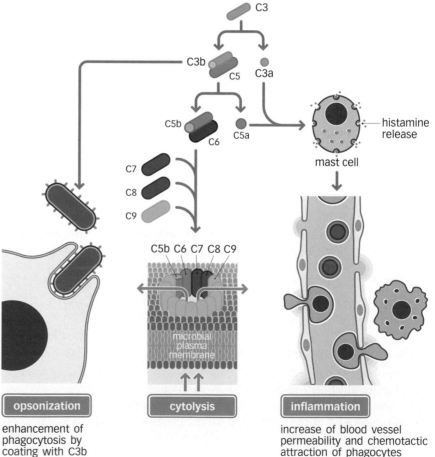

Figure 15.20 The complement cascade from C3 to C9. Note the influence of this part of the complement system on inflammation and phagocytosis.

opsonization
enhancement of phagocytosis by coating with C3b

cytolysis

inflammation
increase of blood vessel permeability and chemotactic attraction of phagocytes

the cell (see Figure 15.17). The same effect can be seen in enveloped viruses: the envelope is destroyed through this membrane attack mechanism. Without its envelope, the virus can no longer infect a host cell.

Some people are genetically deficient and cannot produce complement proteins, a condition that makes them prone to infections (Table 15.5). C3 deficiencies are worst because this protein is the nexus of all the activation pathways. Without C3, the entire complement system and all its beneficial responses are lost. Deficiencies in any of the proteins (C5 to C9) making up the membrane attack complex also result in recurrent infections, especially with *Neisseria* species.

It is important to note that the complement system can kill cells and can adversely affect normal host cells. This potential for damage to the host is controlled by having complement proteins produced in an inactive form and activating them only when an infection occurs. Furthermore, in the antibody–antigen reaction, there is a specificity that prevents random binding and initiation of the classical pathway. In fact, most host cells contain membrane-bound inhibitor proteins that bind to and inactivate complement proteins, thereby interrupting any aberrant cascade. In addition, human cells routinely replace surface membranes at a high rate. Therefore, any membrane attack complex formed inadvertently is shed off the membrane or endocytosed into the cell and degraded.

Evasion of the Complement System

As we have come to expect by now, some organisms have evolved defenses against the complement response. One defense strategy involves encapsulation: bacterial capsules contain large amounts of sialic acid, which discourages the formation of membrane attack complexes. In another approach, certain Gram-negative bacteria lengthen surface glycolipid complexes so as to prevent membrane attack by complement. Some Gram-positive bacteria release enzymes that destroy C5a fragments, to inhibit the amplification of phagocytosis.

Interferon

When viruses are outside a host cell, they are subjected to all the host defenses we have been discussing. However, as we saw in Chapter 12, viruses must enter the host cell as part of the infectious process, and once inside a host cell they are essentially invisible to these host defenses. The production of interferons is one of the host responses to this problem.

Interferon (IFN) is a protein produced by virus-infected host cells and released from those cells. Interferon then moves to uninfected host cells and prompts them to make antiviral proteins that prevent viruses from replicating. The main effect of interferons is that they make a host cell incapable of being infected by a virus. Interferons are not virus-specific and are produced in response to any viral infection.

Different interferon molecules are produced in different tissues, and there are three major forms found in humans: alpha, beta, and gamma (Table 15.6). The alpha and beta forms are very similar and bind to the same receptor on target cells. Consequently, both are referred to as type I. In contrast, the gamma form is distinctly different and is classified as type II. Although there are these various forms, we often speak in terms of the singular interferon, in which case we mean all the forms in general.

Deficiency	Condition
C3	Severe recurrent infections
C1, C2, C5	Less severe recurrent infections
C6, C8	Gonococcal infections
C6	Meningococcal infections
C1, C2, C4, C5, C8	Systemic lupus erythematosus

Table 15.5 Conditions associated with complement deficiency.

Table 15.6 Summary of interferons.

Class	Cell source	Stimulated by	Effects
Alpha interferon (IFN-α)	Leukocytes	Virus infection	Stimulates the production of antiviral proteins in uninfected cells
Beta interferon (IFN-β)	Fibroblasts	Virus infection	Same as those seen with IFN-α
Gamma interferon (IFN-γ)	T lymphocytes and natural killer cells	Virus infection and antigenic stimulation	Activates certain immune cells; kills infected cells and destroys tumors

IFN-α and IFN-β

The alpha forms (of which there are more than a dozen) are produced in monocytes and macrophages, whereas the beta form is made in fibroblasts. Both are synthesized immediately after infection and protect uninfected neighboring cells. This protection occurs as IFN molecules bind to surface receptors on neighboring cells, a process that stimulates these cells to transcribe genes that code for **antiviral proteins (AVPs)** (**Figure 15.21**). These AVPs are inactive until they come into contact with double-stranded RNA. Recall that this form of RNA is not normally seen in host cells: it appears only in cells infected by certain RNA viruses.

IFN-γ

Gamma interferon is produced by T lymphocytes and NK cells. It also appears later in the course of an infection, when the adaptive immune reactions begin. IFN-γ protects against viral infection, as do the other IFNs, but it can also heal macrophages and neutrophils that are infected and re-stimulate their phagocytic activity. For this reason, IFN-γ was originally referred to as macrophage-activating factor. IFN-γ also assists in the destruction of tumors.

Therapeutic Use of Interferon

Initially, the amounts of IFN that could be obtained from normal cells were too small for therapeutic use. This hurdle was overcome with the advent of biotechnology, when the genes for IFN were cloned and large amounts of pure IFN were produced in the laboratory. Interferon was approved in 1986 as a treatment for a rare leukemia called hairy cell leukemia, and it has proved to be effective in keeping this malignancy under control. In fact, if taken off IFN therapy, 90% of these patients relapse. Interferon is also useful in chronic granulomatous disease, a congenital condition in which patients do not produce IFN-γ and have no phagocytic activity in their neutrophils. This condition leads to repetitive serious infections and a life span of about 10 years. Treatment with IFN-γ increases the life expectancy of these patients. Interferon has also been approved for the treatment of several viral diseases, such as genital warts and hepatitis C.

It is important to point out that IFN is primarily a treatment and cannot cure disease by itself. Furthermore, the stability of recombinant IFN is poor, and some adverse side effects are seen with IFN treatment, including nausea, vomiting, fatigue, weight loss, and, in some cases, damage to the central nervous system. IFN can also increase the fever response, and in high concentrations it damages the heart, bone marrow, liver, and kidneys.

Fast Fact Interferons were originally thought to be the magic bullet both for viral diseases and for some malignancies. Sadly, they have not fulfilled this promise.

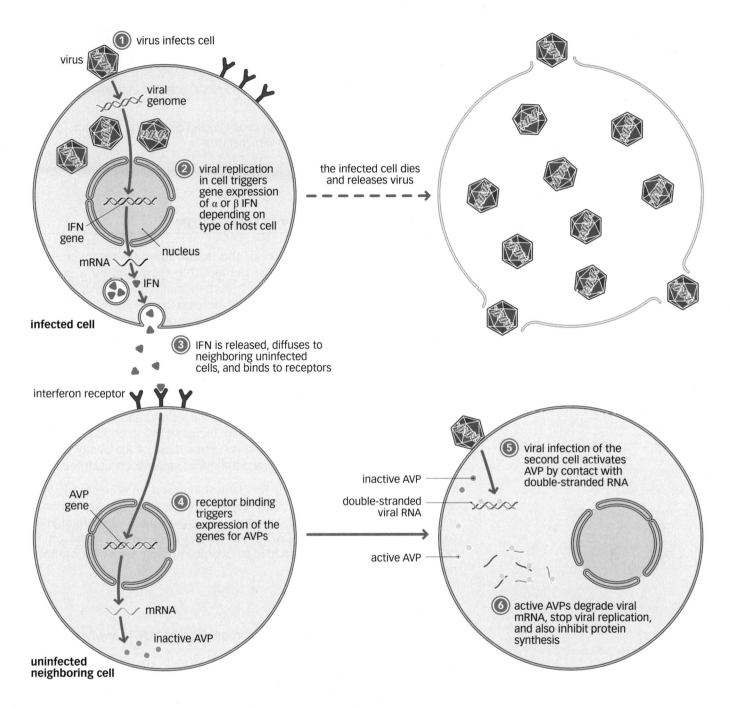

① virus infects cell

virus

viral genome

② viral replication in cell triggers gene expression of α or β IFN depending on type of host cell

IFN gene

nucleus

mRNA

IFN

infected cell

the infected cell dies and releases virus

③ IFN is released, diffuses to neighboring uninfected cells, and binds to receptors

interferon receptor

AVP gene

④ receptor binding triggers expression of the genes for AVPs

mRNA

inactive AVP

uninfected neighboring cell

inactive AVP

double-stranded viral RNA

active AVP

⑤ viral infection of the second cell activates AVP by contact with double-stranded RNA

⑥ active AVPs degrade viral mRNA, stop viral replication, and also inhibit protein synthesis

Figure 15.21 The mechanism of action of α and β interferons. Only uninfected cells neighboring an infected cell are protected. Interferons cannot save cells that are already infected.

Keep in Mind

- Phagocytosis is the cellular response in which microorganisms, damaged host cells, and cellular debris resulting from apoptosis are removed from the body.

- Chemotaxis is the release of chemicals by cells involved in infection that attract phagocytic cells to an infected area.

- Inflammation is a physiological response to body trauma. It involves vasodilation and increased vascular permeability, resulting in redness, pain, heat, swelling, and loss of normal function.

- Fever is a systemic rise in body temperature that often accompanies and augments the effects of inflammation.

- The complement system is part of the innate immune response that destroys bacteria by punching a hole in the bacterial cell membrane. It also enhances other parts of the innate response such as phagocytosis and inflammation.

- Complement can be activated in three ways: the classical, alternative, or lectin-binding pathways.

- Interferon is a protein produced by virus-infected cells that can protect neighboring cells from infection with that virus.

- The five mechanisms of the innate immune response are highly integrated and act together.

GENETIC SUSCEPTIBILITY TO INFECTION

Before we leave our discussion of the innate immune response, it is important to put into perspective the fact that there can be a genetic predisposition to infectious disease. In fact, in some cases, genes that are protective for one bacterial disease may increase susceptibility to another. Genetic abnormalities can involve receptors or other components of the defense network we have been discussing. For example, genetic variability in TLRs can result in serious problems with host defense. Variations in TLR-2 have been shown to be associated with leprosy, and genetic variations in TLR-4 are seen in cases of severe Gram-negative systemic disease. In addition, genetic variability in TLRs can contribute to increased susceptibility to anthrax and tuberculosis. Genetic variation in receptors on T cells and NK cells is associated with increased susceptibility to infection by *Plasmodium*, HIV, and hepatitis C virus. Loss of the ability to produce the cytokine TNF-α can increase the risk of staphylococcal infections.

Variation in the genes that code for chemokines and cytokines are linked to at least 20 infectious diseases and syndromes. At least six variants in chemokine receptors and cytokines are known to contribute significantly to the development of AIDS. Furthermore, variations in the genes coding for IFN production can lead to overproduction, with concomitant host pathogenesis.

SUMMARY

- The innate immune response is a nonspecific response to any type of infection.

- The first line of defense is barriers that can mechanically block the entrance of microbes into the body and/or produce chemicals that kill or affect the microbes' ability to grow and divide.

- White blood cells are heavily involved in the second line of defense.

- Host defense cells use Toll-like receptors to identify nonself antigens.

- A variety of cytokines and chemokines are produced throughout the innate response.

- Diapedesis is the process whereby white blood cells leave the blood vessels and migrate into tissues to reach the site of the infection.

- Phagocytosis is one of the most important parts of the innate response and involves chemotaxis, adherence, ingestion, digestion, and excretion.

- Inflammation involves vasodilation and an increase in vascular permeability and results in redness, pain, heat, swelling, and loss of normal function.

- Complement is a cascade of proteins that results in the destruction of bacterial cells and can enhance other mechanisms of the innate response.

- The complement cascade can be initiated in three ways: the classical pathway, the alternative pathway, and the lectin-binding pathway.

- The mechanisms of the innate response are very powerful and act together in a concerted way.

- There can be a genetic predisposition to infection.

In this chapter we have learned that in the innate immune response there are barriers constituting a first line of innate defense, and also powerful mechanisms in a second line of innate defense. Keep in mind that all these barriers and mechanisms work together to maximize the nonspecific response to any infection. In addition to these powerful innate immune defenses, we also have a pathogen-specific response called the adaptive immune response. Like the innate mechanisms we have discussed here, adaptive mechanisms are also lethal. However, as we will see in the next chapter, the adaptive immune response is an exquisitely sensitive and specific response that also has the property of memory.

 SELF EVALUATION AND CHAPTER CONFIDENCE

Multiple Choice

Answers are given in the back of the book and help can be found in the student resources at:

www.garlandscience.com/micro2

1. Nonspecific defense is
   A. The body's defenses against all pathogens
   B. The body's lack of resistance to infection
   C. The body's defense against a particular pathogen
   D. None of the above

2. The innate response includes all of the following except
   A. Phagocytosis
   B. Inflammation
   C. Production of antibody
   D. Production of interferon
   E. Activation of complement

3. The loss of which barrier makes burns victims most vulnerable to infection?
   A. Lysozyme
   B. Mucociliary escalator
   C. Skin
   D. Saliva
   E. Gastric juice

4. The function of the mucociliary escalator is to
   A. Kill microorganisms
   B. Remove microorganisms from the upper respiratory tract
   C. Remove microorganisms from body cavities
   D. Remove microorganisms from the lower respiratory tract
   E. All of the above

5. Tears contain
   A. Lipocalin
   B. Lysozyme
   C. IgA
   D. All of the above
   E. None of the above

6. Perspiration inhibits bacteria because
   A. It contains mucus
   B. It contains IgA
   C. It contains lysozyme
   D. It flushes them away
   E. Both **C** and **D** are correct

7. Toll-like receptors are used by the immune system to distinguish between
   A. Viruses and bacteria
   B. Alive and dead cells
   C. Toxins and nutrients
   D. Eukaryotic and prokaryotic cells
   E. Self and nonself

8. All of the following cells are involved in the innate response except
   A. Neutrophils
   B. Eosinophils
   C. Basophils
   D. Monocytes
   E. Lymphocytes

9. Neutrophils attach to the vascular linings and move out of the blood and into the tissues in a process known as
   A. Intravascular clotting
   B. Selection
   C. Diapedesis
   D. Phagocytosis
   E. None of the above

10. Margination is the process in which white blood cells
    A. Separate from red blood cells
    B. Leave the blood vessels
    C. Produce selectin
    D. Slow down, stop, and attach to vessel walls
    E. Speed up and attach to vessel walls

11. The phagocytes present in highest proportions in the blood are
    A. Basophils
    B. Eosinophils
    C. Lymphocytes
    D. Monocytes
    E. Neutrophils

12. Which are the first cells to arrive at an infection site?
    A. Neutrophils
    B. Macrophages
    C. Basophils
    D. Monocytes
    E. Lymphocytes

13. Macrophages resident in the liver are called
    A. Alveolar macrophages
    B. Dendrites
    C. Microglial cells
    D. Kupffer cells
    E. None of the above

14. Macrophages resident in the central nervous system are called
    A. Alveolar macrophages
    B. Dendritic macrophages
    C. Microglial cells
    D. Kupffer cells
    E. None of the above

15. The characteristics of cytokines include all of the following except
    A. Regulation of inflammation response
    B. Secretion from white blood cells
    C. Reaction with specific receptors on target cells
    D. Phagocytic activity
    E. Having overlapping functions with other cytokines

16. Mediators released by mast cells include
    A. Histamine
    B. Serotonin
    C. Cytokines
    D. Proteases
    E. All of the above

17. Dendritic cells found in the skin are called
    A. Langerhans cells
    B. Kupffer cells
    C. Microglial cells
    D. None of the above

18. Natural killer cells are
    A. Restricted to immune surveillance of tumors
    B. Part of the adaptive immune response
    C. Involved in both surveillance of tumors and response to pathogens
    D. Restricted to destruction of pathogens

19. The sequence of phases in phagocytosis is
    A. Ingestion, chemotaxis, adherence, digestion, excretion
    B. Digestion, adherence, chemotaxis, ingestion, excretion
    C. Ingestion, digestion, adherence, excretion
    D. Adherence, digestion, ingestion, excretion
    E. Adherence, ingestion, digestion, excretion

20. Phagolysosomes are formed after which phase of phagocytosis?
    A. Ingestion
    B. Digestion
    C. Adherence
    D. Excretion

21. Redness, pain, heat, and swelling are hallmarks of
    A. Phagocytosis
    B. Vasoconstriction
    C. An anti-inflammatory response
    D. An inflammatory response
    E. Intravascular clotting

22. The acute-phase response is
    A. Related to fever
    B. Specific
    C. A type of inflammatory response
    D. Localized
    E. A defense primarily against viruses

23. Fever is caused by chemicals known as
    A. Pyretics
    B. Intravascular clotting factors
    C. Pyrogens
    D. Pyrotechnics
    E. None of the above

24. The complement system can be activated by
    A. The alternative pathway
    B. The restriction pathway
    C. The lectin-binding pathway
    D. **A** and **C**
    E. All of the above

25. The classical pathway for activation of the complement system requires
    A. A phagocytic response
    B. Mannose-binding ligands
    C. Factor D
    D. Antigen–antibody complexes
    E. Properdin

26. The alternative pathway for activation of the complement system is initiated at protein
    A. C1
    B. C2
    C. C6
    D. C3
    E. C1–C2–C4 complex

27. The combination of complement proteins C5 to C9 is known as
    A. The terminal complex
    B. The defense complex
    C. The membrane defense complex
    D. The membrane attack complex
    E. None of the above

28. Gamma interferon is produced by
    A. T lymphocytes
    B. T lymphocytes and NK cells
    C. T lymphocytes, NK cells, and neutrophils
    D. NK cells
    E. NK cells and neutrophils

29–32. Answer **A** if each of the following is involved only in the classical pathway to complement activation; **B** if involved in only the alternate pathway; **C** if involved in both pathways; and **D** if involved in neither pathway. You can use the choices more than once, or not at all.

29. Mannose binding protein

30. Factors B and D

31. C3

32. C2 and C4

33–36. Answer **A** if each of the following is true only for neutrophils; **B** if true only for basophils; **C** if true for both; and **D** if true for neither. You can use the choices more than once, or not at all.

33. Derived from myeloid stem cell

34. Directly involved in inflammation

35. Directly involved in phagocytosis

36. Directly involved in an adaptive immune response

37–39. Answer **A** if each of the following is true only for cells infected with virus; **B** if true only for cells not infected with virus; **C** if true for both; and **D** if true for neither. You can use the choices more than once, or not at all.

37. Expression of antiviral protein genes

38. Expression of Interferon genes

39. Contain interferon receptors

DEPTH OF UNDERSTANDING

Questions listed here require you to bring together the concepts you have learned in this chapter into a discussion format. This helps you to increase your depth of understanding of the material you have learned. Help can be found in the student resources at: www.garlandscience.com/micro2

1. Describe the barriers associated with the innate immune response.

2. The complement system is extremely important in the nonspecific defense response. It can be said that this system is an amplification system. Defend this statement.

3. While chopping an onion, you inadvertently cut your finger. Recalling what you have learned in this chapter, explain the chemical and cellular defense responses that occur at the site of this trauma.

 CLINICAL CORNER

Help can be found in the student resources at: www.garlandscience.com/micro2

1. Your patient is a 35-year-old male who presents with a history of infections. These infections are due primarily to staphylococcal and streptococcal pathogens. His blood work shows a high titer of antibody against staphylococci and streptococci but he is currently infected with a strain of *Neisseria*. Further evaluation of his blood work indicates he has no titer against these organisms and also that he has little to no C3 protein in his blood. As part of his history he has told you that his father was also prone to repeated infections.

   A. How will you explain his situation to him?
   B. Should he be put on broad-spectrum antibiotics?
   C. What do you tell him if he asks whether he will continue to have repeated infections?

2. Mr. Edison is recovering from a kidney infection. He has been receiving antibiotic therapy and his symptoms have diminished markedly. He has returned to the clinic because he has noticed blood in his urine. The doctor tells Mr. Edison that in some cases the response to an infection can have side effects that are as bad as the infection. After the doctor has left, Mr. Edison asks you to explain what the doctor meant.

   A. What would you tell him?
   B. Which of the innate responses would be the most involved in this?

The Adaptive Immune Response

Chapter 16

Why Is This Important?

The adaptive immune response is a very powerful system that protects us from a multitude of infectious organisms. It has the benefit of immunological memory, which provides a more rapid and powerful reaction if the same pathogen is seen again. Without the adaptive immune response we would not survive.

The year is 1970. A little girl named Jane is sick with the chickenpox. Jane is covered with the vesicular rash from head to toe and is itchy and miserable. Although unpleasant, chickenpox was a fairly mild childhood illness and most children contracted it during their elementary school years.

Fast forward to the year 2000. One-year-old Mary is visiting her pediatrician for her immunizations. One immunization she is scheduled to get is her chickenpox vaccine. This vaccine will be boosted again when she is four years old. The vaccine contains weakened chickenpox virus that will induce immunity but not cause infection in healthy children. Mary will have to endure the pain of a needle stick and perhaps soreness at the injection site, but other side effects of the vaccine are rare.

Both Jane and Mary will have produced adaptive immune responses that provided immunological memory to chickenpox that will prevent them from becoming infected by the chickenpox virus in the future. However, Jane had to suffer the unpleasant effects of the chickenpox rash. Jane could also suffer from shingles later in life, because shingles is caused by latent chickenpox virus. Mary will not contract chickenpox but will still have immunity. If enough children are vaccinated, this childhood infection could become part of our past medical history.

OVERVIEW

In this chapter, we look at the adaptive immune response. In Chapter 15, we learned that a host has a formidable array of innate defense mechanisms that in many cases are more than enough to handle potential infections. In addition to innate defenses, the host has the ability to mount a defense that is specific to a particular pathogen. This defense is called the adaptive immune response. The adaptive response has the benefit of immunological memory: when a host is infected by a pathogen, the adaptive immune response not only clears the infection from the body but also remembers the pathogen. When that pathogen invades the host again, even decades after the first infection, the pathogen will be defeated even more quickly. Immunological memory is also the reason that vaccinations work. As we look at the adaptive response, you will see that it is like a carefully choreographed ballet in which the dancers all depend on one another in giving a powerful performance.

We will divide our discussions into the following topics:

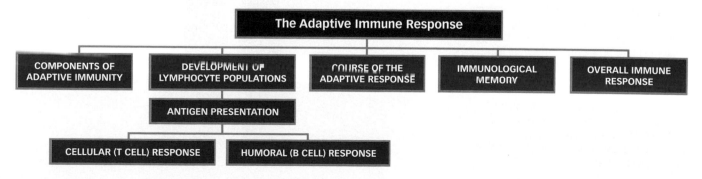

COMPONENTS OF ADAPTIVE IMMUNITY

The adaptive immune response responds to specific antigens and depends on two processes: **humoral immunity** and **cellular immunity**. Humoral immunity involves **B cells** and the production of **antibodies**. The level of antibody in serum can be expressed in units called a **titer** and can be an indicator of the level of protection against a pathogen. Cellular immunity involves **T cells**. The two processes are both separate and interwoven, as we will see as this chapter progresses.

Keep in mind that the innate response is a prerequisite for the adaptive response, and cells that participate in innate immune reactions are also involved in the adaptive response.

Strategic Lymphoid Structures

The adaptive response is associated with the lymphatic system of the body, which is laid out so that it covers the entire body (**Figure 16.1**).

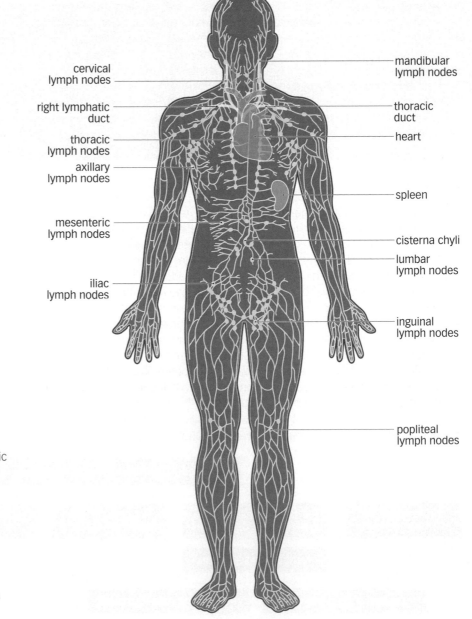

Figure 16.1 An illustration of the human lymphatic system. The lymphatic system is responsible for collecting fluids left in the tissues by the circulatory system and returning them to circulation. Pathogens that enter the body often end up in the lymphatic fluids. Notice how this system covers the entire body. It is the strategic placement of these lymphoid structures that makes it possible for the adaptive immune response to deal with potential pathogens from any place that is involved with infection.

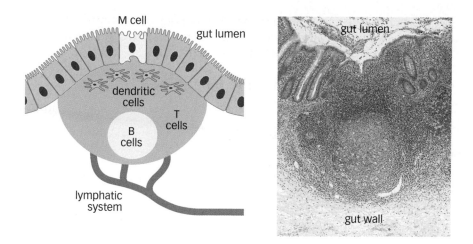

Figure 16.2 Schematic figure and micrograph of a Peyer's patch. Peyer's patches are bundles of lymphatic tissues found in the mucosal epithelium of the small intestine. M cells are distinct from other cells of the epithelium and form pockets to collect antigens. Below the M cells are areas of dendritic cells, T cells, and B cells—all components ready to produce an immune response to pathogens in the digestive tract.

The strategic location of the lymphoid structures, and in particular the lymph nodes, permits the identification of potential invaders from any location in the body. As we saw in previous chapters, there are several areas of the body that are portals of entry for pathogens, where immune protection is essential. In all of these areas we find lymphoid tissue. Lymph nodes can be found in large numbers in the neck (cervical lymph nodes), under the arm (axillary lymph nodes), and in the legs and groin (inguinal lymph nodes), and there are also lymphoid tissues associated with the mucous membranes.

MALT (mucosa-associated lymphoid tissue) is the collective term given to the lymphoid tissues associated with the mucous membranes. In the gastrointestinal tract, the adaptive immune response is focused in the **GALT** (gut-associated lymphoid tissue), which includes the tonsils, adenoids, appendix, and Peyer's patches. Peyer's patches, which are located in the mucosal layer of the small intestine, are the most important part of the GALT because they contain antigen-collecting **M cells**. M cells are unlike any other part of the small intestine because no nutrient absorption takes place there. These cells have no villi; instead they are fenestrated (**Figure 16.2**). Under each M cell, there are germinal centers, regions filled with B cells and surrounded by T-cell areas ready to implement an immune response against pathogens that enter through the intestinal tract.

BALT (bronchus-associated lymphoid tissue) is associated with the mucous membranes of the respiratory system. Recall that both the body's mucosal surfaces and the respiratory system are major portals of entry for infectious organisms and must therefore be well protected by both the innate and adaptive immune responses. In addition to BALT, there are other mucous membranes that contain lymphoid tissues such as NALT (nasal-associated lymphoid tissue) and CALT (conjunctival-associated lymphoid tissue).

Immune mechanisms in all these regions are designed to trap potential antigens (either by dendritic cells or by macrophages) and induce an adaptive response by presenting the antigens to lymphocytes. Of equal significance is the fact that peripheral lymphoid tissues also produce important signals that allow lymphocytes to survive and continue migrating until they encounter their specific antigen.

Cells of the Adaptive Immune Response

The cells involved in adaptive immunity arise in the bone marrow from stem cells (**Figure 16.3**).

Dendritic Cells

Of the cells that participate in the early steps of adaptive immunity, dendritic cells are some of the most important. As we saw in Chapter 15, dendritic cells are also part of the innate response and are a good example of the interrelation between the innate and adaptive immune responses.

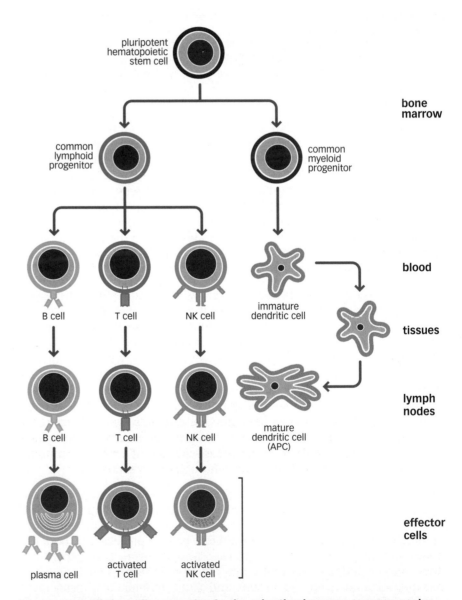

Figure 16.3 All the cells operating in the adaptive immune response arise from stem cells in the bone marrow. These stem cells divide into a common lymphoid precursor that gives rise to T cells, B cells, and natural killer (NK) cells, and a common myeloid precursor that gives rise to dendritic cells (and the cells of the innate immune response as shown in Figure 15.5).

Immature dendritic cells migrate from the bone marrow into the blood and then from blood into tissues. They are resident in most tissues, particularly at epithelial surfaces, and carry Toll-like receptors, making them great watchdogs for potential pathogens. When dendritic cells detect antigens, they use phagocytosis and macropinocytosis (which involves engulfing large amounts of fluids) to capture and process the antigens. Dendritic cells with antigens attached to their surface then migrate from the tissues to the nearest regional lymphoid structure (such as lymph nodes). The antigen recognition and processing causes the dendritic cell to mature into an **antigen-presenting cell** (APC). The migration and maturation of dendritic cells can be seen in **Movie 16.1**. Note the large number of long projecting arms in the dendritic cell in **Figure 16.4**. These projections give the cell an enormous surface area for antigen presentation and make dendritic cells the most efficient APC. Macrophages and B cells also have antigen presentation ability. All three types of APC interact with T cells in lymphoid structures (**Figure 16.5**). APCs also produce the cytokines that influence the adaptive immune response.

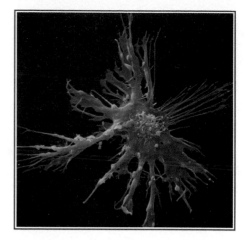

Figure 16.4 A scanning electron micrograph of a dendritic cell, which is the major initiator of the adaptive immune response. Note the extensive elongated projections, which greatly increase the amount of surface area available for antigen presentation.

Lymphocytes

The other cells involved in the adaptive immune response derive from a common lymphoid precursor cell as shown in Figure 16.3 and are therefore known as **lymphocytes**. There are three types of lymphocyte: natural killer cells (part of the innate response and discussed in Chapter 15), T lymphocytes (T cells), and B lymphocytes (B cells).

Although B cells and T cells are both formed in bone marrow, they mature in different places. B cells mature in the bone marrow, whereas T cells mature in the thymus. After maturing, both types of lymphocyte move into other parts of the body. Lymphocytes of the adaptive response circulate through blood, lymph, and peripheral tissue looking for free antigens or presented antigens that fit their receptors. Mature lymphocytes that have not encountered their antigen are known as **naive lymphocytes**. Peripheral lymphoid structures such as lymph nodes are specialized to trap antigen-bearing dendritic cells so that they can be examined by lymphocytes for potential antigens.

In the absence of infection, most lymphocytes circulating in the body are small and featureless, with condensed DNA and few cytoplasmic organelles. This is characteristic of inactive cells; indeed, lymphocytes have no function until they are activated. Activated lymphocytes that have differentiated into their fully effective form are called **effector lymphocytes**. When activated, B cells differentiate into **plasma cells**, which secrete antibody and present antigen to T cells. Activated T cells come in three main classes: **cytotoxic T cells** kill infected cells, **helper T cells** help activate B cells as well as other types of immune cell, and **regulatory T cells** suppress the immune response once the antigen is reduced or gone. Some activated B cells and T cells differentiate into **memory cells**, long-lived lymphocytes that are responsible for **immunological memory**.

The different types of T cell can be differentiated from each other by proteins expressed on their surface. The surface protein called CD4 is found on helper T cells and regulatory T cells, and the surface protein called CD8 is found on cytotoxic T cells.

Both B and T cells can mount a specific response against virtually any antigen because as they mature, they produce a unique antigen receptor. The antigen receptor for B cells is a surface molecule that has two identical antigen recognition sites (**Figure 16.6a**). The antigen receptor

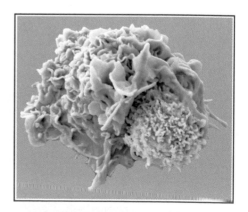

Figure 16.5 A colorized electron micrograph showing a dendritic cell (blue) interacting with a lymphocyte (yellow) during antigen presentation.

(a)   **(b)**

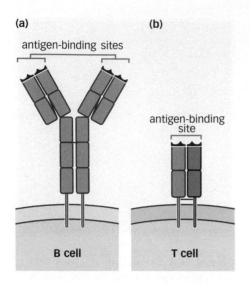

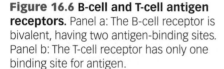

Figure 16.6 B-cell and T-cell antigen receptors. Panel a: The B-cell receptor is bivalent, having two antigen-binding sites. Panel b: The T-cell receptor has only one binding site for antigen.

for a T cell has only one antigen-binding site (**Figure 16.6b**). The main functional difference is that antigen receptors on B cells can recognize free antigen, whereas the antigen receptor on T cells can only recognize antigen especially presented on the surface of APCs.

Both B cells and T cells require not only the signals that result from antigen binding, but also a second **co-stimulatory signal** in order to become fully activated and effective. This signal is generated from, and regulated by, cells of the innate immune response. Contact with antigen without a co-stimulatory signal leads to automatic inactivation of the lymphocyte, a state called **anergy**. This is a safeguard against the adaptive immune system reacting against the body and is known as **peripheral tolerance**.

| **Keep in Mind** |

- The adaptive immune response is specific and involves both a cellular and a humoral (antibody) component.
- Antigens are substances that can elicit an adaptive immune response.
- There are lymphoid structures such as GALT, MALT, and BALT that are strategically located in major portals of entry used by pathogens.
- The adaptive immune response is interrelated with the innate immune response through antigen presentation by cells called antigen-presenting cells (APCs).
- Cells involved in adaptive immunity arise from stem cells in the bone marrow.
- The two lymphocyte types involved in the adaptive response are T cells and B cells.
- Activated B cells differentiate into plasma cells, which produce antibody.
- Activated T cells differentiate into cytotoxic T cells, helper T cells, and regulatory T cells, which have a variety of functions.
- Both T cells and B cells have specific receptors for antigen.

DEVELOPMENT OF LYMPHOCYTE POPULATIONS

As we saw above, B cells mature in the bone marrow, and T cells mature in the thymus. The thymus is prominent in children but atrophies by the time we reach puberty. It was difficult for scientists to accept the fact that the thymus atrophies, even though in many cases the T-cell response lasts essentially for life. As it turns out, the T-cell response is maintained through long-lived T cells that can occasionally divide to maintain the response capability. The situation for B cells is slightly different because the bone marrow remains functional for life. Therefore there is always a supply of mature B cells readily available.

The stages of lymphocyte development are marked by successive rearrangements of antigen receptor genes. In addition, maturation requires signals from the microenvironment in which the lymphocyte develops.

Clonal Selection of Lymphocytes

Clonal selection, illustrated in **Figure 16.7**, is the process by which some lymphocytes are destroyed and others are allowed to mature. This selection process takes place in the bone marrow for B cells and in the thymus for T cells.

A precursor cell produces lymphocytes, each with specificity for a different antigen. This wide range of specificities is the result of segments of the

Figure 16.7 Clonal selection. Each lymphoid progenitor gives rise to a large number of lymphocytes, each bearing a distinct antigen receptor. Lymphocytes bearing receptors for self are lost through the process of clonal deletion before they become fully mature. When a mature naive lymphocyte binds to its particular antigen, the lymphocyte proliferates and differentiates to form a clone of cells all specific for that particular antigen.

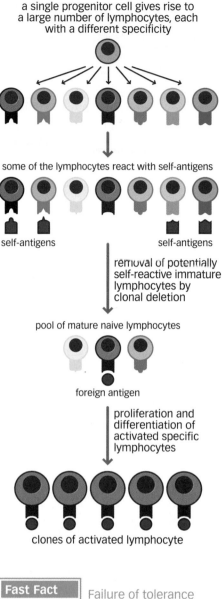

a single progenitor cell gives rise to a large number of lymphocytes, each with a different specificity

some of the lymphocytes react with self-antigens

self-antigens   self-antigens

removal of potentially self-reactive immature lymphocytes by clonal deletion

pool of mature naive lymphocytes

foreign antigen

proliferation and differentiation of activated specific lymphocytes

clones of activated lymphocyte

genes that code for antigen receptors being rearranged during lymphocyte development. This rearrangement generates millions of variants of the antigen receptors on the surface of the lymphocytes. Because each lymphocyte has a unique specificity and because there are billions of lymphocytes in the body, the repertoire of antigen receptors is vast.

Clonal deletion is a part of clonal selection. In this step, those members of the newly formed lymphocyte group that react with self-antigens are eliminated. Since these cells would react against the host, they must be removed. This process is called **central tolerance** and is another safeguard against the immune system attacking the body.

All the lymphocytes remaining after deletion will mature, and each will be specific for a different antigen. When one member of this group encounters its particular antigen, it binds that antigen and becomes activated. Lymphocyte activation gives rise to a clone of identical lymphocytes specific for that antigen. Only those lymphocytes that encounter their specific antigen are activated. This limitation prevents indiscriminate activation of the immune response, because indiscriminate activation can have dangerous consequences for the host. A sustained high concentration of antigen sends strong constant signals through the antigen receptor and acts as a further mechanism of tolerance. Many of the body's normal proteins are produced continuously by many cell types and are abundant throughout the body; this distinguishes them from pathogenic antigens, which are introduced suddenly and are localized in the early stages of infection.

Clonal selection is one of the central principles of adaptive immunity and has three important consequences:

- It enables a limited number of gene segments to rearrange and thereby generates a vast number of different antigen receptors.

- Each receptor is specific for a different antigen.

- Because genetic rearrangement is irreversible, all progeny of a lymphocyte that is a member of a clone will have that same receptor.

Fast Fact   Failure of tolerance leads to autoimmune diseases.

Survival of Lymphocyte Populations

Each day the bone marrow produces millions of new B cells. The survival of these lymphocytes is determined by signals sent out from peripheral lymphoid tissue and received by the antigen receptors on the lymphocytes. These signals cause the B cells to either become activated and proliferate, or die. Eventually, those B cells that never get stimulated undergo apoptosis. This helps to keep the size of the overall B-cell population relatively constant.

For T cells, survival signals come from specialized epithelial cells in the thymus during development. Signals can also come from dendritic cells in the peripheral lymphoid tissues. Once they leave the thymus, T cells migrate to the lymph nodes. After being presented with their antigen, T cells in the lymph nodes stop migrating and become activated. They become larger and increase their numbers up to fourfold every 24 hours for three to five days. Therefore, one naive T cell can give rise to thousands of daughter cells of the same specificity, and each of these can differentiate into an activated effector T cell. These changes also affect the surface adhesion molecules on these cells so that they can either migrate to sites of infection or remain in the lymph nodes. Activated effector T cells have a limited life span and most eventually undergo apoptosis.

As we mentioned above, both B cells and T cells can also differentiate into memory cells, which are long lived and responsible for long-lasting immunity.

Lymphoid Tissues

After lymphocytes leave the thymus or bone marrow, they are carried by the blood into peripheral lymphoid tissues. These tissues, which include the lymph nodes, are organized into distinct areas where T cells or B cells are found. We can use the lymph node to illustrate this organization (**Figure 16.8**). B cells in the lymph node are found in areas called **follicles**, which are located in the outer cortex. T cells are found in zones that surround these follicles in what are called **paracortical areas**.

Normally, a lymphocyte circulates continuously through the lymphoid tissue by way of the blood and lymph fluid until the lymphocyte either encounters its specific antigen or dies. When a lymphocyte dies, it is replaced by a new one.

In peripheral lymphoid tissue, the fate of the lymphocytes is controlled by their antigen receptors. In the absence of an encounter with their specific antigen, B cells die. Although large numbers of B cells leave the bone marrow every day, most die soon after they arrive in the peripheral lymphoid tissue. However, the total number in the body remains constant through replacement.

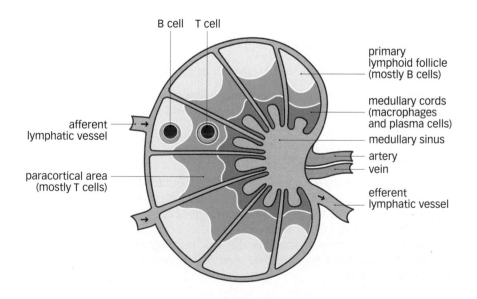

Figure 16.8 A diagrammatic representation of the structure of the lymph node. Notice how B and T cells are located in different regions of the node.

T cells leave the thymus fully mature but in smaller numbers than B cells leaving the bone marrow. T cells generally have a longer life span (measured in years) than B cells, which have a life span measured in days. However, T cells are thought to be self-renewing in peripheral lymphoid tissue.

> **Keep in Mind**
>
> • T cells mature in the thymus; B cells mature in the bone marrow.
>
> • Clonal selection is the process by which some lymphocytes are allowed to mature and others are deleted from the body.
>
> • Lymphocytes continue to circulate through lymphoid tissue by way of the blood or lymph until they either encounter their specific antigens or die.

ANTIGEN PRESENTATION

We have mentioned antigen presentation a few times in this and the previous chapter, and now we'll look at the mechanism. Antigen presentation is the display of antigens on the surface of antigen-presenting cells so that the antigen can be recognized by T cells. B cells do not require antigen to be presented by MHC molecules in order to recognize it as an antigen.

There are three main types of antigen-presenting cell: dendritic cells, macrophages, and B cells. Antigens must be broken into fragments and delivered to the cell surface by specialized glycoproteins known as **major histocompatibility (MHC)** molecules if they are to be recognized by T cells. The genes that code for MHC molecules were first identified because of their effects in tissue transplants; hence their name.

There are two types of MHC molecule: class I and class II. Although both can combine with and display antigen on the surface of cells, they differ in structure, the T cells they activate, the type of pathogen they respond to, and in which cells they are expressed (**Figure 16.9** and **Movie 16.2**).

Figure 16.9 Class I and II MHC molecules present antigens from three different sources. In the first panel, MHC I presents antigens from the cytosol, usually fragments of viruses or bacteria that are infecting the cell. MHC I combined with antigen is presented to cytotoxic T cells, which will be triggered to kill the infected host cell. In the second panel, a phagocytic cell combines phagocytosed antigen with MHC II, which it presents to helper T cells. The helper T cells are activated and stimulate the phagocytic cell to kill the pathogens in their vesicles. In the third panel, antigen from extracellular pathogens (or toxins) is taken into an APC (here a B cell) by endocytosis and returned to the surface bound to an MHC II molecule. The MHC II molecule with bound antigen is presented to helper T cells, which stimulate the B cell to secrete antibodies to help eliminate extracellular pathogens.

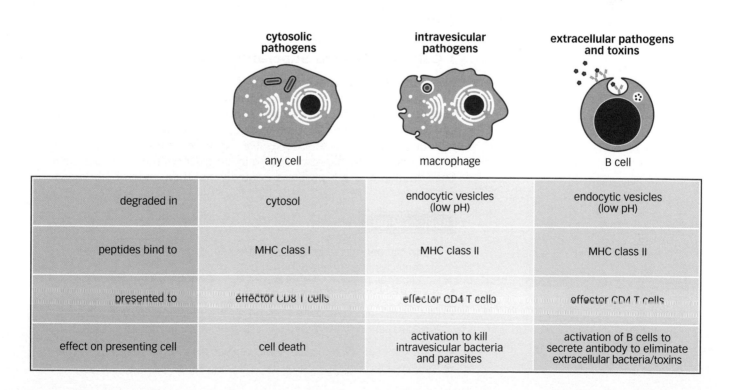

	cytosolic pathogens	intravesicular pathogens	extracellular pathogens and toxins
	any cell	macrophage	B cell
degraded in	cytosol	endocytic vesicles (low pH)	endocytic vesicles (low pH)
peptides bind to	MHC class I	MHC class II	MHC class II
presented to	effector CD8 T cells	effector CD4 T cells	effector CD4 T cells
effect on presenting cell	cell death	activation to kill intravesicular bacteria and parasites	activation of B cells to secrete antibody to eliminate extracellular bacteria/toxins

Class I MHC

Class I MHC molecules (MHC I) are found on all nucleated body cells and help to identify them as self. Antigens from pathogens that multiply in host cell cytoplasm are carried to the cell surface by class I MHC molecules and presented to cytotoxic (CD8) T cells. This results in the proliferation of cytotoxic T cells, which then look for and kill infected host cells expressing that antigen.

Class II MHC

Class II MHC molecules (MHC II) are found exclusively on immune system cells, especially phagocytic antigen-presenting cells (dendritic cells, macrophages, and B cells). MHC II collects antigens generated in the vesicles of infected cells and conveys an entirely different message from that delivered by MHC I. Instead of triggering the death of the presenting cell, MHC II combined with antigen, referred to as the antigen–MHC complex, binds to the CD4 receptors of helper T cells, stimulating them to proliferate and differentiate. The differentiated helper T cells then go on to activate other immune cells and stimulate the phagocytic APC to eliminate the pathogen. MHC II also deals with extracellular pathogens and toxins by stimulating the production of antibodies.

For either class of MHC molecule to function properly, the antigen–MHC complex must be stable at the APC cell's surface. This stability permits long-term display of the complex, which is required for effective presentation of the antigen to a T cell. If the antigen is removed from the MHC molecule while on the surface of the APC cell, or if an MHC molecule shows up on the surface of the APC without an antigen, the entire MHC molecule undergoes a conformational change and is quickly degraded. This is a failsafe mechanism for the adaptive immune response that protects against aberrant reactions that could damage the host.

The T-cell receptor must recognize both the antigen and the MHC molecule in the antigen–MHC complex for proper recognition and production of an immune response. This requirement for dual recognition is called **MHC restriction**.

T-Cell Response to Superantigens

Superantigens are a distinct class of antigen produced by pathogens, such as the exotoxin produced by *Staphylococcus aureus* (see Chapter 5). Superantigens bind to the outside surface of MHC II already bound to an antigen, and can also bind directly to T-cell antigen receptors. This type of antigen presentation causes massive overproduction of cytokines by CD4 helper T cells, which causes severe inflammatory illness and toxic shock. The stimulated T cells proliferate rapidly and then undergo apoptosis.

Fast Fact Superantigens also stimulate the host to produce antibodies and can be used to generate vaccines.

> **Keep in Mind**
>
> - T cells can only recognize antigen if it is bound to MHC molecules and presented on the surface of other cells.
> - There are two types of MHC molecule, class I and class II.
> - MHC class I molecules are expressed on all nucleated cells; MHC class II molecules are only expressed on cells of the immune system, primarily APCs.
> - Cytotoxic T cells bind to class I MHC–antigen via their CD8 receptor, which stimulates the T cell to destroy the antigen-presenting cell.

- Helper T cells bind to class II MHC–antigen via their CD4 receptor, which stimulates the T cell to multiply and activate other immune cells, especially B cells.

- T cells that bind to superantigens produce an immune reaction that damages the host.

CELLULAR (T CELL) RESPONSE

Now that we have examined antigen presentation, let's turn our attention to the cellular response, which is generated by T cells. Naive T cells continually circulate between the blood and peripheral lymphoid tissue, and it is in the lymphoid tissue that they encounter antigens, presented to them as antigen–MHC complexes. Remember that they also need co-stimulatory signals to become fully differentiated effector cells.

Naive cytotoxic T cells are activated by infected cells presenting antigen–MHC I complexes. They also need signals from activated helper T cells.

Production of Activated Effector Helper T Cells

Naive T cells circulating in the lymphoid tissues sample the antigen–MHC II complexes on APCs. As they migrate through a lymph node, naive T cells transiently bind to each APC. This binding allows the T cell to crawl along the APC and sample every antigen present. If the T cell finds an antigen that fits into its antigen receptor, there is an immediate conformational change in the receptor. This change stabilizes the binding of the T cell to that site on the APC. The association of these two cells can last for several days, during which time the T cell proliferates. This proliferation produces around 1000 cells of identical antigen specificity, which will differentiate into activated effector T cells. Activation of T cells can be accomplished by the three types of APC: dendritic cells, macrophages, and B cells (**Figure 16.10**).

Activation by Dendritic Cells

Dendritic cells secrete a chemokine that attracts naive T cells to come and sample antigens that the dendritic cell is presenting (see **Movie 16.3**).

Fast Fact The activation of naive T cells is referred to as the primary cell-mediated immune response. It provides for both the activation of T cells and the development of memory T cells.

Figure 16.10 Activation of T cells by the three groups of antigen-presenting cells.

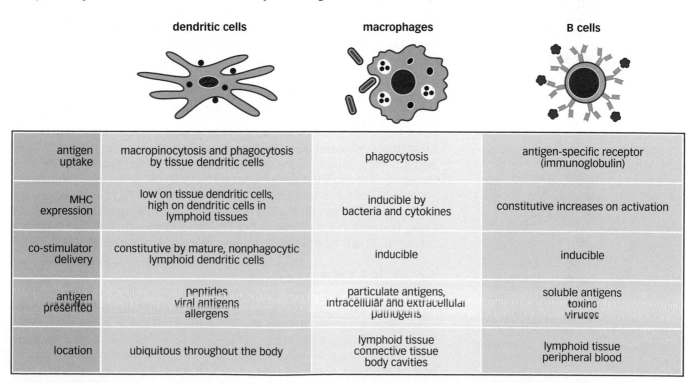

	dendritic cells	macrophages	B cells
antigen uptake	macropinocytosis and phagocytosis by tissue dendritic cells	phagocytosis	antigen-specific receptor (immunoglobulin)
MHC expression	low on tissue dendritic cells, high on dendritic cells in lymphoid tissues	inducible by bacteria and cytokines	constitutive increases on activation
co-stimulator delivery	constitutive by mature, nonphagocytic lymphoid dendritic cells	inducible	inducible
antigen presented	peptides viral antigens allergens	particulate antigens, intracellular and extracellular pathogens	soluble antigens toxins viruses
location	ubiquitous throughout the body	lymphoid tissue connective tissue body cavities	lymphoid tissue peripheral blood

They are the most efficient APCs and present viral, fungal, and bacterial antigens. In addition, they can present antigens either as part of a transplantation rejection process or as part of the allergic response. Contact with a pathogen sends signals via the TLR that activate the dendritic cell to produce an antigen–MHC complex. Dendritic cells produce co-stimulatory molecules all the time so that they can immediately activate naive T cells.

Activation by Macrophages

Macrophages are highly phagocytic but also function as APCs for pathogens that they are unable to digest. Resting macrophages have few or no MHC II molecules on their surface and do not express co-stimulatory signals. When a macrophage engulfs a microorganism, the microorganism is degraded using cellular lysosomes, and antigen–MHC II complexes are generated on the cell surface. This causes the production of co-stimulatory molecules that allow the activation process to continue. Macrophages also digest dead and dying host cells, so they have a variety of receptors such as TLRs, mannose receptors, and complement receptors (see Chapter 15) that ensure they do not activate T cells against self cells.

Activation by B Cells

The B-cell antigen receptor can bind to free antigen; when this binding occurs, the B cell's receptor and its bound antigen are internalized by endocytosis. The antigen is then bound by MHC II molecules and moved back to the surface of the B cell, where it can be presented to T cells. As with macrophages, the production of co-stimulatory molecules is induced by antigen binding.

Functions of Activated Effector T Cells

Effector T cells have three broad functions: killing infected cells, helping to activate other immune cells, and regulating the immune response. Each of these functions is carried out by one of the three types of effector T cell, the function being determined by the effector molecules that the cells produce. Cytotoxic (CD8) T cells release cytotoxins, which kill infected cells. Helper (CD4) T cells release cytokines that help activate other immune cells. Water-soluble cytokines and membrane-associated molecules often act in combination to mediate the effects of helper T cells on their target cells. For example, membrane-bound CD40 ligand is induced on activated helper T cells and delivers activation signals to macrophages, eosinophils, mast cells, B cells, and neutrophils. Regulatory T cells also produce cytokines, which suppress immune responses.

Most effector T cells leave the lymphoid tissue when they are activated, and enter the blood via the thoracic duct. The initial binding of an activated T cell to its target is mediated by nonspecific adhesion molecules in a manner similar to that seen with naive T cells and APCs. However, the number of adhesion molecules is two to four times higher on activated T cells. This allows an activated T cell to bind more tightly to its target and remain bound long enough to release effector molecules.

Cytotoxic T Cells

All viruses and some bacteria and protozoans multiply in the cytoplasm of infected cells. Once inside the host cell, the pathogen is no longer susceptible to antibody, and the elimination of the organism becomes the responsibility of cytotoxic T cells (**Figure 16.11**). Because this elimination must occur without the destruction of any healthy tissue, it must be powerful but accurately targeted. Cytotoxic T cells are stimulated by

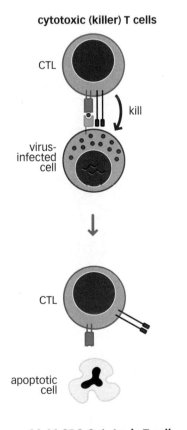

cytotoxic (killer) T cells

CTL

kill

virus-infected cell

CTL

apoptotic cell

Figure 16.11 CD8 Cytotoxic T-cell function. Cytotoxic T cells (CTL) recognize antigen–MHC class I complexes and kill infected cells.

antigen–MHC I complexes and other signals produced by effector helper cells. They can kill an infected cell that is surrounded by healthy tissue without causing any tissue damage.

The principal mechanism of cytotoxic T cells is the calcium-dependent release of preformed specialized granules. These granules are modified lysosomes containing several proteins that are expressed in cytotoxic T cells. Three of these proteins, **perforin**, **granzyme** (a protease enzyme), and the peptide **granulysin** (which induces the onset of apoptosis), are responsible for the death of the target cell. During the killing event, perforin creates a pore in the target cell membrane through which the other proteins gain access to the cytoplasm of the target cell. Programming the onset of apoptosis only takes about five minutes, but the infected cell may take up to several hours to die. Once the cytotoxic T cell has induced apoptosis, it detaches from the infected cell and attaches to another, repeating the process many times. Cytotoxic T cells are therefore selective and repetitive killers of target cells. T-cell action can be seen in **Movie 16.4**.

In another example of cooperation between the innate and adaptive immune responses, cytotoxic T cells also release cytokines, such as interferon and tumor necrosis factor, which are part of the innate immune response.

Helper T Cells

When stimulated (by antigen–MHC II complexes), naive helper T cells proliferate and become immature effector T cells, which can be thought of as T_H0 cells. Upon further stimulation, these cells can mature into several types of helper T cell: T_H1, T_H2, T_H17, or T_{FH} cells, each of which activates different sets of immune cells, to deal with different types of pathogen (**Figure 16.12**).

> **Fast Fact** Cytotoxic T cells also have a role in the destruction of transformed or malignant cells. Because transformation of cells can happen through genetic mutations, there is the possibility that cancerous cells may routinely appear in the body; the cytotoxic T cells keep them from developing into a tumor.

Figure 16.12 Helper T cell functions. T_H1 cells activate macrophages to digest ingested pathogens. T_H2 cells activate mast cells and eosinophils, and stimulate plasma cells to make the anti-parasite antibody. T_H17 cells are helper T cells that activate neutrophils. T_{FH} cells are helper T cells that activate B cells to produce antibodies.

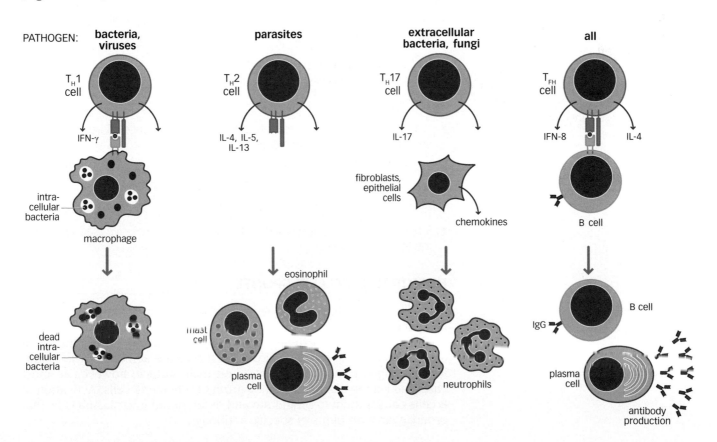

As we saw in Chapter 5, some bacteria can avoid phagocytosis and proliferate inside macrophages. T$_H$1 cells activate these macrophages, increasing their killing power and enabling them to eliminate the pathogen. This is referred to as **macrophage activation**, and T$_H$1 cells must remain in contact with macrophages for long periods for it to occur. Part of the macrophage activation process is the production of oxygen free radicals and nitric oxide, both potent antimicrobial substances. Activation also enhances the fusion of phagosomes to lysosomes. In addition, macrophage activation amplifies the overall immune response by increasing the numbers of MHC II molecules and cytokine receptors on the surface of the macrophage. The presence of these receptors makes macrophages better APCs.

T$_H$2 cells promote anti-parasitic (especially anti-helminthic) responses and allergic responses by activating eosinophils, and mast cells at mucosal surfaces, and by stimulating plasma cells (activated B cells) to produce the antibody associated with anti-parasitic action. T$_H$17 cells protect against extracellular bacteria and fungi by stimulating neutrophils. T$_{FH}$ cells provide signals required to activate B cells that then produce antibody.

Regulatory T cells

Regulatory T (T$_{reg}$) cells are a variable group of T cells that can arise from the thymus or from naive CD4 helper T cells. Regulatory T cells work by producing cytokines that inhibit dendritic cells and T-cell proliferation and suppress the immune response after the antigen has gone. They are also important in preventing autoimmune diseases, a topic that will be discussed in Chapter 17.

Keep in Mind

- The cellular portion of the adaptive immune response is performed by T cells.
- Cytotoxic T cells (CD8 cells) are activated by antigen in association with MHC I molecules.
- Cytotoxic T cells kill infected host cells.
- CD4 helper T cells can be divided into several subpopulations.
- T$_H$1 CD4 helper T cells are involved with pathogens that accumulate inside vesicles in macrophages and dendritic cells.
- T$_H$2 and T$_H$17 CD4 helper T cells deal with extracellular bacteria and parasites by activating B cells, eosinophils, mast cells, and neutrophils.
- T$_{FH}$ cells provide signals to B cells that are required for antibody production.
- Regulatory T cells suppress other T cells and prevent excessive reactions after the antigen has gone.

HUMORAL (B CELL) RESPONSE

Many pathogens that cause infectious disease multiply in the extracellular spaces of the host's body. Attacking this type of problem is the task of the humoral immune response. This response involves the production of antibody and is the province of B cells. B cells are activated by the binding of antigen to antigen receptors on their surface; however, in most cases, activation also requires help from CD4 helper T cells. Activation of B cells causes them to proliferate and differentiate into plasma cells that produce large amounts of specific antibody.

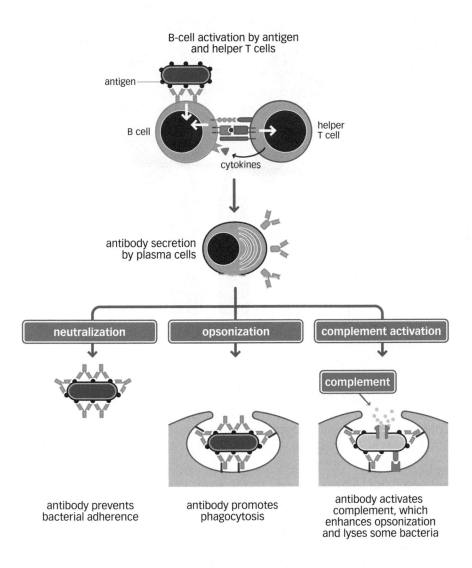

B-cell activation by antigen
and helper T cells

antigen

B cell

helper
T cell

cytokines

antibody secretion
by plasma cells

| neutralization | opsonization | complement activation |

complement

antibody prevents
bacterial adherence

antibody promotes
phagocytosis

antibody activates
complement, which
enhances opsonization
and lyses some bacteria

Figure 16.13 The three mechanisms by which antibody protects a host: neutralization, opsonization, and complement activation. Antibodies that are produced by the plasma cell can prevent attachment and infection by bacteria and viruses (neutralization), assist in phagocytosis through opsonization, and activate the classical complement pathway. Note the complex interactions between surface markers of helper T cells and B cells (antigen presenting) in the top panel: the helper T cell must recognize both parts of the antigen–MHC complex before it stimulates the B cell to proliferate and differentiate into a plasma cell.

Antibody molecules, known as **immunoglobulins** (**Igs**), produced by plasma cells are found in the blood and extracellular spaces, and contribute to immunity in three major ways: neutralization, opsonization, and complement activation (**Figure 16.13**). In the neutralization pathway, antibodies bind to toxins and viruses, preventing them from binding to receptors on host cells, and also prevent bacterial attachment. In the opsonization pathway, antibodies bind to antigens on bacteria or viruses and facilitate their uptake by phagocytic cells by enhancing the attraction of the phagocyte to the antigens. Recall that some pathogens produce capsules that inhibit phagocytosis (Chapter 5), but such capsules can be rendered useless through opsonization. In the complement activation pathway, antibodies activate the classical pathway of the complement system (as we saw in Chapter 15).

Immunoglobulin molecules are also the antigen receptors on B cells and can recognize free antigens. Remember that T-cell antigen receptors can only recognize antigens bound to MHC molecules and presented on the surface of APCs.

The Immunoglobulin Molecule

Every immunoglobulin molecule has a Y shape and is composed of four polypeptide chains, two **light chains** and two **heavy chains** (**Figure 16.14** and **Movie 16.5**). The terms light and heavy refer to the

Figure 16.14 The immunoglobulin molecule is a four-peptide chain structure made up of two light chains bonded by disulfide linkages to two heavy chains. Note the constant and variable regions in the molecule. These regions arise from the three-dimensional structure of the polypeptides. The variable regions of the heavy and light chains combine to form the antigen-binding sites of the molecule. Because there are two binding sites, the molecule is referred to as bivalent.

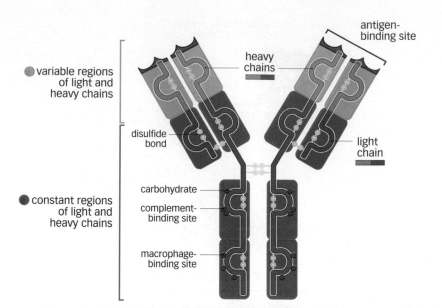

molecular weights of the chains. The four N-terminal ends (one on each of the four chains) make up the **variable region** of the molecule, which contains two identical **antigen-binding sites**. The remainders of the light and heavy chains make up what is referred to as the **constant region**.

Antibodies recognize only a small part of the antigen molecule (the **epitope**), and binding depends on the size and shape of the antigen. The binding of antigen to antibody involves hydrophobic forces and electrostatic forces (in which electrical charges on the immunoglobulin and antigen molecules either attract or repel one another) between the antigen and the amino acids at the antigen-binding site. In addition, hydrogen bonds can help to hold the two molecules together. Because each immunoglobulin molecule has two antigen-binding sites, bound antigens can be joined together in a process known as cross-bridging. The binding resulting from cross-bridging is very stable and also acts as a trigger for B-cell activation.

There are different types of constant region (known as **isotypes**). The isotype of the heavy chain determines the function and class of an antibody. There are five classes of immunoglobulin: IgG, IgA, IgM, IgD, and IgE. The isotypes of the heavy chains are designated with the corresponding Greek letters: γ (gamma) for IgG, α (alpha) for IgA, μ (mu) for IgM, δ (delta) for IgD, and ε (epsilon) for IgE.

The constant region has three roles:

• It is recognized by specialized receptors expressed on immune cells such as macrophages and neutrophils, facilitating opsonization. During opsonization, the binding site of the immunoglobulin molecule binds to the capsule, leaving part of the constant region of the immunoglobulin molecule exposed. This exposed portion is recognized by the phagocytic cell.

• It mediates initiation of the classical pathway of the complement system by interacting with complement protein C1.

• It is involved in delivering antibody to places such as breast milk, mucous secretions, and tears.

Distribution and Function of Immunoglobulins

Because pathogens find their way to most sites in the body, antibodies must be able to do the same. Most antibody molecules are distributed by diffusion, but specialized transport mechanisms are required for bringing antibody to the epithelial surface linings of the intestine and lungs. All five classes of immunoglobulin can occur as membrane-bound antigen receptors or as free antibody.

IgM

IgM is the first antibody to be produced in a humoral response and is a pentamer made of five immunoglobulin molecules (**Figure 16.15a**). Because IgM is the first antibody produced in the humoral response, each individual antigen binding site binds less tightly, but there are 10 binding sites (2 on each immunoglobulin) to add up to a greater binding strength. IgM is one of the best activators of the complement system because its pentameric shape makes five constant regions available for complement binding. IgM is usually found in the blood, in smaller amounts in the lymph, and has a prime role in reactions against bloodborne pathogens. It is also found in the pleural cavity and pleural spaces, where it protects against environmental pathogens. However, the role of IgM is limited because it is too large to leave the blood, lymph, and pleural cavity and spaces to diffuse into the tissues.

IgG

IgG (**Figure 16.15b**) is the principal isotype in the blood and extracellular fluid because it is small (a single molecule) and can easily diffuse out of the blood and into the tissues. IgG is very effective for opsonization, activation of complement, neutralization of toxins, blocking viral infection, and preventing the attachment of pathogens to host tissues. Neutralization occurs when antibody binds to the toxin molecule and thereby blocks the toxin's ability to bind to its cellular receptor site. This prevents the toxin from gaining entry into the cell. IgG is the principal neutralizing antibody. Similarly, IgG can block a virus from attaching to its host cell, or a pathogen from attaching to a host tissue. Newborn babies are born with maternal IgG in their blood and tissues, because this antibody can easily diffuse across the placenta.

IgA

IgA (**Figure 16.15c**) is the principal antibody secreted by the epithelial lining (a mucosal surface) of the intestine and the respiratory tract. IgA is able to neutralize toxins, block infection by viruses, and prevent pathogen attachment to mucosal surfaces. However, IgA is not very efficient at opsonization and only weakly activates the complement cascade. Newborn infants are especially vulnerable to infection, and maternal IgA in colostrum along with IgG (found in breast milk) is transferred to the gut of the infant, where it provides protection from newly encountered bacteria until the infant can synthesize its own antibodies.

Transport proteins can bind to the base of immunoglobulins and carry particular isotypes across epithelial barriers during a process called **transcytosis** (**Figure 16.16**). This form of transport is most frequently seen with the IgA molecule, which can be transported across the epithelial layer of the gut or bronchi. For this to happen, IgA is secreted as a dimer held together with a protein called a **J chain** and a specialized **secretory piece** (see Figure 16.15c). Dimeric IgA uses the J chain to bind to a receptor located on the basal side of the epithelial cell, and is

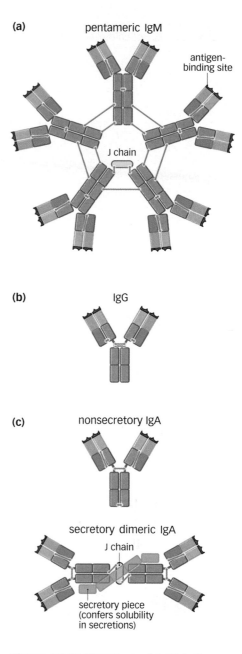

Figure 16.15 Structure of IgM, IgG, and IgA. Panel a: IgM is a pentamer, held together by a J chain, giving it 10 antigen-binding sites and 5 constant regions. Panel b: IgG is a monomer. Panel c: Nonsecretory IgA is a monomer, and secretory IgA is a dimer.

Fast Fact Antibody at the surface of mucosal linings is important because it can neutralize a potential pathogen or promote its elimination before a significant infection can get started.

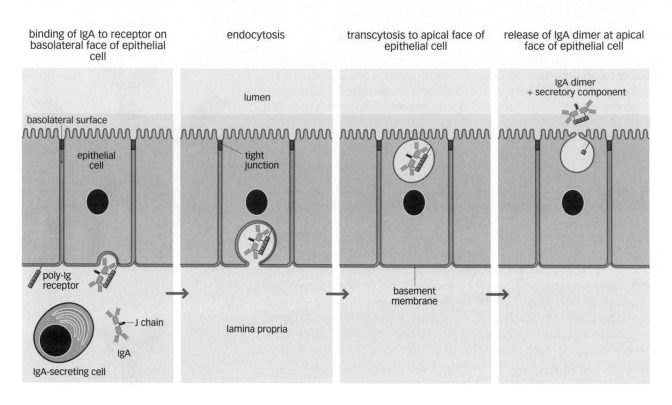

binding of IgA to receptor on basolateral face of epithelial cell

endocytosis

transcytosis to apical face of epithelial cell

release of IgA dimer at apical face of epithelial cell

Figure 16.16 Transcytosis of dimeric IgA across an epithelial layer. This process is mediated by a specialized transport protein known as the poly-Ig receptor and is responsible for the secretion of antibodies into the digestive and respiratory tracts.

Fast Fact Violent physical reactions such as coughing and sneezing are actually beneficial to the host because they can expel infectious agents from the body.

internalized. It then moves through the cytoplasm in a transport vesicle, to the surface of the cell facing the lumen. As the vesicle fuses with the membrane, the IgA is deposited in the mucus. The IgA molecule is held in place in the mucus by mucins that bind to the carbohydrate portions of the secretory piece.

IgE

IgE is found in very low levels in the blood and extracellular fluids, but it is bound tightly by mast cells just below the skin and mucosa, and also along blood vessels in the connective tissue. A substantial amount of IgE is bound to mast cells and basophils. When antigen is bound by IgE on mast cells, the IgE triggers these cells to release powerful chemical mediators, including histamine, that induce reactions such as coughing, sneezing, and vomiting. Because of this, IgE is often called the allergy antibody. We discuss allergic reactions caused by antigen bound to IgE in Chapter 17. Activated mast cells not only release their granules but can also secrete inflammatory mediators and cytokines, and this occurs via specific IgE and IgG receptors, but only when the surface IgE is bound to antigen (**Figure 16.17**). Incredibly, this release occurs in just seconds.

Figure 16.17 Release of inflammatory mediators by mast cells. Notice the dark granules containing these mediators in the resting mast cell at the left. Cross-linking of the IgE molecules on the surface of the cell leads to a rapid release of the mediator molecules, as shown in the activated cell at the right.

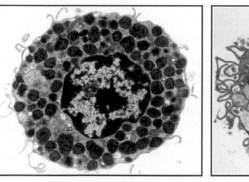
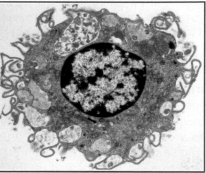

Histamine causes a local increase in blood flow and vascular permeability. This quickly leads to an accumulation of fluid and blood proteins, including antibody, in the surrounding tissue. Shortly thereafter, there is an influx of neutrophils, macrophages, and lymphocytes into the area.

IgD

IgD is found in very small amounts (about 0.2% of the total serum immunoglobulin) and is found in the lymph and blood as well as on the surface of B cells. IgD has no known function in serum, but it is involved as an early antigen receptor on the surface of B cells.

Timing of Immunoglobulin Release

All mature undifferentiated B cells express both IgM and IgD on their surface. After activation, IgD disappears and only IgM remains. When an antigen is encountered for the first time, IgM is always the first immunoglobulin isotype to be produced in large amounts, and this part of the response is known as the **primary immune response**. Days pass while the adaptive immune response develops and IgG is produced. If the antigen is associated with a pathogen, the person will be infected and show symptoms before the level of IgG is sufficient, along with other adaptive immune responses, to kill the pathogen and rid the body of it. The primary response also generates memory cells. When an antigen is encountered again, these memory cells initiate the **secondary immune response**, which is quicker and more powerful than the primary response. IgM is still the first immunoglobulin produced, but in much smaller quantities than in the primary response. A greater amount of IgG is produced sooner and it is sufficient to kill the pathogen before the infected person shows symptoms (**Figure 16.18**). Helper T cells regulate both the production of antibody and also which isotype of immunoglobulin is produced. The secondary immune response is the basis of vaccination.

Antibody Activation of Immune Cells

Another important function of antibody is activating a variety of cells that have receptors for the constant region of the antibody molecule. Nonphagocytic cells, such as NK cells, basophils, and mast cells, are triggered to secrete stored mediators when these receptors for the constant region are engaged. Examples of this receptor binding are illustrated dramatically in parasitic infections in which the parasites are too large to be engulfed. Here, the phagocytic cells attach to the surface of the antibody-coated parasite; the lysosomes of the phagocytic cells move to the cell surface, fuse, and release their contents directly onto the parasite.

Figure 16.18 The primary and secondary response to an antigen shown as the level of antibody produced. Notice the difference in the amounts of the isotopes of antibodies produced.

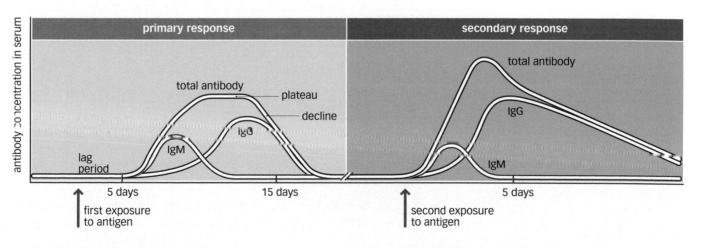

B-Cell Activation and Cooperation with T Cells

The B-cell antigen receptor (recall that it is an immunoglobulin molecule) transmits signals to the interior of the B cell when it binds to antigen. The receptor also brings the bound antigen into the cell by endocytosis. Once inside the B cell, the antigen is degraded, combined with MHC II molecules, and returned to the surface of the B cell. This antigen–MHC complex is recognized by helper T cells, which then activate the B cell to proliferate and differentiate into a plasma cell. The plasma cell will then go on to produce antibodies against the antigen that activated the B cell in the first place.

So, for the humoral response to work, each B cell must find a helper T cell with which to cooperate. This is accomplished by a trapping mechanism that operates in the peripheral lymphoid tissue. The process, which is summarized below, is truly elegant.

- When antigens make their way into the body, they are captured and processed by antigen-presenting cells (APCs), especially dendritic cells, that then migrate to local lymph nodes lodging in the paracortical areas of the nodes.

- Naive T cells circulating in the body are continuously passing through these paracortical areas and sampling the APCs.

- Once a naive T cell passing through the paracortical area of a node is presented with an antigen specific for the T cell's receptor, the T cell becomes activated and remains trapped in the lymph node.

- When B cells enter the lymph nodes, they first pass through the paracortical areas on their way to the follicles.

- If an activated helper T cell is trapped in the paracortical areas, a B cell also becomes trapped. The agents responsible for trapping B and T cells in the paracortical area are adhesion molecules produced by the B and T cells themselves.

- After interacting with helper T cells in the paracortical area of the lymph node, B cells migrate to the follicle region of the node. It is here that B cells proliferate and differentiate into plasma cells, which produce antibody.

The plasma cells created in this process have a variety of life spans. Some survive only days, but others receive signals from stromal cells in the bone marrow that allow them to survive for long periods. The presence of these long-lived plasma cells in the body provides a source of long-lasting antibody protection.

Keep in Mind

- Activation of B cells causes them to proliferate and differentiate into plasma cells, which produce antibody.

- Antibody protects the host through neutralization, opsonization, and complement activation.

- The immunoglobulin molecule is Y-shaped and composed of two heavy and two light chains.

- Antibody specificity is determined by its variable region; antibody function is determined by its constant region.

- There are five isotypes of immunoglobulin molecule (IgG, IgA, IgE, IgD, and IgM), each with different functions in protecting the host.

- The adaptive response can be divided into a primary phase and a secondary phase.

- The primary response is predominantly with IgM and not very powerful. The secondary response is much more powerful than the primary, with IgG being the predominant class of antibody formed.

- B cells must interact with helper T cells to be activated and differentiate.

- Cooperation between T and B cells is carried out using MHC II molecules.

COURSE OF THE ADAPTIVE RESPONSE

The complex set of interactions that make up the adaptive immune response can be summarized as follows and are illustrated in **Figure 16.19** and **Movie 16.6**.

- A crucial effect of infection is the activation of potential APCs that reside in most tissues. These cells take up antigens through interactions with their Toll-like receptors (TLRs). They become activated and mature into APCs. As part of this process, the APCs increase the synthesis rate of MHC II molecules and begin to express co-stimulatory molecules on their surface. APCs carrying antigen move from the site of infection and enter peripheral lymphoid tissue. Here they initiate the adaptive response.

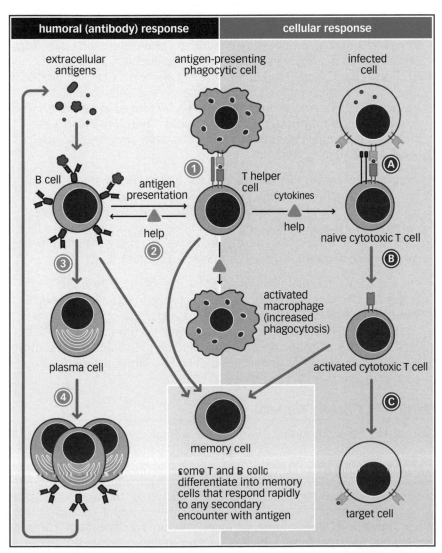

Figure 16.19 Cellular cooperation in the adaptive immune response. This illustration shows the cellular cooperation required for both the humoral (antibody) response and the cellular immune response. The left side of the figure shows the development of an antibody response. In step 1, antigen is presented to the helper T cell by either a B cell or a phagocytic antigen-presenting cell such as a dendritic cell or macrophage. Remember that this interaction is dependent upon MHC class II recognition. In step 2, the helper T cell stimulates the B cell to differentiate into a plasma cell (step 3), which produces antibody against the antigen that was presented (step 4). Helper T cells can also activate macrophages and other immune cells. On the cellular reaction side, helper T cells identify pathogen antigens presented by APCs through antigen–MHC II complexes on their surface and become activated. The helper T cell produces cytokines that cause naive T cells to differentiate into activated cytotoxic T cells if the naive T cell is also recognizing antigen bound to MHC class I from an infected cell (step A). The cytotoxic T cell is activated (step B) and kills the target cell (step C). Note that B cells, helper T cells, and cytotoxic T cells will become memory cells.

- Once APCs arrive in the lymphoid tissue, their only job is to activate antigen-specific naive T cells. These naive T cells are able to recognize the antigen presented on the surface of the APCs and become activated. This activation causes them to divide and mature into activated effector T cells, which then re-enter the circulation.

- The full activation of naive T cells takes four to five days and is accompanied by marked changes in the homing behavior of these cells. Activated cytotoxic T cells must travel from lymph nodes or other peripheral lymphoid tissues, where they became activated so as to attack and destroy infected cells. The same is also true for activated effector helper T cells, which also leave the lymphoid tissue to activate macrophages at the site of infection.

- By the peak of the adaptive response, most of the T cells are specific for the infecting pathogen because of several days of proliferation and differentiation. One or two of these activated effector T cells encountering antigen in tissues can initiate a potent local inflammatory response. This recruits both a greater number of activated effector T cells and many more nonspecific inflammatory cells to the site. Activated effector T cells that enter tissue and do not find antigens there are rapidly lost. Either they leave the tissue immediately and go back into the blood, or they commit suicide through apoptosis. This is important because these cells have the capacity to damage the tissues and must be prevented from doing so.

- The production of antibody is essential in controlling many infections and develops in lymphoid tissues under the direction of helper T cells. This is predominantly the work of the T_{FH} subset of CD4 helper T cells and occurs in the lymphoid tissue. Remember, B cells specific for a protein antigen cannot be activated to proliferate or to differentiate into plasma cells without the help of T cells. Therefore there is not sufficient high-affinity antibody production until after specific helper T cells have been generated.

- Antibody responses are sustained in the lymph nodes and in the bone marrow. In these sites, plasma cells secrete antibody directly into the blood via the efferent lymph flow for distribution to the rest of the body. Plasma cells in the lymph nodes live for only two to four days and then undergo apoptosis. However, plasma cells in the bone marrow can live for long periods (months to years).

- Once the infection has been repelled by the adaptive immune response, two things happen. First, effector cells remove the specific stimulus that recruited them. Second, those effector cells undergo death by neglect once the antigen has gone. Fortunately for us, not all the effector cells disappear. Some of them are retained as memory T and B cells that keep us safe should we encounter the same pathogen again.

IMMUNOLOGICAL MEMORY

One of the most important parts of the adaptive immune response is the development of immunological memory, which is the ability of the adaptive immune system to respond rapidly and effectively to pathogens that have been encountered previously. This occurs because of the persistence of antigen-specific lymphocytes known as memory cells, produced during the primary response in both B-cell and T-cell responses. When an antigen is encountered again, memory cells initiate the secondary immune response, which is quicker and more powerful than the primary response. Immunological memory is long-lived and is the basis for vaccination.

Although most of these memory cells are at rest, a small percentage of them are dividing at all times. The stimulus that causes these resting memory cells to divide remains unclear, but it is known that interleukins help to maintain the memory T-cell population.

After either infection or immunization of a host, the number of host T cells reactive to a given antigen markedly increases as effector T cells are produced. As time goes by and the infection subsides, the number of T cells decreases to a persistent level. This level is 100–1000 times higher than the initial number before activation, with the increase being due to memory T cells. These memory cells are more sensitive to re-stimulation by the same antigen than are naive T cells, and the memory cells produce cytokines more quickly and more vigorously.

As we saw in Figure 16.18, the secondary immune response of B cells produces different antibodies in a different time frame. The powerful IgG antibody is produced sooner and in larger quantity.

- A pathogen elicits an adaptive immune response.

- This stimulates the production of antibody and effector T-cell responses that eliminate the pathogen.

- When the infection is over, most activated effector T cells and antibody levels slowly decline.

- Memory T and B cells remain. These cells are able to initiate a response to a recurrence of infection with the same pathogen.

- This memory response occurs more quickly than the initial response and is more powerful.

Natural and Artificial Immunity

Adaptive immunity can be divided into naturally acquired immunity and artificially acquired immunity. Both of these categories can be further divided into active and passive types.

Naturally acquired immunity is, as the name implies, acquired as part of normal life. Naturally acquired active immunity results from contact with a pathogen, which usually occurs as infection, recovery, and immunity to subsequent infection because of immunological memory. Naturally acquired passive immunity results when maternal antibodies are transferred to the fetus or newborn across the placenta or through breastfeeding. Because the baby did not produce the immune products, the protection against infection is transient and does not produce lasting immunological memory.

Artificially acquired immunity occurs because of some type of medical treatment. Artificially acquired active immunity occurs when a person is given a vaccine. A person is exposed to antigens in a way that does not allow infection but does generate immunological memory and protection from future infection. Artificially acquired passive immunity involves the transfer of immune products, usually antibodies, from one person or animal to another. Again, because the immune products were not produced by that person, there is no immunological memory and the protection is transient.

Notice that in active immunity the patient encounters the antigen, produces an immune response, and then has lasting immunological memory for that antigen. In passive immunity, antibodies are passed from one person to another. The person receiving the antibodies does not produce the immune response that made them and therefore does not have lasting immunological memory. However, this can be very important in treating and preventing immediate infection.

Keep in Mind

- One of the most important properties of the adaptive immune response is the development of memory.

- The adaptive response causes memory to occur for both T and B cells.

- Memory causes a quicker and more powerful response to antigens that have been encountered before.

OVERALL IMMUNE RESPONSE

When we consider how the cells and molecules of the innate and adaptive responses work together, we can see the host defense as an integrated system that eliminates and controls an infectious agent and also provides long-lasting protection. The course of an infection can be broken down into stages as shown in **Figure 16.20**. The innate response is the early response to infection and is a prerequisite for the adaptive response to occur. This is because antigen-specific lymphocytes of the adaptive immune response are activated by the cells of the innate response and the co-stimulatory molecules they produce. However, the interaction between the two responses is not entirely a one-way street: the adaptive response produces signals that stimulate some of the innate response systems (**Figure 16.21**).

Only when pathogens have established a site of infection in the host does disease occur. Most pathogens are repelled by innate immune mechanisms, but if they establish a stable site of infection and disease occurs, the adaptive system is called into action. Damage to the host will remain localized unless the invading pathogen is able either to spread from the original site of infection or to secrete toxins that can spread to other parts of the body. Normally, there is little damage to uninfected tissue during the response to the pathogen, but in some cases collateral

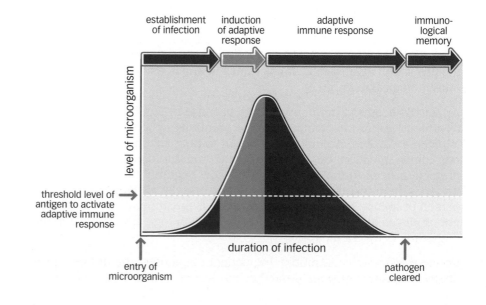

Figure 16.20 The course of a typical acute infection that is cleared by the adaptive immune response.

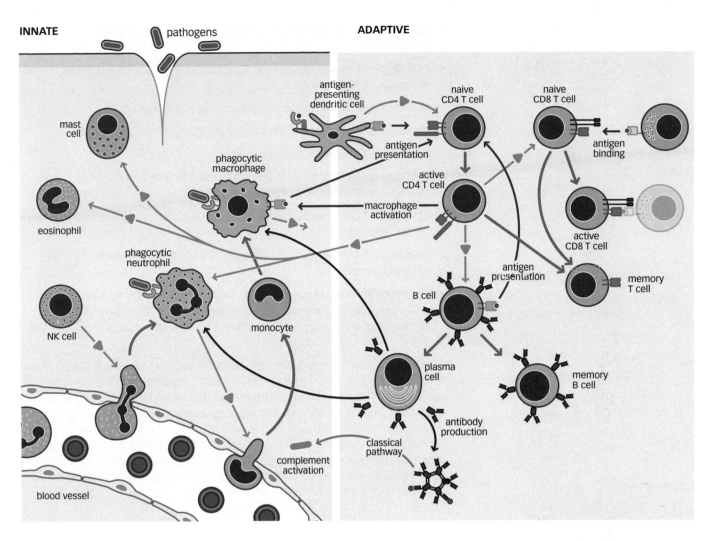

INNATE

pathogens

ADAPTIVE

mast cell

antigen-presenting dendritic cell

naive CD4 T cell

naive CD8 T cell

antigen presentation

antigen binding

phagocytic macrophage

eosinophil

active CD4 T cell

macrophage activation

active CD8 T cell

phagocytic neutrophil

antigen presentation

memory T cell

B cell

NK cell

monocyte

plasma cell

memory B cell

antibody production

blood vessel

complement activation

classical pathway

damage to uninfected tissue can be very serious. Furthermore, some pathogens such as *Mycobacterium tuberculosis* may never be completely cleared and persist in the host in a latent form.

Figure 16.21 Relationship between the innate immune response and the adaptive immune response during infection.

SUMMARY

- The adaptive immune response is connected to the innate immune response.

- The adaptive immune response is specific and involves both cellular and humoral responses.

- There are lymphoid structures strategically placed in major portals of entry.

- T cells and B cells are involved in the adaptive immune response.

- Both T cells and B cells have receptors for antigen.

- T cells mature in the thymus; B cells mature in the bone marrow.

- Clonal selection and deletion are processes that allow some lymphocytes to mature while others are deleted from the body.

- T cells are initially naive and become activated effector cells after encountering their specific antigen.

- There are two classes of MHC molecule, class I (MHC I) and class II (MHC II), that can bind and display antigens.

- Antigen presentation involves combining antigen with MHC I and MHC II molecules.

- CD4 helper T cells recognize MHC II molecules, whereas CD8 cytotoxic T cells recognize MHC I molecules.

- B cells differentiate into plasma cells that produce antibody.

- There are five types of antibody molecule: IgG, IgM, IgA, IgD, and IgE.

- T cells direct the production of antibody.

- CD8 cytotoxic T cells kill specifically identified targets and remember them through the development of memory cells.

- CD4 helper T cells can be divided into several groups, each with a different helper function.

- Regulatory T cells function to suppress the immune response when the antigen has gone.

- The adaptive immune response can be divided into a primary response and a secondary response; the secondary phase is the responsibility of immunological memory and is quicker and more powerful than the primary phase.

In this chapter we have explored the adaptive immune response, noting that it is intimately connected to the innate immune response examined in the previous chapter. The adaptive response is essential when infections overwhelm or evade the capacity of the innate response to deal with them, and get out of hand. This response is specific and results in the development of immunological memory. It is among the most potent and lethal responses in nature and therefore it can cause severe problems if it goes wrong, over-reacts or, worse, is not available at all. In the next chapter, we look at what happens when the immune response is missing or functions incorrectly.

Ⓠ SELF EVALUATION AND CHAPTER CONFIDENCE

Multiple Choice

Answers are given in the back of the book and help can be found in the student resources at:
www.garlandscience.com/micro2

1. The antibody response is part of the

   A. Humoral response
   B. Cellular response
   C. Phagocytic response
   D. Propagation response
   E. None of the above

2. Specificity is seen in each of the following except

   A. The humoral response
   B. The cellular response
   C. The adaptive immune response
   D. The innate response
   E. Antigen–antibody complexes

3. The adaptive response relies upon distinguishing

   A. Complete from incomplete antigens
   B. Proteins from lipid antigens
   C. Carbohydrates from protein antigens
   D. Self from non-self antigens
   E. Carbohydrates from lipids

4. Presentation of antigen is done by

   A. Macrophages
   B. Dendritic cells
   C. Monocytes
   D. **A** and **B**
   E. All of the above

5. Antibody is produced by
   A. T cells
   B. B cells
   C. Plasma cells
   D. Macrophages
   E. Dendritic cells

6. T cells mature in the
   A. Bone marrow
   B. Liver
   C. Lymph nodes
   D. Thymus
   E. Thyroid

7. B cells mature in the
   A. Bone marrow
   B. Liver
   C. Lymph nodes
   D. Thymus
   E. Thyroid

8. Two of the classes of T cells are called
   A. Antigen-presenting and suppressor
   B. Suppressor and killer
   C. Cytotoxic and helper
   D. Suppressor and cytotoxic
   E. None of these pairs is correct

9. M cells are found in all of the following except
   A. The intestines
   B. Peyer's patches
   C. The GALT
   D. The MALT
   E. Lymph nodes

10. A mature antigen-presenting cell
    A. Is older than other cells
    B. Has recognized antigens before
    C. Is able to phagocytose proteins
    D. Has recognized and processed antigens before
    E. Is able to recognize antigens better than an immature cell

11. Clonal selection involves all of the following except
    A. Rearrangement of gene segments
    B. Specific antigen receptors
    C. Reversible genetic rearrangement
    D. Irreversible genetic rearrangement
    E. Passage of genetic rearrangement to progeny

12. After antigen presentation, T cells in the lymph nodes will
    A. Immediately leave the node
    B. Become activated and leave the node
    C. Become activated and remain in the node
    D. Become inactivated and remain in the node
    E. Become inactivated and leave the node

13. Anergy is
    A. Lymphocyte activation by B cells
    B. Lymphocyte inactivation by B cells
    C. Lymphocyte inactivation due to increased co-stimulatory signals
    D. Lymphocyte inactivation due to lack of co-stimulatory signals
    E. Another name for apoptosis

14. The thymus
    A. Grows larger when puberty is reached
    B. Becomes populated with dendritic cells when puberty is reached
    C. Atrophies when puberty is reached
    D. Becomes filled with activated T cells when puberty is reached
    E. None of the above

15. T cells are found in which part of the lymph node?
    A. Stromal
    B. Follicles
    C. Paracortical areas
    D. Capsular areas
    E. Both **A** and **B**

16. T cells that react with self-antigen–MHC have what effect?
    A. Maintenance of the T-cell population
    B. Apoptosis of T cells
    C. Activation of T cells
    D. Inactivation of T cells
    E. None of the above

17. Which of these statements is not true of the B-cell receptor?
    A. Its structure is Y-shaped
    B. It is an immunoglobulin
    C. It has two antigen-binding sites
    D. It can only recognize antigen bound to MHC molecules
    E. It has constant and variable regions

18. The T-cell receptor is specific for
    A. Large antigens
    B. Whole microbes linked to MHC
    C. MHC only
    D. Fragments of antigen linked to MHC
    E. None of the above

19. Class I MHC presents antigen to
    A. Phagocytic cells
    B. B cells
    C. Helper T cells
    D. Cytotoxic T cells
    E. All of the above

20. Class II MHC presents antigen to

    A. Phagocytic cells
    B. B cells
    C. Helper T cells
    D. Cytotoxic T cells
    E. All of the above

21. Superantigens are

    A. Extra-large proteins
    B. Recognized by T cells after being bound to MHC molecules
    C. Recognized by T cells without being bound to MHC molecules
    D. Presented by special antigen-presenting cells
    E. Presented only by dendrites

22. Antibody molecules are bivalent because they have

    A. One binding site
    B. One attachment site for macrophages
    C. Two identical binding sites
    D. Two binding sites that recognize different antigens
    E. Four identical binding sites

23. Which antibody is produced first?

    A. IgA
    B. IgG
    C. IgD
    D. IgE
    E. IgM

24. T cells that have not been presented with antigen are referred to as

    A. Activated
    B. Effector
    C. Cytotoxic
    D. Primed
    E. Naive

25. Immunological memory

    A. Allows protection upon re-exposure to a previous pathogen
    B. Is associated with both T and B cells
    C. Is stimulated by interleukins
    D. Is more powerful than primary exposure to antigen
    E. All of the above

26. Vaccination is an example of

    A. Naturally acquired active immunity
    B. Naturally acquired passive immunity
    C. Artificially acquired active immunity
    D. Artificially acquired passive immunity

(Q) **DEPTH OF UNDERSTANDING**

Questions listed here require you to bring together the concepts you have learned in this chapter into a discussion format. This helps you to increase your depth of understanding of the material you have learned. Help can be found in the student resources at: www.garlandscience.com/micro2

1. Describe the events that lead up to the activation of helper T cells after a pathogen has broken the barrier of the skin.

2. Compare the maturation of T lymphocytes with that of B lymphocytes.

3. Describe the cellular events associated with the adaptive immune response.

(Q) **CLINICAL CORNER**

Help can be found in the student resources at: www.garlandscience.com/micro2

1. Your patient is a 33-year-old previously healthy
   American male who presents to the ER with severe
   and sudden fluid loss and a decrease in blood pressure.
   Fortunately he recovers and wants to know what
   caused his illness. He is surprised to learn that it was
   the result of an infection and even more surprised to
   learn that it was his immune response that caused the
   problem.

   A. What sort of infection could have caused this
   reaction?

   B. What will you explain about his immune response?

Failures of the Immune Response

Chapter 17

Why Is This Important?

In this chapter we will see how the host defense can be either inhibited or lost. When either of these things happens, we no longer have the protection we need to survive the fight against pathogenic organisms.

Julie is a high-school student who is learning to play the flute in her high-school band. After renting a flute for several months, her parents decide to purchase a nickel-plated flute for her birthday. Julie enjoys her new flute but after a few days of practice breaks out in a rash around her mouth. The rash is distressing and does not seem to get better, even with medication. She sees her family doctor and he suspects an allergic reaction. He explains that nickel is a small ion that easily penetrates the skin and can serve as a hapten. Haptens react with self-proteins, creating complexes that can be presented as foreign antigens to the immune system. The reactions require a first exposure to produce memory T cells and then subsequent exposures that cause the inflammation and tissue damage. This abnormal immune response is a type of hypersensitivity or allergy. Many people are allergic to nickel, and Julie is having a reaction to the nickel-plated flute. Her parents return the nickel-plated flute and exchange it for a silver-plated flute. Julie's rash immediately clears and she enjoys four years of playing with her high-school band.

In this chapter we will learn about different types of hypersensitivity as well as other failures of the immune response.

OVERVIEW

In the preceding two chapters, we saw that the host defense relies on two mechanisms to survive against relentless pathogenic assault. The innate (non-specific) immune response protects against most of these infections, and the adaptive (specific) immune response is mobilized against pathogens that persist despite the innate defenses. Recall that both of these systems are elegantly designed to deal with keeping the body safe and that a host with a deficient immune system (an immunocompromised host) is more susceptible to infection. In this chapter, we first examine some of the ways in which host defense can be lost: infection with HIV, infection by other pathogens, and genetic abnormalities. In the second part of the chapter, we look at what happens when the immune system turns on the host in the form of autoimmunity and hypersensitivity reactions.

We will divide our discussions into the following topics:

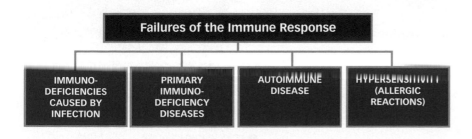

IMMUNODEFICIENCIES CAUSED BY INFECTION

"Pneumocystis Pneumonia—Los Angeles. In the period October 1980 – May 1981, 5 young men, all active homosexuals, were treated for biopsy-confirmed Pneumocystis carinii *pneumonia at three different hospitals in Los Angeles, California. Two of the patients died...."*

Gottlieb MS, Schancker HM, Tan PF et al. (1981)
CDC *Morbidity and Mortality Weekly Report* 30:250–252.

This posting in a CDC weekly mortality report was the first warning of what was to come. There was an editorial note included with this entry that also proved to be prophetic: *"Pneumocystis* pneumonia is almost exclusively limited to immunosuppression patients. ... The occurrence of this disease in these five previously healthy individuals is unusual. ... The fact that these patients are homosexuals suggests an association between some aspects of a homosexual lifestyle or disease acquired through sexual contact."

This editorial was written in 1981, and by 1992 Acquired Immune Deficiency Syndrome (AIDS), had become the leading cause of death in individuals 25–44 years of age in the United States. Recent estimates from the World Health Organization are that 20 million people have died from the epidemic and more than 40 million are currently infected worldwide. Although AIDS was initially identified in homosexual patients, it is by no means restricted to homosexuals.

Acquired Immune Deficiency Syndrome

AIDS is caused by infection with the human immunodeficiency virus (HIV). HIV is an enveloped retrovirus (**Figure 17.1**) that is transmitted through sexual contact or by contact with infected body fluids. Retroviruses contain the enzyme reverse transcriptase, which, as we saw in Chapter 12, allows it to copy its RNA genome into a DNA copy that can then integrate into a host chromosome. There are two major forms of this virus: HIV-1 is the predominant form seen throughout the world, and HIV-2 is the form seen mostly in Africa. HIV debilitates the immune response by attacking cells of the immune system.

Fast Fact Viral entry into target cells can also occur through endocytosis, when the virus is complexed with anti-HIV antibody. Therefore, even when the virus is being attacked by the adaptive immune response it can still infect.

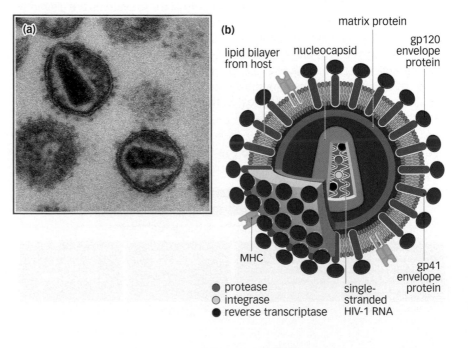

Figure 17.1 The virion of human immunodeficiency virus (HIV-1). Panel a: A transmission electron micrograph of HIV. Panel b: An illustration of the structures of the HIV virion. Notice that the reverse transcriptase, integrase, and viral protease enzymes are packaged in the virion. Although for clarity only one molecule of viral integrase and protease are shown, the virion actually contains many molecules of each enzyme. Note the nucleocapsid structure and envelope proteins gp120 and gp41, which are also found as part of this virus. Also shown (green) are MHC molecules acquired from the host during budding.

Fluid	Estimated quantity of virus (infectious particles per milliliter)
Cerebrospinal fluid	10–10,000
Intestinal mucosal secretions	1–5000
Plasma	1–5000
Semen	10–50
Ear secretions	5–10
Breast milk	<1
Saliva	<1
Tears	<1
Urine	<1
Vaginal–cervical fluid	<1
Feces	None detected
Sweat	None detected

Table 17.1 The estimated quantity of HIV particles in body fluids.

Cellular Targets of HIV

HIV uses a complex of two glycoproteins—**gp120** and **gp41**—in the viral envelope to bind to the CD4 molecule on helper (CD4) T cells, dendritic cells, and macrophages. For the infection to be successful, HIV must also bind to co-receptors, two of which have been identified. As with CD4, they are normal cellular proteins. One is the chemokine receptor known as CXCR4, which usually binds to the chemokine CXCL12. Binding by HIV causes virally infected cells to form syncytia (giant cells), which enable the virus to spread. Some variants of HIV use the co-receptor CCR5, which usually binds to the chemokines CCL3, CCL4, and CCL5. Several other co-receptors have been found in places such as the brain and the thymus, but they are not yet well characterized. People who have mutations in the genes that code for these receptors are less susceptible to infection with HIV.

Modes of HIV Transmission

HIV is transmitted through sexual contact and via the blood and body fluids. The efficiency of transfer is based on the concentration of viral particles in the fluid being exchanged (**Table 17.1**): the highest viral loads are found in peripheral blood monocytes, blood plasma, and cerebrospinal fluid. Semen and genital secretions are also sources of the virus.

Other sexually transmitted diseases can contribute to HIV transmission, probably because infected and ulcerated genital tissue permits direct access of the virus to the blood. Although sexual activity is still the leading mode of transmission, intravenous drug abuse is the next most common form (**Figure 17.2**). Children are also at risk because the virus can be transmitted across the placental barrier and can also be found in breast milk (although transmission rates through breastfeeding are low).

Except for direct access through an injection, a needlestick injury, or a blood transfusion, HIV enters the lamina propria when the mucosal epithelial cell layer is broken. Infection relies on sufficient viral numbers, and infected cells are the richest source of the virus. Spread of virus

Fast Fact Until recently, the transmission of HIV from infected mother to fetus was extremely high. However, the administration of drugs to infected mothers has drastically limited this mode of transmission.

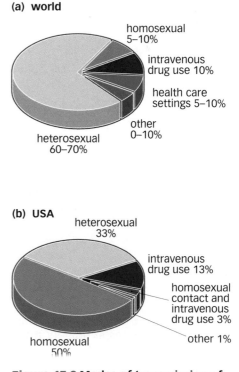

Figure 17.2 Modes of transmission of HIV. Panel a: Worldwide figures. Panel b: Figures for the United States.

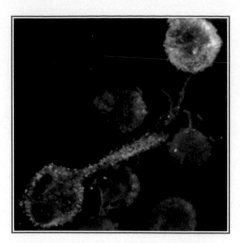

Figure 17.3 Lymphocytes infected with HIV. In this photomicrograph, two lymphocytes are already infected with HIV and producing virus (stained green). Two of the other lymphocytes (colored blue) are beginning to show signs of infection (the green dots).

from infected cells to uninfected cells can be observed in tissue culture (**Figure 17.3** and **Movie 17.1**). Cells of the mucosal immune system, including M cells in the bowel and dendritic antigen-presenting cells in the vagina and cervical epithelium, are prime targets for the initial infection because they function to pick up and process antigens. In fact, it has been shown that dendritic cells can bind to the HIV-1 envelope proteins with high affinity and hold them stably for several days until susceptible T cells come along and are infected (**Figure 17.4**). In sexually active males, cells lining the penis can become infected by contact with virus from infected macrophages or Langerhans cells in the cervix or intestinal mucosa of an infected partner.

HIV Genome

HIV has an RNA genome, and because it is a retrovirus, the RNA must be copied to DNA by reverse transcriptase, which is carried in HIV virions. The DNA copy is integrated into the host DNA using the viral enzyme integrase to form a provirus. The HIV genome comprises only nine genes flanked by long terminal repeat sequences (called LTRs), which are required for the virus to integrate into the host chromosome.

The sequence of cellular events associated with HIV infection is shown in **Figure 17.5** and **Movie 17.2**. Once the provirus has been integrated into a host chromosome, host cell machinery is used to replicate the virus. It is important to note that the reverse transcriptase that converts viral RNA to DNA has no proofreading ability, and therefore many mutations occur. These mutations help HIV develop rapid resistance to antiviral drugs and escape drug therapy.

Dynamics of HIV Replication in Infected Patients

The dynamics of HIV production during infection are truly astonishing. The minimum rate of release into the blood based on one cycle of infection per infected cell per day is an astronomical 10^{10} virions. When this high rate is coupled with the high mutation rate seen in this virus, it can be predicted that every possible mutation at every position in the viral genome can occur numerous times each day! It is estimated that the genetic diversity of HIV produced in a single infected person is greater than all the diversity that would be seen in a worldwide epidemic of influenza (influenza virus is one of the most genetically diverse viruses in the world).

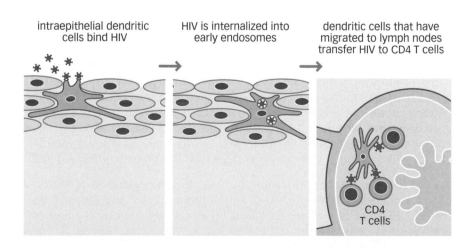

intraepithelial dendritic cells bind HIV

HIV is internalized into early endosomes

dendritic cells that have migrated to lymph nodes transfer HIV to CD4 T cells

CD4 T cells

Figure 17.4 Dendritic cells initiate the HIV infection by transporting HIV from mucosal surfaces to lymphoid tissue. The viral particles gain access to dendritic cells either at the site of mucosal injury or possibly on portions of the dendritic cell that protrude between the epithelial cells.

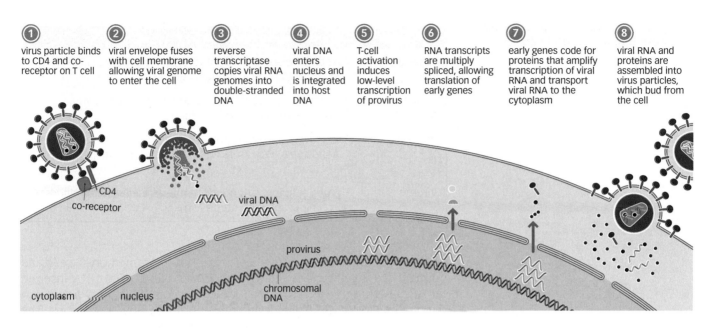

1	2	3	4	5	6	7	8
virus particle binds to CD4 and co-receptor on T cell	viral envelope fuses with cell membrane allowing viral genome to enter the cell	reverse transcriptase copies viral RNA genomes into double-stranded DNA	viral DNA enters nucleus and is integrated into host DNA	T-cell activation induces low-level transcription of provirus	RNA transcripts are multiply spliced, allowing translation of early genes	early genes code for proteins that amplify transcription of viral RNA and transport viral RNA to the cytoplasm	viral RNA and proteins are assembled into virus particles, which bud from the cell

Figure 17.5 The sequence of events associated with HIV infection.

Course of Infection

The pathology of HIV infection consists of three phases (**Figure 17.6**): the acute phase, the asymptomatic phase, and the symptomatic phase with the development of AIDS.

The **acute phase** begins in the first few days after the initial infection. During this time, virus is produced in large quantities by infected lympho-cytes in the lymph nodes, resulting in swollen lymph nodes (lymphade-nopathy) and flu-like symptoms. This initial viremia is greatly reduced by 10 weeks after initial contact. The decrease is most probably a result of the actions of cytotoxic (CD8) T cells, which kill infected targets. Support for this hypothesis is demonstrated by the increase in the number of cytotoxic T cells seen during this period. The helper (CD4) T-cell target population, which is initially decreased, rebounds by the end of this phase to near-normal numbers. During this phase, the viral population is homogeneous, indicating that little mutation is occurring.

The asymptomatic phase of HIV infection usually begins three to four months after the initial infection. By this point, the viremia is very low, with only occasional small bursts of virus being released. Many of the infected

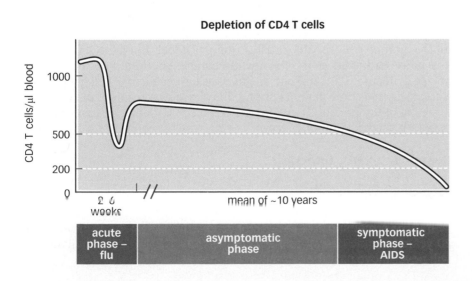

Depletion of CD4 T cells

Figure 17.6 Levels of helper (CD4) T cells during the three phases of an untreated HIV infection. Notice there is an initial decrease in the CD4 (helper) T-cell population in the acute phase of the infection as the patient experiences flu-like symptoms. This corresponds to the increase in the amount of HIV present in the blood. An adaptive immune response causes the helper T-cell counts to rebound. However, the virus has not been eliminated and becomes latent during the asymptomatic phase. Helper T-cell levels decline during this phase until they reach a level of 500 per microliter, when the patient enters the symptomatic phase. When the number drops to 200 per microliter, the patient is said to have AIDS. Death occurs as a result of opportunistic infections or cancers caused by the patient's lack of immunity.

Table 17.2 The types of opportunistic pathogen that infect AIDS patients.

Infections	Opportunistic pathogens
Parasites	*Toxoplasma* species, *Cryptosporidium* species, *Leishmania* species, *Microsporidium* species
Intracellular bacteria	*Mycobacterium tuberculosis, Mycobacterium avium intracellulare, Salmonella* species
Fungi	*Pneumocystis (carinii) jiroveci, Cryptococcus neoformans, Candida* species, *Histoplasma capsulatum, Coccidioides immitis*
Viruses	Herpes simplex, cytomegalovirus, varicella-zoster, Kaposi's sarcoma-associated herpesvirus

cells have been killed by cytotoxic T cells. However, memory T cells and dormant macrophages that contain an HIV provirus (viral DNA inserted into host DNA) serve as reservoirs for the infection. While its genome is in the provirus form, HIV is latent and does not make any new virions. The degree of active viremia seen during this phase is a predictor of how fast the disease will progress in an individual—the greater the viremia seen here, the faster the progression of the disease.

The population of helper T cells declines at a steady rate during the asymptomatic phase. There are three ways in which this may happen: direct killing of infected cells by the virus, increased induction of apoptosis in infected cells, and killing of infected helper T cells by cytotoxic T cells. This is a protracted phase that can last for years and is referred to as **clinical latency**, with virus replication continuing at a low rate, mainly in the lymph nodes. It is thought that cytotoxic T cells are the reason that the infection does not progress more rapidly during this time, but these cytotoxic T cells eventually begin to disappear. In addition, the virus population becomes more heterogeneous as this phase progresses, indicating that mutations are occurring with greater frequency.

The **symptomatic phase** is the end stage of the infection, and it is here that the infected individual develops the clinical symptoms of AIDS. The helper T-cell count drops below 500 per microliter of blood, and the cytotoxic T-cell number is also greatly decreased. When the helper T-cell count drops below 200 per microliter, the patient is said to have AIDS. Viral replication increases in the lymph nodes, and the architecture of the lymph nodes begins to deteriorate. During this phase there is increased susceptibility to opportunistic infections (Table 17.2) as well as generalized lymphadenopathy and the appearance of lesions. The most common lesions are thrush (a fungal infection of the oral cavity) and **hairy leukoplakia**, in which white patches are found on the tongue and oral mucosa. Cancers such as Kaposi's sarcoma and B-cell lymphoma also develop in some patients during this phase as a result of the reduced targeting of tumor cells by cytotoxic T cells. The patient eventually dies of opportunistic infection or cancer.

Response to HIV Infection

Without treatment, approximately 10% of HIV-infected individuals will progress to AIDS in the first three years after the initial infection. More than 80% of infected individuals show signs of clinical disease within 10 years. The remaining 20% are free of disease for long periods (more than 20 years in some cases), and a small percentage of infected individuals will never move past the asymptomatic phase.

The virus replicates best in activated T cells. Because infection activates T cells, the greater the number of opportunistic infections that occur in AIDS patients, the greater the number of T cells activated, and the more T cells that are activated, the larger the quantity of virus particles produced, leading to more infected cells, leading to more opportunistic infections, and so on.

The response to any viral infection is a combination of the humoral (B-cell) portion and the cellular (T-cell) portion of the adaptive response, and this is no different for HIV infection. The antibody response can occur as quickly as a few days after infection, but it is generally seen during the first few months. Antibodies against the gp120 and gp41 proteins of the viral envelope are produced in response to the infection but are unable to clear it. There is some evidence that the antigenic sites on the viral envelope are heavily glycosylated and are therefore inaccessible to the antibodies. Antibody against the virus is secreted into the blood and can be detected in genital and other secretions. IgG is the dominant form of antibody in any HIV infection. This antibody is also involved in neutralizing the virus, in blocking viral receptors, and in complement-mediated reactions.

It has been shown that infected individuals all have a cellular adaptive response to HIV infection, a response in which cytotoxic T cells are programmed to recognize essentially all HIV-1 proteins. This makes sense because there is a good correlation between the cellular response, low viral load, and slower disease progression. The obvious question, therefore, is why the cellular adaptive response does not control and defeat the infection. There is as yet no clear-cut answer to this question.

Major Tissue Effects of HIV Infection

Most HIV is trapped in the germinal centers of lymph nodes, which consist of clusters of B cells, T cells, and follicular dendritic cells. Early in the asymptomatic phase of the infection, the patient develops lymphadenopathy as a result of the rapid growth and proliferation of follicle cells (follicular hyperplasia). Later, in the intermediate stage of the symptomatic phase of the disease, the lymph nodes begin to deteriorate as a result of cell death. In the advanced stage of the symptomatic phase of disease, lymphoid tissue is almost completely destroyed.

About one-third of HIV patients are diagnosed with neurological disorders at some point during the infection. HIV enters the nervous system early in the infection but does not seem to infect neurons. Nearly two-thirds of all patients infected with HIV-1 get sub-acute encephalitis, referred to as **AIDS dementia**. Several opportunistic infections that result from the immunodeficiency associated with AIDS can be neuropathological.

Advanced disease shows damage to the gastrointestinal system manifested in the form of diarrhea and chronic malabsorption of nutrients from food, which results in significant weight loss.

HIV infects the lungs, heart, and kidneys as well as joint fluids. In addition, HIV infection is associated with cancer. The latter finding makes sense because the immune system is responsible for tumor surveillance, and when the immune system is disrupted, there can be an increased incidence of malignancies. HIV infection also leads to high levels of cytokine production, which might activate oncogenic viruses and promote angiogenesis (the formation of blood vessels that is required for tumor

Fast Fact The primary protective mechanism of antibody neutralization of virus is difficult for HIV because the epitopes of the virus are hidden.

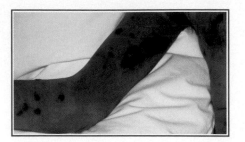

Figure 17.7 Kaposi's sarcoma in a young man infected with HIV-1.

growth). Two malignancies are prevalent in patients with AIDS: Kaposi's sarcoma and B-cell lymphoma.

Kaposi's sarcoma (Figure 17.7) is a tumor that contains many cell types. In immunocompetent individuals, this condition presents as a nonaggressive and rarely lethal malignancy, but in HIV-infected males the sarcoma is very aggressive, affecting both mucocutaneous and visceral areas of the body. Kaposi's sarcoma can occur in HIV-infected men but is seen less commonly in HIV-infected women.

The second form of cancer prevalent in patients with AIDS, **B-cell lymphoma** (a cancer of the B cells, affecting the blood and the lymphatic tissues), is 60–100 times more common in patients with AIDS than in the healthy population. It is found in many areas of the body, including the lymph nodes, intestinal tract, central nervous system, and liver.

Keep in Mind

- Infection with HIV can lead to acquired immune deficiency syndrome (AIDS).

- Helper (CD4) T cells are one target of HIV.

- HIV requires the presence of co-receptors on the T cell to infect the T cell successfully.

- HIV is transmitted sexually or through body fluids.

- HIV uses the enzyme reverse transcriptase to convert its RNA genome into DNA that can be integrated into the host chromosome.

- There are three phases of HIV infection: acute, asymptomatic, and symptomatic with the development of AIDS.

- High rates of mutation by HIV make the infection very hard to treat.

- Patients with AIDS die from secondary infections or cancer, which take hold because the immune system is severely disabled.

- HIV is not the only pathogen that can suppress the immune response.

Other Infections

Many pathogens other than HIV can suppress a host's immune response. One group capable of this suppression is the staphylococcal organisms that produce enterotoxins and toxic shock syndrome (Chapter 5). These toxins also cause the production of cytokines that suppress the immune response.

Some pathogens cause mild or transient immunodeficiency during acute infection. Although the exact mechanisms are not understood, this form of host suppression seems to affect dendritic cells, causing them to become unresponsive and thereby disabling the T-cell response through a lack of antigen presentation. This is important because the patient then becomes more susceptible to secondary infections, which are in many cases more dangerous than the primary infection. In one example, *Neisseria gonorrhoeae* is a pathogen that has been shown to avoid activation of the adaptive immune response by inducing dendritic cells to display proteins on their surface with immunosuppressive properties. Instead of activating T cells, these dendritic cells suppress the immune response. Patients with gonorrhea can be infected multiple times because efficient immunological memory is never produced to this organism.

PRIMARY IMMUNODEFICIENCY DISEASES

Up to now, we have studied immunodeficiency caused by infection. Now let's look at primary immunodeficiencies that result from genetic abnormalities present in the affected individual at birth. The diseases classified as **primary immunodeficiency diseases** are characterized by an immune system that does not function properly. They can occur when parts of the immune response are defective because of mutations in one or more of the genes responsible for coding for these parts. Primary immunodeficiency diseases present as overwhelming infections in young children. The type of infection that occurs can indicate the type of primary immunodeficiency causing the problem. For example, recurrent infections by **pyogenic (pus-forming) bacteria** points to a defect in antibody production, in the complement system, or in the phagocytosis mechanism. In contrast, recurrent fungal or viral infections suggest that the defect is in reactions mediated by T cells.

Inherited immunodeficiency diseases are caused by genetic defects and were first identified in 1952. These diseases can affect both the development and function of T cells and B cells (**Table 17.3**). There can also be defects in phagocytic cells, complement, or cytokine receptors, any of which will result in a lack of proper host defense.

B-Cell Defects

Defects in antibody production can result in severe and repeated infection with encapsulated bacteria. Recall from Chapter 15 that the bacterial capsule is a protective component that defeats phagocytosis. A host deals with the presence of encapsulated bacteria by producing antibody that reacts with components of the capsule. The result is opsonization of the pathogen, which facilitates its elimination. Without the ability to produce capsule-binding antibody, the host is unable to defeat encapsulated pathogens.

Table 17.3 Common primary immunodeficiency syndromes.

Name of deficiency	Specific abnormality	Immune defect	Susceptibility
Severe combined immunodeficiency syndrome (SCID)	X-linked SCID	No T cells	General infections
	Autosomal SCID A DNA repair defect	No T cells or B cells	General infections
DiGeorge syndrome	Thymic aplasia (the thymus does not develop)	Very low T-cell numbers	General infections
MHC class II deficiency	Lack of expression of MHC class II	No CD4 T cells	General infections
Wiskott–Aldrich syndrome	X-linked defective gene	Impaired T-cell activation, defective anti-polysaccharide response	Encapsulated extracellular bacterial infections
X-linked agammaglobulinemia	Loss of specific enzyme activity	No B cells	Bacterial and viral infections
Phagocyte deficiencies	Many different causes	Loss of phagocyte function	Bacterial and fungal infections
Complement deficiencies	Many different causes	Loss of specific complement components	Bacterial infections, especially *Neisseria* species
Ataxia telangiectasia	Defective DNA repair enzyme gene	Reduced numbers of T cells	Respiratory infections

The same is true of neutralizing antibody, which protects against attachment of viruses and also neutralizes toxins. Antibodies can bind to viruses and prevent the virus from binding to the host cell, thus preventing the infection. Defects in antiviral antibody production can result in more viral infections. If a host has a B-cell defect that prevents the formation of neutralizing antibody, the toxins produced by pathogens can cause severe problems.

Many of the immunodeficiency diseases classified as B-cell defects have been identified, and most are due either to defects in the development of B cells or to defects in the activation of the humoral response. There is also a transient humoral deficiency in normal babies that occurs during the first 6–12 months of life. Newborn infants have the same level of antibody as their mother because of the transplacental transfer of maternal IgG to the fetus. However, these maternal molecules are eventually catabolized, and the level of antibody in the newborn infant decreases after birth until the infant can begin to produce its own antibody at about six months of age. Consequently, the infant has low levels of antibody from three months to one year of age and is more susceptible to infection during that time.

T-Cell Defects

Defects in T-cell function can result in **severe combined immunodeficiency syndrome** (**SCID**). In these cases there is no T-cell function and therefore no cellular adaptive immune response. In addition, T-cell defects eventually result in a lack of a humoral (B-cell) adaptive response because, as we saw in the previous chapter, the helper T cell is directly involved in regulating the production of antibody. Some cases also exhibit a class II MHC deficiency in which there is no T-cell development in the thymus. In rare cases there can be deficiency in class I MHC, resulting in chronic respiratory infections and skin ulcerations. For example, in **DiGeorge syndrome**, the thymic epithelium fails to develop properly, and therefore the thymic environment cannot support T-cell maturation. Defects in cytokine production or activation can also result in immunodeficiency disease.

The SCID known as **Wiskott–Aldrich syndrome** causes impaired T-cell function, reduced numbers of T cells, and a failure of B-cell response to encapsulated bacteria. In this disease, the defect is in the actin cytoskeleton of the cell, which is required for cooperation between T cells and B cells. In these cases, there is no T-cell cytotoxic response or T-cell help. Bone marrow transplants or gene therapies can be used to treat these types of immunodeficiency diseases. However, bone marrow transplants have the complication of possible rejection. Success in gene therapy began in the late 1990s, but the treatment was found to induce leukemia in some patients. Clinical trials are currently under way with a new form of gene therapy for some types of SCID; they are showing some success.

Defects in Accessory Cells

Primary immunodeficiency diseases can also be the result of defects in the host defenses other than B-cell and T-cell defects. For instance, defects in the complement cascade can lead to an increase in the occurrence of infectious disease (**Figure 17.8**). Defects in phagocytic cells can cause widespread bacterial infections. Any deficit in the number of phagocytic cells is associated with severe immunodeficiency. In fact, a total absence of neutrophils (neutropenia) is fatal. For inherited neutropenia, the only effective treatment is a successful bone marrow transplant.

Some immunodeficiencies occur because of genetic defects in the production of the adhesion molecules required for the margination (rolling

Fast Fact In people with B-cell defects, infections are usually treated with antibiotics and also by infusion of immunoglobulin collected from large donor pools.

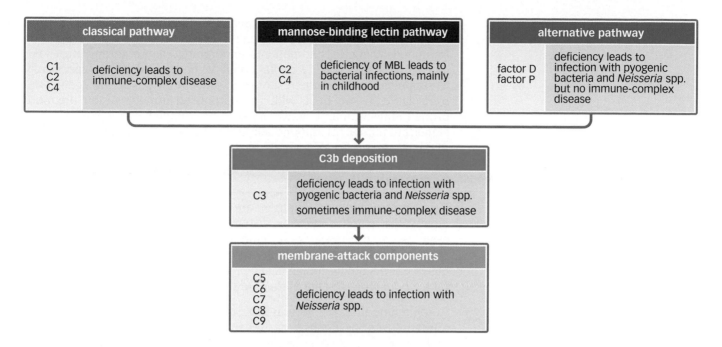

and stopping) of white blood cells in the blood vessels. However, most of the defects in white blood cells affect their ability to kill intracellular bacteria or ingest extracellular bacteria (Table 17.4). Among the best-known of these diseases is **chronic granulomatous disease**, in which the white blood cells cannot produce the superoxide radical, so that their antibacterial activity is seriously impaired. Another immunodeficiency disease involving white blood cells is **Chediak–Higashi syndrome**, which involves a defect in the gene encoding the protein involved in the formation of intracellular vesicles. This problem results in a failure to fuse phagosomes with lysosomes.

Figure 17.8 A summary of primary immunodeficiency diseases resulting from defects in components of the complement system. Deficiencies in the classical pathway lead to immune complex disease. Defects in the mannose-binding lectin (MBL) pathway are associated with increased bacterial infections in early childhood. Deficiencies in the early components of the alternative pathway lead to increased infection with pyogenic pathogens and *Neisseria* species. Perhaps the most problematic deficiencies are associated with defects in C3, which cause inhibition of the formation of the membrane attack complexes and essentially shut down the complement pathway.

Keep in Mind

- Primary immunodeficiency can be caused by mutations in immune response genes.
- Primary immunodeficiency is usually identified by overwhelming infections in young children.
- Inherited immunodeficiency is caused by recessive gene defects.
- Defects in B cells can result in severe and repeated infections with encapsulated bacteria.
- Defects in T-cell function result in severe combined immunodeficiency syndrome (SCID).
- Primary immunodeficiency can also be the result of defects in host defense other than T cells or B cells.

Type of defect/syndrome	Associated infections
Leukocyte adhesion deficiency	Widespread pyogenic bacterial infections
Chronic granulomatous disease	Intracellular and extracellular infections
Myeloperoxidase deficiency	Defective intracellular killing of pathogens
Chediak–Higashi syndrome	Intracellular and extracellular infection

Table 17.4 Defects in phagocytic cells associated with persistent bacterial infections.

As we saw in Chapter 6, host defenses can be compromised in other ways. Drug treatments such as those given for cancer can cause dangerous immunodeficiencies, and patients undergoing chemotherapy must take care to avoid infection. Transplant patients have to take drugs specifically to suppress their immune response to prevent it from acting against the transplanted tissue. Hosts with a compromised immune system are more prone to infection and, as we discussed in Chapter 8, the presence of a significant number of immunodeficient individuals is a contributory factor in the re-emergence of several infectious diseases.

AUTOIMMUNE DISEASE

Recall from Chapter 16 that there are several mechanisms that safeguard against the adaptive immune system reacting against the body. If these systems fail, the resulting autoimmune reactions can be a devastating problem because of the lethality of immune defense. Autoimmune diseases are becoming a significant problem in health care, so we'll now take a brief look at some of them and their causes.

Autoimmune diseases vary in severity and in the tissue affected. Some affect only specific organs or tissues and the reaction against self is confined to that area. A good example is **type 1 insulin-dependent diabetes mellitus**, which is caused by an autoimmune response to insulin-producing cells in the pancreas. Other examples are **Hashimoto's thyroiditis** and **Graves' disease**, both of which attack the thyroid gland. **Rheumatoid arthritis** attacks joints, and **multiple sclerosis** attacks the nervous system. Systemic autoimmunity affects multiple organs, for example **systemic lupus erythematosus** (**SLE**), in which ubiquitous anti-chromatin antibodies attack the skin, kidneys, and brain.

Causes of Autoimmunity

We still do not know the exact events involved in initiating autoimmune reactions, but some triggers are apparent. It is clear that some people are genetically predisposed to autoimmune disease. There is also some preliminary evidence that autoimmunity can be influenced by environmental factors. In addition, drugs and toxins can also cause autoimmunity. In these cases, it is thought that drugs or toxins react with self-antigens in such a way that they form derivatives that are perceived as foreign by the immune system.

Infection can also result in an autoimmune reaction. For example, **rheumatic fever** is an autoimmune disease triggered by infection. In this disease, antibodies against streptococcal species cross-react with host structures that are similar to antigens found on the bacteria. Inflammation from an infection leads to tissue destruction, and self-reactive lymphocytes can become activated as the level of tissue destruction rises. In addition, some pathogens can upset the body's regulation of the immune response either by preventing apoptosis or by secreting their own cytokines. Molecular mimicry can occur when pathogens express proteins or carbohydrate antigens that resemble host molecules, and this can cause autoimmunity. In most of these cases, the autoimmune response is only transitory: when the infection is over, the autoimmunity ends. However, in some individuals, the disease can become chronic.

Autoimmunity results in chronic disease when it is impossible to rid the host of the self-antigen that triggered the response. This results in a vicious cycle in which the disease destroys host tissue containing the self-antigen, which frees more self-antigens into the circulation, which

Fast Fact Procainamide, a medication used to treat cardiac arrhythmias, can induce the production of autoantibodies resembling those present in systemic lupus erythematosus.

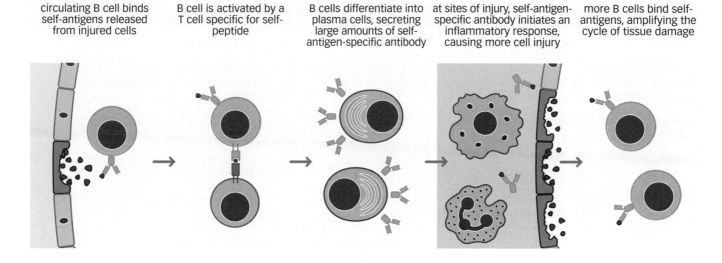

circulating B cell binds self-antigens released from injured cells

B cell is activated by a T cell specific for self-peptide

B cells differentiate into plasma cells, secreting large amounts of self-antigen-specific antibody

at sites of injury, self-antigen-specific antibody initiates an inflammatory response, causing more cell injury

more B cells bind self-antigens, amplifying the cycle of tissue damage

are reacted against, producing more antibody, which destroys more tissue, and so on (**Figure 17.9**).

Mechanisms of Autoimmunity

All parts of the immune system can be involved in autoimmune reactions. The production of autoantibodies (antibodies that react against the host) by B cells requires T-cell help. Autoantibodies are the damaging factor in diseases such as SLE and **myasthenia gravis**. SLE involves autoantibodies against DNA when it is in the form of chromatin in a cell that is not undergoing mitosis, and autoantibodies in myasthenia gravis react against the receptors for the neurotransmitter acetylcholine, a reaction that results in muscle weakness. Cytotoxic T cells can directly damage tissues, as happens in insulin-dependent diabetes mellitus (**Figure 17.10**), rheumatoid arthritis, and multiple sclerosis.

Even though T cells and B cells that react strongly against self are eliminated in the thymus and bone marrow respectively, not all self-antigens are expressed in these two locations when the T cells and B cells are developing. There are antigens associated with the body's immunologically protected sites—brain, eyes, testes, and uterus—that do not induce an immune response but can be the target of autoimmune attack. For example, myelin basic protein from the brain is involved in the autoimmune reaction that causes multiple sclerosis.

Keep in Mind

- In autoimmune reactions, the immune system reacts against the host.

- Tolerance to self-antigens normally prevents the development of autoimmune disease.

- Many autoimmune diseases are the result of production of autoantibodies, but multiple parts of the immune response are usually involved.

- Specialized regulatory T cells (T_{reg} cells) control the development of an autoimmune response.

- Some people are genetically predisposed to autoimmune disease.

- Autoimmunity can be organ-specific, such as type 1 diabetes mellitus, which affects cells in the pancreas, or systemic, affecting many organs.

Figure 17.9 How autoimmune-mediated inflammation leads to the release of self-antigens from damaged tissues, a release that further activates the autoimmune response to self.

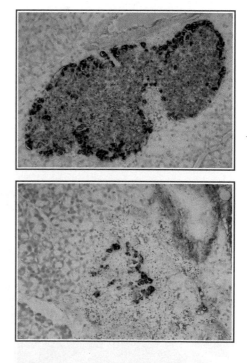

Figure 17.10 Tissue destruction in insulin-dependent diabetes mellitus, an autoimmune disease. Pancreatic cells have been stained to make insulin molecules visible (dark brown). Top: Cells from a healthy person; the many dark brown spots indicate normal levels of insulin-producing cells. Bottom: Cells from a diabetic patient. Notice the absence of dark brown, indicating a loss of insulin-producing cells.

Syndrome	Common allergens	Route of entry	Response
Systemic anaphylaxis	Drugs, serum, venoms, peanuts	Intravenous directly or after oral absorption	Edema, increased vascular permeability, tracheal occlusion, circulatory collapse, death
Acute urticaria (wheal-and-flare)	Animal hair, insect bites, allergy testing	Through the skin	Local increases in blood flow and vascular permeability
Allergic rhinitis (hay fever)	Pollens, dust-mite feces	Inhalation	Edema and irritation of the nasal mucosa
Asthma	Dander (cat), pollens, dust-mite feces	Inhalation	Bronchial constriction, airway inflammation, increased production of mucus
Food allergy	Tree nuts, peanuts, shellfish, milk, eggs, fish	Oral, through the skin, inhalation	Vomiting, diarrhea, itching, hives, anaphylaxis (rare)

Table 17.5 Examples of IgE-mediated allergic reactions.

HYPERSENSITIVITY (ALLERGIC REACTIONS)

An **allergic reaction** to some substance such as pollen, peanuts, dust-mite feces, or animal dander is a form of immune response that, although usually not life-threatening, can sometimes produce serious tissue injury and, in asthma and **anaphylaxis**, even death. Allergies can cause discomfort and distress for the patient as well as lost time from work or school. Allergic responses are the result of a person's becoming hypersensitized to a particular substance. In fact, the terms **hypersensitivity** and allergy mean the same thing. There are four main types of hypersensitivity reaction: type I, type II, type III, and type IV.

Type I hypersensitivities (Table 17.5) occur when the immunoglobulin IgE responds to antigens called **allergens**. IgE antibodies are produced by plasma cells located in lymph nodes that drain the site of allergen entry, and also by plasma cells in inflamed tissue. IgE activates mast cells and basophils located in tissues exposed to these allergens.

Most human allergies are caused by inhaled small, water-soluble proteins carried on dry particles such as pollen grains or dust-mite feces (Figure 17.11). Once it comes in contact with the mucosa of the airways, the allergen is eluted from the carrier particle and diffuses into the mucosa, which allows presentation at a low dose.

It has been estimated that as many as 40% of people in the Western United States show a tendency to mount IgE responses to environmental allergens. This hypersensitivity to environmental allergens is referred to as **atopy**, and these individuals are called atopic. They have higher levels of eosinophils than normal and are more susceptible to such allergic diseases as hay fever, rhinitis, and asthma. There is some evidence that this susceptibility may be genetic. The prevalence of atopic allergies in general, and of asthma in particular, is explained by four environmental factors that predispose a person to atopy:

- Lack of exposure to infectious diseases and allergens in early childhood
- Environmental pollution
- Allergen levels
- Dietary changes

These predisposing factors led to the formulation of what is called the **hygiene hypothesis**, which suggests that a less hygienic environment

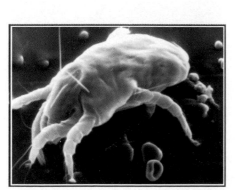

Figure 17.11 Scanning electron micrograph of *Dermatophagoides pteronyssimus* (dust mite) with some of its fecal pellets.

can help protect against atopic allergies. For instance, early population of the gut by commensal bacteria such as lactobacilli or infection by gut pathogens is associated with a reduced prevalence of atopic allergies. In contrast, there is some evidence that children who have had attacks of bronchiolitis associated with viral infection are more prone to later development of asthma.

Type II hypersensitivity reactions involve antibodies that attach to host cells. One example of a type II hypersensitivity is acute hemolytic anemia, in which the antibodies bind to red blood cells and mark them for destruction. Type II hypersensitivities can also be induced by certain drugs. In this case, the drug binds to the surface of cells and serves as a target for anti-drug IgG antibodies that then cause destruction of the cells. This occurs in only a small percentage of the population, and the reason that it occurs is still unclear.

Type III hypersensitivity occurs with soluble antigens that combine with antibodies to form immune complexes. These complexes bind to mast cells and other leukocytes and create a localized inflammatory response with increased vascular permeability at the site. This is called an **Arthus reaction** if the antigen enters the body through a subcutaneous route. Type III reactions are also seen systemically in cases of serum sickness. Serum sickness occurs if large amounts of foreign antigen are injected into the host. Examples of this problem are seen when horse serum containing antibody against a pathogen is used as a treatment for infection. This type of hypersensitivity can also be seen when antigen is not effectively cleared by the host's immune response. For example, in subacute bacterial endocarditis or chronic viral hepatitis, pathogens are constantly generating new antigens and the adaptive immune response fails to clear them completely from the system. This causes immune complex disease, with injury to small blood vessels in many tissues and organs. Antigen entering the body through inhalation can also trigger a type III hypersensitivity reaction. For example, the disease farmer's lung is caused by the inhalation of particles from moldy hay or other moldy crops.

Type IV hypersensitivity reactions are not mediated by antibody but instead result from the reactions of antigen-specific T cells. When stimulated by antigen, these T cells release chemokines and cytokines that result in inflammation and tissue damage. Type IV hypersensitivities are often called delayed hypersensitivities because they commonly take days to develop. A common type IV hypersensitivity occurs in people who are allergic to nickel, a common metal used in cheap jewelry.

Effector Mechanisms in Allergic Reactions

Most IgE is found on the surface of mast cells, basophils, and activated eosinophils and is bound there by high-affinity receptors. Mast cells are highly specialized, prominent residents of mucosal and epithelial tissues. They are located in the vicinity of small blood vessels, where they guard against invading pathogens and continuously express the IgE receptor on their surface. When a person previously sensitized to an allergen is exposed to it again, the allergen cross-links with an IgE molecule bound to a receptor on a mast cell. This causes degranulation of the mast cell, and within seconds a variety of inflammatory mediators are released.

Some of the mediators cause immediate increases in local blood flow, vessel permeability, and enzyme production leading to tissue destruction. Others contribute to smooth muscle contraction, increased secretion of mucus, and the influx and activation of leukocytes.

Fast Fact Both eosinophils and basophils cause inflammation and tissue damage in allergic reactions and induce mast-cell degranulation.

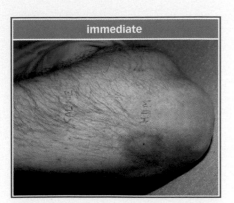

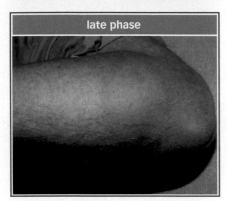

Figure 17.12 Early and late phases of an allergic reaction. The immediate wheal-and-flare reaction (left) is seen within a minute or two after antigen is injected under the skin and lasts for up to 30 minutes. The late-phase edematous reaction (right) develops approximately eight hours later.

Eosinophils are also part of the allergic reaction. They are found mainly in the connective tissue lying immediately under the respiratory, digestive, and genitourinary epithelia. Eosinophils have two functions in allergic reactions. They release highly toxic granules and free radicals that kill microorganisms and cause significant tissue damage, and they produce chemical mediators—prostaglandins, leukotrienes, and cytokines—that enhance and amplify the allergic reaction. Because of their destructive potential, eosinophils are strictly regulated both at the level of origination in the bone marrow (few are produced unless there is an infection) and at the site of infection (the receptor for IgE is never present until the eosinophil has been activated).

Phases of Allergic Reactions

Allergic reactions can be divided into immediate and late-phase responses. IgE-mediated mast-cell activation is an example of the immediate reaction that occurs within seconds after a person has been exposed to an allergen. In contrast, the late-phase reactions can take 8–12 hours to develop. The immediate reaction is due to the release of histamine and preformed mediators. Late-phase reactions are due to elaboration of the mediators (such as leukotrienes, chemokines, and cytokines) associated with smooth muscle contraction, sustained edema, and tissue remodeling. As an example, the wheal-and-flare reaction seen in positive skin tests for allergies has an immediate phase that is seen within a minute and lasts for about 30 minutes, whereas the late-phase reaction occurs about eight hours later in the form of a widespread edematous reaction that can persist for hours (**Figure 17.12**).

Clinical Effects of Allergic Reactions

Symptoms of allergic reactions range from the sniffles seen in hay fever to the life-threatening circulatory collapse seen in systemic anaphylaxis (**Figure 17.13**).

The clinical effects of allergic reactions vary according to the site of mast-cell activation. These clinical effects depend on three variables: (1) the

Figure 17.13 Schematic representation of the effects of mast-cell activation on different tissues during a hypersensitivity reaction.

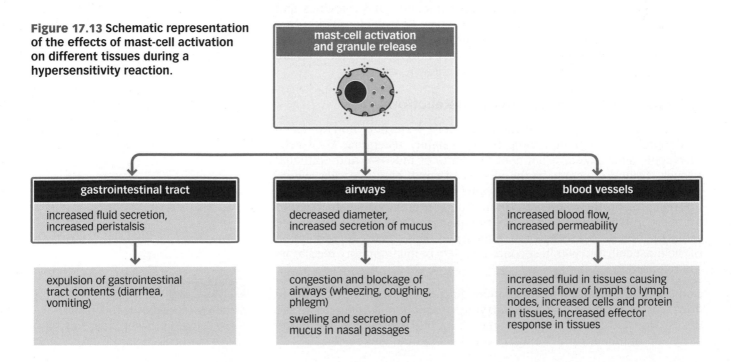

amount of allergen-specific IgE present on the mast cells, (2) the route by which the allergen is introduced into the body, and (3) how much allergen enters the body. If allergen is either deposited directly into the blood or rapidly absorbed from the gut, connective-tissue mast cells associated with the blood vessels become activated, and the result is systemic anaphylaxis. This is a potentially fatal reaction caused by a widespread increase in vascular permeability, leading to a catastrophic loss of blood pressure, airway constriction causing breathing difficulties, and swelling of the epiglottis, which can cause suffocation. This is referred to as **anaphylactic shock** and can be seen in people who are allergic to substances such as penicillin and peanuts. An immediate injection of epinephrine (adrenaline) is required to control this reaction.

Allergic rhinitis (hay fever) is characterized by intense itching, sneezing, and a local edema that leads to blocked nasal passages and nasal discharge. A similar reaction is seen in **allergic conjunctivitis**, the condition that results when allergens enter the eye. In contrast, **allergic asthma**, which is more serious, is triggered by the activation of submucosal mast cells in lower airways. This leads to immediate bronchial constriction (in seconds) and an increased production of fluids and mucus, which makes breathing difficult and can be life-threatening. Asthma is due to a chronic inflammation of the airways that results in the continued presence of increased numbers of T cells, neutrophils, and other white blood cells.

Keep in Mind

- An allergic response is a type of immune response.

- This response is generated against allergens (antigens) such as peanuts, pollen, dust-mite feces, or animal dander.

- Allergic responses are usually not life-threatening but can sometimes cause anaphylaxis, a condition that can be fatal.

- There are four types of hypersensitivity response.

SUMMARY

- The immune system can break down and in some cases fail.

- HIV infection and genetic mutation can cause immunodeficiencies.

- HIV attacks helper (CD4) T cells and without treatment effectively destroys the adaptive immune response.

- Genetic mutations cause primary immunodeficiency diseases.

- Individuals with compromised immune systems are more susceptible to infection by other pathogens, and it is these infections that lead to illness and death.

- Autoimmune disease occurs when the adaptive immune response attacks the host.

- Tolerance to self-antigens is a safeguard against the development of autoimmune disease.

- Autoimmunity can be organ-specific or systemic.

- An allergic response is a type of immune response against antigens that are referred to as allergens.

In this chapter, we have examined ways in which our immune system can be compromised. We have seen that this can result from infections with HIV and other pathogens, and from genetic diseases in which parts of the immune response do not develop. We also have looked at what can happen when the immune response mistakenly reacts against self in the form of autoimmunity, and the damage that can result from the defense mechanisms responsible for allergies. Even in view of the problems that can occur, we are universally better off with an immune system than we would be without one. Given the number of pathogens that attack us on a regular basis and require our host defenses to clear the infections, it is amazing how seldom we have to worry about the failures of our immune system.

(Q) SELF EVALUATION AND CHAPTER CONFIDENCE

Multiple Choice

Answers are given in the back of the book and help can be found in the student resources at:
www.garlandscience.com/micro2

1. All of the following are examples of failures of the immune response except

    A. Autoimmunity
    B. Primary immunodeficiency
    C. Acquired immune deficiency syndrome (AIDS)
    D. Inflammation

2. HIV makes its entry through

    A. The mucosal system
    B. A needle stick
    C. M cells of the digestive system
    D. All of the above

3. The number of T cells drops to below 200 per milliliter in which phase of the HIV infection?

    A. The acute phase
    B. The developmental phase
    C. The symptomatic phase
    D. The asymptomatic phase
    E. None of the above

4. The most frequent mode of HIV transmission in adults is

    A. Kissing
    B. Sexual transmission
    C. Cross-placental transfer
    D. Breast feeding
    E. Blood transfusion

5. The initial symptom of HIV infection is

    A. Loss of T cells
    B. Lymphadenopathy
    C. Appearance of thrush
    D. Significant weight loss
    E. High viral load

6. Lymphadenopathy and flu-like symptoms are characteristics of which phase of HIV infection?

    A. Acute
    B. Latent
    C. Asymptomatic
    D. Symptomatic

7. How many genes does HIV have?

    A. 7
    B. 8
    C. 9
    D. 10

8. Primary immunodeficiencies can affect which parts of the immune system?

    A. Antibody production
    B. Complement
    C. Cytotoxic T cells
    D. Phagocytes
    E. All of the above

9. Persistent fungal infections are indicative of a deficiency in which part of the immune system?

    A. Antibodies
    B. Phagocytic cells
    C. Complement
    D. Infrequent infection with encapsulated bacteria
    E. T cells

10. Which part of the immune system causes the damage in insulin-dependent diabetes mellitus?

    A. Cytotoxic T cells
    B. Autoantibodies
    C. Complement
    D. Cytokines

11. The following are organ specific autoimmune diseases except

    A. Graves' disease
    B. Multiple sclerosis
    C. Type-1 diabetes
    D. Systemic lupus erythematosus
    E. Rheumatoid arthritis

12. Autoimmunity is prevented by which of the following?

    A. The innate response
    B. Tolerance to self-antigens
    C. Amplification
    D. Antibody
    E. Phagocytic cells

13. Which of the following immunoglobulin molecules is associated with hypersensitivity and allergy?

    A. IgG
    B. IgA
    C. IgM
    D. IgE
    E. None of the above

14. Which cells are most prominently involved in hypersensitivity reactions?

    A. Mast cells
    B. Neutrophils
    C. Monocytes
    D. Eosinophils
    E. None of the above

 DEPTH OF UNDERSTANDING

Questions listed here require you to bring together the concepts you have learned in this chapter into a discussion format. This helps you to increase your depth of understanding of the material you have learned. Help can be found in the student resources at: www.garlandscience.com/micro2

1. Compare and contrast acquired immune deficiency syndrome and primary immune deficiency.

2. Discuss the role of T cells during the acute, asymptomatic, and symptomatic phases of HIV infection.

 CLINICAL CORNER

Help can be found in the student resources at: www.garlandscience.com/micro2

1. You are an emergency medical technician and are sitting in a restaurant with your family. The place is crowded and you can't help hearing the woman at the next table asking the waiter about whether there are peanuts in the dish she is ordering. The waiter assures her that there are none. You are about to finish and ask for your check when you notice the woman at the next table is choking and has a bluish tint to her skin. Immediately you sprint for your car and dig out an epinephrine syringe. The man sitting at the table with the woman is trying to lift her up saying that she needs the Heimlich maneuver. You push him out of the way and jab the needle into her thigh. Within minutes she is breathing more easily and her color is returning. The man is grateful but indignant that she would have been okay if he could have used the Heimlich manouvor.

    A. Explain to him why he is wrong.
    B. What do you think caused this to happen?
    C. How could you prove your suspicions?

2. Your patient, Richard Parks, is a 26-year-old white male who is in the clinic because of problems with lethargy and loss of vigor. He works in construction and lately has had no desire to even go to work. Although he eats several meals a day, he is losing weight and he is always thirsty. You have ordered blood work and it shows a high level of triglycerides and a glucose value of 135. You have to tell Mr. Parks that he has type I diabetes.

    A. Can you explain how this happened?
    B. What will you tell him about the immune response and its role in this disease?

In the first four parts of this book, we concentrated on the disease process, the characteristics of organisms that can infect us, and our internal defense against infection. Here in **Part V**, we turn our attention to how humans use antiseptics, disinfectants, and antimicrobial agents to control pathogens.

Chapter 18 describes how we use disinfectants to limit contamination of inanimate surfaces, and antiseptics to limit contamination of the skin. A variety of chemical treatments can be used for these purposes, and we look at the major categories of these chemicals as well as drawbacks that may be associated with their use.

In **Chapter 19**, we look at antibiotics and other antimicrobial drugs. This chapter is important because of our dependence on these chemical treatments for fighting infectious diseases. We progress from the discovery and development of antibiotics to a description of selective toxicity, an important concept in the use of therapeutic drugs. In addition, we examine different categories of antibiotics, their specific structures and targets, and some cutting-edge new antibiotics being developed. We then look at the therapeutic options for viral, fungal, and parasitic infections.

Chapter 20 is one of the most important chapters because it looks at antibiotic resistance, possibly the greatest challenge facing health care today. Here we examine the evolution of resistance and the mechanisms by which pathogens acquire it. We explore new targets and the development of new antimicrobials, and finish with the testing required to bring new products to market.

When you have finished **Part V** you will:

- Have a working knowledge of how the growth of potential pathogens can be inhibited through the use of antiseptics and disinfectants.

- Have learned about the targets used for antibiotic therapy and the different types of antibiotic.

- Know a variety of agents that are used in treating viral, fungal, and parasitic infections.

- Understand the concept of selective toxicity.

- Understand the problems associated with antibiotic resistance, how it occurs, and how to prevent it.

- Be familiar with the process of researching and testing new antibiotic drugs for human use.

Control of Microbial Growth With Disinfectants and Antiseptics

Chapter 18

Why Is This Important?

Infection control through the use of disinfectants and antiseptics is essential to keep infections from spreading, particularly in the hospital setting.

In 1942, British military scientists detonated anthrax-filled bombs on Gruinard Island, located less than a mile off the coast of Scotland, to study its effectiveness as a biological weapon. The island's only inhabitants, a flock of sheep placed there as test subjects, were exposed to aerosolized *Bacillus anthracis* endospores and died within days of exposure. The bodies of the sheep were incinerated, and the island remained uninhabitable for the next four decades.

Then, in 1981, anthrax-contaminated soil from the island was placed outside a UK military science facility, along with an anonymous demand to decontaminate the island. Over the next few years, more than 300 tons of formaldehyde solution was used to disinfect the entire island. After a flock of sheep placed on Gruinard Island remained healthy, the island was finally declared safe in 1990 and the 48-year quarantine was lifted.

OVERVIEW

As we saw in previous chapters, we humans live in constant contact with dangerous and potentially dangerous microorganisms. Because they can produce serious illnesses, it is important to know how we can defeat potential pathogens. In the chapters of Part IV we learned about our defense mechanisms. Here in Part V we will learn about methods we use to rid ourselves of infectious organisms. We begin here in Chapter 18 with a discussion of antiseptics, which are chemicals that we can use to prevent the growth of pathogens on our bodies, and disinfectants, which we commonly use to clean inanimate objects. We use disinfectants to keep down the level of microorganisms on those objects, thereby lowering the chances of our picking up some infection as we handle them. We will see that the targets for these chemicals are the same as those used for antibiotics, which will be discussed in Chapter 19.

We will divide our discussion into the following topics:

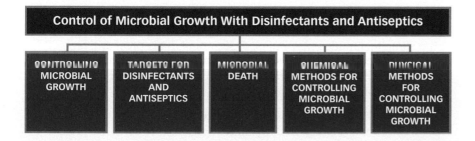

Control of Microbial Growth With Disinfectants and Antiseptics

CONTROLLING MICROBIAL GROWTH	TARGETS FOR DISINFECTANTS AND ANTISEPTICS	MICROBIAL DEATH	CHEMICAL METHODS FOR CONTROLLING MICROBIAL GROWTH	PHYSICAL METHODS FOR CONTROLLING MICROBIAL GROWTH

"Sunlight is said to be the best of disinfectants..."

Justice Louis Brandeis (1856–1941)

Although US Supreme Court Justice Brandeis was speaking about transparency in politics, business, and other societal organizations, his quote is actually quite applicable to our discussion of antimicrobial chemical and physical agents. Sunlight is a form of non-ionizing radiation and includes ultraviolet radiation, which is an effective form of disinfection. For millennia, people have preserved foods under direct sunlight, and, as we will see in this chapter, it is still used today as a form of antimicrobial treatment that kills bacteria, viruses, fungi, and protozoans.

CONTROLLING MICROBIAL GROWTH

The dramatic true story of disinfection on Gruinard Island reminds us of how difficult it can be to kill microbes, even with harshest chemical agents. To enable us to understand how disinfectants and antiseptics can control the growth of microbes, it is important to become familiar with some of the terminology used to describe these substances and their use. Keep in mind that some treatments can be used for both disinfection and antisepsis, with the difference being that disinfection is associated with inanimate objects and antisepsis is associated with human tissue and skin. The following definitions give a basic understanding of the processes used for control of microbial growth.

- **Disinfection** refers to the use of chemical or physical agents to kill microbes on inanimate objects. **Disinfectants** include chemicals, heat, or ultraviolet radiation, which kill or inhibit the growth of microorganisms, including pathogens. It is important to note that some disinfecting agents do not affect endospores or prions, so any organism that has the ability to form an endospore can be very difficult to destroy. Prions are highly resistant to most disinfectants. Some viruses are also unaffected by disinfectants.

- **Antisepsis** refers to the use of chemical or physical agents to kill microbes on the skin and in living tissue. As with disinfectants, endospores may survive **antiseptics**.

- **Sterilization** involves the destruction of all microbes, including bacteria, viruses, fungi, and spores. It should be noted that common sterilization techniques do not destroy prions, which are the infectious proteins we discussed in Chapter 8 that cause spongiform encephalopathies such as Creutzfeldt–Jakob disease.

- **Aseptic** is a term used to describe an environment or procedure that is free from contamination by pathogens. For example, surgeons and laboratory technicians use aseptic techniques to avoid contamination.

- **Degerming** is the removal of microbes from a surface by mechanical means. For example, the act of scrubbing when washing your hands is a degerming mechanism. Another example is the use of an alcohol pad to prepare the skin for injection.

- **Sanitization** refers to the disinfection of places (restrooms) or items (dishes) used by the public. Sanitization is not sterilization, although the same techniques may be used: steaming; high-pressure, high-temperature washing; and scrubbing. Sanitization is used to reduce the number of pathogenic organisms present to a number that meets accepted public health standards.

- **Pasteurization** uses heat to kill pathogens and is most often seen in the food-processing industry. This method does not sterilize but

Fast Fact Pasteurization was originally developed to prevent bottled wine from spoiling. Pasteurization kills *Acetobacter* bacteria, which convert the ethanol in wine into acetic acid (vinegar).

is used to reduce the number of pathogens and also the number of organisms that can cause spoilage of the food. Milk, fruit juices, wine, and beer are routinely pasteurized.

- The two suffixes -**static** and -**cidal** are used on words that describe agents that either kill organisms (-cidal) or inhibit their growth (-static). For example, a **bacteriocidal** agent kills any bacteria exposed to it, but a bacteriostatic agent does not. When bacteria are in a **bacteriostatic** environment, their numbers do not multiply but the organisms do not die. Once the bacteria are taken out of the bacteriostatic environment, they resume their normal growth pattern. These suffixes are used on words describing agents used against bacteria (bacteriostatic, bacteriocidal), viruses (virustatic, **virucidal**), and fungi (fungistatic, fungicidal).

- An **antimicrobial** is defined as a substance that inhibits the growth of a microbial organism.

TARGETS FOR DISINFECTANTS AND ANTISEPTICS

With these terms in mind, let's examine the action of these agents. To do this, we have to think about the vulnerabilities of microorganisms. The discussion in Chapter 9 will be very helpful here. Recall that in that chapter we focused on the anatomy of bacterial cells. Because bacteria are single-celled organisms with a relatively simple anatomy, it is easy to see which targets in the cell could result in their death. For example, the cell wall of a bacterium is crucial for keeping the outside out and the inside in. Thus, damage to this structure could result in the death of the organism. The same is true for the plasma membrane of the bacterium, because this structure is responsible for enclosing the cytoplasm and is involved with DNA replication and ATP production. Therefore, loss or damage to the bacterial plasma membrane is a lethal event. Finally, any inhibition of protein synthesis or alteration in protein structure can result in disastrous consequences for the organism. As we will see, these targets are also some of the same ones that are used for antibiotic therapy (discussed in Chapter 19).

The Cell Wall

As mentioned above, the cell wall of a bacterium maintains the integrity of the cell. Several chemical agents damage this barrier by inhibiting its synthesis, by digesting it, or by breaking down its structure. Any microorganism that loses the integrity of the cell wall becomes very fragile and susceptible to lysis.

The Plasma Membrane

This structure is composed of a phospholipid bilayer that also contains a variety of proteins. As we learned in Chapter 9, this membrane is selectively permeable, which means it has the ability to selectively allow some things, but not others, to enter and leave the cell. When the plasma membrane of a bacterial cell is disrupted, the cell loses its selective permeability, and this change in permeability leads to the death of the cell.

Surfactants are chemicals (commonly called soaps or detergents) that reduce the surface tension of solvents, such as water, by decreasing the attraction between molecules forming the surface layer of the solvent. Surfactants are very effective for disrupting the plasma membrane of cells because, like the phospholipid bilayer, surfactant molecules are also polar molecules that have both hydrophobic and hydrophilic

Fast Fact In the early- and mid-20th Century, Lysol was used as a douche by women for feminine hygiene and birth control. However, it eliminated the normal microbiota in the vagina and masked the odors that indicated vaginal infections.

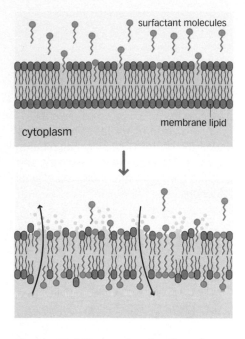

Figure 18.1 The mode of action of surfactants. Surfactants affect the plasma membrane of cells by inserting into the lipid bilayer, disrupting it, and creating abnormal channels that cause leakage from the cell and, ultimately, death.

regions. Surfactant molecules are therefore able to bind to and penetrate the bilayer and cause openings to form in the plasma membrane, resulting in the lysis of the cell (**Figure 18.1**).

In addition, recall that some viruses have envelopes composed of the plasma membranes from host cells they have infected. Damage to this envelope can cause the virus to lose the capacity to infect a host cell.

Protein Structure and Function

Proteins are among the most important molecules in the microbial cell. They are responsible for structural elements of the cell and are also the cell's enzymes, which are important for the physiology and metabolism of the cell. Recall that proteins have a three-dimensional shape and that this shape is directly related to the function of the protein. If this shape is changed in any way, the protein is said to be denatured, and its function may be inhibited or eliminated, resulting in the death of the cell. Denaturation involves breaking the hydrogen bonds and other bonds that hold the three-dimensional shape of the protein together. When these bonds are broken, the protein unfolds and is inactivated.

Heat and strong solvents, such as alcohols and acids, break the hydrogen bonds and can result in total denaturation and coagulation of the protein. In addition, metallic ions can inhibit enzymatic function by blocking the active site on the protein (**Figure 18.2**).

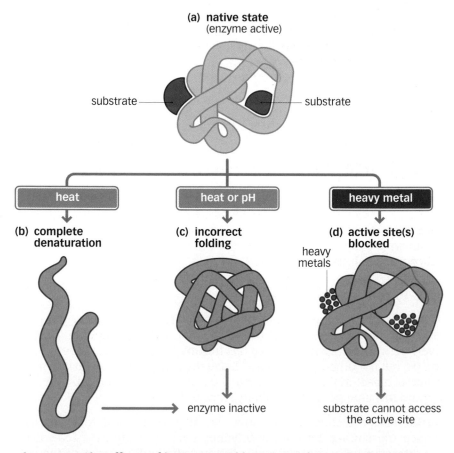

Figure 18.2 The effects of heat, pH, and heavy metals on protein structure. Panel a: The native functional configuration of the protein. Panel b: Denaturation of the protein, involving complete unfolding. Panel c: Incorrect folding of the protein. Panel d: Interference with the binding site.

Nucleic Acids

Like proteins, nucleic acids are required for cell survival. Some chemicals, such as formaldehyde and ethylene oxide, interfere with DNA and RNA function, but the most powerful agent for disrupting nucleic acids is radiation. In fact, irradiation with gamma rays, ultraviolet radiation, or X-rays causes mutations that can result in permanent inactivation of the nucleic acids and the death of the organism.

The synthesis of DNA and RNA (reviewed in Chapter 11) is one of the most important functions of the cell, and any agent that can disturb this synthesis is a powerful antimicrobial compound. Some antimicrobial agents bind irreversibly to DNA, preventing gene expression (transcription and translation), whereas others are mutagenic and cause lethal mutations in gene sequences.

MICROBIAL DEATH

In reality, the death of a microbe such as a bacterium is hard to identify because there are no apparent signs of it. In complex organisms, the failure of a system necessary to the wellbeing of the organism is an indication of death, but in bacteria, these systems do not exist. The difficulty in identifying bacterial death is compounded by the fact that many lethal compounds kill bacteria without changing what they look like under the microscope. Even a lack of movement by the bacteria cannot be a definitive sign of death. Therefore, it is necessary to develop special requirements to define microbial death.

The most efficient way to determine microbial death is to determine whether the organism can reproduce when moved from the antimicrobial environment to an environment that is suitable for growth. If the organism cannot reproduce even in the most suitable growth environment, it is dead. The permanent loss of reproductive capability even under optimum conditions has become the accepted definition of microbial death.

One of the techniques used to evaluate the efficacy of an antimicrobial agent is to calculate the microbial death rate (**Figure 18.3**), which is constant over time and under a particular set of conditions. You can see from the numbers on the vertical axis of the figure that the death rate is logarithmic (just like the logarithmic phase seen in the bacterial growth curve described in Chapter 10).

Factors That Affect the Rate of Microbial Death

There are several factors that can affect the death of microbes, particularly in a clinical setting.

- The greater the number of organisms, the longer it will take to kill all of them. This is a matter of accessibility of the disinfectant or antiseptic to the organisms. If there are large numbers of organisms present, it will take time for the agent to reach each of the organisms.

- Different microbes need different exposure times to the same antimicrobial agent to be killed, and, as above, the exposure time required also depends on the accessibility of the agent to the organism.

- Some antimicrobial agents take longer than others to kill microbes, for example radiation treatment can take a long time to be successfully accomplished.

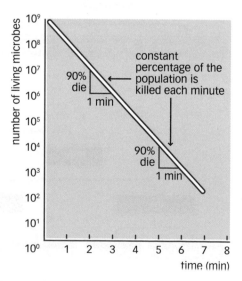

Figure 18.3 Microbial death rate. The killing of organisms involves a constant percentage over time (in this case 90% per minute). Note that the vertical scale is logarithmic.

- The lower the temperature at which the microbes are treated, the longer it will take to kill all of them.

- The environment in which the microbes to be treated exist is particularly important in health care because many pathogens will be associated with organic materials such as blood, saliva, bodily fluids, or even fecal material, which inhibit the accessibility of the antimicrobial agents to the organisms.

- Of all the factors that affect the rate of microbial death, endospore formation may be the most important. Endospores are resistant to many of the agents routinely used to inhibit microbial growth, and spore-forming organisms can evade destruction by these agents. This can be especially important when considered in the context of nosocomial infections (discussed in Chapter 6).

> **Keep in Mind**
>
> - The cell wall, plasma membrane, proteins, and nucleic acids of microbes are targets for disinfectants and antiseptics.
> - The permanent loss of reproductive capability has become the accepted definition of microbial death.
> - Several factors can affect the rate of death, including the number of organisms, the duration of exposure, temperature, environment, and the ability to form a spore.

Figure 18.4 gives an overview of the methods used to control microbial growth. As you can see from this figure, there are three major methods of control: chemical, physical, and mechanical removal.

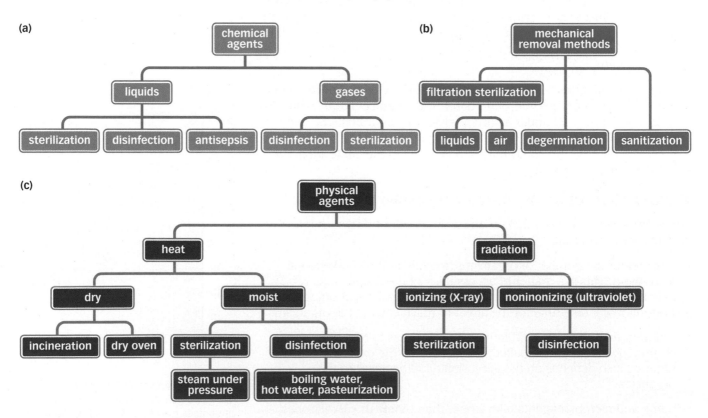

Figure 18.4 An overview of microbial control mechanisms. There are three methods that can be used: chemical (panel a), mechanical (panel b), and physical (panel c).

CHEMICAL METHODS FOR CONTROLLING MICROBIAL GROWTH

In this section we examine the use of chemicals as a means of controlling microbial growth. Many chemicals can kill microbes, but they can also be harmful to humans. Therefore, if they are to be useful as disinfectants or antiseptics, they must also be safe to use.

The Potency of Disinfectants and Antiseptics

There are several factors to consider when evaluating chemical methods for controlling microbial growth, because the effectiveness of antiseptics and disinfectants can be affected by such factors as time, temperature, and concentration. We will use the general term **chemical agent** when discussing a topic that is about both antiseptics and disinfectants. Using this general term, we can say that the death rate of organisms subjected to a chemical agent can be accelerated by, for example, increasing the temperature, as noted above. An increase of 10°C in the temperature at which a microbial specimen is treated will roughly double the rate of the chemical reaction, thereby increasing the potency of the chemical agent. Changes in pH can also either increase or decrease an agent's potency.

For most chemical agents, increases in the concentration of the agent will result in an increase in its potency: higher concentrations will be microbicidal, whereas lower concentrations will be microbistatic. However, this does not hold true for alcohol when used as an antimicrobial. In this case, an increase in the concentration does not increase the killing efficacy but actually hinders it. For alcohol to be effective, it must have some water associated with it in order to penetrate the microbe and denature proteins. Consequently, 70% alcohol is better for microbial control than 99% alcohol.

The use of low concentrations of some disinfectants has been shown to result in bacterial resistance to those disinfectants. Even more worrisome, those microbes also became resistant to antibiotics that they had never been exposed to! For this reason, disinfectants used at home, work, and in hospitals should be used at the correct concentration. Furthermore, varying the disinfectants used will help prevent the target microbes from developing disinfectant and antibiotic resistance.

Evaluation of Disinfectants and Antiseptics

There is no completely satisfactory method for evaluating antimicrobial chemical agents, but there are several tests that can be used, and we will now look at three of them. One is to compare the agent with phenol, a test known as the **phenol coefficient**. The second method is the **disk diffusion method**, which is a rapid test for efficacy that uses tiny disks of filter paper, and the last is called the **use dilution method**.

The Phenol Coefficient

Phenol was first used as a disinfectant by Joseph Lister in 1867 to reduce infection during surgery. It is still the benchmark disinfectant that others are compared with, and the comparison is reported as a phenol coefficient. Usually *Salmonella typhi* and *Staphylococcus aureus* are the two organisms on which the disinfectants are compared. To calculate a phenol coefficient, the concentration of phenol compound that kills the test organism in 10 minutes, but not in 5 minutes, is divided by the concentration of the test compound that kills the organism under the same conditions. A chemical agent with a phenol coefficient of 1.0 has the same effectiveness as phenol. Any number greater than 1.0 indicates an

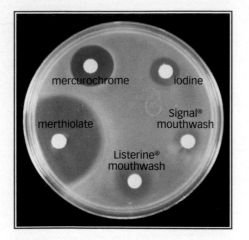

Figure 18.5 The disk diffusion method.
Notice the clear zones (zones of inhibition) around some of the disks. The size of the zone of inhibition around a disk depends on the effectiveness of the agent in killing the organism. The size of the zone also depends on the concentration of the agent and on the rate at which it diffuses away from the disk.

efficiency greater than that of phenol, and any number less than 1.0 indicates efficiency less than that of phenol. Phenol coefficients are reported separately for the two test organisms. As an example, the household disinfectant Lysol® has a coefficient of 5.0 against *Staphylococcus aureus* but only 3.2 against *Salmonella typhi*, whereas ethanol has a phenol coefficient of 6.3 against both.

The Disk Diffusion Method

This simple method for determining the efficacy of a chemical agent against a particular microbe uses tiny disks of filter paper soaked in the agents to be tested. First an agar plate is inoculated with the microbe against which the various agents are being tested. Then the disks, each soaked in a different antimicrobial agent, are placed at various positions on the inoculated agar. Inhibition of growth around each disk (or lack thereof) is easily identifiable by a clear zone called a **zone of inhibition** (**Figure 18.5**). The sizes of the zones seen with different chemicals are not comparable because they may reflect differences in concentration of the chemicals and also differences in the diffusion rates of the molecules (larger chemical molecules on one disk move away from their disk more slowly than smaller chemical molecules on a different disk do from theirs, resulting in a smaller zone of inhibition around the disk containing the larger molecule). This method also cannot distinguish between a microbicidal agent and one that is microbistatic. In both cases there is inhibition of growth during the disk test, but if you were to subculture the organisms from the zone of inhibition surrounding the microbistatic agent, the organisms would resume growing.

The Use Dilution Method

This evaluation method is more time-consuming than the two just described. However, it can tell you whether an antimicrobial agent is bacteriostatic or bacteriocidal, and it uses standardized preparations of certain bacteria. A series of solutions of different concentrations of the disinfectant are prepared, ranging from very dilute to very concentrated. Broth cultures of the test organism are dried down on stainless steel cylinders, and each cylinder is dipped for 10 minutes into one of the solutions. The cylinders are then removed from the solutions, rinsed with water to remove any remaining chemical agents, and placed into a tube of growth medium. The tubes are incubated and observed for the presence or absence of bacterial growth.

An important example of this type of testing is the determination of the **minimal inhibitory concentration** (**MIC**). In this test, various concentrations of an antibiotic are added to the culture media in which bacteria collected from a clinical specimen are grown. The test is repeated with a variety of antibiotics to determine which antibiotic is most effective at treating the bacterial infection, which may be resistant to some antimicrobial chemicals but not to others.

Selecting an Antimicrobial Agent

Many chemicals are tested as possible antimicrobial agents, but it is important to remember that many of them have distinct and inherent potential side effects that would disqualify them from this use. In addition, some chemical agents are better for certain uses than others. In health care facilities, authorities determine which antimicrobial agents should be used. In general, there are three qualities that are considered when deciding which antimicrobial agent to use:

Fast Fact Organic matter (for example blood, dirt, feces, food, or vomit) interferes with the ability of chlorine bleach to work as an effective disinfectant, and therefore must be removed before applying bleach to an inert surface.

- It must be effective against all types of infectious organisms without damaging tissues or inert surfaces.
- It should be effective even in the presence of organic materials, such as bodily fluids, blood, and fecal material.
- It should be stable and, if possible, inexpensive.

As we discussed above, there are several ways to inhibit the growth of microbial organisms, such as destruction of the cell wall or plasma membrane, denaturation of proteins, and destruction of nucleic acids. Antimicrobial chemical agents—both disinfectants and antiseptics—kill by participating in all these types of reaction (Table 18.1).

Table 18.1 Examples of commonly used antiseptics, disinfectants, and sterilants.

Agent	Use	Mode of action
Antiseptics		
Alcohol (60–85% ethanol or isopropanol in water)[a]	Skin	Lipid solvent and protein denaturation
Phenol-containing compounds (hexachlorophene, triclosan, chloroxylenol, chlorhexidine)	Household cleansers, antibacterial soaps, lotions, cosmetics, body deodorants	Disrupts cell membrane
Cationic detergents, especially quaternary ammonium compounds (benzalkonium chloride)	First aid products, antibacterial soaps, lotions	Interact with phospholipids of membrane
Hydrogen peroxide (3% solution)	Skin	Oxidizing agent
Iodine-containing iodophor compounds in solution (Betadine®)	Skin	Iodinates tyrosine residues of proteins
Silver	Wound dressings, catheters, breathing tubes	Protein precipitant
Disinfectants and sterilants		
Alcohol (60–85% ethanol or isopropanol in water)[a]	Disinfectant and sterilant for medical instruments and laboratory surfaces	Lipid solvent and protein denaturant
Cationic detergents (quaternary ammonium compounds)	Disinfectant for medical instruments, food, and dairy equipment; first aid products	Interact with phospholipids
Ethylene oxide (gas)	Sterilant for temperature-sensitive laboratory materials such as plastics	Alkylating agent
Formaldehyde	3–8% solution used as surface disinfectant, 37% (formalin) or vapor used as sterilant	Alkylating agent
Glutaraldehyde	2% solution used as high-level disinfectant or sterilant	Alkylating agent
Hydrogen peroxide	Vapor used as sterilant	Oxidizing agent
Iodine-containing iodophor compounds in solution (Wescodyne®)	Disinfectant for medical instruments and laboratory surfaces and drinking water treatment	Iodinates tyrosine residues
Ozone	Disinfectant for drinking water	Strong oxidizing agent
Peracetic acid	0.2% solution used as high-level disinfectant or sterilant	Strong oxidizing agent
Phenolic compounds	Disinfectant for laboratory surfaces	Protein denaturant

[a]Alcohols, hydrogen peroxide, and iodine-containing iodophor compounds can act as antiseptics, disinfectants, or even sterilants depending on concentration, length of exposure, and form of delivery.

Figure 18.6 Denaturation of proteins can be either permanent or temporary. Panel a: Treatment with concentrated antimicrobial agents usually leads to a permanently denatured protein molecule. Panel b: Treatment with dilute antimicrobial agents can lead to denaturation that is only temporary. Once removed from the antimicrobial environment, the protein molecule reverts to its normal three-dimensional configuration.

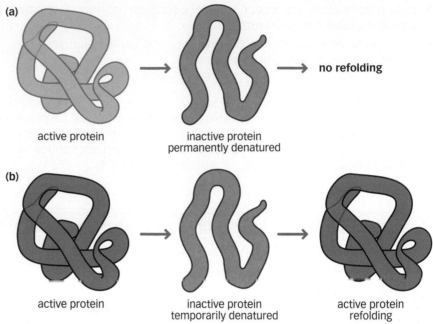

Proteins are the major structural and enzymatic molecules of the cell. They can be denatured by destroying their three-dimensional shape, and this destruction of shape destroys their functional capability. It is important to note that mild treatments, such as low temperature or treatment with dilute acids or alkalis, denature proteins only temporarily, and when the agents are removed (or the temperature is allowed to return to room temperature), the protein will refold to its normal structure and be functional once again (**Figure 18.6**). Consequently, most antimicrobial chemical agents are used in strong enough concentration and for periods that guarantee permanent denaturation of proteins, which is a lethal event for the organism.

Chemical agents can also have a role in the control of viral pathogens by inactivating the ability of the virus to infect or to replicate. This can be accomplished by either destroying the proteins associated with the virion or preventing viral replication (by disrupting viral gene expression). Detergents, alcohols, and other agents that denature proteins can affect the protein found on the capsid of virions, and the envelopes of viruses are susceptible to agents that act on lipids. Furthermore, alkylating agents, such as ethylene oxide and nitrous acid, can act as mutagens for viral nucleic acid, thereby inhibiting the replication and proliferation of virions.

Types of Chemical Agent

Chemical agents are used more than physical means (described below) for disinfection, antisepsis, and preservation. Keep in mind that chemical agents—both disinfectants and antiseptics—affect microbial cell walls, plasma membranes, proteins, or nucleic acids. Remember also that the effect of chemical agents varies with temperature, the length of exposure, the amount of contaminating organic material present, pH, and the concentration and stability (freshness) of the agent.

Chemical agents tend to either destroy or inhibit the growth of enveloped viruses, bacteria, fungi, and protozoans but have trouble dealing with

Fast Fact Some viruses are difficult to inactivate even after being exposed to chemical agents and can remain infective after treatment.

protozoan cysts and bacterial endospores. There are eight major catego-ries of chemical agents used as antiseptics and disinfectants: phenol and phenolic compounds, alcohols, halogens, oxidizing agents, surfactants, heavy metals, aldehydes, and gaseous agents. (Some antimicrobial chemical agents combine one or more of these.) Let's look at these vari-ous categories one at a time.

Phenol and Phenolic Compounds

Phenolic compounds are derived from phenol (**Figure 18.7a**) by adding halogens or organic functional groups to the phenol ring. Many of these derivatives have greater efficacy and fewer side effects than phenol. In fact, natural oils such as pine oil and clove oil are phenolic compounds that can be used as antiseptics.

Bisphenolics (**Figure 18.7b**) are composed of two (the prefix bis- means two) covalently linked phenolic compounds. The bisphenolic *ortho*-phenylphenol is the active ingredient in Lysol, and the bisphenolic tri-closan is found in many household antibacterial products, toothpaste, diapers, garbage bags, and cutting boards. Phenols and phenolic com-pounds are low-level to intermediate-level disinfectants and antiseptics that denature proteins and disrupt the plasma membrane of microbial cells. More importantly, they are very effective in the presence of organic material, such as vomit, pus, saliva, and feces, and can remain active for prolonged periods. They are therefore commonly used as disinfectants in health care settings and laboratories.

Alcohols

Alcohols are bacteriocidal, fungicidal, and virucidal for enveloped viruses but have no effect on fungal spores and bacterial endospores. Alcohols are intermediate-level disinfectants, with the most commonly used being isopropanol (rubbing alcohol) and ethanol (drinking alcohol). Of these two, isopropanol is slightly better than ethanol as an antimicrobial agent. When alcohol is used to carry other antimicrobial chemicals, the solution is referred to as a **tincture**. Tinctures are often more effective than are the ingredients of tinctures by themselves.

Alcohol denatures proteins and also disrupts the plasma membrane of microbial cells. It has the benefit of evaporating, thereby leaving no resi-due, and it is routinely used as a degerming agent to prepare sites for injection. The evaporative effect of alcohol can be a problem if the liquid evaporates before there has been sufficient time to kill all of the organ-isms in the area being treated.

Halogens

Four members of the chemical family known as the halogens have antimicrobial activity: iodine, chlorine, bromine, and fluorine. These are intermediate-level antimicrobial chemical agents that are effective against bacteria and fungal cells, fungal spores, some bacterial endospores, protozoan cysts, and many viruses. They can be used alone or in combination with other elements, but the mechanism of action of halogens is still not completely understood. It is thought that they oxidize and/or denature enzymatic proteins in the microbes they attack.

Iodine is a well-known antiseptic and is used in tablet form to disin-fect water, or medically as a tincture or as an **iodophor**, which is an organic compound that incorporates iodine in such a way that the iodine is slowly released from the organic molecule during a chemical reaction. The advantage of iodophor antiseptics over tincture of iodine is that the

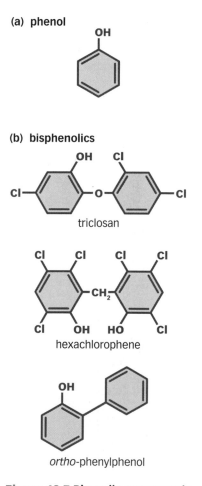

(a) phenol

(b) bisphenolics

triclosan

hexachlorophene

ortho-phenylphenol

Figure 18.7 Phenolic compounds. Panel a: The phenol molecule. Panel b: Bisphenolics are composed of two phenol rings joined together with additional functional groups to make them more effective and safer than phenol.

former are longer lasting and non-irritating to the skin. **Betadine** is an example of an iodophor and is routinely used to prepare skin for surgery and injection as well as to treat burns.

Chlorine is found in drinking water, in swimming pools, and in wastewater from sewage treatment plants. It is also a major ingredient in disinfectants such as chlorine bleach and is used to disinfect kidney dialysis equipment. **Chloramines**, which are combinations of chlorine and ammonia, are used in wound dressings, some skin antiseptics, and some municipal water supplies. Chloramines are less effective than chlorine as disinfectants/antiseptics, but they release their chlorine atoms more slowly and therefore last longer.

Oxidizing Agents

Oxidizing agents are high-level disinfectants and antiseptics that prohibit bacterial metabolism. They are very effective against infections of deep tissues because they release hydroxyl radicals (OH⁻), which kill anaerobic organisms. Consequently, oxidizing agents are routinely used in deep puncture wounds. The three most commonly used are hydrogen peroxide, ozone, and peracetic acid.

Hydrogen peroxide (H_2O_2) is a common household antiseptic. As we learned in Chapter 10, aerobic and facultatively anaerobic organisms can produce catalase, which breaks down hydrogen peroxide, but the amount of peroxide used as a disinfectant is more than sufficient to overwhelm the bacterial production of this enzyme.

Ozone (O_3) is a very reactive form of oxygen generated when O_2 is exposed to electrical discharge. Instead of chlorine, some Canadian and European cities use ozone generated in this way for water treatment. However, ozone is expensive to produce and difficult to maintain at the proper concentration in water.

Peracetic acid (CH_3-CO-O-OH, the peroxide form of acetic acid) is an extremely effective sporicide that can be used to sterilize surfaces. It is not affected by organic contaminants and leaves no toxic residue. Peracetic acid is used by food processors and medical personnel to sterilize equipment.

Surfactants

There are two common surfactants: soap and detergents. One end of a soap molecule is composed of fatty acid and is therefore hydrophobic; the other end is ionic and is therefore hydrophilic. When you wash your hands, the hydrophobic end of the soap molecule dissolves oily deposits into tiny drops that easily mix with water and are washed away. Soaps are good degerming agents but poor antimicrobial agents. However, they can be made more potent by adding antimicrobial chemicals such as triclosan.

Detergents are positively charged organic surfactants that are more soluble in water than soap is. The most popular detergents are **quaternary ammonium compounds** (**QUATS**), which contain the ammonium cation NH_4^+ with the hydrogen atoms replaced by functional groups or hydrocarbon chains (**Figure 18.8**). QUATS are considered low-level disinfectants/antiseptics, but their advantage is that they are odorless, tasteless, and harmless to humans (except at high concentrations). They are therefore used in many industrial and medicinal applications, such as mouthwashes.

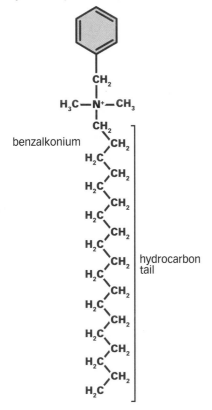

Figure 18.8 Quaternary ammonium compounds (QUATS) are surfactants in which the hydrogen atoms of ammonium ions are replaced by other functional groups.

QUATS function by disrupting the plasma membrane of microbial cells, which causes the cell to lose essential internal ions such as K^+. QUATS are bacteriocidal (especially for Gram-positive bacteria), fungicidal, and virucidal against enveloped viruses. They are not useful for non-enveloped viruses, mycobacteria, or bacterial endospores, and they are inhibited by the presence of organic contaminants.

Fast Fact *Pseudomonas* species grow very well in the presence of QUATS.

Heavy Metals

The ions of heavy metals such as arsenic, zinc, mercury, silver, and copper are inherently antimicrobial, and mercury and silver have been used in clinical situations. However, the use of mercury was abandoned when toxicity was discovered, and only silver is still occasionally used. The mechanism of action is through protein denaturation because these ions interact with the sulfur atoms in the amino acid cysteine. This amino acid forms disulfide linkages in proteins and helps to maintain their three-dimensional shapes (for a reminder see Chapter 2).

Heavy metals are low-level bacteriostatic agents. At one time, it was routine to use silver nitrate cream to treat the eyes of newborn infants as a way to prevent blindness brought on by infection with *Neisseria gonorrhoeae*, which could enter the newborn's eyes as it passes through the birth canal. Today, however, antibiotic creams, which are less irritating and can also kill *Chlamydia trachomatis*, are used instead for this purpose. Nevertheless, silver nitrate is still used in surgical dressings, burn creams, and catheters. A solution or cream of silver sulfadiazine, which releases low concentrations of silver ions over time, is also used on burns to prevent infection.

Aldehydes

Aldehydes are compounds containing a terminal –CHO functional group. There are two highly reactive aldehydes used as antimicrobials: glutaraldehyde, which is used in liquid form, and formaldehyde, which is used in both liquid form and gaseous form. Aldehydes function by cross-linking organic functional groups, and these cross-linking reactions denature proteins and also inactivate nucleic acids. Hospitals and research laboratories use 2% solutions of glutaraldehyde, which effectively kills bacteria, viruses, and fungi. In fact, treatment for 10 minutes will disinfect most objects, and treatment for 10 hours will sterilize. Glutaraldehyde is less toxic than formaldehyde but it is also more expensive.

Health care workers use a 37% formaldehyde solution called **formalin** to disinfect isolation rooms, exhausts, cabinets, surgical instruments, and dialysis machines. However, formaldehyde is an irritant for mucous membranes and has been found to be carcinogenic.

Gaseous Agents

Many items cannot be sterilized with heat or chemicals (some examples are plastics, suture materials, heart–lung machine components, and dried or powdered foods). These items can be sterilized effectively by highly reactive antimicrobial and sporicidal gases such as ethylene oxide, propylene oxide, and β-propiolactone.

Gases rapidly penetrate and diffuse into any space, and over time (14–18 hours) they denature proteins and also DNA by cross-linking organic functional groups. In fact, they kill everything they come into contact with and do so without causing any damage to inanimate objects. Ethylene oxide is the most frequently used gaseous sterilizing agent and is found

Fast Fact Chlorine gas has been used to decontaminate large spaces such as the post offices that were contaminated during the anthrax attack in 2001.

in hospitals and dentists' offices. Large chambers for use with ethylene oxide can be found in hospitals and are used for sterilizing instruments and equipment that might be sensitive to heat.

The primary disadvantages of using gaseous agents are that they are explosive, poisonous, and potentially carcinogenic. Disinfection with gas also takes a considerable time to complete because of the need for continuous cleanup.

Keep in Mind

- The potency of disinfectants and antiseptics can be affected by time, temperature, and concentration.

- The effects of disinfectants and antiseptics can be increased by increasing the temperature.

- Three tests—phenol coefficient test, disk diffusion method, and use dilution—are used to evaluate disinfectants and antiseptics.

- There are eight major categories of chemical agents used as antiseptics and/or disinfectants: phenols, alcohols, halogens, oxidizing agents, surfactants, heavy metals, aldehydes, and gases.

PHYSICAL METHODS FOR CONTROLLING MICROBIAL GROWTH

Physical methods for controlling microbial growth have been used for centuries. Ancient Egyptians and many other ancient civilizations worldwide dried food to preserve it, and Europeans used heat in canning food 50 years before Pasteur worked out why heat was a necessary part of the process. In addition to drying and heating, there are many other physical methods used to control microbial growth, including cold, filtration, osmotic pressure, and radiation. Let's look at all of these in greater detail.

Heat

Elevated temperatures are usually lethal to most pathogenic microbes. Two types of heat are used to control microbial growth: moist and dry. **Moist heat** comes from hot water, boiling water, or steam, and the range of temperatures effective for killing microbes runs from 60°C to 135°C. For example, classic pasteurization occurs when liquids are warmed to 63°C for 30 minutes. Sterilization in an **autoclave** occurs when the steam-based device reaches 121°C for 15 minutes.

High-temperature, short-time pasteurization occurs when the moist or liquid product is warmed to 72°C for 15 seconds. In contrast, **dry heat** is hot air with a low moisture content, such as the air in an oven. Temperatures for dry heat used as an antimicrobial range from 160°C to several thousand degrees. Hot air ovens are used for glassware, metallic instruments, powders, and oils. Exposure to a temperature between 150°C and 180°C for 2–4 hours ensures the destruction of spores as well as vegetative cells.

Moist heat can achieve the same level of effectiveness as dry heat in a much shorter time and at a lower temperature. Moist heat denatures proteins, which halts microbe metabolism and causes death. Dry heat dehydrates microbial cells, and the absence of water then inhibits metabolism. However, the loss of water can also stabilize some proteins present in microbes, and in these cases the object being treated must be exposed to the dry heat for a longer time so that the proteins are denatured.

Type	Temperature (°C)	TDT (min)
Non-spore-forming bacteria	58	28
Vegetative stage of spore-forming bacteria	58	19
Fungal spores	76	22
Yeasts	59	19
Viruses		
Non-enveloped	57	29
Enveloped	54	22
Protozoan trophozoites	46	16
Protozoan cysts	60	6

Table 18.2 Average thermal death times (TDTs) for selected microorganisms.

When the temperature of dry heat is very high, cells are oxidized (burned to ashes). The flame of a Bunsen burner reaches 1870°C at its hottest point, for instance, and furnaces/incinerators operate at 800–6500°C. Direct exposure to such temperatures ignites and reduces microbes to ash and gases. An example of this is flaming the inoculating loop used to inoculate cultures so as to maximize aseptic technique.

Adequate sterilization with heat depends on both the temperature and the length of time for which the heat is used. Normally, high temperatures are associated with short treatment times. These two variables are used to calculate the **thermal death time** (TDT), which is defined as the shortest length of time needed to kill all organisms at a specific temperature (**Tables 18.2** and **18.3** show the TDTs for selected microorganisms and bacterial endospores). Temperature and duration of exposure can also be used to calculate the **thermal death point** (TDP), which is defined as the lowest temperature required for killing all the organisms in a sample in 10 minutes.

Moist Heat Methods

There are three ways of using moist heat for controlling microbial growth: boiling, pressurized steam, and pasteurization (**Table 18.4**). When pressurized steam is used, it is not the pressure that kills, but the

Organism	Temperature (°C)	TDT (min)
Moist heat		
Bacillus subtilis	121	1
Clostridium botulinum	121	10
Clostridium tetani	105	10
Dry heat		
Bacillus subtilis	121	120
Clostridium botulinum	120	120
Clostridium tetani	100	60

Table 18.3 Thermal death times (TDTs) for various endospores.

Table 18.4 Moist heat methods for controlling microbial growth.

Method	Characteristics and typical uses
Pressurized steam	Achieves temperatures above 100°C, which are necessary to destroy endospores. Used to sterilize surgical instruments and other items that can be sterilized by steam. Also used to process canned food
Boiling	Fast, reliable, and inexpensive way to destroy most bacteria and viruses. Does not affect endospores
Pasteurization	Significantly decreases the numbers of heat-sensitive microorganisms, including pathogens and those that spoil food. Does not affect endospores. Routinely used for milk, fruit juices, beer, and wine

high temperatures that result from increased pressure. A good example of this is the autoclave, which is found in any health care facility. The autoclave is like a pressure cooker. It has a complex network of valves, ducts, and gauges that regulate and measure pressure and conduct steam into the pressurized chamber (**Figure 18.9**). Sterilization occurs when steam condenses to liquid water on the objects to be sterilized and the hot water gradually raises their temperature. Autoclaves routinely use pressures of 15 pounds per square inch, which yields a temperature of 121°C for the water, and are superior for sterilizing heat-resistant materials such as glassware, surgical dressings, some types of

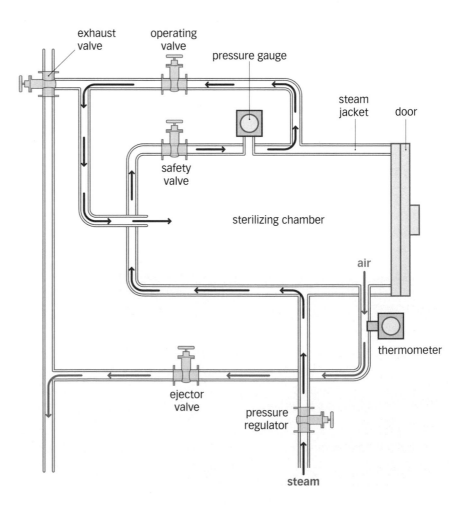

Figure 18.9 An autoclave. Steam (red arrow) is forced round the steam jacket and into the sterilizing chamber, which forces the air out of the chamber (blue arrows). When steam also starts to exit from the chamber, the ejector valve shuts automatically and pressure rises in the sterilizing chamber as more steam enters.

rubber glove, instruments, liquids, paper, some types of culture media (but not all), and heat-resistant plastics. Autoclaves are not useful for substances that repel water, such as waxes, powders, oils, most plastics, and media containing ingredients that break down when heated.

Boiling water is easy to use in homes and clinical settings but it does not kill heat-resistant cells. It is therefore effective only for disinfection, not for sterilization. Immersing an object in boiling water for 30 minutes will kill most non-spore-forming pathogens, including *Mycobacterium tuberculosis* and *Staphylococcus* species. Boiling can also disinfect drinking water, but the disadvantage of boiling is that once the object is removed from the boiling water, re-contamination can occur.

The last heat-associated mechanism that we will discuss is **pasteurization**. As noted earlier in the chapter, pasteurization is used to reduce microbial load and destroy pathogens while preserving flavor and nutritive value, but this process does not sterilize. Pasteurization can be accomplished in two ways: via the flash method, which involves exposure to a temperature of 71.6°C for 15 seconds, or via the batch method, which uses a temperature of 63–66°C for 30 minutes. The flash method is preferable because it is more effective against *Coxiella* and *Mycobacterium* organisms. In addition, longer exposure to a high temperature can alter flavor and nutritive value. These treatments kill 97–99% of bacteria but do not affect endospores or non-pathogenic lactobacilli, micrococci, or yeasts.

Fast Fact Pasteurized milk is not sterile: it still contains more than 20,000 organisms per milliliter.

The primary reason for pasteurization is to prevent the transmission of milk-borne diseases, such as those caused by *Salmonella* species (intestinal food poisoning), *Campylobacter jejunum* (acute intestinal disease), *Listeria monocytogenes* (listeriosis), *Coxiella burnetii* (Q fever), and *Mycobacterium tuberculosis* (tuberculosis).

Refrigeration, Freezing, and Freeze-Drying

Cold temperature retards the growth of microorganisms by slowing the rate of enzymatic reactions, but it does not kill the microorganisms (it is microbistatic). Refrigeration is used to delay the spoilage of food, or of medical supplies such as vaccines, by keeping the temperature at about 5°C. Many bacteria and molds can continue to grow at this temperature, so refrigeration is useful only for a limited period.

Fast Fact Even when stored in a refrigerator, *Clostridium botulinum* has been shown to grow and produce exotoxin while embedded deep in food where anaerobic conditions prevail.

Freezing at −20°C can preserve food but does not sterilize it. This low temperature slows the metabolic rate of microorganisms to such a degree that there is no growth or spoilage of food. Freezing can also be used to preserve microorganisms, but temperatures much lower than those used to freeze food must be used. The organisms to be preserved are usually frozen in glycerol or protein to prevent the formation of ice crystals that could puncture cell membranes, then placed in dry ice (the solid form of carbon dioxide, −78°C) or liquid nitrogen (−180°C). Freezing is, however, an effective method for killing most helminth parasites.

Freeze-drying, also known as **lyophilization**, can also be used to preserve cells. This is the process used to produce instant coffee and requires the removal of water from the organisms. Lyophilization is used for the long-term storage of organisms and also for ease of shipping and handling. Organisms are rapidly frozen in liquid nitrogen and then subjected to high vacuum, which pulls out the water molecules. The containers holding the frozen, dehydrated organisms are then sealed under vacuum, and the organisms are viable in this state for years. The addition of water is all that is required to restart the growth process.

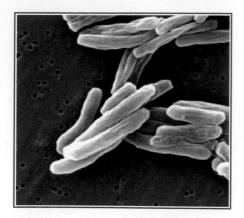

Figure 18.10 Colorized scanning electron micrograph of bacteria on a membrane filter. The diameter of the pores in the filter is smaller than the diameter of the bacterial cells, so the bacteria cannot pass through the filter.

Filtration

Filtration is useful for sterilizing liquids that would be destroyed by heat, or for removing microbes from air. The process involves passing the liquid through membrane filters, which contain pores that are too small to allow the passage of microorganisms (**Figure 18.10**). Membrane filters are usually made of nitrocellulose. These filters can be made with specific pore sizes ranging in diameter from 25 μm to less than 0.025 μm. Membrane filters are relatively inexpensive, do not clog very often, and can be purchased either sterile or non-sterile (the user autoclaves the latter form before use). They can be used for growth media, drugs, and vitamins, and in some cases they are used for commercial food preparation, such as in the filtration of beer. Filtration is used to sample and test water samples for contamination, in particular fecal coliform contamination.

Filters can also be used to purify air. **HEPA** (high-efficiency particulate air) filters are routinely found in ventilation systems, such as operating rooms, burn units, and clean rooms of laboratories, where microbial control is required and sterility is important. These same filters are used in facilities where dangerous organisms are researched, such as the Centers for Disease Control (CDC) in Atlanta. However, in this case the filters are used to keep organisms from escaping into the outside air. Air filters are usually soaked in formalin before disposal, to kill any trapped organisms.

Osmotic Pressure

This technique is another that has been used in food preservation for many decades. As we learned in Chapter 9, high concentrations of salt (or sugar) will create a hypertonic medium that draws water from the organism through osmosis, leading to plasmolysis and death. High concentrations of sugar are used in making preserves, such as jams and jellies (notice that the word preserve fits here), and high concentrations of salt have been used for a long time to 'cure' meats and fish. Keep in mind that some halophilic (salt-loving) organisms thrive in these conditions and can spoil these products.

Radiation

Radiation can be defined as energy that is emitted from atomic activities and dispersed at high velocity through matter or space. There are three types of radiation involved with controlling microbial growth: gamma rays, X-rays, and ultraviolet rays. When a cell is bombarded with radiation, the cell's molecules absorb some of the energy, leading to changes in the structure of the cell's DNA (**Figure 18.11**). The radiation can be one of two types:

- **Ionizing radiation**, which changes atoms in a cell's molecules to ions by causing electrons to be ejected from the atoms. Gamma rays, X-rays, and high-speed electron beams are forms of ionizing radiation. DNA is very sensitive to this type of radiation, which means that in DNA exposed to the radiation there is large-scale mutation and breakdown of chromosomal elements.

- **Non-ionizing radiation**, for example ultraviolet radiation. This excites atoms but does not ionize them, and it leads to abnormal bonds within molecules, such as the formation of thymine dimers (Chapter 11).

Ionizing radiation has become safer to use in recent years and is very useful because it sterilizes without heat or chemicals. All ionizing radiation can penetrate liquids and most solid materials, but gamma rays are the most penetrating. Although irradiation has been used in food preservation for many years, there are still stigmas attached to its use. In fact, any use of radiation in food preservation must be clearly stated on the food label. Flour, meat, fruits, and vegetables are routinely irradiated to kill microorganisms, parasitic worms, and insects.

Although radiation can cause cancer, there are no side effects associated with the irradiation of food. In fact, it has been estimated that the irradiation of just 50% of the meat and poultry in the United States would result in 900,000 fewer cases of infection, 8500 fewer hospitalizations, and 350 fewer deaths per year. Radiation is currently approved for the reduction of *E. coli* and *Salmonella* bacteria in beef and of *Trichinella* worms in pork.

Sterilization of medical products by radiation has become a rapidly expanding area. Drugs, vaccines, plastics, syringes, gloves, and even tissue that is to be used for grafting and heart valves are all sterilized using radiation. The main drawback to this technique is potential radiation poisoning of the operators who perform the sterilization.

Ultraviolet radiation is non-ionizing radiation that disrupts cells by generating free radicals, which then bind to the cells' DNA, RNA, and proteins. This radiation is a powerful killer of fungal cells and spores, bacterial cells, protozoans, and viruses.

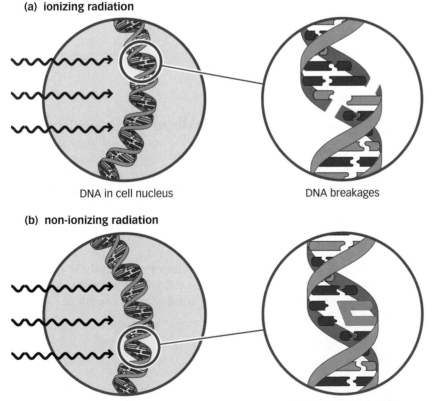

(a) ionizing radiation

DNA in cell nucleus   DNA breakages

(b) non-ionizing radiation

abnormal DNA bonds

Figure 18.11 Disinfection with radiation. Panel a: Ionizing radiation can bombard and penetrate a cell, dislodging electrons from the DNA molecules in the cell and causing the DNA strand to break. Panel b: Non-ionizing radiation enters the cell, strikes the DNA molecules, and excites them, causing mutations and the formation of abnormal bonds in the double helix.

Ultraviolet irradiation is used for disinfection, not sterilization. Germicidal lamps can cut down on microbes by as much as 99%, and the lamps are used in hospital rooms, operating rooms, schools, food preparation areas, and dental offices. In addition, disinfection of air with ultraviolet radiation has been effective in reducing postoperative infection, preventing the droplet transmission of infectious organisms and curtailing the growth of microorganisms in food preparation. It can also inhibit the growth of organisms in water, vaccines, drugs, plasma, and tissues used for transplantation. The major disadvantages of ultraviolet irradiation are poor penetration and damaging effects seen over long exposure to human tissues: retinal damage, cancer, and skin wrinkling.

Keep in Mind

- Physical methods for controlling microbial growth include heat, cold, desiccation, filtration, osmotic pressure, and radiation.

- High temperature usually results in the death of microbes.

- Thermal death time is the shortest time needed to kill all organisms at a specific temperature.

- Thermal death point is the lowest temperature required to kill all organisms in 10 minutes.

- Autoclaves combine temperature and pressure for sterilization.

- Ionizing radiation sterilizes without heat or chemicals and is used on some foods as well as medical products.

- Ultraviolet irradiation is used for disinfection but not for sterilization.

A Word About Hand Washing

It is of interest to note that one of the most important historical discoveries in the field of medicine was that the simple act of washing your hands can inhibit the spread of pathogens. Washing your hands with soap and running water mechanically removes microbes from the skin. Hand washing is required in hospitals and clinics as well as restaurants and bars. Much of the effectiveness of hand washing is related to the type of soap used and the time taken to wash. Liquid soap is recommended over bar soap because it reduces the transmission of microbes from person to person. Many hospitals use bacteriocidal soaps that are very effective at preventing the transmission of pathogens, but even household soap can be effective if enough time is taken to do the job thoroughly. A rule of thumb for many is to simply sing Happy Birthday as you wash and do not stop washing till the song is over.

SUMMARY

- The cell wall, plasma membrane, proteins, and nucleic acids are targets for disinfectants and antiseptics.

- Several factors affect the rate of death of microorganisms, including the number of organisms present, the duration of treatment, temperature, environment, and the ability to form an endospore.

- The potency of antiseptics and disinfectants can be affected by temperature and concentration.

Figure 18.12 Relative susceptibility of selected microbes to antimicrobial agents.

- Three tests are used to evaluate disinfectants and antiseptics: the phenol coefficient test, the disk diffusion method, and the use dilution test.

- There are eight major categories of chemical agent used as antiseptics and/or disinfectants: phenols, alcohols, halogens, oxidizing agents, surfactants, heavy metals, aldehydes, and gases

- Heat, cold, desiccation, filtration, osmotic pressure, and radiation are physical methods used to control microbial growth.

In this chapter, we have seen that there are many ways to control microbial growth, either through chemical or physical means, and that many of these methods are routinely used in commercial applications. It is important to keep in mind that control of microbial growth depends on many factors and that we can establish a hierarchy with regard to the susceptibility of microorganisms to these methods (Figure 18.12). We have also discussed ways to control microbial growth outside the body. In the next chapter we look at how antibiotics are used to control microbial growth inside the body.

most susceptible

enveloped viruses

Gram-positive bacteria

non-enveloped viruses

fungi

Gram-negative bacteria

active-stage protozoans

cysts of protozoans

mycobacteria

bacterial endospores

prions

most resistant

(Q) SELF EVALUATION AND CHAPTER CONFIDENCE

Multiple Choice

Answers are given in the back of the book and help can be found in the student resources at:

www.garlandscience.com/micro2

1. Which of the following best describes microbial death rate?

   A. All the cells in a culture die at once
   B. The cells in a population die at a constant rate
   C. All of the cells in a culture are never killed
   D. The type of antimicrobial agent has no effect on the death of microbes
   E. It is not affected by changes in temperature

2. Bacterial death will result from damage to which of the following structures?

   A. Plasma membrane
   B. Proteins
   C. Nucleic acids
   D. Cell wall
   E. All of the above

3. Sterilization involves

   A. Removal of all microbes including endospores
   B. Removal of all microbes but not endospores
   C. Killing only pathogens
   D. Bacteriostatic agents
   E. None of the above

4. Which of these factors does not affect the rate of bacterial cell death?

   A. The number or organisms present
   B. Temperature
   C. The environment
   D. Endospore formation
   E. All of the above

5. The bactericidal or bacteriostatic nature of a disinfectant can be determined using which of the following test methods?

   A. The phenol coefficient
   B. The disk diffusion method
   C. The thermal death point
   D. The use dilution method
   E. The thermal death time

6. The effectiveness of all chemical disinfectants increases with the following except

   A. Length of exposure
   B. Increasing temperature
   C. Higher concentration
   D. None; they all increase the effectiveness of chemical disinfectants

7. Which of the following substances is the least effective antimicrobial agent?
   A. Phenol
   B. Cationic detergents
   C. Soap
   D. Alcohol
   E. Iodine

8. Which of the following are unaffected by alcohol?
   A. Bacteria
   B. Viruses
   C. Fungi
   D. Endospores
   E. None of the above

9. Oxidizing agents target
   A. Cell membranes
   B. Metabolic pathways
   C. The cell wall
   D. Protein synthesis
   E. None of the above

10. A disk diffusion test using *Staphylococcus* gave the results shown in the following table. Which compound was the most potent killer of *Staphylococcus*?

Disinfectant	Zone of inhibition (mm)
A.	0
B.	2.5
C.	10
D.	5

   A. A
   B. B
   C. C
   D. D
   E. Insufficient information to compare potency

11. QUATS target
   A. Metabolic pathways
   B. The cell wall
   C. The plasma membrane
   D. Protein synthesis
   E. Endospores

12. Which one of the following would indicate the most potent disinfectant?
   A. Kills *Staphylococcus aureus*
   B. Kills lipophilic viruses
   C. Kills *E. coli*
   D. Kills *Pseudomonas*
   E. All are equal

13. All of the following affect microbial growth except
   A. Disinfection
   B. Antisepsis
   C. Neutralization
   D. Pasteurization
   E. Sanitization

14. All of the following are chemicals used in disinfection except
   A. Aldehydes
   B. Ultraviolet irradiation
   C. Phenolic compounds
   D. Halogens
   E. Surfactants

15. The aim of pasteurization is to kill
   A. All microorganisms
   B. Endospores
   C. Pathogens
   D. Gram-positive organisms preferentially
   E. Gram-negative organisms preferentially

16. Refrigeration at −20°C
   A. Sterilizes food
   B. Preserves food for a period of time
   C. Causes bacterial cell death through changes in osmotic pressure
   D. Has no effect on microbial growth

17. Which of the following can sterilize?
   A. Pasteurization
   B. Autoclaving
   C. Refrigeration
   D. Freezing
   E. None of the above

18–21. Compare pasteurization with sterilization and pick the best option from: **A** pasteurization; **B** sterilization; **C** both; **D** neither. Answers can be used more than once or not at all.

18. Kills endospores

19. Kills bacteria

20. Milder treatment

21. Destroys prions

22–25. Compare ionizing with non-ionizing radiation and pick the best option from: **A** ionizing radiation; **B** non-ionizing radiation; **C** both; **D** neither. Answers can be used more than once or not at all.

22. Uses ultraviolet light

23. Uses gamma rays

24. Alters DNA

25. Uses X-rays

DEPTH OF UNDERSTANDING

Questions listed here require you to bring together the concepts you have learned in this chapter into a discussion format. This helps you to increase your depth of understanding of the material you have learned. Help can be found in the student resources at: www.garlandscience.com/micro2

1. Why does the endospore present such a problem when controlling the growth of bacteria?

2. Compare and contrast the effects of dry and moist heat as antimicrobial treatments.

3. Compare the benefits of ionizing and non-ionizing radiation as methods for controlling microbial growth.

CLINICAL CORNER

Help can be found in the student resources at: www.garlandscience.com/micro2

1. It was already a very busy day in the infectious disease ward of the hospital. Patients were being admitted and discharged and others were being sent for tests. It is your responsibility to see that the rooms that are being vacated are cleaned and ready to receive new patients. You have a new nursing assistant to help you with this.

   A. How will you explain the importance of the job at hand?
   B. What procedures should you make your assistant aware of?

2. Hospital standard operating procedures require that you wash your hands regularly even though most of the time you are wearing rubber gloves when working with patients.

   A. Why is washing your hands so important?
   B. What effect will wearing rubber gloves have on the need for washing your hands?
   C. Is hand washing in the hospital the same as at home?

Antibiotics and Antimicrobial Drugs
Chapter 19

Why Is This Important?

Antibiotics have been used for decades and have changed the landscape of health care. Since they were discovered, the number of deaths due to infection has been drastically reduced.

In April 2003, an older gentleman (Mr. Jones) walked into a central Oregon medical clinic with a wound and severe bruising on one of his legs. He had been gored by one of his bulls and after treating himself at home for 48 hours with washing and icing, felt that the wound was becoming infected. After an examination by a doctor, he was sent home on a course of antibiotics and anti-inflammatories that included Augmentin (an attenuated penicillin derivative). Two days later, Mr. Jones returned to the clinic with a significant fever and strong pain radiating from the wound. He had developed a complicating infection with *Streptococcus pyogenes*, a bacteria that causes necrotizing fasciitis, which can be fatal in 30% of cases. Mr. Jones was transferred to a hospital where he underwent surgery to remove the affected tissue as well as a significant portion of the tissue around the original wound. Thankfully, he retained the use of his leg and was successfully discharged from the hospital after several days of intensive antibiotic therapy.

OVERVIEW

The word antibiotic is defined as "a chemical that controls the growth of micro-organisms", which implies that all products used to kill microbes of any kind are antibiotics. However, in common usage, the term antibiotic is associated with products taken internally and designed to kill bacteria only. In this chapter, we use antibiotic in terms of killing bacteria and specify other antimicrobials on the basis of the pathogens they affect.

Perhaps one of the most important scientific findings in history was the discovery that microbes protect themselves from other microbes. This simple principle led to the development of antibiotics, most of which are chemicals that we can take from one organism and use to protect ourselves from other organisms. Infectious diseases that once swept through the human population unabated were suddenly shut down with the simple administration of these chemicals, and people began to think that infectious disease was a thing of the past. We will focus in this chapter on the development and functions of antibiotics and also look at antivirals, antifungals, and antiparasitics (agents used to kill viruses, fungi, and parasites respectively), but keep in mind that overuse and improper use of these important medications has brought about a dangerous rise in the resistance to antibiotics, a topic we discuss in Chapter 20.

We will divide our discussion into the following topics:

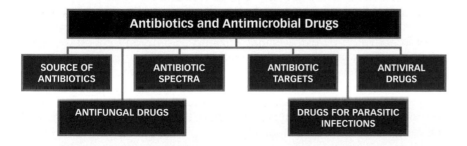

SOURCE OF ANTIBIOTICS

The earliest drugs with antibiotic properties (sulfa drugs) were available in the 1930s, but the first major antibiotic, **penicillin**, came into widespread use during the second world war. It was discovered by accident in 1928 when Alexander Fleming was working with *Staphylococcus aureus*, and while plating this organism he accidentally allowed the fungus *Penicillium* to contaminate his plate. He subsequently observed that the plate had a uniform growth of *S. aureus* except where the fungi were growing. In this area, Fleming saw a clearly defined region where there was no bacterial growth; this was what eventually became referred to as a zone of inhibition. It took more than 10 years and the pressing need to combat infections during the second world war for penicillin to be developed into a useful drug, but this accidental discovery was the beginning of an industry that has become central to the successful care of patients. It is estimated that more than 80 million prescriptions are filled in America each year, with about 12,500 tons of antibiotics being produced annually. Administration of antibiotics has permitted successful outcomes in many diseases for which there were previously no treatments available. In the United States from 1900 to 1980, mortality from infectious diseases dropped from 797 per 100,000 persons to 36 per 100,000 persons. There are now several classes of antibiotic, and the timeline for their introduction into clinical practice is shown in **Figure 19.1**.

We also see from Figure 19.1 that there have been very few major discoveries of natural antibiotic substances for decades. As we will see later in the chapter, efforts have now shifted to modifying existing antibiotics and searching for potential antibiotics in new places.

Antibiotics Are Part of Bacterial Self-Protection

Many antimicrobial compounds are produced by microorganisms as part of their naturally occurring survival mechanisms. These substances keep other organisms away and protect the host's supply of nutrients and oxygen. **Table 19.1** shows microbial sources of some common antibiotics.

The microorganisms that produce antimicrobial compounds use elegant molecular mechanisms to control the production of these toxic molecules, thereby preventing self-destruction. There are several ways in which the organism can do this. For example, some bacteria restrict the production of antibiotics to the stationary phase of bacterial growth, when the bacteria are not actively dividing (Chapter 10). Consequently, bacteria in their stationary phase are protected from the antibiotics they produce which impair cell wall construction and interfere with logarithmic growth of competing organisms. It is also the case that, during the stationary phase, nutrients and other growth requirements are at a low level and competition for them is intense.

Other bacteria keep the intracellular concentrations of the antibiotics they are producing at low levels. This is accomplished by tightly regulating the rates at which these molecules are produced and the rates at which they are exported. In these bacteria, the antibiotic molecules can be exported in an inactive form that will not affect the producing organism and then become activated by extracellular enzymes. Some antibiotic-producing microorganisms modify their own cell walls or polymerase molecules to ensure their safety. Interestingly, the genes responsible for these modifications are clustered with the antibiotic-producing genes so that the functions of these two sets of genes are integrated: when the genes that control antibiotic production are turned on, the modifications to potential self-targets are made automatically.

Fast Fact Alexander Fleming shared the 1945 Nobel Prize with Howard Florey and Ernst Chain, who developed a way to produce enough penicillin for medical use while working at the University of Oxford (UK) in 1940. Florey had to go to the United States to get help with large scale manufacture of the drug and it was first produced on an industrial scale in the United States.

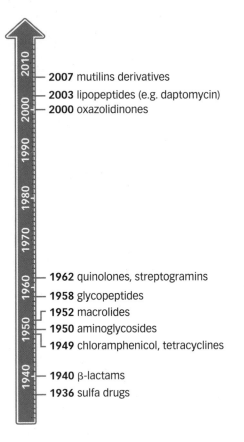

2007 mutilins derivatives
2003 lipopeptides (e.g. daptomycin)
2000 oxazolidinones

1962 quinolones, streptogramins
1958 glycopeptides
1952 macrolides
1950 aminoglycosides
1949 chloramphenicol, tetracyclines

1940 β-lactams
1936 sulfa drugs

Figure 19.1 A timeline for the introduction of new classes of antibiotics into clinical practice. Note that no new naturally occurring antibiotics were discovered between 1962 and 2000.

Microorganism	Antibiotic
Bacteria	
Streptomyces various species	Amphotericin B
	Chloramphenicol
	Erythromycin
	Kanamycin
	Neomycin
	Nystatin
	Rifampin
	Streptomycin
	Tetracyclines
	Vancomycin
Micromonospora various species	Gentamicin
Bacillus various species	Bacitracin
	Polymyxins
Fungi	
Penicillium various species	Griseofulvin
	Penicillin
Cephalosporium various species	Cephalosporins

Table 19.1 Microbial sources of some antibiotics.

| Keep in Mind |

- Penicillin was discovered accidentally.

- There have been very few major discoveries of new natural antibiotics since the 1960s.

- Many microorganisms produce toxic substances as part of their survival mechanisms and have developed methods to protect themselves from the antibiotics they produce.

ANTIBIOTIC SPECTRA

The first medically used antibiotics were natural products isolated from specific microorganisms. As more of these compounds were discovered, they were classified on the basis of their targets, their effects on microbial growth, and on the range of microbes that they affected. Molecules that had an effect on both Gram-positive and Gram-negative bacteria were classified as **broad spectrum**, and those with effects on only one group as **narrow spectrum**. Over time, we have perfected ways to modify the structure of the original molecules chemically or genetically, to broaden their spectrum of activity. This can create some confusion in classifying large groups of antibiotics. For example, an originally narrow-spectrum drug can give rise to several broad spectrum antibiotics.

There are now a number of diverse molecules that inhibit the growth of bacteria. **Table 19.2** lists many of the major groups of antibiotics in clinical use today, their spectra, their cellular targets, and their effect.

Antibiotic class	Drug	Primary effect	Spectrum	Bacterial target
Aminoglycoside	Gentamicin, neomycin, tobramycin, amikacin, kanamycin	Cidal	Gram −, anaerobes	Protein synthesis
Carbapenems	Imipenem, doripenem, meropenem	Cidal	Gram +, Gram −, MRSA resistant bacteria	Cell wall
Cephalosporins[a]	Cefalexina, cefoxitin, ceftiofur, cefepime	Cidal	Gram +, Gram −; each new generation has increased Gram − activity	Cell wall
Glycopeptides	Vancomycin, teicoplanin, televancin	Static/cidal	Gram +	Cell wall
Lincosamides	Clindamycin, lincomycin	Static	Gram +, some protozoan	Protein synthesis
Lipopeptides	Daptomycin	Cidal	Gram +	Protein synthesis
Macrolides[b]	Azithromycin, spiramycin, erythromycin	Static	Gram +, some rickettsial	Protein synthesis
Nitrofurans	Nitrofurantoin, furazolidone	Static/cidal	Gram +, Gram −, some protozoan (dose dependent)	Ribosomal proteins, DNA and metabolism
Oxazolidonones	Linezolid	Static	VRSA Gram −	Protein synthesis
Penicillins	Penicillin, amoxicillin, ampicillin, carbenicillin	Cidal	Gram +, Gram −	Cell wall
Quinolones	Ciprofloxicin, levofloxicin, moxifloxacin	Cidal	Gram +, Gram −	Bacterial nucleic acids
Rifamycins	Rifampicin	Cidal	Gram +, Gram −	Bacterial nucleic acids
Sulfonimides	Sulfamethizole, sulfadiazin, sulfamethoxizole	Static	Gram +, Gram −	Inhibits folate synthesis
Tetracyclines	Oxytetracycline, doxycycline	Static	Gram +, Gram −	Protein synthesis

[a]First-generation cephalosporins have been preferentially spelled with 'ceph-' in the United States, Australia, and New Zealand. However, most European countries have adopted the International nonproprietary spelling using 'cef-', and all subsequent generations use 'cef-' even in the United States.

[b]Often included or closely related to the macrolide group is the ketolide group, which are structurally similar to the macrolides. They are used to treat macrolide-resistant bacteria.

MRSA, methicillin-resistant *Staphylococcus aureus*.

Table 19.2 Properties of some common antibiotics.

Antibiotics can have different effects at different doses and may be listed as cidal and/or static. Many of these classes contain multiple generations of molecules that have been, and are being, continually modified, so only a few specific drug examples in any class are listed.

Antibiotic Structure

Penicillin was the first molecule to be studied structurally in detail for its antimicrobial properties and has been the template for the development of an entire group of antibiotics (more than 50 so far). In its native form, penicillin contains a core four-sided ring structure called the beta (β)-lactam ring (**Figure 19.2**). This structure is also referred to as the

penicillin basic structure

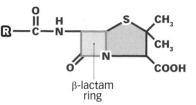

natural penicillins

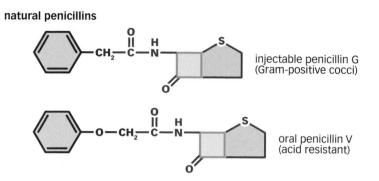

injectable penicillin G
(Gram-positive cocci)

oral penicillin V
(acid resistant)

semi-synthetic penicillins

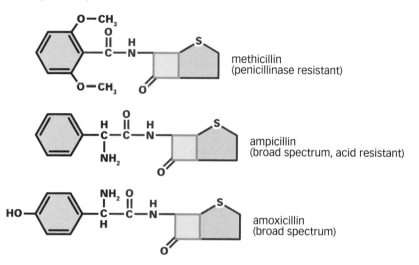

methicillin
(penicillinase resistant)

ampicillin
(broad spectrum, acid resistant)

amoxicillin
(broad spectrum)

nucleus of the penicillin molecule. All derivatives of penicillin contain this ring structure and are therefore called **β-lactam antibiotics**. Derivatives of penicillin such as ampicillin or amoxicillin contain an additional specific structure, referred to as a side chain and designated generically by the letter R, attached to this ring. Changing the side chain changes the antimicrobial activity of the penicillin derivative as well as its resistance to stomach acid and its overall half-life in the body.

Figure 19.2 shows several examples of this kind of manipulation. Here we see that penicillin can be found in two natural forms called **penicillin G** and **V** and that the three synthetic forms of penicillin shown all have the same core structure of the β-lactam ring. However, ampicillin, methicillin, and amoxicillin all have distinctly different side chains attached to the nucleus. These examples are referred to as semi-synthetic forms of penicillin because the modifications can be made in a laboratory. Modifying side chains has become an important part of the search for new antibiotics. Chemists can create and modify side chains that are attached to the core structure, thereby producing new semi-synthetic forms of penicillin. It is important to note that changing a side chain does not guarantee that a functional antibiotic will be formed. In some cases the newly derived molecule may have an increased spectrum of activity (that is, the number of different organisms it affects). In other cases, the synthetic side chain may result in a diminution or loss of activity.

Natural penicillin has a very narrow spectrum and reacts against only a small group of Gram-positive bacteria. Modifying penicillin to form ampicillin broadens the spectrum to include Gram-negative bacteria, whereas modifying it to carbenicillin or ticarcillin can broaden the spectrum to include *Pseudomonas* species. Existing **semi-synthetic penicillins** can be

Figure 19.2 Natural and semi-synthetic forms of penicillin. Natural forms of penicillin (G and V) can be modified by changing the side group (shown in red) on the β-lactam ring. This can change not only the spectrum of activity but also the susceptibility to acids and enzymes.

Fast Fact The penicillin molecule has undergone a number of synthetic modifications, and there are now five categories of the drug based on narrow versus broad spectrum and the reactivity to the organism *Pseudomonas aeruginosa*.

further modified to increase the efficiency of inhibiting bacterial growth. For example, ampicillin can be further modified to mezlocillin or azlocillin.

The same kind of manipulation can be seen with the cephalosporin family of antibiotics, in which modification of the natural molecule has resulted in five generations of semi-synthetics, three of which are shown in **Figure 19.3**. As with penicillin, you will note that modifications result from changing the side chains and leaving the core intact. These modifications also change its reactivity patterns and therefore its spectrum.

Keep in Mind

- Antibiotics are classified by their spectra of reactivity, either broad or narrow, and their effect, cidal or static.

- Penicillin is composed of a core ring structure known as the β-lactam ring.

- Natural penicillin is found in two forms, G and V, both of which have a narrow spectrum of activity.

- Modification or addition of side chains to the core structure of naturally occurring antibiotics can increase their effect and their spectrum of activity.

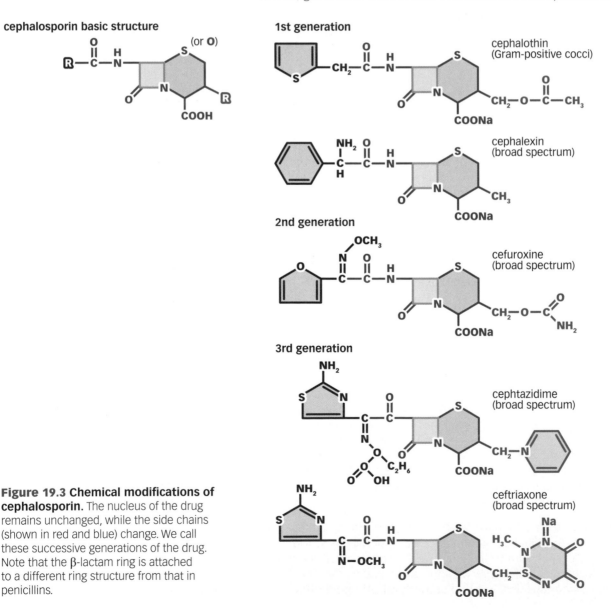

Figure 19.3 Chemical modifications of cephalosporin. The nucleus of the drug remains unchanged, while the side chains (shown in red and blue) change. We call these successive generations of the drug. Note that the β-lactam ring is attached to a different ring structure from that in penicillins.

ANTIBIOTIC TARGETS

One of the fundamental criteria in the selection of an antibiotic for medical use is **selective toxicity**, meaning that the antibiotic should be destructive to the disease-causing organism but have little or no effect on the human host. Achieving such selective toxicity is no trivial matter, and in fact many chemicals that are useful in restricting bacterial growth are inherently powerful toxins and cannot be used therapeutically. The introduction of any chemicals, especially in large quantities, to the human body can disrupt the body's homeostasis (the balance of chemicals and chemical reactions), with potentially harmful results. In addition, many antibiotic molecules can be toxic to the host if administered at concentrations that are too high.

The first antibiotic to be discovered, penicillin, was naturally selectively toxic. This was unfortunate because it gave a false sense of security regarding antibiotic molecules, suggesting that all of them could be used without regard to toxicity. This is sadly not the case and a key point in minimizing toxicity is creating antibiotics with very specific bacterial targets. When we think of potential targets for antibiotic activity, we can use what we learned in Chapter 9 about the structures associated with bacteria. Antibiotic targets can be subdivided into five major groups:

- The bacterial cell wall
- The bacterial plasma membrane
- Synthesis of bacterial proteins
- Bacterial nucleic acids
- Bacterial metabolism

Figure 19.4 shows each of these targets and some of the major antibiotics that affect them.

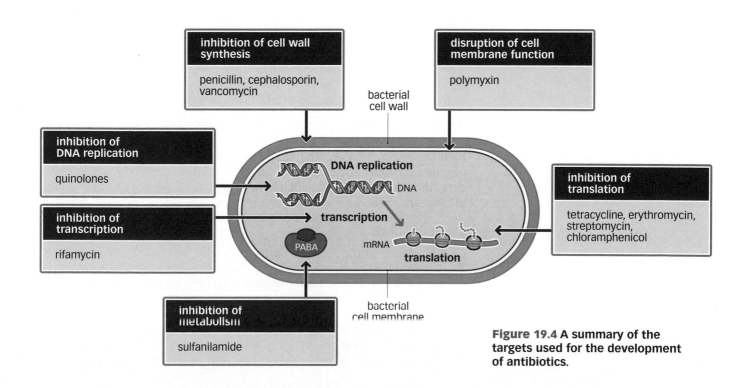

Figure 19.4 A summary of the targets used for the development of antibiotics.

The Bacterial Cell Wall

The most appealing target for antibiotics is the bacterial cell wall, because it is found in bacteria but not in humans. It therefore meets the criterion of selective toxicity. Recall that the bacterial cell wall is a structure built to keep the outside out and the inside in, making it a necessary component for the survival of the bacteria. This wall is found in both Gram-positive and Gram-negative bacteria and is a complex network of linked molecules (Chapter 9). Antibiotics that target the bacterial cell wall include the penicillins, cephalosporins, carbapenems, and monobactams (all β-lactam antibiotics) and glycopeptides. Each has slightly differing mechanisms of action against the synthesis of the bacterial cell wall.

Penicillins

Many enzymatic interactions take place during the construction of cell wall components. The enzymes involved in these processes can therefore be used as targets of antibiotic molecules. As we have learned, the cell wall is composed of layers of peptidoglycan, which is made up of repeating units of N-acetylglucosamine (NAG) and N-acetylmuramic acid (NAM) (see Figure 9.2) The NAG and NAM molecules are cross-linked through the activity of transpeptidase enzymes known as **penicillin-binding proteins** (**PBPs**). They are called penicillin-binding proteins because the β-lactam ring of penicillin binds to these proteins, disrupting bacterial cell wall synthesis by preventing the cross-linking of peptidoglycans (see Figure 9.4). The result of this action is that the cell wall is not properly cross-linked; it is therefore weak and unable to withstand the environmental pressures that are always present.

There can be large numbers of PBPs in a bacterial cell wall. For example, the cell wall in one *Escherichia coli* cell has more than 2500 types of these (although only about 300 can be used by penicillin to kill the cell). New cell wall is continuously being built during the active growth phase, so it is at this time that penicillins are most effective. The more rapidly the bacteria are dividing, the more devastating is the effect of penicillins. Although both Gram-positive and Gram-negative cells contain peptidoglycan, the amounts are markedly less in Gram-negative bacteria. For this reason, Gram-negative bacteria are normally less sensitive to penicillin than are Gram-positive organisms. The semi-synthetic forms of penicillin we looked at earlier have enhanced activity against Gram-negative bacteria. It should be noted that though PBPs are useful as targets of antibiotic activity, mutations in the bacterial genes that code for them could result in the organism's becoming resistant to penicillin.

Fast Fact MRSA is resistant to methicillin because of a mutation in one of its PBP genes.

Another mechanism of resistance to the penicillins involves an enzyme called **β-lactamase** that breaks open the β-lactam ring structure through hydrolysis and nullifies the drug's antibacterial properties. There are augmented forms of penicillin derivatives (for example Augmentin, a derivative that contains clavulanic acid, which blocks the enzyme) which prevent that process.

Cephalosporins

Cephalosporin antibiotics act similarly to the penicillins, by preventing the construction of a stable cell wall. However, the cephalosporins have a much greater effect on Gram-negative bacteria and are naturally broad-spectrum antibiotics. One major reason for their success is that they are less susceptible to the β-lactamase enzymes that inactivate penicillin. There are five generations of cephalosporins, with more than 70 versions now in use. They are one of the most widely prescribed antibiotics, with the fifth generation being used against resistant infections such as MRSA.

The mechanism of action of this antibiotic class is similar to that of the penicillins, but cephalosporins have the capacity to penetrate through the porin molecules (see Figure 9.5) found in the outer layer of Gram-negative bacteria. As the side chains of the cephalosporin molecule continue to be modified, the spectrum of reactivity continues to increase. Cephalosporins have an excellent safety record with regard to adverse reactions and are therefore frequently used both preoperatively and postoperatively. However, the frequent use of these antibiotics in hospital settings has increased the number of bacteria that are resistant to early-generation cephalosporins.

Carbapenems

The carbapenems also contain a β-lactam ring as part of their structure and inhibit the synthesis of bacterial cell walls. However, antibiotics in this class are less susceptible to β-lactamase enzyme activity than are penicillins and cephalosporins, so they can be effective in infections resistant to other β-lactam antibiotics. They reduce the activity of β-lactamase by binding to it. Carbapenems have a very broad spectrum of antibacterial activity, and two have been approved for clinical use in humans: imipenem and meropenem. Both of these antibiotics are useful against *Pseudomonas* species.

Primaxin is a combination of imipenem and a molecule called cilastatin. In the human body, the imipenem is antibacterial and the cilastatin prevents premature destruction of the antibiotic molecule by the kidneys, allowing the drug to remain active longer. Primaxin has become the antibiotic of choice for many nosocomial infections.

Monobactams

Monobactam molecules have a different structure from the penicillins and cephalosporins in that their β-lactam ring is not attached to another ring (see Figures 19.2 and 19.3). They are not susceptible to β-lactamase and can therefore be used against bacteria that have antibiotic resistance related to β-lactamase. Monobactams are only active against Gram-negative organisms, and only one is commercially available: aztrenam. It has good reactivity against *E. coli* and *Pseudomonas* species, both of which are becoming very dangerous hospital infection agents.

Glycopeptide Antibiotics

Glycopeptide antibiotics, such as vancomycin and teicoplanin, affect cell wall production in a different way from the β-lactam antibiotics. Both vancomycin and teicoplanin inhibit cell wall synthesis by binding to the amino acids that make up the wall's peptidoglycan molecules, preventing the addition of new units, but neither can penetrate the porins of Gram-negative cells. They therefore have a very narrow spectrum, being restricted to use against Gram-positive organisms. Glycopeptide antibiotics work on different parts of peptidoglycan synthesis from those affected by penicillin, making the use of these antibiotics in concert with others potentially attractive for enhanced protection. Teicoplanin differs structurally from vancomycin and has not been approved for use in the United States, although it is used in other countries. Both vancomycin and teicoplanin are derived from *Streptomyces* bacteria. When discovered, they were found to have serious side effects, but that toxicity level has been reduced in recent years through improvements in purification.

Fast Fact Vancomycin has achieved a new status as the appearance and magnitude of bacterial antibiotic resistance have increased. It is the last line of antibiotic defense against MRSA as well as against certain streptococci, enterococci, and other pathogens. Some strains have now developed resistance to vancomycin.

Peptide Antibiotics Effective Against Mycobacteria

Isoniazid and ethambutol are used against bacteria that have modified their cell walls for further protection against environmental conditions

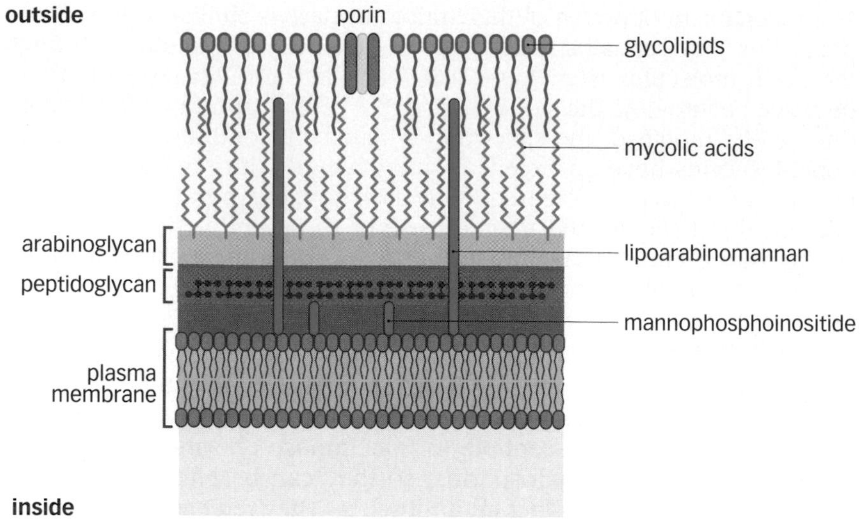

Figure 19.5 Cell wall of a mycobacterium showing the position of mycolic acids.

and host cell defenses. The *Mycobacterium* species, including the pathogens that cause tuberculosis and leprosy, are a good example. As we saw in Chapter 9, the cell walls of these organisms are modified by the incorporation of mycolic acids (see **Figure 19.5**), which are waxy components that add extra protection and make these organisms uncommonly resistant to most antibiotics.

Isoniazid is very effective against mycobacteria because of a mechanism that seems to inhibit the synthesis of mycolic acid. Ethambutol alone is actually not very effective against mycobacterial species, but given in concert with isoniazid it inhibits the incorporation of mycolic acid into the growing bacterial cell wall. The combination of isoniazid, ethambutol, and rifampin is now the treatment of choice for tuberculosis, with the administration of these combinations of drugs also lowering the potential for the development of antibiotic resistance.

Polypeptide Antibiotics

Polypeptide antibiotics such as **bacitracin** and polymixin B are used for superficial infections of Gram-positive or Gram-negative organisms such as *Staphylococcus* and *Streptococcus* species. Bacitracin inhibits the transport of the building blocks of peptidoglycans used to build the cell wall. Polymixin B alters cell wall permeability. These antibiotics are highly toxic and fail to meet the standards of selective toxicity necessary for internal use; they can only be used externally.

The Bacterial Plasma Membrane

The bacterial plasma membrane is involved with many important physiological functions. It is therefore a prime target for antibiotics because any disruption of this membrane will destroy the bacterium's ability to survive. Unfortunately, the structure of a bacterial plasma membrane (the phospholipid bilayer) is remarkably similar to that found in eukaryotic host cells. This similarity makes it difficult for antibiotics that attack this bilayer to be selectively toxic and appropriate for medical use. Consequently, there are few antibiotics in use that target the plasma membrane, and they have restricted use. Polymyxin B is used in topical administrations only, and daptomycin (which has been associated with causing a life-threatening pneumonia) is only used in intravenous dosing regimens in cases of confirmed antibiotic resistance.

Synthesis of Bacterial Proteins

On the basis of what we have discussed about the structure and function of proteins and their importance to all living cells, it is easy to understand how any disruption in the production of these molecules could be devastating to a bacterial cell. As we saw in Chapter 11, proteins are assembled at a ribosome in combination with messenger RNA, and the assembly of a protein begins with the formation of an intact ribosome from two ribosomal subunits (see Figure 11.19). Here, amino acids are linked together through peptide bond formation. Because ribosomes are found in both prokaryotic and eukaryotic cells, the selection of protein synthesis as a target for antibiotic therapy against bacteria would, at first glance, seem to break the rule of selective toxicity. However, the ribosomes of prokaryotes are not the same as those of eukaryotes, so antibiotics that target bacterial ribosomes meet the criterion of selective toxicity. The mitochondria of eukaryotic cells contain ribosomes that are similar to the ribosomes in prokaryotic cells (and different from cytoplasmic eukaryotic ribosomes). Consequently, there can be some antibiotic interference in normal eukaryotic cell function if antibiotics that interrupt protein synthesis are given in excessive amounts.

Recall from Chapter 11 that the intact 70S ribosome is composed of two parts, a small 30S subunit and a larger 50S subunit, which join together when they encounter mRNA. Each subunit contains ribosomal RNA and proteins, with the sites for protein synthesis (called A and P) having a specific three-dimensional configuration. When both sites are occupied by transfer RNA molecules carrying amino acids, the amino acids are in close proximity to each other and will join together through the formation of a peptide bond. Transfer RNA molecules are accompanied to the sites by chaperone proteins that assist in the correct orientation of the RNA molecules, and these chaperone proteins are of interest as possible new targets for antibiotic activity.

Both the 30S and 50S subunits have been examined by X-ray crystallography, and their three-dimensional structures have been determined. Knowing the exact three-dimensional structure has given researchers a new insight into the mechanism of action of antibiotics that target ribosomes. Both subunits of the bacterial ribosome are targets for antibiotic action. For example, streptomycin acts on the 30S subunit and upsets the accuracy of the translation, whereas chloramphenicol acts on the 50S subunit and prevents elongation of the peptide chain (**Figure 19.6a**). Some antibiotics, such as spectinomycin, interfere in the process of peptide elongation by blocking the translocation of the growing peptide chain from the A site to the P site; others, such as paromycin, interfere with the decoding process of the message.

Macrolides

Macrolides such as erythromycin act by blocking the elongation of the peptide chain forming in the 50S subunit, inhibiting ribosomal translocation and potentially dissociating peptidyl tRNA from the ribosome (**Figure 19.6b**). Natural macrolides have a narrow spectrum and are used only in the treatment of Gram-positive infections. The synthetic forms of erythromycin known as azithromycin and clarithromycin have expanded spectra and fewer side effects than erythromycin; they are used to treat respiratory infections.

Tetracyclines

Tetracyclines, which have been used since the late 1940s, are bacteriostatic and block the arrival of tRNA at the A site (**Figure 19.6c**).

Fast Fact Several *Streptomyces* species make a pair of antibiotics called pristinamycin and streptogramin, which work synergistically to block translation at the 50S subunit. These antibiotics have been synthetically modified and make up the antibiotic known as Synercid®, which has recently been approved for the treatment of vancomycin-resistant enterococcal (VRE) infections.

Figure 19.6 Antibiotic targets of the bacterial ribosome. The bacterial 70S ribosome is composed of two subunits, 30S and 50S. Several targets of antibiotics are located on these subunits. Some mechanisms of antibiotic inactivation involve improper orientation of the mRNA, inability to form peptide bonds, or inhibition of peptide elongation.

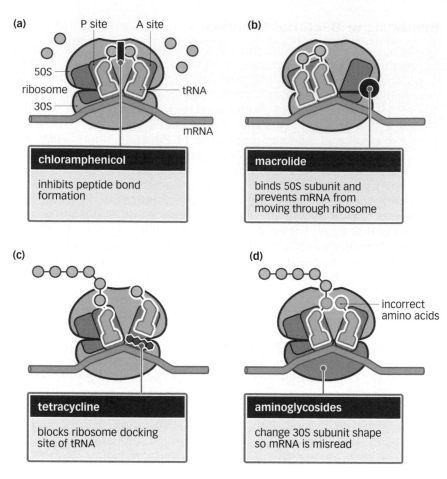

(a)

P site   A site

50S
ribosome
30S

tRNA

mRNA

chloramphenicol

inhibits peptide bond formation

(b)

macrolide

binds 50S subunit and prevents mRNA from moving through ribosome

(c)

tetracycline

blocks ribosome docking site of tRNA

(d)

incorrect amino acids

aminoglycosides

change 30S subunit shape so mRNA is misread

Unfortunately, these antibiotics have been in use for so long that many bacteria have developed resistance to them, and their use has steadily declined over the years. They also cause a number of side effects including phototoxicity and gastrointestinal upset, and are not recommended for use in children under 8 years of age (they can cause permanent discoloration of teeth and nails). It is still the treatment of choice for certain diseases such as those caused by *Chlamydia*, *Rickettsia*, and spirochetes. The long-acting doxycycline is used to treat respiratory infections, and it retains some effect against *Yersinia pestis*.

Aminoglycosides

Aminoglycoside antibiotics, which have also been around since the 1940s, are used in combination with other antibiotics and target the 16S RNA portion of the 30S ribosomal subunit (**Figure 19.6d**). There have been several generations of these compounds, such as gentamicin and tobramycin, and although they are potent against Gram-negative organisms, they are not very effective against Gram-positive bacteria (they are narrow-spectrum drugs). They have been used in combination with β-lactam antibiotics to fight *Pseudomonas* infections, but they produce significant renal toxicity and ototoxicity (damage to hearing). They also require either intramuscular or intravenous injection as the route of administration, which decreases their usefulness in the treatment of common ailments.

Bacterial Nucleic Acids

Perhaps the most obvious target of antibiotic therapy would be the nucleic acids DNA and RNA. Because these molecules have critical roles in the

Fast Fact Oxazolidinones are the newest class of antibiotics that are inhibitors of bacterial protein synthesis. Linezolid was the first in the class to be approved for use clinically and is highly effective against MRSA and VRE infections.

reproduction of bacteria, they would be excellent targets: any disruption in their function will result in the death of the bacteria. Unfortunately, selective toxicity is a major problem. DNA and RNA are universal, and the structure of DNA and RNA in bacteria is no different from the structure of these molecules in humans. Over the years, a variety of potential antibiotics have been developed to target nucleic acids, but they have been found to be unusable. However, two classes of synthetic compound, the quinolones and the rifamycins, have been found to be somewhat effective in attacking bacterial nucleic acids while not harming the same molecules in humans.

Quinolones

Quinolones block either DNA replication or DNA repair. Replication involves a variety of protein molecules, including the enzymes topoisomerase and gyrase. These enzymes are involved with making small cuts in the supercoiled structure of DNA so that it can uncoil, unwind, and separate (Chapter 11). Remember that the three-dimensional structure of enzymes is critical for proper function and that any disruption of this structure will result in a loss of function. This makes them very valuable targets for antibiotics that can disrupt the three-dimensional structure. Perhaps more importantly, the topoisomerase molecules found in bacteria are different from those found in eukaryotic organisms, meaning that antibiotics that attack this enzyme will achieve selective toxicity.

Fluoroquinolones, such as levofloxacin and ciprofloxacin, are used against both Gram-positive and Gram-negative organisms to treat urinary tract infections, osteomyelitis, and community-acquired pneumonia and gastroenteritis. Ciprofloxacin has made the headlines as the primary antibiotic used against infections caused by *Bacillus anthracis* (anthrax). We discuss this disease more in Chapter 21.

Rifamycins

The rifamycins were originally isolated from *Streptomyces*. These antibiotics bind to RNA polymerase and disrupt its three-dimensional shape, rendering it unable to function properly. It is interesting to note that this binding is allosteric: it occurs away from the active site of the polymerase molecule. Once the activity of this molecule has been blocked, there can be no transcription and therefore no protein synthesis, which is lethal. The rifamycin known as rifampin is the only antibiotic with this mechanism of action and is used only in combination therapy because resistance develops rapidly if it is used alone. As mentioned earlier, rifampin is used in combination therapy for tuberculosis. All the rifamycins are considered selectively toxic because the sensitivity of eukaryotic RNA polymerase is 100 times less than that of bacterial RNA polymerase.

Bacterial Metabolism

Two other targets for the inhibition of bacterial growth are the production of nucleic acid precursors and the metabolic pathways that occur at the plasma membrane. Fortunately, there are several metabolic pathways that are exclusive to bacteria, and interruption of these pathways selectively inhibits bacterial growth without affecting eukaryotic organisms. Perhaps the best example of this is the metabolism of folic acid, a molecule needed for the synthesis of nucleic acids (Figure 19.7). One of the intermediates in the pathway is *para*-aminobenzoic acid (PABA). Sulfa drugs competitively inhibit the function of the enzyme that incorporates the PABA molecule into the folic acid metabolic pathway. It is referred to as competitive inhibition because the sulfa molecule is remarkably

Figure 19.7 Competitive inhibition of metabolism. Panel a: The metabolic pathway used to generate nucleic acids and the point at which sulfa drugs inhibit this pathway. Panel b: The reason for this blockade. As you can see, the structure of sulfa drugs is very similar to that of the PABA molecule and will competitively inhibit the pathway.

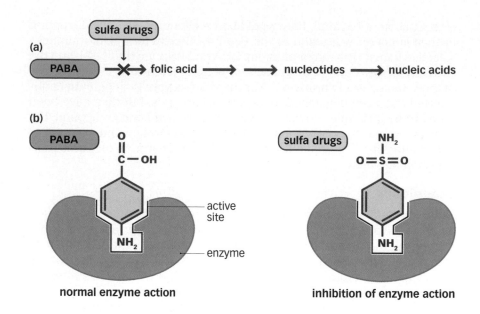

similar in structure to the PABA molecule. X-ray diffraction studies have confirmed this similarity.

The enzyme simply gets fooled into incorporating the sulfa molecule into the folic acid structure instead of PABA. Incorporation of sulfa instead of PABA stops the pathway and is a lethal event for bacteria because they cannot survive without folic acid. This inhibition is selectively toxic because unlike bacteria, which must synthesize folic acid, we obtain folic acid through our diet.

Sulfa drugs have been in use longer than any other form of antibacterial inhibitor. The sulfa drug sulfamethoxazole is usually used in combination with trimethoprim for urinary tract infections and for *Pneumocystis* infections that occur secondarily to AIDS. Both of these drugs block a step in folic acid metabolism. Sulfamethoxazole blocks the biosynthesis of folate, whereas trimethoprim blocks intermediates that are required for the synthesis of DNA. Unfortunately, because these drugs have been used for such a long time, bacterial resistance has continued to increase, decreasing their usefulness.

Keep in Mind

- Antibiotics must satisfy the criterion of selective toxicity: they must react against bacteria but not against the host.

- Most antibiotics have dose-dependent activity and toxicities.

- There are five targets for antibiotics: the bacterial cell wall, the bacterial plasma membrane, protein synthesis, metabolic inhibition, and nucleic acids.

- The bacterial cell wall is the most easily selectively toxic because human cells do not have a cell wall.

- Nucleic acids and the plasma membrane are the least selectively toxic targets because they are very similar in bacteria and in host cells.

ANTIVIRAL DRUGS

Viruses present a different set of problems for therapy because they are obligate intracellular parasites. Many of the drugs that would eliminate the virus would be dangerous, if not lethal, to non-infected cells, so selective toxicity is difficult to maintain when searching for antiviral agents. Because many viruses are difficult, if not impossible, to grow in laboratory conditions, it is challenging to test the efficacy of potential antiviral drugs. This is compounded by the lack of rapid tests that can differentiate between various viral infections. Furthermore, successful antiviral drugs must eliminate all virions because the escape of even one virion from an infected cell into the host's blood could quickly restart the infectious cycle.

The first antiviral to be used therapeutically was the sulfa drug derivative thiosemicarbazone in the 1950s. These compounds are still studied today and have been shown to have not only antiviral properties but also anti-protozoal properties, and some anti-cancer properties through iron and copper binding. The early development of other antiviral drugs was very slow, with fewer than 10 drugs being developed in the next 50 years. A technique called blind screening was employed, which involved looking for any chemical (natural or synthetic) that might inhibit viral replication and testing it in a trial-and-error style process.

The advent of molecular techniques using recombinant DNA, as well as a better understanding of many viral life cycles, has made the use of blind screening obsolete. By identifying specific viral proteins, researchers are able to select chemicals that have potential as antiviral drugs based on their action against that protein. The viral protein is then produced using recombinant DNA techniques and tested against the selected potential drugs. For the drug to be selectively toxic, its target must be a protein that is essential to the virus but not to the eukaryotic host cell. **Figure 19.8** identifies the life-cycle stages and viral proteins that are targets for antiviral drugs. Development of antiviral drugs since the 1990s has been rapid, and there are now more than 40 available. Let's look at a few in more detail.

Fast Fact Thiosemicarbazones were discovered to have good antiviral effect against pox viruses in the 1950s. They showed promise in the area of both prophylaxis and treatment of small pox viral infections, but were abandoned with the success of the smallpox eradication program.

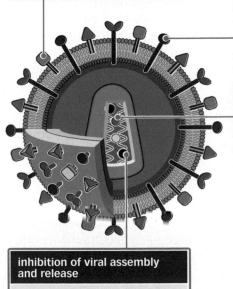

inhibition of viral assembly and release

neuraminidase inhibitors – oseltamivir

inhibition of viral entry and uncoating

amantadine

also potential area for new antiviral drugs

inhibition of viral DNA polymerases (including reserve transcriptase)

nucleotide analogs – cidofovir

nucleotide analogs – acyclovir, AZT

non-nucleoside inhibitors – foscarnet, nevirapine

inhibition of viral assembly and release

protease inhibitors – saquinavir

Figure 19.8 A summary of the targets used for the development of antivirals.

Nucleoside Analogs

Nucleoside analogs are molecules that are structurally similar to the nucleosides used in DNA synthesis. They are converted to a nucleotide analog within the eukaryotic cell and incorporated into the newly synthesized DNA chain, resulting in the premature termination of DNA synthesis. Because the DNA in viruses and host cells is the same molecule there are selective toxicity issues, but they are not as great as you might initially think because viral DNA polymerase enzymes are much more likely to use these analog nucleotides than host cell DNA polymerases. Nucleoside analogs are therefore used to treat infections by viruses that carry their own DNA polymerase enzymes such as the herpesviruses and retroviruses. They are only effective when viral DNA is being synthesized; they cannot treat infections in a latent state.

Acyclovir

Acyclovir is a specific, nontoxic drug that is highly effective against both genital and oral herpes simplex infections and has been used with some success in the treatment of varicella-zoster (chickenpox and shingles). It was discovered in 1974 but was not widely used until the mid 1980s; it can be taken intravenously or orally, or used topically. Acyclovir is a nucleoside analog of guanine and is produced as a **prodrug**, which means a drug that must be activated enzymatically once in the patient's body. This activation is accomplished by enzymes found only in infected cells, making acyclovir more selectively toxic.

Acyclovir is poorly soluble and not very effective when taken orally, which led to the development of analogs such as famciclovir and valaciclovir. These variations of acyclovir (Valtrex® and Famvir®) are very effective and widely used.

AZT

Azidothymidine is also a nucleoside analog and was the first antiretroviral drug used in the treatment of HIV under the name AZT. Remember that HIV is a retrovirus and carries the enzyme reverse transcriptase, which is unique to retroviruses. Reverse transcriptase transcribes RNA to DNA, making it an RNA-dependent DNA polymerase. Any drug that specifically inhibits reverse transcriptase would automatically be highly selectively toxic, making reverse transcriptase an attractive target for drug development. Like acyclovir, AZT must be converted into an active form, but for AZT this is accomplished by cellular enzymes. Activated AZT is structurally similar to thymidine and can be incorporated into the reverse transcription process in place of thymidine, which terminates DNA synthesis (see **Movie 19.1**). Cellular DNA polymerases have proofreading ability (Chapter 11) and can detect the incorrect base, quickly cutting it out and replacing it with the correct one. Reverse transcriptase doesn't have this ability and so cannot tell when an incorrect base has been incorporated, making AZT selectively toxic.

AZT significantly reduces the replication of HIV, creating clinical improvement and lowering transmission. It does not, however, completely stop viral replication and is therefore typically used in combination with other antiretroviral therapies. The use of combination therapy also helps prevent drug resistance from developing. Since the mid 1980s, at least seven nucleoside analogs, including didanosine (ddI) and lamivudine (3TC), have been licensed for the treatment of HIV.

Fast Fact In the long term, high doses of AZT can have significant side effects and were a problem during early treatment regimes.

Other nucleoside analogs include **ganciclovir**, used to treat cytomegalovirus infections in immunocompromised patients, and **ribavirin**, used to treat respiratory syncytial virus (RSV) infections in newborns.

Nucleotide Analogs

Nucleotide analogs work in the same way as nucleoside analogs (selectively inhibiting viral DNA polymerases), but they do not require the initial activation step. Cidofovir is a cytidine analog that is used primarily to treat cytomegalovirus retinitis infections in patients with HIV. However, it has been shown to have *in vivo* effectiveness against a number of other viruses including papillomavirus, herpesvirus, and poxvirus. As it does not require activation by virus-specific enzymes, the virus cannot develop resistance to the activation stage, and cidofovir shows continued activity against acyclovir-resistant viral strains of herpes simplex virus and cytomegalovirus.

Non-nucleoside Inhibitors

Non-nucleoside inhibitors also inhibit viral DNA polymerase enzymes, but they have a different mechanism. They are allosteric inhibitors (Chapter 3) and bind at some distance from the active site, disrupting the shape of the enzyme and therefore inhibiting its function. Foscarnet is used to treat herpesvirus infections that are resistant to nucleoside analogs and also as an HIV reverse transcriptase inhibitor. It has been found to be toxic to both kidney and bone marrow; it is therefore a drug of last resort for life-threatening infections. There are also non-nucleoside inhibitors, such as nevirapine and delavirdine, that specifically inhibit reverse transcriptase and are used in HIV combination therapy.

Inhibiting Viral Assembly and Release

After viral replication in a host cell, virions must be correctly assembled and released into the body of the host. This is a complex process that involves enzymes specific to viruses, and therefore provides potential targets for antiviral therapy that fulfill the need for selective toxicity.

Protease Inhibitors

Protease inhibitors, such as saquinavir and ritonavir used in HIV therapy, are potent anti-retroviral drugs that fulfill the need for selective toxicity because they specifically affect a viral protease enzyme. By inhibiting the viral protease, they interfere with the assembly of virions and result in the production of defective, non-infectious viral particles. Because protease inhibitors are so specific, there is a danger of resistance developing; they are therefore used most effectively in combination therapy with agents such as reverse transcriptase inhibitors. They must also be used at the correct dose over a prolonged period.

Fast Fact Two protease inhibitors are also used in the treatment of hepatitis C infections, which are a now a major burden of health care systems.

Protease inhibitors are also being explored for many viruses that cause chronic diseases, such as the picornaviruses (including the rhinoviruses responsible for the common cold), or are at risk of causing large-scale epidemics such as the coronaviruses (including the viruses that cause SARS and MERS).

Neuraminidase Inhibitors

Neuraminidase inhibitors target the neuraminidase protein (N) of the influenza virus and prevent new virions from budding from the host cell. Drugs in this class, such as zanamivir and oseltamivir (Tamiflu®), are now

Fast Fact There has been much debate about the effectiveness of neuraminidase inhibitors, but they are recommended for use in those at risk of serious complications. They are most effective if taken within 48 hours of symptoms appearing.

the drugs of choice for the treatment of influenza. They act against both influenza A and B viruses.

Inhibiting Viral Uncoating

Amantadine was the first highly specific potent antiviral agent used against influenza A viral infections. It binds a viral protein and inhibits uncoating of the viral particle. Unfortunately, the influenza A virus frequently mutates the protein target of amantadine, thereby becoming resistant to this drug. All of the influenza strains for the 2008/2009 flu season tested resistant to amantadine, and it is no longer recommended for use against influenza A in the United States. Influenza B does not contain the target protein and is therefore unaffected by amantadine.

Inhibiting Viral Entry

An area of exciting research and development right now involves trying to inhibit viral entry to host cells. Viruses attach to the outside of the host cell, penetrate the plasma membrane, and uncoat their genetic material (see Figure 12.8).This is an attractive area for antiviral activity, because the virus is more accessible to drugs before it has gained entry into the host cell and some of the viral proteins involved are common across several groups of viruses. One drug could therefore potentially treat infections by more than one type of virus. New drug development is focusing on small molecules that inhibit one of the proteins that viruses use to fuse with the plasma membrane and is showing very exciting success rates.

DRACO

Another area of development for an antiviral drug that could act against several types of virus is Double-stranded RNA Activated Caspase Oligomerizer (DRACO). These drugs show potential to be effective against many infectious viruses, including some of the most feared: Ebola virus, Marburg virus, and dengue virus. DRACO is reported to be extremely selective for only infected cells and causes rapid apoptosis based on the length of double-stranded RNA (dsRNA) strands detected in the cell, while leaving uninfected cells (which do not contain long dsRNA molecules) unharmed. In the studies published, it was effective against all 15 viruses tested, even when the host was challenged with lethal doses of virus.

Keep in Mind

- Viruses present problems for therapy because they are obligate intracellular microbes; once inside the host cell, selective toxicity is hard to achieve.

- Potential targets for antiviral drugs are specific viral proteins, which are selected on the basis of knowledge of viral life cycles.

- Many currently available antiviral drugs target viral DNA polymerases, including reverse transcriptase, and viral proteases.

- Viral entry and uncoating are also potential targets.

ANTIFUNGAL DRUGS

Fungi have previously been relatively innocuous organisms from the perspective of infectious diseases. However, the emergence of a substantial number of immunocompromised individuals (especially patients with AIDS) has led to an increase in secondary fungal infections. We will concentrate on specific fungal diseases in Part VI of this book; here we limit our discussion to drugs that are currently available for use against fungal infections.

| Fast Fact | Amantadine has side effects associated with the central nervous system and is sometimes used as a therapy in Parkinson's disease, although its effectiveness is disputed.

| Fast Fact | Historical treatment of fungal infection included surgical intervention and the ingestion of toxins such as potassium iodide, phenols, dyes, and oil of turpentine. Most of these were ineffective, although oil of turpentine had a narrow effect against cutaneous sporotrichosis.

The identification of drugs that can be used for fungal infections has been difficult because achieving selective toxicity is difficult. Fungi are eukaryotes, so many of the cellular proteins, structures, and processes are the same as those of the host, meaning that drugs that affect fungi are likely to affect the host in the same way and can have serious side effects.

However, there are some differences that can be exploited as targets for antifungal therapy: the fungal plasma membrane has a different composition of lipids, and fungal cells have a cell wall.

Antifungals That Affect the Plasma Membrane

There are two types of antifungal drug that exploit the different lipid composition of plasma membranes in humans and fungi: the lipid ergosterol is essential in the fungal cell membrane but does not occur in human cell membranes, and polyenes affect the fungal membrane directly whereas azoles and allylamines act indirectly.

Polyenes

Polyenes, such as amphotericin B, are produced by the soil bacterium *Streptomyces* and increase the permeability of the fungal membrane by interacting with ergosterol. For many years, amphotericin B was used for systemic fungal infections such as histoplasmosis and coccidiomycosis. However, it must be used with caution because it has severe and potentially lethal side effects, especially kidney failure. Amphotericin B has historically been used only intravenously, which also limits its use outside health care facilities. There is a lipid-based formulation that has fewer side effects, but it is more expensive and less readily available. There is also an oral formulation in development but it is not yet commercially available. Nystatin is too toxic for systemic use, but it can be used topically.

Azoles

There are two classes of azoles, imidazoles and triazoles, that inhibit the production of ergosterol. Clotrimazole and miconazole, which are derivatives of imidazole, are sold without a prescription and are routinely used topically for the treatment of athlete's foot and vaginal yeast infection. Ketoconazole was developed as a broad-spectrum antifungal, a less toxic alternative to the polyenes. It can be taken orally for systemic fungal infections. However, it was proven to be so hepatotoxic that there was a move to ban it from human use in the European Union. The least toxic forms of azoles are fluconazole and itraconazole, which have become widely used for systemic fungal infections. As a result of resistance developing, second-generation azoles (triazoles) are being developed, including voriconazole and posaconazole. These two offer broad-spectrum activity against many dimorphic fungi, yeasts, molds, and even some resistant strains of *Candida*. Finally, they have the added benefit of oral administration, making them highly accessible for therapeutic use.

Antifungals That Affect the Cell Wall

Fungal cells have a cell wall but animal cells do not, making it a good candidate for antifungal therapy. Echinocandins inhibit the synthesis of glucan in the fungal cell wall. They are very specific and narrow in their spectrum, and are primarily used against organisms in the *Candida* and *Aspergillus* family. With such a narrow spectrum of activity, they may seem to have little clinical relevance. However, the newer analogs (casofungin, micafungin, and anidulafungin) are used as a first line of defense against suspected candidiasis and are especially useful in the prophylactic treatment of hematopoietic transfusion recipients.

Fast Fact Many of the functions of ergosterol in fungal cell membranes are performed by cholesterol in animal cells.

Antifungals That Affect Nucleic Acid Synthesis

The antifungal drug **flucytosine** interferes with DNA and RNA synthesis and is used to treat systemic infections. It is taken up preferentially by fungi but has a high level of toxicity toward kidney and bone marrow that limits its use.

The exact mechanism of **pentamidine** is unknown, but it appears to disrupt DNA synthesis; it is used in the treatment of *Pneumocystis* pneumonia, a fungal pneumonia that is commonly seen in immunocompromised individuals, particularly patients with AIDS.

Fast Fact Pentamidine is also an effective treatment for protozoal infections such as leishmaniasis, trypanosomiasis, toxoplasmosis, and *Acanthamoeba* infections.

Keep in Mind

- Fungal infections have become more prevalent since the appearance of immunocompromised individuals, especially in those infected with HIV.

- Because fungi and host cells are eukaryotes, selective toxicity is hard to achieve and side effects of antifungal drugs can be serious.

- Targets for antifungals are ergosterol in the fungal cell membrane, and the fungal cell wall, neither of which is present in animal cells.

DRUGS FOR PARASITIC INFECTIONS

As we saw in Chapter 14, many of our most serious diseases are caused by protozoal organisms. One of the most widespread and devastating diseases, malaria, was overlooked in developed nations because the incidence of this disease was low in those places. And what was true for malaria has been true for many other parasitic infections: the development of drugs for them has lagged behind the development of antibacterial and antiviral drugs because parasitic infections do not occur at high frequency in developed nations. In other words, there was no money to be made in the development of these types of drug. International agencies and governments are now making funding available to develop drugs against parasitic infections, which has led to the advent of new synthetic drugs.

Many protozoans act in a similar fashion to viruses in that they are intracellular parasites, and once they are inside host cells it is very difficult to find effective therapies that will not harm the host and therefore achieve selective toxicity. Many antiparasitic drugs can have severe side effects on patients; some can even be lethal.

Anti-protozoan Drugs

Plasmodia are found in red blood cells, *Leishmania* is found in macrophages, and trypanosomes are found in many different cell types. Many of these organisms can even produce cytokines that down-regulate host immune responses. The two main targets for most anti-protozoan drugs are the asexual reproduction stage and folate metabolism.

Antimalarials

The first widely used antiparasitic drug was quinine. Quinine as a treatment for malaria has been used since the 1600s, and over the centuries it has been chemically modified into several synthetic forms, such as chloroquine and mefloquine. Another derivative, diiodohydroxyquin, is used for the treatment of several intestinal amebic diseases but has been found to be toxic to the optic nerve. All these drugs act against the asexual stage of *Plasmodium* when it is resident in red blood cells. The drugs enter infected red blood cells and cause them to digest themselves, destroying the parasite in the process. Chloroquine also affects internal pH, causes a buildup of toxic metabolites, and interferes with nucleic acid synthesis in the parasite. It is a cheap and effective antimalarial drug that is widely used as a **prophylactic** as well as a treatment, but it can have significant side effects.

As with other pathogens, *Plasmodium* can develop resistance to drugs, and many areas in the world have widespread chloroquine-resistant strains of *Plasmodium*. In these areas, other quinine derivatives such as mefloquine must be substituted for effective action, but chloroquine is still in heavy use in sub-Saharan Africa. As a result of resistance to quinine-based drugs, artemisinin is now widely prescribed for malarial infection. The WHO has specifically recommended against the use of artemisinin as a sole agent to prevent resistance from developing, but strongly recommends its use in combination with others such as mefloquine and chloroquine as a first-line treatment for malaria worldwide. Artemisinin is considered the most effective drug against *Plasmodium* species, but there is no consensus regarding its mechanism of action. There have also been reports of resistance developing in Southeast Asia in 2008 and Thailand in 2012.

Primaquine is another quinine-based antimalarial, but it targets a different life-cycle stage of *Plasmodium*. Primaquine acts against the liver stage of the parasite and must be used to treat relapsing forms of malaria.

Quinacrine is related to mefloquine and was approved as an antimalarial agent in the 1930s (it was extensively used by US troops in the second world war). It is now used mainly to treat giardiasis; its mechanism of action is uncertain.

Folate Agonists

Drugs that disrupt folic acid metabolism are known as folate agonists. They disrupt nucleic acid synthesis and repair; they gain their selective toxicity by being able to selectively inhibit microbial enzymes. Nevertheless they have adverse effects on rapidly dividing host cells such as bone marrow, and are not recommended for use during the first trimester of pregnancy.

Pyrimethamine and proquanil are antimalarial agents but are typically not used in solo treatment; they are used in combination with others to increase their effectiveness. They are also used to treat toxoplasmosis.

Drugs for Anaerobic Protozoa

Metronidazole is one of the most widely used anti-protozoan drugs. It is sold under the name Flagyl® and is the drug of choice for diseases such as vaginitis resulting from infection with *Trichomonas vaginalis*, giardiasis, and amebic dysentery. Metronidazole interferes with DNA synthesis in anaerobic organisms and also acts against anaerobic bacteria such as *Clostridium difficile*. Metronidazole is typically also used in combination therapies. For example, when treating amebiasis caused by *Entamoeba histolytica*, metronidazole is used against the invasive form, and diloxanide is used against the noninvasive form.

Iodoquinol is also used to treat amebiasis caused by infections with *Entamoeba histolytica*. Although poorly absorbed, it is active against both cysts and trophozoites. Its mechanism of action is unknown.

Fast Fact Quinine occurs naturally in the bark of the *Chinchona* tree, which is native to South America and was used by native people as a treatment for fevers. Artemisinin is found in the bark of the sweet wormwood tree and has been used in traditional Chinese medicine for thousands of years.

Fast Fact Halofantrine is related to quinine and is effective against all species of malarial parasite including multi-drug resistant *P. falciparum*. Unfortunately it can cause major cardiac side effects.

Drugs for *Trypanosoma* and *Leishmania*

Eflornithine was initially developed as a treatment for cancer, but although it was not very effective in this capacity it was found to be highly effective in treating African sleeping sickness, caused by *Trypanosoma bruceii*. It inhibits an enzyme that is very stable in *Trypanosoma* but has a high turnover in humans, meaning that the parasite enzyme suffers much more from inhibition and the drug has a remarkable degree of selective toxicity. Unfortunately, production of the drug ceased in the 1990s because treating African sleeping sickness did not make a profit. Production of a cream version of eflornithine was started in 2001 as a treatment for facial hair, and after international pressure and media attention, the production of an intravenous version that could be used to treat African sleeping sickness was started. Since then, the use of eflornithine has greatly contributed to reducing the burden caused by African sleeping sickness.

Drugs based on heavy metals, such as melarsoprol, which is a compound of arsenic, are also used to treat both forms of trypanosomiasis (African sleeping sickness and Chagas' disease, which is found in the Americas). Arsenic is highly toxic to humans, and melarsoprol can have severe side effects, even death. Sodium stibogluconate, which contains antimony, is the primary treatment for leishmaniasis but can cause serious damage to the veins and pancreas. These drugs should only be used under close medical supervision.

Anti-helminthic Drugs

Anti-helminthic drugs, like anti-protozoan drugs, have been largely ignored until recently because the affected populations were not found in developed countries. However, the popularity of sushi in the developed world has led to an increase in tapeworm infestations, and increased world travel has led to an increased occurrence of helminth infections in travelers. Niclosamide is based on phenol and inhibits the production of ATP. ATP production is essentially the same in helminths and in humans, but niclosamide is selectively toxic because it is absorbed by cestodes (tapeworms) and not intestinal cells. It is the treatment of first choice for tapeworm infections but is ineffective against roundworms and flukes.

The broad-spectrum anti-helminthic praziquantel is the drug of choice for trematode (fluke) infections, such as schistosomiasis. It is thought to act by increasing the permeability of the plasma membrane and inducing muscle spasms, which apparently expose antigenic sites for attack by the host immune system. It is also effective against tapeworms.

Mebendazole is used for the treatment of several common intestinal helminthic infections, such as pinworm (*Enterobius vermicularis*) and ascariasis (*Ascaris lumbricoides*). Mebendazole disrupts microtubule formation, which affects the motility of the worm, and blocks glucose uptake. It is poorly absorbed from the intestine and so cannot be used to treat tissue infections. Albendazole also interferes with microtubule formation and can be used against intestinal and tissue infections.

Ivermectin is an important and influential anti-helminthic drug. It was the world's first endectocide and is effective against a wide variety of internal and external parasites, for example internal nematodes, cutaneous larvae, and ectoparasites such as lice and scabies. Ivermectin was first used in humans in 1988 to treat onchocerciasis and proved to be safe, highly effective, well tolerated, and easily administered. It is the main component in the global disease eradication program against onchocerciasis and lymphatic filariasis.

Piperazine and tetrahydropyrimidine compounds paralyze intestinal worms and are used to treat infections with *Ascaris* and *Trichuris*.

Keep in Mind

- The development of drugs that are useful against parasitic protozoans and helminths has been slow because diseases caused by these organisms do not often occur in developed countries.

- Selective toxicity is hard to achieve because parasites are eukaryotes, like humans; many antiparasitic drugs have serious side effects.

- There are several antimalarial drugs based on quinine, but they have similar resistance issues to antibiotics.

- Metronidazole is one of the most widely used anti-protozoan drugs and is also used as an antibiotic; it has good selective toxicity.

- Ivermectin is used to treat several types of intestinal worm infection.

SUMMARY

- Microorganisms produce toxic chemicals as part of their natural defense to protect themselves from other microorganisms.

- The term antibiotic is generally used for chemicals that act against bacteria.

- Antibiotics can be broad or narrow spectrum and have cidal or static activity.

- Chemical modification of natural antibiotics can broaden their spectrum.

- Antibiotics and all antimicrobial compounds must be selectively toxic to be used therapeutically—they must harm the pathogen, but not the host.

- The five targets for antibiotics are the cell wall, the plasma membrane, the ribosome, nucleic acids, and metabolic synthesis pathways.

- Viruses present problems for treatment because they are obligate intracellular parasites, so selective toxicity is hard to achieve once they are inside the host cell.

- Targets for antiviral drugs are viral DNA polymerases such as reverse transcriptase and viral entry into the host cell.

- Selective toxicity is hard to achieve for antifungal drugs because fungi are eukaryotes.

- The development of drugs against parasitic protozoans and helminths has been slow because these infections occur mostly in underdeveloped countries.

- Protozoans and helminths are also eukaryotes and present selective toxicity issues.

- Many antifungal and antiparasitic drugs have significant side effects.

- There are many more antibiotics than there are antiviral, antifungal, or antiparasitic drugs.

- The choice of drug with which to treat an infection depends on several factors, including the route of administration, side effects, and cost.

With all the improved methods for drug discovery and testing, one fundamental problem remains: all pathogenic organisms can very quickly become resistant to new drugs. The problem of drug resistance has been referred to as one of today's greatest medical threats. We discuss this in the next chapter.

Ⓠ **SELF EVALUATION AND CHAPTER CONFIDENCE**

Multiple Choice

Answers are given in the back of the book and help can be found in the student resources at:

www.garlandscience.com/micro2

1. The first antibiotic discovered was

   A. Streptomycin
   B. Quinine
   C. Sulfa drugs
   D. Penicillin

2. Most of the available antimicrobial agents are effective against

   A. Viruses
   B. Fungi
   C. Bacteria
   D. Protozoa
   E. All of the above

3. Which of the following does not target the same bacterial component as the others?

   A. Monobactam
   B. Cephalosporin
   C. Bacitracin
   D. Streptomycin
   E. Penicillin

4. Which of these antibiotics has the fewest side effects?

   A. Penicillin
   B. Chloramphenicol
   C. Tetracycline
   D. Erythromycin
   E. Streptomycin

5. Which of the following methods of action would be bacteriostatic?

   A. Inhibition of cell wall synthesis
   B. Inhibition of RNA synthesis
   C. Inhibition of DNA synthesis
   D. Injury to plasma membrane
   E. None of the above

6. Which of the following antimicrobial agents is recommended for use against fungal infections?

   A. Amphotericin B
   B. Penicillin
   C. Bacitracin
   D. Cephalosporin
   E. Polymyxin

7. More than half of our antibiotics are

   A. Produced by Fleming
   B. Produced by bacteria
   C. Produced by fungi
   D. Synthesized in laboratories
   E. None of the above

8. Which of the following drugs is not used primarily to treat tuberculosis?

   A. Sulfonamide
   B. Rifampin
   C. Isoniazid
   D. Ethambutol
   E. None of the above.

9. Which of the following organisms would most probably be sensitive to natural penicillin?

   A. *Streptococcus pyogenes*
   B. *Penicillium*
   C. Penicillinase-producing *Neisseria gonorrhoeae*
   D. *Mycoplasma*

10. *Streptomyces* bacteria produce all of the following antibiotics except

    A. Erythromycin
    B. Nystatin
    C. Kanamycin
    D. Rifampin
    E. Bacitracin

11. Broad-spectrum antibiotics

    A. React only with Gram-positive bacteria
    B. React only with Gram-negative bacteria
    C. React only with *Pseudomonas*
    D. React with Gram-positive bacteria, Gram-negative bacteria, and *Pseudomonas*
    E. React with only large bacteria

12. A bacteriostatic antibiotic

    A. Inhibits bacterial growth by killing the organism
    B. Increases the electrical charge of the organism
    C. Damages the bacterial plasma membrane
    D. Inhibits bacterial growth but does not kill the organism
    E. None of the above

13. The difference between penicillin and ampicillin is

   A. The β-lactam ring
   B. The type of carbohydrates associated with the drug
   C. The side chains affixed to the core ring structure
   D. All of the above
   E. None of the above

14. β-Lactamase is

   A. A ring structure seen in semi-synthetic penicillin
   B. Found only on cephalosporin
   C. A chemical that enhances the effect of antibiotics
   D. An enzyme that cleaves the ring structure of penicillin
   E. None of the above

15. All of the following are targets for antibiotics except

   A. The cell wall
   B. Bacterial ribosomes
   C. The glycocalyx
   D. The plasma membrane of the bacteria
   E. Nucleic acids

16. Monobactam antibiotics have

   A. Different side chains from penicillin
   B. The same ring structure as penicillin
   C. A different ring structure to penicillin
   D. No core structure

17. Protein synthesis is

   A. Not a target for antibiotics
   B. Not a selectively toxic target for antibiotics
   C. A selective target
   D. A target of last resort for antibiotic therapy
   E. Used only for viral infections

18. All of the following target the bacterial ribosome except

   A. Streptomycin
   B. Tetracycline
   C. Penicillin
   D. Chloramphenicol
   E. Erythromycin

19. The first target of sulpha drugs in bacteria is _____ synthesis

   A. Folate
   B. Cell wall
   C. Plasma membrane
   D. Ribosome
   E. DNA

20. Antiviral drugs

   A. Are all selectively toxic
   B. Only affect infected cells
   C. Only affect free viral particles
   D. Must eliminate all viral particles to be effective
   E. Only effect the uncoating of viral particles

21. Acyclovir

   A. Is effective against HIV infections
   B. Blocks all DNA replication
   C. Blocks viral DNA replication
   D. Is an analog of cytosine
   E. Blocks viral transcription

22. All of the following are antifungal drugs except

   A. Polyenes
   B. Bacitracin
   C. Azoles
   D. Griseofulvin
   E. Flucytosine

23. Which of the following is used in treating protozoan infections?

   A. Griseofulvin
   B. Bacitracin
   C. Polyenes
   D. All of the above
   E. None of the above

24–28. Answer **A** if the reason is correct; answer **B** if the reason is incorrect.

24. The activity of penicillin is most effective during cell growth because penicillin prevents the cross-linking of the NAG and NAM units and thereby prevents the formation of an intact cell wall.

25. Isoniazid is very effective against mycobacteria because it inhibits the synthesis of protein.

26. The plasma membrane in bacteria is a prime target for antibiotics because any disruption of this membrane will destroy the bacteria's ability to survive.

27. The synthetic forms of erythromycin known as azithromycin and clarithromycin have expanded spectra and are used for respiratory infections because they are more potent.

28. Rifampin is used only in combination therapy because resistance develops rapidly if the antibiotic is used alone.

29–32. Each of the following drugs has a different mechanism. Match the mechanism with the following antibiotics: **A** flucytosine; **B** chloroquine; **C** mebendazole; **D** azoles.

29. Destroys red blood cells

30. Inhibits the production of sterols

31. Inhibits microtubule formation

32. Interferes with nucleic acid synthesis

DEPTH OF UNDERSTANDING

Questions listed here require you to bring together the concepts you have learned in this chapter into a discussion format. This helps you to increase your depth of understanding of the material you have learned. Help can be found in the student resources at: www.garlandscience.com/micro2

1. Discuss the differences between natural and semisynthetic penicillin, using structure and function as the foundation of your answer.

2. Evaluate the main targets used for antibiotics with regard to selective toxicity.

3. Most infectious diseases are caused by viruses, but there are few effective antiviral drugs. Explain why.

(Q) CLINICAL CORNER

Help can be found in the student resources at: www.garlandscience.com/micro2

1. Your patient has an infection with Gram-positive staphylococci. She has been on cephalosporin for seven days and has shown little improvement. The doctor is switching her to a combination of streptomycin and penicillin. She does not understand why this is necessary.

   A. How do you explain to her why the initial treatment did not seem to work?
   B. How do you explain to her what the benefits of the new drug therapy will be?
   C. Do you think that this switch will make a difference?

2. A patient has come to the surgery with a heavy cold that they have been suffering from for several days and is upset when they are not prescribed an antibiotic.

   A. How do you explain why antibiotics are not appropriate for their illness?
   B. How do you explain why there is no effective drug therapy?
   C. How do you explain what steps they should take?

Antibiotic Resistance

Chapter 20

Why Is This Important?

It is now clear that the most important problem associated with infectious disease today is the rapid development of resistance to antibiotics. This resistance will force us to change the way we view disease and the way we treat patients.

One Sunday evening in a busy emergency room in Chicago, Dr. Smith saw a young 6-year-old boy for a severe ear ache. Upon examination, the ear showed signs of otitis media. In addition, the child had a mild to moderate sinus infection. The doctor discharged the family with a prescription for Augmentin (a derivative of penicillin) and instructions to see their family physician if the symptoms worsened. The family was not financially secure and so did not immediately follow up with a pediatrician. Over the next 48 hours, the child's symptoms persisted with a low-grade fever starting and the child complained of a headache. The mother called the ER and was told to give the child 200 mg of Ibuprofen for the fever and pain. In the next 12 hours, the child developed a high fever and began to have mini-seizures. The parents rushed the child back to the ER where he was started on fluids and an injection of penicillin while blood tests were taken and analyzed. The condition worsened steadily, with the child becoming unresponsive. Finally, labs returned with a diagnosis of resistant *Streptococcus pneumoniae* causing meningitis. Expensive additional antibiotics were added and after a prolonged hospitalization, he was released alive, but permanently hearing impaired and prone to seizures for the rest of his life.

S. pneumoniae is one example of a drug-resistant organism of growing concern. According to a report done by the CDC, it causes 1.2 million drug-resistant infections per year, resulting in 96 million US dollars in excess medical costs, and 7000 deaths.

OVERVIEW

This chapter is primarily dedicated to bacterial resistance to antibiotics. Even though modifying existing antibiotics can increase the life spans of those anti-biotics, resistance is inevitable. In other words, resistance to antibiotics is a matter not of if but of when. Resistance has become such a widespread and alarming problem it is important that health care professionals understand why and how this resistance occurs. The development of resistance to antiviral and antiparasitic drugs will be briefly covered. We will also look at the development of new antibiotics here, since resistance has become the driving force behind the need to develop new antibiotics.

We will divide our discussions into the following topics:

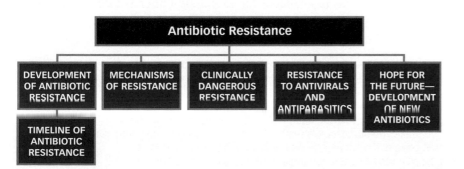

"The message on this World Health Day is loud and clear. The world is on the brink of losing these miracle cures. In the absence of urgent corrective and protective actions, the world is heading towards a post-antibiotic era, in which many common infections will no longer have a cure and, once again, kill unabated."

WHO statement on World Health Day 2011

As the WHO statement says, antibiotics were miracle cures when they were introduced. Since the discovery of these drugs, the world health care landscape has changed drastically, not only for the patient but also for the health care worker. Antibiotics gave the physician powerful weapons to treat bacterial infections and allowed surgical procedures to be conducted safely. All this is now under threat.

DEVELOPMENT OF ANTIBIOTIC RESISTANCE

As we learned in the previous chapter, microbes naturally produce antibiotics, so it isn't surprising that bacteria naturally have mechanisms to resist the actions of antibiotics. However, being able to resist the action of antibiotics is only an advantage if there is antibiotic present: the more antibiotic there is, the more of an advantage it is to be resistant to it. Changing the environment by increasing the amount of antibiotic present creates an evolutionary pressure for bacteria to be resistant to that antibiotic. It's a matter of survival. Like everything else, the ability to resist antibiotics ultimately comes from genes and relies on mutations to generate new mechanisms of resistance. Let's now look at some of the factors involved in the development of antibiotic resistance.

Bacterial Growth and Mutation Rates

Because bacteria have very short generation times, they can quickly grow into large populations. Consequently, the potential for mutation and transfer of genetic material from one generation to the next is considerable. Among the mutations that occur are those that allow the survival of an organism exposed to otherwise lethal antibiotics, and these mutations will be selected for in the presence of such antibiotics (**Figure 20.1**). As you can see from the figure, those bacterial cells that have developed resistance to a given antibiotic are not killed off when treated with the drug. These resistant cells continue to divide, and the resulting population will be resistant by means of vertical gene transfer.

Plasmids and Conjugation

As well as a main chromosome, bacteria have genetic material on plasmids and, as we saw in Chapter 11, these plasmids can be transferred between individual bacteria and between bacterial species via conjugation. This mechanism is also known as **horizontal gene transfer** (HGT). If a gene for antibiotic resistance is coded for on a plasmid, or part of the chromosome, that is available for exchange either between organisms

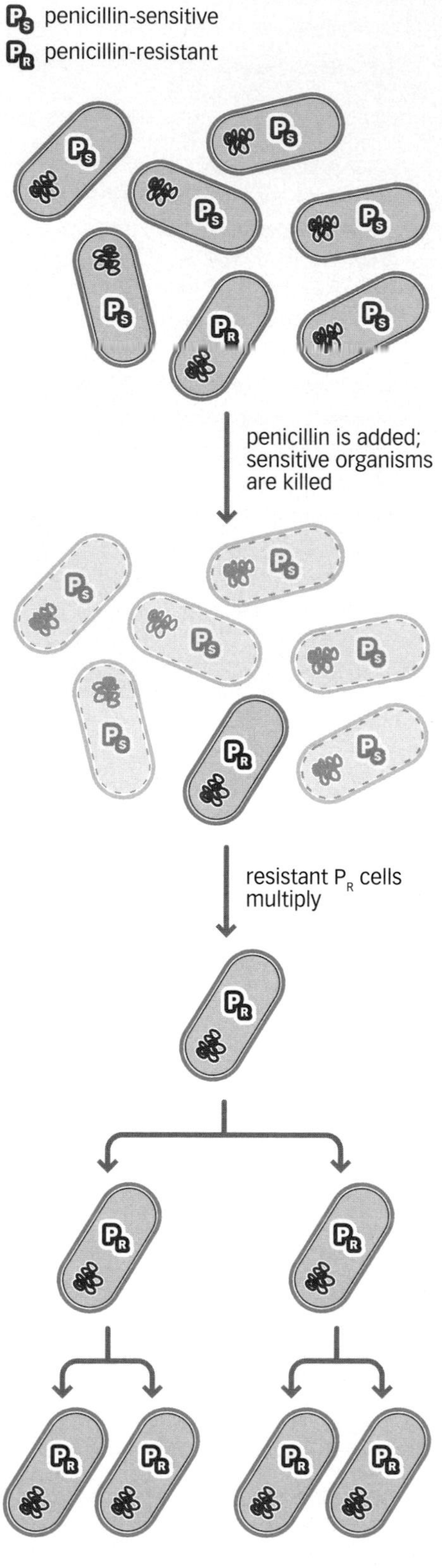

P_S penicillin-sensitive

P_R penicillin-resistant

penicillin is added; sensitive organisms are killed

resistant P_R cells multiply

Figure 20.1 Development of antibiotic resistance. Resistance is an evolutionarily favorable mutation. In the initial population, most of the bacterial cells are sensitive (P_S) to the antibiotic being used for treatment, but one is resistant (P_R) to the drug. When the antibiotic is applied, the P_S cells die but the P_R cell survives and reproduces. Eventually the population is made up entirely of P_R cells. Any mutation that permits the survival of the organism will be selected for and passed on from generation to generation, as indicated here.

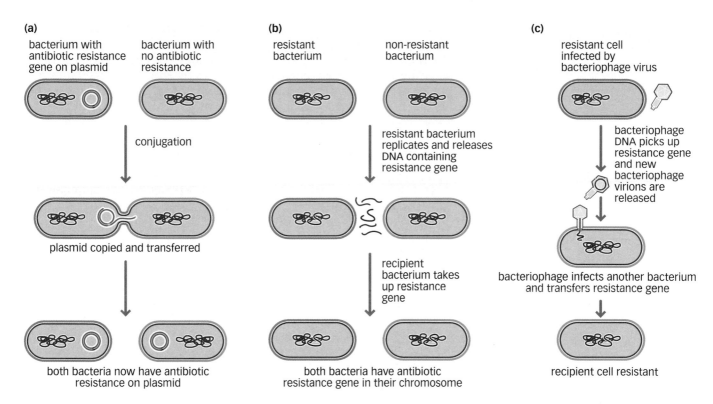

(a)

bacterium with antibiotic resistance gene on plasmid

bacterium with no antibiotic resistance

conjugation

plasmid copied and transferred

both bacteria now have antibiotic resistance on plasmid

(b)

resistant bacterium

non-resistant bacterium

resistant bacterium replicates and releases DNA containing resistance gene

recipient bacterium takes up resistance gene

both bacteria have antibiotic resistance gene in their chromosome

(c)

resistant cell infected by bacteriophage virus

bacteriophage DNA picks up resistance gene and new bacteriophage virions are released

bacteriophage infects another bacterium and transfers resistance gene

recipient cell resistant

Figure 20.2 Horizontal gene transfer.
Panel a: Genetic material is encoded on a plasmid and transferred via conjugation to a recipient cell. Panel b: Genetic material from a donor cell is released and taken up directly by a recipient cell via transformation. Panel c: Genetic material is transferred via the use of an intermediary or carrier, such as a viral phage.

or species, the recipient bacteria will gain resistance (**Figure 20.2**). Examples of this transfer have been seen between several species important in infectious disease, for example *Bacillus*, *Streptococcus*, *Salmonella*, *Neisseria*, *Staphylococcus*, *Shigella*, and *Enterococcus*. Bacterial plasmids containing genes for resistance can integrate into the chromosome of a recipient bacterium at specific sites, and these sites can be referred to as **resistance islands** where resistance genes accumulate and are consistently maintained.

Inappropriate Clinical Use of Antibiotics

Perhaps the most important contributing factor for the development of antibiotic resistance is inappropriate clinical use. Take as an example upper respiratory infections, including the common cold, which represent one of the most prevalent clinical problems we face. It has been estimated that upward of 60% of all upper respiratory infections are of viral etiology, but many patients seeing a doctor for flu-like symptoms will be given a prescription for an antibiotic. In fact, many people wonder about a visit to their doctor that did not result in a prescription. Because antibiotics are antibacterial and therefore are not effective against viruses, taking the prescribed antibiotic does nothing to treat the actual underlying cause of disease, which (given a normal immune system) should subside naturally within a week.

Patients who, feeling better, decide not to finish the course of treatment, patients who use outdated drugs, and patients who use antibiotics that were prescribed for something else all contribute to the development of resistance. Sharing antibiotics between family members and friends leads to not one but two cases treated with incomplete regimens. There are countries where antibiotics are available without prescription, leading to improper dosing or use in many cases. Finally, some patients will discard their unused portions into water sources and landfills, introducing antibiotics into the environment. All these situations create the necessary evolutionary pressure for antibiotic resistance to develop.

Inappropriate use of antibiotics not only increases the amount of antibiotic present, it can also have a further negative effect by destroying the normal flora of our bodies. Using broad-spectrum antibiotics where a narrow-spectrum antibiotic would be just as effective exacerbates this situation by killing more of the normal flora. As we have seen, the bacteria that make up these resident populations have a critical role in keeping under control organisms that can become opportunistic pathogens.

Use of Antibiotics in the Food Chain

The use of antibiotics as feed additives for the promotion of growth in food animals is a common practice in the United States, although the practice has been banned in the European Union. The antibiotics add pounds per day of average weight gain when used in high-concentration feedlots, and add value to commercial operations such that there is cheaper, more readily available meat for the consumer. However, such use can also result in contamination of the surrounding environment and ground water with low levels of antibiotics. This subtherapeutic level is enough to create an evolutionary pressure for bacterial populations to develop resistance and has been implicated in a decrease in effectiveness of many classes of antibiotics over time. Antibiotic resistance can develop by mutation of a plasmid in a non-pathogenic bacterium and then be passed to a pathogenic bacterium by HGT.

It has also been hypothesized that both hormones and antibiotics in food have affected the normal microbial flora of the human body in such a way as to add to conditions including abnormal digestion, weight gain in children, and diabetes, as well as contributing to chronic infections.

Immunocompromised Population

An important social change that affects bacterial resistance is the increase in the number of people who are immunocompromised. In Chapter 17, we saw that there are individuals who are immunocompromised for a variety of reasons such as genetic abnormalities, AIDS, cancer and cancer therapies, organ transplants, and age. These individuals often have an increased need for antibiotics, which has in turn increased the likelihood of the development of resistance.

Health Care Facilities

Hospitals and other health care facilities are outstanding settings for the acquisition of resistance. They have all the necessary conditions, including a high concentration of bacteria, many of which can be extremely pathogenic, and a population of people who for one reason or another have had their health compromised. The hospital is also the place where large amounts of antibiotics are constantly in use. Depending on the patient, these may include drugs such as vancomycin, which is considered an antibiotic of last resort. Because increased use leads to resistance, the hospital is a place where resistance can develop rapidly, and resistance to antibiotics of last resort can be clinically devastating.

Lifestyle

There are more large cities in the world today than ever before. The presence of large numbers of people in relatively small areas means that passing a contagious disease or antibiotic-resistant pathogen from one person to another is more easily accomplished than in areas where populations are spread out. This can be further compounded by the fact that many of these large urban populations occur in developing countries, where sanitation is poor, health care is less widely available, and poverty affects treatment possibilities. In these situations, organisms such as the bacterium *Vibrio cholerae* can easily cause epidemics of cholera. *Vibrio* organisms have become resistant to many antibiotics, making these outbreaks even more dangerous. A larger population of older people in care homes also allows people with poorly functioning immune systems and on long-term antibiotics to live in close proximity, which are excellent conditions for the development and spread of antibiotic resistance.

A person can travel anywhere in the world within 24 hours and bring resistant bacteria with them. That travel usually takes place in an enclosed space (plane, train, or car) in association with several or many other people. In a plane, for instance, an individual infected with a resistant bacterium will be seated in an enclosed space, in which the air is continuously recirculated, for several hours. The problem increases because the individuals who may become secondarily infected by the resistant bacterial strain are usually making connections to other planes, where the process can be repeated.

Keep in Mind

- Like the production of antibiotics, antibiotic resistance is a natural phenomenon.

- Mutations that confer resistance to antibiotics are selected for if there is an evolutionary pressure that makes possessing them an advantage by substantially increasing the chances of survival.

- The more an antibiotic is used, the greater is the chance of resistance developing.

- The genes for antibacterial resistance can be passed from one generation of bacteria to the next (vertically) and between bacteria of the same generation and bacteria of different species (horizontally).

- Inappropriate use of antibiotics, in clinical and non-clinical settings, is the major cause of the rapid increase in resistant populations of bacteria.

- An increase in the number of immunocompromised individuals and modern lifestyle are also contributory factors.

- Hospitals are ideal settings for the development and spread of antibiotic resistance.

TIMELINE OF ANTIBIOTIC RESISTANCE

In retrospect, it is not surprising that resistance to penicillin in some strains of staphylococci was recognized almost immediately after introduction of the drug in 1946. Likewise, very soon after their introduction in the late 1940s, resistance to streptomycin, chloramphenicol, and tetracycline was noted. By 1953, the first example of multi-drug resistant bacteria (*Shigella dysenteriae*) was isolated, exhibiting resistance to chloramphenicol, tetracycline, streptomycin, and the sulfonamides. Over the years almost every known bacterial pathogen has developed resistance to one or more antibiotics in clinical use.

The clinical success of early antibiotics led to ever-increasing efforts to discover new antibiotics, coupled with a great emphasis on the modification of existing drugs. The intent was to find or develop antibiotics with broader spectra of reactivity. More recently, that effort has been replaced by the need to overcome bacterial strains that are resistant to treatment

Table 20.1 Evolution of resistance to antibiotics.

Antibiotic	Year deployed	Resistance observed
Sulfonamides	1930s	1940s
Penicillin	1943	1946
Streptomycin	1943	1959
Chloramphenicol	1947	1959
Tetracycline	1948	1953
Erythromycin	1952	1988
Vancomycin	1956	1988
Methicillin	1960	1961
Ampicillin	1961	1973
Cephalosporins	1960s	Late 1960s

with antibiotics. It is important to note how quickly the development of resistance can occur (**Table 20.1**). Several groups of antibiotics—the sulfonamides, streptomycin, erythromycin, and vancomycin—were in use for many years before resistance to them was observed. In contrast, resistance to penicillin was observed only three years after it became widely used.

For some semi-synthetic forms of penicillin, such as ampicillin, the length of time before resistance was observed was relatively long (deployed in 1961, resistance observed in 1973). Other semi-synthetic forms, such as methicillin, lasted only a year before resistance was observed. In some cases, the short interval for some antibiotics is directly related to increased use. Methicillin is a good example of this. This modified penicillin had a narrow spectrum of activity and was so effective that it became overprescribed.

Remember, the more an antibiotic is used, the greater is the chance of resistance developing through genetic mutation and the selection of resistant strains. Listed below are six guidelines that can be used to increase the useful life span of drugs:

- Optimal use of all antibacterial drugs
- Control of certain antibiotic classes and selective use of broad-spectrum antibiotics
- Use of antibiotics in rotation or cyclic patterns
- Use of combination therapy
- Evaluation of routes of resistance
- Implementation of global changes

Susceptibility Testing

The successful restriction of broad-spectrum antibiotics in treating an infection requires the use of susceptibility testing. Testing involves culturing the infective agent and then isolating and identifying each organism involved in the clinical disease presentation. Then, a list of likely drugs is created on the basis of the identity of the pathogen and tested using the Kirby–Bauer method (**Figure 20.3**) or disk diffusion

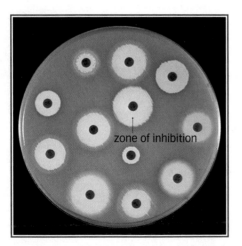

Figure 20.3 The Kirby–Bauer test for testing potential antibiotic drugs. This plate shows an agar layer with continuous bacterial growth. When tiny disks containing either different compounds or different concentrations of the same compound are placed on the agar, the compounds diffuse out. If they are pharmacologically active, they form a zone of inhibition in which bacteria do not grow.

method (Chapter 18) to determine the extent to which a drug affects the plated pathogen. As with the evaluation of disinfectants, any disk that inhibits pathogen growth is easily identifiable by the zone of inhibition surrounding the disk (see Figure 20.3).

In addition to testing various concentrations for a single new compound, the disk method is also used to compare the relative effectiveness of different compounds. In these tests, it is important to recall from Chapter 18 that a larger zone of inhibition does not necessarily indicate a more powerful compound, because there can be differences in diffusion rates. Comparison of similar compounds can be done by measuring the zones and referring to a standardized table for that type of drug, and that concentration in relation to a particular species. When testing specific organisms by the disk method, the organisms can be classified as sensitive, intermediate, or resistant to the compound or compounds being tested. Although this is a simple and inexpensive test, it is inadequate for most clinical purposes based on time constraints in treating active infections.

A more advanced diffusion test, known as the **E test**, permits the determination of the minimal inhibitory concentration (MIC), the lowest antibiotic concentration that prevents visible pathogen growth. This test employs plastic-coated strips that contain gradients of antibiotic concentrations. After incubation of pathogen-coated plates, the MIC can be read from the scale printed on each strip (see **Figure 20.4**).

As we saw in Chapter 18, a dilution test is needed to determine whether the antibiotic is microbicidal or microbistatic. The **broth dilution test** is used for this purpose and determines the **minimum bactericidal concentration (MBC)** of a compound. This procedure involves incubating a specific organism in a sequence of wells containing decreasing amounts of the antibiotic compound being tested. Microbes from wells that do not show growth can be recultured in nutrient broth medium containing none of the test compound. Growth in this medium indicates that the compound being tested inhibited the growth of the organism but did not kill it; the compound is therefore microbistatic. If no growth occurs in the reculture, the test compound killed the organism and is classified as microbicidal.

These types of dilution test are highly automated, and additional testing can use colorimetric methods to indicate reactivity and can determine serum concentrations of the test compound. This is important because many drugs are toxic at high concentrations, and the use of a higher concentration than necessary will increase the speed at which antibiotic resistance develops. Higher concentration equals more antibiotic, which means that resistance will develop faster.

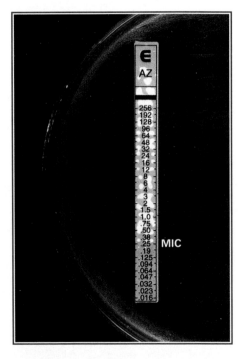

Figure 20.4 The E test uses a quantitative scale that can be used to determine the MIC (minimum inhibitory concentration) of the compound being tested. This picture shows the drug azithromycin; the strip contains a gradation of antibiotic with the strongest at the top and the weakest at the bottom. The minimum inhibitory concentration of antibiotic that will inhibit the bacteria is determined by the point at which the growth of the bacteria starts, or the area of the growth where the ellipse meets the strip (in this case 0.25).

MECHANISMS OF RESISTANCE

Now that we have looked at the factors involved in the development of bacterial resistance to antibiotics, let's look at the ways in which they protect themselves. As we discuss them, you will notice that they are the same mechanisms that antibiotic-producing bacteria use to protect themselves:

- Inactivation of the antibiotic
- Efflux pumping of the antibiotic
- Modification of the antibiotic target
- Alteration of the pathway

In some ways, these protective mechanisms represent countermeasures to the methods we use in our search for potential antibiotics. For instance, we choose targets that would be lethal for the organism, while the bacteria develop ways to change those targets so they are no longer identifiable. Let's look at each of these mechanisms in more detail.

Inactivation of Antibiotic

If a bacteria can inactivate an antibiotic molecule, it will be resistant to that antibiotic. Inactivation usually involves the enzymatic breakdown of the antibiotic molecules. We learned in Chapter 19 about β-lactamase, the enzyme produced by bacteria in response to antibiotics that have a β-lactam ring as their core. There are more than 190 forms of β-lactamase, and these enzymes are usually secreted into the bacterial periplasmic space, where they attack the antibiotic as it approaches its target (the bacterial cell wall). Lactamases hydrolyze the ring structures in penicillins, cephalosporins, and carbapenems (**Figure 20.5**). Efforts to obtain antibiotics that are not affected by β-lactamases led to the development of cefotaxime and ceftazidime (next-generation cephalosporins). Another way to inhibit the destruction of the β-lactam ring in an antibiotic molecule is to augment the antibiotic with lactamase inhibitors such as potassium clavulanate. Potassium clavulanate has no antibacterial activity itself, but it can inhibit the activity of β-lactamase, thereby depriving the bacteria of this major resistance mechanism and allowing the antibiotic to function normally.

Several of these combinations have been developed, including:

Augmentin® (Timentin®) = amoxicillin + potassium clavulanate

Primaxin® = imipenem + cilastatin

Unasyn® = ampicillin + sulbactam

AmpC β-lactamase

Many enteric bacteria carry the *AmpC* gene on their chromosome, where its expression is inducible: it is usually turned off, but is turned on (induced) by the presence of antibiotics containing a β-lactam ring. *AmpC*

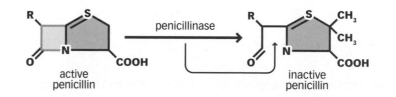

Figure 20.5 Action of β-lactamase. A β-lactamase enzyme, in this case penicillinase, cleaves the β-lactam ring and renders the antibiotic inactive.

codes for a β-lactamase that can inactivate broad-spectrum cephalosporins and is not inhibited by lactamase inhibitors. It is therefore a powerful enzyme and its expression is controlled by several other genes in the *Amp* group. The *AmpG* gene codes for a bacterial protein that binds to fragments of peptidoglycan created when the antibiotic destroys the bacterial cell wall. The AmpG protein transports these fragments through the cytoplasm and induces transcription of the *AmpC* gene. Therefore, when the bacterial cell wall begins to be destroyed, the gene for β-lactamase is turned on to stop the destruction.

Mutations in *AmpG*, as well as in the *AmpR* and *AmpD* genes, cause *AmpC* to be overexpressed, which confers resistance to broad-spectrum cephalosporins. The *AmpC* gene has also been acquired by plasmids and transferred to bacteria such as *Escherichia coli* and *Klebsiella pneumoniae* that otherwise lack *AmpC*, meaning they, too, can develop resistance to cephalosporins. Antibiotic resistance based on *AmpC* is a major clinical challenge because once it has developed, the only option is treatment with carbapenems.

Aminoglycoside-Inactivating Enzymes

The most common form of resistance to aminoglycoside antibiotics, such as gentamicin, is conferred by enzymes that modify the antibiotic molecule and render it ineffective. There are many aminoglycoside-inactivating enzymes that act through different mechanisms, meaning that many parts of the aminoglycoside can be attacked. These enzymes may have a physiologic function that does not involve antibiotic inactivation, but the presence of large quantities of aminoglycosides causes the enzyme-producing genes to be overexpressed and confers high levels of resistance. The genes encoding these enzymes are usually found on plasmids and transposons, meaning that resistance is easily passed to other bacteria.

Efflux Pumping of Antibiotic

A major route of drug resistance is through the use of bacterial pumps, known as **efflux pumps**, that are located in the plasma membranes of all bacteria. Many species of bacteria use them not only for protection against antibiotics but also for everyday physiological functions. Efflux pumps in Gram-positive bacteria function independently, but those in Gram-negative organisms are often associated with two other proteins that eject the antibiotic through the outer membrane as well. The pumps work by keeping the concentration of antibiotic in the bacterial cell below levels that would destroy the cell.

Like the antibiotics they protect against, efflux pumps can be classified as narrow spectrum or broad spectrum, with the broad-spectrum pumps conferring resistance against more than one type of antibiotic. Although efflux pumps are active against β-lactams and fluoroquinolones, their greatest activity is seen against tetracyclines. The three-protein efflux pumps seen in Gram-negative bacteria work on tetracycline, ciprofloxacin, chloramphenicol, and β-lactams. *Pseudomonas* has four families of these pumps and they all overlap functionally, making it one of the most antibiotic-resistant organisms. Three-protein pumps are not restricted to Gram-negative bacteria and have been implicated in the appearance of multi-drug resistant tuberculosis.

Fast Fact *E. coli* has more than 35 different genes that code for drug efflux pumps.

The genes that code for efflux pumps can be located on the bacterial chromosome or on plasmids and transposons. They can therefore be readily transferred to non-resistant bacteria, transforming them into resistant species.

Figure 20.6 Efflux pumps. Panels a, b, and d: show efflux pumps for Gram-positive bacteria. Three types of pump (panels a, b, c) exchange the antibiotic molecule for a cation, either Na⁺ or H⁺, while a fourth (panel d) uses ATP to generate energy. Panel c: shows the efflux pump for Gram-negative bacteria, which has to cross the outer membrane as well.

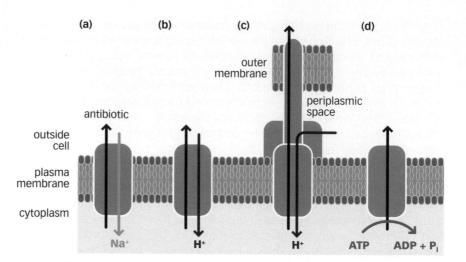

Mechanism of Efflux Pumps

Efflux pumps are active transport systems, ones that require energy, because the concentration of antibiotic is high outside the cell so that movement of antibiotic out of the bacterial cells is movement against a concentration gradient. There are four mechanisms by which the pumps operate (**Figure 20.6**). Three of them use the principle of counterflow, in which the antibiotic is pumped out of the bacterial cell at the same time cations such as H⁺ and Na⁺ are pumped in. Bacteria also use these pumps to remove antiseptic and disinfectant substances (in particular the quaternary ammonium compounds). The fourth pump mechanism is a one-way transport that depends solely on ATP expenditure and does not simultaneously import cations as antibiotic molecules are moved out of the cell. This form of efflux pump is relatively rare, and it is thought that although this ATP-consuming pump is required for normal physiology, it is co-opted into antibiotic efflux activity by the bacteria when necessary.

Tetracycline Efflux Pumps

The efflux pumps coded for and regulated by the *Tet* gene family are used for ejecting tetracycline by both Gram-positive and Gram-negative organisms. As with regulation of the β-lactamase enzymes, it is an inducible system in which the protein (pump) is only made when tetracycline is present. The *TetR* gene codes for a repressor protein that binds to the *Tet* operon. This restricts transcription of the *TetA* gene, which codes for the efflux pump protein that pumps out tetracycline. When no tetracycline is present in a bacterial cell, the protein is not made. However, when tetracycline is present, it binds to the TetR protein, which is then inhibited from repressing the *TetA* gene. The efflux pump protein is then made in large quantities and sent to the bacterial plasma membrane, where it pumps tetracycline out of the cell, simultaneously importing needed cations. This same type of regulation occurs in *Pseudomonas*, *Neisseria gonorrhoeae*, and *E. coli*. The bacterium *Bacillus subtilis* uses the same system for the efflux of fluoroquinolones, chloramphenicol, and doxorubicin antibiotics.

It should be noted that in addition to membrane efflux pumping, some bacteria reduce the permeability of their membranes as a way of keeping antibiotics out. This is accomplished by turning off the production of porin and other membrane channel proteins. This mechanism is seen in *E. coli* O157:H7 and in *Pseudomonas*, which use this decreased membrane permeability for resistance to streptomycin, tetracycline, and sulfa drugs.

Modification of Antibiotic Target

Bacteria can also escape antibiotic activity by modifying a cellular component that is the target of an antibiotic. This is a very interesting mechanism because the modified target must be insensitive to drug activity but still be able to function. This can be achieved in two ways: mutation of the gene that codes for the target protein, or importing a gene that codes for a modified target that performs the required function but is not recognized by the attacking antibiotic.

Penicillin-Binding Proteins

Recall from Chapter 19 that bacteria have penicillin-binding proteins (PBPs) in their plasma membranes and that these proteins are targets for the penicillin family of antibiotics. Methicillin-resistant *Staphylococcus aureus* (MRSA) has acquired the *mecA* gene, which codes for a PBP that has a different three-dimensional structure from the PBP usually found in *S. aureus* cells and is less sensitive to penicillin antibiotics. Because of this, MRSA is essentially resistant to all β-lactam antibiotics, cephalosporins, and carbapenems, making it a very dangerous pathogen, particularly in burn patients.

The production of these modified PBPs is another example of operon function at the genetic level. The gene coding for a modified PBP is always kept switched off by a repressor protein. Through a series of reactions, the bacterial enzymes cut the repressor protein into fragments, and the absence of the intact repressor protein then allows the modified PBP to be made. Because the PBP does not attach to any circulating penicillin molecules, the bacterial cell wall can be constructed correctly even in the presence of the antibiotic. Through natural selection of generations of bacterial growth, this type of antibiotic resistance can accumulate to very high levels. For example, when MRSA was treated with the fluoroquinolone ciprofloxacin, resistance to the drug increased from 5% to more than 85% in one year.

Another example of PBP target modification can be seen with *Streptococcus pneumoniae*, an organism that can make as many as five different types of PBP. Apparently this is not the product of simple mutation but rather the ability of the bacterium to rearrange or shuffle these genes. This is referred to as genetic plasticity and permits increased resistance.

Modification of Target Ribosomes

The ribosomes in bacterial cells are a primary target for antibiotics, and several antibiotics affect this target in different ways. For example, erythromycin and its descendants azithromycin and clarithromycin attack the 23S ribosomal RNA of the 50S subunit. This family of antibiotics has been routinely used for respiratory infections. However, resistance to these drugs has increased markedly in the past 20 years. This resistance is the result of modification of the 23S RNA, a modification that makes the RNA no longer sensitive to the antibiotics. Some organisms use target modification in conjunction with the production of efflux pumps, making the resistance even more effective.

Modifying ribosomes to make them insensitive to antibiotics is also a suicide-prevention mechanism used by organisms that produce antibiotics. For example, erythromycin-producing bacteria have a gene called *ErmE*, which is expressed constitutively (it is always on and actively producing protein) to protect these bacteria from the erythromycin they produce. This same type of gene has been found in pathogenic bacteria that were

Fast Fact The incidence of MRSA in hospitals in the United States has risen to levels of 20–40%. In addition, it is being recognized in veterinary hospitals at an alarming rate. The isolation of methicillin-resistant strains of *Staphylococcus* are between 12% and 36%, and in addition to resistance to β-lactam antibiotics, many are also resistant to one or more macrolides, lincosamides, and trimethoprim-sulfa drugs.

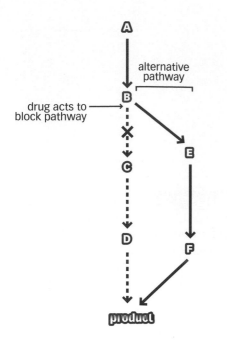

alternative
pathway

drug acts to
block pathway

product

Figure 20.7 Alternative metabolic pathway. Bacteria can use alternative metabolic pathways to circumvent a pathway that has been blocked by the action of a drug.

originally not resistant to erythromycin but have developed resistance over the years. In those bacteria, however, the gene is inducible. The presence of low levels of erythromycin causes the gene to be expressed, which produces a protein that changes the three-dimensional shape of the ribosome. This change in shape closes off the site that erythromycin attacks, making the antibiotic ineffective.

Alteration of a Metabolic Pathway

Some of the drugs used to control the growth of bacteria focus on competitive inhibition of metabolic pathways as a mechanism of action (recall our discussion of the sulfa drugs in Chapter 19). Bacteria can overcome this method of control by changing to an alternative pathway. This is illustrated in **Figure 20.7**. Because metabolic pathways can parallel one another, it is easy to circumvent the blockade of an established pathway by a drug. This alternative pathway still achieves the required outcome.

Table 20.2 lists some of the resistance mechanisms used by bacteria against commonly prescribed antibiotics. It should be noted that some antibiotics elicit multiple resistance mechanisms. This table also lists the antibiotic target and the association of drug resistance with mutation and plasmid localization. Most resistance mechanisms result from mutation, and in more than 75% of those listed, resistance is located on a plasmid. Remember, any resistance capability that is located on a transmissible element, such as a plasmid, can efficiently and in some cases rapidly transfer that resistance to other bacteria.

Keep in Mind

- Bacteria have natural mechanisms to protect themselves from antibiotic action.

- There are four mechanisms: inactivation of antibiotic, efflux pumping, target alteration, and alteration of metabolic pathways.

- Many of the enzymes involved in antibiotic resistance are coded for by inducible genes that are only switched on in the presence of antibiotic.

- When the genes involved in antibiotic resistance mechanisms are on plasmids, they can be easily transferred to another bacterium, which then acquires resistance.

CLINICALLY DANGEROUS RESISTANCE

There are currently several antibiotic-resistant bacteria (sometimes known as superbugs) that are considered especially clinically dangerous. It is important that we discuss a few in some detail because they are rapidly becoming what can only be described as a health care nightmare.

MRSA

Methicillin-resistant *Staphylococcus aureus* (MRSA) causes several infections in humans that are very difficult to treat and can be fatal. It has three or four resistance islands located on its chromosome, and more than 20 additional resistance conferring gene clusters located on mobile genetic

Antibiotic	Structural class	Target	Mutant/ plasmid	Resistance mechanism			
				Inactivation	**Efflux**	**Porin**	**Target alteration**
Ampicillin	Penicillin	E	+/+	Yes	Yes	Yes	Yes
Ceftriaxone	Cephalosporin	E	+/+	Yes	Yes	Yes	Yes
Imipenem	Carbapenems	E	+/+	Yes	Yes	Yes	Yes
Fosfomycin	Phosphonic acid	E	+/+	Yes		Yes	
Gentamicin	Aminoglycoside	R	+/+	Yes	Yes		Yes
Chloramphenicol	Phenylpropanoid	R	+/+	Yes	Yes		Yes
Tetracycline	Polyketide (II)	R	+/+	?	Yes		Yes
Erythromycin	Macrolide	R	+/+	Yes	Yes		Yes
Clindamycin	Lincosamide	R	+/+	Yes			Yes
Synercid	Streptogramin	R	+/+	Yes	Yes		Yes
Telithromycin	Ketolide	R	+/+	Yes	Yes		Yes
Ciprofloxacin	Fluoroquinolone	D	+/+		Yes		Yes
Vancomycin	Glycopeptide	E	+/+				Yes
Sulfisoxazole	Sulfonamide	M	+/+				
Trimethoprim	–	M	+/+				
Rifampin	Ansamycin	P	+/+	Yes			Yes
Fusidic acid	Steroid	T	+/+		Yes		Yes
Linezolid	Oxazolidinone	R	+/–				Yes
Novobiocin	Coumarin	D	+/+				Yes
Isoniazid	–	M	+/–				
Pyrazinamide	–	M	+/–				
Nitrofurantoin	Nitrofuran	M	+/–	Yes			
Polymyxin	Peptide	E	+/–		Yes		Yes
Capreomycin	Peptide	R	+/–	Yes			Yes
Mupirocin	Pseudomonic acid	T	–/+				Yes

D, replication; E, cell wall; M, metabolism; P, RNA polymerase; R, ribosome; T, translation.

elements, such as plasmids, which can easily transfer to other bacteria. In fact, genes for antibiotic resistance make up approximately 7% of the total *S. aureus* genome. The methicillin resistance that MRSA is named after comes from the *mecA* gene we discussed above, which also renders it essentially resistant to all β-lactam antibiotics (penicillins, cephalosporins, and carbapenems). MRSA is also resistant to other classes of antibiotics through different mechanisms such as aminoglycosides by means of antibiotic-altering enzymes, erythromycin via ribosome modification, and tetracyclines through the operation of efflux pumps. MRSA infections can be treated with powerful antibiotics such as vancomycin (a glycopeptide),

Table 20.2 Bacterial resistances to various classes of clinically used antibiotic.

but there are now also vancomycin-resistant strains of *S. aureus* (VRSA). By 2013, isolates of VRSA had been reported in the United States, Iran, India, Europe, and Latin America. MRSA and VRSA strains develop resistance to new antibiotics very quickly, and in some cases, there are no longer any single antibiotic treatments effective for some strains.

MRSA infections were initially seen only in hospitals or other health care settings, but now they are common in the general community (community-associated MRSA). MRSA is spread by skin contact and by the bacteria then entering the body through a break in the skin. Anyone can be infected with MRSA, but some people are at greater risk. In health care settings, patients with surgical wounds or long-term intravenous lines are particularly vulnerable. MRSA can be spread from one patient to another by a health care worker or visitor, so it is essential that hands are thoroughly washed before and after touching patients and clean gloves are worn for each patient. These procedures have now become adopted in hospitals in most developed countries resulting in a significant decrease in infection rates.

VREs

VREs are **vancomycin-resistant enterococci**. One species, *Enterococcus faecalis*, accounts for more than 90% of all vancomycin-resistant bacteria. Resistant enterococcal strains are the leading cause of endocarditis and are common pathogens in patients with indwelling catheters. Very few antibiotics are effective against VRE, and the current treatment for VRE infection involves the administration of Synercid® or a combination of oxazolidinone and linezolid.

Genetic resistance to vancomycin involves five tandem genes (meaning that they are located together) that work in sequence to change the structure of peptidoglycan such that it is no longer affected by the antibiotic. One of the major problems with these resistance genes is that they are easily transferred by plasmids or transposons; therefore, resistance to vancomycin can rapidly spread. VRSA gained its vancomycin resistance from VRE in this way.

E. coli

Bacteria that are part of the normal flora of the body are becoming clinically dangerous as they acquire antibiotic resistance. The best example is *E. coli*, which is part of the normal flora of the large intestine but is now a frequent component of potentially fatal systemic infections (entero-hemorrhagic *E. coli*) and common localized infections such as urinary tract infections. Most of the drug-resistant strains are acquired through the food chain. Antibiotic-resistant *E. coli* infections are now being seen in countries throughout the world, and this bacterium has become more and more resistant to antibiotics, including third-generation and fourth-generation cephalosporins.

Re-emerging Diseases

The situations we have discussed so far in this section are emerging diseases, infectious diseases not seen before in humans. Antibiotic resistance is also a major issue in re-emerging diseases: diseases that were previously under control but are posing serious clinical issues again. Perhaps the best example of a re-emerging bacterial disease with antibiotic resistance issues is tuberculosis (caused by *Mycobacterium tuberculosis*), which we discussed in Chapter 8. Use of antibiotics developed in the 1950s

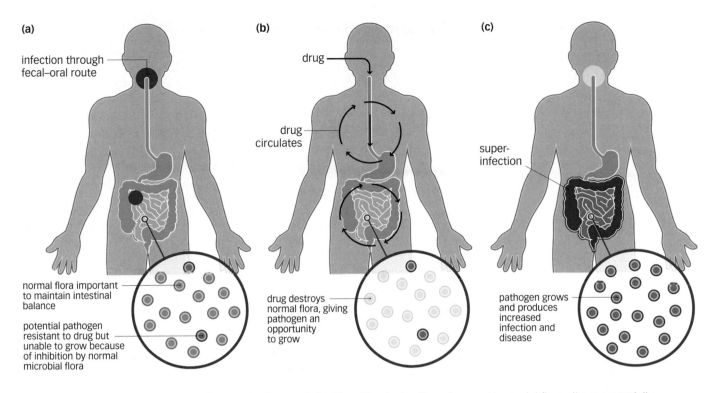

(a)

infection through fecal–oral route

normal flora important to maintain intestinal balance

potential pathogen resistant to drug but unable to grow because of inhibition by normal microbial flora

(b)

drug

drug circulates

drug destroys normal flora, giving pathogen an opportunity to grow

(c)

super-infection

pathogen grows and produces increased infection and disease

Figure 20.8 Development of superinfection. The destruction of normal bacterial flora allows potentially pathogenic bacteria that would normally be prohibited from growing to take over, resulting in disease.

decreased the spread and death rate of TB significantly, but by 2000 TB was becoming a significant problem again. Inappropriate treatment of TB is the major cause of the emergence of drug-resistant TB, and more than 450,000 new cases of multi-drug resistant TB (MDR-TB) were reported in 2012. Extensively drug-resistant TB (XDR-TB), which is even harder to treat, was first reported in 2006 and has now spread to 92 countries.

Superinfections

The overuse of broad-spectrum antibiotics, such as penicillins and cephalosporins, permits the development of a **superinfection** in which pathogens occupy areas where normal microbes have been killed. In these cases, the antibiotics have essentially compromised the patient by eliminating the natural bacteria present in the gut and allowing colonization by opportunistic pathogens. An example of this is colonization by the pathogen *Clostridium difficile*, which can establish itself in the intestinal tract as part of a superinfection. This organism is very resistant to antibiotics, and patients with this infection are difficult to treat. **Figure 20.8** illustrates how superinfections like this occur.

Unless an infection is life-threatening, clinicians should always take the time to test for sensitivity and specificity of treatment options, that would permit the use of more specific antibiotics.

Keep in Mind

- Antibiotic resistance is a major clinical problem.
- Antibiotic resistance has a major role in emerging (MRSA, VRE, and *E. coli* infections) and re-emerging infections (TB).
- Inappropriate use of antibiotics can lead to the development of superinfections.

RESISTANCE TO ANTIVIRALS AND ANTIPARASITICS

Up to now, this chapter has been about bacterial resistance to antibiotics, and there is no doubt that it is a serious issue in health care today. However, viruses and parasites also develop resistance to drug treatment, and we will now briefly look at some of the issues involved. The underlying premise is essentially the same: the use of drugs to kill organisms creates an evolutionary pressure on those organisms to become resistant to the actions of those drugs.

Antiviral Resistance

Antiviral resistance develops mainly through mutation. Because viral DNA polymerases have no proofreading activity they make mistakes during replication, meaning that progeny viruses are not exactly the same. Some of the progeny will be less susceptible to the antiviral drug and will therefore have a survival advantage. Further mutations will confer additional resistance, and eventually all viruses will be resistant.

Antiviral resistance is therefore more likely to occur in patients being treated with long-term antiviral therapy, especially those who are significantly immunocompromised such as organ transplant patients and those with bone marrow dysplasia.

The use of combination therapy means that many more mutations must be acquired for the virus population to become resistant, which is why combination therapies are used for HIV, HBV, and influenza treatments. However, resistance can develop if the incorrect dose is used, so the proper use of drugs is of utmost importance. Because there are few options for viral therapy, resistance becomes a large clinical problem when it occurs. Just as antivirals are hard to develop, viral resistance is hard to detect and a considerable amount of work is going into developing rapid and reliable assays.

Once resistance has developed, it can spread quickly. Viruses such as influenza undergo recombination as well as mutation, and a combination of these events has resulted in the widespread resistance of both seasonal and pandemic flu strains to adamantine antivirals. These antivirals are no longer recommended for use in the United States. Other antivirals such as oseltamivir (Tamiflu®) are still effective, although the 2009 H1N1 strain showed some resistance to oseltamivir. Resistance in this strain is caused by a single mutation in the gene for neuraminidase. The mass movement of people in today's world also contributes to the rapid spread of resistant virus strains.

Parasitic Resistance

Malaria affects approximately 500 million people, so the specter of resistance to all current antimalarials is frightening. Resistance has been reported to every available single treatment developed within 5–10 years of its introduction into clinical use. As of the mid 2000s, significant resistance to chloroquine has arisen in all endemic areas of Asia and Africa, and it is no longer considered the first-line treatment of malarial protozoa. Resistance has also been reported to sulfadoxine–pyrimethamine therapy. These two have historically been the least expensive and most readily available treatment options in developing countries. With their loss of efficacy, the best treatments are now combination therapies including artemisinin derivatives. Although effective, these treatments are also 20–30 times more expensive. Resistance to artemisinin is now spreading through South East Asia, and with no new class of drug expected for at least five years, recent gains in controlling malaria are at risk.

HOPE FOR THE FUTURE — DEVELOPMENT OF NEW ANTIBIOTICS

Resistance is the natural response of bacteria to the use of antibiotics. The response of the pharmaceutical industry to resistance is to modify existing drugs and develop new drug classes to combat the resistant strains. In this continual game of one-upmanship, which can be likened to an arms race, the stakes are very high.

As we have seen, antibiotics are natural products produced by microbes that can inhibit the growth of bacteria. It is safe to assume that all of the naturally occurring, readily isolatable antimicrobial products produced by known organisms have already been found. These compounds have also been synthetically modified to give new generations of antibiotics. This strategy works well until disease-causing microbes inevitably develop resistance to the modified molecules, which can happen very quickly. There are few, if any, modifications left to be made, so we need to find or develop new antibiotics from somewhere else. One approach is to search for antibiotics produced by previously unknown or poorly under-stood microorganisms, such as those that live in the soil, in the sea, and in extreme environments such as thermal vents. Another is to identify novel microbial structures or functions that can be used as potential new targets.

DNA and RNA Analysis

Bacterial chromosomes and plasmids can be **sequenced** quickly and eas-ily, and by comparing the sequences with others, DNA sequences that are seen only in prokaryotes and are conserved over generations can be identified. These sequences are likely to be genes that code for essential products that can be used as targets.

Messenger RNA can be investigated using **microarray analysis** to look for the most abundant copies of specific messages. If these copies are abundantly produced over a variety of growth conditions, it is possible that the products they code for are required for the organism to survive, and they could be new targets for antibiotic therapy.

Structural Analysis

Detailed **X-ray crystallography** studies of bacterial structures can pro-vide new targets for antibiotic action. The bacterial ribosome has been studied in this way and has been shown to be composed of a series of channels through which mRNA and newly formed proteins are constantly moving. Blockade of these channels would be devastating to the bacte-ria. The structure of bacterial efflux pumps is also being investigated for antibiotic potential. Proteins specific to the activation or creation of the pumps may be potential targets for novel antibiotics.

Auxiliary Targets

Molecular biological techniques can also be used to look for auxiliary targets, targets that can help antibiotics attack previously unavailable targets. As an example, *Mycobacterium tuberculosis* produces a waxy outer coat that protects it from many drugs. If it were possible to identify and inactivate the proteins that assemble that waxy coat, *M. tuberculosis* would be susceptible to antibiotics that could then enter the cell. In a similar fashion, RNA helicase proteins required for proper folding of the RNA molecule can be targeted such that the folding of the RNA molecule is absent or incorrect, thereby destroying normal RNA function.

Fast Fact So far, hundreds of gene products have been identified as potential targets for antibiotic therapy.

Automated Synthesis and Screening

We can couple our understanding of genetically defined potential targets with new techniques in chemistry to rapidly and efficiently produce new synthetic molecules with antibacterial potential. Computer analysis of bacterial genomes can be coupled with computer-generated construction of chemical fragments to produce potential antibiotic compounds without initially working on the bacteria themselves. Combinatorial chemistry uses computer programs to produce all possible combinations of a basic set of modular components and can generate thousands of compounds in a day. These compounds are then screened by automated techniques that can perform the initial evaluation of as many as 50,000 compounds in a single day. This is known as high-throughput screening. In addition, all the compounds can be placed into libraries for future testing when new targets are identified. Some companies have libraries with as many as 500,000 compounds in them.

Virulence Factors

Other new targets could be virulence factors or microbial survival mechanisms. One example could be targeting the production of the lipid A component of the lipopolysaccharide (LPS) layer in the outer membrane of Gram-negative organisms (Chapter 9). Because lipid A is intimately associated with the toxicity of these organisms, elimination of this component could negate toxicity. A second example would be the targeting and inactivation of proteins that organisms such as *Listeria*, *Salmonella*, and *Yersinia* use to avoid destruction by phagocytic enzymes.

Further Investigation of Known Antibiotic Compounds

As well as investigating new compounds, there are existing compounds with antibiotic activity whose mechanisms are not well understood. A good example is the **lantibiotics,** peptides produced by Gram-positive bacteria. The best-known lantibiotic is nisin, which has been used as a preservative in the dairy industry for more than 40 years, but the molecular mechanism is not fully understood. Knowing the mechanism could lead the way for the development of new antibiotics. **Defensins** are a large and diverse class of peptides, produced as part of the innate immune response by both vertebrates and invertebrates, that have antibacterial activity. They act by disrupting the phospholipid plasma membrane and have good potential for development into therapeutic agents. The peptides drosocin and apidaecin have antibacterial activity and are produced by insects.

Testing of Antibiotics

Many compounds have antibiotic activity, but for an antibiotic to be used therapeutically it must be selectively toxic (see Chapter 19). Although high-throughput techniques permit the rapid identification of potentially useful drugs, they do not give information on the potential toxicity or side effects of these compounds. Intensive and expensive testing is required before any compound can be considered for therapeutic use in humans.

Rules and regulations have been put in place to protect us from inadequately researched and poorly manufactured drugs that could be harmful. However, the need for product safety means that a great deal of time and money is needed to produce an effective and safe antibiotic. A compound is first tested against a panel of pathogen strains, some of which are resistant to antibiotics. If positive against these strains, the new compound moves on to the next step, which involves testing for toxicity at the cellular level,

Fast Fact Bacteriophage viruses, viruses that attack only bacteria, are currently being investigated for their therapeutic potential. All these techniques can also be used to develop new antiviral, antifungal, and antiparasitic compounds. The use of these viruses is called "phage therapy" or "phage mediated biocontrol of bacteria" and has been in use in the former Soviet Union for years. Bacteriophages are far more specific than antibiotics and have been used in Western countries in the dairy industry, as well as targeted destruction of *Listeria* associated with food poisoning. There are, however, no approved therapeutic regimens at this time in Western countries.

using cultures of cells. The next step is testing on infected animals to see whether they are cured by the compound, and the side effects, toxicities, and lethal levels are also evaluated. Results are compared with standard antibiotics that are currently in use to determine whether they are more effective. Finally, clinical trials must be performed on humans. This is the most expensive and time-consuming part of the procedure. It is estimated that the cost of a new drug can range from US$100 million to US$500 million and take as long as five to ten years to reach the clinic.

Keep in Mind

- New targets for antibiotics are being investigated using several techniques.
- There are also known antibacterial compounds that are being investigated further.
- Compounds must undergo stringent safety checks to ensure that they are suitably selectively toxic and will not harm the patient.

SUMMARY

- Antibiotic resistance is a natural phenomenon and will occur wherever antibiotics are used.
- Mutations that confer resistance are selected for by bacteria and can be transferred between generations (vertically) or between organisms of different species (horizontally).
- The most important factor in the development of antibiotic resistance is inappropriate or excessive use.
- The time it takes to develop resistance to an antibiotic can be short, but it can be extended by using antibiotics in combination or rotation.
- Antibiotic susceptibility testing can reduce the development of resistant strains.
- Resistance to an antibiotic can occur through inactivation of the antibiotic, pumping the antibiotic out of the cell, modifying the target of the antibiotic, or using alternative metabolic pathways.
- Antibiotic resistance is a major factor in emerging and re-emerging infectious diseases.
- Resistance to antiviral therapies is mainly due to their high mutation rate.
- New antibiotics are under development, but developing new drugs is costly and time consuming.

This chapter focuses on the variety of ways in which pathogenic bacteria and viruses have become resistant to antibiotics and antivirals. We must keep in mind (1) that resistance develops through evolutionary pressure that comes from the use of antibiotics, and (2) that this pressure will continue for as long as the bacteria are in danger of extinction from these drugs. Our overuse and misuse of antibiotics have placed them in danger of becoming useless. Given that we may have exhausted the identification of easily accessible new sources of antibacterial drugs, it is critically important to give careful consideration to how we continue to use antibiotics.

(Q) **SELF EVALUATION AND CHAPTER CONFIDENCE**

Multiple Choice

Answers are given in the back of the book and help can be found in the student resources at:
www.garlandscience.com/micro2

1. All of the following are involved in the spread of antibiotic resistance except
   A. Travel
   B. Overuse of antibiotics
   C. Specific prescriptions for antibiotics
   D. Improper use of antibiotics
   E. Immunocompromised population

2. Resistance to antibiotics is facilitated by which of the following?
   A. The antibody response
   B. Host immunity
   C. Frequency of use
   D. The inflammatory response

3. VRSA stands for which of the following?
   A. Very resistant *Streptococcus aureus*
   B. Vancomycin-resistant *Streptococcus aureus*
   C. Variably resistant *Staphylococcus aureus*
   D. Vancomycin-resistant *Staphylococcus aureus*

4. MRSA organisms are resistant to which of the following?
   A. Penicillin
   B. Cephalosporins
   C. Carbapenems
   D. None of the above
   E. All of **A**, **B**, and **C**

5. Resistance to antibiotics seen at the level of the ribosome is caused by
   A. Efflux pumping
   B. Destruction of anti-ribosomal antibiotics
   C. Changes in the shape of the ribosome
   D. None of the above

6. Increased use of antibiotics can be attributed to
   A. Inappropriate prescriptions
   B. Food production
   C. Increase in the number of immunodeficient people
   D. All of the above
   E. None of **A**, **B**, and **C**

7. The best way to prevent antibiotic resistance is to
   A. Prescribe antibiotics for all infections no matter what the cause
   B. Tell patients to stop taking antibiotics as soon as they feel better
   C. Immediately prescribe the most powerful antibiotic available
   D. Use antibiotics as sparingly and precisely as possible

8. The useful life of antibiotics can be increased by
   A. Increasing the doses
   B. Using more broad-spectrum antibiotics
   C. Using combinations of antibiotics
   D. None of the above

9–17. Answer **A** if the reason is correct, answer **B** if the reason is incorrect.

9. Active transport of penicillin out of the bacterial cell is required because the concentration of antibiotic is lower outside than inside the cell.

10. Genes for antibiotic resistance are more of a problem on plasmids because they are expressed at a higher rate.

11. Bacteria can resist the action of β-lactam antibiotics by producing penicillin-binding proteins that do not bind penicillin.

12. Potassium clavulanate augments the action of antibiotics by inhibiting the action of β-lactamase enzymes.

13. Antibiotics promote superinfections because they make the bacteria normally present in the body stronger.

14. Because antibiotics are not effective against viruses, taking antibiotics does not prevent or cure viral infections.

15. The main cause of viral resistance to antiviral drugs is because they produce many proteins.

16. *E. coli* has become a clinical problem because it has acquired antibiotic resistance and can cause dangerous systemic infections.

17. Because metabolic pathways can parallel one another, circumvention of the blockade of an established pathway can occur.

18–22. Answer **A** if true or **B** if false.

18. Mutants that are resistant to antibiotics are selected for in the presence of antibiotics.

19. Antibiotic resistance does not spread easily in health care facilities.

20. Handwashing is important to combat the spread of antibiotic resistance.

21. The time taken for antibiotic resistance to develop can be lengthened by using more broad-spectrum antibiotics.

22. Resistance to an antibiotic can only occur through destruction of the antibiotic.

 DEPTH OF UNDERSTANDING

Questions listed here require you to bring together the concepts you have learned in this chapter into a discussion format. This helps you to increase your depth of understanding of the material you have learned. Help can be found in the student resources at: www.garlandscience.com/micro2

1. It has been said that antibiotic resistance may be the greatest threat in medicine today. Defend this statement.

2. Using your knowledge of microbial genetics, describe the rise of antibiotic resistance.

3. Discuss the mechanisms involved in the development of bacterial resistance, bearing in mind the changes in bacterial function required.

 CLINICAL CORNER

Help can be found in the student resources at: www.garlandscience.com/micro2

1. One of the attending physicians who see patients on your floor is well known for prescribing very broad-spectrum antibiotics for all of the patients he sees. He has told the patients that it is the quickest and easiest way to get them well.

A. What is wrong with this approach?
B. What would be the best way to prescribe antibiotics?

2. The neighbor that lives next door received a flu shot and then developed a runny nose coupled with sneezing and coughing. She went to see her doctor and he prescribed a regimen of penicillin to be taken for seven days. After three days, she felt better and stopped taking the drugs. She mentions to you that she stopped taking them so if she got sick again she could use the rest of the medicine and would not have to pay to see the doctor and get a new prescription.

A. What should you tell her about her idea of saving what was left of the prescription for the next time she felt sick?
B. Do you think the prescription was appropriate in the circumstances?

As we have progressed through this book, we have built a strong foundation for understanding microbiology and infectious disease. Part I gave us a basic understanding of the infectious process and the chemistry we need. In Part II, we looked at the infection process. In Part III, we had a detailed discussion of the organisms that can infect us. There we divided our discussions into bacterial pathogens, viruses, fungi, and parasitic organisms, and included chapters on growth requirements and bacterial genetics. Part IV examined ways in which the body can defend itself against infection. Part V showed us how we can control microbial growth by using disinfectants, antiseptics, and antibiotics, as well as ways in which pathogens can defeat our attempts to control their growth.

In **Part VI**, we look at infections of body systems, examining each system in turn. In addition to the importance of these topics, these chapters can also be used as a form of ready reference. In each chapter, we discuss the infection process for a body system and then divide our discussion into bacterial, viral, fungal, and parasitic infections that affect that system. The coverage of each infection includes a brief discussion of the pathogenesis of the infection and the treatments recommended for it. Although we briefly discuss the anatomy of these systems, it will be helpful for you to review your anatomy and physiology textbooks so that you can more easily understand the mechanisms that we are looking at. To get the most out of this part of the book it is important, when you look at the various infections, to incorporate the basic concepts of getting in, staying in, defeating the host defenses, damaging the host, and being transmissible. In addition, think about control mechanisms and the consequences of antibiotic resistance in the infections discussed in these chapters.

In **Chapters 21–23**, we look at the three most often infected systems: the respiratory, digestive, and genitourinary systems. Because these systems are open to the outside world, they are very disposed to infection, and many of the infections seen in these three systems are routinely seen in clinical settings. **Chapters 24–26** give us an understanding of the infections we see in the nervous system, the blood, and the skin and eyes.

When you have finished **Part VI** you will:

- See that each system of the body can be involved in infection.

- Understand how the respiratory, digestive, and genitourinary systems are both major portals of entry for pathogens and major portals of exit that can transmit pathogens from one person to another.

- Understand the mechanisms and pathogenesis of some of the more common bacterial, viral, fungal, and parasitic infections seen in clinical settings, as well as some of the treatments used for these infections

Infections of the Respiratory System

Chapter 21

Why Is This Important?

The respiratory system is the most commonly infected of all the systems of the body. As a health care provider, you will see more respiratory infections than any other form of routine infection.

In 1882, Robert Koch announced his discovery of the rod-shaped bacteria that cause tuberculosis, a respiratory disease also known as consumption, phthisis, and white plague. Tuberculosis has been infecting humans for at least 4000 years. The ancient Greek physician, Hippocrates, described it as the most devastating disease of his time. From 1700 to 1900 AD, tuberculosis killed approximately 1 billion human beings. When Koch made his discovery, 7 million people a year died from it.

Today, tuberculosis has not been eradicated (there were more than 1 million deaths from it in 2010), and some worrisome antibiotic-resistant strains have evolved. Still, Koch's discovery of the bacteria that cause it, and other researchers' cures, vaccines, diagnostics, and preventative measures, have saved innumerable human lives. For his discovery, Koch was awarded the Nobel Prize in Physiology or Medicine in 1905.

OVERVIEW

In this chapter, we discuss infections of the respiratory system. As we learned in Part II, this is one of the major portals of entry into the body for infectious organisms. We can divide the respiratory system into two tracts, upper and lower, on the basis of the structures and functions found in each part. However, the dichotomy between the upper and lower respiratory tracts can also be seen in the types of infection that occur in each region. We will see that infections of the upper respiratory tract, which consists of the nasal cavity, sinuses, pharynx, and larynx, are fairly common and are usually nothing more than an irritation. In contrast, infections of the lower respiratory tract are more dangerous and can be very difficult to treat.

In this chapter, we organize our discussion of respiratory infections in two ways: (1) according to whether they occur in the upper or lower tract, and (2) according to whether they are bacterial, viral, or fungal.

We will base our discussions on the following topics:

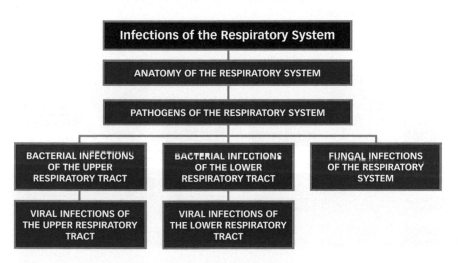

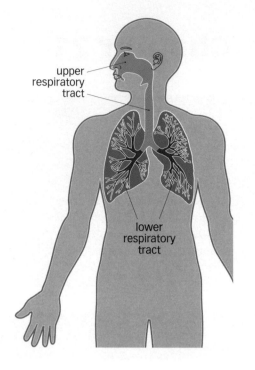

upper
respiratory
tract

lower
respiratory
tract

Figure 21.1 The upper and lower respiratory tracts.

"If the importance of a disease for mankind is measured by the number of fatalities it causes, then tuberculosis must be considered much more important than those most feared infectious diseases, plague, cholera and the like. One in seven of all human beings dies from tuberculosis."

Robert Koch (1843–1910)

ANATOMY OF THE RESPIRATORY SYSTEM

The respiratory system is the most accessible system inside the body. The simple and required act of breathing brings in clouds of potentially infectious bacteria, viruses, and even opportunistic fungi. As we examine some of the infections caused by these pathogens, remember that there are a variety of host defense mechanisms associated with the respiratory tract, including physical barriers, chemical barriers, and the innate and adaptive immune responses.

We can divide the respiratory system into an upper tract and a lower tract (**Figure 21.1**). The upper tract, the primary entryway into the system, is continuously exposed to potential pathogens drifting around in the external environment. In contrast, the lower respiratory tract, which begins with the bronchi and continues into the lung, is essentially a sterile environment. When this area is infected, pneumonia can result.

PATHOGENS OF THE RESPIRATORY SYSTEM

The list of bacterial organisms that infect the respiratory system is extensive and includes *Streptococcus pneumoniae*, *Haemophilus*, group A streptococci, *Mycoplasma pneumoniae*, *Chlamydophila pneumoniae*, *Bordetella pertussis*, and *Mycobacterium tuberculosis*. The upper respiratory tract is also a portal of entry for such viral pathogens as influenza, parainfluenza, respiratory syncytial virus, and rhinovirus (a cause of the common cold). Vaccination has eliminated many infections of the respiratory tract, but some, such as diphtheria and pertussis (whooping cough), are still seen in underdeveloped parts of the world. Some of the microbial sources of respiratory infection are shown in **Figure 21.2**.

Many respiratory infections are caused by pathogens that move from one human to another, with the infection spreading as the pathogens circulate within a community. Other respiratory pathogens are acquired from animal sources, and so the infections they cause are classified as zoonotic infections. For example, Q fever stems from farm animals that spread the organism *Coxiella burnetii*, and the respiratory infection called psittacosis, caused by *Chlamydophila psittaci*, is spread from parrots and other birds to humans.

Figure 21.2 Common infections of the respiratory system. Panel a: Some of the more common infections and symptoms associated with the upper respiratory tract. Panels b and c: Infections seen in the lower respiratory tract.

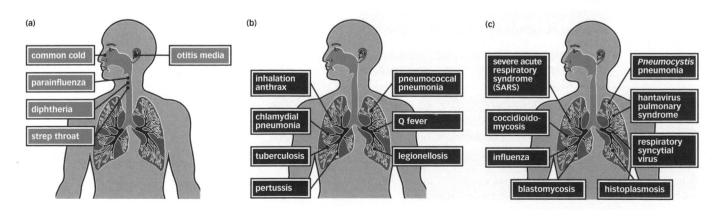

(a)

common cold

parainfluenza

diphtheria

strep throat

otitis media

(b)

inhalation anthrax

chlamydial pneumonia

tuberculosis

pertussis

pneumococcal pneumonia

Q fever

legionellosis

(c)

severe acute respiratory syndrome (SARS)

coccidioido-mycosis

influenza

blastomycosis

Pneumocystis pneumonia

hantavirus pulmonary syndrome

respiratory syncytial virus

histoplasmosis

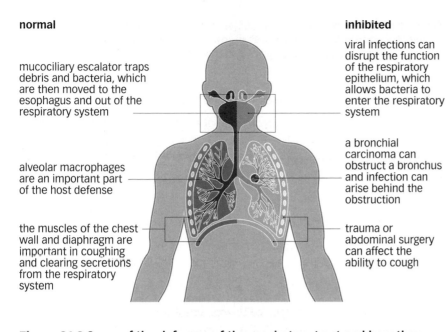

normal

mucociliary escalator traps debris and bacteria, which are then moved to the esophagus and out of the respiratory system

alveolar macrophages are an important part of the host defense

the muscles of the chest wall and diaphragm are important in coughing and clearing secretions from the respiratory system

inhibited

viral infections can disrupt the function of the respiratory epithelium, which allows bacteria to enter the respiratory system

a bronchial carcinoma can obstruct a bronchus and infection can arise behind the obstruction

trauma or abdominal surgery can affect the ability to cough

Figure 21.3 Some of the defenses of the respiratory tract and how they can be inhibited. Loss of these defenses is directly correlated with increased pathogen activity.

Water can also be a source of respiratory infections. The prime example is legionellosis (legionnaires' disease), a form of pneumonia caused by *Legionella pneumophila*. This organism can survive in water over the range of temperatures from 25°C to 45°C. If water contaminated with this organism is aerosolized as a mist (as is done in some air-conditioning systems), droplets can be inhaled, and an infection can result.

Fungi, which are ubiquitous in the environment, are also a source of respiratory infection. However, these organisms do not normally cause infection unless the patient is in some way immunocompromised. Some of the more dangerous respiratory fungal infections are caused by *Aspergillus* species and *Pneumocystis (carinii) jiroveci*.

Some pathogens are restricted to certain sites. *Legionella* infecting only the lungs is one example. Other pathogens can cause infection in multiple sites. For example, *Streptococcus* can cause middle ear infections and **sinusitis** (both infections of the upper respiratory tract) as well as pneumonia (an infection of the lower respiratory tract). Four areas of the upper respiratory tract—the middle ear, mastoid cavity, nasal sinuses, and nasopharynx—are frequent sites of infection.

Keep in mind that there are significant defenses in the respiratory system (**Figure 21.3**). For example, the mucociliary escalator of the upper respiratory tract traps pathogens in mucus, and the rhythmic movement of ciliated epithelia cells (**Figure 21.4**) moves these pathogens up and out of the system. In addition, the act of coughing is a mechanical process that forcefully eliminates organisms from the respiratory system, making it a good portal of exit. In the lower respiratory system, alveolar macrophages in the alveoli of the lungs help protect against infection. It is important to remember that compromise of any of these defenses can lead to a predisposition to respiratory system infection. A breach of these defense systems can also be caused by primary infections, resulting in secondary infections caused by normal flora.

Figure 21.4 A colorized scanning electron micrograph showing ciliated cells. These are important barriers used to keep pathogens from entering the lower respiratory tract.

otitis media, sinusitis, mastoiditis		
	organism	**treatment**
	Streptococcus pneumoniae	amoxycillin, erythromycin
	Haemophilus influenzae	amoxycillin, ciprofloxacin
	Staphylococcus aureus	flucloxacillin, erythromycin, ciprofloxacin
	group A streptococci	amoxycillin, benzylpenicillin
pharyngitis		
	group A streptococci	benzylpenicillin, amoxycillin
	group C streptococci	benzylpenicillin, amoxycillin
	Corynebacterium diphtheriae	benzylpenicillin, erythromycin (antitoxin)
	Chlamydophila pneumoniae	erythromycin, tetracycline
	Mycoplasma pneumoniae	erythromycin, tetracycline
community-acquired pneumonia – typical		
	Streptococcus pneumoniae	amoxycillin, erythromycin, cefuroxime
	Haemophilus influenzae	amoxycillin, cefuroxime, ciprofloxacin
	Staphylococcus aureus	flucloxacillin, erythromycin
community-acquired pneumonia – atypical		
	Chlamydophila pneumoniae	erythromycin, tetracycline
	Chlamydophila psittaci	erythromycin, tetracycline
	Mycoplasma pneumoniae	erythromycin
	Coxiella burnetii	erythromycin, tetracycline
	Legionella pneumophila	erythromycin (and rifampicin)
hospital-acquired pneumonia		
	Citrobacter spp.	gentamicin, ciprofloxacin, imipenem
	Enterobacter spp.	gentamicin, ciprofloxacin, imipenem
	Pseudomonas aeruginosa	gentamicin, ciprofloxacin, imipenem
	Staphylococcus aureus	flucloxacillin (and gentamicin)
	MRSA	vancomycin (and gentamicin)

Figure 21.5 Bacteria that can infect the respiratory system and antibiotics commonly used against them. Gram-positive bacteria are shown in purple and Gram-negative bacteria in pink.

Bacteria That Infect the Respiratory System

Figure 21.5 provides an extensive list of bacteria that are involved with infections of the respiratory system. They can be divided into several groups as follows:

- Those that cause otitis media (middle ear infections), sinusitis, and mastoiditis

- Those that cause **pharyngitis**

- Those responsible for typical and atypical community-acquired pneumonia

- Those that cause hospital-acquired (nosocomial) pneumonia

We have included an example of the morphology and Gram reaction for each of the organisms listed, as well as commonly used antibiotics for infections caused by these organisms.

Keep in Mind

- The respiratory system is the most accessible system in the body and is therefore a major portal of entry for pathogens.

- The upper respiratory tract is continuously exposed to pathogens, whereas the lower respiratory tract is essentially sterile.

- Many infections of the lower respiratory tract result in pneumonia.

- The respiratory system is also a good portal of exit, and so it is easy for a pathogen to use this system to spread infections from person to person.

- Fungal infections of the respiratory system are usually restricted to individuals who are immunocompromised.

BACTERIAL INFECTIONS OF THE UPPER RESPIRATORY TRACT

Otitis Media, Mastoiditis, and Sinusitis

The middle ear, mastoid cavity, and sinuses are all connected either directly or indirectly to the nasopharynx. The ciliated epithelial cells that line both the sinuses and the eustachian tube push out bacteria that are trapped in mucus. In middle ear and sinus infections, it is likely that a virus such as respiratory syncytial virus initially invades the ciliated epithelium and destroys the cells involved with the mucociliary escalator, thereby allowing bacteria to invade and remain in the respiratory tract.

Mastoiditis is an uncommon problem but is very dangerous because of the proximity of this cavity to the nervous system and large blood vessels.

Pharyngitis

A variety of bacteria can cause infection in the pharynx. The classic form of pharyngitis is called **strep throat** (Figure 21.6), caused by *Streptococcus pyogenes*. Recall from earlier discussions that this organism has an M protein protruding from its cell wall that inhibits phagocytosis. In addition, proteins such as hemolysin O and S, DNase, streptokinase,

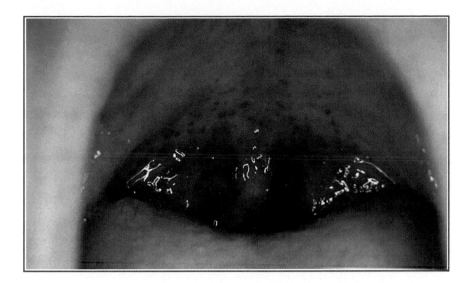

Figure 21.6 An example of strep throat, which is caused by *Streptococcus pyogenes*. Notice the redness and edema of the oropharynx, and the petechiae (small red spots) on the soft palate.

and pyrogenic toxins (**Figure 21.7**) produced by this pathogen account for the symptoms seen with pharyngitis. Group A streptococci (such as *S. pyogenes*) can also cause abscesses on the tonsils, a development that may require the removal of these lymphoid organs. *S. pyogenes* can also cause other infections that are potentially dangerous, in particular scarlet fever and toxic shock syndrome.

Scarlet Fever

Scarlet fever is a disease that is caused by group A streptococci and can sometimes occur in people who have strep throat. The disease is usually seen in children under the age of 18 years.

Pathogenesis of Scarlet Fever

The symptoms of scarlet fever usually begin with a rash that appears as tiny bumps on the chest and abdomen. This rash can spread over the entire body and is usually redder in the armpits and groin areas. The rash normally lasts two to five days. In addition, the patient may experience flushing and a very sore throat, which can be accompanied by yellow or white papules. There can be a fever of 101°F or higher and lymphadenopathy in the neck region. Headache, body aches, and nausea can also occur.

Treatment of Scarlet Fever

Treatment of scarlet fever involves the use of antibiotic therapy.

group-specific carbohydrate (e.g. A)

M protein is a major virulence factor, with antiphagocytic properties

extracellular products

streptococcal pyrogenic toxin which causes:
- scarlet fever rash
- toxic shock syndrome

streptolysins (hemolysins)
- streptolysin O
- streptolysin S

streptokinase

some strains produce a capsule which makes them resistant to phagocytosis

Figure 21.7 The virulence factors of streptococcus organisms. Different groups of streptococci are defined by the carbohydrates in the cell wall.

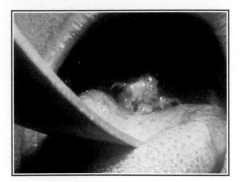

Figure 21.8 The pseudomembrane seen in diphtheria infections. In severe cases, this membrane can cover the trachea.

Fast Fact The removal of tonsils and adenoids was once a common procedure until medical workers realized that these organs served an important protective function.

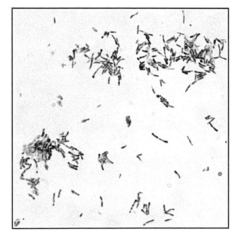

Figure 21.9 A photomicrograph showing the organization of *Corynebacterium diphtheriae* in the characteristic Chinese-letter pattern. This pattern results from snapping, which is the unique form of cell division seen in this organism.

Diphtheria

Diphtheria is caused by diphtheria toxin, which is produced by *Corynebacterium diphtheria* and is a potent inhibitor of protein synthesis. The infection is localized and initially presents as severe pharyngitis accompanied by a plaque-like pseudomembrane in the throat (**Figure 21.8**). The life-threatening aspects of the infection are due to toxemia, which can involve multiple organ systems, including acute myocarditis. Diphtheria is highly contagious and is transmitted by droplet aerosol, by direct contact with the skin of an infected person, or, to a lesser degree, by fomites.

Vaccination against diphtheria, part of the DTaP (diphtheria, tetanus, pertussis) vaccine protocol, is very effective, and infection is rare when vaccination is in place. In fact, fewer than 10 cases are reported in the United States each year, with the highest incidence seen in migrant workers or immigrants who have not been vaccinated. Diphtheria still occurs frequently in some parts of the world, particularly where socioeconomic conditions do not permit vaccination. For example, when a vaccination program existed in the Soviet Union, there were normally fewer than 200 reported cases of diphtheria per year. However, when the numbers of children vaccinated decreased, adults who had been exposed passed the infection to unvaccinated children and the number of cases increased enormously, with more than 47,000 cases and 1700 deaths between 1990 and 1995.

Pathogenesis of Diphtheria

Corynebacterium diphtheriae is a small bacillus that appears in V and L forms resembling Chinese letters (**Figure 21.9**). This arrangement occurs because of a unique process of cell division referred to as snapping. The process causes the cells to arrange themselves both parallel and perpendicular to one another. These bacilli have an unusual cell envelope similar to that of *Mycobacterium* (Chapter 9), and show weak Gram-positive staining. This organism is poorly invasive, and the effects of the infection are mainly due to the exotoxin produced. As we saw in Chapter 5, this toxin has a classical configuration of two polypeptide chains, with the B chain used for entry into the target cell and the A chain used for inhibiting protein synthesis.

Local effects are seen as epithelial cell necrosis accompanied by inflammation, and the pseudomembrane that forms is composed of a mixture of fibrin, leukocytes, and cell debris. The size of this membrane varies from small and localized to extensive, in which case it can cover the trachea. Diphtheria can also be systemic, causing acute myocarditis. Iron seems to have a role in how much toxin is produced, and the *Tox* genes that code for this toxin are regulated by operons like those we learned about in Chapter 11.

Incubation takes two to four days and usually presents as pharyngitis or **tonsillitis** accompanied by fever, sore throat, and malaise. Patches of pseudomembrane can be seen on tonsils, uvula, soft palate, or pharyngeal walls and may extend downward into the larynx and trachea. In uncomplicated cases, the infection will resolve and the membrane will be coughed up after five to ten days. However, complications caused by respiratory obstruction can result in suffocation, and systemic infection can result in myocarditis during the second or third week of the infection. Diphtheria can also cause nonrespiratory infections, particularly of the skin, demonstrated by the formation of simple pustules or chronic non-healing ulcerations.

Treatment of Diphtheria

The most important treatment is neutralization of the toxin and elimination of the organism. Toxin neutralization is the more critical and should be done as quickly as possible. Antitoxin can be used to neutralize free toxin but has no effect on toxin that has become fixed to target cells. *Corynebacterium diphtheriae* is sensitive to many antibiotics, including penicillin, cephalosporin, erythromycin, and tetracycline.

VIRAL INFECTIONS OF THE UPPER RESPIRATORY TRACT

Rhinovirus Infection (the Common Cold)

There are several hundred serotypes of rhinovirus, and fewer than half of them have been well characterized. Those that have been characterized are all picornaviruses (Figure 21.10), which are extremely small, single-stranded RNA viruses that do not have envelopes. The optimum temperature for the growth of picornaviruses is 33°C, which is the temperature in the nasopharynx. In fact, there is some thought that this lower temperature is why these viruses localize there.

Pathogenesis of Rhinovirus Infection

Rhinovirus uses the glycoprotein ICAM, which is an adhesion molecule, as a receptor to infect host cells. It is known as the common cold virus because it is the major cause of the mild upper respiratory infections that affect people of all ages but especially older children and adults. This infection is seen at any time of the year and is usually epidemic in the early fall and spring. Rhinovirus infections are rarely seen in the lower respiratory tract.

The incubation period for rhinovirus infection is approximately two to three days, and the acute symptoms can last for three to seven days. There is little damage to the mucosal layer of the upper respiratory tract during this infection, but some data do suggest that the infection causes an increase in the production of bradykinin, which may cause excessive secretion, vasodilation, and sore throat.

Treatment of Rhinovirus Infection

There is no specific therapy or treatment for these infections. Because of the large number of serotypes of this virus, it is unlikely that an effective vaccine will ever be developed. However, there is a possibility that soluble ICAM receptors (discussed in Chapter 12) may be a way of inhibiting viral attachment.

Parainfluenza

There are four types of parainfluenza virus: 1, 2, 3, and 4. This virus is enveloped and belongs to the paramyxovirus group. It is a single-stranded RNA virus, but the viral genome is not segmented as it is in the influenza virus. Like the influenza virus, the parainfluenza virus contains hemagglutinin and neuraminidase, and it is this similarity that makes the transmission and pathology of the parainfluenza viruses similar to those of the influenza virus.

There are also differences between these two viruses, one being that parainfluenza virus replicates in the cytoplasm of the host cell, whereas influenza virus replicates in the nucleus. In addition, parainfluenza is genetically more stable, and very little mutation is seen with this virus. Consequently there is little antigenic drift and no antigenic shift.

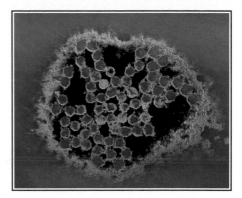

Figure 21.10 A colorized electron micrograph of a cluster of rhinoviruses. Rhinovirus is the etiologic agent of the upper respiratory infection known as the common cold.

The parainfluenza virus can cause serious problems in infants and small children, especially between the ages of one and three years. Overall, it is responsible for 15–20% of nonbacterial respiratory infection that requires hospitalization in infancy and childhood. There is only a transitory immunity to re-infection, but the infection becomes milder as the child ages.

Pathogenesis of Parainfluenza Infection

The onset of infection with parainfluenza may be abrupt and appear as an acute spasmodic **croup** (hoarseness and a barking cough). This usually begins as mild and progresses over one to three days to involve the lower respiratory tract. The duration of the illness can be between four and twenty-one days but is usually seven to ten days. Of the four types of parainfluenza virus, types 1 and 3 are the most clinically relevant. Type 1 is the major cause of laryngotracheitis (**acute croup**) in infants and young children. It can also cause severe upper respiratory illness, pharyngitis, and **tracheobronchitis** in all age groups. Outbreaks usually occur in the fall of the year.

Parainfluenza type 3 is the major cause of severe lower respiratory infection in infants and young children, often causing bronchitis and pneumonia in children less than one year of age. Infections with this virus can occur throughout the year, and it is estimated that 50% of all children are exposed to this virus during their first year of life.

Treatment of Parainfluenza Infection

There is currently no method of treatment or control for this viral infection except supportive care.

Keep in Mind

- Bacterial infections of the upper respiratory tract can involve otitis media, mastoiditis, sinusitis, and pharyngitis.

- Viral infections of the upper respiratory tract include the common cold (rhinovirus) and parainfluenza.

- There are no specific treatments for viral infections of the upper respiratory tract.

BACTERIAL INFECTIONS OF THE LOWER RESPIRATORY TRACT

Bacterial Pneumonia

Of all of the infections of the lower respiratory tract, pneumonia is one of the most serious. This infection of the lung parenchyma can be divided into two types: community-acquired and nosocomial, and each type can be caused by a variety of organisms. Nosocomial pneumonia occurs approximately 48 hours after a patient has been admitted to the hospital and is usually associated with *Staphylococcus aureus* or with Gram-negative bacteria such as *Pseudomonas aeruginosa*. This infection can be particularly difficult to deal with if the organism is resistant to antibiotics.

In contrast, community-acquired pneumonia usually presents as a lobar pneumonia accompanied by fever, chest pain, and the production of purulent sputum. It is important to note that there can also be cases of atypical pneumonia characterized by coughing without the production of sputum. Atypical pneumonia can be caused by a variety of organisms, including

Mycoplasma pneumoniae, *Chlamydophila pneumoniae*, *Chlamydophila psittaci*, *Coxiella burnetii*, and *Legionella pneumophila*. In some cases, bacterial pneumonia can progress to the production of lung abscesses.

Pathogenesis of Nosocomial Bacterial Pneumonia

Ironically, a hospital setting is one of the most clinically dangerous, mainly because of the large number of infected patients in one location (many of whom are immunocompromised), each carrying different pathogens, many of which have developed antibiotic resistance (Chapter 20). Among these resistant bacteria are *Enterobacter* species, *Pseudomonas aeruginosa*, and methicillin-resistant *Staphylococcus aureus* (MRSA).

The problem is compounded by the fact that the hospital is filled with debilitated patients. Debilitation causes an increase in proteolytic enzyme activity in the saliva of these patients, which contributes to the rapid turnover of the fibronectin layer that covers the epithelium of the pharynx. This layer is associated with normal bacterial flora; when it is lost, the area can become colonized with opportunistic pathogens that can then be aspirated into the lungs and cause pneumonia (**Figure 21.11**).

Pathogenesis of Typical Community-Acquired Bacterial Pneumonia

This infection usually occurs after the aspiration of pathogens such as *Pneumococcus* in great enough numbers to overwhelm the resident defenses present in the lungs. The establishment of an infection in the lungs depends not only on the number of pathogens entering the lungs, but also on the competence of the mucociliary escalator to keep them out. The classical lobar pneumonia resulting from infection with *Pneumococcus* has four stages:

- **Acute congestion**: local capillaries become engorged with neutrophils that are part of the host defense response.

- Red hepatization: red blood cells from the capillaries flow into the alveolar spaces.

- Gray hepatization: large numbers of dead and dying neutrophils are present, and degenerating red cells are seen.

- Resolution: the adaptive immune response begins to produce antibodies that control the infection.

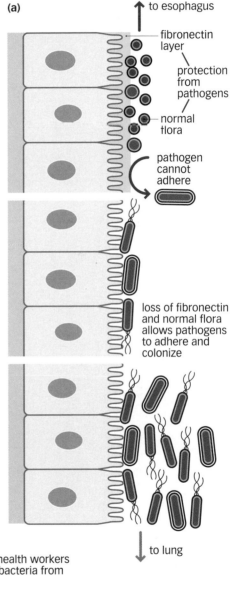

Figure 21.11 Mechanisms associated with nosocomial pneumonia. Panel a: Mucosal surfaces are coated with fibronectin and normal flora. Removal of this layer allows Gram-negative bacteria to colonize the oropharynx in significant numbers. Panel b: Some of the contributory factors for nosocomial pneumonia.

(a) to esophagus

fibronectin layer

protection from pathogens

normal flora

pathogen cannot adhere

loss of fibronectin and normal flora allows pathogens to adhere and colonize

to lung

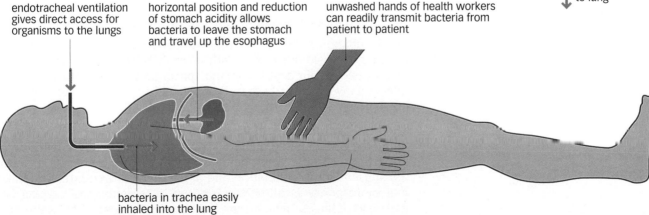

(b) endotracheal ventilation gives direct access for organisms to the lungs

horizontal position and reduction of stomach acidity allows bacteria to leave the stomach and travel up the esophagus

unwashed hands of health workers can readily transmit bacteria from patient to patient

bacteria in trachea easily inhaled into the lung

Treatment of Nosocomial and Typical Pneumonia

Treatment depends on the severity of the infection and the type of organism causing the infection. The leading cause of bacterial pneumonia is *Streptococcus pneumoniae*, which is treated with penicillin, amoxicillin-clavulanate (Augmentin®), and erythromycin. Other antibiotics used to treat bacterial pneumonia include cefuroxime (Ceftin®), ofloxacin (Floxin®), and trimethoprim-sulfamethoxazole.

Chlamydophila Pneumonia

This infection is caused by *Chlamydophila pneumoniae*, which is found throughout the world and is responsible for 10% of pneumonias and 5% of bronchitis cases. The infection occurs throughout the year and is spread through person-to-person contact. There have been reports of this form of pneumonia in both community-acquired and nosocomial forms, and there are more infections with this organism in the elderly.

Pathogenesis of *Chlamydophila* Pneumonia

The infection can present as either pharyngitis or lower respiratory tract infection, or both, and is clinically similar to *Mycoplasma* pneumonia (see below). An initial pharyngitis lasting one to three weeks is replaced by a persistent cough that can last for weeks. There may be an association between *Chlamydophila* pneumonia and vascular endothelial disease such as atherosclerosis.

Treatment of *Chlamydophila* Pneumonia

Tetracycline or erythromycin will alleviate the symptoms of this infection.

Mycoplasma Pneumonia

This is a mild form of pneumonia that accounts for about 10% of all pneumonias. It is often referred to as walking pneumonia because there is no need for hospitalization during this infection. The most common age for patients with this infection is between five and fifteen years, and it causes approximately 30% of all teenage pneumonias. It is caused by *Mycoplasma pneumoniae*, an unusual bacterium that lacks a cell wall. As a consequence, this bacterium has no rigid shape, cannot be treated with drugs targeting the peptidoglycan cell wall, and is able to penetrate normal sterilization filters. The infection is acquired by droplet transmission, and the infectious dose required to cause infection is fewer than 100 cells, making it very easy to contract. *Mycoplasma pneumoniae* is found throughout the world, especially in temperate climates.

Pathogenesis of *Mycoplasma* Pneumonia

The incubation period is between two and fifteen days, and the infection has an insidious onset, with fever, headache, and malaise occurring for two to four days before the appearance of respiratory symptoms. Infection involves the trachea, bronchi, and bronchioles and may extend down to the alveoli.

The organism initially attaches to the cilia and microvilli of the cells lining the bronchial epithelium. This attaching interferes with the ciliary action, with the result being detachment of the mucosal layer and the subsequent inflammation and appearance of **exudate**. The inflammatory response is initially composed of lymphocytes, plasma cells, and macrophages, which may infiltrate and thicken the walls of the bronchioles and alveoli. The organism can be shed in upper respiratory secretions for two to eight days before symptoms appear and for up to fourteen weeks afterward.

This infection causes a mild tracheobronchitis with fever, cough, headache, and malaise. It can occur with a sore throat, and some patients also experience symptoms of otitis media.

Treatment of *Mycoplasma* Pneumonia

Erythromycin or tetracycline is the usual treatment. The use of either drug can shorten the clinical symptoms, but the organism may be in the nasopharynx for long periods after the symptoms have subsided. Azithromycin, clarithromycin, and most quinolones are also effective.

Tuberculosis

It is estimated that 2 billion people worldwide are infected with tuberculosis. That amounts to about one-third of the population of the world, and 1.5 million people die of tuberculosis each year. In fact, *Mycobacterium tuberculosis* has been responsible for more deaths than any other infectious agent. The development of effective antibiotics in the 1950s slowed the spread of tuberculosis for some years, but by the year 2000 the incidence of tuberculosis had begun to rise. The appearance of AIDS and HIV infection has had a significant role in the increase of tuberculosis, because these infections increased the efficiency of the tuberculosis transmission cycle.

Poverty and poor socioeconomic conditions are breeding grounds for tuberculosis, and drug resistance has become an increasingly dangerous problem. One of the major reasons is non-compliance with the therapy regimen. Because the treatment of tuberculosis can require the daily administration of antibiotics for long periods (at least six months), many patients will stop taking the drugs. This can lead to the evolution of drug-resistant strains of *Mycobacterium tuberculosis*.

Early detection is critical to prevent the spread of tuberculosis and to increase the chances of successful treatment. Because the initial symptoms are similar to those seen in other respiratory infections, it is important to look for symptoms such as fever, fatigue, weight loss, chest pain, and shortness of breath in conjunction with coughing as an indication of potential tuberculosis.

M. tuberculosis is an obligate aerobic rod-shaped bacillus that is acid-fast and forms no spores (**Figure 21.12**). Recall from Chapter 9 that *Mycobacterium* produces mycolic acid as part of its cell wall, and it is this component that makes it difficult to Gram stain and also protects the pathogen from antibiotic therapy and host defenses.

Pathogenesis of Tuberculosis

Mycobacterium is a problem for the host defense because of its unique cell wall, which interferes with macrophage function and with T-cell activation. When *Mycobacterium* is ingested by macrophages, it inhibits the formation of the phagolysosome and eventually escapes into the cytoplasm of the macrophage. Here the bacterium will increase in number and eventually spread to the lymph nodes, where it will enter the blood and distribute throughout the body. The cell wall components of *Mycobacterium* attract T cells and macrophages to the site of the infection, and there is an uncontrollable release of enzymes and metabolites that destroy tissues, leading to

> **Fast Fact** In sub-Saharan Africa it is estimated that 50% of HIV-infected patients also have tuberculosis.

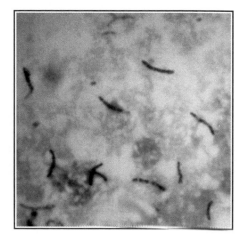

Figure 21.12 An acid-fast stain of *Mycobacterium* bacteria. Because of the mycolic acid that is part of the cell wall of this bacterium, heat is required in staining it.

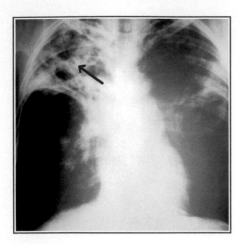

Figure 21.13 An X-ray of the lungs of a patient with advanced tuberculosis. The arrow shows infected areas containing tubercles filled with live *Mycobacterium*. These areas of the lungs are no longer functional.

necrosis. Necrosis in the lung liquefies and spreads to adjacent areas of the lung, a migration that causes the cycle to continue (**Figure 21.13**).

There are two basic types of tuberculosis: primary, which follows initial exposure to the pathogen, and secondary, which can occur years later. Let's briefly examine the difference between primary and secondary tuberculosis. Primary tuberculosis occurs when the host encounters *M. tuberculosis* for the first time. Organisms find their way to the alveoli and a localized inflammatory response develops. This involves the phagocytosis of the bacilli by macrophages and neutrophils; however, the pathogens are not killed but are transported by these phagocytes to the regional lymph nodes and continue to divide intracellularly. A cell-mediated immune response now begins, including a delayed-type hypersensitivity reaction to tuberculin protein. This can lead to a positive tuberculin (PPD) skin test reaction.

If the infection is not contained by host defenses, tubercles (or granulomas) will form. A tubercle is composed of aggregates of enlarged macrophages, some of which are infected with bacteria. This tubercle is surrounded by fibroblasts and lymphocytes. Frequently, the center of the tubercle will undergo caseous necrosis, which may later calcify. When this occurs, these calcifications are referred to as **Ghon complexes**; they are readily seen on X-ray.

At this point the primary infection becomes quiescent and asymptomatic in about 90% of patients. In the remaining 10%, the infection can evolve into clinical disease with bacilli spreading through the lymphatic channels and bloodstream and via the bronchi and gastrointestinal tract. These events result in tuberculous meningitis, miliary (disseminated) tuberculosis or both. If the localized tubercles discharge their contents into the bronchus, they can be aspirated and distributed to other parts of the lungs.

Secondary tuberculosis is usually due to the breakdown of old tubercles and the release of reactivated bacteria that have been there since the primary infection. While in the tubercle, the bacteria are semi-dormant because there is no oxygen (they are obligate aerobes), but release into the oxygenated environment of the lungs means they can grow again. This recurrence of disease occurs in a small percentage of patients whose initial symptoms have subsided. Secondary tuberculosis usually manifests itself in the top of the lungs and occurs within two years of the primary infection. However, secondary tuberculosis may evolve many decades later if innate resistance begins to diminish. Patients whose immune systems become compromised are therefore vulnerable to developing secondary tuberculosis.

Treatment of Tuberculosis

Treatment usually consists of a triple therapy involving isoniazid (INH), pyrazinamide (PZA), and rifampicin/rifampin (RFP). These three drugs are taken once a day for two months, followed by a regimen of INH and RFP for a further six to nine months. If the tuberculosis strain is drug resistant, the initial regimen will also include ethambutol. Both isoniazid and ethambutol target the cell wall of *Mycobacterium* and are therefore specific for this family of bacteria. Resistance to isoniazid is most common, and in difficult cases, ciprofloxacin and clarithromycin have also been used. Multi-drug resistant tuberculosis (MDR-TB) is defined as resistance to isoniazid and rifampicin and must be treated with more powerful second-line drugs. Extensively drug resistant tuberculosis (XDR-TB) has recently been seen and is resistant to first-line and second-line tuberculosis drugs.

DOT is required for the these patients	DOT is recommended for these patients
HIV patients co-infected with tuberculosis	Congregate living
Homeless	Children
A history of criminal incarceration	Adolescents
Psychiatric disorders	Elderly with cognitive impairment
Cognitive dysfunction	Recently immigrated individuals
Drug abuse or misuse	Patients who seem to have difficulty in understanding
History of non-adherence to medication therapies	
Physiological resistance to one or more anti-tuberculosis drugs	
Persistently positive test results	
History of refusal of medical care	

Patients must be monitored carefully because compliance with the drug therapy is very important. Compliance by patients can be difficult because of the toxicity of the drugs and the consequent side effects, the most serious of which is liver toxicity. **Directly observed therapy** (**DOT**) is used as a way of preventing MDR-TB. It involves the delivery of scheduled doses of tuberculosis medication by a health care worker. The health care worker directly administers, observes, and documents the patient's ingestion or injection of the medication. The purpose of DOT for tuberculosis treatment is to ensure that patients receive the medication required to prevent the spread of tuberculosis and to prevent MDR-TB. Patients who are required to receive DOT or are recommended for this treatment are listed in Table 21.1.

Table 21.1 Directly observed therapy.

Pertussis (Whooping Cough)

This infection of the lower respiratory tract is spread by airborne droplets produced by patients in the early stages of illness. It is highly contagious and infects 80–100% of exposed susceptible individuals. Consequently, the spread of pertussis is rapid in schools, hospitals, offices, and homes— in other words, just about anywhere. The pathogen *Bordetella pertussis* is not found in animals and does not survive in the environment. Therefore, the reservoir for this pathogen is humans, and it has been shown that previously immunized individuals are becoming an important reservoir for this infection. Because the symptoms can be similar to those of a cold, many infected adults will spread the infection to places such as schools and nurseries without knowing it. Mortality is highest in infants (70%) and in children under one year old.

Immunization against pertussis was begun in the 1940s and continues today as part of DTaP vaccination. Once use of this vaccine became widespread, the incidence of pertussis in the USA dropped from as many as 147,000 cases a year in the 1940s, to around 1000 cases a year in the 1970s. Unfortunately, pertussis seems to be making a comeback of sorts since 1980, with epidemics occurring every three to four years. In addition, the greatest numbers of infections are being seen in 10- to 20-year olds, which may be due to people who were not immunized because of fears associated with vaccine side effects. There is a clear relationship between lack of vaccination and infection.

Pathogenesis of Pertussis

Bordetella pertussis, a Gram-negative coccobacillus, is a strictly human pathogen that has an affinity for ciliated bronchial epithelium. Once attached, the organism produces a **tracheal toxin** that immobilizes and progressively destroys the ciliated cells. The persistent coughing associated with this infection is caused by the inability to move the mucus that builds up in the respiratory tract up and out.

Pertussis does not invade cells of the respiratory tract or deeper tissues. After an incubation period of seven to ten days, the infection follows three stages:

- The catarrhal stage features persistent perfuse and mucoid **rhinorrhea** (runny nose) for one to two weeks. There may also be sneezing, malaise, and anorexia. The infection is most communicable during this stage.

- The paroxysmal stage starts with persistent coughing that can build to 50 times a day for two to four weeks. This is when the characteristic whooping sound is heard. It is the result of the patient's trying to catch his or her breath during a coughing episode. In addition, apnea may follow the coughing, especially in infants. There is also a significant increase in lymphocytes during this period, with counts sometimes reaching 40,000 per mm^3 of blood.

- In the convalescent stage, the frequency and severity of coughing and other symptoms gradually decrease.

The most common complications of pertussis are superinfection with *Streptococcus pneumoniae* and convulsions. However, subconjunctival and cerebral bleeding as well as anoxia can occur as a result of the persistent coughing.

Treatment of Pertussis

Antibiotics such as erythromycin and clarithromycin can be used in early stages of the infection, and this treatment limits the spread of the infection to others. However, once the paroxysmal stage is reached, therapy is only supportive.

Inhalation Anthrax

This infection produces a fulminate pneumonia (one that comes on suddenly with great severity) leading to respiratory failure and death. Anthrax is primarily a disease of herbivores such as cattle, horses, and sheep and is acquired from spores found in pastures. If these spores are inhaled, anthrax can occur in the respiratory tract. Infection is rare, and when it does occur, it usually affects farmers, veterinarians, and meat handlers; it usually presents as localized lesions. However, it has in recent years become of interest as a biological weapon, a topic we discussed in Chapter 8. In the 1980s there was an unconfirmed report of an explosion at a biological weapons plant in the Soviet Union that killed 50 workers and released anthrax spores into the environment. More recently, in October 2001, letters contaminated with powdered anthrax spores were mailed to various locations in the United States.

Pathogenesis of Inhalation Anthrax

Bacillus anthracis is a Gram-positive spore-forming rod (Figure 21.14) commonly found in the soil of pastures. When they reach the rich environment of human tissues, the spores germinate into the bacilli. The

Figure 21.14 A photomicrograph of *Bacillus anthracis* in the lung tissue of a case of fatal inhalation anthrax.

anti-phagocytic properties of the capsule that surrounds the organism aid in its survival and growth in large numbers. Pathology is the result of the exotoxin produced by the organism.

Symptoms of pulmonary anthrax are one to five days of nonspecific malaise, mild fever, and a nonproductive cough that leads to progressive respiratory distress and **cyanosis**. There is rapid and massive spread to the central nervous system and bloodstream, followed by death.

Treatment of Inhalation Anthrax

Antibiotic therapy can be successful, and *B. anthracis* is susceptible to penicillin. Doxycycline and ciprofloxacin are alternatives recommended for prophylaxis.

Legionella Pneumonia (Legionnaires' Disease)

This infection is caused by *Legionella pneumophila*, a Gram-negative rod that cannot be stained or grown using normal techniques (**Figure 21.15**). Unrecognized before 1976, it was the cause of a widely publicized outbreak at an American Legion convention held in Philadelphia that year. There is evidence that the bacterium may have been around since the 1950s but not identified properly.

In nature, *Legionella* is found ubiquitously in fresh water, particularly in warm weather. In water, *Legionella* is found living within *Acanthamoeba* organisms, and these are the infectious reservoirs. The bacterium is transmitted to humans as a humidified aerosol from contaminated water systems; person-to-person transmission has never been seen. Most outbreaks occur in large buildings that use cooling towers for their air-conditioning system.

Healthy people do not seem to be affected very often, and many cases of infection may go undetected. In fact, *Legionella* seems to have a low virulence for humans, with infections occurring in less than 5% of the population; these people are usually immunocompromised in some way. It is estimated that there are about 25,000 cases per year in the United States.

Pathogenesis of *Legionella* Pneumonia

Legionella is a facultative intracellular parasite that aggressively attacks the lungs, producing a necrotizing multifocal pneumonia involving the alveoli and terminal bronchioles, but not usually the larger bronchioles. An inflammatory response occurs during which an exudate is formed containing fibrin, polymorphonuclear leukocytes, and red blood cells. *Legionella* organisms are inhaled and enter the alveoli, where they infect the alveolar macrophages through the production of an endocytic vesicle. Inside this vesicle, the bacteria continue to replicate and prevent the fusion of the vesicle with lysosomes. The infected macrophage takes on a coiled morphology (**Figure 21.16**) and eventually dies, releasing the infectious bacteria.

Legionnaires' disease is a severe toxic pneumonia that begins with myalgia and headache followed by a rapidly rising fever. There are chills, pleuritic chest pain, vomiting, diarrhea, and occasional confusion and delirium. Patchy or interstitial infiltrates in the lung can be seen on X-ray, and there can also be hepatic dysfunction. Serious cases show

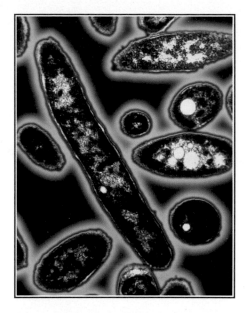

Figure 21.15 A transmission electron micrograph of *Legionella pneumophila*, the etiologic agent of Legionnaires' disease.

Fast Fact *Legionella* can persist in the water systems of air-conditioning cooling towers even in the presence of chlorine.

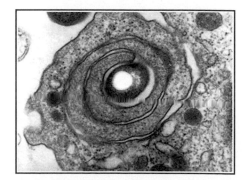

Figure 21.16 The coiled vesicle seen in macrophages that have been infected by *Legionella*.

progressive illness for three to six days and end in shock or respiratory failure, or both. The mortality rate for *Legionella* infection is about 15% but can be as high as 50% in hospital outbreaks because in these cases the infected population is immunocompromised or immunosuppressed.

Treatment of *Legionella* Pneumonia

Erythromycin is better than penicillin because *Legionella* produces a β-lactamase enzyme that destroys penicillin. Although tetracycline, rifampin, and quinolones are good, azithromycin and clarithromycin are the agents of choice.

Q Fever

Q fever is a zoonotic infection seen throughout the world, with cattle, sheep, and goats being the primary reservoirs for humans. Q fever stands for query fever, because the infectious agent was unknown during the first outbreak. We now know that Q fever is caused by *Coxiella burnetii*, which is the only Gram-negative spore-forming bacterium. This organism is an obligate intracellular pathogen that grows well in the placenta of animals so that, as birth occurs, large numbers of *Coxiella* are liberated onto the ground. The organism can survive in the soil for very long periods (even years) in its spore stage, and workers involved with animal births or in slaughtering animals are the most prone to infection. *Coxiella* can be transmitted by inhalation as well as by the ingestion of unpasteurized milk.

Pathogenesis of Q Fever

The pathology of this infection is not clearly understood. It usually begins nine to twenty days after inhalation, with the abrupt onset of fever, chills, and headache. There can also be a mild, dry, hacking cough, and a patchy interstitial pneumonia may develop. In some cases, there can also be abnormal liver function.

Treatment of Q Fever

Most cases of Q fever resolve spontaneously without therapy, but tetracycline can be given to shorten the fever and reduce the risk of a rare chronic infection. Death can occur in 1–2% of patients with acute infection.

Psittacosis (Ornithosis)

This infection is a zoonotic pneumonia contracted by the inhalation of bird droppings infected with the bacterium *Chlamydophila psittaci*. Infection was originally seen in parrots and parakeets, but it is now found in many birds, including turkeys. The infection is usually latent in birds, but the stress of confinement in cages may cause large numbers of *C. psittaci* bacteria to be shed into the feces. Some strains of *C. psittaci* are extremely contagious.

Pathogenesis of Psittacosis

In humans, psittacosis presents as an acute infection of the lower respiratory tract. There is acute onset of fever, headache, malaise, and muscle aches, accompanied by a dry, hacking cough and bilateral pneumonia. There can also be occasional systemic complications, such as myocarditis, endocarditis, and hepatitis. In some cases, there may be **splenomegaly** and **hepatomegaly**.

Treatment of Psittacosis

Both tetracycline and erythromycin are effective, but only if given early in the infection.

Keep in Mind

- Bacterial infections of the lower respiratory tract include both nosocomial and community-acquired bacterial pneumonia.

- *Chlamydophila* and *Mycoplasma* can also cause pneumonia.

- Tuberculosis is a serious lower respiratory tract infection caused by *Mycobacterium tuberculosis*, an organism that is becoming more resistant to antibiotic treatment.

- Pertussis (whooping cough), inhalation anthrax, legionellosis, Q fever, and psittacosis are serious infections of the lower respiratory tract.

VIRAL INFECTIONS OF THE LOWER RESPIRATORY TRACT

When both upper and lower tracts are considered, respiratory infections account for 75–80% of all the acute infections in the United States, and most of these are of viral origin. In fact, there are three or four viral infections per person per year in this country. The incidence of these infections varies inversely with age and is greatest in young children. Most viral infections are also seasonal, with the lowest rates in summer and the highest in winter.

The two viruses that cause the majority of acute viral infections in the lower respiratory tract are influenza and respiratory syncytial virus. A common characteristic of infections with these viruses is a short incubation period of about one to four days and transmission from person to person. This transmission can be either direct (through droplets) or indirect (through hand transfer of contaminated secretions to conjunctival or nasal epithelium). Influenza and syncytial infections are seen worldwide. We have already mentioned both of these in earlier discussions (Chapters 8, 12, and 13). Please make sure to refer to those sections to broaden your understanding.

Influenza

The influenza virus is a member of the *orthomyxovirus* group, and its virions are surrounded by an envelope (**Figure 21.17**). The virions contain single-stranded RNA in eight segments, allowing the virus to undergo a high rate of mutation, and it is this ability to mutate so easily that makes the virus a repetitive problem for humans. There are three major serotypes—A, B, and C—with the differences being based on the antigens associated with the nucleoprotein. Because influenza A is the best documented of the three serotypes, and the most virulent, our discussions will center on infection with this virus.

Influenza is a significant health concern, and there is currently concern about its potential to recombine with an avian influenza strain to produce a significant pathogenic influenza virus (Chapter 8). Recall from Chapter 13 that the ability of a virus to mutate gives it the ability to undergo antigenic drift and antigenic shift, both of which have a role in our lack of ability to defend against it. Humans are the hosts for influenza, but the reservoir for this virus is birds. Severe respiratory problems are the primary manifestation of influenza infections, and outbreaks have been described since the sixteenth century. Outbreaks of differing severity occur nearly every year; the most severe pandemics occurred in 1743, 1889, 1918 (Spanish flu), 1957 (Asian flu), 1968 (Hong Kong flu), and 2009 (swine flu).

(a)

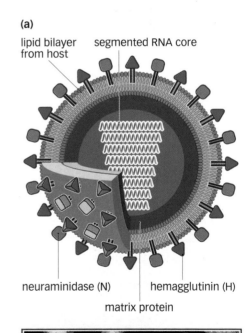

lipid bilayer from host
segmented RNA core
neuraminidase (N)
hemagglutinin (H)
matrix protein

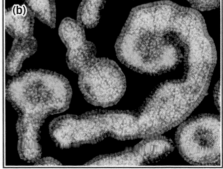

(b)

Figure 21.17 The influenza virus. Panel a: The influenza A virion. Panel b: Colorized electron micrograph of influenza virus.

Direct droplet transmission is the most common method by which influenza spreads, with outbreaks occurring more frequently in the winter months. There seems to be an interval of two to three years between major outbreaks; the typical epidemic lasts for three to six weeks and involves as many as 10% of the population. However, illness rates may exceed 30% in school-aged children and in residents of closed institutions such as prisons and convalescent homes. The identification of an influenza outbreak as an epidemic is based on excessive mortality rates (in other words, more deaths from the infection than expected).

Pathogenesis of Influenza

The influenza virus has a predilection for the respiratory epithelium, and viremia is not a feature of infection. The virus multiplies in the ciliated cells of the lower respiratory tract, resulting in functional and structural abnormalities in these cells. As part of the infection, cellular synthesis of nucleic acids and proteins is shut down, and both the ciliated and mucus-producing epithelial cells are shed, resulting in substantial interference with the mechanical clearance mechanisms of the lower respiratory tract. In addition, there is localized inflammation associated with the death of these cells. The respiratory epithelium may not be restored to normal for two to ten weeks after the initial infection. There is also viral destruction of tissues accompanied by inflammation and impaired phagocytic and chemotactic responses, which can cause superinfection by bacteria.

Recovery from influenza infection starts with the production of interferon, which limits the spread of the infection. This is accompanied by the rapid generation of natural killer cells, which reverse the infection. Shortly thereafter, cytotoxic T cells and specific antibodies, which are part of the adaptive immune response, appear in large numbers and control the infection so that tissue repair can begin.

In some cases, patients develop **acute influenzal syndrome**, which differs from the common infection course just described. This acute syndrome has a short incubation time of about two days, and symptoms can develop in a matter of hours. These include fever, myalgia, headache, and occasional shaking chills. The infection reaches maximum severity in just six to twelve hours, and a nonproductive cough develops. These acute symptoms can last three to five days and are usually followed by gradual improvement. However, occasionally these patients develop a progressive infection that involves the tracheobronchial tree and lungs, resulting in lethal pneumonia.

As mentioned above, one of the most common and important complications of an influenza infection are bacterial superinfections. This usually involves the lungs and can develop during the convalescent stage of the viral infection, when the patient is debilitated (recall that these secondary infections can be more dangerous than the primary infection). Superinfection is identified by the development of an abrupt worsening of the patient's condition after an initial stable period. Most often, the secondary superinfection involves *Streptococcus pneumoniae*, *Haemophilus influenzae*, or *Staphylococcus aureus*.

It is important to note that there are three ways in which an influenza infection can cause death:

- Underlying disease can occur in people with limited cardiovascular activity or pulmonary function. This progression usually occurs in the elderly.

- Superinfection can lead to bacterial pneumonia or, in some cases, disseminated bacterial disease.

Feature	Amantadine	Rimantadine	Zanamivir	Oseltamivir
Susceptible viruses	Influenza A only	Influenza A only	Influenza A and B	Influenza A and B
Emergent resistant strains	Yes	Yes	Not known	Not known
Administration	Oral	Oral	Inhalation	Oral

Table 21.2 Comparison of antiviral drugs for influenza.

- Direct rapid progression of the infection can lead to overwhelming viral pneumonia and asphyxia.

Treatment of Influenza

There are two basic approaches for managing influenza: symptomatic care and anticipation of potential complications. The best treatments for influenza are rest, adequate fluid intake, conservative use of analgesics for myalgia and headache, and cough suppressants.

When influenza is diagnosed, four to five days of administration of amantidine or rimantadine may be considered, but this is useful only if the infection is diagnosed within 12–24 hours of onset (Table 21.2).

Respiratory Syncytial Virus (RSV)

This virus is so named because of the syncytia associated with it. (Recall from Chapter 5 that a syncytium is a large multinucleated cell produced during a viral infection.) Community outbreaks of RSV occur annually in the late fall to early spring, with the usual outbreak lasting about eight to twelve weeks. These outbreaks can involve 50% of families with small children. It is usually an older sibling that brings the virus into the home, but it infects young children or infants most often. Virus is normally shed for five to seven days but can be shed for up to twenty days in infants. This virus is a major cause of nosocomial infections, and control of these infections in a hospital setting is difficult. However, attention to handwashing as well as the exclusion of staff and visitors with respiratory symptoms helps to hold down the level of infection. There is no vaccine for RSV.

Fast Fact Wearing a surgical mask is of no use in controlling RSV infection because the virus is small enough to move through the material of the mask easily.

Pathogenesis of RSV Infection

RSV spreads to the upper respiratory tract by contact with infectious secretions, and infection is usually confined to the respiratory epithelium with progression to the middle and lower airways. Viremia is rare, and the effect of the virus on the respiratory epithelium is similar to that seen in influenza infections. Cytotoxic T cells seem to have a significant role in controlling acute RSV infection.

Major pathological consequences involve the bronchi, bronchioles, and alveoli, including necrosis of the epithelial cells, interstitial mononuclear cell infiltration, and inflammation that can involve the alveoli and the alveolar ducts. These developments can result in the plugging of the small airways with mucus, necrotic cells, and fibrin. The incubation period is usually two to four days, followed by the onset of **rhinitis**. The severity of the infection peaks within three days.

Clinical signs can include hyperexpansion of the lungs, hypoxia (low oxygenation of the blood), and hypercapnia (retention of CO_2). There can also be pulmonary collapse seen on X-ray, but the normal duration of acute signs is only ten to fourteen days. The fatality rate among hospitalized infants is about 1%, but it can be as high as 15% in compromised children. RSV infection is usually mild in adults and older children.

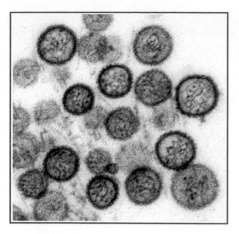

Figure 21.18 An electron micrograph of the hantavirus (Sin Nombre).

Treatment of RSV Infection

Treatment is primarily directed at the underlying clinical pathology and includes adequate oxygenation and ventilation. Additionally, close observation to deal with potential bacterial superinfections is indicated.

Hantavirus Pulmonary Syndrome (HPS)

It has been known for some time that rodents in the United States are infected with hantavirus (**Figure 21.18**), but there was no recognized infection in humans until an outbreak in 1993. This infection usually occurs in the southwestern United States (where the outbreak in 1993 occurred), but cases have been reported in 21 states. The virus causes a fulminant respiratory infection with high (50–70%) mortality.

Three types of hantavirus cause HPS, with the type known as Sin Nombre being the most common. Hantavirus infections are associated with increases in the rodent population.

Pathogenesis of Hantavirus Pulmonary Syndrome

Transmission is by the inhalation of dried rodent excreta, by the conjunctival route, or by direct contact through breaks in the skin. Human-to-human transmission has not been observed. The early symptoms of this infection include fatigue, fever, and muscle aches especially in the large muscle groups, thighs, hips, and back. There can also be headache, dizziness, and abdominal problems. Four to ten days after the initial phase of the illness, late symptoms appear. These include coughing and shortness of breath as the lungs fill with fluid.

Treatment of Hantavirus Pulmonary Syndrome

The major control of HPS infection involves controlling the rodent population. Once a person is infected, aggressive respiratory support is required, although intravenous ribavirin has been used with some success.

Keep in Mind

- Influenza and RSV cause the most common viral infections of the lower respiratory tract.

- They are serious infections in themselves and also make the patient more susceptible to secondary bacterial infections.

- There are vaccines and antiviral therapies for influenza, though they are of limited effectiveness.

- There is no vaccine or specific treatment for RSV.

- HPS is an emerging infectious disease.

FUNGAL INFECTIONS OF THE RESPIRATORY SYSTEM

As you read this discussion of fungal infections of the respiratory system, it is important to remember that fungal infections are for the most part seen only in patients who are in some way compromised. There are two major factors governing the incidence and spread of fungal infections: (1) the ubiquity of the infectious organisms, which are found not only in the soil but also as resident flora of the body; and (2) the fact that the adaptive immune response usually keeps these infections under control. Consequently, in patients who are compromised in some way, the risk for fungal infection rises drastically. Although a variety of fungal organisms can cause respiratory infections in compromised hosts, we will focus on only the most common.

Pneumocystis Pneumonia (PCP)

In immunocompromised patients, particularly those with AIDS, the fungus *Pneumocystis (carinii) jiroveci* (**Figure 21.19**) causes a lethal pneumonia called *Pneumocystis* pneumonia (PCP). This fungus has never been grown in culture, and most of what is known about it comes from clinical information obtained from patients with infected lungs. For a long time, the infection was thought to be caused by a protozoan because of the shape, nucleus, and spores of *Pneumocystis*, which resembled structures seen in protozoans. In addition, this organism has cholesterol in the plasma membrane rather than the ergosterol seen in other fungal cells. Eventually, RNA typing was used to confirm that *P. (carinii) jiroveci* is indeed a fungus.

Pulmonary infection occurs in humans and animals throughout the world, and antibodies against it are found in almost all children by the age of four years. Although the reservoirs and modes of transmission have yet to be defined, it has been shown in animal models that the fungus can be transmitted by aerosol.

Before the AIDS pandemic, *Pneumocystis* pneumonia was seen only sporadically in infants with congenital immunodeficiencies and in some compromised adults. AIDS is now the most common predisposing factor for PCP. In fact, before aggressive therapy for AIDS was introduced, PCP was seen in more than 50% of patients presenting with AIDS, and most patients with AIDS develop this form of pneumonia as the infection progresses.

Pathogenesis of *Pneumocystis* Pneumonia

Pneumocystis (carinii) jiroveci has a low level of virulence and therefore rarely affects immunocompetent hosts who have normal T-cell function. However, in AIDS the risk of PCP increases as CD4 T cells disappear. Little is known about the early stages of the infection, but it is thought that the organism may attach by way of a surface glycoprotein that binds to host cell proteins or fibronectin. This fungal glycoprotein also seems to undergo antigenic variation, which may be the reason for the persistence of the infection.

Pneumocystis pneumonia is characterized by alveoli filled with sloughed-off alveolar cells, monocytes, and fluid, producing a distinct foamy honeycombed appearance. In compromised hosts, this infection presents as progressive diffuse **pneumonitis**. For patients with AIDS and infants, the onset of the infection is insidious, and the infection can be present for three to four weeks before it is discovered.

The principal manifestation of infection is progressive **dyspnea** and tracheal pneumonia, with eventual cyanosis and hypoxia. A non-productive cough appears in 50% of patients, with X-rays showing some alveolar infiltrates that spread out from the hili of the lung, eventually affecting the entire lung. Radiographic abnormalities are accompanied by decreased O_2 saturation of arterial blood as well as decreased lung vital capacity. Death occurs through progressive asphyxiation. There can also be lesions in the lymph nodes, bone marrow, spleen, liver, eyes, thyroid, adrenal glands, and kidneys.

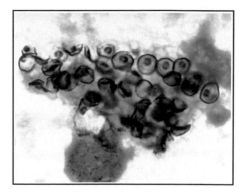

Figure 21.19 A photomicrograph of *Pneumocystis (carinii) jiroveci* in the sputum of a patient with *Pneumocystis* pneumonia. This organism cannot yet be grown in a laboratory setting, and consequently all images of it have been made from clinical samples.

Treatment of *Pneumocystis* Pneumonia

In patients who do not have AIDS, a combination of trimethoprim and sulfamethoxazole (TMP-SMX) for 14–21 days is the treatment of choice. Most AIDS patients have severe side effects with TMP-SMX therapy, and the preferred alternative therapy is with pentamidine and trimetrexate. In addition, patients with AIDS usually receive treatment for longer than 21 days because in most cases they present with a more advanced infection, and respond more slowly and relapse more often.

Blastomycosis

Blastomycosis (also known as Gilchrist's disease) is an infection caused by the fungus *Blastomyces dermatitidis*. The spores of the fungi enter the body through the respiratory system and primarily affect the lungs. However, the disease can occasionally spread through the bloodstream and affect other parts of the body, including the skin. Most infections occur in the United States but they have also been seen widely spread in Africa. Men between the ages of 20 and 40 years are the most commonly infected with this fungal disease. Unlike most fungal infections, blastomycosis is not seen more often in people who have AIDS.

Pathogenesis of Blastomycosis

Infection of the lungs begins gradually with fever, chills, and drenching sweats. Chest pain, difficulty in breathing, and a cough may also develop. Infection in the lungs develops slowly and can sometimes heal without treatment. When the infection spreads, it can affect many areas of the body, but the skin, bones, and genitourinary tract are most often affected. In the skin, the infection begins as papules, which may contain pus. Warty patches then develop and are surrounded by tiny painless abscesses. Painful swelling of the bones can also occur. Men may experience prostatitis or painful swelling of the epididymis.

Treatment of Blastomycosis

Blastomycosis can be treated with intravenous amphotericin B or oral itraconazole. With treatment the patient begins to feel better quickly, but the drug therapy must be continued for months. Without treatment, the infection slowly worsens and leads to death.

Histoplasmosis

This fungal infection is caused by *Histoplasma capsulatum*, a fungus commonly found in temperate, subtropical, and tropical zones in soil contaminated with bat or bird droppings. In the United States, *H. capsulatum* is most often found in areas around the Ohio and Mississippi rivers.

Between 50% and 90% of residents in areas where this organism is found test positive for it, indicating that they have been exposed. People who live and work in the vicinity of bat or bird droppings are at increased risk of developing this infection.

Pathogenesis of Histoplasmosis

Transmission is through the inhalation of conidia, which are small enough to reach the bronchioles and alveoli of the lung. Because of their minute size, these spores are usually referred to as microconidia (**Figure 21.20**).

Most cases of histoplasmosis are asymptomatic or present with only fever and mild cough. Initial infection is pulmonary, but the lymph nodes,

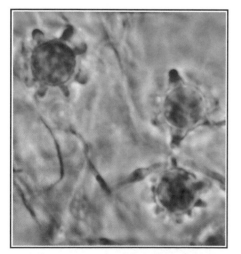

Figure 21.20 The microconidia of *Histoplasma capsulatum*, the causative agent of the fungal infection histoplasmosis. These structures are small enough to bypass some of the defenses of the upper respiratory tract and enter directly into the lung, causing pneumonia-like symptoms.

spleen, bone marrow, and other elements of the mononuclear phago-cytic system can become involved. During the infection, microconidia are inhaled and converted to the yeast forms, which are phagocytosed by macrophages and polymorphonuclear leukocytes. The yeast forms sur-vive the formation of the phagolysosome in phagocytic cells by capturing iron and lowering the pH of the phagolysosome. They then continue to divide in the cytoplasm of the phagocytic cell.

As growth continues, there is formation of a tubercle similar to that seen in tuberculosis. The vast majority of cases never go further in the infec-tious process, although some patients may develop fever and cough for a few days or even weeks. In severe cases there may be chills, malaise, chest pain, and extensive pulmonary infiltration, but even these severe infections usually resolve spontaneously.

Treatment of Histoplasmosis

The infection usually resolves spontaneously and there is no need for treatment. If necessary, amphotericin B is the treatment of choice, but it is toxic and used only for short times, and then only in severe cases. Itraconazole and ketoconazole have been used in patients with AIDS who have histoplasmosis.

Coccidioidomycosis

This infection is caused by *Coccidioides immitis*. Infection can occur either in an asymptomatic form or as **valley fever**. Like histoplasmosis, coc-cidioidomycosis is restricted to certain geographical areas. This fungus grows only in alkaline soil and semi-arid climates known as the lower Sonoran life zone, usually in places with hot, dry summers, mild win-ters, and an annual rainfall of 10 inches. These locations are scattered throughout the United States (Arizona, New Mexico, western Texas, and California), and 50–90% of long-term residents in these areas test posi-tive for exposure to *Coccidioides*.

Pathogenesis of Coccidioidomycosis

The arthroconidia of the fungus are inhaled and are small enough to bypass defenses of the upper respiratory tract and lodge directly in the bronchioles. The outer wall of *Coccidioides immitis* has antiphagocytic properties that prevent elimination of the organism, and the arthroconidia convert to spherules (**Figure 21.21**), which slowly grow and completely inhibit phagocytosis. It is thought that, in the spherule stage, cell wall pro-teases may contain virulence factors, but they have not yet been identified.

More than half of infected individuals show no signs of infection, but the remainder progress to valley fever and present with malaise, cough, chest pain, fever, and arthralgia (joint pain), all of which occur one to three weeks after the infection begins. The signs can last for up to six weeks; however, most patients spontaneously resolve the infection, and only 10% of patients ever experience pulmonary symptoms. Disseminated coccidioidomycosis has been seen in patients with AIDS and in individu-als on immunosuppressive therapy. In addition, a form of coccidioidal meningitis can occur, which can be fatal if not treated aggressively.

Treatment of Coccidioidomycosis

Because this infection is in most cases self-limiting, no treatment is indi-cated. However, in cases of progressive pulmonary infection or infection of the central nervous system, amphotericin B can be used. Fluconazole and itraconazole are also effective.

Fast Fact There is no person-to-person transmission of histoplasmosis, and for reasons that are not clear it seems to be more prevalent in men than in women.

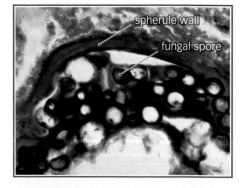

Figure 21.21 The spherule structure seen in coccidioidomycosis. The fungal spores are contained inside the spherule.

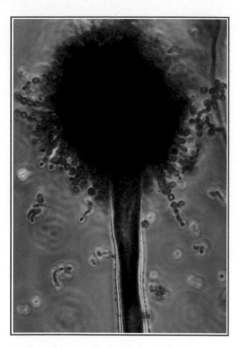

Figure 21.22 A photomicrograph showing the appearance of a conidiophore of the fungus *Aspergillus.*

Fast Fact *Aspergillus* can be present in older homes and become aerosolized when disturbed as the homes are remodeled.

Aspergillosis

Invasive aspergillosis is distinguished in immunocompromised individuals by rapid progression to death. This infection is typically seen in immunocompromised patients, in particular patients with leukemia or AIDS and those undergoing bone marrow transplantation. The fungus *Aspergillus* (**Figure 21.22**) is widely distributed in nature and is found throughout the world.

Dispersal occurs through the inhalation of resistant conidia and has been seen more and more in nosocomial infections associated with air-conditioning systems.

Pathogenesis of Aspergillosis

The conidia of *Aspergillus* are small enough to reach the alveoli of the lung when inhaled. Infection is rare when good immune potential is present. The mechanism by which the conidia attach to host cells is not well understood, but it has been shown that, after attachment, the fungus produces extracellular proteases and phospholipases. However, the involvement of these extracellular products in the infection process is not yet understood. Most *Aspergillus* species also produce toxic metabolites, but their involvement in the infection process is also not clear.

Aspergillosis usually occurs in individuals with a preexisting pulmonary disease, such as chronic bronchitis, asthma, or tuberculosis, and also in immunocompromised hosts. Colonization with *Aspergillus* leads to the invasion of tissues by the branching septate hyphae. Invasion of lung tissue can lead to the penetration of blood vessels, causing **hemoptysis** (coughing up blood) and/or acute pneumonia in immunocompromised patients. This is accompanied by multifocal pulmonary infiltrates that consolidate and present with high fever. The prognosis of this infection is grave, and the mortality for invasive aspergillosis is 100%.

Treatment of Aspergillosis

Amphotericin B and itraconazole for the invasive form of the infection can be used but are usually ineffective.

Keep in Mind

- The adaptive immune response usually keeps fungal infections under control.
- Fungal infections of the respiratory tract are usually opportunistic infections seen in immunocompromised individuals.
- Once developed, fungal infections of the respiratory tract can be serious, and treatment options are limited.
- Fungal infections of the respiratory tract include *Pneumocystis* pneumonia, histoplasmosis, coccidioidomycosis, and aspergillosis.

SUMMARY

- The respiratory system is a major portal of entry and exit for pathogens.
- The upper respiratory tract is continuously exposed to pathogens, whereas the lower respiratory tract is essentially sterile.
- Infections of the upper respiratory tract are not usually serious (diphtheria is an exception).

- Infections of the lower respiratory tract are serious and can be life threatening.

- Pneumonia and tuberculosis are serious bacterial infections of the lower respiratory tract.

- The most common viral infections of the lower respiratory tract are caused by influenza virus and respiratory syncytial virus.

- Viral infections of the lower respiratory tract also leave the patient more susceptible to bacterial infections.

- Fungal infections of the respiratory tract are usually opportunistic infections seen in immunocompromised individuals.

In this chapter, we have seen that bacteria, viruses, and fungi can all use the respiratory tract as a portal of entry and cause infection. This picture will be repeated in all of the body systems that we explore in the following chapters. It is important to remember that although many pathogens can gain entry to the respiratory system, the impressive defense mechanisms associated with the system work to control the occurrence of infections, especially in the lower respiratory tract.

 SELF EVALUATION AND CHAPTER CONFIDENCE

Multiple Choice

Answers are given in the back of the book and help can be found in the student resources at:

www.garlandscience.com/micro2

1. Penicillin is used to treat all of the following except

    A. Pneumococcal pneumonia
    B. *Mycoplasma* pneumonia
    C. Scarlet fever
    D. Diphtheria
    E. Streptococcal sore throat

2. Pneumonia can be caused by which of the following?

    A. *Haemophilus*
    B. *Mycoplasma*
    C. *Streptococcus*
    D. *Legionella*
    E. All the above

3. Which of the following does not produce exotoxin?

    A. *Bordetella pertussis*
    B. *Streptococcus pyogenes*
    C. *Corynebacterium diphtheriae*
    D. *Mycobacterium tuberculosis*
    E. All of the above

4. Which of the following infections can result in the formation of a pseudomembrane in the throat?

    A. *Corynebacterium diphtheriae*
    D. *Streptococcus pyogenes*
    C. *Bordetella pertussis*
    D. *Mycobacterium tuberculosis*
    E. None of the above

5. Which of the following is a leading cause of nosocomial pneumonia?
    A. *Streptococcus pyogenes*
    B. *Chlamydophila pneumoniae*
    C. *Corynebacterium diphtheriae*
    D. *Pseudomonas aeruginosa*
    E. *Mycobacterium tuberculosis*

6. Fungal infections of the respiratory tract are usually associated with which of the following?
    A. Compromised immune system
    B. Young children
    C. International travel
    D. Vaccination
    E. All of the above

7. *Legionella* is transmitted by
    A. Fomites
    B. Airborne droplets
    C. Vectors
    D. Foodborne transmission
    E. Person-to-person contact

8. The patient has a sore throat. Which of the following could be involved?
    A. *Mycobacterium*
    B. *Haemophilus*
    C. *Corynebacterium*
    D. *Bordetella*
    E. All of the above

9–15. Answer **A** if the reason is correct; answer **B** if the reason is incorrect.

9. *Mycobacterium* is a problem for the host defense because it has no cell wall.

10. Compliance by patients can be difficult because they are too weak to concentrate.

11. Because the symptoms of pertussis can be similar to those of a cold, many infected adults can spread the infection to places such as schools and nurseries without knowing it.

12. Erythromycin is better than penicillin for treating legionellosis because *Legionella* has rapid protein synthesis.

13. For a long time, the *Pneumocystis* infection was thought to be caused by a protozoan because of the shape of its nucleus and its spores.

14. In patients with AIDS, the preferred therapy for tuberculosis is with pentamidine and trimetrexate because most of them have severe side effects with TMP-SMX therapy.

15. Because of their cone-like shape, *Histoplasmosis* spores are usually referred to as microconidia.

(Q) DEPTH OF UNDERSTANDING

Questions listed here require you to bring together the concepts you have learned in this chapter into a discussion format. This helps you to increase your depth of understanding of the material you have learned. Help can be found in the student resources at: www.garlandscience.com/micro2

1. Discuss the differences between the upper and the lower respiratory tract with regard to infections and their causative agents.

2. You are informed by your hospital supervisor that you need to get a flu shot. Your co-worker had one last year and didn't enjoy the experience. On the basis of what you have learned in this chapter, explain to her why she will have to continue getting these shots.

3. Describe how viruses that attack the endothelial cells of the respiratory tract can facilitate the onset of pneumonia. Make sure to include the defense mechanisms associated with the respiratory tract in your discussion.

4. Discuss the pathogenesis of tuberculosis, including the host defense or lack thereof.

(Q) CLINICAL CORNER

Help can be found in the student resources at: www.garlandscience.com/micro2

1. A new office building opened near a hospital and has been touted as being a state-of-the-art, energy-efficient structure. On a late Thursday afternoon, the emergency room is overwhelmed with patients complaining of different levels of respiratory distress ranging from mild coughing to severe pneumonia. Two have died of pneumonia shortly after being admitted. Most of the patients work at the new office building. Their offices are in all floors of the building.

   A. What is the most likely cause?
   B. What would be your recommendation for treatment?

2. You work in the hospital nursery, where you take care of newborn babies. You have been having flu-like symptoms the last few days but they are nowhere near strong enough to make you stay home from work, but you really need the money and don't want to miss days because you don't feel that great. A co-worker has noticed that you are ill and has told you to stay home until you feel better. To placate her you have decided to wear a mask while at work but even this step has not made her happy. She has threatened to report you to the nursing supervisor if you do not stay home until you feel better.

   A. Why is she so insistent about your not coming to work?
   B. Why is wearing the mask not enough to overcome her objections?

Infections of the Digestive System

Chapter 22

Why Is This Important?

The digestive system is the second leading portal by which infectious organisms enter the body. As a health care professional, you will see many patients with infections of the digestive system.

Mary was a cook. She loved cooking and worked for a very wealthy family in New York in 1906. The family had visitors one day, and one of them infected Mary with typhoid fever (caused by the bacterium *Salmonella enterica* serotype Typhi). Interestingly, Mary did not get sick but she became a carrier. Within a short time several people in the house where she worked developed typhoid fever, so Mary moved to another home. The same thing happened there and at two other places where Mary worked, and she was finally forced to seek employment under assumed names. After six years, the New York Board of Health

caught up with Mary and, although she was unwilling to cooperate, forced her to be examined in a hospital. The examination showed that Mary's gall bladder was infected with the typhoid-causing bacteria and that the only way to stop the infections was to have the gall bladder removed. Mary refused to agree to the operation and was therefore sentenced to isolation in the hospital. For three years, Mary was 'imprisoned' in the isolation ward of the hospital, but finally public sympathy forced the authorities to release her with a promise she would never be a cook again.

Five years later, 25 cases of typhoid fever occurred in a woman's hospital, and 8 of the infected people died. Coincidentally, the cook for the hospital suddenly disappeared. It did not take long for the authorities to discover that the cook was Mary. When she was arrested, Mary defended her actions by saying she had a right to be a cook and that she would continue to cook. She was sent back to the quarantine ward of the hospital and was forced to remain there for the rest of her life (23 more years). Remarkably, Mary died of a stroke, not typhoid fever.

OVERVIEW

The digestive system is a major portal of entry for bacteria, viruses, and parasites. Many infectious diseases occur through the ingestion of water and food, but it is important to keep in mind that the defenses of the digestive system are very strong and keep many infections from ever happening. Remember also that the mouth and large intestine are crowded with microorganisms that are part of the body's normal flora (the microbiome), and the presence of this normal flora is part of the body's defense against opportunistic infection.

We will divide our discussions into the following topics:

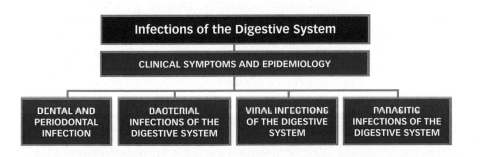

"A nationwide study published by the USDA found that ... 78.6 percent of the ground beef contained microbes that are spread primarily by fecal matter."

Eric Schlosser from
Fast Food Nation: The Dark Side of the All-American Meal

CLINICAL SYMPTOMS AND EPIDEMIOLOGY

The most common symptoms of gastrointestinal infections are fever, vomiting, abdominal pain, and diarrhea. These symptoms vary with the specific infection and with the stages of infection, but they can be used to classify gastrointestinal infections.

Watery Diarrhea

Watery diarrhea is the most common result of gastrointestinal infection, caused by pathogenic mechanisms that attack the intestines. It develops rapidly and results in frequent voiding. There may also be accompanying vomiting, fever, and abdominal pain. The purest form of watery diarrhea is caused by enterotoxin-secreting bacteria, such as *Vibrio cholerae* and enterotoxigenic *Escherichia coli*. These pathogens cause diarrhea without destroying the cells of the intestines, whereas other pathogens, such as rotavirus, cause diarrhea and in the process destroy intestinal cells. Watery diarrhea is usually acute but brief (lasting for one to three days) and self-limiting. However, watery diarrhea caused by either *V. cholerae* or the parasite *Giardia* can last for weeks and be very severe.

Dysentery

Dysentery has a rapid onset with frequent intestinal evacuations that include blood and mucus but are smaller in volume than those seen in watery diarrhea. There can be accompanying cramps and abdominal pain, but there is very little vomiting. The pathologic consequences of dysentery center in the colon. Organisms that cause dysentery damage the colonic mucosa either directly or through the production of toxins. This tissue damage results in the blood and mucus seen in the stool. There is not as much fluid loss in dysentery as in watery diarrhea, but dysentery lasts longer. Even so, most cases resolve in three to seven days. There are two main types of dysentery: **bacillary dysentery** (shigellosis), caused by *Shigella* bacteria, and **amebic dysentery**, caused by the protozoan (ameba) *Entamoeba histolytica*.

Enteric Fever

Enteric fever is a systemic infection with a focus in the gastrointestinal tract. The prominent features of this condition are a fever and abdominal pain that take days to develop. In this condition, diarrhea is mild and does not usually appear until late in the infection. The pathogenesis of enteric fever is more complex than that of either watery diarrhea or dysentery and involves the penetration of **enterocytes**, which are epithelial cells of the small and large intestine. Pathogens spread from here to the biliary (bile) tract, liver, and organs of the reticuloendothelial system. The most investigated form of enteric fever is typhoid fever, which is caused by *Salmonella enterica* serotype Typhi. Although it is usually self-limiting, enteric fever can be serious and result in significant mortality.

Treatment and Management Options for Gastrointestinal Infections

Treatment for most gastrointestinal infections involves supportive care with liquid replacement and rest. In some cases, such as with cholera,

Fast Fact Gastrointestinal diseases are a major cause of death of children in developing countries, killing approximately a million children under five years of age each year.

where liquid loss can be substantial, intravenous replacement of liquids may be required. Chronic bacterial infections can be treated with antibiotics, and infections resulting in renal failure require dialysis or in some cases a transplant. There are also some antiparasitic drugs available.

Endemic and Pandemic Gastrointestinal Infections

Many gastrointestinal infections are endemic: they are constantly in the population but do not pose a public health risk. Some are seen throughout the world and others are geographically restricted, for example by the need for a warm climate. In addition, some organisms show a seasonal variation and age-related predisposition. In developed countries, the most common causes of endemic gastrointestinal infections are the bacteria *Campylobacter*, *Salmonella*, and *Shigella*. These infections are more commonly seen in children because of their underdeveloped immunity and the prevalence of fecal–oral contact.

Outbreaks of gastrointestinal infections are usually associated with poor hygiene and contaminated food or water, and in the right circumstances can spread quickly to become epidemic or even pandemic. Natural disasters, such as earthquakes and hurricanes, and wars cause disruption to water and sanitation and bring about the displacement of a large number of people into camps; these conditions are therefore ideal for the spread of gastrointestinal infections. Cholera is now pandemic in Asia, Latin America, and Africa.

Traveler's Diarrhea

A significant proportion (20–50%) of travelers to underdeveloped countries will get traveler's diarrhea (TD) in the first week of their visit. Usually, the problem is brief and self-limiting, but in some cases it can be serious. Bacterial pathogens cause approximately 80% of TD, with enterotoxigenic strains of *E. coli* being the most common. Improperly cooked food is the major source of transmission; some toxigenic *E. coli* are also found in salads and vegetables. Viral and parasitic infections can also cause TD.

Nosocomial Infections

Employees who are ill but come to work anyway and contaminate food that has been prepared outside the hospital and then brought into hospital are common causes of gastrointestinal infections acquired in hospitals. The close proximity of other patients with infections is also a risk. Two pathogens are responsible for the majority of nosocomial gastrointestinal infections: *E. coli*, which is usually seen in infants and small children; and *Clostridium difficile* (**Figure 22.1**), which accounts for 90% of cases. Nosocomial *C. difficile* infections can be mild but are serious in patients who have had their normal gut bacteria decimated by antibiotics, causing fulminant pseudomembranous colitis. Overuse of antibiotics allows the development of opportunistic infections (superinfections) that are normally prevented by the microbial flora present in our bodies (**Figure 22.2**). Gloves, gowns, and handwashing in the hospital prevent the spread of *C. difficile*.

Food Poisoning

Many gastrointestinal infections involve food; food poisoning is usually connected to one single meal as the source and almost always involves improper food handling. The spread of this type of foodborne infection has become greater with the popularity of fast food. Food poisoning is typically seen in multiple patients connected to a single source of contamination

Fast Fact *Vibrio cholerae* infection is now rare in developed countries but was common in the 1800s. Proper water and sewage treatment have virtually eliminated the spread of cholera in industrialized countries.

Figure 22.1 Photomicrograph of Gram-positive *Clostridium difficile* bacteria isolated from a stool sample.

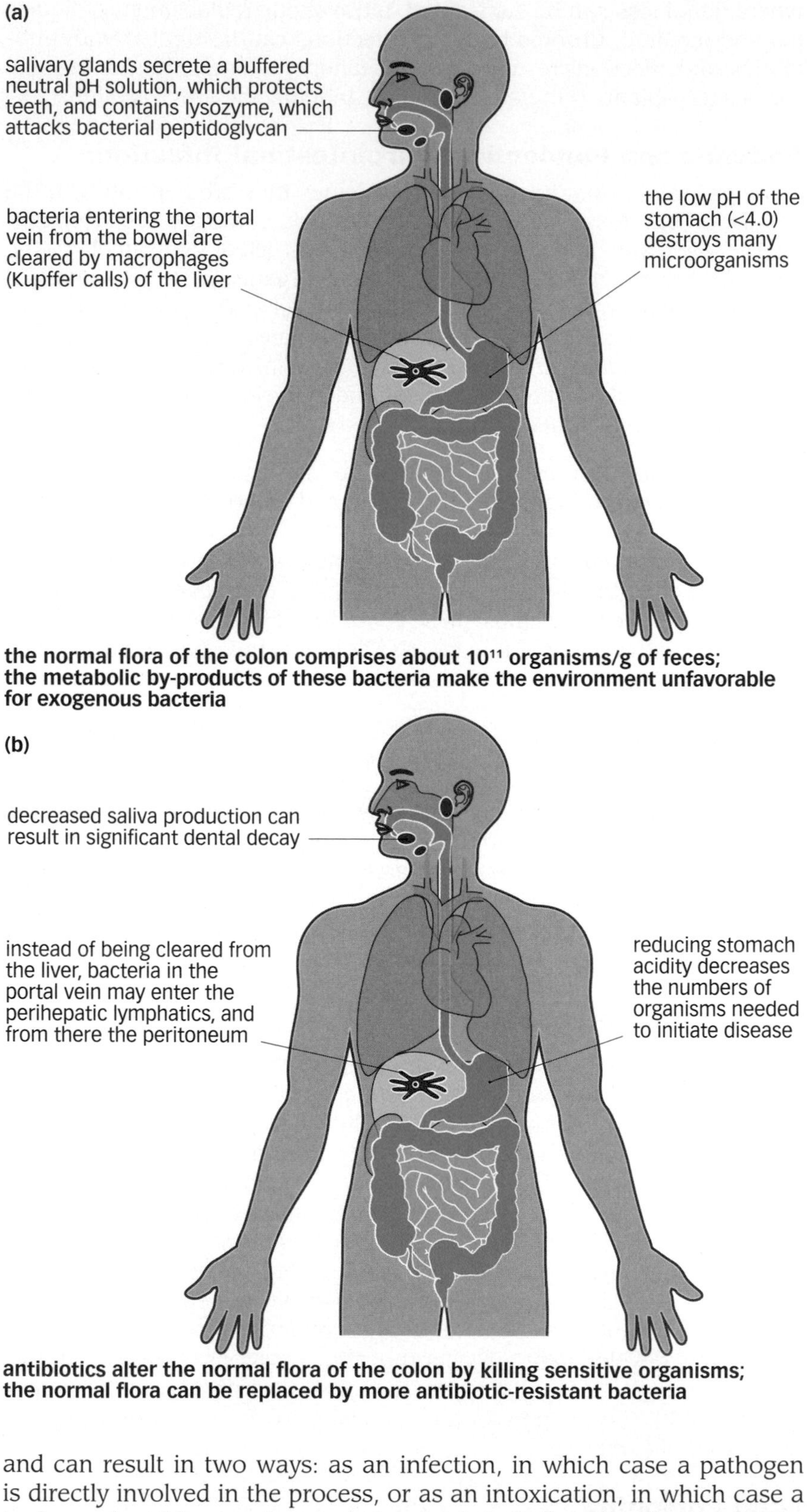

Figure 22.2 The body's defenses against gastrointestinal infection. Panel a: Anatomical features and normal bacterial flora that are part of the human body's defense against gastrointestinal infection. Panel b: How changes to these anatomical features or defenses can predispose to gastrointestinal disease.

(a)

salivary glands secrete a buffered neutral pH solution, which protects teeth, and contains lysozyme, which attacks bacterial peptidoglycan

the low pH of the stomach (<4.0) destroys many microorganisms

bacteria entering the portal vein from the bowel are cleared by macrophages (Kupffer calls) of the liver

the normal flora of the colon comprises about 10^{11} organisms/g of feces; the metabolic by-products of these bacteria make the environment unfavorable for exogenous bacteria

(b)

decreased saliva production can result in significant dental decay

reducing stomach acidity decreases the numbers of organisms needed to initiate disease

instead of being cleared from the liver, bacteria in the portal vein may enter the perihepatic lymphatics, and from there the peritoneum

antibiotics alter the normal flora of the colon by killing sensitive organisms; the normal flora can be replaced by more antibiotic-resistant bacteria

and can result in two ways: as an infection, in which case a pathogen is directly involved in the process, or as an intoxication, in which case a toxin produced by a pathogen is involved. In both types, the incubation time and severity of the illness usually depend on the number of pathogens ingested (in infection) or the amount of toxin present (in intoxication). In general, the incubation time is shorter in intoxication than in infection, because organisms need time to reproduce and colonize before causing symptoms. Intoxication may also involve organs outside the

Table 22.1 Features of microbial food poisoning.

Etiology	Percentage of cases[a]	Typical incubation period	Clinical findings	Foods
Intoxication[b]				
Clostridium botulinum	5–15	12–72 hours	Vomiting, paralysis, weakness	Honey and improperly preserved vegetables, meat, fish
Staphylococcus aureus	5–25	2–4 hours	Vomiting	Meats, custard, salads
Bacillus cereus	1–2	1–6 hours	Vomiting, diarrhea	Rice, meat, vegetables
Infections[c]				
Clostridium perfringens	5–15	9–15 hours	Watery diarrhea	Meat, poultry
Salmonella	10–30	6–48 hours	Dysentery	Poultry, eggs, meat
Shigella	2–5	12–48 hours	Dysentery	Variable
Vibrio parahaemolyticus	1–2	10–24 hours	Watery diarrhea	Shellfish
Trichinella spiralis	1–3	3–30 days	Fever, myalgia	Meat, especially pork
Hepatitis A	1–3	10–45 days	Hepatitis	Shellfish

[a]Figures based on documented outbreaks reported to the Centers for Disease Control.
[b]Disease caused by toxin in the food at time of ingestion.
[c]Disease caused by infection after ingestion.

digestive tract, as in botulism, which affects the central nervous system. The most common features of both types of food poisoning are shown in **Table 22.1**.

Most cases of both types of food poisoning involve failing to cook food adequately and then allowing the undercooked food to sit warm for some length of time. The latter serves as an incubation period for the pathogen, allowing it either to multiply or else to produce a sufficient amount of toxin to cause illness. In about 80% of investigated cases, an additional contributing factor is the improper storage of food. Of the 400–500 reported outbreaks of food poisoning in the United States each year (which involved more than 15,000 people), fewer than 200 were solved. However, *Salmonella*, *Clostridium perfringens*, and *Staphylococcus aureus* account for more than 70% of cases in which a microbial source was identified.

Koop in Mind

- The digestive system is a major portal by which pathogens may enter the body.

- The main symptom of a digestive system infection is diarrhea, but there can also be fever, vomiting, and abdominal pain.

- Dysentery differs from diarrhea in that in dysentery the stool contains mucus and blood.

- Gastrointestinal diseases are a major cause of death of children in developing countries.

- Treatment for most gastrointestinal infections is supportive, replacing the liquids lost as a result of diarrhea and/or vomiting.

- Most infections of the digestive system are endemic, but epidemics and pandemics occur and are usually caused by poor public health management and spread by overcrowding.

- Food poisoning is a common cause of gastrointestinal infection.

DENTAL AND PERIODONTAL INFECTIONS

The mouth is the portal of entry for many pathogens of the digestive tract. Many opportunistic pathogens reside here, but it also contains many microorganisms that make up the normal microbial flora. The most commonly seen infections in the mouth are dental caries (also called cavities) and infections of the gum tissue (gingivitis). In both cases, the major source of the problem is *plaque*, which is a soft, adherent dental deposit that forms as a result of bacterial colonization of the surface of the teeth. Plaque is insoluble in aqueous media, including saliva, and resists removal by all but the most vigorous brushing and flossing. The formation of caries in a tooth results from the progressive destruction of the mineralized tissue of the tooth, destruction that occurs because of the organic acids produced as a part of the metabolic activity of bacteria located in the plaque coating the tooth.

Formation of Dental Plaque

The tooth surface is normally covered by a thin organic film, the pellicle, that results from the absorption and binding of specific molecules (mainly proteins and glycoproteins) found in saliva. The pellicle forms around the tooth and because of it bacteria never interact directly with the surface of the tooth but instead adhere to the pellicle.

This adherence is facilitated through the interaction of bacterial adhesion molecules, and the initial adhesion of bacteria (usually *Streptococcus mutans*; **Figure 22.3**) to the pellicle is followed by growth of the bacteria and additional aggregation of other organisms onto the primary layer of bacteria. Primary among these additional organisms are Gram-positive cocci such as streptococcal species and short, Gram-positive rods such as actinomycetes. After two to four days, new layers of organisms have piled on to the growing plaque, and they are followed by Gram-negative motile anaerobic organisms. By this time, there can be as many as 400 species of bacteria affixed to the now mature plaque adhering to the pellicle covering the tooth (**Figure 22.4**).

Dental plaque is considered a biofilm (Chapter 4) because it is permeated with channels that transport nutrients to the bacteria that form the plaque and remove the waste products of bacterial metabolism. It also acts as a mediator for many types of interaction between the organisms making up the plaque. It is important to remember that antiseptic substances such as biguanides, chlorhexidine, fluorides, triclosan, and quaternary ammonium compounds can inhibit plaque formation and reduce the level of plaque buildup.

Dental Caries

Caries are the single greatest cause of tooth loss. There are several factors involved in the development of caries, including tooth structure,

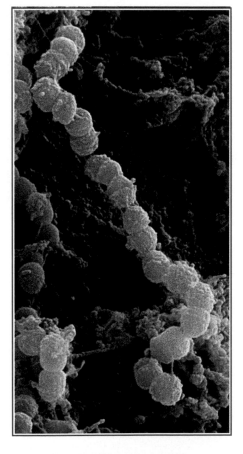

Figure 22.3 Colorized scanning electron micrograph of a biofilm of *Streptococcus mutans*. This organism is one of the first to colonize the pellicle around a tooth.

types of microflora residing on the tooth, and types of substrate available to the microflora to produce the organic acids that destroy a tooth. Normally, saliva will protect against the establishment of many bacteria because it contains lysozyme, IgA, and other antibacterial products. However, when organisms that produce acid by-products colonize the pellicle, they can cause dental caries. The most common acid-producing organisms are *Streptococcus mutans*, *Streptococcus salivarius*, *Lactobacillus acidophilus*, and *Actinomyces* species.

Bacteria that cause tooth decay must continue to have an appropriate substrate to metabolize if they are to continually produce the acids that destroy the teeth. The most readily usable substrates are sugars, and when sugars are available in the mouth, they are absorbed and metabolized by the bacteria so quickly that the acid by-products begin to accumulate. The longer this acid accumulation lasts, the more damage is done to the structure of the tooth.

Studies in humans have shown that *S. mutans* is a major cause of dental caries, and this organism initiates the problem, but other organisms also contribute to the destruction of the tooth structure. Carbohydrates easily enter the plaque where all these organisms are located, and are readily metabolized by the bacteria. The frequency of application of substrate is very important in the overall process, and repeated snacking on sugar or carbohydrates keeps the acid level high and continues the demineralization of the tooth.

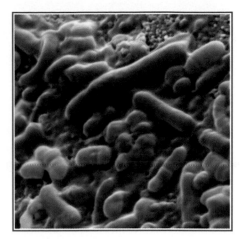

Figure 22.4 Colorized scanning electron micrograph of dental plaque, which can be made up of as many as 400 different species of bacteria.

Gingivitis and Periodontitis

There are two forms of plaque-induced periodontal disease:

- **Gingivitis** is an inflammatory condition limited to the marginal surfaces of the gums (**gingiva**). This condition does not involve the loss of bone and, depending on the degree of severity, can be corrected.

- **Periodontitis** is an infection of the gingiva that results in the loss of supportive bone and ligaments, and is responsible for most of the tooth loss seen in people over 35 years of age. There are very aggressive forms of periodontitis in which the loss of bone is very rapid.

Both of these problems are thought to be caused by certain bacteria in the dental plaque that lie next to the gingival surfaces. Gingivitis is an inflammation of the connective tissue attached to the tooth and causes a loss of collagen. The tissue destruction is caused by enzymes such as hyaluronidase and collagenase, as well as by bacterial toxins such as leukotoxin and endotoxin. In mild forms of gingivitis, there is no bacterial invasion of tissues, but such invasion does occur in aggressive forms of periodontitis.

Gingivitis will continue as long as dental plaque remains on a tooth. If it is removed and the tooth is kept plaque-free, gingivitis can completely resolve, and the tissue will return to normal. However, when the supporting bone begins to be resorbed, the condition goes from gingivitis to periodontitis. In this case, the bone is not replaced. Gingivitis can begin in as little as two weeks if teeth are not cleaned effectively.

Necrotizing Periodontal Disease

The condition known as **necrotizing periodontal disease**, previously referred to as Vincent's disease or **trench mouth**, is a spectrum of acute inflammatory diseases, starting with destruction of the mouth's soft tissue as well as the bone and ligaments associated with the teeth.

The onset of necrotizing periodontal disease can be acute, and there is an association of this condition with emotional stress and poor oral hygiene. Necrotizing periodontal disease causes rapid ulceration of the tissues and pronounced bone loss, with spirochete bacteria in direct contact with and invading the tissues. The condition is very painful and causes extremely bad breath. Both systemic and topical administration of antibiotics relieves the symptoms, but resolution depends on professional cleaning of the teeth.

It is important to note that the dental plaque we have discussed in this section must be viewed as a hazard for immunocompromised patients and patients with heart valve malfunctions, in whom there can be an increased incidence of endocarditis. In immunocompromised patients, plaque can give rise to serious systemic infections. In fact, one of the most frequent sources of lethal infections in leukemia patients is the oral cavity.

Keep in Mind

- Many opportunistically pathogenic organisms are found in the mouth.
- Most infections in the oral cavity involve the formation of dental caries, which destroy the teeth, and gum disease, with dental plaque being the source of the infections.
- Dental plaque is a biofilm made up of hundreds of different organisms.
- Gum diseases include gingivitis, chronic periodontitis, and necrotizing periodontal disease.

BACTERIAL INFECTIONS OF THE DIGESTIVE SYSTEM

In the developed world, the bacterium *Campylobacter* is the most common cause of gastrointestinal infections, with the bacteria *Salmonella* and *Shigella* being the next most common causes of infections. In the developing world and the tropics, *Vibrio cholerae* is the most common. Sources and transmission routes for several bacterial infections in the digestive tract are shown in **Figure 22.5**.

Infections in the digestive system can be classified in two groups: **exogenous infections**, which are caused by pathogens that come into

Figure 22.5 Examples of the sources and modes of transmission of some common bacterial causes of gastrointestinal infections.

organism	source	transmission	result
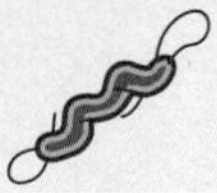 ***Campylobacter* spp.** part of the normal flora of chicken intestine	contaminated chicken carcasses	consumption of contaminated and undercooked chicken	gastroenteritis
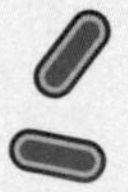 ***Escherichia coli* O157:H7** flora of farm animals	foodborne, educational farm visit, contamination of agricultural products	contaminated hand to mouth	hemorrhagic colitis, hemolytic uremic syndrome, renal failure
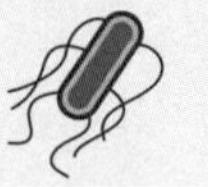 ***Salmonella enteritidis*** flora associated with hen eggs	home-made mayonnaise made using raw eggs	storage at room temperature promotes bacterial growth	gastroenteritis
 ***Shigella* spp.** in a child with diarrhea	preschool	contaminated hand to mouth	gastroenteritis
 ***Staphylococcus aureus*** (enterotoxin producer)	nasal carriage in cook, food contaminated during preparation	storage at room temperature promotes bacterial growth, with toxin production	toxin consumed, results in acute food poisoning

exogenous			
	organism	**disease**	**mech.**
	Campylobacter spp.	gastroenteritis	In/Inf/P
	Salmonella enteritidis	gastroenteritis	In/Inf
	Salmonella typhimurium	gastroenteritis	In/Inf
	Shigella dysenteriae	bacillary dysentery	In/Inf
	Shigella sonnei	gastroenteritis	In/Inf
	Escherichia coli O157	hemorrhagic colitis hemolytic uremic syndrome acute and chronic renal failure	T
	Salmonella typhi	typhoid/enteric fever	P
	Salmonella paratyphi	enteric fever	P
	Vibrio cholerae	cholera	T
	Staphylococcus aureus	vomiting and/or diarrhea	T
	Bacillus cereus	vomiting and/or diarrhea	T
	Helicobacter pylori	peptic ulcer/gastric malignancies	T/Inf
	Clostridium difficile	antibiotic-associated diarrhea	T

endogenous		
	organism	**disease**
mouth		
	Streptococcus sanguinis	dental caries, dental abscess
	Streptococcus mutans	dental caries, dental abscess
	Prevotella spp.	gingivitis, periodontitis
	Fusobacterium spp.	gingivitis, periodontitis
intestine		
	Streptococcus spp.	diverticulitis, appendix abscess, hepatobiliary sepsis, periodontitis
	Enterococcus spp.	diverticulitis, appendix abscess, hepatobiliary sepsis, periodontitis
	Coliforms (e.g. *Escherichia coli*)	diverticulitis, appendix abscess, hepatobiliary sepsis, periodontitis
	Klebsiella spp.	diverticulitis, appendix abscess, hepatobiliary sepsis, periodontitis
	Bacteroides spp.	diverticulitis, appendix abscess, hepatobiliary sepsis, periodontitis
	Peptostreptococcus Peptococcus spp.	diverticulitis, appendix abscess, hepatobiliary sepsis, periodontitis
	Clostridium spp.	diverticulitis, appendix abscess, hepatobiliary sepsis, periodontitis

Figure 22.6 Some bacteria that cause infections of the digestive system. For exogenous organisms the main mechanisms of disease are indicated as follows: In, invasion; Inf, inflammation; P, penetration; T, toxin.

the body from the outside, and **endogenous infections**, which are caused by organisms that are part of the normal microbial flora of the body (**Figure 22.6**). Exogenous organisms are brought into the digestive system through contaminated food or water. This group includes *C. difficile*, which is frequently brought into the body from the hospital environment, and *Helicobacter pylori*, which spreads from human to human through oral–oral or fecal–oral contact. Infection with these exogenous pathogens can cause nausea and vomiting within eight hours of ingestion. Endogenous organisms, such as the bacteria *Streptococcus* and *Enterococcus* and anaerobes of the intestinal tract, are found as part of the normal flora. However, given the right set of circumstances, these organisms can cause dental diseases, infections of the bowel, appendix, and liver, and **diverticular abscesses**.

Enterobacteriaceae is the name of a diverse family of Gram-negative rod-shaped bacteria, some of which are free-living and some of which are part of the indigenous microflora of the human body. These bacteria grow rapidly under both aerobic and anaerobic conditions, and a small number of them are important etiologic agents of diarrheal diseases. In addition, spread of these organisms to the blood can cause endotoxic shock, which can be fatal. *Entero-* means pertaining to the intestines, and as the family name indicates, all these bacteria damage the host intestines.

Members of the family Enterobacteriaceae have a variety of morphologies, ranging from coccobacilli to elongated filamentous rods. They do not form spores, and the cell wall, plasma membrane, and internal structures are morphologically similar in all species of the family. Some of the cell wall components and other bacterial structures are antigenic, and, as we learned in Chapter 9, these structures have been used to divide species into serotypes. The lipopolysaccharide in the outer membrane is referred to as the **O antigen** (**Figure 22.7**). Cell surface polysaccharides may form a well-defined capsule that is referred to as the **K antigen**. The motile strains use flagella, and the proteins that make up the flagella can be distinguished antigenically and are collectively called the **H antigens**.

Most members of the family Enterobacteriaceae are colonizers of the lower gastrointestinal tract of humans. However, some readily survive in nature and live freely anywhere that water and minimal energy sources are available. In humans, many of these organisms are components of the normal colonic flora. It should be noted that the enterobacteria *Shigella* and *Salmonella* are not part of the normal flora, but members of these genera are not free-living: they are strictly animal and human pathogens.

Any enterobacteria that are part of the resident flora in the human digestive system always have the potential to cause infection, and it is important to remember that many infections are simply cases of these resident organisms finding their way out of the digestive system to other systems in the body. It is known that surface structures on the bacterial cells, such as fimbriae, aid these organisms in causing infections, but once they get inside the deep tissue of a human host, their ability to persist and cause injury is not really understood except for the action of exotoxins, endotoxins, and capsules.

Salmonella, *Shigella*, *Yersinia enterocolitica* and some strains of *E. coli*, all members of the family Enterobacteriaceae, produce disease in the gastrointestinal tract, and these pathogens have invasive properties. In addition, some produce virulence factors, such as cytotoxins and enterotoxins, that correlate with the type of diarrhea they produce. Enterotoxin-producing bacteria normally cause a watery diarrhea, whereas invasive and cytotoxic strains produce dysentery. For some species, such as *Salmonella*, the gastrointestinal tract is just the portal of entry but the actual disease is systemic.

In addition to adherence by means of fimbriae and the production of endotoxins and exotoxins, enterobacteria can also produce a variety

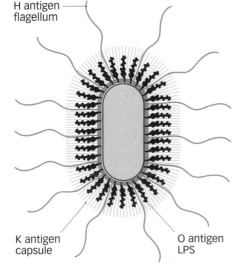

H antigen
flagellum

K antigen
capsule

O antigen
LPS

Figure 22.7 Antigens associated with the outer layer of members of the family Enterobacteriaceae. Depending on the organism, there can be a variety of combinations of these antigens. LPS, lipopolysaccharide.

of virulence factors that cause disease. Some of these bacteria use a contact secretion system to introduce virulence factors into the cytoplasm of the host cell. The genes for these factors are located either on the chromosome of the bacteria or in a plasmid (sometimes in both) and are organized into pathogenicity islands. Enterobacteria produce the widest variety of infections of any group of microbial organisms. Let's look at several of the most clinically important members of this group in more detail.

Escherichia coli

The first enterobacterium we consider is *E. coli*. This Gram-negative bacterium ferments lactose and can be distinguished from other organisms that harm the human digestive system by the biochemical reactions they perform. There are over 150 different types of *E. coli* classified according to their O, K, and H antigens. All these distinct groups are defined by numbers (for example *E. coli* O157:H7). Both fimbriae and pili are often found on these bacteria, and these structures, which are how the bacteria attach to host cells, have an important role in the virulence of the organism.

Escherichia coli can produce every toxin seen in the Enterobacteriaceae family, including a pore-forming cytotoxin, protein inhibition toxins, and toxins that affect cellular signaling pathways in infected cells. Some of the major toxins produced by pathogenic *E. coli* are:

- Pore-forming toxin, a form of hemolysin that inserts itself into the membrane of the host cell, destroying the integrity of the cell, which eventually dies.

- **Shiga toxin**, which was originally thought to be exclusive to *Shigella* bacteria but is now known to be produced by *E. coli* as well. It is a two-chain AB toxin. The B chain attaches to host cells, and this attachment allows the A chain to move into the cell cytoplasm, where it enzymatically modifies the host ribosomal RNA, thereby inhibiting the binding of tRNA to the ribosome. This blocks protein synthesis and causes the death of the host cell.

- Heat-labile toxin is also a two-chain AB toxin; it is called heat-labile because of its sensitivity to heat. The B chain binds to the membrane of the host cell and allows entry of the A chain, which causes a breakdown in the signaling mechanisms of the cell. The result is an accumulation of water and other liquids in the lumen of the intestine, an accumulation that becomes a basis for the diarrhea seen in infections involving these bacteria.

- Heat-stable toxin is a small peptide that is more resistant to heat than is heat-labile toxin. It binds directly to glycoprotein receptors on target cells. This results in the secretion of liquids and electrolytes into the lumen of the intestine (in a similar manner to that seen with heat-labile toxin).

With virulence as a method of classification, **diarrheagenic** (diarrhea-causing) *E. coli* can be divided into the following categories: enterotoxigenic (ETEC), enteropathogenic (EPEC), enteroinvasive (EIEC), enterohemorrhagic (EHEC), and enteroaggretive (EAEC). Each type causes disease by a different mechanism, and the diseases differ both clinically and epidemiologically. A summary of the pathogenesis, clinical syndromes, and epidemiology for these five categories of *E. coli* is shown in **Table 22.2**, but we will look at the enterotoxigenic, enteropathogenic, and enterohemorrhagic forms in more detail here.

Organism	Toxin	Lesion	Pathogenic gene location	Transmission	Disease
Common	Hemolysin	Inflammation	None	Adjacent flora	Opportunistic
ETEC	Heat-labile and heat-stable	Hypersecretion	Plasmid	Fecal–oral	Watery diarrhea
EPEC		Effacement (covering) of the small intestine	Pathogenicity island	Fecal–oral	Watery diarrhea
EIEC		Invasion and inflammation	Large plasmid, pathogenicity island	Fecal–oral	Dysentery
EHEC	Shiga toxin	Effacement (covering) of colon and hemorrhage	Pathogenicity island	Fecal–oral, cattle	Bloody diarrhea
EAEC		Adherent biofilm			Mucoid watery diarrhea

Table 22.2 Characteristics of *Escherichia coli* strains.

Enterotoxigenic *E. coli*

As the name indicates, these pathogens release a toxin that is the cause of the disease, and they are the most important cause of traveler's diarrhea. These organisms also produce diarrhea in infants and are the leading cause of morbidity and mortality in the first two years of life. In developing countries, they are a leading cause of intellectual development disorder and malnutrition, but these symptoms are rarely seen in developed countries.

Transmission of enterotoxigenic *E. coli* is by the consumption of food or water that has been contaminated either by actively infected individuals or by carriers of the organism. Person-to-person transmission is very rare. Uncooked foods are the greatest risk.

Pathogenesis of enterotoxigenic *E. coli* infections

The organism adheres to the cells of the small and large intestine, and the diarrhea seen in this infection is caused by the heat-labile or heat-stable enterotoxins (described above), which are secreted into the small intestine. Genes for these toxins are found on plasmids. Recall our discussions of transfer of genetic information in Chapter 11 and keep in mind that these toxin genes can be readily moved from organism to organism. The bacteria remain attached to the surface of the host cell, and the toxin causes water and electrolytes to flow into the intestine. There is no invasion of the host cells, no damage to those cells, and no inflammatory response in these infections.

Enteropathogenic *E. coli*

Enteropathogenic *E. coli* were first seen in hospital outbreaks of diarrhea in the 1950s but have essentially disappeared in developed countries. They account for about 20% of diarrhea seen in bottle-fed infants less than one year of age, and transmission is through the fecal–oral route, with infants being the reservoir.

Pathogenesis of enteropathogenic *E. coli* infections

These pathogens initially attach to cells in the intestine, using fimbriae to form clustered colonies on the surface of the cells. The lesions resulting from attachment cause effacement of the microvilli of the intestinal cells and change the overall morphology of the cells. The secretion system of these *E. coli* then delivers at least five different proteins into the target cell cytoplasm, and these proteins then inhibit cell signaling and induce modifications to cytoskeletal proteins. The cause of the diarrhea seen in enteropathogenic *E. coli* infections is not understood but may have something to do with morphological changes to the microvilli.

Enterohemorrhagic *E. coli*

These *E. coli* organisms are so named because they involve the production of Shiga toxin, which causes capillary thrombosis and blood in the stool. They are associated with diseases in which a host consumes products from animals colonized with these pathogens. However, person-to-person transmission can occur. Interestingly, these infections are seen more in developed industrialized countries than in underdeveloped nations. The most talked-about **enterohemorrhagic *E. coli*** infections are linked to the serotype *E. coli* O157:H7, which causes bloody diarrhea and is associated with ground meat, contaminated produce, and unpasteurized juices.

Pathogenesis of enterohemorrhagic *E. coli* infections

These pathogens have a very low ID_{50} and a common reservoir (cattle), both of which are factors that increase the possibility of human infection. In the food-processing industry in the United States and Europe, accidental contamination of meat with intestinal contents and the subsequent mixing of intestinal bacteria into ground beef have caused outbreaks of disease. Because ground meat cooked only to the rare stage can harbor live pathogens, most states have mandated that all ground meat must be thoroughly cooked. Fruits and vegetables can also be contaminated with enterohemorrhagic *E. coli* and should be thoroughly washed before eating.

The distinguishing clinical factors for enterohemorrhagic *E. coli* are the production of Shiga toxin and the intestinal microvilli effacement seen with enteropathogenic strains. Enterohemorrhagic strains attack the colon by adhering through attachment proteins and using the secretion injection system to deliver proteins into the target cells. The proteins then radically alter the cytoskeletal components of the cells. Attachment and effacement cause the diarrhea, while the Shiga toxin produces capillary thrombosis and inflammation of the colonic mucosa, leading to **hemorrhagic colitis**. The Shiga toxin can also circulate in the blood and bind to renal tissue, causing glomerular swelling and the deposition of fibrin and platelets in the blood vessels.

Treatment of all five forms of *E. coli* infection

Most *E. coli* diarrhea is mild and self-limiting, and therefore treatment is not required. When diarrhea is severe, replacement of liquid may be required. In enterohemorrhagic infections, further measures such as dialysis may be necessary. Treatment with trimethoprim/sulfamethoxazole or quinolones can reduce the duration of the diarrhea, but antibiotics will have no effect if hemorrhagic colitis has occurred.

Shigella

Shigella species are members of the family Enterobacteriaceae and are closely related to *E. coli*, but they do not ferment lactose. They also lack flagella and cannot be identified by H antigens. There are four species of *Shigella*—*S. dysenteriae*, *S. flexneri*, *S. boydii*, and *S. sonnei*—all of which are able to invade and multiply inside a wide variety of cells, including enterocytes. All species also produce the Shiga toxin, with *S. dysenteriae* producing the most (see **Table 22.3**).

Shigella causes dysentery and is a strictly human pathogen. There are 8–12 cases of shigellosis per 100,000 people per year in the United States,

Fast Fact The *E. coli* contamination of California spinach in the summer of 2006 caused several deaths and enormous financial losses for farmers. The strain responsible was enterohemorrhagic *E. coli* O157:H7.

Organism	Toxin	Lesion	Pathogenic gene location	Transmission	Disease
Shigella dysenteriae	Shiga toxin	Invasion, inflammation	Large plasmid, pathogenicity island	Fecal–oral	Severe dysentery
Shigella flexneri	Shiga toxin	Invasion, inflammation	Large plasmid, pathogenicity island	Fecal–oral	Dysentery
Shigella boydii, Shigella sonnei	Shiga toxin	Invasion, inflammation	Large plasmid, pathogenicity island	Fecal–oral	Dysentery

Table 22.3 Characteristics of *Shigella* species.

but shigellosis is one of the most common causes of diarrhea worldwide and is responsible for more than 600,000 deaths each year. Transmission of the infection can occur by the fecal–oral route, by person-to-person transmission, and by the consumption of contaminated food or water. The ID_{50} of *Shigella* is fewer than 200 organisms, making the infection easily transmissible. In fact, 40% of patients get this infection from a family member. There is also a direct connection between the incidence of *Shigella* infections and community sanitary practices.

The most common *Shigella* species are *S. flexneri* and *S. sonnei*. *Shigella dysenteriae* is found mainly in underdeveloped countries, where it causes the most severe form of infection, bacillary dysentery (shigellosis).

Pathogenesis of Shigellosis

Shigella species are acid-resistant and survive passage through the stomach and small intestine into the large intestine. Once there, the bacteria invade the cells of the colonic mucosa, triggering an intense acute inflammatory response that causes mucosal ulcerations and abscess formation. This multi-step process is illustrated in **Figure 22.8**.

Shigella cells cross the mucosal membrane by entering the M cells of the intestine. The bacteria selectively adhere to M cells and move through them and into the underlying phagocytic cells of the host, where the bacteria induce apoptosis and kill both the M cells and the phagocytes. Any *Shigella* cells released from the M cells contact the basolateral side of the neighboring enterocytes and then initiate a sequence of steps to invade these cells. *Shigella* is nonmotile, and during this process each bacterial cell creates an actin tail to be used as a means of transport through the cytoplasm of the infected host cell.

Figure 22.8 Sequence of events in *Shigella* infection. Notice that the initial invasion of the target cell is by way of the apical surface, but subsequent movement of the pathogen is by way of the basolateral surfaces. Also note the projection of the fingerlike structure into the neighboring cell.

Shigella moves to the membrane of the host cell, and here some rebound back into the cytoplasm while the rest push into the adjacent cell. This pushing causes the formation of a fingerlike projection into the next cell (the fourth step in Figure 22.8), and eventually this pinches off, forming a vacuole that contains *Shigella* and is surrounded by a double membrane that protects it from the host immune response. *Shigella* then lyses

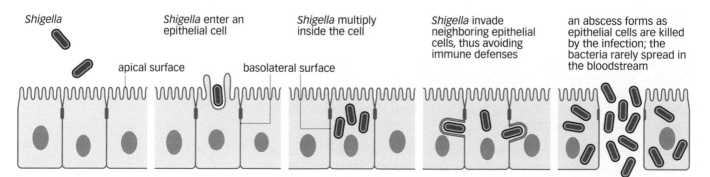

Shigella   *Shigella* enter an epithelial cell   *Shigella* multiply inside the cell   *Shigella* invade neighboring epithelial cells, thus avoiding immune defenses   an abscess forms as epithelial cells are killed by the infection; the bacteria rarely spread in the bloodstream

apical surface   basolateral surface

Organism	Lesion	Pathogenic gene location	Transmission	Disease
Salmonella enterica	Ruffles, invasion, inflammation	Pathogenicity island	Fecal–oral, animals and humans	Gastroenteritis, sepsis
Salmonella enterica serotype Typhi	Macrophage survival, mononuclear phagocytic system growth	Pathogenicity island	Fecal–oral	Typhoid fever

both membranes and is released into the cytoplasm of the newly invaded host cell, and the process starts again. This cell-to-cell extension creates localized ulcers in the mucosa, particularly in the colon, and adds to the infection a hemorrhagic component that allows *Shigella* to enter the lamina propria. An intense inflammatory response results and diarrhea begins. The stool is small and contains white blood cells, red blood cells, and bacteria.

Shigella cells produce the Shiga toxin, which also contributes to the overall severity of the illness, resulting in an acute inflammatory colitis and bloody diarrhea that presents as dysentery with cramps and bloody mucoid discharges. The symptoms of fever, malaise, and anorexia are the initial indications, and these are followed by the dysentery. Most shigellosis cases resolve spontaneously after two to five days, but the mortality in shigellosis epidemics in Asia, Latin America, and Africa has been as high as 20%.

Treatment of Shigellosis

Several antibiotics have been effective at shortening the period of illness by limiting excretion of the bacteria, with trimethoprim and sulfamethoxazole being the antibiotics of choice. Standard sanitation disposal and water chlorination are important in preventing the spread of *Shigella*.

Salmonella

There are many types of *Salmonella* and they used to be named in a variety of ways. However, they are now classified as two species, *S. enterica* and *S. bongori*. These can be subdivided into serotypes on the basis of the different lipopolysaccharide O antigens found in the cell wall and on the variety of K (capsule) and H (flagellar) antigens. *Salmonella* serotypes can also be distinguished by host range; some, such as *Salmonella enterica* serotype Typhi, are strictly adapted to humans (**Table 22.4**).

Salmonella cells possess multiple pili that bind to mannose receptors on various eukaryotic cells, and most of these bacteria are very motile (**Figure 22.9**). Using common clinical patterns of pathogenesis, we can divide *Salmonella* infections into five groups: gastroenteritis, bacteremia, enteric fever, chronic infections, and typhoid fever.

Bacteremia occurs when bacteria get into the blood. Approximately 70% of AIDS patients get *Salmonella* bacteremia, which can lead to septic shock and death. *Salmonella* bacteremia also leads to the spread of the pathogens to the meninges, the bones, and sites with preexisting abnormalities, such as atherosclerosis plaque.

Enteric fever is a multiorgan *Salmonella* infection with sustained bacteremia and profound infection of organs in the mononuclear phagocytic system (in particular the lymph nodes, liver, and spleen). Incubation takes about 13 days, with the first symptoms being fever and headache. The fever can increase over the first 72 hours and if untreated can last for

Table 22.4 Characteristics of *Salmonella*.

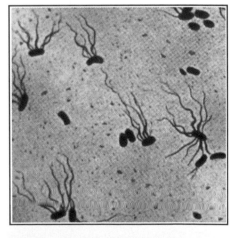

Figure 22.9 A photomicrograph of flagella-stained *Salmonella* bacteria. The flagella make *Salmonella* very motile.

weeks. In addition, some patients will exhibit constipation whereas others will have diarrhea.

Chronic infection by *Salmonella* is a very serious problem if the bacteria enter the blood of the host continuously, because the continuous release of endotoxin can lead to myocarditis, encephalopathy, intravascular coagulation, or infection of distal sites in the body. This is particularly true of the biliary tract, which can continue to harbor organisms that re-infect the intestines and cause diarrhea late in the disease.

Let's now look in more detail at gastroenteritis caused by *Salmonella* and typhoid fever.

Salmonella Gastroenteritis

Gastroenteritis typically begins 24–48 hours after ingestion of the pathogen, with nausea and vomiting being the initial symptoms. This stage of the infection is followed by cramps and diarrhea that persist for three to four days and then resolve spontaneously. Fever occurs in about 50% of patients, and diarrhea varies from loose stool to severe.

As the name indicates, *Salmonella* gastroenteritis occurs both in the stomach (gastro-) and in the intestines (entero-). Transmission is from animal or human reservoirs to humans, but the ID_{50} is higher than that seen with *Shigella*, making *Salmonella* less infectious: 1000 or more organisms are required to cause infection.

Salmonella is the leading cause of foodborne gastrointestinal infections, with poultry and contaminated eggs being the most common transmission vehicle. In addition, poor food handling and preparation are implicated, and the infection can also be transmitted by exotic pets such as turtles. *Salmonella* gastroenteritis is predominantly a disease of industrialized societies. Incidence in the United States is double that of *Shigella* infections: 40,000–50,000 *Salmonella* cases are reported each year. It is important to note that this may represent only 1–4% of the total cases that occur, because many people do not report it. Nearly 30% of *Salmonella* infections occur in nursing homes, hospitals, and mental health facilities. Approximately 5% of patients recovering from salmonellosis will shed the organism in their feces for up to 20 weeks, and chronic carriers who are food handlers can be an important reservoir for these bacteria.

Fast Fact The geographical distribution of salmonellosis has changed with the advent of long-distance distribution systems that can deliver large amounts of contaminated food to many different places in a relatively short time.

Pathogenesis of *Salmonella* gastroenteritis

Ingested *Salmonella* bacteria pass through the stomach acid and swim through the intestinal mucus layer. Eventually organisms reach the enterocytes and M cells of the large and small intestine, and here they use their pili to mediate adherence to M cells, causing the surface of the M cells to form ruffles (**Figure 22.10**). These ruffles are specialized plasma membrane sites where there has been rearrangement of the filamentous actin cytoskeleton. This rearrangement of the host-cell cytoskeleton is stimulated by 12 or more proteins that are produced by the bacteria and are coded for by genes located on pathogenicity islands in the *Salmonella* chromosome.

The ruffled surface of the host M cells engulfs the adherent bacteria into an endocytic vacuole that transcytoses from the apical surface of the cell to the basolateral surface. From here the bacteria enter the lamina propria, initiating a powerful inflammatory response. *Salmonella* cells can withstand the phagocytic defense of the host cell by inducing apoptosis after being taken up by a host phagocytic cell. It seems that the combination of transcytosis of the pathogen and the vascular permeability

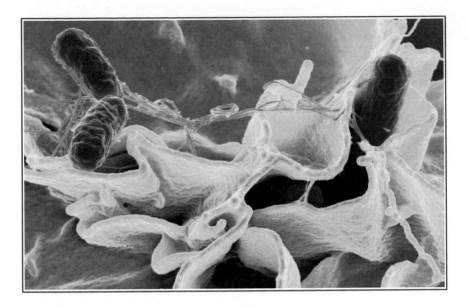

Figure 22.10 Colorized scanning electron micrograph of *Salmonella* (red) invading the epithelial cells of an intestine (yellow). Notice the ruffling of the intestinal cell membrane.

associated with the inflammatory response causes the onset of diarrhea. With *Salmonella*, however, this process remains localized to the mucosa and submucosa of the host intestinal cells.

Typhoid Fever

Typhoid fever, the result of infection by *Salmonella enterica* serotype Typhi, is a strictly human disease, with chronic carriers being the primary reservoirs. Some patients carry the disease for years, one example being the infamous 'Typhoid Mary' described at the start of this chapter. Long-term carrying occurs because bacteria become sequestered in the gall bladder and biliary tract when stones are present. The bacteria are transmitted to the water supply when sewage contaminates drinking water, and are passed from person to person by the fecal–oral route. The ID_{50} of *S. enterica* serotype Typhi is not as low as that of *Shigella* (fewer than 200 organisms) but can become lower if the organism is encapsulated. Although the incidence of typhoid fever is low in the United States, there is still significant morbidity and mortality from this organism in Latin America, Asia, and India.

Pathogenesis of typhoid fever

Because there is no animal model in which we can study this infection, it is difficult to show positively the events that occur. It is presumed that there is a killing of intestinal M cells and macrophages similar to that seen in *Salmonella* gastroenteritis infections. However, unlike other *Salmonella* serotypes, *S. enterica* serotype Typhi can survive for long periods inside viable host macrophages by inhibiting the release of the oxidative poisons used by macrophages to kill invading bacteria. This allows *S. enterica* serotype Typhi to multiply and infect new macrophages, which eventually leads to spilling of these bacteria into the lymphatic circulation, allowing them to migrate to the lymph nodes, spleen, liver, and bone marrow of the host.

This systemic infection is exacerbated by the release of lipopolysaccharide endotoxin, which causes a fever that increases and persists. The bacteria can also spread to the host urinary system and other organs, eventually coming full circle and re-infecting the intestine. The most important complication of typhoid fever is hemorrhaging that causes perforation of the wall of the colon or the ileum at the site of Peyer's patches that have

Organism	Growth	Urease	Epidemiology	Pathogenesis	Disease
Vibrio cholerae	Facultative	–	Fecal–oral, water	Cholera toxin	Watery diarrhea
Campylobacter jejuni	Microaerophilic	–	Animals, unpasteurized dairy products	Unknown	Dysentery, watery diarrhea
Helicobacter pylori	Microaerophilic	+	Transmission not understood	Vacuolating cytotoxin, urease	Chronic gastritis, ulcers

**Table 22.5 Features of *Vibrio,
Campylobacter,* and *Helicobacter*.**

become necrotic. The entire cycle from intestine back to intestine takes only two weeks.

General Treatment of *Salmonella* Infections

The primary therapy for all forms of *Salmonella* infections is the replacement of fluid and electrolytes and the control of nausea and vomiting. Antibiotic therapy is not appropriate because it increases the duration and frequency of the carrier state. However, therapy with chloramphenicol, ampicillin, trimethoprim sulfamethoxazole, or some of the cephalosporins can be used prophylactically to prevent the spread of the disease. A vaccine for typhoid fever has been available for many years in both the oral and injectable forms. It is also essential to provide clean water and treatment for those carrying the disease.

This ends our discussion of bacteria belonging to the family Enterobacteriaceae. Next let's look at some other bacteria that also infect the digestive system (**Table 22.5**). We will look at *Vibrio cholerae*, *Campylobacter jejuni*, and *Helicobacter pylori*, all of which are Gram-negative rods with a similar morphology.

Vibrio

Members of the genus *Vibrio* are Gram-negative, non-spore-forming bacteria commonly found in salt water. They have a unique morphology in that they form S shapes and half-spirals (comma shapes), and they are highly motile by means of a single polar flagellum (**Figure 22.11**). *Vibrio* can grow either aerobically or anaerobically and have a cell structure similar to that of other Gram-negative bacteria. They have a low tolerance for acidic conditions but grow well in mildly alkaline environments.

The species responsible for the disease cholera, *Vibrio cholerae*, produces a toxin that causes a devastating intestinal infection. Cholera is endemic in many countries and we are currently in the seventh cholera pandemic: it started in South Asia in 1961, reached Africa in 1971, and the Americas in 1991.

Pathogenesis of *V. cholerae*

Epidemic cholera is spread primarily by contaminated water and poor sanitation. Its short incubation period, about two days, ensures a rapid infection cycle. To produce disease, *V. cholerae* must reach the small intestine in sufficient numbers to multiply and colonize. The bacteria possess long filamentous pili that form bundles on the bacterial surface, and these are used for colonization. *V. cholerae* will colonize the entire intestinal tract, from the jejunum to the colon, and liquid loss resulting from colonization depends on a balance between bacterial growth, toxin production, and host liquid secretion and absorption. The loss of liquids and electrolytes, which can amount to multiple liters a day, is greatest in the small intestine, and the results of liquid

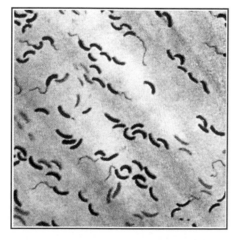

**Figure 22.11 A photomicrograph of
Vibrio cholerae.**

loss are dehydration, hypokalemia (loss of potassium), and metabolic acidosis resulting from loss of bicarbonate.

Vibrio cholerae does not invade or damage the enterocytes of the digestive system but instead uses its toxin plus a variety of virulence factors to cause disease. These factors, which are part of a remarkable controlled, coordinated system involving environmental sensors, are all coded for by genes located on pathogenicity islands in *V. cholerae*.

Cholera has a rapid onset, beginning with abdominal fullness and discomfort, rushes of peristalsis and loose stools. The stools quickly become watery, voluminous, and almost odorless and can progress to what is referred to as **rice stool** because it contains bits of mucus. There is no fever with cholera and no blood in the stool.

Treatment of Cholera

The outcome of cholera depends on balancing liquid and electrolyte loss. This balance can be accomplished by oral or intravenous liquid replacement, and this is all that is required except in the most severe cases. Tetracycline can shorten the duration of diarrhea and the magnitude of liquid loss. It is important to note that epidemic cholera does not exist in areas where waste disposal is adequately dealt with.

Campylobacter Enteritis

The digestive infection *Campylobacter* enteritis is caused by *Campylobacter jejuni* (**Figure 22.12**), which was not recognized as a human pathogen until 1973 but is now one of the most common causes of diarrhea. It is found in 4–30% of diarrheal stools, making it the leading cause of gastrointestinal infection in developed countries, and has an ID_{50} of only a few hundred, making it easily transmissible. In fact, there are more than 2 million cases of *C. jejuni* infection in the United States each year.

The primary reservoir of *C. jejuni* is animals, and it is transmitted to humans either by the ingestion of contaminated food or by direct contact with pets that harbor the organism. The most common source of human infection is undercooked poultry, but transmission can also occur via contaminated water or unpasteurized milk. It should be noted that *Campylobacter* is commonly found as part of the gastrointestinal and genitourinary tract flora of warm-blooded animals, and for this reason domestic pets may have a significant role in transmission to humans.

Pathogenesis of *Campylobacter* Enteritis

Oral ingestion of the pathogen is followed by colonization of the intestinal mucosa, where bacteria adhere to and then enter cells in endocytic vacuoles. Once inside the cells, the bacteria move in association with microtubule structures and produce a lethal **distending cytotoxin** that arrests cell division while the cytoplasm continues to increase. How this leads to the common symptom of diarrhea is not known. Illness with *C. jejuni* begins about seven days after ingestion of the bacteria, with fever and lower abdominal pain that may be severe enough to mimic appendicitis. This is followed within hours by dysenteric stools containing blood, mucus, and pus.

There is an association between *C. jejuni* infection and Guillain–Barré syndrome. Up to 40% of patients with this acute demyelinating disease have evidence of infection with *Campylobacter* at the time that neurologic symptoms arise. This is probably due to an autoimmune reaction

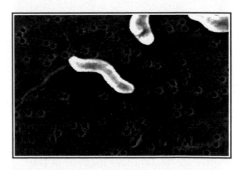

Figure 22.12 A scanning electron micrograph of *Campylobacter jejuni*, which has a corkscrew appearance.

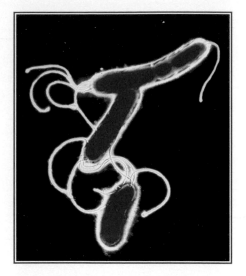

Figure 22.13 A colorized transmission electron micrograph of *Helicobacter pylori*, which is a causative agent of ulcers. Notice the multiple flagella, which help to propel this organism through the acidic stomach.

Fast Fact *Helicobacter pylori* was one of the first bacteria to be classified as a carcinogen by the World Health Organization.

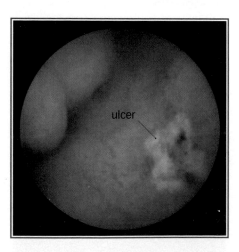

ulcer

Figure 22.14 An ulcer located in the duodenum of a patient. This lesion was caused by *Helicobacter* infection.

in which antibody against the bacteria cross-reacts with the myelin of the neurons.

Treatment of *Campylobacter* Enteritis

This infection is self-limiting, and signs and symptoms last for only three to five days (though they can last for up to two weeks in severe cases). Fewer than 50% of patients benefit from antibiotic treatment; however, if the infection is severe, ciprofloxacin is the treatment of choice.

Helicobacter pylori

This organism is similar to *Campylobacter* in morphology and growth characteristics. It is a slender, microaerophilic, Gram-negative curved rod with polar flagella (**Figure 22.13**) and a cell wall structure similar to that seen in most other Gram-negative organisms.

One of the unique features of *H. pylori* is the production of a urease enzyme that allows it to survive in very acidic environments by generating ammonia. Another circulating protein associated with this pathogen, **vacuolating cytotoxin**, causes apoptosis in infected eukaryotic cells. These infected cells are easily identified under the microscope because they have large vacuoles throughout their cytoplasm. Like many other intestinal pathogens, *H. pylori* cells possess a contact injection system that introduces proteins that disrupt many of the proteins in the infected cell. As in other bacteria, these virulence factors are coded for by genes located in pathogenicity islands.

Helicobacter pylori infection causes what may be the most prevalent disease in the world, ulcers (**Figure 22.14**). The organism is found in 30–50% of all adults in developed countries and in practically 100% of adults in developing countries. The mode of transmission for the pathogen is not known but is presumed to be from person to person by the fecal–oral or oral–oral route. It has been shown that colonization increases with a patient's age and persists for decades. *Helicobacter* is the most common cause of **gastritis** and of gastric and duodenal ulcers and is also the main predisposing factor for gastric adenocarcinoma, which is one of the most common causes of cancer deaths in the world.

Pathogenesis of *H. pylori* Gastritis

Helicobacter pylori uses multiple mechanisms to adhere to a host's gastric mucosa and survive the acidic environment. The bacterium is highly motile and swims to less acidic areas located below the mucus layer surrounding the lining of the stomach. Once in these less acidic areas, the bacterium adheres by using surface proteins, some of which bind to blood group antigens. Colonization is almost always accompanied by cellular infiltration as part of the host's inflammatory response to the infection. The inflammatory response can be extensive and can cause the formation of microabscesses that contribute to the ulceration and are also accompanied by the release of virulence factors that enhance cellular erosion.

The primary infection with *H. pylori* is either without symptoms or may cause some nausea and mild upper abdominal pain that lasts for about two weeks. Years later, however, there can be gastritis, or peptic ulcer disease with nausea, anorexia, vomiting, and pain. Many patients are asymptomatic for decades even up to perforation of the ulcer, which leads to extensive internal bleeding.

Treatment of *H. pylori* Gastritis

Helicobacter pylori is sensitive to a wide variety of antibiotics. First-line therapy is amoxicillin, clarithromycin, and the acid secretion inhibitor omeprazole. Treatment with bismuth salts and a combination of metronidazole and tetracycline is recommended for patients that fail initial treatment.

Keep in Mind

- The family Enterobacteriaceae comprises Gram-negative rods and contains a variety of organisms involved in infections of the digestive system.

- Many bacteria of the family Enterobacteriaceae are resident flora of the human digestive system.

- O antigens are part of the lipopolysaccharide in the outer membrane, whereas H antigens are found on the flagella, and K antigens are part of the capsule.

- The enterobacteria *Escherichia coli*, *Salmonella*, *Shigella*, and *Yersinia* cause infections of the digestive system through the production of toxin as well as the invasion of tissue.

- Several toxins are produced by *E. coli*, with one of the most dangerous being the Shiga toxin.

- Virulence properties can be used to subdivide *E. coli* into five groups: enterotoxigenic, enteropathogenic, enteroinvasive, enterohemorrhagic, and enteroaggretive.

- Four species of *Shigella* are involved in digestive infections, and all produce the Shiga toxin: *S. dysenteriae*, *S. flexneri*, *S. boydii*, and *S. sonnei*.

- Two prominent infections of the digestive system that are caused by *Salmonella* are gastroenteritis and typhoid fever.

- *Vibrio cholerae*, a member of the family Spirillaceae, produces a powerful toxin that causes devastating infections of the digestive system.

- *Campylobacter* enteritis is one of the most common causes of diarrhea.

- *Helicobacter pylori* is the most common cause of gastric and duodenal ulcers.

- Gastrointestinal infections can be exogenous (coming from the outside) or endogenous (caused by organisms found in a person's normal intestinal flora).

VIRAL INFECTIONS OF THE DIGESTIVE SYSTEM

As you might expect, the digestive system is an important portal of entry for viruses. Some viruses—poliovirus is one example—use the digestive system only as an entry point. These viruses infect intestinal epithelial cells, but the disease caused by the virus occurs somewhere else in the body. In this section, we discuss some of the most common viruses that cause infections of the digestive tract.

As with bacteria, the most common sign associated with these viral infections is diarrhea. Usually the diarrhea has a rapid onset (within hours after the virus enters the body) but lasts for less than three weeks. There is abundant excretion of virions in the stool, with amounts in excess of 10^8 per gram of stool. These viral particles are easily identified in stool samples by electron microscopy.

Specific criteria are required to implicate viruses in digestive infections, and these are similar to the Koch's postulates used for bacteria (Chapter 7).

- The virus must be detected in ill patients, and viral shedding must correlate with the onset of symptoms.

- There must be a significant antibody response in patients who are shedding virus.

- The disease must be reproduced through experimental inoculation of nonimmune humans or animals with the virus in question. This is a very difficult requirement to satisfy, because many viruses cannot be grown in culture.

- Other causes of the signs and symptoms of the infection (mainly diarrhea) must be excluded.

On the basis of these criteria, several groups of viruses that cause gastrointestinal infections have been established including rotavirus, calicivirus (including norovirus), astrovirus, some serotypes of adenovirus, and enterovirus. The infections caused by all these viruses have similar characteristics, including brief incubation periods, fecal–oral routes of transmission, and vomiting that either precedes or accompanies diarrhea. The characteristics of some of the viruses that cause digestive disease are shown in Table 22.6, but because of space limitations we will only discuss rotavirus, norovirus, and enterovirus in detail.

Rotavirus

The rotaviruses belong to the family *Reoviridae* and are non-enveloped, spherical, double-stranded RNA viruses. Under an electron microscope, a rotavirus looks like a wheel (*rota* is Latin for wheel) and has a double-capsid structure in which the outer capsid is attached to the inner one by short, spokelike structures. Rotaviruses were not discovered until 1973,

Table 22.6 Characteristics of viruses that cause gastrointestinal infections.

Feature	Rotavirus	Calicivirus	Astrovirus	Adenovirus
Biological features				
Nucleic acid	dsRNA	+ssRNA	+ssRNA	dsDNA
Shape	Naked, double-shelled capsid	Naked, round	Naked, star-shaped	Naked, icosahedral
Number of serotypes	Four that infect humans	More than four	At least five	Unknown
Pathogenic features				
Site of infection	Duodenum, jejunum	Jejunum	Small intestine	Small intestine
Epidemiologic features				
Seasonality	Usually winter	None known	None known	None known
Ages primarily affected	Infants, children <2 years old	Older children and adults	Infants, children	Infants, children
Transmission	Fecal–oral	Fecal–oral; contaminated water and shellfish	Fecal–oral	Fecal–oral
Incubation	1–3 days	12 hours to 2 days	1–2 days	8–10 days
ds, double-stranded; ss, single-stranded.				

when electron microscopy was used to examine biopsy specimens from infants with diarrhea. Since then, members of the rotavirus genus have been found around the world and are believed to account for 40–60% of cases of acute gastroenteritis. Rotavirus causes acute gastrointestinal disease in a variety of species, and it has the ability to undergo genetic reassortment, a property that makes this virus difficult to deal with immunologically (recall our discussion of influenza virus in Chapters 12 and 13).

Outbreaks of rotavirus infections are common in infants and children under two years of age, but adults are usually only minimally affected. These infections can affect elderly institutionalized people. Usually, rotavirus infection in infants results in little or no clinical illness, and by the age of four years more than 90% of individuals have developed antibody against the virus. It is estimated that rotavirus infections kill almost half a million infants worldwide each year, but in the United States there are fewer than 100 deaths attributed to rotavirus each year. However, rotavirus infection is still a major cause of hospitalization early in life.

Pathogenesis of Rotavirus Infection

The virus localizes primarily in the duodenum and proximal jejunum, where it blunts (shortens) the microvilli of the epithelial cells. This change in the microvilli causes a mild infiltration of mononuclear and polymorphonuclear leukocytes into them. The primary effect of this is a decreased absorptive surface on the microvilli coupled with decreased enzymatic function. This results in malabsorption and defective handling of fats and carbohydrates. Interestingly, the gastric and colonic mucosa are not affected. It can take eight weeks to restore normal histology and function to the damaged area after the infection is over.

The incubation period is between one and three days, and symptoms begin with the abrupt onset of vomiting, followed within hours by frequent copious watery brown stools. There can often be low-grade fever as well, and the vomiting and diarrhea can last for several days. As in many other intestinal infections, the major complication of these effects is dehydration.

Treatment of Rotavirus Infection

There is no specific treatment, but severe cases require vigorous replacement of fluid and electrolytes. Rotavirus is highly infectious and can spread rapidly in institutional settings. Therefore, control measures involving hygienic practices are important to inhibit the spread of the infection. A live attenuated vaccine has been developed.

Norovirus

Noroviruses are members of the *Caliciviridae*. All noroviruses are considered strains of the Norwalk virus. Worldwide, these viruses cause 90% of nonbacterial gastroenteritis, affect more than 250 million people, and kill more than 200,000 people each year. In the UK, it is often referred to as 'winter vomiting bug' because it causes vomiting and is spread more often in winter, when people spend more time indoors and in close contact with others. Outbreaks often occur in hospitals, schools, prisons, dormitories, cruise ships, and nursing homes, where crowded indoor conditions exist.

These viruses are spread by direct or indirect (for example, via contaminated food) fecal–oral transmission from person to person. Symptoms associated with gastroenteritis begin after one or two days, and last for 24–60 hours. The virus can remain infective for 7–12 days on food preparation surfaces, fabrics, and other surfaces, and for months in contaminated water.

Pathogenesis of Norovirus Infection

Noroviruses are replicated within the small intestine. They are highly infectious; fewer than 20 virions can cause an infection. The viruses are still shed from the host for weeks after symptoms have disappeared. Vomiting, coughing, and flushing toilets can aerosolize noroviruses. The main symptoms of infection include forceful vomiting, watery diarrhea, and abdominal pain. Lethargy, weakness, muscle aches, headaches, coughing, loss of taste, and fever may also occur.

Treatment of Norovirus Infection

Norovirus infection is usually self-limiting and will resolve on its own, but life-threatening symptoms such as dehydration require treatment. None of the currently available antiviral agents have been shown to be effective against the *Caliciviridae* noroviruses.

Enterovirus

Enteroviruses are members of the *Picornaviridae*. They are very resistant to acidic environments, which helps them survive passage through the stomach on their way to the intestines. They also resist common disinfectants and various detergents, especially if embedded in organic material. Humans are the natural hosts for these enteroviruses, and asymptomatic infection is common. However, there are some enteroviruses that do not infect humans. Although whether infection is symptomatic or not depends on the species of enterovirus in question and on the age of the host, all the enteroviruses that infect the digestive system show a seasonal infection pattern and are predisposed to temperate climates.

Direct or indirect fecal–oral transmission from person to person is the most common way for these infections to be spread, and the virus will normally spend one to four weeks in the oropharynx after infection. The virus can be shed for 18 weeks in the feces.

Pathogenesis of Enterovirus Infection

About 60% of infections with enteroviruses occur in children aged nine years or less, and the incubation time is two to ten days. Attachment usually occurs between attachment proteins on the surface of the virion that are specific for receptors on the host cell. The virus is brought into the host cell by envelopment in the host membrane, and viral RNA is released into the cytoplasm. Here it binds directly to ribosomes and begins the synthesis of viral proteins. Enteroviruses are lytic, and the end result of the infectious cycle is destruction of the host cell and the release of new virions that infect other cells. The primary infection occurs in the digestive system but then spreads to other sites in the body.

Treatment of Enterovirus Infection

None of the currently available antiviral agents have been shown to be effective against the *Picornaviridae* enteroviruses.

Hepatitis Viruses

The name **hepatitis** describes any disease that affects the hepatocytes of the liver, and these diseases can be caused by a variety of agents,

Feature	Hepatitis A	Hepatitis B	Hepatitis C
Virus type	+ssRNA	dsDNA	+ssRNA
Percentage of viral hepatitis	50	41	5
Incubation period (days)	15–45	7–160	15–160
Onset	Usually sudden	Usually slow	Insidious
Age preference	Children, young adults	All ages	All ages
Transmission			
Fecal–oral	+++	+/–	–
Sexual	+	++	+
Transfusion	–	++	+++
Severity	Usually mild	Moderate	Mild
Chronicity (%)	None	10	>50
Carrier state	None	Yes	Yes
Protection by immune serum globulin	Yes	Yes	Yes

Table 22.7 Comparison of hepatitis A, B, and C viruses.

including bacteria, protozoans, viruses, toxins, and drugs. There are at least six different viruses that cause hepatitis, and we will look at each of these in detail. It is important to note that the hepatitis viruses are distinctly different from one another. Three of the most frequently encountered types are summarized in **Table 22.7**.

Hepatitis A

This is a non-enveloped, single-stranded RNA virus with cubic symmetry. It resists inactivation and is stable at –20°C and low pH. Hepatitis A is classified as a member of the family *Picornaviridae*. There is only one serotype of this virus, and humans are the most common natural host for it. The major form of transmission is the fecal–oral route, and hepatitis A infections are commonly seen in situations where there is crowding and poor hygiene. The rates of hepatitis A infection are higher in lower socio-economic groups. The rates of infection in the United States have been declining since 1970. In developing nations, however, as many as 90% of the population show evidence of previous infection with hepatitis A virus.

Pathogenesis of hepatitis A infection

This infection often results from poor personal hygiene in food handlers. Patients are most contagious one to two weeks before the onset of clinical symptoms of infection. The virus is believed to replicate initially in the intestinal mucosa and can be seen in feces by electron microscopy 10–14 days before the onset of symptoms. It is interesting to note that when symptoms begin, virus particles are no longer being shed in feces. Multiplication in the intestines is followed by spread to the liver. This causes lymphoid infiltration into the liver, which leads to necrosis of the parenchymal cells and to the proliferation of Kupffer cells. The extent of the necrosis coincides with the severity of the infection. Patients with anti-hepatitis A antibodies cannot be re-infected with this virus, indicating that the immune response to this virus is a protective one.

Incubation times for hepatitis A vary from 10 to 50 days, followed by the onset of fever, anorexia, nausea, pain in the upper right abdominal quadrant, and jaundice. The urine of infected patients with jaundice can become dark, and their stool can become clay-colored one to five days before onset of the jaundice. Many people infected with hepatitis A will be asymptomatic or only mildly affected and do not develop jaundice. This form of the infection, referred to as anicteric hepatitis A, is a function of the patient's age. It is seen more in children and less in adults.

Treatment of hepatitis A infection

There is no treatment for hepatitis A infection, and supportive measures such as rest and adequate nutrition are the only recommendation. A passive prophylactic treatment with human immune serum globulin obtained from pools of donors can give protection during the period of incubation and is 80–90% effective. There is also an active immunization protocol that uses either live attenuated virions or killed virions for those who are repeatedly exposed to hepatitis A virus.

Hepatitis B

Hepatitis B belongs to the family *Hepadnaviridae* and is unrelated to any other human virus. It has a spherical shape with a surrounding envelope and a unique DNA genome (**Figure 22.15**). The DNA genome is only partly double-stranded and contains a short stretch that is single-stranded. It also carries with it its own viral DNA polymerase. The envelope of hepatitis B virus contains viral surface antigens called HBsAg, and aggregates of these antigens are often found in abundance during the infection. Hepatitis B DNA can be found in the nucleus of infected hepatocytes, and viral HBsAg can be found in the cytoplasm.

Hepatitis B virus has a unique replication cycle as a result of its incompletely double-stranded DNA. During viral replication, full-length positive viral RNA transcripts are inserted into the core of the virus, and these RNAs are used to form a template for reverse transcription in which a negative DNA strand is made. Then a positive DNA strand is made from this negative DNA strand, but it never gets finished before the virus is released. Therefore the new virions have a double-stranded DNA plus a stretch of single-stranded DNA.

Hepatitis B is found worldwide, though its prevalence varies from one country to another. Chronic carriers are the main reservoir for the virus. In the United States, it is estimated that 1.5 million people are infected with hepatitis B, and there are 300,000 new cases each year. Five to ten percent of these people will become chronic carriers of the virus, and 300 others will die of acute viral infection. One striking statistic is that up to 4000 of the 300,000 Americans who contract the infection in a given year will develop hepatitis B cirrhosis, and 1000 will get hepatocarcinoma as a result of the infection. Fifty percent of infections are sexually transmitted, but screening of blood donors has markedly reduced the incidence of transfusion transmission.

Pathogenesis of hepatitis B infection

The major mode of transmission is through close contact with body liquids from infected individuals. Hepatitis B antigens have been found in most body liquids, including saliva, semen, and cervical secretions, and

Fast Fact There is a high risk of hepatitis B infection in health care workers, as a result of exposure to patients' bodily fluids.

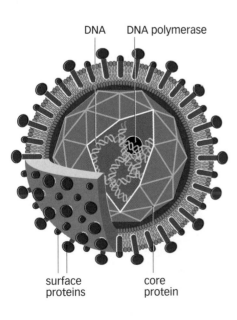

DNA   DNA polymerase

surface   core
proteins   protein

Figure 22.15 A diagrammatic representation of the hepatitis B virus. This DNA virus carries its own viral DNA polymerase.

it has been shown that as little as 0.0001 milliliters of infected blood is able to produce an infection. Consequently, inadequately sterilized hypodermic needles or instruments used in tattooing or piercing can easily transmit this viral infection.

Factors that determine the appearance of clinical symptoms are not yet completely understood, but some seem to involve the immune response of the host. For example, there is an arthritis that sometimes precedes the jaundice, with the arthritis being mediated by the complement system of the host. Lesions in acute hepatitis B infections resemble those seen with other hepatitis viruses in that chronic active infection causes a continued inflammation resulting in necrosis of hepatocytes in the liver. This destruction of liver cells can lead to collapse of the reticular framework of the liver and progressive fibrosis.

The clinical symptoms are variable, and the incubation time is anywhere between 7 and 160 days. Acute infection is usually manifested by a gradual onset of fatigue, loss of appetite, nausea, and pain in the upper right abdominal quadrant. Early in the infection, painful swollen joints and arthritis may appear, and some patients will develop a rash during this period. Jaundice can also be a complication, and these symptoms can last for months. The ratio of infection to disease varies with the age of the host, but fulminant hepatitis is seen in less than 1% of infections. Approximately 10% of infected individuals will develop chronic hepatitis, and this risk is higher in children and immunocompromised persons. Chronic hepatitis B can lead to **cirrhosis**, liver failure, or **hepatocellular carcinoma**.

Treatment of hepatitis B infection

There is no effective treatment for acute hepatitis B infection, but it can be prevented by using safe sex and avoiding needle sticks. Hepatitis B serum globulin is available, and administration of this soon after exposure can reduce development of the disease. A vaccine that is recombinant and made in yeast provides excellent protection. Health care workers are required to receive it.

Hepatitis C

This RNA virus is in the *Flaviviridae* family (as are yellow fever and dengue fever) and has a very simple genome consisting of only eight genes. There are six major genotypes and multiple subtypes of hepatitis C virus, and the genotypes are related to the geographic distribution of the virus and the severity of the disease produced. Transmission of hepatitis C via blood transfusions is well known, but it can also be transmitted sexually. Needle sharing accounts for more than 40% of infections, and hemodialysis patients are also at risk. More than 3.5 million people in the United States are infected with hepatitis C.

Pathogenesis of hepatitis C infection

The incubation time for this infection varies between 6 and 12 weeks, and the infection is usually mild or asymptomatic. However, 85% of those infected will become carriers of the infection and progress to chronic hepatitis (this can take 10–18 years). Cirrhosis and hepatocellular carcinoma are late consequences of chronic hepatitis C infection (Figure 22.16). This condition is the leading cause of liver transplants in the United States.

Treatment of hepatitis C infection

Combination therapy with interferon-α and ribavirin is the treatment of choice for hepatitis C infection.

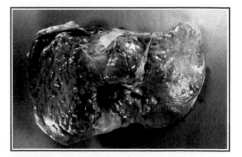

Figure 22.16 A liver damaged by hepatitis C virus infection.

In addition to the three main strains of hepatitis—A, B, and C—there are three others worth mentioning here: D, E, and G.

Hepatitis D

Hepatitis D is a small single-stranded RNA virus that is currently referred to as a satellite virus because it requires the presence of hepatitis B antigens to complete its replication cycle and become infective. It is seen only in people infected with hepatitis B. How hepatitis D virions replicate is unknown, but it has been shown that the capsid of hepatitis D is made up of hepatitis B proteins. This infection is seen most often in intravenous drug abusers, a group that also happens to be at high risk for hepatitis B infection.

Pathogenesis of hepatitis D infection

There are two types of infection with this virus. The first involves co-infection with hepatitis B, and the second presents as a superinfection of people already infected with hepatitis B. In simultaneous infection, the clinical picture mirrors that seen with hepatitis B. In superinfection, the hepatitis D infection can cause a relapse, recurrence of jaundice, and increased risk of cirrhosis. A rapid progression of liver disease and death is also seen in 20% of these cases.

Treatment of hepatitis D infection

Interferon-α is given to patients doubly infected, but it is not as effective as with hepatitis B alone, and only about 15–25% of patients improve. Prevention methods include safe sex and no sharing of needles.

Hepatitis E

This virus, which is in the family *Caliciviridae*, causes a form of hepatitis that is transmitted by the fecal–oral route. Hepatitis E resembles hepatitis A virus, although the two belong to different families, and is a non-enveloped RNA virus that has pronounced spikes on the capsid. Infection with hepatitis E is frequently subclinical; it causes acute disease only in pregnant women. The highest attack rates for this virus are seen in young adults and are associated with contaminated drinking water. However, the virus does not seem to be transmitted by person-to-person contact. Incubation takes about 40 days, and there is no treatment.

Hepatitis G

This RNA virus was discovered in 1995. It is a member of the family *Flaviviridae* and is similar to hepatitis C virus. Approximately 2% of blood donors are found to be positive for hepatitis G RNA, but the pathogenesis and disease process associated with this virus are not yet understood. There is no treatment for infection with hepatitis G.

Keep in Mind

- The digestive tract is an important portal of entry for viruses.

- The most common symptom of viral infection of the digestive system is diarrhea.

- Viral pathogens of the digestive system include rotavirus, calicivirus, astrovirus, and adenovirus.

- Hepatitis describes any disease that affects hepatocytes of the liver, and there are six different viruses that are classified as hepatitis viruses.

PARASITIC INFECTIONS OF THE DIGESTIVE SYSTEM

There are several protozoan and helminthic infections associated with the human digestive system. In Chapter 14, we looked at amebiasis (caused by *Entamoeba histolytica*), enterobiasis (caused by the pinworm *Enterobius vermicularis*), and tapeworm infections (such as *Taenia saginata*). Here we discuss other common parasitic infections that affect the digestive system: the protozoa *Giardia* and *Cryptosporidium*, the whipworm (helminth) *Trichuris trichiura*, and the hookworms *Necator americanus* and *Ancylostoma duodenale*.

Giardiasis (*Giardia duodenalis*)

Giardia duodenalis is a flagellate protozoan parasite found throughout the world, and the infection of the digestive system that they cause is **giardiasis**. *Giardia* exists in both a trophozoite form and a cyst form, and the organism is quite large. The trophozoite form resembles a stingray and has four pairs of flagella located ventrally, laterally, anteriorly, and posteriorly (**Figure 22.17**). These flagellates reside in the duodenum and jejunum of the human small intestine, where they absorb nutrients from the host digestive tract. They move about and through the mucus layer at the base of the intestinal microvilli in two ways, either by using a tumbling motion or through a large ventral sucker that attaches to the epithelium. Unattached *Giardia* is carried in the fecal stream to the large intestine.

While *Giardia* is in the descending colon, the flagella are retracted into cytoplasmic sheaths and a clear cyst wall is secreted and encloses the organism. While in the cyst, *Giardia* divides, producing a quadrinucleate organism with two sucking disks. This mature cyst is the infectious form of the parasite. It can survive in cold water for months and is also resistant to the chlorine used in most municipal water supplies. Transmission of the infection is by the fecal–oral route. Once inside a host, the cyst divides into two trophozoites.

Giardia is one of the most widely distributed intestinal protozoans and is found in fish, amphibians, reptiles, birds, and mammals. There are three morphologically distinct groups of *Giardia*, and the prevalence of the infection in young children and young adults is highest in areas of poor hygiene and sanitation. Travelers and hikers can become infected through contaminated water or food, and there have been more than 20 waterborne outbreaks in the United States associated with the contamination of municipal water supplies by sewage.

Pathogenesis of Giardiasis

The disease is associated with malabsorption by the intestinal tract, in particular fats and carbohydrates. The precise pathogenic mechanisms are unknown, but there can be blockage of the intestine by large numbers of *Giardia*. Destruction of microvilli by the parasite's sucking disk is also possible, as are damage to bile-production pathways and altered intestinal motility. The parasite may also cause an accelerated turnover of the mucosal epithelium and invasion of the mucosal tissue. Because none of these events correlates with the symptoms normally seen, pathogenesis remains a mystery.

In endemic situations, more than 60% of the infected patients are asymptomatic; however, in acute outbreaks, most of those infected will show symptoms. When these symptoms occur, they begin one to three weeks

Fast Fact Although the risk of human infection with *Giardia* from cats and dogs is small (domestic animals get infected with a different genetic type of *Giardia*), avoiding contact with infected animals' feces and handwashing after cleaning their litter pans can reduce the chances of transmission.

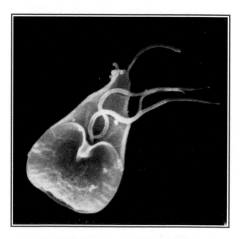

Figure 22.17 Micrograph of one species of *Giardia*. This image shows the trophozoite stage, with the characteristic four pairs of flagella.

after exposure to the parasite and include diarrhea that is sudden in onset and explosive. The stool is foul-smelling and greasy in appearance, but there is no blood or mucus. Abdominal cramping is also common, and large amounts of gas are produced, resulting in abdominal distension and abundant flatulence. Nausea and low-grade fever are also possible.

Acute illness usually resolves itself in one to four weeks but may persist in children and lead to significant weight loss and malnutrition. In many adults, the acute phase may give way to a subacute or chronic stage consisting of heartburn, weight loss, and flatulence that can last for weeks to months.

Treatment of Giardiasis

There are four drugs currently available for this infection in the United States: quinacrine hydrochloride, metronidazole, furazolidone, and paromomycin. These drugs require five to seven days of therapy and should not be used during pregnancy.

Cryptosporidiosis (*Cryptosporidium*)

Cryptosporidium is a small protozoan parasite that infects the intestinal tract of both humans and other animals, causing the infection called **cryptosporidiosis**. It is an obligate intracellular parasite that alternates between sexual and asexual reproduction cycles, both of which are completed in the gastrointestinal tract. This parasite was not identified as a human pathogen until 1976, when it was shown that the species *Cryptosporidium parvum* could infect humans. The organism is small and spherical and arranges itself along the microvilli of the intestinal tract (**Figure 22.18**). It is interesting to note that the parasites remain external to the cell cytoplasm but eventually become covered over by the membrane of the host cell, which is why they are referred to as intracellular.

Infectious *C. parvum* oocysts are excreted in the stool of infected persons, and the oocysts are fully mature and infectious on passage into the feces. On ingestion of these oocysts by a new host, sporozoites are released from the oocyst and attach to the microvilli of the epithelial cells in the small intestine. Here they transform into trophozoites, which divide by schizogony (fission) to form schizonts containing eight daughter cells. When released from the schizont, each daughter cell attaches to another epithelial cell, where the schizogony cycle is repeated.

After several rounds of schizogony, the trophozoites develop into male and female forms, and the sexual reproductive cycle takes place. The resulting zygote develops into an oocyst that is shed into the lumen of the small intestine. The zygote has a thick protective wall that ensures safe passage both in the fecal stream and in the external environment. Approximately 20% of the zygotes formed will not develop protective walls, and the resulting oocysts will rupture, releasing infectious sporozoites back into the lumen.

Various species of *Cryptosporidium* can infect most vertebrates, with infection usually occurring in the young or immature. Domestic animals are reservoirs for this parasite, and most human infections of *C. parvum* result from person-to-person transmission. In developed countries, 1–4% of children harbor oocysts, but this number doubles in underdeveloped nations, with the highest rate of infection being seen in children, families of infected children, medical workers, and travelers to countries where the disease is endemic. The principal route of transmission is fecal–oral, but there can also be transmission by contamination of food and water.

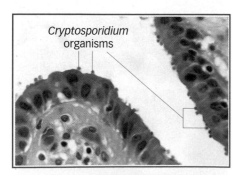

Figure 22.18 A histological section of the gall bladder of an AIDS patient, showing numerous *Cryptosporidium* organisms along the luminal surfaces of the epithelial cells.

Pathogenesis of Cryptosporidiosis

The jejunum is the most heavily involved area for these parasites, but in some severe cases they can be seen throughout the entire digestive system. Bowel changes are minimal for this infection, with mild to moderate destruction of microvilli and some mononuclear cell infiltration of the lamina propria. The pathology of the resulting diarrhea is not understood but may be similar to that seen in cholera. Antibodies against *C. parvum* are protective against infection, and it has been shown that CD4 T cells and interferon can also have a role in clearance of the parasites.

In immunocompetent patients, there is onset of profuse, explosive watery diarrhea one to two weeks after exposure to the parasite. The illness lasts for about five days and then clears rapidly, although there may be continued mild weight loss for up to a month. In contrast, for those patients who are immunodeficient, the diarrhea can be more severe with a liquid loss of 25 liters per day. Unless the deficiency is reversed, the disease can last for life. Half of the cryptosporidiosis patients who have AIDS will die within six months.

Treatment of Cryptosporidiosis

In the immunocompetent patient, the disease is self-limiting, and therefore no treatment is required, although it may be necessary to rehydrate small children. In immunocompromised patients, the diarrhea is so pronounced that rehydration is essential. There is no uniformly effective therapy for this infection.

Whipworm (*Trichuris trichiura*)

The adult whipworm is 30–50 mm in length, with the first two-thirds thin and the last one-third bulbous (**Figure 22.19**). The female worm can produce 3000–10,000 eggs per day, and the eggs have a brown shell with translucent knobs at each end. Whipworms infect about 1 billion people worldwide, and infection is usually concentrated where there is indiscriminate defecation and a warm, humid environment. The infection rate in tropical environments can be 80%. This infection affects more than 2 million people in rural areas of the southeastern United States. The adult worm can live for eight years.

Pathogenesis of Whipworm Infection

Adult worms live with their anterior end attached to the host colonic mucosa. While attached in the cecum, the female releases eggs into the lumen of the intestine. These eggs are passed out of the body with the feces and deposited in the soil. At this stage, the eggs are immature and not infective, and they must incubate in the soil for 10 days before they embryonate and become infectious. The infectious eggs are picked up on hands, passed into the mouth and swallowed. They move into the duodenum, where the larvae mature for about one month before migrating to the cecum.

Attachment of adult worms to the colonic mucosa produces hemorrhaging and localized ulcerations that can be used as portals of entry for opportunistic bacterial infections. Consequently, concomitant bacteremias can often be seen with this parasitic infection. Although light infections are usually asymptomatic, higher worm loads result in damage to the intestinal mucosa, accompanied by nausea, diarrhea, and abdominal pain. Some children can have worm loads as high as 800, and in these cases there is significant mucosal damage, blood loss, and anemia. When

Figure 22.19 The whipworm *Trichuris trichiura*, which infects more than a billion people worldwide. Notice how the first two-thirds of the worm are quite thin, whereas the last part is thick and bulbous. This example is a female, which is able to shed between 3000 and 10,000 eggs per day.

these children strain to defecate, the sheer force of the fecal stream on the worm bodies can cause prolapse of the colonic or rectal mucosa out through the anus.

Treatment of Whipworm Infection

There is no need for treatment of asymptomatic infections; however, for more severe disease, mebendazole is very effective and is the treatment of choice. The cure rate is 60–70%, with 90% of the worms being expelled with this treatment. However, even though the patient becomes asymptomatic, worms may still be present. Prevention of the infection involves good sanitation.

Hookworms

There are two species of this parasite that infect humans: *Necator americanus* and *Ancylostoma duodenale*. Adult worms are pinkish and about 10 mm long. Because the direction in which the anterior end curves is often opposite to the direction in which the rest of the body curves, the organism takes on the appearance of a hook. The hookworm is found worldwide, with transmission through the deposition of eggs into shady, well-drained soil. Infection with these worms can be significant in densely populated communities. In the early twentieth century, about 40% of people in the southern United States were infected as a result of poor sanitation and fecal contamination of soil. Hookworm infection was especially common in impoverished communities. Eradication and education programs, including the installation of outhouses, were tremendously successful in reducing the infection rates.

The hookworm *Necator americanus* is found in southern Asia, Africa, and the Americas; *Ancylostoma duodenale* is found in the Mediterranean, the Middle East, northern India, China, and Japan.

Fast Fact It has been estimated that together *Necator americanus* and *Ancylostoma duodenale* extract 7 million liters of blood each day from more than 700 million infected people.

Pathogenesis of Hookworm Infection

Ancylostoma duodenale uses four sharp toothlike structures (**Figure 22.20**), whereas *Necator americanus* has dorsal and ventral cutting plates for attachment to the mucosa of the small intestine. Fertilized females can release 10,000–20,000 eggs per day, which are passed in the feces. These eggs are passed in a four-cell or eight-cell stage of development and on reaching the soil hatch within 48 hours, releasing larvae that feed on soil bacteria. They double in size and molt to become infectious larvae that can survive four to six weeks in the soil.

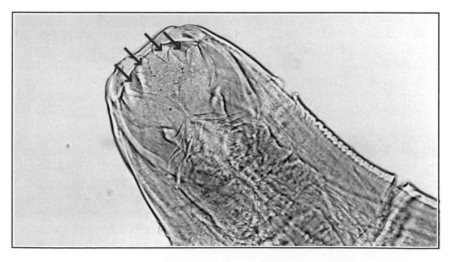

Figure 22.20 This unstained micrograph of the hookworm *Ancylostoma duodenale* shows the four sharp toothlike structures. These structures aid the worm in attaching to the mucous membranes of the host's small intestine.

On contact with human skin, the larvae penetrate the epidermis and move into the blood and lymph. Using the blood system, they move to the heart and eventually the lungs, where they rupture in the alveolar spaces and are coughed up and swallowed. They then move into the small intestine and mature into adult worms. As the worms migrate in the intestine, they leave behind bleeding points at the former attachment site, and because the worms can live up to 14 years, the blood loss can be enormous.

In the overwhelming majority of patients, the worm load is small and the infection is asymptomatic. However, in symptomatic cases there can be rash and swelling where the worm makes initial entry (usually between the toes), and this can persist for several days. Pulmonary problems are infrequent and less severe, but in the gut, the worms can cause abdominal pain and abnormal peristalsis. The major clinical manifestation of this infection is blood loss and concomitant anemia. The severity of the anemia depends on worm load and on the patient's dietary iron intake. Usually the development of severe anemia can take months or even years, but in children there can be earlier problems, including heart failure and retardation of mental and physical development.

Treatment of Hookworm Infection

The anemia that results from this infection is the primary consideration and must be corrected. The three most widely used anti-helminthic drugs for this infection are pyrantel pamoate, mebendazole, and albendazole, which are all highly effective. Prevention of this infection is tied to proper sanitation.

> **Keep in Mind**

- There are several protozoa and helminths that infect the human digestive tract.
- Parasitic protozoan pathogens of the digestive tract include *Giardia* and *Cryptosporidia*.
- Infection of the digestive tract by helminths can be caused by whipworms or hookworms.

SUMMARY

- The digestive system is a major portal of entry for pathogens.
- The primary symptom of an infection of the digestive system is diarrhea, and there can also be fever, vomiting, and abdominal pain.
- Many gastrointestinal infections are endemic and can cause epidemic and pandemic outbreaks.
- Treatment for most infections of the digestive system includes supportive care and the replacement of fluids.
- Most infections in the oral cavity involve dental caries or diseases of the gums.
- Many of the bacteria that cause infections in the digestive system are Gram-negative rods.
- Enterobacteria are classified by their antigens: O antigens are part of the lipopolysaccharides in the outer membrane, H antigens are found on the flagella, and K antigens are part of the capsule.

- Many bacteria that produce gastrointestinal infections do so through the production of enterotoxins such as the Shiga toxin.

- *Campylobacter* infection is the most common cause of diarrhea in the developed world.

- *Helicobacter pylori* is the most common cause of gastric and duodenal ulcers.

- Rotavirus, norovirus, and enterovirus all infect the digestive system, causing diarrhea and vomiting.

- There are six different (unrelated) viruses that infect the liver and cause hepatitis.

- Protozoan parasites and helminths infect the human digestive system and can cause serious health problems such as amebic dysentery and giardiasis.

In this chapter, we have seen that the digestive system is a major portal of entry for infectious organisms. It is also a portal of exit through the fecal–oral route of infection. As with other systems, there are bacteria, viruses, and parasites that can produce an array of symptoms; the most prominent is diarrhea. It is important to recall continually that although this system seems to be overwhelmed by potential pathogens, there are significant defenses that are located in and around the digestive tract that protect us from many of these diseases. In the next chapter, we look at the third of the systems that provides a good portal of entry for pathogens, the genitourinary tract.

Ⓠ SELF EVALUATION AND CHAPTER CONFIDENCE

Multiple Choice

Answers are given in the back of the book and help can be found in the student resources at:
www.garlandscience.com/micro2

1. The most frequent cause of gastrointestinal infections in the developed world is
    A. *Salmonella*
    B. *Shigella*
    C. *E. coli*
    D. *Campylobacter*
    E. *Bacillus*

2. The causative agent for gastric ulcers is
    A. *Clostridium difficile*
    B. *Helicobacter pylori*
    C. *E. coli*
    D. *Campylobacter*
    E. None of the above

3. The primary cause of nosocomial gastrointestinal infections is
    A. *Clostridium difficile*
    B. *Helicobacter pylori*
    C. *E. coli*
    D. *Campylobacter jejuni*
    E. MRSA

4. Dental plaque is an example of
    A. A pellicle
    B. A biofilm
    C. Gingivitis
    D. Trench mouth
    E. All of the above

5. Dental caries are initiated by
    A. *E. coli*
    B. Hyaluronidase
    C. *Streptococcus mutans*
    D. *Actinomyces*

6. Which of the following are common features of the Enterobacteriaceae?
    A. Gram-negative
    B. Rod shaped
    C. Colonize the intestines
    D. Do not form spores
    E. All the above

7. Which of the following is produced by *E. coli*?
   A. Pore-forming toxin
   B. Shiga toxin
   C. Heat-labile toxin
   D. Heat-stable toxin
   E. All of the above

8. Transmission of traveler's diarrhea occurs through
   A. Sexual contact with infected individuals
   B. The respiratory route
   C. The skin
   D. Contaminated food or water
   E. None of the above

9. Bacillary dysentery is caused by
   A. *Salmonella*
   B. *Shigella*
   C. *E. coli*
   D. *Campylobacter*
   E. All of the above

10. Serotypes of *Salmonella* are identified primarily by
    A. H antigens
    B. O antigens
    C. Capsules
    D. **A** and **B**
    E. All of the above

11. The most serious complication of typhoid fever is
    A. High fever
    B. Diarrhea
    C. Perforation of the colon wall
    D. Dementia
    E. None of the above

12. Which of the below is not a feature of cholera?
    A. Colonization of the entire intestinal tract
    B. Pathogenesis caused by a toxin
    C. Long incubation period
    D. Retention of fluid electrolytes
    E. Production of rice stool

13. Which of the following are features of viral infections of the digestive system?
    A. Vomiting and diarrhea
    B. Enteric fever
    C. Long incubation time
    D. Transmission by contaminated food
    E. All of the above

14. Chronic carriers are the main reservoirs for
    A. Hepatitis A
    B. Hepatitis C
    C. Hepatitis E
    D. Hepatitis B
    E. None of the above

15. The intestinal parasite *Giardia* is found in
    A. Fish
    B. Mammals
    C. Reptiles
    D. Birds
    E. All of the above

16. Rice stools are characteristic of
    A. Amebic dysentery
    B. Bacillary dysentery
    C. Typhoid fever
    D. Cholera

17. Most of the normal microbial flora of the digestive system are found in the
    A. Stomach
    B. Mouth
    C. Small intestines
    D. Large intestine

18. Poultry products are a likely source of infection by
    A. *Salmonella*
    B. *Shigella*
    C. *Vibrio*
    D. Streptococci

19. Most gastrointestinal infections should be treated with
    A. Water and electrolytes
    B. Quinacrine
    C. Penicillin
    D. None of the above

20–23. Answer **A** if the reason given is correct, answer **B** if the reason given is incorrect.

20. Antibiotic therapy of *Salmonella* infections is not appropriate because it decreases the duration and frequency of the carrier state.

21. Hepatitis D is a small single-stranded RNA virus that is currently referred to as a satellite virus because it requires the presence of HIV antigens to complete its replication cycle and become infective.

22. Hookworms have the shape of a hook because the anterior end of the body curves in the opposite direction to the rest of the body.

23. As hookworms migrate in the intestine, they leave behind bleeding points that can lead to significant blood loss.

Ⓠ **DEPTH OF UNDERSTANDING**

Questions listed here require you to bring together the concepts you have learned in this chapter into a discussion format. This helps you to increase your depth of understanding of the material you have learned. Help can be found in the student resources at: www.garlandscience.com/micro2

1. Describe how infectious diseases of the digestive system can be a hazard after flooding.

2. Describe the pathogenic features of enterohemorrhagic *E. coli*.

3. Summarize the pathogenesis of whipworm infections.

Ⓠ **CLINICAL CORNER**

Help can be found in the student resources at: www.garlandscience.com/micro2

1. A patient comes into the clinic showing signs of dehydration and explains that she has had serious diarrhea for several days. The patient reports that she and some friends went out to dinner a couple of days before and ordered a hamburger cooked rare. One of her friends ordered a steak also cooked rare and the other friend had a salad. She suspects that she may have a contracted food poisoning. She says her friend who ate salad is also experiencing diarrhea, but the friend who had steak is not ill.

   A. Would you consider this a consequence of food poisoning?
   B. What would you tell the patient about possible causes for her problem?
   C. What tests would you recommend be done to confirm your suspicions?

2. Your patient is dehydrated and has metabolic acidosis. She has had severe diarrhea with intermittent bouts of cramping and a feeling of fullness. Her stools are voluminous but odorless and look to be composed mainly of fluids. These symptoms began about 24 hours after she attended a party at the yacht club where she consumed a number of raw oysters that were served as appetizers. She delayed coming to the doctor because she thought that she had a simple case of food poisoning.

   A. Does she have a simple case of food poisoning?
   B. How does the case history help in your formulating possible explanations for her problems?

Infections of the Genitourinary System

Chapter 23

Why Is This Important?

The genitourinary system is the third of the body's systems that are open to the outside world, and many pathogens use this system as a portal of entry into the body.

Aaron started having flu-like symptoms, tiredness, and muscle soreness. He had a small open sore in his inner, upper thigh region that he first thought was an ingrown hair. But after checking online he found that his symptoms matched those of genital herpes, a condition caused by a sexually transmitted virus. He found that, although genital herpes commonly causes blisters on the penis, sores can also occur on the skin around the groin, buttocks, and anus. He made an appointment with his doctor and had to wait an anxious nine days to get the test results. When he finally got a call from his doctor, the results were positive. He was devastated by the news and every time he thought about it, it made him want to cry.

He has not had any new sexual partners, but if and when he does, he realizes he will need to be honest about his infection.

He has started taking an antiviral medication called valacyclovir that decreases the frequency of outbreaks and reduces the possibility of passing it on to partners. Even though it will be with him forever, he realizes it's not the end of the world. He has learned that approximately one in five men in America also live with genital herpes. Now he views life in a different way and realizes he is not alone.

OVERVIEW

In this chapter, we look at diseases that occur in the urinary and reproductive systems, collectively called the genitourinary system. This system, which is open to the outside environment, is the third of the three major portals by which pathogens enter the body. As we will see, some urinary tract infections (UTIs) begin in the urethra and can travel up to the bladder and in severe situations even to the kidneys. The outcome of these infections can be severe and potentially life-threatening. We will also look at a variety of reproductive system infections, many of which are designated sexually transmitted infections (STIs). Infections of the genitourinary system can be caused by bacteria, viruses, yeast, and protozoan organisms. Our discussions will be divided into the urinary and reproductive systems, keeping in mind that the anatomy of these systems has a role in the kinds of infection that occur. We discussed protozoan genitourinary infection in Chapter 14 and won't repeat it here.

We will divide our discussions into the following topics:

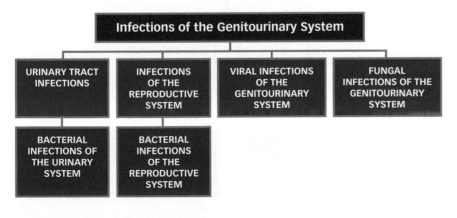

"... through midnight streets I hear how the youthful harlot's curse blasts the new-born infant's tear ..."

William Blake (1757–1827)

The above excerpt, from Blake's poem *London*, is a reference to **congenital syphilis** in a baby born at a time when venereal diseases were widespread and devastating. As we will see in this chapter, **syphilis** is a sexually transmitted disease of the genitourinary system that has plagued people worldwide for centuries. To this day, sexually transmitted diseases remain a hidden epidemic that carries tremendous health, economic, and psychological burdens to millions of people.

URINARY TRACT INFECTIONS

Because urine is essentially sterile, the presence of pathogens or inflammatory cells in the urine is an indication of a urinary tract infection. UTIs are more common in women than in men, because of several factors, including anatomical differences (the urethra is shorter in women, and the urethral opening is closer to the anus, allowing the transfer of fecal bacteria) and contraceptive use (diaphragms and spermicides in the vagina increase susceptibility to UTIs). The pathogen involved is usually a bacterium or yeast.

In hospitals, UTI is a serious problem that is usually seen associated with indwelling catheters. These catheters compromise the structure and physical barriers of the urethra and the bladder, with the result that bacteria or yeast can ascend either the outside or the lumen of the catheter to reach the bladder. It should be noted that after removal of the catheter, a 14-day course of antibiotics should be given to halt the infection.

Looking at the anatomy of the urinary tract, we see that urine flow is ideally in one direction: from kidney to bladder to urethra and then out of the body (**Figure 23.1a**). However, there can be a reflux action, allowing pathogens to enter the urinary tract from the outside and spread throughout the urinary tract (**Figure 23.1b**). A UTI is called **urethritis**

Fast Fact A single catheterization has a 1% risk of causing a UTI, and 10% of people with catheters get a UTI.

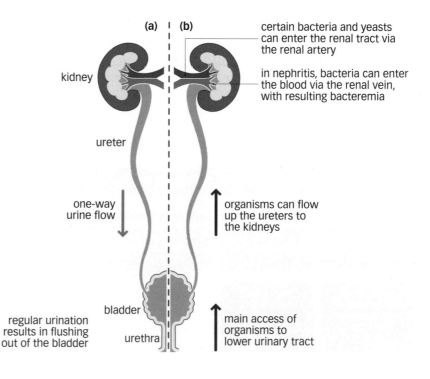

Figure 23.1 Generic version of the human urinary tract. Panel a: Normal structure and physiology permit descending flow and a sterile environment for urine. Panel b: Pathogens can enter the urinary tract from the external environment and produce infection that may spread all the way up the urinary tract. In an alternative pathway, pathogens in the blood enter the urinary system from the renal arteries, as indicated at the top of the illustration.

when it occurs in the urethra, **cystitis** when it occurs in the bladder, and **nephritis** when it occurs in the kidneys. In males, infection of the urinary tract can lead to **prostatitis**, infection of the prostate.

Keep in Mind

- The presence of pathogens or inflammatory cells in the urine is an indication of a urinary tract infection.
- UTIs are very serious in hospital settings and are usually related to indwelling catheters.
- UTIs are usually caused by bacteria.
- Infection in the urethra is called urethritis; in the bladder, cystitis; in the prostate, prostatitis; and in the kidney, nephritis.

BACTERIAL INFECTIONS OF THE URINARY SYSTEM

It is likely that a few bacteria routinely enter the urinary tract, either from the external environment or from blood passing through the renal artery, but these are normally flushed out during urination. The prevalence of bacterial UTIs varies with age (**Figure 23.2**). In the first three months of life, bacterial UTIs are more common in males, but by preschool age they are more common in females as a result of refluxing of urine from the bladder into the ureters. In older males and females, anatomical changes associated with aging predispose to chronic bacteria in the urine (**bacteriuria**), but this condition is often asymptomatic. In males, enlargement of the prostate can increase the incidence of these infections, and in the elderly of both sexes, gynecological or prostatic surgery, incontinence, or chronic catheterization increases the rates of bacterial UTIs to 30–40%.

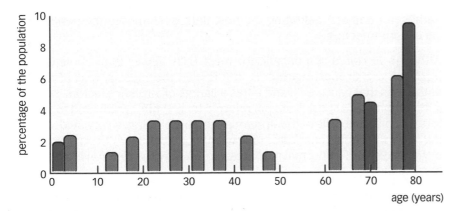

Figure 23.2 The prevalence of urinary tract infections in females (orange) and males (blue), according to age.

The problem of nosocomial bacterial UTIs is complicated by the increasing number of antibiotic-resistant bacteria found in hospitals, and in particular resistant bacteria such as *Pseudomonas*, *Enterococcus*, and MRSA (methicillin-resistant *Staphylococcus aureus*). *Escherichia coli* accounts for more than 90% of the 7 million or more cases of cystitis and the 250,000 cases of nephritis estimated to occur in otherwise healthy people in the United States each year. A list of the most common causes of bacterial UTIs is shown in **Figure 23.3**.

Pathogenesis of Bacterial UTIs

Infection occurs when bacteria are able to get into the urine and remain there. Because all portions of the urinary tract are connected to one

Figure 23.3 Some of the bacteria that commonly cause urinary tract infections and the antibiotics that can be used to treat them.

	organism	treatment
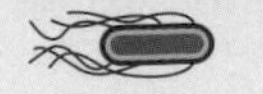	*Escherichia coli*	trimethoprim, cephalexin, gentamicin
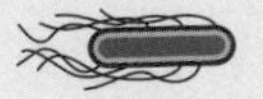	*Proteus* spp.	trimethoprim, cephalexin, gentamicin
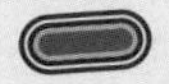	*Klebsiella* spp.	trimethoprim, cephalexin, gentamicin
	*Pseudomonas aeruginosa*	ciprofloxacin, gentamicin
	*Enterococcus* spp.	amoxicillin, vancomycin
	*Staphylococcus aureus*	trimethoprim, cephalexin, gentamicin
	coagulase-negative staphylococci	trimethoprim, cephalexin, gentamicin

another by a liquid medium, infection is spread easily. Much of the pathogenesis seen in these infections is the result of anatomy. Males are protected from bacterial infections to some degree because the male urethra is longer than in the female, and the shorter female urethra means that bacteria invading from the external environment have a shorter route to the bladder. In addition, the female urethra ends in the vaginal area, which is colonized by a variety of bacteria as part of the normal flora, many of which can initiate a UTI. Furthermore, sexual intercourse and manipulation of the female anatomy are likely to increase the chances of bacteria entering the bladder. Any process that interferes with the complete emptying of the bladder can allow some bacteria to remain and increase in numbers. As we have seen throughout our discussions, both the ability to remain in one location (staying in) and increases in pathogen numbers (defeating the host defenses) are requirements for a successful infection.

Although *E. coli* is responsible for most UTIs, fewer than 10 serotypes of *E. coli* are **uropathogenic**. *E. coli* is also the most potent of all the pathogens that cause UTI, and it has a variety of virulence factors, such as α hemolysins and specialized pili referred to as **P pili**, which bind to receptors on epithelial cells of the urinary tract. This binding is very avid, allowing successful adherence and subsequent colonization of the urinary tract by the bacteria. The normal flushing associated with urination is unable to remove them.

Infections cause an inflammatory response in which neutrophils migrate to the infection site and, along with bacterial toxins and enzymes, irritate the lining of the bladder and urethra. This causes the symptoms of increased frequency of urination, urgency, and dysuria (painful urination), which are commonly seen in these infections. In males, the infection can reach the prostate from the urethra, lymphatics, and blood to cause prostatitis. The inflammation caused by prostatitis can lead to compression of the urethra lumen, a compression that can obstruct or retard the flow of urine. Prostatitis can be either acute or chronic.

Symptoms and Diagnosis of Bacterial UTIs

The clinical sequelae associated with UTIs can vary, and more than 50% of these infections do not produce recognizable illness. The four types of UTI have different symptoms.

Urethritis and Cystitis

With bacterial urethritis and cystitis, the symptoms are dysuria, frequency, and urgency. There can also be low back pain and abdominal pain or tenderness over the bladder area. In addition, the urine may be cloudy (**Figure 23.4**). Cystitis can be distinguished from urethritis by the fact that the former has a more acute onset, more severe symptoms, and the presence of bacteria and blood in the urine.

Nephritis

Bacterial kidney infection usually presents with pain in the flanks of the body and a fever above 38.3°C. It may be preceded or accompanied by symptoms of cystitis and in more severe cases can present with diarrhea, vomiting, and tachycardia. Nephritis can occasionally result in septic shock. Usually, the symptoms resolve themselves and there is no damage to kidney function. In 20–50% of pregnant women, however, the infection causes premature birth. Some people develop chronic nephritis without ever showing any symptoms of a UTI.

Prostatitis

Prostate bacterial infection usually presents with lower back pain and pain in the perirectal area and testicles. In acute prostatitis, there can be high fever, chills, and symptoms of bacterial cystitis. Inflammatory swelling during bacterial prostatitis can lead to obstruction of the urethra and the retention of urine. In addition, there can be abscess formation, epididymitis, and seminal vesiculitis. Acute prostatitis is usually seen in young men, whereas the chronic form is associated with the elderly and usually with catheterization.

Diagnosis

The diagnosis of a bacterial UTI is based on the examination of urine for evidence of bacteria or accompanying inflammation. This requires the collection of a clean voided midstream urine sample. For about 90% of patients, UTIs are identified as **pyuria** (more than 10 white blood cells per cubic millimeter of urine) or, more specifically, by the presence of white blood cell casts in the urine. The most positive way to confirm a UTI is to Gram-stain a urine sample. The presence of at least one bacterium per microscopic oil-immersion field is an indication of infection.

Treatment of Urinary System Bacterial Infections

The antibiotics used to treat some of the most common organisms involved in bacterial UTIs are shown in Figure 23.3. The choice of antibiotic is best guided by results of cultures and antimicrobial susceptibility tests, although trimethoprim (alone or in combination with sulfamethoxazole or a fluoroquinolone) is the most commonly used antibiotic treatment for UTIs. The duration of treatment depends on the severity of the bacterial infection and on the risk to the patient. The success of the treatment is determined by a culture of urine two weeks after therapy.

It is important to keep in mind that antibiotic resistance is becoming an increasing problem, especially in hospital and other institutional settings, where coliform bacteria account for 85% of the bacteria isolated from urine specimens. At least 50% of these isolated bacteria are resistant to amoxicillin, and 20% are resistant to trimethoprim. UTIs caused by *E. coli* that produce the enzyme extended-spectrum β-lactamase (EBSL) are increasing and are a serious public health problem because these infections are resistant to penicillins and cephalosporins.

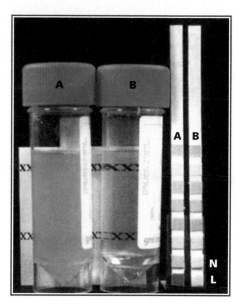

Figure 23.4 Urine specimens and diagnostic dipsticks. Specimen A is cloudy, and both the leukocyte (L, purple) and nitrate (N, deep pink) results shown on the dipstick are positive, indicating a bacterial infection in the genitourinary tract. Specimen B is clear, and the dipstick for it is negative (the colors in the L and N regions are unchanged).

Keep in Mind

- Bacterial UTIs are seen more in women than in men because of the difference in the lengths of the urethra.
- *Escherichia coli* is responsible for the majority of UTIs and is the most potent of all pathogens that cause UTIs.
- Bacterial UTIs are routinely treated with antibiotics, but resistance is now a serious issue.

INFECTIONS OF THE REPRODUCTIVE SYSTEM

Most infections of the reproductive system are sexually transmitted. Even though many STIs have been well studied and successful treatments are available for them, in many cases the infected individuals will not seek medical help because they are asymptomatic.

Many STIs have a different impact on women than men. Anatomically, the vaginal lining is thinner, more delicate, and moister than the skin of a penis, so it is easier for microbes to penetrate and grow in the vagina. Furthermore, women may be less likely than men to notice the signs and symptoms of an STI. For example, herpes and syphilis lesions are not as visible inside the vagina as they are on the penis. Discharge, discomfort, or pain may be confused with menstrual discharge or pain in women, but are unusual signs and symptoms in men and are more likely to be alarming. As we will see, there are more serious health complications from STIs in women than in men.

STIs can cause urethritis, **cervicitis** (inflammation of the cervix), prostatitis, and pharyngitis. Although the pharyngitis is usually asymptomatic, the infected individual is a carrier and can readily infect sexual partners. In addition, **pelvic inflammatory disease** (PID) can result from gonococcal or chlamydial infection and can cause infertility and ectopic pregnancy. Infections can also infect the fetus and newborn, sometimes with devastating consequences. Expectant mothers are therefore routinely screened during prenatal visits for exposure to sexually transmitted disease.

Sexually transmitted infections were first identified in the 1600s. They affect all populations and social strata, and interest in them has been rekindled by the appearance of HIV and AIDS. The most common pathogens that cause STIs are listed in **Figure 23.5**. Depending on the pathogen, a sexually transmitted infection can be either localized or systemic. For localized infection, the most common manifestations are inflammatory (for example urethritis or cervicitis) and may not be noticed by the patient. In some cases, the infection can involve deeper tissues and structures, causing epididymitis and **salpingitis** (inflammation of the fallopian tubes). It is important to understand that these latter types of infection can become systemic.

Keep in Mind

- Most infections of the reproductive system are sexually transmitted infections (STIs).
- A wide variety of pathogens can cause STIs, and infection can become systemic.
- Women are more affected than men by STIs because of anatomical differences in the reproductive systems.
- STIs can be passed to fetuses and newborns.

pathogen		infections and disease
bacteria		
	Neisseria gonorrhoeae	urethritis, cervicitis, prostatitis, pharyngitis, pelvic inflammatory disease, disseminated gonococcal infection, salpingitis
	Chlamydia trachomatis	non-gonococcal urethritis, epididymitis, cervicitis, pelvic inflammatory disease, salpingitis, and lymphogranuloma venereum
	Treponema pallidum	primary, secondary, latent, tertiary, and congenital syphilis
	Haemophilus ducreyi	chancroid
	Ureaplasma urealyticum	non-gonococcal urethritis
	Gardnerella vaginitis	bacterial vaginitis
viruses		
	HIV	AIDS, AIDS-defining illnesses
	herpes simplex virus 2	primary and recurrent genital herpes
	human papillomavirus	genital warts and cervical carcinoma
	hepatitis B and C virus	chronic liver disease, cirrhosis, carcinoma
fungi		
	Candida albicans	vaginal and penile candidiasis, cystitis, nephritis
protozoan parasites		
	Trichomonas vaginalis	trichomonal vaginitis

Figure 23.5 The major sexually transmitted pathogens and the infections they cause.

BACTERIAL INFECTIONS OF THE REPRODUCTIVE SYSTEM

The major bacterial infections seen in the reproductive system are sexually transmitted. These infections are caused by a wide range of organisms, including group B streptococci, *Neisseria gonorrhoeae*, *Treponema pallidum* (the bacterium that causes syphilis), and *Chlamydia trachomatis* (the bacterium that causes non-gonococcal urethritis).

Figure 23.6a shows the bacterial species found in the microflora present in the vagina of the female genital tract. The anatomy of the female genital tract (**Figure 23.6b**) means that these vaginal organisms can easily move up to the uterus and fallopian tubes. In addition, the close proximity of the anus to the vagina makes it easy for some of the organisms normally found in the digestive tract to move into and infect the female

Figure 23.6 The microbial flora of the female reproductive tract. Panel a: Bacteria that make up the normal vaginal flora. Panel b: The female reproductive tract.

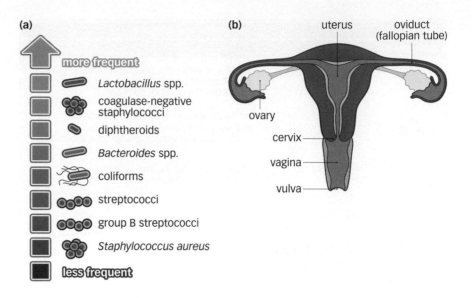

reproductive tract. Furthermore, the vaginal epithelium in prepubescent females is not yet keratinized and can easily support the growth of gonococcal organisms. Changes in the vaginal flora that occur as part of the menstrual cycle can be involved in potential infection.

Common Clinical Conditions Associated With STIs

We will now look briefly at some of the common clinical syndromes seen with bacterial STIs and then move on to a more in-depth look at three bacteria responsible for reproductive system infections: *Treponema*, *Neisseria*, and *Chlamydia*.

Genital ulcers are lesions (either single or multiple) that arise on the genitalia. They begin as either a papule (solid skin bump) or a pustule (skin bump containing pus) and evolve into ulcers. The features of these ulcers are described in Table 23.1. Some of these differences can be significant. For example, the syphilis ulcer is single and painless, whereas genital herpes ulcers are often multiple and very painful.

The condition known clinically as **sexually transmitted urethritis** usually presents as dysuria or urethral discharge, or both. The discharge may be prominent or may have to be milked from the urethra. The major causes of sexually transmitted urethritis are *N. gonorrhoeae* and *C. trachomatis*, and in many cases the infection will involve both pathogens. Diagnosis of

Infection	Type of ulcer	Involvement of inguinal lymph nodes
Genital herpes	Multiple grouped vesicles and painful coalesced ulcers	Tender and nonsuppurative
Syphilis	Non-tender indurated ulcer that occurs singly	Rubbery consistency
Chancroid	Tender, shallow, painful ulcer	Suppurative
Lymphogranuloma venereum	Painless, small ulcer or papule usually healed by the time of presentation	Discrete progressing to suppurative

Table 23.1 Causes and types of genital ulcers.

gonococcal urethritis usually requires culture of the organism, but it can also be done by direct microscopic examination of Gram-stained samples as well as by DNA analysis. However, for detection of *C. trachomatis*, DNA amplification analysis is required. Successful treatment of sexually transmitted urethritis depends on the agent causing the infection and also on whether the infection has spread beyond the local area.

Epididymitis, which is unilateral swelling of the epididymis, is commonly seen in sexually active men and is usually quite painful, with fever and swelling of the testicles. Two bacteria are usually implicated in this infection, *N. gonorrhoeae* and *C. trachomatis*. In men over 35 years of age and in homosexual men, Enterobacteriaceae and coagulase-negative staphylococci may also cause this condition.

The microbial etiology of cervicitis can vary, but it is usually caused by *N. gonorrhoeae* and *C. trachomatis* infecting the stratified squamous epithelium of the cervix. This infection usually involves a **mucopurulent vaginal discharge** as the cervix becomes inflamed, and phagocytic polymorphonuclear leukocytes will be found in the discharge.

Bacterial vaginitis is the most common form of **vaginitis** and is associated with overgrowth of multiple members of the vaginal anaerobic flora and the Gram-negative rod *Gardnerella vaginalis*. In bacterial vaginitis, there can be a yellowish discharge that is homogeneous and stays adhered to the vaginal wall. Vaginal epithelial cells called clue cells (**Figure 23.7**) found in the discharge are covered with bacteria. Vaginal discharge can occur alone or in connection with salpingitis, **endometritis**, or cervicitis, and the clinical findings of this condition vary with the etiological agent.

The clinical manifestations of PID vary, but they generally include abdominal pain. Approximately 50% of PID is caused by *N. gonorrhoeae* infection, but there can be non-gonococcal infections caused by a combination of bacteria including *C. trachomatis*, *Bacteroides*, and anaerobic streptococci. The infections caused by these non-gonococcal PID agents are more complex than the gonococcal variety, but they are usually milder than that caused by *Neisseria*.

The condition **lymphadenitis**—inflammation of lymph nodes—is seen in several sexually transmitted infections, especially in herpes infections and **lymphogranuloma venereum**, which is caused by *C. trachomatis*. Lymphadenitis usually begins as a small genital ulcer that is frequently unnoticed. Usually the first evidence is a tender swollen lymph node in the groin (**Figure 23.8**).

Now let us look at three of the most common bacterial sexually transmitted infections: syphilis, gonorrhea, and non-gonococcal urethritis.

Syphilis (*Treponema pallidum*)

Syphilis may be the earliest recorded sexually transmitted infection, first described in the 1600s. It is caused by the bacterium *T. pallidum* (**Figure 23.9**), a slim spirochete with regular spirals that resemble a corkscrew. This pathogen shows a characteristic slow rotating motility with occasional 90° flexion (trembling a person bowing from the waist). *Treponema* cannot be grown on bacterial media but can be grown in mammalian cell cultures at a low oxygen tension. It is extremely susceptible to any changes in its environment and to any deviation in physical conditions. *Treponema* dies rapidly if dehydrated or heated and is very sensitive to detergents and disinfectants. The transmission method for this pathogen is restricted to direct contact.

Figure 23.7 A clue cell found in a vaginal smear. Notice the bacteria (the small dark dots) clinging to the surface of the clue cell.

Fast Fact The incidence of PID is slightly higher in women using intrauterine devices than in women not using this form of contraception. Condom use, however, significantly reduces the incidence of PID by 30–60%.

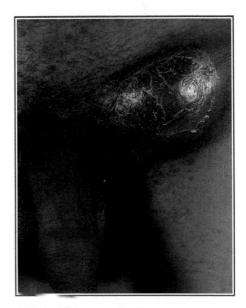

Figure 23.8 Early development of bilateral enlarged lymph nodes called buboes, which are seen in lymphogranuloma venereum caused by *Chlamydia trachomatis*.

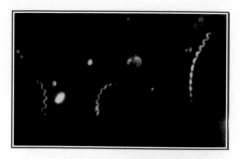

Figure 23.9 A darkfield preparation of the spirochete *Treponema pallidum*.

Treponema is exclusively a human pathogen, and infection is acquired by sexual contact with someone that has active **primary** or **secondary syphilis**. However, it has been shown that there is a possibility of transmission through the sharing of contaminated needles and also transplacentally. Since 1990, the number of syphilis cases in the United States has steadily declined and is now below 40,000 per year. Of these 40,000 patients, 20% have primary or secondary disease, and the rest are either latent or in the tertiary stage of the disease. Syphilis is still a major health problem worldwide, with more than 12 million cases reported each year.

Pathogenesis of Syphilis

Because there is no animal model for this disease, the pathogenesis has been extrapolated from observations of patients with syphilis. The spirochetes reach the subepithelial tissues either by moving through breaks in the skin or by passing between the epithelial cells of the mucous membranes of the reproductive tract. As the organism multiplies slowly, there is little or no inflammatory response during the initial stage of the infection. As a lesion develops, small arterioles begin to swell, and the endothelial cells of these vessels proliferate. This increase in the number of endothelial cells reduces the blood flow through the arterioles, leading to the necrotic ulceration seen at the primary infection site. This is followed by an influx of granulocytes, lymphocytes, monocytes, and plasma cells that surround the affected blood vessels. The primary lesion heals spontaneously, but by this time the bacteria have spread to other locations by way of the blood and lymph. For reasons that are not yet understood, syphilis then goes silent for a period before the secondary stage develops. The disease also undergoes this silent period before the onset of the **tertiary** stage. *Treponema* binds to host proteins such as immunoglobulins and complement components, and it has been suggested that this binding is a type of camouflage to protect the bacterium. It does not produce any virulence factors during the progression of the disease, which has several clinically defined stages.

Clinical Stages of Syphilis

Primary syphilis is associated with the appearance of the primary syphilitic lesion, which is a papule that evolves into an ulcer. This ulcer is usually located on the external genitalia or on the cervix but can also be found in the oral cavity or anus, depending on the type of sexual interaction. The ulcer remains painless and is referred to as a **chancre** (**Figure 23.10**). The incubation time from contact to chancre is about three weeks, with lymphadenopathy arising within one week of the appearance of the initial lesion. The lymphadenopathy can persist for months, but the chancre will disappear in four to six weeks. Primary syphilis can also be associated with unilateral or bilateral enlargement of the lymph nodes of the groin.

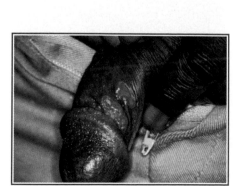

Figure 23.10 The chancre seen in primary syphilis on the genital area of a male.

Secondary syphilis is also known as disseminated syphilis and develops two to eight weeks after the chancre formed in the primary stage has disappeared. It is characterized by generalized lymphadenopathy and the appearance of a symmetric mucocutaneous maculopapular rash accompanied by fever, malaise, and lymphadenitis. The skin rash is seen on the face, trunk, and extremities, including the palms of the hands (**Figure 23.11**) and soles of the feet. The lesions associated with this rash are teeming with spirochetes and are extremely infectious. The lesions resolve in a few days, but in one-third of cases, resolution takes many weeks. In the two-thirds of cases where the lesions are gone in a few days, the next step is development of **latent syphilis**.

Latent syphilis can last for years. There are no clinical signs or symptoms, but the infection is continuing. During the first few years, latency can be interrupted by progressively less severe bouts of secondary syphilis. When the infection is in the latent stage, transmission of the disease to others is possible only during these relapse periods, but transmission from mother to fetus is possible throughout the latent period.

Tertiary syphilis is reached by about one-third of untreated patients. Signs and symptoms can appear as soon as five years after the initial infection but are usually not seen for 15–20 years. The clinical findings seen in tertiary syphilis depend on whether the infection spreads to the cardiovascular system (cardiovascular syphilis) or to the nervous system (neurosyphilis). In cardiovascular syphilis, the bacteria move to the vaso vasorum of the aorta, causing necrosis and destruction of elasticity. This can lead to the development of aneurisms and aortic valvular incompetency. During this process **gummas** appear. These are localized granulomatous lesions seen in the skin, bones, joints, and internal organs (**Figure 23.12**). Neurosyphilis is characterized by a mixture of meningovasculitis and degenerative changes in the parenchymal tissue, changes that can occur in any area of the body. The most common symptoms of neurosyphilis are chronic meningitis, fever, headache, and increased numbers of cells and protein in the cerebrospinal fluid. There can also be cortical degeneration of the brain, causing such mental changes as decreased memory, hallucinations, and psychoses. These findings of the central nervous system are classified as PARESIS (**p**ersonality, **a**ffect, **r**eflexes, **e**yes, **s**enses, **i**ntellect, and **s**peech).

Congenital syphilis is the form passed from mother to fetus. The fetus is susceptible to infection with *T. pallidum* only after the fourth month of gestation, but the infection can have devastating consequences. If the mother is treated for the disease before the fourth month of pregnancy, the fetus will show no signs of infection. However, if she is not treated she can lose the baby or the baby can be born with congenital syphilis, which is clinically similar to secondary syphilis. The condition can result in changes to the entire skeletal structure of the newborn, anemia, thrombocytopenia, and liver failure.

Treatment of Syphilis

Treponema pallidum has remained very sensitive to penicillin, which is the treatment of choice. If there are allergies to penicillin, treatment with tetracycline, azithromycin, or cephalosporin is successful. Safe sex is effective for prevention of this infection.

Gonorrhea (*Neisseria gonorrhoeae*)

The STI gonorrhea is caused by *N. gonorrhoeae*, which is a Gram-negative diplococcus that possesses numerous pili extending through the outer membrane of the cell. This outer membrane also contains phospholipids, lipopolysaccharides, and several distinct outer membrane proteins, including porins and adherence proteins.

Neisseria grows well on chocolate agar and requires supplementation with CO_2. The organism can also change the antigens associated with its surface from generation to generation, and this antigenic variability is also seen in the pili, which undergo multiple changes through recombinant gene exchanges. Because *Neisseria* is easily transformed, there can be extensive genetic changes in this pathogen. These genetic changes are important because they allow the pathogen to escape the host defensive

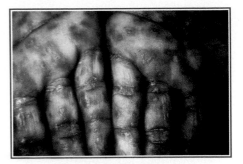

Figure 23.11 Palm lesions in secondary syphilis. As well as on the palms, these lesions can be seen on any other surface of the body.

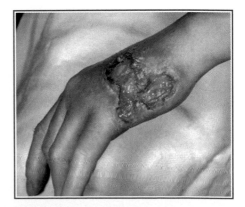

Figure 23.12 Gummas seen in tertiary syphilis.

response and also make it possible for the organism to bind to a variety of different receptors, thereby maximizing the potential for infection.

Although reported cases of gonorrhea have been decreasing in the United States for the past 20 years, the reported cases probably represent only 50% of the actual number of cases. Therefore, gonorrhea is still a major public health problem. The overall incidence in the United States is about 130 cases per 100,000 people, but the rate of infection among adolescents is very high and increasing by 10% per year, with the highest rates of infection being seen in women aged between 15 and 19 years and men between 20 and 24 years.

The major reservoir for *N. gonorrhoeae* is asymptomatic patients, with almost 50% of these individuals being infectious. In fact, the infection rate can be as high as 20–50% if there is sexual intercourse with an asymptomatic individual.

Fast Fact There is no truly effective means of controlling gonorrhea because it is difficult to detect asymptomatic cases (which are still infectious) and also because of increased resistance to antibiotic treatment.

Pathogenesis of Gonorrhea

The infectious process can be divided into three parts: attachment and invasion, survival, and dissemination.

Neisseria gonorrhoeae is not part of the normal microbial flora found in the body. This bacterium contains pili and adherence proteins that are used to attach to the urethral and vaginal epithelium, as well as to sperm and to parts of the fallopian tubes. After attachment, the pathogen invades host epithelial cells through a unique process in which the microvilli of the epithelial cells surround the organism and escort it into the cytoplasm. This process has been termed **parasite-directed endocytosis** because it seems to be initiated by the bacterium and not by the host cell. This entry involves nonphagocytic cells. Once inside the target cell, the bacterium transcytoses through the cytoplasm and exits through the basal membrane of the cell to enter the submucosa.

In the submucosa, the bacterium is immediately exposed to host defenses and must evade them in order to survive. This evasion occurs in a variety of ways. *Neisseria* can block the deposition of the C3 component of complement, effectively shutting down the complement pathway. Additionally, the antibody response is inhibited by *Neisseria* surface proteins that bind to host antibodies in such a way that the bacteriocidal activity of the antibodies is blocked. These blocked antibodies are readily found in patients who have repeated gonococcal infections. *Neisseria* can also survive phagocytosis by interfering with the attachment of phagocytic cells and also by producing excessive amounts of catalase, the enzyme that neutralizes the oxidative killing that is part of phagocytosis.

Neisseria organisms tend to stay localized in the genital structures, an immobility that facilitates transmission of the infection and also causes increased inflammation and localized tissue injury. Purulent exudates containing sticky clusters of *Neisseria* held together with bacterial proteins can be found, and these are probably the primary infectious units. However, infection may spread to deeper structures by progressive extension to adjacent mucosal glandular epithelial cells of the prostate, cervical glands, and fallopian tubes. In women, this spread can be facilitated by bacteria adhering to sperm. In addition, small numbers of infectious *Neisseria* can reach the blood and produce a systemic infection.

Clinical Manifestations of Gonorrhea

There are several clinical manifestations seen in gonorrhea, including genital gonorrhea, PID, and disseminated gonoccocal infection.

Genital gonorrhea in men occurs primarily in the urethra. Symptoms appear two to seven days after infection and consist of purulent urethral discharge (**Figure 23.13**) and dysuria. Infection can spread to the epididymis and the prostate. In women, symptoms include increased vaginal discharge, urinary frequency, dysuria, abdominal pain, and menstrual abnormalities. It should be noted that all these symptoms can be mild or completely absent, particularly in women.

PID develops in about 10–20% of women infected with *Neisseria*. Symptoms include fever, bilateral abdominal tenderness, and leukocytosis and are caused by the spread of the bacteria along the fallopian tubes, producing salpingitis (**Figure 23.14**), and into the pelvic cavity, causing pelvic peritonitis and abscess formation. The most serious complications of PID resulting from gonorrhea are infertility and ectopic pregnancy.

Disseminated gonococcal infection is the result of either localized gonorrhea or PID. The clinical features are fever, polyarthralgia, and petechial maculopapular or pustular rash. In fact, some of these signs may be caused by the host response to the bacteremia. The spread of the bacteria can lead to endocarditis or meningitis but it most commonly appears in the form of purulent arthritis.

Treatment of Gonorrhea

Both individual patient issues and public health concerns must be considered in treating gonorrhea. Patients who discontinue treatment early risk continuing to transmit the disease and also developing antibiotic-resistant strains of *Neisseria*. Resistance to penicillin has rendered this drug useless for the treatment of gonorrhea. Resistance is due to alterations in the penicillin-binding protein, alterations that require administration of such large amounts of penicillin they would be toxic. In addition, *Neisseria* acquired a new β-lactamase (penicillinase) during the Vietnam War that is now found worldwide.

The development of resistant *Neisseria* strains has caused treatment options to shift to third-generation cephalosporins, which are not affected by penicillinase. In addition, these drugs have such high activity that treatment involves only one dose of antibiotic. Antibiotics such as fluoroquinolones, azithromycin, and doxycycline are also effective.

Non-gonococcal Urethritis (*Chlamydia trachomatis*)

Non-gonococcal urethritis (NGU) is the most common form of sexually transmitted disease, with the number of people contracting it being twice as high as the number contracting gonorrhea. More than 700,000 cases are seen each year in the United States. NGU is caused by a unique form of bacteria called *Chlamydia*, which are obligate intracellular bacteria. *Chlamydia trachomatis* is the most common cause of NGU. In addition to being an agent of sexually transmitted disease, this organism also causes the eye infection called trachoma (discussed in Chapter 26).

Chlamydia is a round cell surrounded by an envelope. The envelope is a trilaminar outer membrane containing lipopolysaccharides and proteins similar to those seen in Gram-negative bacteria. However, *Chlamydia* does not contain peptidoglycan. These organisms cannot be grown outside host cells, and they contain one of the smallest genomes of all the prokaryotes. Although they contain ribosomes and can produce energy in the same way as other bacteria, they lack the genes necessary to synthesize amino acids. *Chlamydia trachomatis* contains

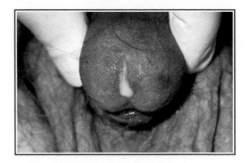

Figure 23.13 Purulent discharge from the male urethra of a patient with gonorrhea.

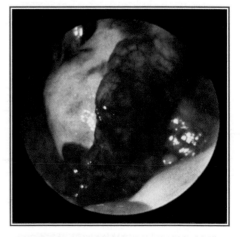

Figure 23.14 Endoscopic view of an infected and inflamed fallopian tube (bright red). Such tubal inflammation is called salpingitis. The ovary (white) can also be seen.

multiple outer-membrane proteins that further divide the members of this species into strains.

Replication of *Chlamydia*

Chlamydia has a unique replication cycle that involves two forms of the bacterium. The first is a small, hardy, infectious form, the **elementary body** (EB), and the second is a larger, more fragile, replicative form called the **reticulate body** (RB; sometimes called the initial body). The EB form is metabolically inert, and the proteins of this form contain a large number of disulfide bonds, making the EB structurally strong.

The replication cycle (**Figure 23.15**) begins when an EB attaches to unknown receptors on the plasma membrane of susceptible host cells (usually columnar epithelial cells) and enters the cell through endocytosis. While in the endocytic vesicle, the EB converts to an RB. An interesting observation is that endocytic vesicles carrying *C. trachomatis* do not fuse with lysosomes in the way you would expect them to. Instead, they fuse with other endocytic vesicles carrying the pathogen, and as the number of *Chlamydia* cells in each vacuole increases, the endosome membrane expands by fusing with the lipids of the Golgi apparatus to form a large inclusion body. After 24 to 72 hours, the process reverses and the RB form re-organizes and condenses to the EB form. The endosome membrane then either disintegrates or fuses with the host cell membrane, releasing the EBs, which go on to infect new targets. *Chlamydia trachomatis* inhibits apoptosis of epithelial cells, an inhibition that allows it to complete its replication cycle.

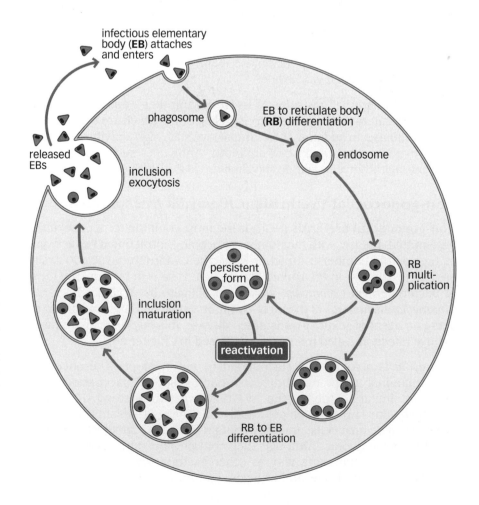

Figure 23.15 The replication cycle of *Chlamydia*. The elementary bodies (EB) do not replicate but are involved in the initial infection of the host cells. The elementary bodies transform into replicative reticulate bodies (RB).

Pathogenesis of *C. trachomatis* Genital Infections

Humans are the only reservoir for this pathogen, and the prevalence of chlamydial urethritis in men and women ranges from 5% in the general population to as high as 20% in populations receiving medical treatment for STIs. Approximately one-third of men who have sexual contact with women infected with *C. trachomatis* develop NGU after an incubation period of two to six weeks. It is important to note that many of these infected men will show no symptoms of the disease. Re-infection with NGU is a common occurrence.

Chlamydia has an affinity for epithelial cells of the endocervix and upper genital tract of women, and for the urethra and rectum of both men and women. Once an NGU infection is established, there is also a release of pro-inflammatory cytokines by the infected epithelial cells. It is thought that this inflammatory response may be generated by the lipopolysaccharides of the *Chlamydia* outer membrane. The inflammatory response results in tissue infiltration by polymorphonuclear leukocytes followed by lymphocytes, macrophages, plasma cells, and eosinophils. If the infection is not treated or if there is a failure in the immune response, aggregates of lymphocytes and macrophages may form in the submucosa, causing necrosis, fibrosis, and scarring. These symptoms can become chronic.

The same sorts of clinical sequelae seen in gonorrhea—urethritis and epididymitis in men, and cervicitis, salpingitis, and urethritis in women—can also be seen in *C. trachomatis* NGU infections, which usually present with dysuria and a thin urethral discharge. Infection of the cervix is usually asymptomatic but can also manifest as a vaginal discharge. In addition, ascending infections resulting in salpingitis and PID can occur in 5–30% of infected women. Chlamydial infections are considered as a major cause of female infertility.

Fast Fact More than 50% of all infants born to mothers infected with *Chlamydia trachomatis* who are excreting the organism during labor will be infected, usually with the eye condition known as conjunctivitis.

Treatment of *C. trachomatis* Genital Infections

Chlamydia organisms are sensitive to doxycycline, azithromycin, and some fluoroquinolones. Topical administration of erythromycin or silver nitrate has little effect on the conjunctivitis seen in newborn infants of infected mothers, and 15–20% of these infants will develop conjunctivitis. There is no vaccine against NGU.

Keep in Mind

- Bacterial infections of the reproductive system are usually sexually transmitted.
- Bacterial STIs can cause ulcers, sexually transmitted urethritis, epididymitis, cervicitis, vaginitis, and pelvic inflammatory disease.
- The most common bacterial STIs are syphilis (caused by *Treponema pallidum*), gonorrhea (*Neisseria gonorrhoeae*), and non-gonococcal urethritis (*Chlamydia trachomatis*).
- *Treponema pallidum* is an exclusively human pathogen that does not produce virulence factors.
- Syphilis has distinct clinical stages: primary, secondary, latent, and tertiary (which is systemic and can affect the nervous system).
- Gonorrhea is a public health problem, with the reservoir being asymptomatic carriers.
- Infection by *Neisseria gonorrhoeae* has three stages: attachment and invasion of the epithelia, survival in the submucosa, and dissemination to reproductive glands (and sometimes blood).
- *Chlamydia* is a unique form of bacteria: it is an obligate intracellular bacterium with a very small genome.

VIRAL INFECTIONS OF THE GENITOURINARY SYSTEM

The most important viral infection of the genitourinary system is HIV, which results in acquired immune deficiency syndrome (AIDS). We discussed HIV in detail in Chapter 17 and will use this section to discuss two other prominent viruses, herpes simplex virus type 2 and **human papillomavirus** (HPV).

Herpes Simplex Virus Type 2

There are two distinct epidemiological and antigenic types of herpes simplex virus (HSV), HSV type 1 and HSV type 2. These are DNA viruses with linear double-stranded DNA molecules, and the two types share many of the same antigens and have 50% homology in their genomes. HSV-1 is described as an above-the-waist virus because it causes cold sores in the area of the mouth, and HSV-2 is a below-the-waist virus because it causes the infection known as genital herpes. HSV-1 can also cause genital herpes but on a much smaller scale than HSV-2 so we will only consider HSV-2 here.

HSV-2 is distributed throughout the world, and humans seem to be the only reservoir for it. Transmission of infection is through direct contact with infected secretions. Interestingly, antibodies against HSV-1 can be found in a large portion of the population, but antibodies against HSV-2 are rarely seen before puberty. In adults, 15–30% of sexually active persons in Western industrialized countries carry antibodies against HSV-2.

Many patients infected with HSV-2 either are asymptomatic or else have small lesions on the penile or vulvar skin that go unnoticed. It is important to note that even though these people are asymptomatic, shedding of the virus can still occur. This accounts for transmission of the virus by individuals who have no active genital lesions and often no history of genital herpes. Although genital herpes is not a reportable disease in the United States, it is estimated that there are 1 million new cases each year.

Pathogenesis of HSV-2

There are two types of infection seen with genital herpes: acute and latent. In the acute infection, the obvious pathology is the appearance of multinucleate giant cells, ballooning degeneration of epithelial cells, focal necrosis, eosinophilic intraneural inclusion bodies, and an inflammatory response. The virus can spread intraneuronally, interneuronally, or through supporting cellular networks of axons or nerves. Spread of virus can also be through cell-to-cell transfer, a pathway that inhibits the effects of circulating antibody.

In the latent infection, the virus has been demonstrated in the sacral region (S2–S3). Although infection does not result in the death of the infected neuron, the effects of the virus on the host cell are not understood. It is known that there can be several copies of the viral genome in each infected host cell, and these are found in a circular form. Because latent infections do not require the synthesis of viral proteins, most antiviral drugs do not eradicate the virus when in its latent state.

Reactivation of latent virus from ganglionic cells with subsequent release of infectious virions accounts for most recurrent genital infections. The mechanisms associated with the reactivation are not yet known, but several precipitating factors, such as exposure to ultraviolet radiation, fever, and trauma, are known to initiate reactivation.

More than 70% of first episodes of genital herpes in the United States are caused by HSV-2. However, 90% of patients who test positive for HSV-2 antibody have never had a clinical genital episode, and in many cases the first episode does not occur for years after the initial exposure and infection.

Clinical Manifestations of Genital Herpes

Herpes infections can be primary, recurrent, or neonatal. Relatively few people develop clinically evident primary genital herpes. For those that do, the incubation time from sexual contact to the onset of lesions is about five days. The lesions begin as small erythematous papules that form vesicles and then pustules on the mucosal tissues (**Figure 23.16**). Within three to five days, these pustular lesions break to form painful coalesced ulcers. Some of these ulcers may crust over before healing, but they all eventually do so without scarring.

In primary genital herpes, the lesions are usually multiple (about 20), bilateral, and extensive. The urethra and cervix can also be involved, and bilateral enlarged and tender lymph nodes in the groin may persist for weeks or even months. About one-third of patients with primary genital herpes show systemic symptoms such as fever, malaise, myalgia, and, in some cases, aseptic meningitis. First episodes of primary genital herpes last an average of 12 days.

In contrast to primary genital herpes, recurrent genital herpes has a shorter duration, is usually localized in the genital region and usually has symptoms, such as a burning or prickly sensation in the pelvic area, that occur 12–24 hours before the appearance of grouped vesicular lesions in the external genitalia. There is accompanying pain and itching that can last four to five days, but the lesions usually disappear in two to five days. At least 80% of patients with primary genital HSV-2 infection develop recurrent episodes within 12 months, and in these patients the median number of recurrences is four or five per year. These are not evenly spaced in time, however; sometimes flare-ups repeatedly occur for several months and are followed by long periods with no symptoms. Recurrent viral shedding from the genital tract may occur without clinically evident disease.

Herpes infection in newborn infants usually results from transmission of the virus during delivery as the infant is exposed to infected maternal genital secretions. Most cases of severe neonatal herpes infection are associated with a woman who has a primary infection at or near the time of delivery. This results in an intense viral exposure to the infant during the birth. The incidence rate of neonatal herpes in the United States is approximately 1 in every 2500 live births; this infection is very serious, with an overall mortality rate of approximately 60%. Even those infants that survive will experience abnormal nervous system function and can show disseminated vesicular lesions with widespread internal organ involvement, necrosis of the liver and adrenal glands, listlessness, or seizures.

Treatment of Genital Herpes

Several antiviral drugs inhibit HSV but the most effective and most commonly used is the nucleoside analog acyclovir. This drug inhibits the function of viral DNA polymerase and significantly decreases the duration of a primary infection. If taken daily, it can also suppress recurrent infections. Resistant HSV virions have been recovered from immunocompromised patients who have persistent lesions, and in these cases the drug foscarnet has been effective. In 1996, the FDA approved valacyclovir and famciclovir for the treatment of recurrent genital herpes.

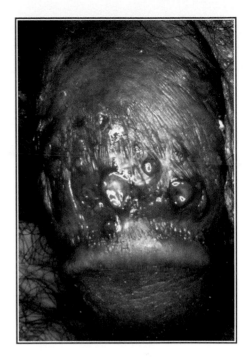

Figure 23.16 Vesicular lesions associated with genital herpes.

Prevention can be accomplished by avoiding contact with infected individuals who are expressing lesions. Although this strategy can limit the spread of the disease, it is important to remember that virus is still being shed in asymptomatic individuals and can be transmitted not only via urethral and cervical secretions but also via saliva.

Human Papillomavirus

Papillomaviruses are small, non-enveloped, double-stranded DNA viruses with an icosahedral symmetry. They cause papillomas (which are benign tumors) or warts in a wide range of higher vertebrates, and the infections are species specific. In some cases, tumors caused by these viruses can be malignant. Because papillomavirus has not been grown in culture, most of what we know about them comes from molecular studies. There is wide genetic diversity among human papillomaviruses and this diversity is indicated by using numbers to identify different genotypes. Currently more than 70 genotypes of HPV have been identified, and some are associated with specific lesions.

HPVs have been identified in a variety of genital hyperplastic epithelial lesions, including cervical, vulvar, and penile warts. They are also strongly associated with premalignant and malignant cervical cancer. Twelve of the HPV genotypes have been identified in human genital lesions, but many of the other genotypes of HPV may be associated with silent infections. It is possible to be infected with more than one genotype of HPV simultaneously.

The incidence of HPV infection is rising, and today between 20% and 60% of women in the United States are infected with one or another of the genotypes. It is estimated that there are nearly 1 million new cases of genital warts in the United States each year. HPV types 6 and 11 are associated with benign genital warts in males and females, and types 16, 18, 31, and 45 cause warty lesions of the vulva, cervix, and penis. Infections with any of the latter four, especially type 16, may progress to malignancy, and the genomes of these four types have been found in cases of dysplastic (dysplasia is abnormal tissue development) uterine cervical cells and in malignant lesions. The increase in HPV infections and the association of HPV with cervical cancer led to the development of a vaccine, which is now part of childhood vaccination programs in the United States, Europe, Canada, and Australia.

Pathogenesis of HPV Infection

Papillomaviruses have a predilection for infection at the junction of squamous and columnar epithelium (for instance, in the cervix). The mechanism of malignant transformation is not understood and is difficult to study because HPV is difficult to grow in culture. However, it has been shown that part of the viral genome can be found integrated into the host cell chromosome, and this integration does not seem to be site-specific. Host cells normally produce a protein that inhibits the expression of papillomavirus-transforming genes, but the virus seems to inactivate that protein.

External genital HPV infection presents in the form of genital warts (**Figure 23.17**). This is most often caused by genotypes 6 or 11, and these lesions may grow to a cauliflower-like appearance during pregnancy or immunosuppression. Genital HPV infections are usually benign, and many lesions reverse spontaneously. However, they may become dysplastic and proceed to severe dysplasia or carcinoma. Paradoxically, the most common malignant lesion is caused by HPV type 16, but the lesions of this genotype are also those that reverse most quickly.

Fast Fact HPV infection is now considered to be the cause of the majority of carcinomas of the cervix, and HPV DNA has been found in 95% of cervical carcinoma specimens.

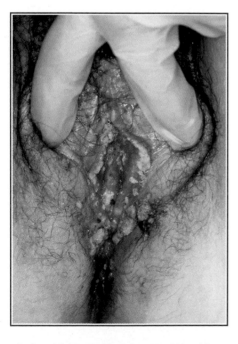

Figure 23.17 Genital warts caused by human papillomavirus occurring on the vulva.

Treatment of HPV Infection

Currently, the only treatments for HPV infection are surgical, cytotoxic drugs, and cryotherapy with liquid nitrogen. The most commonly used cytotoxic drugs are podophyllin, podophyllotoxin, 5-fluoroacetic acid, and trichloroacetic acid. Recurrence of the infection is common after cessation of the treatment, because the virus is able to survive in the basal layers of the epithelium.

Fast Fact The HPV vaccine is recommended for girls and boys in most countries. Men can get other cancers and genital warts from HPV, and spread the virus to their partners.

Keep in Mind

- Several viruses cause STIs, including HIV, HSV-2 (herpes simplex virus 2), and HPV (human papillomavirus).

- Humans seem to be the only reservoir for HSV-2, and many people infected with genital herpes are asymptomatic but still infectious.

- There is no cure for genital herpes infection, and transmission to newborns causes a serious infection that can be fatal.

- Human papillomavirus (HPV) can cause genital warts and is also the primary cause of cervical cancer.

- The incidence of HPV infection is rising, and it is estimated that as many as 60% of women in the United States are infected with one or another of the genotypes of HPV.

- A vaccine is now available for the prevention of HPV infection.

FUNGAL INFECTIONS OF THE GENITOURINARY SYSTEM

Because it is so open to the external environment, the genitourinary tract can be exposed to many fungal organisms. However, as we have discussed before, fungal infections are usually opportunistic and caused by part of the normal microbial flora. This is true of one of the most prominent fungal infections, **vaginal candidiasis**, caused by *Candida albicans*.

Vaginal Candidiasis (*Candida albicans*)

Candida albicans is part of the normal microbial flora of the oropharyngeal and gastrointestinal regions in both males and females, and the genital tract of females. It can grow in multiple morphological forms but is most often seen as yeast. The general name for infections caused by *C. albicans* is **candidiasis** (or vaginal thrush), and the infection can be either local or systemic. In vaginal candidiasis, the main symptoms are itching and a thick white discharge. Infections with *Candida* are normally endogenous except in cases of direct mucosal contact with a person expressing lesions, such as would occur during sexual intercourse. Indwelling catheters and overuse of antibiotics provide additional opportunities for *Candida* to become opportunistically pathogenic.

Pathogenesis of *C. albicans* Infection

Because this organism is a part of the normal flora found on mucosal surfaces, it must undergo changes to become pathogenic. One of the changes is the appearance of hyphae, which are seen when *Candida* invades tissues. It is not known what causes the hyphae to form, but their appearance is accompanied by the production of several factors that permit strong attachment of the organism to human epithelial cells. This attachment involves the usurping of host cell enzymes.

The hyphae excrete proteases and phospholipases that digest epithelial cells and further facilitate the invasion of tissues. *Candida* can also protect itself by binding to the C3 fragment of complement in such a way that the complement is not available to opsonize the yeast. In point of fact, any factors that inhibit T-cell function or compromise the host's immune response enhance the ability of *Candida* to invade tissue. Infection of the vagina by *C. albicans* produces a thick discharge with the consistency of cottage cheese, and accompanying itching of the vulva. Most women will have at least one episode of vaginal candidiasis in their lifetime. In addition, a small percentage of women may become chronically infected and experience recurrent symptoms. *Candida* can also infect the urinary tract, leading to cystitis, nephritis, abscesses, and expanding fungus ball lesions in the renal pelvis.

Treatment of *C. albicans* Infection

This fungus is usually susceptible to azole drugs, amphotericin B, nystatin, and flucytosine. In many cases, the lesions seen in this infection will resolve spontaneously on the elimination of predisposing conditions (mainly the removal of a catheter).

Keep in Mind

- The most prominent form of fungal infection of the genitourinary tract is vaginal candidiasis caused by *Candida albicans*.

- Indwelling catheters and the overuse of antibiotics provide additional opportunities for opportunistic Candida infection.

SUMMARY

- The genitourinary system is open to the outside environment and is a major portal of entry for pathogens.

- Infections of the genitourinary system are split into urinary tract infections (UTIs) and sexually transmitted infections (STIs).

- UTIs in hospitals are a serious problem and are usually related to indwelling catheters.

- UTIs in the community are seen in women more than in men because of the difference in the lengths of the urethra.

- *E. coli* is the major cause of UTIs.

- Bacterial UTIs can be treated with antibiotics, but resistance is an increasing problem.

- The majority of infections seen in the reproductive system are sexually transmitted and pose a serious public health issue.

- STIs can be caused by bacteria, viruses, fungi, and protozoan parasites.

- Bacterial STIs can be treated by antibiotics: *Treponema pallidum* (syphilis) is still sensitive to penicillin, but *Neisseria gonorrhoeae* (gonorrhea) is now resistant.

- Viral STIs can be contained by antivirals (such as acyclovir for HSV-2 and podophyllin for HPV), but there is no cure.

- There is now a vaccine for HPV which is recommended as part of the standard vaccination program.

- The most common form of fungal infection is candidiasis caused by indwelling catheters coupled with the overuse of antibiotics, bringing about a superinfection.

This chapter completes the discussion of the three systems that are the most open portals of entry for microorganisms. As we have seen, the genitourinary tract, like the respiratory and digestive tracts, is susceptible to a variety of infections. We saw that in the urinary system there can be an ascending infection going from the urethra to the bladder, to the prostate in males, and on up to the kidneys. We also saw that most infections of the reproductive system were classified as sexually transmitted infections. The same ascending sequence seen in urinary tract infection can be seen with STIs, which can go from the urethra to the reproductive organs.

 SELF EVALUATION AND CHAPTER CONFIDENCE

Multiple Choice

Answers are given in the back of the book and help can be found in the student resources at:

www.garlandscience.com/micro2

1. Cystitis can be distinguished from urethritis by all the following except

    A. Cystitis has a more acute onset
    B. Cystitis has more severe symptoms
    C. Cystitis often exhibits bacteremia
    D. Cystitis often produces flank pain

2. UTIs are more common in women because

    A. The urethra in men is nearer the anus
    B. The urethra is shorter in women
    C. The urethra is longer in women
    D. Men have a prostate gland but women do not

3. Which of the following is treated with penicillin?

    A. Candidiasis
    B. Syphilis
    C. Genital herpes
    D. Papillomas
    E. Gonorrhea

4. Lymphogranuloma venereum is caused by

    A. *Candida albicans*
    B. *Neisseria gonorrhoeae*
    C. *Chlamydia trachomatis*
    D. *Treponema pallidum*

5. Which of the following recurs at the site of the infection?

    A. Gonorrhea
    B. Genital herpes
    C. Syphilis
    D. Chancroid

6. Which of the following can be an opportunistic infection?

    A. Gonorrhea
    B. Candidiasis
    C. Syphilis
    D. Genital herpes

7. Most nosocomial urinary tract infections are caused by

    A. *E. coli*
    B. *Enterococcus*
    C. *Pseudomonas*
    D. *Staphylococcus*

8. The major cause of cervical cancer is infection with

    A. Herpes simplex virus 2
    B. Herpes simplex virus 1
    C. Human papillomavirus
    D. Human immunodeficiency virus

9. Which of the following is not a mechanism used by *Neisseria gonorrhoeae* to evade the immune system?

    A. Blocking the complement cascade
    B. Using multiple host cell receptors
    C. Interfering with phagocytosis
    D. Inhibiting the antibody response

10. Which of the following is not true about *Chlamydia* replication?

    A. The bacterium attaches to membrane receptor
    B. It enters the host cell through endocytosis
    C. Once inside the cell, the elementary body converts to a reticulate body
    D. Vesicles containing reticulate bodies then fuse with lysosome

11–15. Answer **A** if the reason is correct, answer **B** if the reason is incorrect.

11. Infections spread easily through the urinary tract because all portions of the urinary tract are connected to one another by a liquid medium.

12. In many cases the individuals infected with HSV-2 will not seek medical help because they are frightened of the results.

13. The pathogenesis of *Treponema pallidum* has been extrapolated from observations of patients with syphilis because there is no animal model for syphilis.

14. Because *Treponema pallidum* multiplies rapidly, there is little or no inflammatory response during the initial stage of the infection.

15. HSV-1 is described as an above-the-waist virus because it causes colds.

 DEPTH OF UNDERSTANDING

Questions listed here require you to bring together the concepts you have learned in this chapter into a discussion format. This helps you to increase your depth of understanding of the material you have learned. Help can be found in the student resources at: www.garlandscience.com/micro2

1. Describe how urinary tract infections can begin as urethritis and develop into nephritis.

2. Discuss the phases of a syphilis infection and the symptoms seen as the infection progress.

3. Compare the different forms of herpes infections.

 CLINICAL CORNER

Help can be found in the student resources at: www.garlandscience.com/micro2

1. Your best friend comes to you with great news; she has found a new boyfriend. You can see that she is really excited and the romance continues for several dates and then stops. When you ask her what happened, she breaks down and tells you that when she tried to take the relationship to the next level her boyfriend told her that he had contracted genital herpes several months ago from a former girlfriend. Though he assured her that they could still be intimate by his wearing a condom, she is still very fearful that she will be infected if they have intercourse. She really cares for him and wants to continue the relationship and asks your advice.

   A. What do you tell her about the possible risks of becoming infected?
   B. Will his wearing a condom be able to protect her?
   C. Can using the available prescription medicines available for genital herpes help them be together?

2. Your patient was admitted to the hospital for elective cosmetic surgery. After the surgery, which went very well, it was discovered that she had a urinary tract infection with *Pseudomonas*. She was placed on broad-spectrum antibiotics for two days, during which she developed vulvovaginal candidiasis. By day four she showed signs of pyelonephritis and was put on dialysis. Obviously her family is distressed and does not understand how she has gone from having a simple 'tummy tuck' to dialysis.

   A. How can you explain what has happened?
   B. What is the significance of the candidiasis?
   C. What are the major concerns you have regarding this patient?

Infections of the Central Nervous System

Chapter 24

Why Is This Important?

Because the central nervous system controls the function of the body, infections that affect the central nervous system can be catastrophic and potentially lethal for the host.

Simon was only three months old when he started exhibiting muscle weakness, reduced appetite, and excessive drooling. His mother became worried and sought medical attention immediately. Although he was severely constipated, doctors managed to obtain and analyze a stool sample. They also performed an electromyogram. On the basis of these tests, they determined that he had a disease called infant botulism.

Infant botulism most often occurs after a baby (under one year of age) eats honey contaminated with the endospores of *Clostridium botulinum.* Adults are generally not susceptible to this because of their stronger immune systems and more mature gastrointestinal tracts. After the baby consumes the spores, the bacteria germinate and

reproduce rapidly, releasing dangerous toxins that cause paralysis by blocking the chemical interactions between nerves and muscles. The doctors gave Simon intravenous fluids containing human antibodies directed against botulism that act as an antitoxin. Because of the early diagnosis and medical attention, Simon made a full recovery and was soon released from the hospital.

OVERVIEW

In this chapter, we look at some of the infections that occur in the central nervous system. We will find that in some cases this system is the main target of the infection, whereas in others the effect on the central nervous system is secondary. As in all the other systems of the body, infections of the central nervous system can be caused by bacteria, viruses, fungi, and parasites. In addition, in the central nervous system we will see infections caused by infectious proteins called prions. Once again, we must think about the anatomy of the nervous system to put these infection agents into perspective.

We will divide our discussions into the following topics:

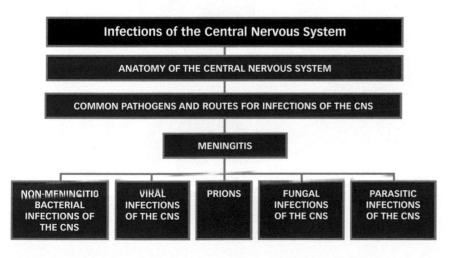

"...no one foresaw the possible entry of a lethal agent far more virulent than any conventional pathogen or virus."

David Brown

In the above quote, a journalist provides a disturbing description of the "lethal agent" (an infectious protein called a prion) that caused more than 170 deaths in Britain between 1987 and 2014. This prion causes a central nervous system disease commonly called mad cow disease in cattle, and variant Creutzfeldt–Jakob disease in people. As we will see in this chapter, a variety of pathogens affect the nervous system, but none are as enigmatic or difficult to inactivate as prions. The central nervous system is perhaps the most critical part of the body; it controls everything else, and when it is compromised, the results can be devastating.

ANATOMY OF THE CENTRAL NERVOUS SYSTEM

The central nervous system (CNS) is divided into two major parts, the brain and the spinal cord. Both of these structures reside in the dorsal body cavity and are protected by the skull and vertebral bones. Both the brain and the spinal cord are surrounded by three layers of connective tissue called the **meninges** (Figure 24.1). The dura mater, which is the most superficial meningeal layer, has the consistency of wax paper. The middle layer is called the arachnoid mater because it has spidery extensions that connect it to the deepest layer, the pia mater.

Between the arachnoid mater and pia mater is a space referred to as the subarachnoid space, and it is here that cerebrospinal fluid is found. This fluid, which is produced by the choroid plexus and bathes the brain and spinal cord, is the intermediary for nutrients required by nervous tissue and is also a vehicle for carrying cellular waste away from the CNS. The brain and spinal cord are protected from the rest of the body by the **blood–brain barrier**, a structural arrangement in which the capillary endothelial cells rest on basement membranes that do not allow

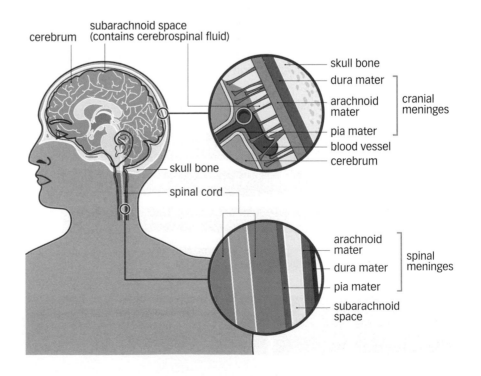

Figure 24.1 The meninges and the cerebrospinal fluid. Three layers of meninges surround the brain and the spinal cord: the dura mater, which is the most superficial; the arachnoid mater, which is the middle layer; and the pia mater, which is the deepest layer. Between the arachnoid and pia maters there is the subarachnoid space, through which the cerebrospinal fluid circulates.

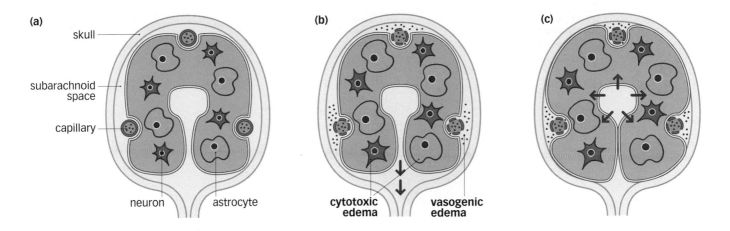

Figure 24.2 The effects of edema (swelling) on the brain. Panel a: The normal brain. Panel b: Vasogenic and cytotoxic edema caused by inflammation in the meninges. Panel c: The effects of these types of edema include swelling of the brain, and oxygen and nutrient deficiencies.

the passage of certain materials from the blood to the brain. This barrier protects the body against infectious disease by preventing pathogens that might find their way into the blood from infecting nervous tissue of the brain. As we will see, some pathogens can pass through the blood–brain barrier.

Because there is no room for swelling of the nervous tissue enclosed by the skull, any pathology that occurs in the brain and results in increased intracranial pressure can have disastrous and potentially lethal consequences, all the result of compression, herniation, and brain cell death. In addition, as we learned throughout our discussions of infectious disease, inflammation is one of the first and most formidable responses to infection. Inflammation causes increased vascular permeability and therefore swelling. An inflammatory response in the subarachnoid space, with the requisite release of cytokines, results in loosening of the tight junctions between vascular endothelial cells. The loosened junctions allow albumin into the cerebrospinal fluid and the result is the form of swelling known as **vasogenic edema** (Figure 24.2). Swelling can also result from the toxic substances produced by bacteria and from neutrophil invasion, both of which result in oxygen and nutrient deficiencies. This produces another kind of swelling, the kind known as **cytotoxic edema**. Infection can also affect the biochemical CNS reactions required for proper brain function through acidosis, hypoxia, and the destruction of neurons.

The effects of increased intracranial pressure, biochemical abnormalities, and neural tissue necrosis can be profound and sometimes irreversible. Furthermore, the blood–brain barrier, which is supposed to protect the brain, can make it difficult to treat and control infections of the CNS, because many drugs cannot cross it.

Keep in Mind

- The central nervous system (CNS) comprises the brain and spinal cord.
- The blood–brain barrier protects the CNS from infection but can prevent drugs getting to the CNS if an infection does occur.
- Swelling and inflammation can have disastrous results for the CNS.

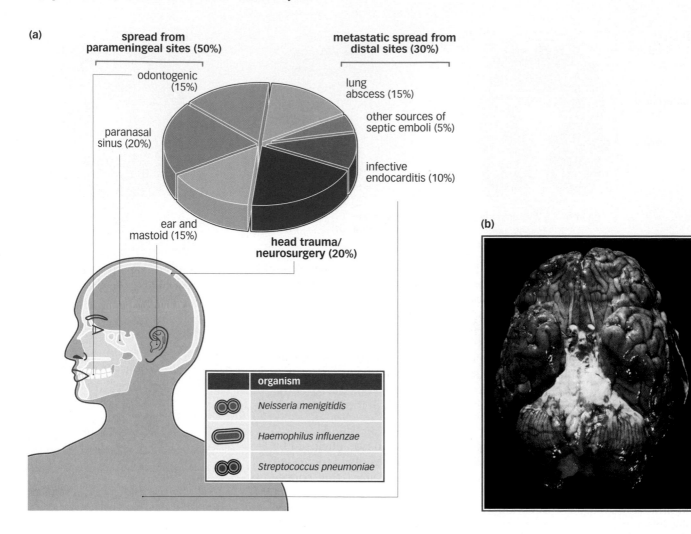

(a)

spread from
parameningeal sites (50%)

metastatic spread from
distal sites (30%)

odontogenic
(15%)

lung
abscess (15%)

paranasal
sinus (20%)

other sources of
septic emboli (5%)

infective
endocarditis (10%)

ear and
mastoid (15%)

head trauma/
neurosurgery (20%)

(b)

	organism
⬤⬤	*Neisseria menigitidis*
⬤	*Haemophilus influenzae*
⬤⬤	*Streptococcus pneumoniae*

Figure 24.3 Meningitis and brain abscesses. Panel a: The main routes by which organisms reach the central nervous system to cause meningitis and brain abscesses. Panel b: Purulent basilar meningitis due to infection with *Streptococcus pneumoniae*.

COMMON PATHOGENS AND ROUTES FOR INFECTIONS OF THE CNS

Microorganisms reach the brain and spinal cord in a variety of ways (**Figure 24.3a**), with blood being one of the most common because there is extensive vascularization in the CNS. Any microbe that circulates in the blood and can enter the cerebrospinal fluid can cause **meningitis** (infection of the meninges). In addition, the skull contains various sinuses and mastoid air spaces that are separated from the brain by the bone of the skull. Infections in these areas can eventually cause the erosion of the bone, and then organisms can enter the brain and cause abscesses to form (**Figure 24.3b**).

Most infections of the CNS result from passage across the blood–brain barrier of bacteria or viruses that are circulating in the blood as a result of some non-CNS infection. A list of bacteria that infect the CNS is shown in **Figure 24.4**. Some of these, such as *Neisseria meningitidis*, *Haemophilus influenzae* type b and *Streptococcus pneumoniae*, can be found as normal flora in the human body, whereas other organisms, such as *Listeria monocytogenes*, are ingested. Group B *Streptococcus* can also be acquired during the birthing process. It is important to note that many infections of the CNS occur through the contamination of shunts, in particular with coagulase-negative *Staphylococcus* species. Deep fungal mycoses are caused by *Cryptococcus neoformans* and *Coccidioides immitis*.

organism		acquisition	treatment
bacterial meningitis			
	Streptococcus pneumoniae	acquired member of the normal flora	benzylpenicillin, cefotaxime, chloramphenicol
	Haemophilus influenzae	acquired member of the normal flora	cefotaxime, chloramphenicol
	Neisseria meningitidis	acquired member of the normal flora	benzylpenicillin, cefotaxime, chloramphenicol
	Escherichia coli	neonatal meningitis acquired from maternal vaginal flora	cefotaxime, gentamicin, chloramphenicol, meropenem
	Streptococcus agalactiae (group B)	neonatal meningitis acquired from maternal vaginal flora	benzylpenicillin + gentamicin, cefotaxime
	Listeria monocytogenes	neonatal meningitis acquired from maternal vaginal flora	benzylpenicillin + gentamicin, cefotaxime
	Staphylococcus aureus	a member of the normal flora	floxacillin + rifampicin
brain abscess			
	streptococci	members of the normal flora	benzylpenicillin, chloramphenicol, clindamycin
	coliforms	fecal flora	cefotaxime, chloramphenicol
	Staphylococcus aureus	a member of the normal flora	floxacillin, chloramphenicol, clindamycin
ventriculo-peritoneal shunt infections			
	coagulase-negative staphylococci	skin commensals	vancomycin + rifampicin
	Staphylococcus aureus	normal flora	floxacillin, vancomycin + rifampicin
	Pseudomonas aeruginosa	hospital-acquired	ceftazidime, gentamicin, meropenem
toxin and immune-mediated disorders			
	Clostridium botulinum	contaminated food products	ventilatory support
	Clostridium tetani	contaminated wounds, intravenous drug use	ventilatory support, surgery, wound debridement, benzylpenicillin
	Campylobacter spp.	contaminated food, gastroenteritis	ventilatory support

Figure 24.4 Brain infections and treatments.

Some of the viruses that can infect the nervous system are shown in Table 24.1. The most common viral causes of acute infections of the CNS are enteroviruses, herpes simplex virus, HIV, Epstein–Barr virus, and several arthropod-borne viruses. Viral infections of the CNS manifest as aseptic meningitis, encephalitis, or poliomyelitis. The viral infections of the CNS referred to as subacute sclerosing panencephalitis usually result from measles or rubella infections.

The initial source of a CNS infection can be either occult (infection of mononuclear phagocytic system cells being one example) or overt, through complications of, for example, pneumonia, pharyngitis, skin abscesses, or **infectious endocarditis** (an infection in the heart). In some cases

Table 24.1 Primary acute viral infections of the CNS.

Agent	Major age group affected	Seasonal predominance
Enteroviruses	Infants and children	Summer–fall
Mumps	Children	Winter–spring
Herpes simplex		
Type 1	Adults	None
Type 2	Newborns, young children	None
Arboviruses		
Western equine encephalitis	Infants and children	Summer–fall
Eastern equine encephalitis	Infants and children	Summer–fall
West Nile meningoencephalitis	Adults	Summer–fall
Rabies	All ages	Summer–fall
Measles	Infants and children	Spring
Varicella-zoster	Infants and children	Spring
Epstein–Barr virus	Children and young adults	None
HIV and cytomegalovirus	All ages	Variable

the infection site is either close to or in direct contact with the CNS. For example, infections in the middle ear, mastoiditis, or sinusitis can easily make the short trip into the CNS under the proper set of circumstances.

Anatomical defects in the structures that encase the CNS can also permit infection. Surgical, traumatic, or congenital developmental abnormalities can leave openings into the CNS that are used by pathogens. For example, a fracture at the base of the skull is in close proximity to the body's respiratory system, which contains many organisms and presents them with ready access to the CNS. It is interesting to note that the least common route to the CNS is by intraneural pathways. However, there are exceptions to this statement. The rabies virus, for example, uses the peripheral sensory nerves to gain access to the CNS, and herpesvirus uses the trigeminal nerve to gain access to the CNS.

Although brain abscesses are relatively rare, they present a special problem. They can be found in the subdural space (between the dura mater and the arachnoid mater), in the epidural space (the region superficial to the dura mater), or directly in the brain tissue, and are commonly formed by bacteria or fungi from a distant site. Brain abscesses can also result from extensions of pathogens located at the site of mastoiditis or sinusitis, or from complications of surgery.

General Treatment of Infections of the CNS

Bacterial and fungal infections of the CNS require prompt and aggressive treatment, and treatment periods vary depending on the type of infection. For uncomplicated bacterial meningitis the treatments can last from 10 days to 12 months, and even longer if the infection is caused by *Mycobacterium tuberculosis*. For fungal infections, treatment can last for years. In contrast, treatment of viral infections of the CNS is mostly supportive.

Now let's look at some of the more common infections of the CNS. As in previous chapters, we will separate our discussions into bacterial, viral, fungal, and parasitic infections, and we will also include a review of prion diseases, which we learned about in Chapter 8.

| Keep in Mind |

- The CNS is well protected structurally and employs a blood–brain barrier to prevent infections.

- Organisms that move from the blood to the cerebrospinal fluid can cause meningitis.

- Most infections of the CNS result from bacteremia or viremia from distal infections.

- In some cases, infection comes from locations close to or in direct contact with the CNS, such as the middle ear or the sinuses.

- Bacterial and fungal infections of the CNS require immediate and aggressive treatment. Viral infections of the CNS are more difficult to treat, and therapy includes mostly supportive care.

MENINGITIS

Meningitis is an infection of the fluid surrounding the spinal cord and brain, usually caused by either a viral or a bacterial infection. It is important to know which type of infection it is, because the severity of the illness as well as the treatment will differ.

Bacterial Meningitis

Bacterial meningitis is a severe infection with acute progression that is usually fatal if untreated and should be treated as a medical emergency. With treatment there can still be brain damage, hearing loss, or learning disability, so treatment must be started immediately if bacterial meningitis is suspected. The inflammatory response to the presence of bacteria in the CNS causes vasogenic edema and cytotoxic edema, leading to an acute increase in cranial pressure.

It is important to know what type of bacterium is causing the meningitis, because antibiotic therapy can prevent the spread of the infection. Before 1990, *H. influenzae* type b (Hib) was the leading cause of bacterial meningitis, but vaccines now given to all children as part of their routine immunizations have reduced the occurrence of *H. influenzae* meningitis. Today, *Streptococcus pneumoniae* and *Neisseria meningitidis* are the leading cause of bacterial meningitis. There is an effective vaccine against group C *N. meningitidis*, and a vaccine for group B strains is under development.

Some forms of bacterial meningitis are contagious, with bacteria being spread through exchange of respiratory and throat secretions. However, meningitis is not as contagious as the common cold or flu. Meningitis caused by *N. meningitidis* can spread to other people who have had close or prolonged contact with infected individuals. This is why cases of meningitis are causes of concern for daycare centers and schools.

In newborns, group B streptococci and *Escherichia coli* are frequently involved in meningitis acquired through the birthing process.

Viral Meningitis

Viral meningitis, also known as aseptic meningitis, is more common, generally less severe and usually resolves without treatment. It is caused by infection with one of several types of virus. About 90% of cases are caused by enterovirus, although herpesvirus and mumps virus can also cause the disease. Aseptic meningitis can also be seen in syphilis and other spirochete infections. This type of meningitis is associated with meningeal inflammation evidenced by increased numbers of lymphocytes and mononuclear cells in the cerebrospinal fluid, with an absence of culturable bacteria or fungi. The primary site of inflammation is in the meninges, with no clinical involvement of neural tissue.

Although there are 25,000 to 50,000 hospitalizations due to viral meningitis in the United States each year, viral meningitis is rarely fatal if the patient has a competent immune system. The symptoms last from 7 to 10 days and the patient usually recovers completely.

Chronic Meningitis

Chronic meningitis has a slow onset, with signs and symptoms developing over weeks. It is usually caused by *Mycobacterium tuberculosis*, by fungi, or by protozoan parasites (Table 24.2).

Symptoms of Meningitis

Symptoms of meningitis include high fever, stiff neck, and headache. These symptoms can develop over several hours or may take one to two days. Other accompanying symptoms can include nausea, vomiting, sensitivity to bright light, confusion, and sleepiness. As the disease progresses, patients may develop seizures. It is important to note that symptoms in newborns and infants may be difficult to detect.

Diagnosis and Treatment of Meningitis

If symptoms of meningitis occur, the patient should see a doctor immediately. Bacterial meningitis is diagnosed by microscopic analysis of, or growing bacteria from, a sample of spinal fluid. The identification of the specific type of bacterium is then made so that effective antibiotic therapy can begin. If antibiotic therapy is started early, it can limit the risk of death to 15%, although this may be higher in the elderly.

Viral meningitis is also diagnosed using spinal fluid, but no antibiotic therapy is necessary. Patients are given bed rest, plenty of fluids, and medicines to relieve fever and headache.

Form of chronic meningitis	Agents
Chronic granulomatous infection	*Mycobacterium tuberculosis, Coccidioides immitis, Cryptococcus neoformans, Histoplasma capsulatum*
Protozoan parasitic infection	*Toxoplasma gondii, Trypanosoma, Acanthamoeba* species
Nematode parasitic infection	*Trichinella spiralis*
Other infections	*Leptospira* species, *Treponema pallidum, Borrelia burgdorferi*

Table 24.2 Causes of chronic meningitis.

- Meningitis is an infection of the meninges and can be caused by bacteria, viruses, fungi, or parasitic pathogens.

- Bacterial meningitis is more severe than viral meningitis, but can be treated with antibiotics.

NON-MENINGITIS BACTERIAL INFECTIONS OF THE CNS

We will now look at two bacterial infections, tetanus and botulism, that affect the CNS in a different way. These infections cause damage through the production of exotoxins that have an affinity for CNS tissue. Recall from Chapter 5 that exotoxins are extremely toxic proteins that are produced by some bacteria and usually have a two-chain configuration, with one chain binding to the target cell and the other chain causing the damage. It is also important to remember that once an exotoxin is produced, it no longer matters whether the organism that produced it survives. Therefore, antibiotic therapy against exotoxin-producing organisms will be ineffective once the exotoxin has been produced. In addition, remember that the binding of exotoxins is irreversible.

Tetanus (*Clostridium tetani*)

Tetanus is caused by the bacterium *Clostridium tetani*, which is a Gram-positive, anaerobic, spore-forming rod. It yields a terminal spore that gives it a drumstick appearance (**Figure 24.5**). This organism is a strict anaerobe and cannot survive in the presence of oxygen. It is commonly found in the soil, particularly soil that has been contaminated by manure, and its spores can survive in the soil for years. These spores, which are very resistant to disinfectants and can withstand boiling, are introduced into wounds contaminated with soil.

The toxin produced by *C. tetani* is a neurogenic toxin, meaning that it has an affinity for and targets nervous tissue. Called either **tetanospasmin** or **tetanus toxin**, it acts by enzymatically degrading proteins required for normal physiology in nervous tissue.

Pathogenesis of Tetanus

Tetanus spores require areas of low oxygen content if they are to germinate, and the area of necrosis around a tissue injury is perfect for the initiation of this process. The spores germinate, and *Clostridium* begins to grow in that location. In newborns, the remainder of the umbilical cord forms an ideal environment for growth of *C. tetani*, and hygienic treatment during birth is the most important prevention measure for neonatal tetanus. The bacteria do not cause any damage to the tissue where they reside; their role is to produce the neurogenic toxin, which then enters the presynaptic terminals of the lower motor neurons and from there they gain access to the CNS. In the spinal cord, the toxin acts at the level of the anterior horn cells and blocks postsynaptic inhibition of the spinal motor reflexes. This produces spasmodic contraction of the muscles, contractions that occur locally at first but then may extend up and down the spinal cord.

The incubation period for *Clostridium* can vary between four days and several weeks. Generally, the shorter the incubation period, the more severe the consequences will be. The tetanus toxin is systemic for muscles and the nerves affecting them; it therefore also affects the peripheral nervous system. The masseter muscle of the jaw is usually the first to

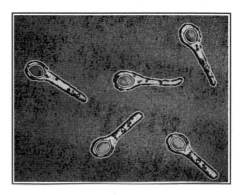

Figure 24.5 False-color transmission electron micrograph of *Clostridium tetani*, showing the drumstick shape of an enclosed terminal endospore.

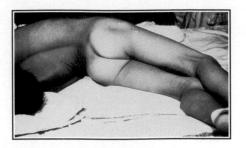

Figure 24.6 A patient with opisthotonos resulting from advanced tetanus. The characteristic sign is a curvature of the body that draws the heels toward the head.

be affected, which means that the mouth cannot be opened (tetanus is sometimes referred to as **lockjaw.**) Eventually the muscles responsible for respiration and swallowing can be compromised, and in severe cases the patient can also suffer from **opisthotonos**, a condition in which the head and heels move toward each other and the body bows out dramatically (**Figure 24.6**). Death can result from exhaustion and respiratory failure, and the mortality for untreated tetanus can vary from 15% to 60%. Several factors affect the mortality rate, including location of the lesion, incubation period, and the patient's age. Mortality rates are highest in infants and the elderly.

Treatment and Prevention of Tetanus

Because the toxin is the problem, once it is expressed, the death of the *C. tetani* organism becomes secondary. Therefore, neutralization of the toxin with large doses of human tetanus immunoglobulin (derived from volunteers who are hyperimmunized with the toxoid of this toxin) is important for initial therapy. In addition, supportive measures—including maintenance of a dark and quiet environment, sedation, and provision of an adequate airway for breathing—should be provided if necessary. Antibiotics are not effective once the toxin has been produced. As discussed in Chapter 5, the DTaP vaccination is effective in preventing tetanus.

Botulism (*Clostridium botulinum*)

The etiologic agent of botulism is *Clostridium botulinum*, a Gram-positive, anaerobic, spore-forming rod. The spores are found in the soil as well as in the sediments of ponds and lakes worldwide. If they contaminate food under anaerobic conditions, these spores can convert to the vegetative state and begin to produce toxin. This toxin is among the most poisonous in the world, and contamination of food with botulinum toxin can occur without affecting the smell, taste, or color of the contaminated food. Botulism is commonly seen in cases of home canning in which the temperatures used in the process are not high enough to destroy the endospores.

Pathogenesis of Botulism

Botulism begins with cranial nerve palsy and develops into a descending symmetrical motor paralysis that may involve the respiratory muscles (and can therefore also affect the peripheral nervous system). There is no fever or inflammation and no obvious sign of infection. The time course of the infection depends on the amount of toxin produced and on whether the toxin was ingested in a preformed state or was produced in the intestinal tract.

Foodborne botulism is classified as an intoxication (Chapter 22), not an infection. The toxin is ingested as a preformed component and is absorbed directly through the intestinal tract, reaching the neuromuscular junction target via the bloodstream. Once bound there, the toxin inhibits the release of acetylcholine, causing muscular paralysis. The symptoms observed depend on which nerves are bound by the toxin, and damage to the nerves after the toxin has bound is permanent. Any recovery of function requires the formation of new synapses.

Foodborne botulism usually starts 12–36 hours after ingestion of the toxin. The first symptoms are nausea, dry mouth, and in some cases diarrhea. Symptoms of nervous system dysfunction start later and include blurred vision, pupillary dilation, and rapid eye movements. Symmetrical paralysis begins with ocular, laryngeal, and respiratory muscles, and spreads to the trunk and extremities. The most serious complication is complete respiratory paralysis, with the mortality for botulism varying from 10% to 20%.

There are two other forms of botulism, the infant form and the wound form. Infant botulism occurs in infants between the ages of three weeks and eight months and is the most commonly diagnosed form of botulism. The organism is introduced either on weaning or through dietary supplements, particularly honey. It multiplies in the infant's colon, and the botulinum toxin is then absorbed into the blood. The symptoms of infant botulism include constipation, poor muscle tone, lethargy, and feeding problems. In severe cases, vision problems and paralysis can also occur. One of the difficulties in diagnosing infant botulism is that it mimics sudden infant death syndrome.

Wound botulism occurs very rarely and is usually seen in intravenous drug users. The symptoms are similar to those of food poisoning and usually begin with muscle weakness in the extremities used for injection.

Treatment of Botulism

The single most important determinant is the availability of intensive support measures, in particular mechanical ventilation. If proper ventilation support is provided, mortality is less than 10%. Because botulism is caused by an exotoxin, antibiotic therapy is given only to patients with the wound form. As described in the chapter opening, an antitoxin called botulism immune globulin intravenous can also be used to combat the botulinum toxin.

Keep in Mind

- Tetanus and botulism are serious infections (or intoxications) that can be devastating.

- Both tetanus and botulism are caused by exotoxins produced by the pathogens.

- Tetanus toxin is a neurogenic toxin that blocks postsynaptic inhibition of the spinal motor reflexes, which leads to spasmodic contraction of muscles.

- Botulism toxin moves from the intestinal tract to the CNS via the blood and inhibits the release of acetylcholine, causing muscular paralysis.

VIRAL INFECTIONS OF THE CNS

Viruses with an affinity for the CNS cause symptoms principally as a result of increased intracranial pressure and inflammation. **Encephalitis** is a term used to describe patients who have no symptoms of aseptic meningitis but who show obvious signs of CNS dysfunction, such as seizures, paralysis, or defective mental faculties. In this case, the problem is associated not with the meninges but with the actual nervous tissue. We will look at two types of viral infections of the CNS: acute and persistent (slow).

Rabies

An acute and fatal viral infection of the CNS, rabies was first reported more than 3000 years ago, and the term rabies is derived from the Latin word meaning rage. This is fitting, because part of the infection process is an overproduction of saliva and an inability to swallow that make

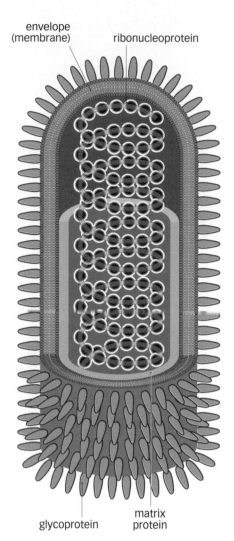

envelope
(membrane)   ribonucleoprotein

glycoprotein   matrix
protein

Figure 24.7 A diagrammatic illustration of the rabies virus.

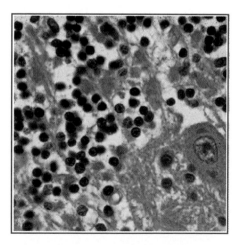

Figure 24.8 Using Sellers stain, the Negri bodies will appear as dark purple inclusions.

patients seem to be foaming at the mouth. Rabies can affect all mammals and is transmitted by infected secretions, usually through a bite. The virus is large and bullet-shaped (**Figure 24.7**), with glycoproteins that cover the entire virion.

Pathogenesis of Rabies

Rabies involves severe neurological symptoms and signs in a patient bitten by a rabid (infected) animal. The CNS abnormalities are characteristic and include a relentless progression of excess motor activity, agitation, hallucinations, and excessive salivation. There can also be severe throat contraction when swallowing is attempted.

Rabies exists in two forms, urban and sylvatic, with the urban form being associated with unimmunized dogs and cats, and the sylvatic form being seen in wild animals. Infection in humans is incidental and does not contribute to the maintenance or transmission of the infection. In the United States, 75% of cases occur in wild animals, and the occurrence of human rabies worldwide is about 15,000 cases per year. Approximately two cases of rabies infection in humans are reported in the United States each year.

The first event of rabies infection is introduction of the virus, usually through the epidermis via an animal bite, but inhalation of a heavily contaminated material such as bat droppings can also introduce the virus. Rabies virus first replicates in the striated muscle at the site of the bite (or in the lungs if the virus was inhaled), and immunization at this time will keep the virus from migrating into the nervous tissue. Without intervention, however, the virus moves into the peripheral nervous system at the neuromuscular junction and spreads into the CNS, where it replicates exclusively in the gray matter. After replication, the virus moves into other tissue, such as the adrenal medulla, kidneys, lungs, and salivary glands. At the same time, lymphocytes and plasma cells are infiltrating into the CNS tissue, and these agents destroy nerve cells. **Negri bodies**, which are inclusions found in rabies virus-infected nerve cells, can be seen in the brain cells of infected hosts (**Figure 24.8**).

The incubation period for rabies can vary from 10 days to as long as a year, depending on the amount of virus initially deposited with the bite or inhalation, the amount of tissue infected, the host's immune response, innervation at the site, and the distance that the virus must travel to reach the CNS.

Rabies presents as acute, fatal encephalitis, and only a handful of infected people have ever recovered without immediate treatment. Once the symptoms have started, the infection is irreversibly fatal. The illness begins with nonspecific fever, headache, malaise, nausea, and vomiting. The onset of encephalitis is marked by periods of excessive motor activity and agitation accompanied by hallucinations, combativeness, muscle spasms, and seizures followed by coma. There can also be excessive salivation, dysfunction of the brain stem and cranial nerves, double vision, facial palsy, and difficulty in swallowing. Involvement of the respiratory centers of the brain causes respiratory paralysis, which is the major cause of death. The median survival after the onset of symptoms is 4 days, and the maximum is 20 days.

Treatment and Prevention of Rabies

Prevention is the main way to control this infection. Treatment consists of a course of injections that are beneficial only if administered before the onset of symptoms. Intensive supportive care can result in longer-term survival. However, even with supportive care, the mortality for rabies is 90%. As discussed in Chapter 13, a vaccination provided before or after exposure to the virus is effective in preventing rabies.

Polio

Poliovirus is one of the most important enteroviruses in the world; the infection it causes first emerged as an epidemic during the latter half of the twentieth century. Although this condition was first known as infantile paralysis, the risk of paralysis from infection with this virus actually increases with age. In most modern countries, poliomyelitis is essentially nonexistent because an effective vaccine exists, but in under-developed countries this infection is still a major problem.

Pathogenesis of Polio

The poliovirus has an affinity for the CNS and normally reaches it by crossing the blood–brain barrier. The virus can also use the axons or perineural sheath of the peripheral nervous system to gain access. Motor neurons are particularly vulnerable to infection, and various levels of neuronal destruction cause necrosis of the neural tissue and infiltration by mononuclear cells, primarily lymphocytes. About 90% of poliomyelitis infections are very mild and are subclinical, with incubation times varying from 4 to 35 days but averaging about 10 days.

Three types of infection are caused by the polio virus:

- Abortive poliomyelitis is a nonspecific febrile illness lasting two to three days and having no signs or symptoms of CNS involvement.

- Aseptic meningitis (nonparalytic poliomyelitis) is characterized by meningeal irritation, a stiff neck, back pain, and back stiffness. Recovery from this form of poliomyelitis is rapid and complete.

- Paralytic poliomyelitis occurs in approximately 2% of persons infected by the virus, which targets and destroys the cells associated with the anterior horn of the spinal cord and brain stem. The hallmark of this form of the infection is asymmetric paralysis, with the extent of the paralysis varying from case to case. The muscle groups responsible for breathing can be affected, leading to respiratory difficulties. Temporarily damaged neurons can regain their function, but this recovery can take six months. Paralysis persisting after this recovery period is permanent.

Prevention of Polio

As mentioned above, the polio vaccine has essentially wiped out this infection in developed countries. Two types of vaccine are licensed for use in the United States, both developed in the 1950s: an inactive form, and a live attenuated vaccine. There are three serotypes of the virus, and both vaccines contain all three serotypes, a design that effectively inhibits infection. In fact, no cases of poliomyelitis attributed to indigenously acquired wild poliovirus have been reported in the United Sates since 1979. (Recall from our discussion of influenza and rhinovirus infections in Chapter 21 that there is no way of producing a completely effective vaccine because of the multiple numbers of serotypes of these viruses.)

Fast Fact There have been cases of polio that could be attributed to reversion of the virus back to the infectious form in the live attenuated vaccine.

Viral Encephalitis

The family of neurological infections classified as viral encephalitis are caused by mosquito-borne viruses called arboviruses (arthropod-borne), which is not a microbial taxonomic group. These viruses are common in the United States, and there is an increased occurrence of these infections in the summer months as the number of mosquitoes increases. A variety of clinical types of arbovirus cause infections that range in severity from subclinical symptoms to rapid death. These infections are all usually characterized by chills, headache, and a fever that can lead to mental confusion and coma. Survivors of viral encephalitis infections can subsequently develop permanent neurological disease.

Both horses and people are affected by arboviruses, and the names of some of the infections reflect this fact. For example, **eastern equine encephalitis** (**EEE**) and **western equine encephalitis** (**WEE**) both cause severe infection in humans. In fact, EEE has a 35% mortality rate in humans, and survivors normally suffer brain damage, deafness, and other neurological problems. **Saint Louis encephalitis** is the most common form of arbovirus encephalitis, but fewer than 1% of cases show clinical symptoms.

West Nile virus, an emerging encephalitis infection, was first seen in New York in 1999. It mostly affects birds but can also infect humans and horses. Although most human cases of West Nile disease are subclinical, there can be severe infection and rapid death in the elderly.

Persistent Viral Infections of the CNS

A variety of progressive neurological diseases in both humans and other animals are caused by viruses (Table 24.3). These are termed slow viral diseases, but a better term would be persistent viral infections, both because of the long period between infection and the onset of disease and because of the prolonged period of illness.

Subacute sclerosing panencephalitis is a rare chronic measles infection that occurs in children and produces progressive neurological disease. It is characterized by an insidious onset of personality change, progressive intellectual deterioration, and dysfunction of the autonomic nervous system. Progressive panencephalitis is also seen after rubella infection.

AIDS dementia complex is part of the pathology of HIV infection and presents as a persistent infection of the CNS in asymptomatic AIDS patients. The clinical course of this infection can vary from mild to very severe progressive dementia.

Persistent enterovirus infection is seen in patients with congenital or acquired immunodeficiency. It is a chronic infection of the CNS characterized by headache, confusion, lethargy, seizures, and increased numbers

Infection	Agent
Subacute sclerosing panencephalitis	Measles virus
Progressive panencephalitis after rubella infection	Rubella virus
AIDS dementia complex	Human immunodeficiency virus
Persistent enterovirus infection of the immunodeficient	Enterovirus

Table 24.3 Conventional viruses that cause persistent viral infections of the CNS.

of mononuclear cells in the CNS. This infection is caused by both echo-viruses and enteroviruses. Therapy with human hyperimmune globu-lin can cause clinical improvement, but relapses occur if the therapy is discontinued.

Keep in Mind

- There are two types of viral infection of the CNS: acute and persistent.
- Acute viral infections of the CNS include rabies, polio, and viral encephalitis.
- There are no treatments for acute viral infections but there are effective vaccines.
- Some persistent infections of the CNS are caused by conventional agents such as measles, rubella, and enterovirus.

PRIONS

As we learned in Chapter 8, prions have been proved to cause bovine spongiform encephalopathy in cattle, scrapie in sheep, and five fatal infections of the CNS in humans. Prions are transmissible infectious pro-teins, with the following properties:

- A diameter of 5–100 nm
- No nucleic acids
- Unusual resistance to ultraviolet irradiation, alcohol, formalin, boiling, proteases, and nucleases
- Can be inactivated by prolonged exposure to steam autoclaving or to 1 M or 2 M NaOH
- Replication to high titers in susceptible tissue
- Transmissible to experimental animals
- No initiation of host inflammatory response
- Do not elicit interferon production
- No alteration in pathogenesis by immunosuppression
- Chronic progressive pathology without remission or recovery

The pathogenesis of these infections is not well understood, but the path-ological features of each of the human infections are similar, including loss of neurons and proliferation of astrocytes. They are called spongi-form because of vacuoles seen in the brain cortex and cerebellum of infected patients. The incubation period can be from months to years, with the course of the infection being protracted and always fatal. Prions can remain infectious in brain tissue even after years of being immersed in formalin, and they are very resistant to ionizing radiation and many common disinfectants.

A prion is a protein coded for by a normal gene. This protein is desig-nated **PrP**^{c} and is converted into the infectious form called **PrP**^{sc} (the sc stands for scrapie) by a change in conformation (**Figure 24.9**). Contact of normal PrP^c with the infectious PrP^{sc} form causes the normal form to reconfigure into the infectious form, and this is the way in which the infectious form multiplies. It is the proliferation of the PrP^{sc} form that causes the pathology seen in prion infections (**Figure 24.10**).

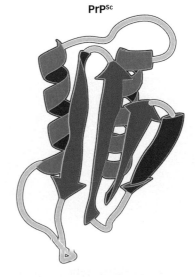

PrP^c

PrP^{sc}

Figure 24.9 Change in conformation that converts the PrP^c protein into the infectious PrP^{sc} prion.

Figure 24.10 Changes in brain tissue after infection with prion disease. Light micrographs of a section of normal brain (panel a) and a section of brain infected with prions (panel b). Note the large number of vacuoles visible in the latter.

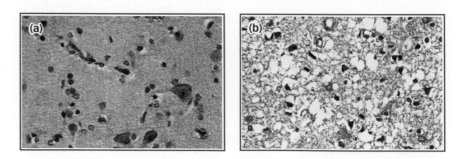

There are several prion infections that affect humans. Kuru is a subacute progressive neurological disease originally discovered in 1957 in the Fore people of New Guinea. Epidemiological studies done at that time showed that kuru usually affected adult women and children of both sexes. Symptoms were a failure of muscular coordination (ataxia), hyperactive reflexes, and muscular spasms that led to progressive dementia and death. There was diffuse neuronal degeneration and spongiform change of the cerebral cortex and basal ganglia. The studies showed that kuru was transmitted through the ingestion of soup made from human brains. (The Fore people cannibalized their dead relatives as a way of celebrating their lives, rather than burying or cremating them.) Incubation could take as long as 20 years after initial exposure. When the practice of cannibalism stopped, kuru disappeared.

Fatal familial insomnia presents as a difficulty in sleeping followed by progressive dementia. It occurs in people between the ages of 35 and 61 years and is fatal, with death occurring between 13 and 25 months after diagnosis.

Creutzfeldt–Jakob disease (CJD) is a progressive fatal infection of the CNS, most often seen in patients who are 60–70 years old. The infection initially presents as a change in cerebral function that can be mistaken for a psychiatric disorder. The patient exhibits forgetfulness and disorientation, and there is a progression to overt dementia that can involve changes in gait, involuntary movements, and seizures. This progression can last from four to seven months, with eventual paralysis, wasting, pneumonia, and death. The infection is seen throughout the world, and approximately one case per million people is reported each year.

The mode of transmission is unknown but has been attributed to contaminated dura mater grafts, corneal transplants, and contact with contaminated electrodes or instruments used in neurosurgical procedures. Transmission has also been linked to contaminated growth hormone, but there is no evidence of transmission by direct contact or airborne spread.

The incubation period for Creutzfeldt–Jakob disease is anywhere from 3 to 20 years, and the pathology is identical to that seen in kuru, with high levels of infectious prions found in the brain. In fact, brains of individuals with Creutzfeldt–Jakob disease show the same fibrils as those seen in sheep that have died of scrapie (**Figure 24.11**). Examination of brain tissue is the only way to confirm Creutzfeldt–Jakob disease, and there is no therapy.

Bovine spongiform encephalopathy ('mad cow' disease) was first identified in the UK in 1986. The source of the prions that caused infection was traced to cattle feed that contained meat and bonemeal from sheep that had scrapie. The cows that ate the feed became infected. Humans who ate products from the infected cows developed an infection known as variant Creutzfeldt–Jakob disease (vCJD). vCJD was first reported in 1996, and 229 deaths worldwide (177 in the UK) have been reported up to June 2014. Cases present in young adults as psychiatric problems that

Figure 24.11 False-color electron micrograph showing the characteristic fibrils seen in brain tissue of patients with prion-caused infections.

progress to neurological changes and eventually dementia. The average life expectancy after diagnosis is 14 months.

FUNGAL INFECTIONS OF THE CNS

As we discussed in Chapter 14, fungal infections are primarily opportunistic and are usually seen in immunocompromised patients. In our discussions of infections of the CNS, the most important fungal infection is cryptococcosis.

Cryptococcosis

The fungal infection called cryptococcosis is caused by *Cryptococcus neoformans*, which is an encapsulated form of yeast. Capsule production varies with the strain and is associated with environmental conditions. This fungus grows best at 35–37°C and in culture produces colonies within two to three days. *C. neoformans* can be found throughout the world, especially in soil contaminated with bird droppings (even though birds are never sick from this fungus).

Pathogenesis of Cryptococcosis

C. neoformans causes chronic meningitis, with a slow, insidious onset of infection. Symptoms include low-grade fever and headache, progressing to altered mental status and seizures. This infection is usually seen in patients who are immunocompromised and is common in AIDS patients.

The infection begins with inhalation of the yeast cells. Once inhaled, each yeast cell begins to overproduce its capsule (**Figure 24.12**). As we saw for encapsulated bacteria, this capsule is anti-phagocytic and can bind to complement components, thereby reducing the opsonization defense of the host. The capsule can also interfere with the presentation of antigens to T cells, thereby inhibiting the adaptive immune response. After inhalation, the yeast cells multiply outside the lungs and move into the nervous system. This results in intermittent headache, dizziness, and difficulty with complex cerebral functions, symptoms that continue over a period of weeks or months. Seizures, cranial nerve damage, and papilledema (edema of the optic nerve) appear in the later stages of the infection, accompanied by dementia and decreased levels of consciousness. These symptoms are accelerated in patients with AIDS.

Treatment of Cryptococcosis

Amphotericin B or fluconazole is used to treat cryptococcosis, and 75% of patients with cryptococcal meningitis initially respond to treatment. However, a significant portion of these patients relapse when the therapy is stopped, and many patients with chronic infection require repeated courses of therapy. There is residual neurological damage in more than half of cured patients.

Fast Fact Cryptococcosis can occur in individuals in whom there is no evidence of the immune system being compromised.

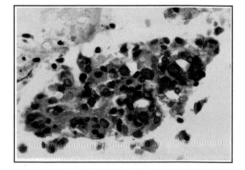

Figure 24.12 Micrograph depicting the histopathology associated with cryptococcosis of the lung. The stain used was mucicarmine.

PARASITIC INFECTIONS OF THE CNS

The most important parasitic infection of the CNS is trypanosomiasis (African sleeping sickness) caused by the protozoa of the *Trypanosoma* species, which we covered in Chapter 14. In this chapter we look at infections of the CNS caused by ameba.

Primary Amebic Meningoencephalitis

The infection of the CNS known as **primary amebic meningoencephalitis** is caused by free-living amebas belonging to the genera *Naegleria* and *Acanthamoeba*. *Naegleria* infections affect children and young adults and are acquired by swimming in fresh water (it enters the nasal passages). This infection occurs infrequently but is almost always fatal, and *Naegleria* organisms are found in large numbers in shallow freshwater ponds, particularly during warm weather.

Primary amebic meningoencephalitis resulting from infection by *Acanthamoeba* species is a subacute or chronic illness that is almost always fatal. *Acanthamoeba* is found in soil and also in fresh brackish water and has been found in the oropharynx of asymptomatic humans. Most *Acanthamoeba* infections occur in the southeastern United States, and infected patients typically fall ill during the summer months after swimming or water skiing in small, shallow freshwater lakes.

Pathogenesis of Amebic Meningoencephalitis

Histologic evidence suggests that *Naegleria* organisms enter the CNS by traversing the nasal mucosa and the cribriform plate. Once in the CNS, the organisms produce a purulent, hemorrhagic inflammatory reaction that extends perivascularly from the olfactory bulb to other regions of the brain. The infection is characterized by rapid onset of severe bifrontal headache, seizures, and occasionally abnormal taste and smell, progressing to coma and death within days. Examination of the cerebrospinal fluid shows blood and an intense neutrophil response, and wet mounts of cerebrospinal fluid reveal trophozoite forms of the parasite.

The epidemiology of *Acanthamoeba* encephalitis has not been clearly defined, but it is known that infection involves the elderly and immunocompromised. It is thought that the ameba reaches the brain by hematogenous dissemination from an unknown site, possibly the respiratory system, eye, or skin. The infection produces diffuse, necrotizing granulomatous encephalitis (**Figure 24.13**), frequently involving the midbrain, with both cysts and trophozoites being found in the lesions. In patients with AIDS, there can also be cutaneous ulcers and hard nodules containing amebas; in these patients, amebas are also seen in the cerebrospinal fluid. The clinical course of *Acanthamoeba* infection is more prolonged than that of *Naegleria* infection, and the former can occasionally end in spontaneous recovery.

Treatment of Amebic Meningoencephalitis

So far, very few patients have ever survived infection with *Naegleria*, and all of these individuals were diagnosed early and treated with high doses of amphotericin B together with rifampin. Studies on drug therapy for *Acanthamoeba* infections have not yet been completed.

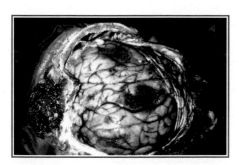

Figure 24.13 Autopsied brain from a patient with *Acanthamoeba* encephalitis. Notice the necrotizing granulocytic lesions.

Keep in Mind

• Infections of the CNS caused by the free-living amebas *Naegleria* and *Acanthamoeba* are rare but almost always fatal.

SUMMARY

- Infections of the CNS can be very dangerous and require immediate and aggressive therapy.

- The blood–brain barrier protects the CNS from bloodborne pathogens.

- Some pathogens can cross the blood–brain barrier; most infections of the CNS result from bacteremia or viremia.

- Any swelling or inflammation of brain tissue has serious consequences because the brain is encased by the skull, which cannot expand.

- Pathogens that can move from the blood to the cerebrospinal fluid cause meningitis.

- Bacterial meningitis is more serious than viral meningitis, but there are treatments and vaccines.

- Viral infection of the CNS can be acute or persistent.

- Acute viral infections of the CNS include rabies, polio, and encephalitis.

- Measles and rubella viruses can cause persistent infections of the CNS.

- Prions cause persistent diseases of the CNS.

- Fungal infections of the CNS are primarily opportunistic and occur in immunocompromised individuals.

- Parasitic infections of the CNS are usually caused by free-living amebas.

In this chapter, we have looked at infections that involve the CNS. These infections can be the result of bacterial, viral, fungal, or parasitic organisms, all of which can take advantage of the anatomy of the CNS to invade and cause infection. In addition, there are brain infections caused by infectious proteins called prions. When reviewing these infections, remember the anatomy of the CNS and how it is implicated in the infectious process. Also keep in mind that the condition referred to as meningitis, which is an infection of the meninges of the CNS, can be caused by a variety of organisms that all produce the same symptoms.

 SELF EVALUATION AND CHAPTER CONFIDENCE

Multiple Choice

Answers are given in the back of the book and help can be found in the student resources at:

www.garlandscience.com/micro2

1. One of the most dangerous results of infection of the central nervous system is
   A. Production of cerebrospinal fluid
   B. Inflammation
   C. Lack of nutrients to brain tissue
   D. All of the above

2. Which of the following infect the central nervous system?
   A. *Neisseria meningitidis*
   B. *Listeria monocytogenes*
   C. *Streptococcus pneumoniae*
   D. *Haemophilus influenzae*
   E. All of the above

3. Infections of the CNS rarely result from which of the following?
   A. Infections of the middle ear
   B. Sinusitis
   C. Mastoiditis
   D. Tuberculosis

4. Treatment of viral meningitis involves all of the following except
   A. Antibiotics
   B. Bed rest
   C. Fluids
   D. Medicines to control fever

5. The clinical effects of botulism are caused by

    A. Bacteria entering the brain
    B. Viruses entering the brain
    C. An endotoxin
    D. An exotoxin

6. Antibiotic therapy for tetanus would be

    A. Very effective at any time during the infection
    B. Moderately effective late in the infection
    C. Not effective at any time
    D. Effective only if given early in the infection

7. The symptoms of infant botulism include

    A. Constipation
    B. Bleeding from the rectum
    C. Increased activity and restlessness
    D. Diarrhea
    E. None of the above

8. The rabies virus is transmitted by

    A. The fecal–oral route of contamination
    B. The *Anopheles* mosquito
    C. The saliva of an infected animal
    D. As a co-infection with the influenza virus

9. The urban form of rabies is associated with

    A. Wild animals
    B. Domesticated cats
    C. Bats
    D. Unimmunized dogs and cats

10. The polio virus binds to receptors on

    A. Astrocytes
    B. Lymphocytes
    C. Motor neurons
    D. Neutrophils

11. Prions are

    A. Viruses
    B. Parasites
    C. Proteins
    D. Bacteria

12. All of the following are prion infections except

    A. Kuru
    B. Bovine spongiform encephalopathy
    C. Creutzfeldt–Jakob disease
    D. Subacute sclerosing panencephalitis
    E. Fatal familial insomnia

13. Cryptococcosis is a

    A. Bacterial infection of the CNS
    B. Viral infection of the CNS
    C. Parasitic infection of the CNS
    D. Fungal infection of the CNS
    E. Prion infection

14. *Naegleria* infection is acquired through

    A. Contaminated food
    B. Swimming
    C. Drinking water containing fungal spores
    D. Working with infected pigs

15–19. Answer **A** for true statements and **B** for false statements.

15. Poliovirus is the only virus with an affinity for the CNS.

16. There are three types of viral infection of the CNS: acute, persistent, and chronic.

17. Acute viral infections of the CNS include rabies, polio, and viral encephalitis.

18. Some persistent viral infections of the CNS are caused by measles, rubella, and enterovirus.

19. Prions become infective by associating with nucleic acids.

20–23. Each of the following questions lists two phenomena followed by how they might be differentiated. If the mode of differentiation is correct, answer **A**. Answer **B** if it is incorrect.

20. Vasogenic versus cytotoxic edema—cytotoxic edema involves neutrophil invasion.

21. Herpes simplex type 1 virus versus enterovirus infections of the CNS—enterovirus infections mainly affect adults.

22. Occult versus overt infections—occult infections can occur from bacterial endocarditis.

23. Bacterial versus aseptic meningitis—aseptic meningitis is always caused by viral infections.

DEPTH OF UNDERSTANDING

Questions listed here require you to bring together the concepts you have learned in this chapter into a discussion format. This helps you to increase your depth of understanding of the material you have learned. Help can be found in the student resources at: www.garlandscience.com/micro2

1. Describe the various methods in which pathogens can gain entry to the central nervous system.

2. Many of the host defenses against infections can cause damage to the host. Describe how this happens in respect of the CNS.

3. Describe the pathogenesis of tetanus from initial infection to a fatal outcome.

CLINICAL CORNER

Help can be found in the student resources at: www.garlandscience.com/micro2

1. You are an avid jogger and living in New York, and Central Park is the place you love to run. It has great scenery and you enjoy watching the horse-drawn carriages taking tourists on sightseeing trips through the park. One morning as you enter the park you are told that it is closed for mosquito spraying and will not be open again for several days because of West Nile virus. Although this ruins your jogging plans, you understand the problem. Several other joggers and drivers of the carriages are not so understanding and demand to know what the problem really is. You want to be helpful.

   A. What do you tell them about West Nile virus?
   B. Why are they spraying for mosquitoes?
   C. Who are the people most at risk of this infection?
   D. When one of the joggers points out that the Nile River is thousands of miles away, how do you explain to him about West Nile virus being found in New York City's Central Park?

2. Your patient presents complaining of headache and dizziness, which have been going on for more than a month. He has also become forgetful to the point that he can no longer manage the automotive parts store he owns. His medical history shows that he received a kidney transplant two years ago but he has no signs of an infection of the transplanted kidney. He is not taking any medication other than the immunosuppressants required to prevent the rejection of his transplanted kidney. Tests have shown no indication of cardiovascular abnormalities.

   A. What are some of the possible explanations for his problems?
   B. Does the fact that he is a transplant recipient mean anything?

Infections of the Blood and Lymph

Chapter 25

Why Is This Important?

Infections that occur in the blood can easily become systemic—that is, affecting multiple organs throughout the body—and have devastating effects.

Lia was a patient who recently suffered from Lyme disease, a blood infection caused by the bacterium *Borrelia burgdorferi* and spread by the bite of a tick. The insidious illness initially presented as a skin rash, but ultimately caused her painful joints, recurring headaches, chills, body aches, and low-grade fevers. As she described it, Lyme disease drained her of her *joie de vivre* and made her extremely ill for an extended period of time. Fortunately, the prescribed course of antibiotics relieved most of her symptoms and the disease never progressed to more damaging neurological, liver, or heart problems.

For a while afterward, she was afraid to venture into the woods behind her New England home for fear of contracting the disease again. She now enjoys the woods again, but checks herself thoroughly for ticks and uses insect repellent. These precautions should help ensure that it won't happen again.

OVERVIEW

In this chapter, we look at infections that occur in the blood and lymph. We will first look at pathogens in the circulatory system and why the danger of systemic infection requires that they be dealt with quickly and aggressively. We will then move on to bacterial, viral, and parasitic pathogens that always use the blood and lymph in the course of their infections, and we will see that several of these infections involve transmission by vectors.

We will divide our discussions into the following topics:

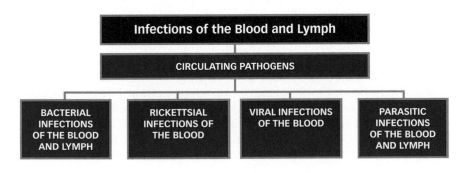

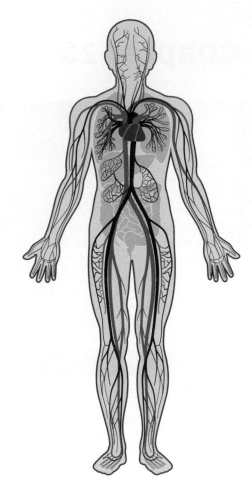

Figure 25.1 The human circulatory system. Notice that this system has access to every area of the body, making it the perfect highway for pathogens moving through the body.

"The story [of Lyme disease] is an interesting one—with deer, mice, and acorns in starring roles."

Robert Buckman (1948–2011)

The story of the blood infection called Lyme disease is not just limited to woodland mammals and seeds; it is a complex tale that also includes ticks and, unfortunately, humans. As we will see in this chapter, blood infections are varied and can affect the entire body. Because blood travels to all the systems of the body (**Figure 25.1**), it can easily carry pathogens from one location to another.

CIRCULATING PATHOGENS

In many ways, the presence of circulating pathogens in the blood is part of the natural progression of an infection, and these organisms are usually quickly removed from the blood by the host defenses. However, in some cases, pathogens in the blood are a reflection of a serious, uncontrollable infection that can lead to **sepsis** or **septic shock**. Depending on the type of organism, such infections are classified as one of the following:

- Bacteremia—bacteria in the blood (**Figure 25.2**)

- Viremia—viruses in the blood

- Fungemia—fungi in the blood

- Parasitemia—parasites in the blood

Bacteremia and fungemia may also be caused by pathogens growing on the inside or the outside of intravenous devices. The infection may start out as a minor problem but can quickly become serious. This is particularly true in patients who are debilitated from long stays in hospital.

Bacteremia

Transient bacteremia occurs if bacteria that are part of the normal flora are exposed to the blood through manipulation or trauma to the body. It is common after some medical, dental, and surgical procedures and is usually of no importance clinically. However, if these bacteria are not cleared from the blood, they can travel round the body and cause a serious infection.

Most cases of significant bacteremia result from **extravascular infections**. Pathogens move into the lymphatics from infected tissues and from there into the blood. Recall that lymph has to pass through the lymph nodes, and many pathogens are captured in the nodes. However, if there are overwhelming numbers of pathogens present, some of them can reach the blood. With staphylococcal pneumonia, for example, there can be thousands of pathogens per milliliter of blood. **Table 25.1** shows the frequency with which certain bacteria produce bacteremias. The most common sources of bacteremia are urinary tract infections (Chapter 23), respiratory infections (Chapter 21), and skin infections (Chapter 26). Organisms that produce meningitis (Chapter 24) usually also produce bacteremia at the same time.

Intravenous Line and Catheter Bacteremia

We have seen throughout the previous chapters that nosocomial infections are the cause of serious problems for patients who are hospitalized because these patients are in many cases debilitated or immunocompromised. Additionally, there are always large numbers of antibiotic-resistant

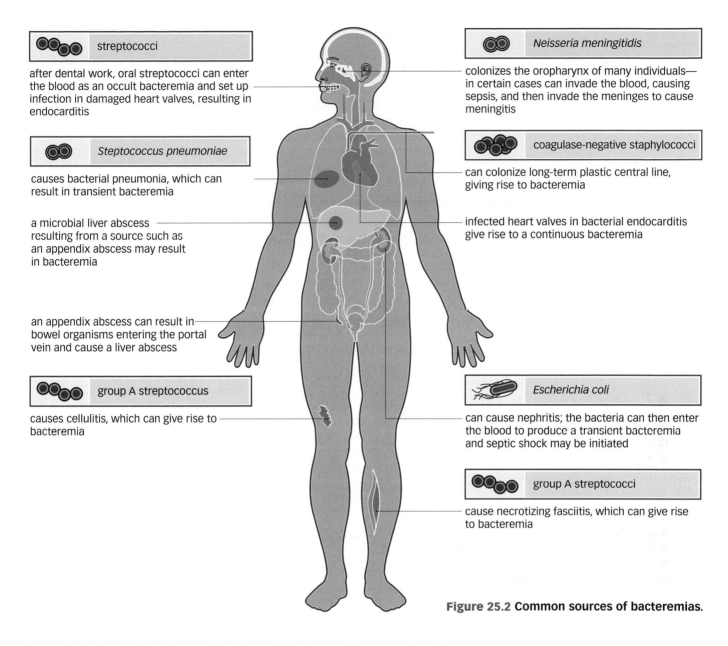

streptococci

after dental work, oral streptococci can enter the blood as an occult bacteremia and set up infection in damaged heart valves, resulting in endocarditis

Steptococcus pneumoniae

causes bacterial pneumonia, which can result in transient bacteremia

a microbial liver abscess resulting from a source such as an appendix abscess may result in bacteremia

an appendix abscess can result in bowel organisms entering the portal vein and cause a liver abscess

group A streptococcus

causes cellulitis, which can give rise to bacteremia

Neisseria meningitidis

colonizes the oropharynx of many individuals— in certain cases can invade the blood, causing sepsis, and then invade the meninges to cause meningitis

coagulase-negative staphylococci

can colonize long-term plastic central line, giving rise to bacteremia

infected heart valves in bacterial endocarditis give rise to a continuous bacteremia

Escherichia coli

can cause nephritis; the bacteria can then enter the blood to produce a transient bacteremia and septic shock may be initiated

group A streptococci

cause necrotizing fasciitis, which can give rise to bacteremia

Figure 25.2 Common sources of bacteremias.

bacteria in hospitals. The combination of these two factors makes the implantation of intravenous lines and catheters a potentially serious matter.

The insertion of either an intravenous line or a catheter can cause the blood to become colonized with organisms normally found on the skin. If this were a transient bacteremia in a healthy host, the infection would quickly be dealt with by the immune response. In debilitated hosts, however, the bacteremia can persist and increase the chances of such secondary complications as **intravascular infections**. Pathogens most commonly involved in this type of secondary infection include *Staphylococcus epidermidis*, *Staphylococcus aureus,* and *Corynebacterium* species.

In some cases, it is the intravenous fluid that is contaminated rather than the apparatus, and in these cases the organisms involved are usually Gram-negative rods such as *Pseudomonas* species. In cases of catheter bacteremia, removal of the contaminated device and treatment with antibiotics usually remedies the situation.

More than 90% of cases
Haemophilus influenzae type b
Neisseria meningitidis
Streptococcus pneumoniae
Brucella
Salmonella enterica serotype Typhi
Listeria
Variable frequency (10–90% of cases) depending on stage and severity of infection
Beta-hemolytic streptococci
Staphylococcus aureus
Neisseria gonorrhoeae
Leptospira
Borrelia
Acinetobacter
Shigella dysenteriae
Enterobacteriaceae
Pseudomonas species

Table 25.1 Bacteria most often involved in bacteremia.

Fast Fact Before antibiotics, death was inevitable for all forms of endocarditis.

Sepsis and Septic Shock

Sepsis and septic shock can result from any infection, but they are most commonly seen as a progression from bacteremia. As we discussed in Chapter 7, severe sepsis occurs when an infection causes a massive response from the host immune system, with systemic inflammation, fever, chills, tachycardia, organ dysfunction, and changes in mental status. If this syndrome continues, it leads to septic shock, which is characterized by the development of severe hypotension. If not treated, septic shock leads to **refractory septic shock**, in which the hypotension cannot be resolved by standard treatments. Continuation of septic shock leads to multiorgan failure, disseminating intravascular coagulation (DIC), and death.

The initial events in severe sepsis are vasodilation with decreased peripheral resistance and increased cardiac output. This is followed by capillary leakage, reduced blood volumes, and multiorgan failure. Early recognition of this problem is critical because resolution depends on more than treatment with antibiotics. Treatment of sepsis conditions also requires adequate maintenance of tissue perfusion, and careful fluid and electrolyte management.

Intravascular Infections

Intravascular infections arise when pathogens gain entrance to the blood and damage the structures of the cardiovascular system. Infectious **endocarditis** (infection of the heart), **thrombophlebitis** (infection in the veins), and **endoarteritis** (infection of the arteries) are most commonly caused by bacteria, but fungi can sometimes be involved. Infections of the cardiovascular system are very dangerous and can be fatal if not promptly and adequately treated. These infections commonly produce constant shedding of organisms into the blood, a condition that causes persistent low-grade fever.

Infectious Endocarditis

Infectious (or infective) endocarditis was once referred to as bacterial endocarditis, but the more general name was adopted once it was realized that organisms other than bacteria can cause it. Table 25.2 lists some of the more common bacterial causes. The infection can be either acute, which presents with high fever and toxicity and can result in death within a few days or weeks, or subacute, which presents with low-grade fever, weight loss, and night sweats, with death taking weeks to months.

If bacteria from the normal flora get into the blood, they can cause endocarditis if conditions are correct. Dental procedures can be dangerous for people who are predisposed to infectious endocarditis,

Organism	Approximate percentage of cases caused
Viridans streptococci	30–40
Other streptococci	15–25
Staphylococcus aureus	15–40
Enterococci	5–18
Coagulase-negative staphylococci	4–30
Gram-negative bacilli	2–13

Table 25.2 Common bacterial sources of infectious endocarditis.

because dislodged plaque-forming bacteria at the gum line can enter the bloodstream and infect other parts of the body, including the valves of the heart. These are people who already have cardiac valvular lesions, such as those seen in heart disease, and they need to take high doses of antibiotics before dental procedures.

Pathogenesis of infectious endocarditis

Infectious endocarditis affects the heart valves, either the body's natural valves or synthetic ones in patients who have received replacement valves. The infection can also develop on the septa of the heart and on cardiac shunts. The endothelium of the heart is altered by the infecting pathogens, and this alteration facilitates the colonization of the area by the pathogens. This in turn causes the deposition of platelets and fibrin at the site (**Figure 25.3**). The turbulence of the blood flow around these deposits can lead to further irregularities of the endothelial surfaces, irregularities that facilitate the further deposition of platelets and fibrin. Eddies caused by slower blood flow in these areas cause even more pathogen growth here.

Circulating pathogens adhere to the fibrin and platelets, which causes an inflammatory response that includes both activation of complement and further damage to the endothelial surfaces. As this process continues, a thrombotic mesh composed of platelets, fibrin, and inflammatory cells forms and leads to the formation of a structure known as **mature vegetation** (**Figure 25.4**). This structure protects the pathogens from the host defense and also helps keep out antibiotics. The mature vegetation causes alterations in the cardiac endothelium, which obstructs blood flow and increases turbulence. The increased turbulence can cause part of the mature vegetation to fall off and form an **embolus** (plug; plural emboli) that may move into the smaller vessels and also obstruct blood flow to other parts of the body (**Figure 25.5a**). Transport of these emboli to the brain or the coronary arteries can be a lethal event.

Colonization of the heart valves can also lead to the formation of immune complexes by the host (**Figure 25.5b**). These complexes form as antibodies against the infecting organisms bind to them and form large aggregates. These aggregates activate the complement system, which then causes such peripheral problems as nephritis, arthritis, and cutaneous vascular lesions.

Complications from infectious endocarditis include a risk of congestive heart failure, rupture of the chordae tendineae of the heart valves, and perforation of the valves (**Figure 25.5c**). In addition, the kidneys are usually affected, and blood can be found in the urine. If the infection occurs on the left side of the heart, coronary emboli can form. Infection of the right side of the heart can lead to infection of the lung in addition to the formation of emboli.

Treatment of infectious endocarditis

In addition to treatment, management of infectious endocarditis is important. The nature of the infection means that it is difficult to cure, and response to therapy is slow. Therefore, the antibiotic therapy must be aggressive and include bacteriocidal drugs that can be given in high enough concentration to guarantee that levels remain high in the blood without causing toxicity, a balancing act that can be difficult to achieve. The course of therapy can last more than four weeks. Removal or replacement of the infected valve may be necessary, and this involves its own set of potential consequences.

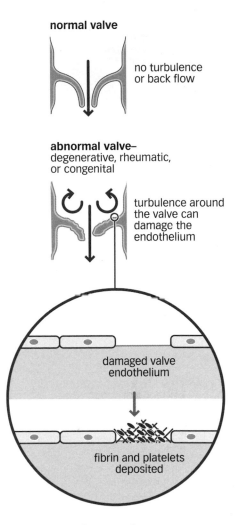

Figure 25.3 Changes that occur in heart valves that predispose to infectious endocarditis. The turbulence that occurs around the heart valves gives the infecting organisms an opportunity for additional growth.

Figure 25.4 Mature vegetation (shown by arrows) in a heart that has been subjected to subacute infectious endocarditis.

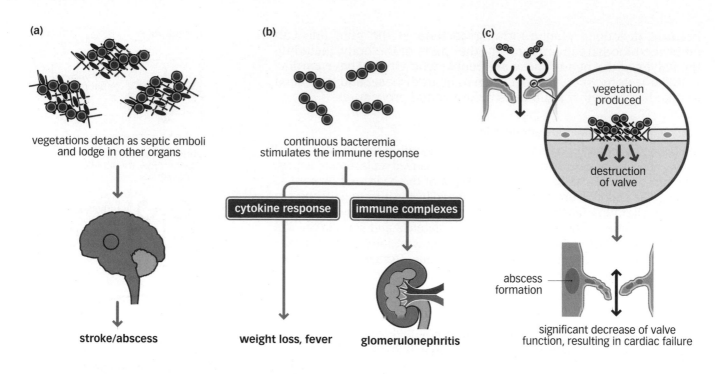

(a)

vegetations detach as septic emboli
and lodge in other organs

↓

stroke/abscess

(b)

continuous bacteremia
stimulates the immune response

cytokine response | immune complexes

weight loss, fever | glomerulonephritis

(c)

vegetation
produced

destruction
of valve

abscess
formation

significant decrease of valve
function, resulting in cardiac failure

Figure 25.5 Effects of infectious endocarditis. Panel a: The mature vegetation formed during the infection can detach from an infected heart valve and become an embolus that can then lodge in some distal organ. Panel b: Stimulation of the immune system can cause immune complexes, which cause distal problems. Panel c: Perforation and destruction of the valve.

Keep in Mind

- Blood and lymph travel throughout the body, and an infection in any part of the body can be spread throughout the body if pathogens gain entry to the blood or lymph.

- All types of pathogenic organism can enter the blood.

- Pathogens, especially bacteria, in the blood can lead to sepsis and septic shock, which can result in multiorgan failure and death.

- Pathogens in the blood can cause damage to cardiovascular structures, for example endocarditis.

BACTERIAL INFECTIONS OF THE BLOOD AND LYMPH

Now let us examine some bacterial infections that we can consider to be major infections of the circulatory system: plague, tularemia, brucellosis, Lyme disease, and relapsing fever. Bacteria causing these infections are always seen in the blood or lymph.

Plague

Plague is an infection that is vector-transmitted to humans and may be the most explosively virulent bacterial infection ever known. The infection spreads from the lymph nodes to the blood and can spread from there to the lungs. Once spread to the lungs, the infection is referred to as **pneumonic plague** and can be easily spread from person to person. This is the form of plague that was referred to as the **Black Death** in the Middle Ages. All forms of plague can lead to toxic shock and death in days. Few other bacterial infections kill previously healthy people so quickly.

Plague is caused by the bacterium *Yersinia pestis* and is an infection of rodents that can be transferred to humans by the bite of the rat flea (*Xenopsylla cheopis*). Plague exists in two forms: sylvatic, which is seen in wild rodents and is the primary reservoir of the organism, and urban, which is seen in cities.

Fast Fact In the fourteenth century, the population of Europe was about 105 million. In less than four years (1346–1350), the Black Death killed 25 million people.

The plague life cycle begins when fleas become infected after feeding on infected rats. The *Y. pestis* organisms multiply in the intestine of the flea and are regurgitated into a host as the flea bites. The bite of the infected fleas leads to **bubonic plague**, which is not normally contagious. However, some infected individuals will develop the pneumonic form of the infection as a result of bacteria spreading to the lungs. This form is highly contagious and can spread between people.

Pathogenesis of Plague

Once *Y. pestis* has been injected past the barrier of the skin, the difference between flea body temperature and human body temperature and the difference in ionic environment cause the bacterium to produce virulence factors. One of these factors, the **F1 protein**, forms a gel-like capsule that prevents phagocytosis and allows the bacteria to multiply in the submucosa. The bacteria eventually reach the lymph nodes and multiply there very rapidly. This produces a **bubo**, which is a suppurative, hemorrhagic, bulging lymph node (**Figure 25.6**) and is the source of the name bubonic plague. From the bubo, the organism spreads rapidly into the blood.

In the blood, *Yersinia* produces extensive systemic toxicity resulting from the release of lipopolysaccharides and virulence enzymes. The bacteria spread to other organs, most notably the lungs, where the result is necrotizing hemorrhagic pneumonia. The incubation period, which is the length of time between when the patient is bitten and when the first symptoms appear, is two to seven days. The main signs are fever and painful buboes, most often in the groin. Without treatment, 50–75% of patients get bacteremia and die of septic shock within days. If plague localizes in the lungs, death can occur within two to three days. Plague is almost always fatal if treatment is delayed more than a day after the onset of symptoms. Approximately 5% of patients develop the pneumonic form of the infection.

Treatment of Plague

Streptomycin is the treatment of choice for both the bubonic and pneumonic forms of plague, but doxycycline, ciprofloxacin, gentamicin, and chloramphenicol can also be used. If treatment is timely, the mortality from plague is less than 10%.

Tularemia

Tularemia is a zoonotic infection that moves from animals to humans and is caused by *Francisella tularensis*, a small Gram-negative facultative intracellular coccobacillus. This organism grows only on a specialized medium that contains sulfhydryl compounds, and even in this supportive environment it takes up to 10 days of incubation before tiny colonies appear. Tularemia is an infection of wild animals, most often ground squirrels and rabbits in the United States. Transmission can be by inhalation, tick bite, ingestion of contaminated meat or water, or directly by contact with an abrasion or cut while skinning an infected animal. Infected animals may not show any signs of this infection. Because the ID$_{50}$ is small (less than 100 organisms) and because humans can become infected from a bite, there are many routes to tularemia infection. It has been shown that tularemia can be acquired through the inhalation of *F. tularensis* organisms as an infected animal is being skinned. Tularemia is distributed throughout the Northern Hemisphere, but distribution patterns vary widely from one region to another. In the United States, between 100 and 200 cases are reported each year.

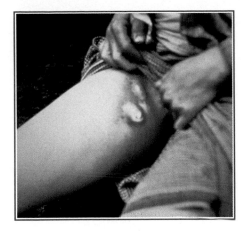

Figure 25.6 A bubo, the salient feature of bubonic plague, on the thigh of a patient.

Fast Fact Tularemia is considered to have potential as a bioterrorist weapon (discussed in Chapter 8).

Pathogenesis of Tularemia

The incubation phase for tularemia lasts between two and five days and usually results in an ulcerated lesion at the inoculation site. From this site, the infecting organisms move to the organs of the mononuclear phagocytic system and form granulomas. *Francisella* multiplies in macrophages, using acidification to disrupt the fusion of the phagosome and the lysosome, and also multiplies in hepatocytes and endothelial cells. The presence of the pathogens in the latter cells can result in abscess formation.

Tularemia can follow a number of courses depending on the inoculation site and on how far into the body the pathogens spread. All possible courses, however, begin with the acute onset of fever, chills, and malaise. In **ulceroglandular tularemia**, a localized **papule** (a small, solid elevation on the skin) forms at the inoculation site and becomes ulcerated and necrotic. This leads to swelling of the regional lymph nodes, which become very painful.

The infection can also be acquired through the eyes, in which case it is referred to as **oculoglandular tularemia** and produces a painful purulent conjunctivitis. A person ingesting a large number of *Francisella* bacilli can develop **typhoidal tularemia**, which presents with symptoms and signs similar to those seen in typhoid fever.

Any form of tularemia can progress to a systemic infection in which lesions are found in multiple organs. The mortality of untreated tularemia can be as high as 30%.

Treatment of Tularemia

Streptomycin is the drug of choice for all forms of this infection, but gentamicin, doxycycline, and ciprofloxacin are also effective. Two important factors for prevention are the use of rubber gloves, a mask, and eye protection when in contact with potentially infected animals, and the prompt removal of any ticks found on a person's body.

Brucellosis

The zoonotic infection brucellosis is seen in sheep, cattle, pigs, goats, dogs, and other animals where there is infection of the reproductive tract. The infecting organisms are various Gram-negative bacteria of the genus *Brucella*. Humans become infected by occupational contact or by ingesting contaminated animal products. In humans, brucellosis is a chronic illness characterized by fever, night sweats, and weight loss that can last for weeks or months. In some cases, the patient develops a cycling pattern of symptoms referred to as **undulant fever**.

Pathogenesis of Brucellosis

Brucella organisms gain access through cuts in the skin, contact with mucous membranes, inhalation, or ingestion. After they have penetrated the skin or mucous membranes, the organisms multiply in macrophages in the host's liver, spleen, bone marrow, and other components of the mononuclear phagocytic system. Their ability to keep host phagosomes and lysosomes from fusing allows the organisms to survive and multiply. In addition, *Brucella* impairs the host's ability to produce and release cytokines as part of the immune response. If not controlled locally, a *Brucella* infection causes the formation of small granulomas at sites throughout the mononuclear phagocytic system. At these sites, bacteria multiply and are released back into the circulation, intermittently causing recurrent bouts of chills and fever (thus the name undulant fever).

Symptoms start with malaise, chills, and fever, continuing for 7–21 days after initial infection. During this period, drenching night sweats and fever reaching 40°C are common. This nocturnal fever can continue for weeks, months, or even years. Other symptoms include headaches, body aches, and weight loss. Fewer than 25% of infected individuals show any signs in the mononuclear phagocytic system, but such individuals exhibit splenomegaly (enlargement of the spleen), hepatomegaly (enlargement of the liver), and lymphadenopathy.

Treatment of Brucellosis

Doxycycline and gentamicin are the primary antibiotics used for brucellosis, and doxycycline is preferred because of its pharmacologic characteristics. In seriously ill patients, these drugs can be supplemented with streptomycin, gentamicin, or rifampin.

Lyme Disease

Lyme disease is caused by *Borrelia burgdorferi*, which is transmitted to humans by the *Ixodes* tick. This bacterium is a Gram-negative spirochete with properties similar to *Treponema pallidum* (the causative agent of syphilis). It requires a specialized medium for growth, and even in that medium the doubling time is between 8 and 24 hours. Consequently, isolation can be difficult. In addition, there are at least 10 subspecies of *B. burgdorferi*, all of which are geographically localized. *B. burgdorferi* exists as part of a complex life cycle involving ticks, mice, and deer (**Figure 25.7**). Humans are incidentally involved when ticks feed on people who enter their habitat.

Figure 25.7 The life cycle of *Ixodes*, the tick vector of Lyme disease. Notice that the tick depends on the availability of deer to complete its life cycle.

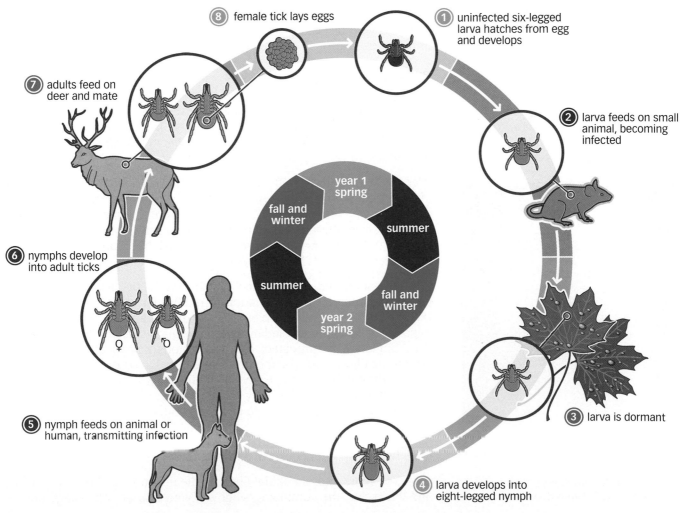

8 female tick lays eggs

1 uninfected six-legged larva hatches from egg and develops

7 adults feed on deer and mate

2 larva feeds on small animal, becoming infected

6 nymphs develop into adult ticks

year 1 spring
summer
fall and winter
summer
year 2 spring
fall and winter

♀ ♂

5 nymph feeds on animal or human, transmitting infection

3 larva is dormant

4 larva develops into eight-legged nymph

Lyme disease is endemic in several parts of the United States, Canada, and temperate areas of Europe and Asia. Approximately 90% of the 10,000–15,000 cases of Lyme disease reported worldwide are reported in the United States.

The primary reservoir for *B. burgdorferi* is mice. Mice serve as the host for the early stages of the *Ixodes* life cycle, and deer serve as the host for the final stages. In spring, fertile adult female ticks living on the deer become engorged on blood, fall from the deer, and lay their eggs in the soil. When hatched, the tick larvae seek out mice for blood meals. Because mice are the host for *B. burgdorferi*, the bacterium is picked up by the tick larvae feeding on the mice and then remains with the ticks throughout their development. During the following spring or summer, the larvae mature to adulthood and parasitize deer. This complex life cycle takes two years, and the deer host is essential to the existence of Lyme disease because this is where *Ixodes* matures and mates.

Humans contract Lyme disease primarily through tick larvae, but the pathogenic mechanisms of acquisition are not yet understood. It is known that the outer membrane of *B. burgdorferi* contains proteins and a toxic form of lipopolysaccharide that differs from the usual Gram-negative endotoxin. These bacteria also contain peptidoglycan, which has inflammatory properties that can survive for long periods in tissues and may cause arthritis if deposited in joint tissues.

Patients with Lyme disease exhibit modulations in their immune response, including the inhibition of mononuclear and natural killer cell function, the proliferation of lymphocytes, and the production of cytokines. In fact, chronic Lyme disease has aspects very similar to those seen in auto-immune disease.

Pathogenesis of Lyme Disease

Acute Lyme disease is characterized by fever, a migratory bull's-eye rash, and muscular and joint pain. Additionally, there is often meningeal irritation associated with the acute form. Chronic Lyme disease is character-ized by the evolution of meningoencephalitis, myocarditis, and disabling recurrent arthritis.

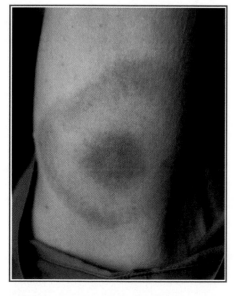

Figure 25.8 The bull's-eye rash associated with Lyme disease.

Both the acute and chronic forms of Lyme disease are highly variable and involve multiple body systems. Signs and symptoms occur in multiple overlapping patterns that come and go at different times, but skin lesions spreading from the site of the tick bite and relapsing arthritis are the most persistent findings. Lyme disease is rarely fatal, but if untreated it can be the source of chronic ill health.

Primary lesions appear in the first month after the tick bite and expand to become annular lesions with a raised red border and central clear area (**Figure 25.8**). The ring of the bull's-eye expands and forms an **erythema** migrans lesion accompanied by fever, myalgia, headache, and joint pain. If the infection is untreated, the skin lesions disappear but the other symptoms can persist for months. Days, weeks, or even months later, a secondary infection stage involving the central nervous system and the cardiovascular system may develop. Neurologic abnormalities can include meningitis, facial nerve palsy, and peripheral nerve destruction; cardiovascular involvement can lead to acute myocarditis and enlarge-ment of the heart. Normally both neurologic and cardiovascular symp-toms resolve spontaneously in a matter of weeks.

Weeks to years after the initial onset of the infection, arthritis can begin, and this marks the continuing stage of Lyme disease. This stage occurs in two-thirds of untreated patients, fluctuates intermittently, and involves the large joints, in particular the knee. In serious cases, this arthritic condition may cause the erosion of bone, but less frequently there can be chronic involvement of the central nervous system, affecting memory, mood, and sleep.

Treatment of Lyme Disease

Prevention involves prompt removal of the tick. Doxycycline and amoxicillin are the preferred antibiotics for the treatment of early-stage Lyme disease, but intravenous penicillin G is used for patients with neurologic and cardiovascular involvement. The response to therapy is very slow and requires the continuous use of antibiotics for 30–60 days.

Relapsing Fever

The infection known as **relapsing fever** is caused by *Borrelia* species other than *B. burgdorferi* and is transmitted to humans by the bite of either ticks or body lice. There are two forms of relapsing fever, and which of them a patient contracts depends on whether the vector was a tick or a louse, and also on the *Borrelia* species involved. The louse-borne form is usually seen in epidemics, whereas the tick-borne form is not.

The occurrence and distribution of the tick-borne form are determined by the biology of the tick and by the relationship between the tick and the *Borrelia* reservoir, which can be rodents, rabbits, birds, or lizards. Ticks that harbor *Borrelia* can remain infectious for several years. In addition, an infected tick is able to transfer the spirochete it carries to its progeny, making the offspring carriers even if the offspring do not feed on infected rodents or other *Borrelia* vectors.

The cycle is different for the louse-borne form of relapsing fever, because lice have only human hosts. In addition, lice live for only about two months and do not pass the pathogenic spirochete to progeny. The louse-borne form is caused by *Borrelia recurrentis*, and lice become infected with the spirochete after biting an infected human. *B. recurrentis* multiplies in the endolymph of the louse but never moves to the salivary glands or the feces. Therefore, to infect an uninfected human, the louse must be crushed and scratched into the bite wound.

Pathogenesis of Relapsing Fever

This illness presents with fever, headache, muscle pain, and weakness. The symptoms last about one week, disappear, and then return a few days later. During the relapse, the spirochetes can be found in the patient's blood. Relapsing fever develops when thousands of spirochetes are circulating per milliliter of blood, and although the exact mechanisms of the infection are not known, it has been shown that these circulating organisms disappear during the periods between relapses. It is thought that they may sequester in the organs, and at each relapse the spirochetes appear with new antigenic markers, causing the synthesis of new antibodies against the new antigens. The periods of relapse correspond to the development of new antibodies.

The incubation period is approximately seven days and is characterized by a huge number of spirochetes in the blood (this condition is called **spirochetemia**). For tick-borne relapsing fever there are only two relapses, but for the louse-borne form there can be as many as ten relapses. Fatalities are rare in the tick-borne form, but mortality in the louse-borne form can reach 40% in untreated individuals, usually from myocarditis, cerebral hemorrhage, and liver failure.

Treatment for Relapsing Fever

A single dose of doxycycline or erythromycin is sufficient to deal with this infection.

Keep in Mind

- Bacterial infections of the blood include plague, tularemia, brucellosis, Lyme disease, and relapsing fever.

- Plague is one of the most virulent bacterial infections ever known.

- Plague can present as either pneumonic plague or bubonic plague.

- Tularemia and brucellosis are zoonotic infections.

- Lyme disease and relapsing fever are caused by the spirochete *Borrelia*.

RICKETTSIAL INFECTIONS OF THE BLOOD

Rickettsia species are coccobacilli, but we treat them here in a section separate from the section on bacterial infections because they have characteristics of both bacteria and viruses. These coccobacilli divide by binary fission and are very small (0.3–0.5 μm). Although they are Gram-negative organisms, they stain very poorly and are therefore better resolved with Giemsa stain. They have a peptidoglycan cell wall with a Gram-negative outer layer containing lipopolysaccharides and outer membranes that extend to the cell surface. All of these traits are similar to those of bacteria, but *Rickettsia*, like viruses, are obligate intracellular parasites.

Rickettsia organisms cause spotted fevers and typhus-related illnesses (**Table 25.3**). They use animal reservoirs and are transmitted by arthropod vectors. Infections produced by *Rickettsia* pathogens are typically fevers that are often accompanied by vasculitis. The most common of these is **Rocky Mountain spotted fever** (**RMSF**), which, despite its name, is seen throughout the world. There are several types of typhus, but we will restrict our discussion to the epidemic and endemic forms.

Rickettsia grows freely in the cytoplasm of infected eukaryotic cells and can be cultured only in cell cultures and fertile chicken eggs. These pathogens grow in cytoplasmic vacuoles of the host and then escape the vacuole and begin growing in the cytoplasm. They also use directional actin polymerization like that seen with some viruses to move through the cell and from cell to cell. Eventually their numbers become so large that they rupture the host cell, which is another similarity to viruses. *Rickettsia* cannot survive outside host cells, and if they do not find a host

Infection	Pathogen	Distribution	Vector	Reservoir
Spotted fever group—Rocky Mountain spotted fever	*Rickettsia rickettsii*	North and South America	Tick	Rodents and dogs
Rickettsial pox	*Rickettsia akari*	USA, Russia, Korea, and Africa	Mite	Mouse
Typhus group	*Rickettsia prowazekii*	Africa, Asia, and South America	Body louse	Humans

Table 25.3 Some pathogenic rickettsiae.

they cease metabolic activity and begin to leak proteins, nucleic acids, and other essential molecules. This instability leads to a rapid loss of potential infectivity.

The classic example of a rickettsial infection is **epidemic typhus**, but the most prevalent rickettsial infection in the United States is RMSF. Both of these infections are characterized by fever, rash, and muscle aches. Both diseases may be fatal as a result of vascular collapse.

Rocky Mountain Spotted Fever

The etiologic agent for RMSF is *Rickettsia rickettsii*. The infection is an acute febrile illness that occurs in association with residential and recreational exposure to wooded areas infested with ticks, which are the vectors for the *Rickettsia* species that cause RMSF.

Different RMSF vectors are found in different geographic locations: the wood tick (*Dermacentor andersoni*) in the western United States, the dog tick (*Dermacentor variabilis*) in the eastern part of the country, and the lone star tick (*Amblyomma americanum*) in the southwestern and midwestern states of the country. It is interesting to note that the *Rickettsia* pathogens do not harm the tick but live in the endolymph and are passed from one vector generation to the next. More than two-thirds of RMSF cases occur in children younger than 15 years of age between April and September.

Pathogenesis of Rocky Mountain Spotted Fever

The incubation period from time of bite to onset of symptoms is usually two to six days. The symptoms include fever, headache, muscle aches, mental confusion, and a rash that appears first on the soles, palms, wrists, and ankles (**Figure 25.9**) and then moves toward the trunk. The rash is the most characteristic feature of the infection. It usually develops on the third day of illness, and its appearance makes it easy to distinguish RMSF from viral infections in which a rash appears. Muscle tenderness, which can become extreme, particularly in the calf muscles, is also a feature of this infection. If untreated and in some cases, even with treatment, RMSF can include complications such as disseminated vascular collapse and renal and heart failure, resulting in death.

In most cases, infection with *Rickettsia* causes vascular lesions, and the pathogens multiply in the endothelial cells lining the patient's small blood vessels. This leads to thrombocytosis (development of clots) and leakage of blood into the surrounding tissue (causing the rash). Although these vascular lesions occur throughout the body, they are most apparent in the skin and most serious in the adrenal glands.

Treatment of Rocky Mountain Spotted Fever

Antibiotic therapy is highly effective if given in the first week of illness. However, if treatment is delayed, it is much less effective. Doxycycline is the antibiotic of choice. The mortality associated with untreated infection can be as high as 25%, but with treatment the mortality is only 5–7%.

It is important to note that sulfonamides contribute to the infection process and are therefore strongly contraindicated.

Epidemic Typhus

Epidemic typhus is caused by *Rickettsia prowazekii*, which is transmitted by the human louse *Pediculus humanus corporis*. It is historically seen in

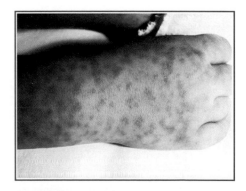

Figure 25.9 The rash caused by Rocky Mountain spotted fever.

times of war or famine, in situations—such as crowding and infrequent bathing—that favor body lice.

Pathogenesis of Epidemic Typhus

Rickettsia circulates through the blood during acute febrile illness, and lice feeding on the body of an infected human become infected. After five to ten days, the number of *Rickettsia* organisms in the infected lice increases and the pathogen is found in the louse feces. The lice defecate while feeding, and the *Rickettsia* organisms are rubbed into the bite wounds when they are scratched. Dried louse feces can also be infectious by entering a human through the eyes, respiratory tract, and mucous membranes.

Within two weeks of being bitten, a patient with epidemic typhus gets fever, headache, and a rash that begins on the trunk of the body and moves to the extremities. Complications of epidemic typhus are myocarditis and central nervous system dysfunction.

As with RMSF, infection causes vascular lesions, and the pathogens multiply in the endothelial cells lining the patient's small blood vessels, leading to thrombocytosis (development of clots) and leakage of blood into the surrounding tissue (causing the rash). Although these vascular lesions occur throughout the body, they are most apparent in the skin and most serious in the adrenal glands.

Treatment of Epidemic Typhus

Doxycycline and chloramphenicol are very effective in treating this infection. In addition, control of lice is very important for preventing infection. If untreated, the fatality rate increases with age to as high as 60%.

Endemic Typhus

Endemic typhus is caused by *Rickettsia typhi* and is transmitted to humans by the rat flea (*Xenopsylla cheopis*). Human infection is incidental, and the primary infection is from rodent to rodent. About 30–60 cases of endemic typhus are reported in the United States each year, with more than half being along the Gulf Coast and southern California.

Pathogenesis of Endemic Typhus

The pathogenesis of endemic typhus is similar to that of epidemic typhus, except for the vector. The rat flea defecates into the bite wound, and the symptoms of infection—headache, muscle aches, and fever—appear one to two weeks later. A **maculopapular rash**, which is one made up of broad lesions that slope away from a central papule, also forms. If untreated, the fever may last 12–14 days, but mortality and clinical complications are rare in this infection even if untreated.

Treatment of Endemic Typhus

Doxycycline or chloramphenicol can reduce the fever period from two weeks to two or three days. In addition, control of rats helps prevent the development of this infection.

Keep in Mind

- *Rickettsia* are bacteria but they have some properties usually associated with viruses.

- Rickettsial infections of the blood include Rocky Mountain spotted fever, epidemic typhus, and endemic typhus.

VIRAL INFECTIONS OF THE BLOOD

Like other microbes, viruses can be found in the blood of an infected person (viremia), and they use the blood to travel throughout the host's body. In this section we discuss viruses whose cellular hosts are found in the blood. We will not include HIV here, because it was discussed in detail in Chapter 17.

Cytomegalovirus

Cytomegalovirus (**CMV**) causes the formation of perinuclear cytoplasmic inclusions and enlargement of the host cell. There are innumerable strains of this ubiquitous virus, which has a very large genome. In developed countries, more than half of the population have antibodies against it, indicating that they have been exposed, and 10–15% of children are infected in the first five years of life. CMV can be isolated from saliva, cervical secretions, semen, urine, and white blood cells for years after the initial infection.

The rate of congenital infection, which means that the fetus is infected *in utero*, is 1% worldwide (about 40,000 a year in the United States). These infants excrete CMV either in urine or in nasopharyngeal secretions. Most of these infections are asymptomatic, but about 20% of infants born infected can have neurological impairment, either sensory-nerve hearing loss or psychomotor retardation, or both. Infants with systemic infection can develop **hepatosplenomegaly**, jaundice, anemia, low birth weight, microencephaly, and chorioretinitis.

In contrast to congenital infection, neonatal infection, which is infection acquired during or shortly after birth, rarely has consequences. CMV can also be efficiently transmitted by breast milk, but infections acquired via this route are usually asymptomatic.

Pathogenesis of Cytomegalovirus Infection

Cytomegalovirus causes a latent infection in leukocytes and bone marrow cells and is responsible for infections associated with blood transfusions and organ transplants. Although CMV is a herpesvirus (human herpesvirus 5), it differs from herpes simplex in that it does not cause skin infections. Instead, it produces a visceral infection that affects the organs and triggers a mononucleosis syndrome. The virus infects both epithelial cells and leukocytes and can cause both tissue damage and immunological damage.

In healthy young adults, CMV can cause a mononucleosis syndrome. In immunosuppressed patients, both primary infections and reactivation of latent infections can be quite severe. For example, in patients receiving bone marrow transplants, CMV causes interstitial pneumonia, which is the leading cause of death in these patients (50–90% mortality). In patients with AIDS, CMV often disseminates to visceral organs, causing gastroenteritis, chorioretinitis, and infection of the central nervous system.

Treatment of Cytomegalovirus Infection

CMV does not respond well to any antiviral drugs. However, ganciclovir, a nucleoside analog of acyclovir, has been shown to inhibit CMV replication and prevent infection in AIDS and transplant patients, and to reduce the retinitis caused by the virus. Combinations of immunoglobulin and ganciclovir have been shown to reduce the high mortality rate seen with CMV pneumonia infections in patients with a bone marrow transplant, but the long-term survival of these patients is not good.

Epstein–Barr Virus

Epstein–Barr virus (EBV) is the major etiologic agent of infectious mononucleosis and of Burkitt's lymphoma. This virus is human herpesvirus 4; it has a small genome, which has been completely mapped. EBV can be grown in culture with long-term lymphoblastoid cell lines (these are cells that are transformed and grow continually) derived from humans and has an affinity for human B lymphocytes and epithelial cells. The infection is less productive in B cells than in epithelial cells.

EBV can be cultured from the saliva of 20–25% of healthy adults; the infection is not highly contagious. Approximately 90–95% of adults worldwide are seropositive, indicating that they have been exposed to this virus. It is important to note that if a primary EBV infection does not occur until the second decade of life or later, it is usually accompanied by the signs and symptoms of infectious mononucleosis.

Pathogenesis of Epstein–Barr Virus Infection

EBV is transmitted only after repeated contact with infected individuals. One of the most important findings about EBV infection is the role of the virus in the development of malignant infections, including Burkitt's lymphoma, nasopharyngeal carcinoma, and lymphoproliferative infections in immunocompromised patients.

EBV does not produce cytopathic effects or inclusion bodies like those seen with herpes viruses, and the major consequence for infected B cells is transformation. When this occurs, only a small amount of the viral DNA integrates into the host chromosome; most of the viral DNA stays in a separate circular **episome** form. After infection, viral proteins called EBNAs (Epstein–Barr nuclear antigens) appear in the nucleus of the infected cell just before the initiation of virus-directed protein synthesis.

The virus enters a human B lymphocyte by means of glycoproteins located on its envelope, and the glycoproteins bind to CD21 receptors on the lymphocyte. These receptors are normally used by the B cell as a complement receptor, but recall from Chapter 12 that viruses often usurp normal receptors to gain access to target cells. After about 18 hours, EBNA proteins are detectable in the nucleus of the infected cell, and the infected B cells begin to express these proteins. It is these viral proteins that mark the cell as infected, and they are the targets for cell-mediated killing by the host defense. During the acute phase of mononucleosis, more than 20% of B lymphocytes express EBNA proteins.

Patients with infectious mononucleosis experience fever, malaise, pharyngitis, tender lymphadenitis, and splenomegaly. These symptoms persist for days to weeks and then slowly resolve spontaneously. In 1–5% of cases, however, there can be complications such as laryngeal obstruction, meningitis, encephalitis, hemolytic anemia, thrombocytopenia, and splenic rupture.

The Burkitt's lymphoma caused by the EB virus is a common malignancy in children in sub-Saharan Africa (**Figure 25.10**). The highest number of lymphomas occurs in areas where malaria is prevalent, and there is some thought that malaria acts as an infectious cofactor or a predisposing factor for Burkitt's lymphoma.

Treatment of Epstein–Barr Virus Infection

The treatment for infectious mononucleosis is mostly supportive, and more than 95% of those infected make a full recovery. In a small minority,

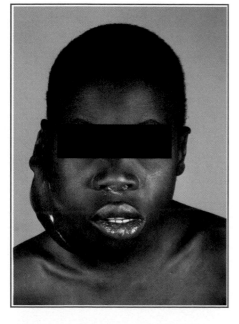

Figure 25.10 A child with Burkitt's lymphoma, which is caused by the Epstein–Barr virus.

the spleen becomes at risk of rupture, and the patient is restricted from taking part in contact sports. In laboratory tests, the EB virus has been shown to be sensitive to acyclovir, but systemic administration of this drug has little effect on clinical illness.

Arbovirus Infections

A variety of infections are seen with arboviruses (Table 25.4), and some of them affect the blood. In the blood, **arbovirus** infections can be classified as fever infections because they always cause fever. We will briefly look at two of these infections, dengue fever (covered in more detail in our discussion of emerging diseases in Chapter 8) and yellow fever.

Arbovirus infections are usually transmitted by mosquitoes, and in some cases the virus can be transferred from one generation of mosquitoes to the next. The vector suffers no ill effects from the virus it harbors. The virus multiplies in the vector during what is called the extrinsic incubation period, which increases its numbers and enhances the chance of causing infection when a host is bitten. Transient viremia that can last for a week or more is a feature of dengue and yellow fever infection in humans.

Pathogenesis of Arbovirus Infections

There are three major manifestations of arbovirus infection in humans that are associated with the affinity of the virus for various target organs. Some arboviruses target the CNS, others attack major organs, particularly the liver (the yellow fever virus works in this way), and the virus that causes hemorrhagic fever damages small blood vessels (which causes hemorrhaging). In addition, all arboviruses produce a cellular necrosis that instigates inflammation and leads to fever.

Organism and infection	Distribution	Vector	Infection expression
Togaviruses			
Western equine encephalitis	North America	Mosquito	Encephalitis
Eastern equine encephalitis	North America	Mosquito	Encephalitis
Flaviviruses			
St Louis encephalitis	North America	Mosquito	Encephalitis
Dengue fever	All tropical zones	Mosquito	Febrile illness
Yellow fever	Africa, South America, and the Caribbean	Mosquito	Hepatic necrosis
West Nile fever	Africa, Eastern Europe, Middle East, Asia, and North America	Mosquito	Febrile illness and encephalitis

Table 25.4 Selected arboviruses that cause infection in humans.

Fast Fact Yellow fever had plagued the southeastern United States and Caribbean for 200 years and had caused the French to stop work on the Panama canal. Walter Reed, an army medical officer, found that the infection was transmitted by mosquitoes. When steps were taken to control the mosquito population, the Panama Canal was able to be completed. Today, the Walter Reed Army Medical Center in Washington is named after this medical pioneer.

Yellow fever

In yellow fever, the arbovirus attacks the liver and causes necrosis of the hepatocytes. The virus can also cause brain hemorrhaging, destroy the myocardium, and affect the urinary system by destroying the renal tubules. Hemorrhage is the major complication in yellow fever, which is distributed throughout the Caribbean, Central and South America, and Africa. Yellow fever is also a potential threat to the southeastern United States because the vector for this virus (the *Aedes aegypti* mosquito) has migrated to this area. The clinical symptoms associated with yellow fever include the abrupt onset of fever, chills, headache, and hemorrhaging that may become severe and cause bradycardia (a slow heart rate) and shock.

Dengue fever

There are four related serotypes of the virus responsible for dengue fever. The serotypes are spread throughout the world, particularly in the Middle East, Africa, the Far East, and the Caribbean. The vector for dengue fever is the same as for yellow fever, the *Aedes aegypti* mosquito, but the clinical infection seen in dengue is different from that seen in yellow fever. In dengue fever, the infection brings about fever, rash, and severe pain in the back, head, muscles, and joints. Some of the more severe forms of the infection are characterized by shock, pleural effusion, hemorrhage, and death.

Treatment of Arbovirus Infections

There is no specific treatment for arbovirus infections other than supportive care. Prevention can be enhanced by control of the vector population, but this is not easy. There is a live attenuated vaccine for yellow fever, and many countries where it occurs require travelers to these countries to have been vaccinated.

Filovirus Fevers

Filoviruses are filamentous viruses that occur in branched, fishhook, and circular configurations. They vary in length and are negative single-stranded RNA viruses. Two of the best-known are Ebola and Marburg, the only two filoviruses that infect humans. There are both classed as emerging diseases, and more detail about Ebola is given in Chapter 8.

Ebola caused outbreaks of hemorrhagic fever in 1976, with a mortality rate of 88% in Zaire and 50% in Sudan. In 1995, an outbreak in Zaire made headlines with more than 200 cases and a 75% mortality rate. The 2014 outbreak in Liberia, Sierra Leone, and Guinea was declared an international emergency by the WHO in August of that year; by March 2015, more than 9000 deaths had been recorded. Before the most recent outbreak, about 10% of the population in rural Central Africa carried antibodies against the virus, indicating that they had been exposed.

Marburg was first recognized in Germany when technicians working with monkey kidney cells began dying of hemorrhagic infection. This virus has also been seen in nosocomial settings, with 25% mortality.

Pathogenesis of Ebola and Marburg Infection

Both Ebola and Marburg cause hemorrhaging in the skin, mucous membranes, and internal organs. Liver cells, lymphoid tissue, kidneys, and gonads are all destroyed, and there can also be brain edema.

The reasons for such rapid lethal hemorrhaging are still not clear, but there is evidence that Marburg replicates in vascular endothelial cells, causing necrosis and bleeding. Ebola may destroy cells by secreting a glycoprotein that interacts with neutrophils and inhibits the inflammatory defense reaction of the host. Ebola causes symptoms in as little as four to six days after infection, and the mortality rate is extremely high: 30–80%.

Treatment of Ebola and Marburg Infection

There is no approved treatment for either Ebola infection or Marburg infection. Experimental treatments for Ebola were used in the 2014 outbreak with some success, and clinical trials of three potential treatments were fast-tracked.

Keep in Mind

- Cytomegalovirus and Epstein–Barr virus both infect white blood cells.
- Cytomegalovirus causes a latent infection.
- Epstein–Barr virus causes mononucleosis and Burkitt's lymphoma.
- Three important infections caused by arboviruses are yellow fever, hemorrhagic fever, and dengue fever.
- Ebola and Marburg are the only filoviruses that infect humans.

PARASITIC INFECTIONS OF THE BLOOD AND LYMPH

We discussed three of the most important and dangerous parasitic infections of the blood in detail in Chapter 14: malaria, toxoplasmosis, and African trypanosomiasis (sleeping sickness). In this chapter we look at two other infections of the blood and lymph: Chagas' disease and filariasis.

Chagas' Disease

Chagas' disease, the American form of trypanosomiasis, is caused by the flagellate protozoan *Trypanosoma cruzi* (**Figure 25.11**). Recall from Chapter 14 that the African form of trypanosomiasis is caused by *Trypanosoma brucei*.

The trypomastigote forms of *T. cruzi* resemble those of *T. brucei*. *Trypanosoma cruzi* trypomastigotes are disseminated in the fecal material of the transmitting vector, which is the **reduviid**, a large winged insect that feeds on sleeping hosts in the evening hours. Once a host has been bitten by an infected reduviid, the trypomastigotes spread from the site of the bite by circulating in the host's blood.

Unlike *T. brucei*, *T. cruzi* does not multiply extracellularly. Instead, the trypomastigotes must invade host tissue cells to assume the **amastigote**

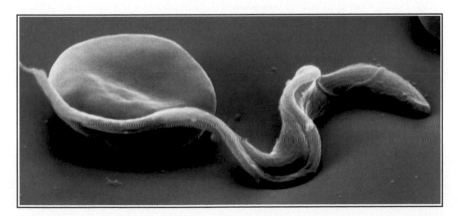

Figure 25.11 A false-color scanning electron micrograph of *Trypanosoma cruzi* next to a red blood cell.

form. This is the form in which they can begin to multiply by binary fission. The large numbers of amastigotes created by this fission soon cause the host cell to rupture. The parasites are then released back into the blood, where they revert to the trypomastigote form and move on to invade other host cells.

When an uninfected reduviid bites an infected host, some of the trypomastigotes circulating in the host blood are taken up by the reduviid. When that infected insect moves on to bite a new host, the infection cycle continues. The trypomastigotes in the reduviid move to the hindgut and are deposited in the feces of the insect. The infected feces are then deposited on a new host during a blood meal. This process can recur with every bite for up to two years.

Pathogenesis of Chagas' Disease

From North to South America, Chagas' disease affects between 16 million and 18 million people, with approximately 50,000 deaths each year. In these areas, it is the leading cause of heart infection and accounts for 25% of all deaths among people between the ages of 25 and 44 years.

The reduviid is referred to as the kissing bug because it preferentially bites near the lips and eyes. Most infections are in children, but infection can also be acquired *in utero* and through breast feeding. Chagas' disease also affects dogs, cats, rats, and opossums, which can then become reservoirs for *T. cruzi*. It is estimated that there are approximately 50,000 infected Latin American immigrants currently in the United States, and *T. cruzi* has now been isolated from vertebrates and invertebrates throughout the southwestern United States, indicating that the infection may be gaining a foothold there.

Multiplication of the parasite at the site of the bite stimulates the accumulation of neutrophils, lymphocytes, and tissue fluids, resulting in the formation of a local chancre called a **chagoma**. Dissemination of the parasite in the host body and tissue invasion produce a febrile illness that can persist for up to three months and cause widespread organ damage. Although any cells can be infected, cells of the heart, skeletal muscle cells, and glial nerve cells are the most susceptible.

After penetration of the cell, the trypomastigote transforms to the amastigote form and multiplies freely to produce a pseudocyst, which is a greatly enlarged and distorted host cell packed with parasites. When the pseudocysts rupture, many of the parasites disintegrate—an event initiating a powerful inflammatory response that destroys the surrounding tissue. The eventual development of an adaptive response will destroy the *T. cruzi*, and the acute infection will be terminated.

Only about one-third of newly infected individuals ever develop clinical symptoms, and these are primarily children. The onset of parasitemia is signaled by the development of a sustained fever and enlargement of the spleen and lymph nodes, and there can be a transient rash. A small number of patients have heart involvement and the potential for congestive heart failure. Clinical symptoms can last for weeks to months, and 5–10% of untreated patients can develop heart or brain problems that are lethal.

There are also chronic forms of Chagas' disease. They are seen only in adults, and usually result in coronary dysfunction.

Treatment of Chagas' Disease

Effective treatment for Chagas' disease has still not been worked out. Nifurtimox and benznidazole can reduce the severity of the acute form, but they are ineffective in the chronic forms. In addition, if these drugs are taken for long periods there can be serious side effects.

Filariasis

The infection called **filariasis** is an umbrella term for a group of infections caused by certain members of the superfamily Filarioidea. These parasites inhabit the lymphatic systems of humans, and their presence induces an acute inflammatory response, chronic lymphatic blockade, and, in some cases, the swelling of the extremities and genitalia known as **elephantiasis** (Figure 25.12).

There are two parasites most commonly involved in filariasis infections: *Wuchereria bancrofti* and *Brugia malayi*. Both are threadlike roundworms (nematodes) that lie coiled up for decades in the lymphatic vessels of a human host. Gravid females produce large numbers of fertile eggs, and once the eggs have been laid, the embryos uncoil to their full length of between 200 and 300 μm. At this point they are referred to as **microfilariae**. The shell of each egg elongates and becomes a flexible sheath around its microfilaria.

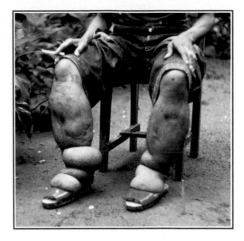

Figure 25.12 Elephantiasis of the leg caused by the parasite *Wuchereria bancrofti*. The swelling is the result of a massive blockade of the lymphatics by the adult worms.

The microfilariae eventually leave the lymph and enter the host's blood, where most of them accumulate in the pulmonary vessels during the day. This movement between the pulmonary circulation and the systemic circulation is called periodicity, and it is important because it determines which type of mosquito will serve as the vector and intermediate host.

A mosquito feeding on an infected human ingests the microfilariae along with the human's blood. The microfilariae are transformed into the larval form in the thoracic muscles of the mosquito, and when the mosquito bites an uninfected human, the larvae penetrate the feeding site. In the new host, the larvae migrate to the lymphatic vessels and go through a series of molts until they reach adulthood, a process that can take 6–12 months.

Pathogenesis of Filariasis

Currently there are about 120 million people infected with either *W. bancrofti* or *B. malayi*, mostly in Africa, Latin America, the Pacific islands, and Asia, with more than 75% of the cases being in Asia. Humans are the only known vertebrate hosts for most strains of *Brugia* and *Wuchereria*.

The pathology of filariasis infections is confined primarily to the lymphatic system, and there are two types of infection: acute and chronic. In the acute form, the presence of molting adolescent and dying adult worms stimulates dilatation of the lymphatics and hyperplastic changes to the vessel endothelium. This brings about the infiltration of lymphocytes, plasma cells, and eosinophils, and this infiltration results in the formation of a granuloma, fibrosis, and permanent lymphatic blockade. Repeated infections eventually result in massive lymphatic blockade, which causes the skin and subcutaneous tissues to fill with edematous fluids and fibrous tissue. At this point, there can be bacterial and fungal superinfections of the skin, which contribute to further tissue damage.

The chronic form of filariasis usually develops 10–15 years after the onset of the first acute attack. The incidence and severity of chronic clinical manifestations tend to increase with age. The main characteristic features of chronic filariasis are chronic **lymphangitis**, thickened lymphatic trunk, and chronic lymphedema.

People indigenous to areas where *W. bancrofti* and *B. malayi* are found usually remain asymptomatic after infection; however, some can experience filarial fevers and lymphadenitis for 8–12 months. These fevers are usually low-grade and are accompanied by chills and muscle aches. Lymphadenitis will normally first be noted in the femoral areas as an enlarged, red, tender lump. Inflammation then spreads down the lymphatic channel of the leg, and vessels become enlarged and tender, with the overlying skin red and edematous.

Acute manifestations last a few days, resolve spontaneously, and then reoccur periodically for weeks or even months. If infection occurs repeatedly, however, there can be permanent lymphatic obstruction involving edema, ascites, and pleural and joint effusion. If the lymphadenopathy persists, lymphatic channels can rupture and cause the formation of abscesses. In patients heavily and repeatedly infected over periods of decades, the infection will result in elephantiasis.

Treatment of Filariasis

Diethyl carbimazine eliminates the microfilariae from the blood and kills or injures the adult worms, resulting in long-term suppression of the infection and a potential cure. However, the dying worms can elicit an allergic reaction in the host. This reaction is occasionally severe and requires the use of antihistamines and corticosteroids.

Tissue changes seen in elephantiasis are often irreversible, but enlarging of the extremities may be ameliorated through the use of pressure bandages. As with any vector-transmitted infection, control of the vector (in this case, mosquitoes) will help prevent infection.

Keep in Mind

- Chagas' disease (American trypanosomiasis) and filariasis are important parasitic blood infections.

- Chagas' disease is caused by the flagellate protozoan *Trypanosoma cruzi*, which is vector-transmitted by reduviids, flying insects known as kissing bugs.

- Filariasis is caused by the parasites *Wuchereria bancrofti* and *Brugia malayi*, which infect the human lymphatic system and can cause acute inflammatory responses and, in severe cases, elephantiasis.

SUMMARY

- Blood and lymph travel to all parts of the body, so they are good ways to spread infection.

- The presence of any pathogen in the blood, especially bacteria, can develop into sepsis or septic shock, which can be lethal.

- Pathogens that have gained access to the blood can infect and damage the structures of the cardiovascular system.

- Some bacterial infections always involve the blood or lymph, including plague, tularemia, brucellosis, Lyme disease, and relapsing fever.

- Plague is one of the most virulent bacterial infections known.

- *Rickettsia* are bacteria but have some viral properties and cause Rocky Mountain spotted fever, epidemic typhus, and endemic typhus.

- Several viruses infect blood cells, including HIV, cytomegalovirus (CMV), Epstein–Barr virus (EBV), several arboviruses (yellow fever and dengue fever viruses), and two filoviruses (Ebola and Marburg).

- EBV infection is associated with several cancers.

- Hemorrhagic fevers caused by the filoviruses are some of the most virulent viral infections known.

- Chagas' disease and filariasis are parasitic infections of the blood and lymph (malaria, trypanosomiasis, and toxoplasmosis are also important).

- Pathogens of the blood are transmitted by vectors.

In this chapter, we have seen that, like the body's other systems, the blood and lymph can be targets for a variety of infections caused by all of the usual suspects (bacteria, viruses, and parasites). Notice that we did not discuss fungal infections in this chapter, but keep in mind that systemic fungal infections make use of the blood to move to distal locations in the body.

 SELF EVALUATION AND CHAPTER CONFIDENCE

Multiple Choice

Answers are given in the back of the book and help can be found in the student resources at:

www.garlandscience.com/micro2

1. A condition in which bacteria are circulating in the blood is called
   A. Bacteremia
   B. Sepsis
   C. Intoxication
   D. Parasitemia
   E. None of the above

2. Septic shock is characterized by
   A. A rise in blood pressure
   B. A fall in blood pressure
   C. Thrombophlebitis
   D. Inflammation
   E. Fever

3. Endocarditis involves all of the following except
   A. Colonization of the heart by pathogens
   B. Minimal platelet deposition
   C. Deposition of fibrin
   D. Formation of mature vegetation

4. Which of the organisms below is most frequently found in endocarditis?
   A. Enterococci
   B. *Candida albicans*
   C. *Staphylococcus epidermidis*
   D. *Corynebacterium*
   E. Viridans streptococci

5. Plague is transmitted to humans by
   A. The exchange of bodily fluids
   B. The use of contaminated needles
   C. The bite of the rat flea
   D. The bite of a mosquito
   E. The bite of a tick

6. Plague is caused by
   A. *Streptococcus mutans*
   B. *Yersinia pestis*
   C. *Neisseria plagus*
   D. *Xenopsylla cheopis*
   E. Filovirus

7. The most contagious form of plague is
   A. The cutaneous form
   B. The neurologic form
   C. The pneumonic form
   D. The oral form
   E. The genitourinary form

8. Organisms of the genus *Francisella* causes
   A. Plague
   B. Typhoid fever
   C. Tularemia
   D. Relapsing fever
   E. None of the above

9. Lyme disease is caused by
   A. *Staphylococcus*
   B. *Borrelia*
   C. *Candida*
   D. *Chlamydia*
   E. *Francisella*

10. Lyme disease is transmitted from
    A. Human to human
    B. Tick to human
    C. Deer to human
    D. Sand flea to human

11. Lyme disease involves all of the following except

    A. Humans
    B. Deer
    C. Ticks
    D. Rats

12. All of the following are rickettsial infections of the blood except

    A. Endemic typhus
    B. Rocky Mountain spotted fever
    C. Yellow fever
    D. Relapsing fever

13. Epidemic typhus is transmitted by

    A. Ticks
    B. Rat fleas
    C. Body lice
    D. *Rickettsia typhi*
    E. None of the above

14. The etiologic agent for infectious mononucleosis is

    A. Variola major
    B. Epstein–Barr virus
    C. Cytomegalovirus
    D. Herpes simplex type 1
    E. Herpes simplex type 2

15. Arboviruses cause all of the following except

    A. Dengue fever
    B. St Louis encephalitis
    C. Yellow fever
    D. Relapsing fever
    E. West Nile fever

16. The protozoan *Trypanosoma cruzi* causes

    A. Sleeping sickness
    B. Malaria
    C. Relapsing fever
    D. Chagas' disease
    E. Dengue fever

17. *Wuchereria bancrofti* is

    A. An ameba
    B. A fungus
    C. A flatworm
    D. A roundworm

18–20. Answer **A** for true statements and **B** for false statements.

18. Infection in any part of the body can be spread throughout the body if pathogens gain entry to the blood or lymph.

19. Pathogenic organisms in the blood are restricted to viruses.

20. Although the Black Death was feared, it killed very few people.

21. Extravascular infections can result in either sepsis or septic shock.

22. Bacterial infections of the blood include yellow fever and dengue fever.

23. EBV infections are associated with malignancies.

24. Tularemia and brucellosis are zoonotic infections.

Q DEPTH OF UNDERSTANDING

Questions listed here require you to bring together the concepts you have learned in this chapter into a discussion format. This helps you to increase your depth of understanding of the material you have learned. Help can be found in the student resources at: www.garlandscience.com/micro2

1. Describe the events associated with the development of endocarditis and why this condition is so dangerous.

2. Discuss the different aspects of infection with *Borrelia burgdorferi* and how this infection can be controlled environmentally.

3. Compare and contrast endemic and epidemic typhus and the impact of socioeconomic conditions on each.

(Q) CLINICAL CORNER

Help can be found in the student resources at: www.garlandscience.com/micro2

1. A patient comes into the clinic for follow-up after having his spleen removed three months previously because of trauma resulting from an automobile accident. After checking to make sure that he is feeling well and healing properly, the doctor has given him a prescription for antibiotics and told him it is important that he take these before and after he visits the dentist, even if he is only going to have his teeth cleaned. He asks you why this has been ordered.

   A. What do you tell him?

   B. What are the potential consequences if he neglects to follow the doctor's orders?

2. You are working as a physician's assistant in a Doctors Without Borders/Médecins Sans Frontières clinic in the Democratic Republic of the Congo. Your patient is a 60-year-old white female Catholic nun who has been on a religious mission to the interior areas. She is complaining of difficulty in breathing, pain, and cramping in the groin that has been going on for about four days. Upon examination, you discover several bulging lymph nodes in the groin, some of which are suppurative and bleeding.

   A. What are your initial thoughts about this patient?

   B. What do you do, and how would you prevent the spread of this infection?

Infections of the Skin and Eyes

Chapter 26

Why Is This Important?

The skin and eyes are in contact with potentially pathogenic organisms all the time.

In 1994, while preparing for a family Christmas vacation, David contracted a *Streptococcus pyogenes* (group A *Streptococcus*) infection via a small cut on his finger. The bacterium, commonly known as the flesh-eating bacterium, caused necrotizing fasciitis, a disease that spread rapidly and caused a devastating and prolonged illness. David was given a very low chance of survival by his doctors. They performed surgery on him multiple times, removing large amounts of infected muscle mass from his arm, back, and buttock regions. After repeated surgeries, powerful antibiotics, blood transfusions, skin grafts, and frequent exposure to a hyperbaric chamber, he narrowly escaped his brush with death. He coped with financial difficulties and media attention for months after his release from the hospital. As an English professor, David wrote about this intense and dramatic story in his book Miracle Victory Over The Flesh-eating Bacteria.

OVERVIEW

In this chapter, we conclude our discussions of infections that affect different parts of the body with a look at infections of the skin and eyes. For the skin, we will see that many infections can occur here and that there are two main reasons for this: first, the skin is always exposed to pathogenic organisms, and second, the soft tissue just below the skin is an excellent breeding ground for infection. Eyes are also open to the outside world, and infections here can be extremely dangerous because of the proximity to the nervous system and the potential for loss of vision.

We will divide our discussions into the following topics:

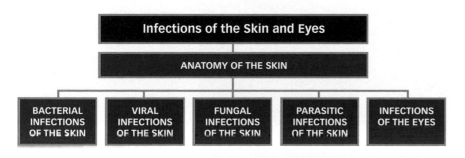

Infections of the Skin and Eyes
ANATOMY OF THE SKIN

BACTERIAL INFECTIONS OF THE SKIN | VIRAL INFECTIONS OF THE SKIN | FUNGAL INFECTIONS OF THE SKIN | PARASITIC INFECTIONS OF THE SKIN | INFECTIONS OF THE EYES

"As they discussed my condition privately the first few days, they couldn't figure out how I could possibly even be alive. ... I looked like a cadaver with its head bantering wittily with the therapists and nurses."

David L. Cowles

ANATOMY OF THE SKIN

Before we look at the kinds of infection that occur in the skin, it would be wise to revisit the anatomy of the skin briefly. The skin is the largest organ in the body and accounts for a significant portion of our weight. It is the barrier between our body and the outside world and, as we learned in Chapter 15, is the first line of defense against invading microorganisms. When you look closely at the skin, it is apparent that this organ is very well engineered to fit the role of protective barrier (**Figure 26.1**). The outer layer (the epidermis), which comes into direct contact with the environment, is essentially contiguous and made up of several layers of dead or dying cells. There is constant shedding of these cells, and this shedding helps prevent pathogens from successfully attaching to the skin.

There are several other mechanical mechanisms that discourage pathogens from colonizing the skin. For example, the production of perspiration has a flushing action that constantly moves pathogens off the surface of the skin. The surface of the skin usually has a pH of 5.0–6.0, which is too acidic for many microorganisms, but the skin is home to a selection of microbes (bacteria and yeasts) that are part of the normal human microbiota. The number and type of microbes found in a particular area of the skin depend on the local environment. Dense populations of microbes are found in moist areas such as the groin and underarms, but there are few on dry, flat areas such as the limbs. The most common species found naturally on the skin are the coagulase-negative *Staphylococcus epidermidis*, *Propionibacterium acnes*, *Candida* species, and *Corynebacterium* species, which act as additional barriers to opportunistic pathogens. *Staphylococcus aureus* is also found commonly on the skin, but as we have seen in previous chapters, it is a potential pathogen with serious consequences. Sebum, which is produced by the **sebaceous glands**, contains free fatty acids and lipids that are broken

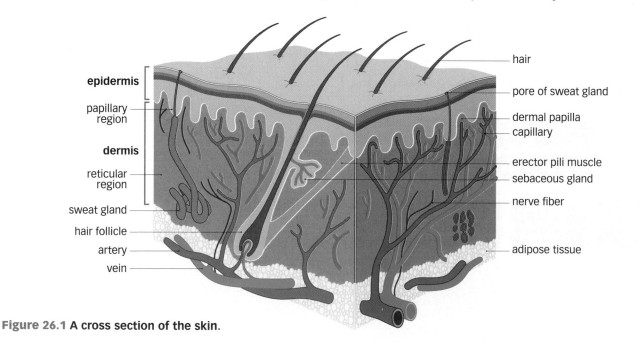

Figure 26.1 A cross section of the skin.

Infection	Organism causing the infection	Characteristics of the infection
Bacterial		
Folliculitis	*Staphylococcus aureus*	Skin abscess
Scalded skin syndrome	*Staphylococcus aureus*	Vesicular lesions over the skin. Seen mostly in infants
Erysipelas	*Streptococcus pyogenes*	Skin lesions that can spread to systemic infection
Acne	*Propionibacterium acnes*	Skin lesions caused by excess of hormones, which is common in teens
Impetigo	Staphylococci and streptococci	Skin lesions seen in children spread by hands and fomites
Viral		
Rubella	Rubella virus	Mild red rash on the face, trunk, and limbs. Dangerous in pregnancy
Measles	Rubeola virus	Severe infection with fever, conjunctivitis, cough, and rash
Chickenpox	Varicella-zoster virus	Generalized itchy rash that develops into vesicles
Shingles	Varicella-zoster virus	Pain and skin lesions usually on the trunk
Smallpox	Smallpox virus	Raised, fluid-filled bumps that are dimpled in the middle
Warts	Human papillomavirus	Small growths on the skin or mucous membranes
Fungal		
Dermatomycoses	Dermatophytes	Dry scaly lesions on various parts of the skin
Candidiasis	*Candida albicans*	Patchy inflammation of the mucous membranes of the mouth or vagina

Table 26.1 Infections of the skin.

down by the normal microbiota into more free fatty acids. These fatty acids are naturally antibacterial and inhibit the growth of such pathogens as group A streptococci.

The surface of the skin is normally penetrated by ducts, hairs, and sweat glands. These structures serve specific functions that are part of the normal physiology of the body. Microbial invasion can occur along any of these routes, especially through any ducts that are obstructed for some reason. The skin is also often breached by trauma, such as abrasions, wounds, punctures, surgical procedures, and bites from insects or animals. When this occurs, pathogens can get into the underlying tissue. Pathogens can also get into the underlying layers of the skin from the blood and lymph.

Infections of the skin can be caused by bacteria, viruses, and fungi (**Table 26.1**). Parasitic infections of the skin also occur. As we have done in other chapters, we will look at each of these types of infection separately.

Infections of the skin often present as skin lesions, of which there are four main types: **macules**, **papules**, **vesicles**, and **pustules** (**Figure 26.2**).

Keep in Mind

- Skin is a barrier impenetrable to pathogens and must be broken for infection to occur from the outside.
- Pathogens can enter through natural openings in the skin such as ducts and glands.
- Skin can also be accessed from the blood and lymph.

macules

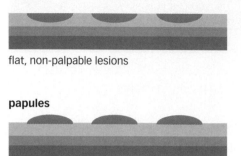

flat, non-palpable lesions

papules

palpable lesions

vesicles

palpable, fluid-filled lesions

pustules

palpable lesions containing pus

Figure 26.2 Differences between skin macules, papules, vesicles, and pustules, the four types of skin lesion.

BACTERIAL INFECTIONS OF THE SKIN

In this section, we look at both bacterial infections of the skin and bacterial infections of the soft tissue just under the skin (**Figure 26.3**). Most of these latter infections accompany a break in the skin. Because any surgical procedure breaches the skin, wound infections are an important problem in hospital settings. *Staphylococcus aureus* and group A streptococci are major concerns for nosocomial infections. The bite of a dog or cat can introduce organisms from their oral flora into the soft tissues, and this can result in serious infection.

Once the skin has been breached, organisms can enter the soft tissue below, which contains the blood supply and is a rich environment for pathogen growth. In addition, the nearness of the body's blood supply to this soft tissue increases the possibility of systemic infection.

Fasciitis

One of the most difficult-to-treat skin infections is **fasciitis**. Fasciae are sheets of fibrous tissue lying just below the subcutaneous layer of the skin, and fasciitis is any infection in a fascia. **Necrotizing fasciitis** (**Figure 26.4**), an infection in which fascia is destroyed, can be caused by an organism such as group A streptococci acting either alone or in combination with other bacteria through a synergistic effect. Usually, this synergy comes from bowel flora, which can cause fasciitis in the abdominal wall after surgery. After trauma or surgery, bacteria enter the fascia. In patients with bacteremia, group A streptococci can reach the fascia and settle in small hematomas or in bruised areas.

Once bacteria enter the fascia, they spread rapidly, and the resulting inflammatory response then affects the neurovascular bundles found in the fascia. Thrombosis of these vessels compromises the blood supply and nerves to that area, and the area rapidly becomes necrotic. In fact, this infection can be so rapid that only surgical removal of the tissue can resolve the spread.

	organism	disease
infections of the skin		
	group A streptococci	impetigo, ecthyma, erysipelas, cellulitis, paronychia
	Staphylococcus aureus	impetigo, folliculitis, carbuncles, cellulitis, paronychia
	Pseudomonas aeruginosa	folliculitis, paronychia, cellulitis
	Propionibacterium acnes	acne
soft tissue infections		
	group A streptococci	necrotizing fasciitis, myositis
	coliforms	synergistic necrotizing fasciitis, Fournier's gangrene
	anaerobes of the bowel	synergistic necrotizing fasciitis, Fournier's gangrene
	Clostridium perfringens	gas gangrene

Figure 26.3 Some of the bacteria involved in infections of the skin and underlying soft tissue.

Erysipelas

The skin condition known as **erysipelas** (sometimes known as cellulitis) is characterized by a rapidly spreading infection of the deeper layers of the dermis. It is an acute infection caused by group A streptococci. Symptoms include edema of the skin marked by erythema (redness), pain, and systemic infection, including lymphadenopathy and fever. Erysipelas can progress to septicemia and local necrosis of the skin and can be serious, requiring immediate treatment. Either penicillin or streptomycin is effective in dealing with this infection.

Folliculitis

This minor infection of hair follicles, usually caused by *S. aureus*, is associated with sweat gland activity. Infections are most often seen in areas where these glands predominate, such as the neck, face, axillae, and buttocks. Folliculitis can also be caused by *Pseudomonas aeruginosa*. In fact there has been an increase in these infections associated with hot tubs and whirlpools. At the elevated temperatures used in these devices, *Pseudomonas* can grow in large numbers, causing large areas of folliculitis in those parts of the body that are immersed. The symptoms normally subside when the insult is discontinued.

Pathogenesis of Folliculitis

Blockage of the sweat gland ducts predisposes to folliculitis, and serious infections can result in the formation of **boils** (furuncles), which are a type of **abscess**. In this case, there is a localized region of pus surrounded by tissue inflammation. If the body defenses do not wall off this localized infection, neighboring tissues become infected and there is usually extensive damage and the onset of fever. The enlarged infected region, made up of several furuncles, is referred to as a **carbuncle**.

Treatment of Folliculitis

Antibiotics cannot penetrate a boil very well, so folliculitis is usually treated locally by draining the abscess.

Acne

Acne is the most common skin infection in humans, affecting 17 million people in the United States alone, with 85% being teenagers. There are three categories of this skin infection: comedonal acne, inflammatory acne, and nodular cystic acne.

Pathogenesis of Acne

Comedonal acne results from inflammation of the hair follicles and associated sebaceous glands of the face, which become plugged by a mixture of shedding skin cells and sebum. As the sebum backs up and accumulates, whiteheads, called *comedos*, appear on the skin. If the blockage protrudes through the skin, blackheads, called *comedones*, appear on the skin.

Inflammatory acne (**Figure 26.5**) is caused by the bacterium *Propionibacterium acnes*, which is part of the natural skin microbiota. It is the predominant anaerobe of the skin and metabolizes the glycerol component of sebum. This glycerol metabolism causes free fatty acids to form, and the presence of the fatty acids initiates an inflammatory response. Neutrophils, which are part of the inflammation process, secrete enzymes that damage the wall of the hair follicles, causing pustules and papules to form. The primary cause of inflammatory acne is hormonal influence

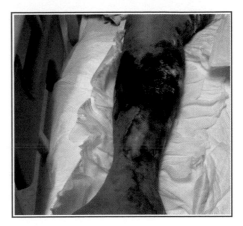

Figure 26.4 Necrotizing fasciitis arising from infection with group A streptococci.

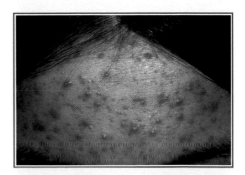

Figure 26.5 Inflammatory acne, a skin infection seen in young people, is caused by the bacterium *Propionibacterium acnes*.

on the secretion of sebum, which is increased during puberty. Usually, inflammatory acne resolves spontaneously in adulthood.

The nodular cystic form of acne is characterized by the formation of **cysts**, which are inflamed lesions that lie deep in the skin and are filled with pus. These lesions leave prominent scars on the skin.

Treatment of Acne

All three types of acne can usually be effectively treated with topical drying agents.

Impetigo

Impetigo is a common, sometimes epidemic, skin lesion that is caused primarily by group A streptococci or staphylococci. The infection is usually found in children and is associated with heat, humidity, and poor hygiene. It is highly contagious and can be spread by fomite transmission when people share contaminated towels or clothing.

Pathogenesis of Impetigo

The initial lesion is often a small vesicle that develops at the site where the bacteria have entered the skin. The vesicle ruptures and causes superficial spread of the bacteria, with the spread being characterized by skin erosion and serous exudate. This area dries to a honey-colored crust that contains numerous infectious *Streptococcus* or *Staphylococcus* species (**Figure 26.6**).

Treatment of Impetigo

The usual treatment involves penicillin or erythromycin taken orally and topical skin antiseptics to limit the spread of the infection.

Scalded Skin Syndrome

The bacterial infection **scalded skin syndrome** derives its name from the fact that the salient sign is the blistering and peeling off of large sheets of skin (**Figure 26.7**). The infection is caused by two exotoxins secreted by certain strains of *S. aureus*. The exotoxins are referred to as **exfoliatins**. The gene for one exfoliatin is located on the bacterial chromosome; the gene for the other is located on a plasmid. It is therefore possible for some strains of *S. aureus* to have only one of the genes and for other strains to have both genes. Scalded skin syndrome is normally restricted to infants but can occur in adults, especially in the late stages of toxic shock syndrome.

Pathogenesis of Scalded Skin Syndrome

Exfoliatins are transported through the blood to distal sites, where they cause the upper layers of the skin to separate and peel off. The first sign is a reddened area of the skin, usually around the mouth. The area soon spreads out, forming large, soft vesicles over the whole body. As the top layer of skin peels away, it leaves the exposed dermal layer looking scalded. This condition is only temporary, because the skin will regenerate in 7–10 days. There is an accompanying high fever throughout the infection.

Treatment of Scalded Skin Syndrome

There is a good immune response to this infection, so that recurrence is unlikely. Most of the bacteria responsible for the infection are sensitive to penicillin, and in cases of penicillin-resistant

Fast Fact The use of antibiotics for minor comedonal acne can be a problem: the acne will resolve naturally, but the use of these drugs contributes to antibiotic resistance.

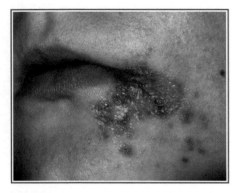

Figure 26.6 Impetigo infection caused by staphylococci. This infection is characterized by isolated pustules that become crusted over.

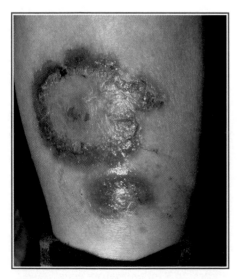

Figure 26.7 Scalded skin syndrome, an infection caused by *Staphylococcus aureus*.

strains, cephalosporins are effective. Alternatives include vancomycin, clindamycin, and erythromycin.

Gas Gangrene (*Clostridium perfringens*)

Gangrene is tissue necrosis resulting from an obstructed blood supply, and the skin infection known as **gas gangrene** gets its name from the fact that the bacteria responsible for the infection release gases as part of their metabolic activity. The infection is caused by *Clostridium perfringens* and is usually associated with deep tissue wounds often seen in battle-field injuries. However, car and motorcycle accidents can also result in this type of infection.

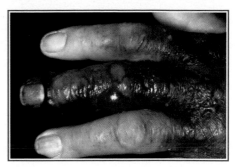

Figure 26.8 Gas gangrene of the hand.

Pathogenesis of Gas Gangrene

Clostridium perfringens is a Gram-positive, anaerobic, spore-forming rod. The spores are introduced into tissue where blood circulation has been impaired and the tissue is dead. Because this environment is perfect for anaerobic growth, the spores germinate and the bacteria multiply. They produce toxins, proteinase, lipase, hyaluronidase, and collagenase, all of which destroy the tissue surrounding the already-dead tissue that is the focus of the infection. The destruction of the surrounding tissue expands the anaerobic environment, and the infection spreads.

The onset of gas gangrene is sudden, appearing anywhere from 12 to 48 hours after the initial injury to the tissue. As the bacteria grow, they ferment and produce hydrogen gas, which causes breaks in the tissue. (This tissue is referred to as crepitant tissue.) Movement of the affected area causes snap, crackle, and popping sounds. There is also a foul smell associated with the tissue destruction, making it obvious that infection has set in. The infection is accompanied by a high fever, massive tissue destruction, shock, and blackened skin (**Figure 26.8**). If not treated quickly, a gas gangrene infection is lethal.

Treatment of Gas Gangrene

Excision of the affected tissue opens the area to oxygen, which kills the anaerobic pathogens. Treatment in hyperbaric chambers is effective in reducing the spread of infection. Penicillin and clindamycin should also be given.

Cutaneous Anthrax

As we saw in Chapter 21, *Bacillus anthracis* can cause serious respiratory infection. It can also cause less harmful infections in the skin, and the one we look at here is **cutaneous anthrax**.

Pathogenesis of Cutaneous Anthrax

The first signs of a cutaneous anthrax infection usually appear two to five days after anthrax spores have been inoculated into an opening in the skin, most often the forearm or hand. The initial lesion is a papule that looks like an insect bite. This papule progresses through vesicular and ulcerative stages in seven to ten days to form a black **eschar** (scab) surrounded by edema. Although the eschar is sometimes referred to as a malignant pustule, it is neither.

The symptoms of cutaneous anthrax are normally mild, and the lesions typically heal slowly after the scab has fallen off. In some cases, however, the infection can become systemic, progressing to a massive edema and toxemia that can be fatal.

Treatment of Cutaneous Anthrax

Antibiotics have little effect on this infection except to protect against dissemination. *Bacillus anthracis* is susceptible to ciprofloxacin, which is the treatment of choice.

| Keep in Mind |

- A wide variety of bacteria cause skin infections.

- Nosocomial infections are a serious hazard, with most being caused by *Staphylococcus aureus* and group A streptococci.

- Group A streptococci can cause necrotizing fasciitis, which is a very serious infection.

- Bacteria can also infect the hair follicles, sebaceous glands, and sweat glands, causing infections such as acne, which is the most common skin infection in humans.

- Gas gangrene causes necrosis of tissues and can be rapidly fatal.

- Cutaneous anthrax is usually a mild infection unless it becomes systemic, in which case it can be fatal.

VIRAL INFECTIONS OF THE SKIN

The skin is also a barrier to viral pathogens, which, as we have seen throughout our discussions, are constantly surrounding us. The same barrier constraints that apply to bacteria also apply to viruses, which means that there needs to be an entry point for the pathogens. Some viruses enter through broken skin, but several common viral infections manifest their signs on the skin after systemic infection, entering the dermis through the blood or lymph (see Figure 26.1).

Measles

Measles is an extremely contagious infection caused by a single-stranded RNA virus of the paramyxovirus family. It is the leading cause of vaccine-preventable disease worldwide. Common forms of measles include **rubeola** (five-day measles) and **hard measles**, for which the infections last 7–18 days. The measles virus can produce severe infection in children, with accompanying high fever, widespread rash, and transient immunosuppression.

Measles usually occurs in preschool children who have not yet received the MMR (measles, mumps, **rubella**) vaccination. Although there is only one serotype of the measles virus and it is restricted to humans, there may be some antigenic drift associated with the virus.

Pathogenesis of Measles

Between 9 and 11 days after exposure to the virus, the infection begins in the respiratory tract with cough, runny nose, and fever. These initial signs are followed by viremia and lymphatic spread of the virus throughout the body, including the lymph tissue, bone marrow, and skin. Virus can be found in the blood during the first week of illness, and viruria can persist for up to four days after the rash appears. There is also a depressed immunity and a susceptibility to bacterial superinfections during this time. One to three days after the onset of the respiratory signs but before the skin rash breaks out, small red spots known as **Koplik's spots** appear on the mucous membrane inside the cheeks (**Figure 26.9**).

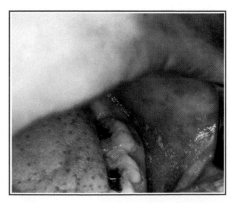

Figure 26.9 A red skin rash plus Koplik's spots on the inner surface of the cheek are early signs of a measles infection.

One day after the Koplik's spots comes the characteristic red skin rash. Significant numbers of virions can be found in the Koplik's spots as well as in the areas around the skin rash once it appears. Lymphadenopathy is common during this infection.

Measles can be very severe in individuals who are immunosuppressed and can result in death. In developing countries, a 15–25% mortality is associated with this infection. There can also be complications in up to 15% of cases, including otitis media, sinusitis, mastoiditis, pneumonia, and sepsis. In addition, 1 in every 1000 cases will develop encephalitis, which can cause permanent nerve damage or death.

Treatment of Measles

There is no therapy other than supportive care and close observation for potential complications. There is a very effective vaccine that is routinely given to children as part of their vaccination schedule.

Rubella (German Measles)

Rubella is a very mild or asymptomatic infection that can involve low-grade fever, lymphadenopathy, and faint macular rash. However, this infection is very serious in pregnant women and can cause congenital abnormalities in the developing fetus. Rubella is usually seen in the spring, and 30–60% of susceptible individuals can develop a clinical infection. An infected individual is contagious for seven days before and seven days after the appearance of the rash.

Fast Fact Many states in the United States require a blood test for rubella as part of the procedure to obtain a marriage license.

Pathogenesis of Rubella

The virus enters through the respiratory tract and spreads to the blood, lymph, organs, and skin. The viremia is seen up to eight days before the rash appears, and viral shedding can be seen in the oropharynx up to eight days after the onset of infection. As mentioned above, in a pregnant woman there can be transplacental transfer of virions to the fetus.

Treatment of Rubella

There is no specific therapy for rubella infection or for the congenital infection that sometimes accompanies it. Since 1969, the administration of live attenuated vaccine (MMR) has caused a marked decrease (90%) in the number of rubella cases, and this vaccination is recommended in the first year of life.

Smallpox (Variola)

Smallpox, an infection caused by a DNA poxvirus, has been known since the Middle Ages, when 80% of the population of Europe suffered from this infection. It was introduced to the Americas by European colonists.

There are two forms of smallpox infection: **variola major**, for which the mortality rate can be 20% or higher, and **variola minor**, which has a very low mortality rate (approximately 1%). Technically speaking, smallpox has been eradicated from the entire world, with the last victim being seen in Somalia in 1977. Because the only reservoir for this virus is humans, there should be no more smallpox cases ever. However, there are stocks of the smallpox virus in both Russia and the United States, a situation that leaves open the possibility of further infections. This is further compounded by the decrease in herd immunity to smallpox. The dominant feature of smallpox infection is the appearance of a papulovesicular rash involving pustules that form within the first two weeks of the onset of infection.

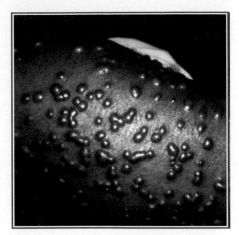

Figure 26.10 Photograph (from 1968) showing the right arm of a person with smallpox in the late pustular stage. The disease manifested itself as the classical maculopapular rash.

Pathogenesis of Smallpox

Smallpox virus is transmitted from person to person via the inhalation of small droplets of virus-containing saliva. The incubation period is 12–14 days but can be as short as 4–5 days, accompanied by the abrupt onset of fever, chills, and muscle aches. A rash appears 3–4 days later and evolves to papulovesicles, which are seen most prominently on the head and extremities and become pustular over 10–12 days (Figure 26.10). These pustules appear only once, and they crust over and slowly heal. Death from smallpox results from either overwhelming virus infection or bacterial superinfection.

Treatment of Smallpox

Because vaccination for smallpox was so effective, the infection was wiped off the planet. However, the potential use of this virus as a bio-weapon has caused many countries to begin stockpiling the vaccine. Because there is no infected population, it is difficult to test potential antiviral drugs, but several candidate drugs, such as cidofovir, are being evaluated.

Chickenpox and Shingles

Before the development of a vaccine against the varicella-zoster virus in 1995, approximately 4 million Americans a year were newly infected with the virus. Since then, varicella-zoster infections, and the resultant hospitalizations and deaths, have declined dramatically. The infection has two clinical manifestations: **chickenpox** in children and **shingles** in adults.

Pathogenesis of Chickenpox and Shingles

The virus is spread through secretions of the respiratory tract, and the infection occurs in the upper respiratory tract and the lymph nodes. Varicella causes a generalized vesicular rash, usually found initially on the back of the head and ears and then appearing on the face, trunk, and proximal extremities. There is commonly involvement of mucous membranes and fever early in the infection. As few as 10 or as many as several hundred irritating, itchy lesions can appear during the course of the infection. Secondary viremia includes infection of the skin.

In the early days of medicine, varicella and zoster were considered separate entities, the former the agent for chickenpox and the latter the agent for shingles. The possibility that varicella and zoster were clinical manifestations of the same virus was shown as early as 1892 and was confirmed in 1954. A latent form of the virus resides in the dorsal root ganglia of adults who had chickenpox as children. When this latent virus is reactivated, it multiplies in a sensory neuron and then travels down that neuron to the skin. The shingles rash comprises vesicles similar to those seen in chickenpox, but in shingles the vesicles are localized in distinct areas of the body, usually the waist and in some cases the upper chest and back (Figure 26.11). These areas are usually the areas of skin innervated by the sensory neuron harboring the latent virus, and it seems that reactivation increases in frequency with advancing age. The shingles lesions are very painful; they appear several days to several weeks after pain is experienced in the affected area and can persist for months. In immunocompromised patients, there can be multiple organ involvement and a significant mortality rate (about 17%).

Treatment of Chickenpox and Shingles

Acyclovir and famciclovir can reduce fever and skin lesions and are the recommended treatments for patients over 18 years old and patients with immunodeficiency.

Fast Fact A vaccine to prevent shingles is available and is recommended for people aged 60 years and over in the United States.

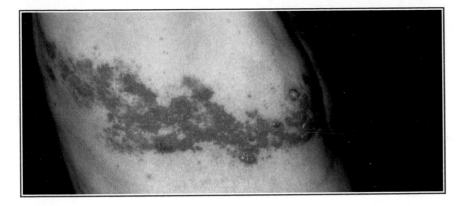

Figure 26.11 The lesions associated with shingles, which occurs when a latent form of the varicella-zoster virus is activated in an adult.

Herpes Simplex Type 1

Infections caused by the herpes simplex type 1 virus (HSV-1) are referred to as above-the-waist infections to distinguish them from the genital infections caused by herpes simplex type 2. HSV-1 causes a latent infection, with signs appearing only when the virus is reactivated. This virus is found throughout the world, and humans are the only reservoir. In developing countries, 90% of the population has antibodies against HSV-1 by the age of 30 years, whereas in the United States antibodies against this virus are found in 60–70% of adults.

Pathogenesis of Herpes Simplex Type 1

During this acute infection, syncytia develop, and there is a degeneration of epithelial cells that brings about necrosis at the infection focus. The inflammatory response occurs, with infiltration by neutrophils followed by mononuclear cells. The virus spreads either interneuronally or intraneuronally, and intraneural spread means that the virus can hide from the immune response and lie latent, sometimes for years. The latent virus resides on the trigeminal, superior cervical, and vagus nerve ganglia, but the reactivation stimuli are not yet understood.

HSV-1 infection is often asymptomatic, but when that is not the case the principal sign is grouped or single vesicular lesions that become pustular and coalesce to form multiple ulcers. There can be painful ulcerative lesions on the tongue, gums, and pharynx. These lesions usually persist for 5–12 days, and after the primary infection, latent reactivation can occur in the form of cold sores (also called fever blisters) that appear on or near the lips and can last for as long as 7 days (Figure 26.12).

HSV-1 sometimes infects the fingers in the area of the nails. This usually occurs because of a break in the skin and causes the formation of painful

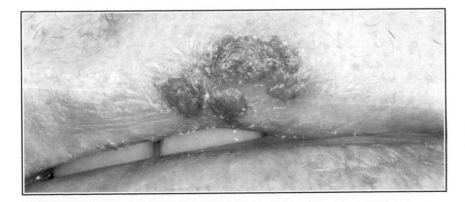

Figure 26.12 Cold sores (fever blisters) occur around the mouth because of reactivation of latent herpes simplex type 1 virus.

pustular lesions on the fingers. This virus can also infect the eye (ophthalmic herpes) and is the most common cause of corneal damage and blindness in the developed world. The damage occurs because of dendritic ulcerations in the conjunctiva and the cornea, which cause scarring.

Treatment of Herpes Simplex Type 1

The most effective treatment for HSV-1 infection is the nucleoside analog acyclovir. This antiviral significantly reduces the duration of the primary infection and can suppress recurrence. Immunocompromised patients may harbor resistant HSV, in which case foscarnet can be used.

Warts

Warts (papillomas) are small growths that appear either on the skin or on the mucous membrane of the respiratory tract, genital tract, and interior of the mouth. They are caused by human papillomavirus (HPV), which we discussed in Chapter 23. HPV infection is a lifelong infection, and warts can return even after removal. This is because the virus is still associated with the tissue adjacent to where a wart was removed.

Warts vary in appearance, location, and pathogenicity. Some are small and self-limiting, others are large but benign, and others are malignant. HPV infection causes more than 95% of cervical cancers, but genital warts do not generally become malignant. Dermal warts are more common in children than in adults. Warts are larger and occur more frequently in people who are immunodeficient.

Pathogenesis of Warts

HPV is one of the few viruses infecting the skin that is transmitted by direct contact between humans and by fomites. In dermal warts, which incubate for one to four weeks, the virus gains entry through broken skin. Genital warts (**Figure 26.13**) can be sexually transmitted and require 8–20 months of incubation before signs are expressed.

In dermal warts, the virus infects epithelial cells and causes them to proliferate, resulting in warts. This occurs at the boundary between the dermis and the epidermis (see Figure 26.1). Usually only one or at most a few warts occur during an outbreak. The removal of one wart in a cluster causes the others in the cluster to regress, and in some cases there is spontaneous regression without a first wart being removed. It is thought that this may be the result of an immune response.

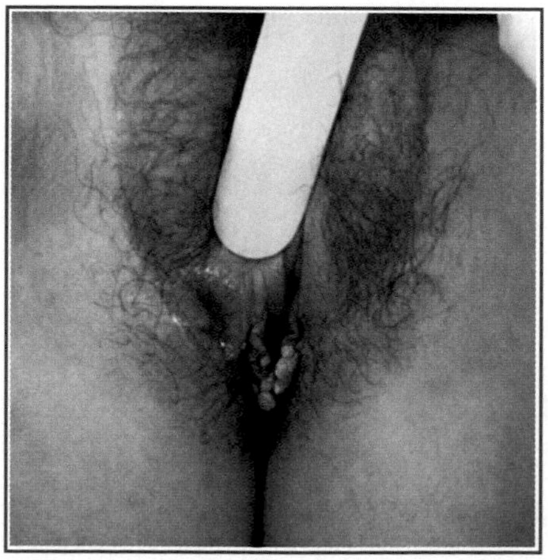

Treatment of Warts

There is no satisfactory drug treatment for warts. However, in many cases the growth can be removed by cryotherapy with liquid nitrogen, accompanied by the removal of adjacent infected tissue. It has been shown that antimetabolites such as 5-fluorouracil and interferon can block HPV infection.

Figure 26.13 Genital warts.

> **Keep in Mind**
>
> - Viruses can enter through broken skin or through the blood and lymph.
> - Several systemic viral infections, including measles, rubella, and smallpox, manifest themselves through skin lesions.
> - Varicella-zoster virus and HSV-1 cause latent infections that reside in the nervous system and cause skin lesions on reactivation.
> - Warts are caused by the human papillomavirus.

FUNGAL INFECTIONS OF THE SKIN

Remember that fungi are always present on the skin, and they rarely bother us as long as we have a competent immune system. However, if there is some compromise of an individual's health, fungi can be opportunistically infectious. The unbroken skin is a barrier to fungi, just as it is to bacteria and viruses.

Cutaneous Candidiasis

The fungal infection known as **cutaneous candidiasis** is caused by *Candida albicans*, which we discussed in Chapter 14. This organism is a member of the normal flora in the oropharyngeal, gastrointestinal, and genitourinary tracts. *Candida albicans* can grow in multiple morphological forms but is most often seen as a yeast. It does have the capacity to form hyphae, and the change from the single-celled yeast form to the formation of hyphae is strongly associated with pathogenicity.

Pathogenesis of Candidiasis

Candida albicans organisms use their hyphae to invade deep into tissues. The stimuli for this change in morphology have not yet been identified, but it is known that the change is associated with the appearance of factors that increase adherence to and destruction of tissues.

The hyphae form strong attachments to human epithelial cells and secrete proteinases and phospholipases that digest epithelial cells and further facilitate tissue invasion by the hyphae. *Candida albicans* contains surface proteins that bind to C3 receptors, thereby preventing opsonization. The compromise of T-cell function or the overuse of antibiotics also allows *C. albicans* to increase in numbers, resulting in local and invasive infection. Indwelling catheters and chemotherapy may also advance the invasion.

Candida albicans infections of the skin usually occur in folds of skin and other areas in which two wet skin surfaces are opposed to each other (on the buttocks of infants, for example, causing diaper rash). The initial lesions are erythematous papules or areas of tenderness and fissured skin. Infants, whose normal flora has not yet been established, can get the infection called **thrush** (**Figure 26.14**), which is the development of a whitish overgrowth in the oral cavity. This same condition can be seen in patients with immunodeficiencies.

A *C. albicans* infection usually remains confined to chronically irritated areas. However, in rare cases in which the infected individual is immunocompromised, the condition called **mucocutaneous candidiasis** is seen. In this case, infections of the hair and skin fail to heal and require therapy. This form of the infection can result in considerable discomfort and in some cases disfigurement with extensive areas of lesions.

Treatment of Candidiasis

Candida albicans is usually susceptible to nystatin, flucytosol, the azole antifungals, and, in cases of deep tissue involvement, amphotericin B. In addition, measures should be taken to decrease moisture and chronic trauma. Fluconazole is the most effective treatment for mucocutaneous candidiasis.

Dermatophytosis

Dermatophytes are fungi that are pathogenic to the skin, and the cutaneous mycoses they cause result in slow, progressive eruptions of the

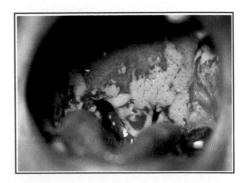

Figure 26.14 Oral candidiasis, also known as thrush.

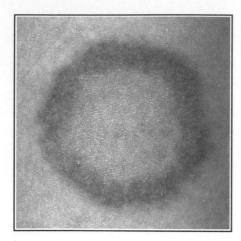

Figure 26.15 The classic lesions of tinea capitis, also known as scalp ringworm.

skin that are unsightly but not painful or life-threatening. There are several forms of dermatophyte-caused infection, classified according to the inflammatory response in the skin, but all forms typically involve erythema, induration, itching, and scaling. The umbrella term for all these conditions is **dermatophytosis**.

The most familiar dermatophyte infection is **tinea capitis**, more commonly known as scalp ringworm (**Figure 26.15**), which gets its name from the shape of the creeping margin at the edge of the dermatophyte growth. Dermatophytosis in the groin is referred to as tinea cruris or jock itch, and dermatophytosis on the feet is tinea pedis or athlete's foot. There are ecological and geographic differences in the occurrence of different dermatophyte infections, and many domestic cats and dogs as well as cows and horses are infected and act as reservoirs for these fungi.

Human-to-human transfer requires close contact because dermatophytes are poorly infective and have very low virulence. The infection is usually seen in families, barber shops, or locker rooms.

Pathogenesis of Dermatophytosis

All three forms of tinea infection begin when minor traumatic skin lesions come into contact with the dermatophyte hyphae from an ongoing infection. Once the stratum corneum, which is the most superficial layer of the epidermis, has been penetrated by the hyphae, the dermatophytes proliferate in the keratinized layers of the skin, aided by the production of protease enzymes. The course of the infection depends on anatomical location, moisture, and the rate of skin-cell shedding. The speed and strength of the inflammatory response also have a role in the infection. It is interesting to note that the faster the skin is shed, the less time it takes to get over the infection. Inflammation can increase the shedding rate and therefore affect the time course of the infection. In contrast, immunosuppression increases the length of the infectious period. In any case, invasion of deep tissues during a tinea infection is extremely rare. Most of these infections are self-limiting, but in cases where the inflammatory response is poor, the infection can become chronic. Dermatophyte infections can also affect the nails and hair follicles, in the latter case by plugging them up and causing the hair to become brittle.

Treatment of Dermatophyte Infections

Most skin dermatophyte infections resolve without therapy. Those that do not can be treated with topical agents such as tolnaftate, allylamines, and azoles. More extensive infections in the nail beds require systemic therapy for weeks or months with griseofulvin or itraconazole. Tinea capitis also requires systemic therapy.

Keep in Mind

- Fungi are always present on the skin and are rarely infectious.

- The most common fungal infection of the skin is candidiasis, caused by *Candida albicans*.

- Several forms of dermatophyte-caused infection are classified according to the inflammatory response seen in the skin.

PARASITIC INFECTIONS OF THE SKIN

Cutaneous Leishmaniasis

The zoonotic parasitic infection **cutaneous leishmaniasis** is caused by protozoa of the genus *Leishmania*. It is common in central Asia, India, the Middle East, and South and Central America and affects as many as 12 million people worldwide. Humans contract this infection when they enter areas inhabited by infected mammals (often rodents). The infection is vector-transmitted to human by sandflies.

Pathogenesis of Cutaneous Leishmaniasis

The lesions associated with cutaneous leishmaniasis appear on the extremities or the face, weeks to months after the bite of the sand fly. They appear as itchy pustules accompanied by lymphadenopathy. Within a few months, the pustules ulcerate (**Figure 26.16**), and there can be several of these lesions on the body. There is spontaneous healing in 5–12 months, leaving depigmented scars on the skin. If the lesions occur on the ear, the infection can cause destruction of the pinna, and in patients with AIDS this infection causes multiple nonhealing lesions.

Treatment of Cutaneous Leishmaniasis

If there is no involvement of the mucous membranes, no treatment is required. Amphotericin B, ketoconazole, and itraconazole can be used for more severe lesions.

Pediculosis (Lice)

The parasitic infection caused by lice on the human body is called **pediculosis**. Infection can occur on the head (caused by the head louse, *Pediculus humanus capitis*) or on the body (caused by the body louse or clothes louse, *Pediculus humanus corporis*). Infection on the body is often in the genital region, and this condition is referred to colloquially as crabs. Body lice can transmit infections such as epidemic typhus. Lice have been around for centuries as infecting agents, and outbreaks of lice infections, usually head lice, are frequently seen in schools, where the transfer of lice occurs easily.

Pathogenesis of Pediculosis

Lice require blood and feed several times a day. The itching that accompanies a pediculosis infection is due to a reaction against the saliva of the biting louse. Scratching can result in secondary bacterial infection. The head louse has specialized legs with which to grasp the hair shaft, usually at the base (**Figure 26.17**).

Treatment of Pediculosis

Treatment usually involves combing out the lice and nits (louse eggs) with a fine-toothed comb in combination with nonprescription drugs such as Nix and Rid. However, there has been increasing resistance to these drugs in recent years. Other topical ointments, such as lindane and malathion, can be effective but are very toxic.

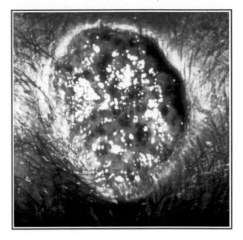

Figure 26.16 A lesion caused by cutaneous leishmaniasis.

Keep in Mind

- Cutaneous leishmaniasis is one of the most common parasitic infections of the skin.
- Lice cause pediculosis and can also transmit other infections such as epidemic typhus.

Figure 26.17 A human head louse.

INFECTIONS OF THE EYES

We are including the eyes in this chapter because they are affected by infections in the same manner as the skin. It is important to remember that the eyes are an immunologically protected site, which means that they are not protected by the immune system in the same way as most other areas of the body. This makes sense because it would be devastating if a defensive response such as inflammation were triggered every time a foreign particle entered the eye. As we saw in Chapter 15, the lacrimal apparatus produces tears, which constantly flush foreign particles from the eye and contain antimicrobial substances to protect the eye from infecting organisms.

Eye infection usually involves pain and the potential for vision loss. A summary of some common eye infections is shown in **Table 26.2**.

Conjunctivitis is the main infection of the eye; it can occur in all age groups. Conjunctivitis is inflammation of the inner membranes of the eyelids with an associated discharge. The infection is easily spread if a contaminated hand rubs the eyes.

Parts of the eye other than the conjunctiva can also become infected, including the cornea—in which case the infection is called **keratitis**—and the anterior and posterior chambers. Structures such as the orbital sinuses can also be involved, and because of the close proximity to the central nervous system, these infections can become life-threatening. The use of contact lenses has led to an increased number of cases of keratitis, which can be caused by bacteria (usually *Pseudomonas aeruginosa*) or protozoa (*Acanthamoeba*). The condition is painful, and loss of vision can occur. In most instances topical eye drops or ointments containing erythromycin or gentamicin are used to treat acute bacterial conjunctivitis, whereas fluoroquinolones can be used for eye infections caused by *Pseudomonas*. Quinolones such as ciprofloxacin are also useful for all types of bacterial eye infection.

Eyelid abscesses such as styes are fairly common, and infection of the entire eyelid is also possible. In addition, the infections of the lacrimal gland (**dacryocystitis**) and lacrimal duct (canaliculitis) can occur.

Infection	Causative organism	Infection characteristics
Bacterial		
Neonatal gonorrheal ophthalmia	*Neisseria gonorrhoeae*	An infection acquired during passage through the birth canal; can lead to blindness
Conjunctivitis	Several bacterial species	Highly contagious inflammation of the conjunctiva of the eye
Trachoma	*Chlamydia trachomatis*	Infection and eventual destruction of the cornea and conjunctiva; can cause blindness
Keratitis	Bacteria, viruses, and fungi	Ulceration of the cornea; occurs mainly in immunodeficient and debilitated patients
Parasitic		
Onchocerciasis	*Onchocerca volvulus*	Blackflies transmit the microfilariae through bite, microfilariae infect other tissues; causes blindness
Loaiasis	*Loa loa*	Transmitted by the bite of the deer fly; causes inflammation of the conjunctiva and the cornea
Acanthamoeba keratitis	*Acanthamoeba*	Typically occurs in healthy people and is often associated with contact lenses; can result in permanent visual impairment or blindness

Table 26.2 Infections of the eye.

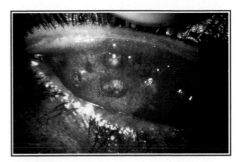

Figure 26.18 Inflammation of the eyelid in trachoma, a chronic and contagious eye disease caused by *Chlamydia trachomatis*.

In developing countries, a leading eye infection is **trachoma** (**Figure 26.18**), caused by *Chlamydia trachomatis*. It is estimated that 500 million people are affected by this condition, with 10 million having been blinded by it. The pathogen is spread to the eyes by hands, by such fomites as towels and clothing, and by flies. Trachoma, which is essentially a chronic conjunctivitis, leads to scarring, corneal ulceration, and eventual vision loss.

The parasite *Onchocerca volvulus* causes the eye infection known as **river blindness**, which affects 20 million people worldwide. It is spread by the blood-sucking blackfly, which transmits the parasite to the eye. The parasite invades the anterior chamber of the eye, causing corneal ulceration, fibrosis, and blindness.

The infection **loaiasis** is caused by the parasitic worm *Loa loa*, found in the African rain forests. The worm is transmitted to humans by the deer fly, which feeds during the day and acquires the parasite from infected humans.

The *Loa loa* larvae migrate to the mouth of the deer fly and are then transmitted to a human when the fly bites. The larvae mature into worms, which migrate through the infected person's subcutaneous tissue, causing inflammation as they go, and settle in the cornea and conjunctiva. The worms can grow to more than an inch in length and are easily seen on ophthalmic examination (**Figure 26.19**). There is a

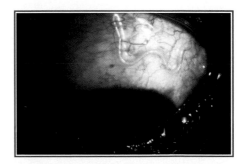

Figure 26.19 The *Loa loa* worm, which infects the eyes and can grow to an inch long.

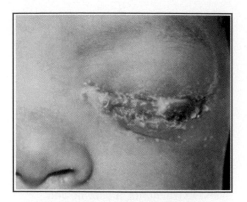

Figure 26.20 Neonatal gonorrheal ophthalmia caused by *Neisseria gonorrhoeae.*

very elevated eosinophil count throughout this infection, and the drug albendazole can reduce the problem.

Neonatal Eye Infections

A serious conjunctivitis caused by the bacterium *Neisseria gonorrhoeae* is **neonatal gonorrheal ophthalmia**, which can be contracted as an infant is passing down the birth canal of an infected mother. The infant acquires large numbers of *N. gonorrhoeae* organisms during the passage.

In many cases, mothers infected with *N. gonorrhoeae* are also infected with *Chlamydia trachomatis*, which is involved in non-gonococcal urethritis, and this organism can also make its way into the eyes of newborn infants. Both neonatal gonorrheal ophthalmia and the infection caused by *C. trachomatis* cause large amounts of pus to form in the eyes (**Figure 26.20**) and without treatment will lead to ulceration and scarring of the cornea.

It is common practice to treat the eyes of newborn infants with erythromycin to prevent the onset of both neonatal gonorrheal ophthalmia and the problems caused by *C. trachomatis* infections. This practice originally involved the application of silver nitrate to the eyes, but this has no effect on *Chlamydia*, so antibiotics are used instead.

Keep in Mind

- Like the skin, the eyes are affected by infections through direct exposure to pathogens.

- The most common eye infection is conjunctivitis.

- People who wear contact lenses are prone to keratitis.

- Trachoma, which is a form of chronic conjunctivitis, is caused by *Chlamydia trachomatis* and is a serious problem in developing countries.

- *Neisseria* pathogens in infected mothers can cause eye infection in babies as they move down the birth canal.

SUMMARY

- Skin is a barrier to pathogens.

- Pathogens must enter through a break in the skin or via the blood and lymph.

- A wide variety of bacteria can cause infection of the skin, with necrotizing fasciitis being one of the worst infections.

- Bacteria can infect hair follicles, sebaceous glands, and sweat glands.

- Several systemic viral infections manifest themselves with skin lesions.

- Viral infections that cause lesions on the skin include measles, rubella, smallpox, chickenpox, herpes simplex type 1, and human papillomavirus.

- Fungi are always present on the skin but rarely cause infection.

- The most common fungal infection of the skin is candidiasis.

- One of the most common parasitic infections of the skin is leishmaniasis.

- Eyes are infected through direct exposure to pathogens.

- Trachoma and river blindness are parasitic infections of the eye, affecting hundreds of millions of people.

In this chapter we have examined the skin and eyes as targets for infectious pathogens. Keep in mind that the barrier nature of the skin is very effective in limiting the number of these types of infection, and only after that barrier has been breached do we see the development of infection. Remember that inflammation, which is a very important defense against infection, can also cause damaging side effects that are clearly seen in infections of the skin and eyes.

 SELF EVALUATION AND CHAPTER CONFIDENCE

Multiple Choice

Answers are given in the back of the book and help can be found in the student resources at:

www.garlandscience.com/micro2

1. The skin is important in protecting against infectious pathogens because

   A. It is made of layers
   B. It is a barrier to invading organisms
   C. It is easily broken
   D. It is sterile

2. All of the following act as barriers to invasion of the skin by pathogens except

   A. Sebum
   B. Perspiration
   C. Acidic pH of the skin
   D. The hair follicles

3. Necrotizing fasciitis is caused by

   A. *Neisseria*
   B. *Staphylococcus*
   C. Group A *Streptococcus*
   D. *Borrelia* organisms

4. Inflammatory acne is caused by

   A. *Borrelia*
   B. Group A streptococci
   C. *Pseudomonas aeruginosa*
   D. *Propionibacterium*
   E. *E. coli*

5. Scalded skin syndrome is caused by

   A. *Staphylococcus aureus* endotoxin
   B. *Streptococcus pyogenes*
   C. *Streptococcus* exotoxin
   D. *Staphylococcus aureus* exotoxin

6. The etiologic agent of gas gangrene is

   A. *Clostridium botulinum*
   B. *Clostridium tetani*
   C. *Bacillus anthracis*
   D. *Clostridium perfringens*
   E. None of the above

7. The characteristic signs of smallpox are

   A. Pimples that form only on the face
   B. Papulovesicular rash on the head and extremities
   C. Open sores on the face and neck
   D. None of the above

8. The etiologic agent of chickenpox is

   A. Variola minor
   B. Variola major
   C. Paramyxovirus
   D. Varicella-zoster
   E. None of the above

9. Cold sores are formed as a result of infection with

   A. *Streptococcus pyogenes*
   B. Variola major
   C. Herpes simplex type 1
   D. Herpes simplex type 2
   E. Paramyxovirus

10. Thrush results from infection with

    A. *Cryptococcus neoformans*
    B. *Streptococcus mutans*
    C. *Candida albicans*
    D. *Staphylococcus aureus*
    E. None of the above

11. Humans acquire cutaneous leishmaniasis through

    A. Contaminated food
    B. Contaminated water
    C. The bite of the tsetse fly
    D. The bite of the *Anopheles* mosquito
    E. The bite of the sand fly

12. Conjunctivitis is an

    A. Infection of the epidermis
    B. Infection of the eye
    C. Infection of the dermis
    D. Infection of the hair follicle

13. The current treatment for neonatal gonorrheal ophthalmia is with

    A. Mercury
    B. Silver nitrate
    C. Antibiotics
    D. Eye wash

14–18. Answer **A** if the statement is correct; answer **B** if it is incorrect.

14. Most skin infections are caused by only a few types of pathogens.

15. Folliculitis is treated with antibiotics.

16. Gas gangrene causes necrosis of tissues and can be rapidly fatal.

17. Cutaneous anthrax is usually a fatal infection.

18. The eye is protected from infection by the lacrimal apparatus.

(Q) **DEPTH OF UNDERSTANDING**

Questions listed here require you to bring together the concepts you have learned in this chapter into a discussion format. This helps you to increase your depth of understanding of the material you have learned. Help can be found in the student resources at: www.garlandscience.com/micro2

1. Describe the physical mechanisms that control the level of infection through the skin.

2. Alice has a chlamydial infection and is about to give birth. Describe the precautions that will be taken at delivery and why they are necessary.

(Q) **CLINICAL CORNER**

Help can be found in the student resources at: www.garlandscience.com/micro2

1. A patient is brought into the emergency room with a very swollen arm. He is in a lot of pain; as you examine him, you notice what looks like a surgical scar that has not healed properly and is now surrounded with dark purple skin. He explains that he had surgery to reattach his bicep, which had been hurt lifting weights, and the site of surgery has continued to hurt. You suspect that this might be a case of gas gangrene, but there is no smell associated with wound. After examination, the patient is moved from the emergency room; the next day, you hear that he had his arm amputated and he is on long-term heavy antibiotic therapy and will be for some time. In addition, the perioperative nurse says he will be hospitalized and closely monitored for some time.

    A. What condition did he have?
    B. Why did you think it was gas gangrene and how would you have proven it was or wasn't?
    C. Why is the long-term antibiotic therapy necessary?

2. Curtis Worthy is a 13-year-old white male who has come to the clinic with his mother. After your initial workup, Curtis seems to be in good health and seems to be well developed both physically and emotionally. He has a moderate case of facial acne with several pustules and papules on his face and neck. His skin also has an oily feel to it. He seems to be handling his skin condition very well, but his mother is very concerned. She is adamant that Curtis be given a prescription for antibiotics that will alleviate her son's condition.

    A. What can you tell her about her son's condition?
    B. Would a prescription benefit Curtis?
    C. What would be your overall concern about giving her a prescription?

Epilogue

You have now reached the end of this exciting learning journey.

The route started by gradually building and establishing a sound knowledge foundation in **Part I**, which we used to understand the more specific and advanced concepts of infection and disease covered in **Parts II** and **III**. We then turned our attention to our inbuilt defense mechanisms in **Part IV**, and how we can use external chemical agents as therapy in **Part V**. The final leg of the journey, **Part VI**, put all the knowledge into a clinical context to help you apply your skills to solve some of the problems you will meet in your professional career.

On the way we have discussed principles of chemistry, molecular biology, and genetics, and the relationship between host and pathogen including their respective structure and function, diversity, and classification. We have encountered concepts of virulence and pathogenicity, infectious disease mechanisms and processes, transmission and epidemiology, the link of prevention and cure with detecting infections, and new potential therapeutic modalities and development of vaccines. Emerging and re-emerging diseases as major problems for public health, microbial growth and control, antibiotics and antibiotic resistance, the immune response, and the catastrophic consequences of an impaired immune system were further stops on the way.

You have learned many things about microbiology and infection, some of it possibly unexpected. We are able to do amazing things using microbiology and the rapidly growing field of biotechnology. Biotechnology is now almost ubiquitous in our day-to-day existence: from food to reproduction, religion and politics, there is no area that cannot somehow be related to new biological technology. Some of the tools available to us have incredible beneficial uses for health care professionals throughout their career. But health care workers must also guard against complacency, because technological advances can be used to cause deliberate harm (such as war and bioterrorism) as well as inadvertently providing situations that allow infectious diseases to thrive. The ease of travel helps infections spread; climate change means that areas are being affected by infections where they were previously not a problem; modern living has seen the emergence and re-emergence of infectious disease; but perhaps the most serious challenge is the rise of antibiotic resistance.

This brings to a close our study of microbiology. Without doubt there is a great deal of information for you to master. As a health care worker (or consumer), you will face problems addressed by microbiology concepts, but they won't always present themselves in the same way in which you learned them. Reaching for the highest levels of understanding is crucial so you can generalize and recognize concepts in the novel ways in which they will occur in health care. While learning, you might have needed to re-read the Learning Skills sections, and you might find it useful again in the future. Learning is a skill for the journey of life and part of the actual journey.

The authors hope that this textbook and its ancillary materials have made the journey easier for you and wish you well in your future career!

Multiple Choice Answers

Chapter 1
1. A
2. D
3. C
4. A
5. B
6. E
7. D
8. C
9. C

Chapter 2
1. B
2. E
3. B
4. E
5. A
6. C
7. D
8. A
9. A
10. E

Chapter 3
1. D
2. E
3. C
4. B
5. A
6. A
7. E
8. D
9. B
10. C
11. E
12. B
13. B
14. A
15. D
16. D
17. D
18. D
19. C
20. C
21. D
22. A
23. A
24. E
25. C
26. E
27. E
28. E
29. B
30. A
31. A
32. B
33. D
34. B
35. A
36. B
37. A

Chapter 4
1. A
2. E
3. A
4. A
5. E
6. B
7. E
8. D
9. D
10. E
11. D
12. D
13. A
14. C
15. B

Chapter 5
1. B
2. E
3. A
4. C
5. A
6. A
7. B
8. D
9. C
10. C
11. A
12. D
13. C
14. C

Chapter 6
1. C
2. A
3. A
4. B
5. C
6. C
7. C
8. C
9. A
10. E
11. D
12. D
13. A
14. D
15. A
16. C
17. D
18. B
19. D
20. D
21. B
22. B
23. A
24. C
25. A
26. A
27. A
28. B
29. D

Chapter 7
1. B
2. D
3. A
4. C
5. B
6. E
7. B
8. E
9. E
10. B
11. B
12. A
13. D
14. D
15. E

Chapter 8
1. C
2. A
3. A
4. A
5. B
6. E
7. C
8. D
9. D
10. C
11. D
12. B
13. E
14. E
15. E
16. E
17. E
18. C
19. C
20. A
21. B
22. A
23. D
24. A
25. C
26. D
27. E
28. A

Chapter 9
1. B
2. E
3. E
4. E
5. A
6. B
7. A
8. B
9. B
10. A
11. C
12. C
13. C
14. A
15. C
16. C
17. A
18. C
19. D
20. B
21. D
22. D
23. C
24. B
25. B
26. D
27. D
28. E
29. E
30. B
31. C
32. A

Chapter 10
1. A
2. E
3. B
4. C
5. B
6. D
7. C
8. A
9. A
10. C
11. A
12. C
13. A
14. A
15. C
16. C
17. D
18. D
19. B
20. A
21. A
22. B
23. A
24. B

Chapter 11
1. E
2. E
3. D
4. D
5. C
6. A
7. D
8. C
9. B
10. A
11. C
12. B
13. E
14. D
15. A
16. C
17. C
18. A
19. A
20. D
21. B
22. E
23. B
24. E
25. D
26. C
27. C
28. C
29. D
30. C
31. C
32. D
33. B
34. A
35. C
36. A
37. A
38. C
39. E
40. E
41. A
42. D
43. D

Chapter 12
1. D
2. A
3. B
4. C
5. D
6. A
7. C
8. A
9. D
10. C
11. D
12. A
13. D
14. A
15. B
16. C
17. E
18. A
19. C
20. D
21. E
22. A
23. A
24. A
25. A
26. B
27. A

Chapter 13
1. D
2. C
3. C
4. B
5. A
6. D
7. C
8. D
9. C
10. E
11. E
12. C
13. C
14. D
15. B
16. D
17. D
18. D
19. B
20. A
21. C
22. C
23. B
24. C
25. B
26. C
27. A
28. B
29. D

Chapter 14
1. D
2. D
3. E
4. C
5. C
6. C
7. B
8. D
9. C
10. C
11. C
12. D
13. D
14. C
15. C
16. C
17. B
18. E
19. B
20. D
21. C
22. E
23. D
24. C
25. C
26. A
27. B
28. B
29. B
30. A
31. B
32. B
33. D
34. A
35. D
36. B
37. A
38. C

Chapter 15
1. A
2. A
3. C
4. C
5. D
6. E
7. E
8. E
9. C
10. D
11. D
12. E
13. D
14. C
15. E
16. E
17. A
18. C
19. E
20. A
21. D
22. C
23. C
24. D
25. D
26. D
27. B
28. B
29. D
30. B
31. C
32. A
33. C
34. B
35. A
36. D
37. B
38. A
39. C

Chapter 16
1. A
2. D
3. D
4. D
5. C
6. D
7. A
8. C
9. E
10. D
11. C
12. C
13. D
14. C
15. C
16. B
17. D
18. D
19. D
20. C
21. C
22. C
23. E
24. E
25. E
26. C

Chapter 17
1. D
2. D
3. C
4. B
5. B
6. A
7. C
8. E
9. E
10. A
11. D
12. B
13. D
14. A

Chapter 18
1. B
2. E
3. A
4. A
5. E
6. D
7. C
8. C
9. B
10. E
11. C
12. D
13. D
14. B
15. C
16. B
17. A
18. B
19. C
20. A
21. B
22. B
23. A
24. C
25. A

Chapter 19
1. D
2. C
3. C
4. A
5. E
6. A
7. B
8. A
9. A
10. E
11. D
12. D
13. C
14. D
15. D
16. C
17. C
18. C
19. A
20. A
21. C
22. B
23. E
24. B
25. B
26. A
27. B
28. A
29. B
30. D
31. C
32. A

Chapter 20
1. C
2. C
3. C
4. D
5. C
6. D
7. C
8. C
9. B
10. B
11. A
12. A
13. B
14. A
15. A
16. A
17. A
18. A
19. B
20. A
21. B
22. B

Chapter 21
1. B
2. E
3. D
4. A
5. A
6. A
7. B
8. E
9. A
10. A
11. A
12. B
13. A
14. A
15. B

Chapter 22
1. D
2. B
3. A
4. B
5. C
6. E
7. E
8. D
9. B
10. B
11. C
12. C
13. A
14. D
15. E
16. D
17. D
18. A
19. A
20. B
21. B
22. A
23. A

Chapter 23
1. D
2. B
3. B
4. C
5. B
6. B
7. A
8. C
9. B
10. D
11. A
12. B
13. A
14. B
15. B

Chapter 24
1. B
2. E
3. D
4. A
5. D
6. D
7. A
8. C
9. C
10. C
11. C
12. D
13. D
14. B
15. B
16. B
17. A
18. A
19. A
20. A
21. B
22. B
23. A

Chapter 25
1. A
2. B
3. B
4. E
5. C
6. B
7. C
8. C
9. B
10. B
11. D
12. C
13. C
14. B
15. D
16. D
17. D
18. A
19. B
20. B
21. A
22. B
23. A
24. A

Chapter 26
1. B
2. D
3. C
4. D
5. D
6. D
7. D
8. D
9. C
10. C
11. E
12. C
13. C
14. B
15. B
16. A
17. B
18. A

Glossary

A site A site on the ribosome occupied by aminoacyl-tRNA.

ABC transport system A type of active transport including several proteins and several steps in which the molecule being transported is handed off from one protein to the next.

Abscess A lesion that contains pus in a cavity hollowed out by tissue damage.

Acidophiles Organisms that grow best in an environment with a pH of 4.0 to 5.4.

Active agglutination Type of agglutination where the antigen occurs naturally on a particle.

Active immunization An immunization in which antigen representing the infectious agent is administered and confers immunity. This is used to increase herd immunity.

Active site The location on an enzyme where reactants are made into products.

Active transport The mechanism used to transport things from one side of a membrane to the other in a process that requires ATP.

Active viremia Viruses replicating in the blood.

Acute congestion A phenomenon that occurs where local capillaries become engorged with neutrophils.

Acute croup Also known as laryngotracheitis. A severe infection of the larynx and trachea.

Acute disease The opposite of chronic disease; one in which infection occurs quickly and lasts a short time.

Acute inflammatory colitis Rapid inflammation of the colon.

Acute influenzal syndrome An acute influenza in which symptoms can develop in a matter of hours.

Acute pancreatitis The rapid onset of serious inflammation of the pancreas.

Acute phase of HIV infection This phase occurs in the first few days after initial infection: the virus is reproducing in the lymphocytes of the lymph nodes, resulting in lymphadenopathy and flu like symptoms.

Acute-phase protein A protein such as C-reactive protein or mannose-binding protein that is related to the development of the inflammatory response but is seen only in acutely ill patients.

Acute septic shock The rapid onset of hypotension associated with overwhelming infection. It is thought to result from the action of endotoxins.

Acute urticaria Sudden onset of vascular reactions in the upper dermis marked by transient appearance of slightly elevated patches that are redder than the surrounding tissue.

Acyclovir An antibiotic used for viral infection. Acyclovir is a synthetic purine nucleoside with activity against herpes simplex virus.

Adaptive immune response A specific host defense composed of the humoral (antibody) and cellular immune responses. It has memory and takes several days to get started.

Adhesin A protein or glycoprotein found on attachment pili or in capsules that helps microorganisms attach to host cells.

Aerobe An organism that uses oxygen.

Aerobic respiration Metabolism that uses oxygen as the final electron acceptor.

Aerotolerant bacteria Bacteria that grow in the presence of oxygen but do not use oxygen for metabolism.

African trypanosomiasis A neurological disease caused by the parasite *Trypanosoma* otherwise known as African sleeping sickness.

Agar A low-melting-point complex polysaccharide derived from marine algae and used to solidify media for the growth of microorganisms.

AIDS (acquired immune deficiency syndrome) A viral infection caused by the human immunodeficiency virus (HIV), which destroys a patient's helper T cells and thereby his or her adaptive immune response.

AIDS dementia A metabolic encephalopathy induced by HIV infection manifested by cognitive, behavioral, and motor abnormalities.

Alcoholic fermentation Fermentation in which pyruvic acid is reduced to ethanol.

Alkaline phosphatase Enzyme that catalyzes the removal of phosphate groups from molecules.

Allergen An ordinarily innocuous foreign substance that can elicit an adverse immunological response in a sensitized individual.

Allergic asthma Recurrent attacks of difficulty in breathing and wheezing due to spasmodic contractions of the bronchi.

Allergic conjunctivitis Inflammation of the conjunctiva caused by allergic reaction and characterized by itching, tearing, and redness.

Allergic reaction Also called hypersensitivity. It is a disorder in which the immune system reacts to antigens that it would normally ignore.

Allergic rhinitis Allergic reaction of the nasal mucosa.

Allolactose A fragment of the sugar lactose that acts as an inducer molecule and allows the lactose operon to be turned on.

Allosteric inhibition Also known as the allosteric effect. It involves the binding of a non-competitive inhibitor to a site on the enzyme molecule that causes a change in the shape of the active site and inhibits the binding of the substrate in the active site.

Alpha hemolysis Partial lysis of red blood cells, leaving a greenish ring in the blood agar medium around the colonies.

Alternative pathway of complement One of the sequences of reactions seen in the complement system of innate immune response initiated by factors B, D, or P.

Alveolar macrophage A macrophage that is found in the alveoli of the lungs.

Amantadine An antiviral agent that prevents the penetration of host cells by the influenza virus.

Amastigote The intracellular morphologic stage in the development of certain hemoflagellates.

Amebiasis Infection with ameboid protozoan parasites, in particular *Entamoeba histolytica*.

Amebic dysentery A gastrointestinal infection characterized by bloody diarrhea, caused by *Entamoeba histolytica*.

Ameboids Cells that move or change shape by means of protoplasmic flow.

American trypanosomiasis The American form of trypanosomiasis, also known as Chagas' disease.

Aminoacyl-tRNA synthetase The enzyme required for binding the amino acid to the transfer RNA molecule.

Aminoglycoside antibiotics Chemical agents that block bacterial protein synthesis.

Amphitrichous flagella One of the four forms of flagellar arrangement. In this case there are flagella at both ends of the bacterial cell.

Anabolic Relating to anabolism, the creation of large molecules from small molecules.

Anabolism Chemical reactions in which energy is used to synthesize large molecules by using smaller molecules (this process involves synthesis reactions).

Anaerobic respiration Metabolism that takes place without oxygen. In this case the final electron acceptor is an inorganic molecule other than oxygen.

Analytical epidemiology A form of bioinformatics that focuses on establishing the cause-and-effect relationship and can use prospective or retrospective data.

Anaphylactic shock Condition resulting from a sudden extreme decrease in blood pressure due to an allergic reaction.

Anaphylaxis An immediate exaggerated allergic response to antigens.

Anergy Automatic inactivation of a lymphocyte due to lack of co-stimulation.

Angiogenesis Development of blood vessels.

Animal virus A virus that infects animal cells. Human cells are considered animal for this definition.

Anion A negatively charged ion.

Anthrax A fatal disease of ruminants contracted by humans through contact with contaminated wool or animal products, or inhalation of spores.

Anthrax toxin A powerful cytotoxin produced by *Bacillus anthracis*, which increases vascular permeability. It has three components: edema factor, protective antigen, and lethal factor.

Antibiotics Chemicals that are taken internally to control the growth of microorganisms in a host.

Antibody Also known as an immunoglobulin. A class of proteins that are produced by the humoral immune response and respond to a specific antigen.

Antibody titer The quantity of antibody found in serum.

Anticodon loop A portion of the transfer RNA molecule that associates with the codon of the messenger RNA molecule at the ribosome.

Antigen Also known as an immunogen. Any molecule that will elicit an immune response.

Antigen binding site The site on the antibody molecule that interacts with epitopes on the antigen. It is made up of a portion of the variable region of the heavy chain and a portion of the variable region of the light chain.

Antigen-presenting cell A cell that processes antigens and places them on their surface for presentation to helper T cells. The predominant ones are macrophages or dendritic cells, but the B cell can also present antigen.

Antigenic drift The process of antigenic variation that results from mutations.

Antigenic shift The process of antigenic variation caused by the re-assortment of genes.

Antigenic variation Mechanisms by which pathogens change their antigenic surface proteins in order to evade the adaptive immune system.

Antimicrobial An agent that will inhibit the growth of a microbial organism.

Antisepsis The application of an antimicrobial agent to living tissue or skin.

Antiseptic A chemical substance that can be used on tissues to control the growth of microorganisms.

Antiviral proteins Proteins produced through stimulation by interferon that become active in the presence of double-stranded RNA and can protect cells from being infected with virus.

Apoptosis Genetically programmed cell death; a process in which the cell commits a kind of suicide. This occurs when cells are worn out or if specific signals are given to the cell, as in cytotoxic reactions.

Arbovirus A group of viruses that are transmitted to humans by mosquitoes or ticks, including agents of yellow fever and viral encephalitis.

Arthroconidia (singular arthroconidium) Specialized stalk structures that hold conidia found in some fungi.

Arthus reaction A localized inflammatory response with increased vascular permeability at the site.

Aseptic The use of techniques that minimize the chances of cultures becoming contaminated.

Asian influenza The term used to describe an influenza pandemic that occurred in 1957 and caused more than 70,000 deaths in the United States.

Aspergillosis Disease caused by *Aspergillus* fungi which causes the development of inflammatory granulomatous lesions in the skin, nasal passages, lungs, bones, and meninges.

Ataxia Lack of voluntary coordination of muscle movements.

Ataxia telangiectasia Primary immunodeficiency syndrome involving reduced numbers of T cells.

Atopy Hypersensitivity to environmental allergens.

ATP The biological energy molecule consisting of adenosine coupled with three phosphate atoms. Breaking the third phosphate off releases energy and ADP, that can be converted back to ATP with the addition of a phosphate and energy.

Atrichous The condition of nonmotile bacteria, in which flagella are absent.

Autoclave An instrument for sterilizing by means of moist heat and pressure.

Autotrophy Obtaining carbon atoms from CO_2.

Avian influenza Also known as bird flu. A type of influenza seen in avian species that has been able to infect humans. It is similar to the influenza virus that caused the Spanish flu pandemic of 1918.

Axial filament Also known as an endoflagellum. It is a subsurface filament found in spirochete organisms that causes the spirochete to rotate in a corkscrew fashion.

Axostyle A microtubule used for attachment and tissue damage found on the parasite *Trichomonas vaginalis*.

β-lactam antibiotics Any natural or semisynthetic antibiotic containing a β-lactam ring structure.

β-lactamase An enzyme produced by penicillin-resistant bacteria that destroys the β-lactam ring of penicillin.

B cell (B lymphocyte) A lymphocyte that is part of the adaptive immune response. This cell is formed in the bone marrow and matures there; it is responsible for the humoral (antibody) response of the adaptive response and differentiates into a plasma cell which produces antibody.

B-cell lymphoma Non-Hodgkin's form of neoplasm seen in lymphoid tissue.

Bacillary dysentery A gastrointestinal infection characterized by bloody diarrhea, caused by *Shigella* organisms.

Bacillus (plural bacilli) A rod-shaped bacterium

Bacitracin An antibacterial polypeptide produced by *Bacillus subtilis* that prevents cell wall synthesis.

Bacteremia Bacteria in the bloodstream.

Bacterial hemorrhagic disease Loss of blood from vessels caused by certain bacterial species.

Bacteriocidal Term describing an agent that kills bacteria.

Bacteriocins Antimicrobial peptides with narrow or broad host ranges produced by bacteria. They have no effect on the organism producing them.

Bacteriophage Viruses that infect bacteria.

Bacteriostatic Term used for chemicals that inhibit microbial growth but do not kill the organisms.

Bacteriuria Bacteria in the urine.

BALT Bronchus-associated lymphoid tissue.

Basophil A type of white blood cell containing large numbers of granules. This cell releases histamine and other molecules as part of the inflammatory response.

Bed sore A decubitus ulcer on the skin.

Beta hemolysis Complete destruction of red blood cells by bacterial enzymes.

Betadine An iodophor routinely used to prepare human skin for surgery and injection.

Binary fission The process in which a bacterial cell duplicates its components and divides into two identical daughter cells.

Biofilm A structure composed of a variety of bacterial species that coexist. It is often seen as the beginning layer of plaque, which builds up on teeth.

Biosensors Analytical tool using a biological component in the detection principle.

Biotechnology An industry that uses molecular biology and microorganisms for specialized purposes, including genetic engineering.

Bioterrorism The use of microorganisms as weapons.

Bioweapon A harmful biological agent that is used as a weapon.

Bisphenolics Chemical disinfectants consisting of two covalently linked phenolic compounds.

Black Death Also known as pneumonic plague and bubonic plague. A noncontagious form of plague caused by *Yersinia pestis* and transmitted to humans by the bite of a flea. It is a systemic disease that spreads through the blood and lymph fluid.

Blastoconidia The products of the budding form of reproduction seen in fungi.

Blood agar Bacterial growth medium that contains blood; used to identify organisms that produce hemolysins, which break down red blood cells.

Blood–brain barrier A selectively permeable barrier separating the circulating blood from the cerebrospinal fluid of the brain.

Boil A massive pus-filled lesion resulting from an infection in which the invading pathogens have been walled off. It is usually seen in the neck and back.

Botulinum toxin A neurotoxin associated with botulism, a digestive intoxication resulting from food poisoning.

Broad spectrum antibiotics Chemicals that attack a wide variety of microorganisms.

Bronchitis Acute or chronic inflammation of one or more of the bronchi.

Broth dilution test A test used to determine the minimal microbicidal concentration.

Bubo Enlargement of infected lymph nodes, especially in the groin and armpit, due to the accumulation of pus. This is seen in bubonic plague and other diseases.

Bubonic plague Also known as black death and pneumonic plague. A noncontagious form of plague caused by *Yersinia pestis* and transmitted to humans by the bite of a flea. It is a systemic disease that spreads through the blood and lymph fluid.

Cadherin A transmembrane protein that has a role in cell adhesion.

Candidiasis A yeast infection caused by *Candida albicans*. It can appear as thrush in the oral cavity or as vulvovaginitis.

Cannula (plural cannulas) A tube with a retractable inner core to be inserted into a blood vessel or other body cavity.

Capsid A protein outer coat of a virus. It is made up of repeating protein subunits known as capsomeres.

Capsomere A protein subunit that, along with other identical subunits, makes up the capsid (protein coat) of a virus.

Capsule A protective structure found around the outside of a bacterial cell. It can be made up of polysaccharides, polypeptides, or a combination of both.

Carbapenem antibiotics Antibiotics that attack the cell wall of bacteria.

Carbon source Compound that is used by an organism to source carbon from to be integrated in the organism's biomass.

Carbuncle A massive pus-filled lesion resulting from an infection in which the invading pathogens have been walled off. It is usually seen in the neck and back.

Carrier molecule Usually a co-factor or coenzyme used in metabolic pathways.

Catabolic Relating to catabolism, the breakdown of large molecules into smaller molecules.

Catabolism The chemical breakdown of organic molecules in which there is a release of energy.

Catalase An enzyme that converts hydrogen peroxide to water and oxygen.

Catalyst A molecule that speeds up a chemical reaction but is not used up or changed in the process.

Cation A positively charged ion.

Cell line A population of cells from a multicellular organism that are kept in culture media and used for research.

Cellular immunity Part of the adaptive immune response. It comprises the activity of T cells in response to antigens.

Central tolerance A part of clonal selection where newly formed lymphocytes that react with self antigens are eliminated.

Cephalosporin family A group of antibiotics derived from the fungus *Cephalosporium*. These chemicals attack the cell wall of bacteria.

Cercaria A tail-bearing larval form of trematodes.

Cervicitis Infection of the cervix.

Cestode A tapeworm.

Cestodiasis Infection with parasitic flatworms (cestodes).

Chagas' disease The American form of trypanosomiasis.

Chagoma A local chancre associated with Chagas' disease, filled with neutrophils, lymphocytes, and tissue fluids.

Chancre An ulcer located on the external genitalia or cervix seen in cases of syphilis.

Chediak–Higashi syndrome Primary immunodeficiency disease leading to lethal progressive systemic disorders and recurrent or chronic bacterial infections.

Chemical agent A general term that includes both antiseptics and disinfectants.

Chemically defined growth medium Growth medium in which each of the ingredients is precisely defined.

Chemiosmosis The process of capturing energy in which a proton gradient is created in the electron transport chain and is used to provide energy for the formation of ATP.

Chemoautotroph An autotroph that obtains energy by oxidizing inorganic substances such as nitrites and sulfides.

Chemoheterotroph Organism that obtains energy from breaking down already-formed organic molecules.

Chemokine A class of cytokines that attract additional phagocytes to the site of infection.

Chemostat A device used to maintain the logarithmic growth of bacteria by the continuous addition of fresh growth medium.

Chemotaxis A nonrandom movement of an organism toward or away from a chemical.

Chickenpox A highly contagious disease characterized by skin lesions and caused by the varicella-zoster virus. It is usually seen in children.

Chloramines Combinations of chlorine and ammonia used in wound dressings.

Chocolate agar Type of medium made with heated blood and used for growing organisms that require heme for growth.

Cholera An acute infectious disease caused by *Vibrio cholerae*. It causes severe diarrhea with extreme fluid and electrolyte depletion.

Cholera toxin An exotoxin produced by *Vibrio cholerae* pathogens. It is an enterotoxin that causes increased permeability in the intestinal tract and a consequent loss of fluids.

Chromatin The form of DNA seen in cells that are not ready to divide. It has the appearance of threads.

Chromosome The structure containing DNA that is seen in cells that are ready to divide. It is derived from the condensation of chromatin.

Chronic disease The opposite of acute disease. Such a disease starts slowly and lasts a long time.

Chronic granulomatous disease Primary immunodeficiency disease causing frequent severe infections of the skin, oral mucosa, intestinal tract, bones, lungs, and genitourinary tract.

Chronic infection A type of persistent infection in which there is continuous and long-term production of virions.

Cilia (singular cilium) Short cellular projections seen in eukaryotic cells, used for movement that result from the beating of these projections in coordinated waves.

Ciliates Organisms that move by the use of cilia; they are rarely parasitic to humans.

Circulatory shock The profound hemodynamic disturbance caused by the failure of the circulatory system to maintain adequate perfusion of organs.

Cirrhosis A general term for a group of liver diseases marked by interstitial inflammation and fibrosis of the liver.

Class I MHC A group of cell-surface proteins that are essential to immune recognition reactions. Class I MHC molecules are involved in natural killing functions and are found on all cells except those involved in the adaptive immune response.

Class II MHC A group of proteins found only on the surface of cells, involved in the adaptive immune response. These proteins are absent until an immune response is generated and they have an important role in antigen presentation.

Classical pathway of complement One of the ways in which the innate immune response works to protect the body. This pathway responds to antigens that have been seen previously and involves antibody against those antigens.

Clinical latency The effect seen in HIV infections in which virus replication is very low and is mainly in the lymph nodes.

Clinical trial A research study that uses human participants in health-related interventions to evaluate the effects of e.g. a new drug or vaccine on the health outcomes.

Coccodiomycosis Disease caused by members of the genus *Coccidioides*.

Clonal selection A theory that explains how exposure to an antigen selects and stimulates a specific lymphocyte to proliferate, giving rise to a clone of identical responsive cells.

Clonorchiasis Infection of the biliary passages of the liver by the liver fluke *Clonorchis sinensis*.

Coccus (plural cocci) A spherical bacterium.

Codon A sequence of three bases in messenger RNA that code for a particular amino acid.

Coenzyme An organic molecule bound to or associated with an enzyme.

Cofactor (1) An inorganic ion required for the function of an enzyme. (2) A component necessary for viral attachment to the host cell.

Cold sore A lesion caused by herpes simplex type 1 virus that is usually found on the lips or gingiva of the oral cavity.

Colony morphology The physical appearance (e.g. color, shape) of a bacterial colony.

Commensalism A symbiotic relationship in which one organism benefits and the other is unaffected.

Common-source outbreak An outbreak arising from contact with contaminated substances such as water.

Communicable disease An infection that can be spread from person to person.

Competitive inhibition A reaction in which a molecule that is similar in structure to a substance competes with that substance by binding to the active site of an enzyme.

Complement protein Serum protein that is part of the innate immune response.

Complement system A set of more than 20 proteins found in the blood that when activated can destroy bacteria by making holes in the bacterial cell wall. This system also amplifies the inflammatory response to infection.

Complete blood count (CBC) A laboratory procedure in which the formed elements (cells) of the blood are counted. This information can be of great value in determining the stage of infection seen in a patient.

Complex growth medium A growth medium that contains ingredients such as beef extract or blood, whose exact composition is not known.

Complex viruses Viruses without either helical or icosahedral symmetry.

Concentration gradient A condition in which the amount of a substance differs between locations, for example, between the two sides of a membrane.

Condyloma Also known as a genital wart. It is an often malignant lesion that arises on the genitalia and is usually associated with sexually transmitted viral diseases.

Congenital syphilis Syphilis that is present at birth. It occurs when a child is born to a mother with syphilis.

Conidia The asexual reproductive elements found in fungi.

Conjugation The transfer of genetic information from one bacterial cell to another. In Gram-negative bacteria this occurs through the use of a pilus, whereas in Gram-positive cells it occurs when two cells stick together.

Conjunctivitis An infection of the conjunctiva of the eye.

Constant region of the antibody molecule The part of the immunoglobulin molecule with a relatively constant amino acid sequence. It determines the effector function of the immunoglobulin molecule.

Constitutive gene A gene that is always turned on, in contrast with inducible or repressible genes.

Contact transmission A mode of disease transmission; it can be direct, indirect, or through droplets.

Contagious A disease that is communicable on contact and spreads quickly.

Continuous cell line A cell line that grows for long periods of time or indefinitely.

Co-stimulatory signal A signal generated from, and regulated by, cells of the innate immune system.

Covalent bond A bond between atoms in which electrons are shared.

C-reactive protein (CRP) An acute-phase protein that binds to phospholipids and is seen in acute infections.

Croup Acute obstruction of the larynx that causes a hoarseness and barking cough.

Cryptosporidiosis Disease caused by protozoans of the genus *Cryptosporidium*, commonly seen in patients who are immunocompromised.

Cultures Term used for microorganisms grown on purpose e.g. in the laboratory.

Cutaneous anthrax Infection with *Bacillus anthracis* that appears on the skin two to five days after endospores enter the epithelial layers of the skin.

Cutaneous candidiasis A fungal infection of the skin caused by *Candida albicans*.

Cutaneous leishmaniasis An infection of the skin caused by the parasitic protozoan *Leishmania* and transmitted to humans by the bite of a sand flea.

Cyanosis Bluish discoloration of the skin and mucous membranes caused by a lack of oxygen in the blood.

Cyst A spherical thick-walled structure that resembles an endospore; formed by certain parasites.

Cysticercus An embedded form of tapeworm found in meat, giving the meat a "mealy" appearance.

Cystitis Infection of the bladder.

Cytocidal effect Part of the pathology seen when host defenses kill virally infected cells.

Cytokines Low-molecular-weight proteins that are released by a variety of cells in the body. There are two types: the hematopoietin family, which includes growth hormones and interleukins, and the tumor necrosis factor family.

Cytomegalovirus (CMV) One of a widespread and diverse group of herpesviruses that often produce severe effects in immunodeficient patients.

Cytopathic effect (CPE) The destruction of host cells through the lytic cycle of viruses.

Cytoplasm The semifluid substance inside cells, excluding the nucleus of eukaryotic cells.

Cytoskeletal structures Structures found inside cells made up of protein fibers that give rigidity and support to eukaryotic cells and permit cell movement.

Cytotoxic chemotherapy The use of cytotoxic drugs to kill malignant cells or depress the immune system to allow the survival of transplants. These drugs cause leucopenia, which is a decrease in the number of white blood cells.

Cytotoxic edema Swelling in the brain caused by the production of toxic substances by bacteria and from neutrophil invasion.

Cytotoxic T cell (CTL) A specific thymus-derived T lymphocyte that kills other cells and has memory.

Cytotoxins Toxins produced by cytotoxic cells that kill infected host cells.

Dacryocystitis Infection of the tear ducts.

Death phase Also known as logarithmic decline phase. The fourth phase of bacterial growth, in which bacteria die faster than they divide.

Deep sequencing Nucleotide sequence analysis with increased scale of readings of parts of the sequence to increase accuracy of the reconstructed sequence.

Defensins Antibacterial peptides produced by humans.

Definitive host An organism that harbors the adult sexually reproducing form of a parasite.

Degenerate genetic code The genetic code is said to be degenerate because more than one codon can code for the same amino acid. This allows for mistakes that can take place in the DNA sequence: the appropriate amino acid can still be placed in the primary protein sequence.

Degerming The method of moving organisms away from a place on the body; an example is the application of alcohol before injections.

Dehydration synthesis A chemical reaction in which water is removed so as to build a complex organic molecule.

Dendritic cell A phagocytic cell found in the dermis; responsible for antigen presentation to helper T cells in the adaptive immune response.

Deoxyribose The sugar found in DNA.

Dermatophytosis The general term for a variety of dermatophyte infections involving erythema, induration, itching, and scaling.

Descriptive epidemiology The study of the physical aspects of patients and the spread of disease.

Detergents Positively charged organic surfactants that are more soluble in water than soap.

Diapedesis The process by which white blood cells move through unbroken capillary walls to the site of an infection.

Diarrhea Abnormally frequent evacuation of watery feces.

Diarrheagenic Causing diarrhea.

Differential blood analysis A routine lab test in which the percentage of each of the white blood cell populations is determined.

Differential media A growth medium that includes components that cause an observable change in color or pH when a particular chemical reaction occurs, making it possible to distinguish between organisms.

Differential stain The use of two or more dyes to differentiate between bacterial species or distinguish different structures of an organism.

DiGeorge syndrome Primary immune deficiency disease in which the thymus does not develop properly, causing a deficiency of T-cell functions.

Dimorphism The term used to describe the ability of some fungi to grow in either the mold or the yeast form.

Diphtheria An acute bacterial infection of the nose, throat, or larynx caused by *Corynebacterium diphtheriae*. It is marked by formation of a gray/white pseudomembrane.

Diphtheria toxin An exotoxin produced by *Corynebacterium diphtheriae* that affects the membranes of the nose, throat, and larynx. It can also affect the heart and the central nervous system.

Diplococcus (plural diplococci) The arrangement of bacteria that grow in pairs. *Streptococcus pneumoniae*, the leading cause of pneumonia, is an important example.

Direct method A measurement of bacterial growth in which cells or colonies are observed.

Directly observed therapy (DOT) Observation of the patient taking the prescribed medication by a health care worker to ensure compliance with the medication schedule.

Disease A negative disturbance in the state of health during which the body does not function properly.

Disinfectants Chemical agents used on inanimate objects to destroy microorganisms.

Disinfection The application of an antimicrobial agent to inert surfaces, such as floors or tabletops.

Disk diffusion method Used to test the efficacy of an antimicrobial chemical. The chemical is applied to a small filter paper disk and placed on a bacteria-inoculated petri plate, to determine how well it inhibits bacterial growth.

Disseminated Spread throughout an organ or the body.

Disseminated intravascular coagulation (DIC) Widespread coagulation of the blood in different areas of the body.

Dissimilation plasmid An extrachromosomal piece of genetic material that contains genes that enable organisms to be resistant to disinfectants and environmental pressure.

Distending cytotoxin A toxin produced by *Campylobacter* that arrests cell division while the cytoplasm continues to increase.

Disulfide bridge A bond that forms between sulfur-containing amino acids. It is one of the primary ways in which the three dimensional folded shapes of proteins are maintained.

Diverticular abscess A pus-containing lesion occurring in the diverticulum of the intestinal tract.

DNA ligase An enzyme that is used to connect sections (fragments) of DNA together by filling in gaps. It is seen on the lagging strand of DNA during replication.

DNA polymerase The enzyme used to match and bond complementary base pairs during the process of DNA replication.

Double-stranded DNA (dsDNA) virus A virus that possesses two complementary strands of DNA.

Dry heat Hot air with a low moisture content. For example, in an oven.

Dual immunological response A vaccine response in which both specific antibodies and T cells are produced within the vaccinated host.

Dysentery A severe diarrhea in which the fecal material contains mucus and blood.

Dyspnea Labored or difficult breathing.

Dysuria Pain and burning sensations during urination.

E site A site on the ribosome where the tRNA exits after the removal of its amino acid.

E test A diffusion method used to determine the MIC (minimal inhibitory concentration) of a drug.

Eastern equine encephalitis (EEE) Type of viral encephalitis seen most often in the eastern United States; it primarily infects horses but can also infect humans.

Ecchymosis (plural ecchymoses) A small hemorrhagic spot in the skin or mucous membranes forming a nonelevated irregular blue or purplish patch.

Effector lymphocytes Activated lymphocytes that have differentiated into their fully effective form.

Efflux pumping A mechanism used by bacteria to resist attack by antibiotics. It takes advantage of pumps in the plasma membrane that pump out the antibiotics.

Elective media Growth media that include components that only support growth of the organism of interest.

Elementary body The infectious form of *Chlamydia*.

Elephantiasis Swelling of the extremities and genitalia as a result of an intense inflammatory response and lymphatic blockade by the parasite *Wuchereria bancrofti*.

Embolus (plural emboli) A mass of clotted blood or other material that is brought by the blood and forced into a smaller vessel, obstructing the circulation.

Emerging infectious disease Diseases that have not been seen before. They arise from movement into areas where humans have not ventured before. They can also be diseases that move from certain areas of the world to places where the diseases have not been before.

Encephalitis Inflammation of the brain.

Endemic disease Refers to a disease that is constantly present in the population.

Endemic typhus A flea-borne typhus caused by *Rickettsia typhi*.

Endoarteritis Inflammatory lesions that occur in the arteries.

Endocarditis Exudative and proliferative inflammatory lesions found on the endocardium and valves of the heart.

Endocytosis The process in which vesicles form by invagination of the plasma membrane of the cell and move substances into the cell.

Endogenous infection An infection caused by organisms that are part of the normal microbial flora of the body.

Endometritis Infection of the endometrium.

Endoplasmic reticulum An extensive network of membranes that form tubes and plates in the cytoplasm of eukaryotic cells. They are involved in the synthesis and transport of proteins and lipids.

Endosome A membrane enclosure seen during the virus penetration step of the infection process.

Endospore A highly resistant dormant structure that is formed by certain bacteria.

Endosymbiotic theory Evolutionary theory holding that some organelles of eukaryotic cells arose from bacteria that came to live in a symbiotic relationship inside eukaryotic cells.

Endotoxin A toxin incorporated into the outer layer of Gram-negative cells that is released when the cell dies.

Energy of activation The energy needed to start a chemical reaction.

Enteric Having to do with the intestine.

Enteric fever A systemic infection with a focus on the intestinal tract.

Enterobacteriaceae A family of bacteria, many of which are intestinal Gram-negative facultative anaerobes that have flagella.

Enterocytes Epithelial cells of the large and small intestine.

Enterohemorrhagic E. coli A type of *Escherichia coli* that causes destruction of the blood vessels; the causative agent in infections seen with contaminated ground beef.

Enterotoxin An exotoxin that attacks the tissues of the intestinal tract.

Enveloped virus A virus with a lipid bilayer surrounding its capsid.

Enzyme immunoassay Analytical approach where an enzyme bound to an antigen or antibody is used to produce a signal for detection.

Enzyme–substrate complex The association of the enzyme with its substrate.

Eosin methylene blue (EMB) A blue dye used for a variety of bacterial stains, both simple and differential.

Eosinophil A white blood cell normally found in very low numbers in the blood but in very high numbers during a parasitic infection.

Eosinophilia An abnormal increase of eosinophils in the blood.

Epidemic disease A disease that has a higher than normal incidence in the population over a short period.

Epidemic typhus A louse-borne rickettsial disease caused by *Rickettsia prowazekii*; seen most often in conditions of overcrowding and poor sanitation.

Epidemiology The study of factors and mechanisms involved in the spread of disease within a population.

Epididymitis Infection of the epididymis.

Episome A DNA segment in bacterial cells that can exist as part of a chromosome, or separate from the chromosome.

Epithelial hyperplasia An increase in epithelial cell number in response to a specific stimulus, such as infection with a pathogen.

Epitope Area of an antigen molecule to which specific antibodies bind; also called an antigenic determinant.

Epstein–Barr virus A virus that causes Burkitt's lymphoma and mononucleosis.

Erysipelas A rapidly spreading infection of the deeper layers of the dermis.

Erythema Redness of the skin due to congestion of the capillaries.

Eschar The thick crust or scab that forms, for example, over a severe burn.

Etiology The cause of a disease.

Eukaryote A cell that has a distinct nucleus and other membrane-enclosed organelles.

Exfoliatin An exotoxin produced by *Staphylococcus aureus* and seen in scalded skin syndrome.

Exocytosis Process in which vesicles inside a cell fuse with the plasma membrane and release their contents to the outside of the cell.

Exogenous infection An infection caused by organisms that enter the body from the outside.

Exotoxin A soluble protein toxin produced by living bacteria. It is seen in many types of systemic infection.

Extravascular infection An infection outside of the bloodstream.

Extreme halophile A bacterium that thrives in extremely high concentrations of salt.

Extreme thermophile A bacterium that grows at temperatures above 80°C.

Exudate Fluid containing high levels of protein and cellular debris that is deposited in tissues or on tissue surfaces. It is usually the result of inflammation.

F1 protein A factor seen in plague that forms a gel-like capsule preventing phagocytosis and allowing the bacteria to multiply in the submucosa.

Facilitated diffusion Diffusion across a membrane that is carried out by a nonspecific carrier molecule and does not require ATP.

Factors B, D, and P (properdin) Chemical molecules found in the blood that can bind to antigens on the surface of microbes and initiate the alternative complement pathway.

Facultative anaerobe A bacterium that uses oxygen for metabolism but shifts to anaerobic metabolism when oxygen is no longer available.

Facultative halophile A bacterium that can grow in high concentrations of salt as well as normal concentrations of salt.

FAD (flavin adenine dinucleotide) A coenzyme that carries hydrogen atoms and electrons.

Fasciitis Infection of the fascia of the body.

Fastidious bacteria Bacteria that do not grow well without specific supplements added to the medium. Even then, these bacteria can take longer to grow than other organisms.

Fatal familial insomnia A very rare prion disease of the brain causing complete sleeplessness. It is untreatable and fatal.

Fatty acid A long chain of carbon and hydrogen atoms with a carboxyl group at one end.

Favus A disease causing hair loss due to permanent destruction of the hair follicle, resulting in bald spots associated with crusty scarred skin.

Fecal coliform A Gram-negative, rod-shaped, facultatively anaerobic bacterium associated with fecal contamination.

Fecal coliform count A test done to determine the level of fecal coliform contamination in water.

Fecal–oral route of contamination A major route of infection associated with poor sanitation and hygiene.

Feedback inhibition A regulatory mechanism in which a product in a reaction inactivates an enzyme necessary for the reaction to proceed.

Fever A body temperature that is abnormally high.

Filariasis Disease of the blood and lymph caused by any of several different roundworms.

Fimbria (plural fimbriae) Also known as an attachment pilus. A short hairlike appendage found exterior to the cell wall. It is used as a mechanism for staying in the host during infection.

Flagella (singular flagellum) Long thin helical appendages found on certain cells. They provide a means of locomotion.

Flagella stain A technique used for observing flagella by coating the surface of the flagella with multiple layers of dye or metal such as silver.

Flagellates Parasitic protozoan organisms that use flagella to move.

Flagellin protein Globular protein that forms the flagella.

Flucytosine An antifungal antibiotic used in the treatment of severe candidal and cryptococcal infections.

Fluid mosaic model of the membrane Model incorporating all of the structures found in the cell membrane and based on the phospholipid bilayer structure of the plasma membrane.

Flukes Adult trematodes that can live for decades in human tissue and blood vessels.

Fluorescent molecules Molecules that emit light after absorbing light of a different wavelength.

Fluorescent proteins Proteins that emit light after absorbing light of a different wavelength.

Flushing Transient redness of the face and neck.

Focus of infection An area where the infection is localized and from which it can spread to other parts of the body.

Follicles Areas of the lymph node where B cells are found.

Folliculitis Inflammation of the follicles.

Fomites Nonliving substances (such as clothing, dishes, or paper money) that are capable of transmitting disease.

Forespore The structure formed during the development of a spore in which the plasma membrane wraps around the developing spore.

Formalin A 37% solution of formaldehyde used in clinical settings for disinfection.

Fosfomycin Antibiotic that targets the *mur*A gene and prevents peptidoglycan subunits from being produced.

Frameshift mutation Mutation resulting from an insertion or deletion of one or more bases in DNA.

Freeze-drying Also referred to as lyophilization. It is a method of extracting water from a frozen state.

GALT (gut-associated lymphoid tissue) A collective name for tissues of lymphoid nodules, especially those in the digestive tract.

Gametocyte A male or female sex cell.

Gametogony The development of merozoites into male and female gametes, that later fuse to form a zygote.

Gamma hemolysis A type of hemolysis in which there is no destruction of red blood cells.

Ganciclovir A derivative of acyclovir used to treat retinitis caused by cytomegalovirus infection.

Gas gangrene A deep wound infection in which tissue is destroyed, often caused by species of the genus *Clostridium*.

GasPak™ jar A container used to incubate obligate anaerobic organisms.

Gastritis Inflammation of the stomach.

Gene constellations The term used to describe clusters of genes that determine virulence.

Gene expression The process of transcription and translation.

Generalized transduction Type of transduction in which a fragment of DNA from the degraded chromosome of an infected bacterial cell is accidentally incorporated into a new bacterial virus particle, which can then be transferred to another cell.

Generation time The time required for a population of organisms to double in number.

Genetic code The one-to-one relationship between each codon and a specific amino acid.

Genetic engineering Process of changing the genetic information of an organism.

Genital herpes Primarily sexually transmitted disease of the genital region which can give rise to confluent ulcerations.

Genomics Discipline in genetics considering entire genomes rather than individual genes.

Genotyping Determination of the genotype of an organism in comparison to a reference organism.

Genus A taxonomic term consisting of one or more species.

Germ theory of disease The theory that diseases can be caused by microorganisms.

German measles Also known as rubella. A viral infection characterized by a short-term skin rash.

Germination The mechanism by which organisms develop from the endospore state to the vegetative (growing) state.

Gerstmann–Sträussler–Scheinker syndrome A group of rare prion diseases having the common characteristic of cognitive and motor disturbances.

Ghon complex Calcified necroses seen in the tubercles, associated with tuberculosis.

Giardiasis A gastrointestinal disorder caused by the flagellated protozoan *Giardia*.

Gingiva The soft tissue (gums) in the mouth.

Gingivitis The mildest form of periodontal disease characterized by inflammation of the gingiva.

Glycocalyx A term used to refer to all substances containing polysaccharides that are found external to the cell wall.

Glycolipid A lipid molecule that contains carbohydrate.

Glycolysis An anaerobic metabolic pathway used to break down glucose into pyruvate; it produces some ATP.

Golgi apparatus An organelle found in eukaryotic cells that receives, modifies, and transports substances coming from the endoplasmic reticulum.

Gonorrhea A sexually transmitted infection caused by *Neisseria gonorrhoeae*, commonly asymptomatic in females but marked by a painful purulent discharge from the urethra in males.

gp41 A protein found on HIV.

gp120 A protein found on HIV.

Gram stain A differential stain that differentiates bacteria into either Gram-positive or Gram-negative groups.

Granulocyte A leukocyte (white blood cell) with granular cytoplasm and an irregularly shaped nucleus.

Granulocyte–macrophage colony-stimulating factor (GM-CSF) A substance that recruits large numbers of phagocytic cells to the site of infection.

Granuloma A collection of epithelial cells, macrophages, lymphocytes, and fibers seen in chronic inflammation.

Granulysin A peptide secreted by killer cells that causes the onset of apoptosis.

Granzyme A protease enzyme secreted by killer cells that causes the death of a target cell by triggering apoptosis.

Graves' disease Autoimmune disease in which antibodies attack thyroid stimulating hormone causing it to be overproduced, leading to hyperthyroidism.

Group translocation One of three types of active transport seen in bacteria, in which chemical substances brought into the cell are chemically modified so they cannot diffuse back out of the cell.

Gruber reaction Identifying an unknown antigen by using a known antibody.

Gummas Localized granulomatous lesions seen in the skin, bones, joints, and internal organs in tertiary syphilis.

H antigen Refers to the proteins that make up the flagella of bacteria; they are antigenic and can initiate an immune response.

Hairy leukoplakia White patch on the tongue or buccal mucosa caused by Epstein–Barr viral infection and associated with HIV infection.

Halophile A salt-loving organism that requires a moderate to extreme level of salt so as to grow. There are three categories: obligate, facultative, and extreme.

Hard measles Common form of measles; also called rubeola.

Hashimoto's thyroiditis Autoimmune disease causing the destruction of thyroid tissue.

Heavy chain of the antibody molecule Two of the four polypeptide chains that make up the antibody molecule. The heavy chains have a greater molecular weight than the other two chains which are called the light chains.

Helical symmetry Also known as symmetry. The "spiral staircase" arrangement of the DNA molecule, which has a precise symmetry.

Helicase An enzyme that unwinds the helical structure of the DNA during replication.

Helminth A worm with bilateral symmetry; includes roundworms and flatworms.

Helper T cell The T lymphocyte that stimulates the function of other immune cells such as B cells, cytotoxic T cells, and macrophages.

Hemagglutinin A glycoprotein on the envelope of some viruses, which assists in attachment to and entering the host cell.

Hematogenous dissemination Movement through the blood; the most efficient way for systemic viruses to disseminate throughout the body.

Hemolysin An enzyme that destroys red blood cells.

Hemoptysis The spitting of blood or bloodstained sputum.

Hemorrhagic colitis Infection of the colon with bloody discharge.

HEPA filter A high-efficiency particulate air filter containing holes that are too small for microorganisms to pass through. It is used to keep rooms sterile.

Hepatitis An inflammation of the liver usually caused by viral infection but can also be caused by an ameba or by toxic chemical damage to the liver.

Hepatocellular carcinoma Malignancy involving cells of the liver.

Hepatomegaly Enlargement of the liver.

Hepatosplenomegaly Enlargement of both the liver and the spleen.

Herd immunity The proportion of individuals in a population who are immune to a particular disease.

Hermaphrodite Term used to describe the reproductive system of some flukes that have both male and female reproductive systems.

Heterotrophy The use of carbon atoms from preexisting organic molecules to produce new biomolecules.

Hfr (high frequency of recombination) cell A bacterial cell that harbors an F plasmid incorporated in its chromosome.

Highly virulent The condition in which a microbe is extremely harmful and damaging to the host.

Histamine A chemical that is produced and stored by mast cells and basophils that can cause vasodilation in the inflammatory response.

Histatins Antimicrobial proteins found in saliva that also enhance the migration of epithelial cells and fibroblasts to speed up wound healing.

Histoplasmosis A fungal respiratory disease endemic to the central and eastern United States, caused by *Histoplasma capsulatum*.

Homolactic fermentation Fermentation in which pyruvate is converted into lactate.

Hong Kong influenza One of the three major influenza epidemics in the United States. It occurred in 1968 and caused more than 30,000 deaths.

Horizontal gene transfer (HGT) The transfer of genetic material between organisms not involving a reproductive cellular division.

Human papillomavirus (HPV) Virus that attacks skin and mucous membranes, causing papillomas or warts. It has been proved to cause cervical carcinoma.

Humoral immunity A response to infection involving the production of antibodies.

Hybridization The binding of oligonucleotides to a complementary sequence of nucleotides.

Hydrogen bond A relatively weak bond between a hydrogen carrying a partial positive charge and an oxygen or nitrogen molecule carrying a partial negative charge.

Hydrolysis A chemical reaction that produces smaller products by using water to break down larger molecules.

Hydrophilic Having an affinity for water.

Hydrophobic Lacking an affinity for water.

Hygiene hypothesis The theory that a less hygienic environment can help protect against certain allergens.

Hypersensitivity Also called allergy. It is a disorder in which the immune system reacts to antigens that it would normally ignore.

Hypertonic A solution containing a concentration of dissolved material greater than that within a cell.

Hyphae Long threadlike extensions of the cytoplasm seen in molds.

Hypotonic A solution containing a concentration of dissolved materials lower than that within the cell.

Hypoxia Condition resulting from reduced levels of oxygen being supplied to the tissues.

Iatrogenic Transmission of infection because of poor techniques employed by health care workers.

ICAM-1 A cell receptor used for rhinovirus infection.

Icosahedral symmetry Also known as symmetry. Capsid arrangement seen in some viruses in which there are 20 geometric sides that make up the virus protein coat.

ID$_{50}$ Infectious dose 50%. The number of organisms required to infect 50% of the subject population.

Immune escape Refers to the ability of some pathogens to evade the host immune response.

Immunocompromised Refers to an individual whose immune defenses are weakened. It can result from a variety of causes such as HIV infection, drug therapy for malignancy or transplantation as well as old age or debilitation.

Immunofluorescence Analytical approach where fluorescent molecules are bound to an antigen or antibody to produce a signal for detection.

Immunoglobulin Also known as an antibody. A class of proteins that are produced by the humoral immune response and respond to a specific antigen.

Immunological memory The ability of the adaptive immune response to immediately recognize and respond against antigens it has previously been exposed to.

Impetigo A highly contagious pyoderma caused by staphylococci or streptococci, or both.

Inactivated vaccine Also known as a killed vaccine. A vaccine composed of virus that is either dead or inactivated.

Incidence of disease The number of new cases of a particular disease seen in a population over a specific period of time.

Inclusion bodies (1) Form of cytopathic effect in which viral particles, viral components, or remnants of virus aggregate in the cytoplasm of infected cells. (2) The aggregation of reticulate bodies seen in *Chlamydia*. (3) In bacteria, used for storage of materials.

Incubation period The time between exposure to the pathogen and the first appearance of signs and symptoms of the infection.

Index case The first person to have been infected with the disease.

Indirect method A measurement of bacterial growth in which cells or colonies are estimated, not observed.

Infectious endocarditis Infection of the heart.

Inflammation The body's defensive response to any trauma or infection of the body.

Influenza An acute viral infection of the respiratory tract marked by inflammation of the nasal mucosa, pharynx, and conjunctiva.

Inhalation anthrax One of several types of anthrax infection caused by *Bacillus anthracis*, resulting in highly fatal pneumonia.

Initiation codon The sequence of messenger RNA that sets the reading frame.

Initiator protein The protein that recognizes the replicator sequence and begins transcription.

Innate immune response The nonspecific immune response that is present at birth.

Inoculate To add microorganisms to growth medium to start a culture.

Integral protein Membrane protein that extends through the entire membrane.

Interferon A group of small protein molecules often released in response to a viral infection that bind to noninfected cells, causing them to produce antiviral proteins that protect against viral infection of the cell.

Interleukin 1 (IL-1) The best-known endogenous pyrogen, which causes the onset of fever during an infection.

Interleukin 6 (IL-6) A cytokine produced during an infection that causes the liver to produce acute-phase proteins.

Interleukins A class of cytokine produced by leukocytes.

Intermediate host An organism in which a parasite develops but does not undergo sexual reproduction.

Interstitial macrophages Cells found in the stroma of the lung. They are smaller and less phagocytic than alveolar macrophages.

Intravascular infection An infection within the bloodstream.

Invasin A virulence factor.

Iodophor An organic compound that incorporates iodine in such a way that the iodine is released slowly.

Ionic bond The bond that forms between cations and anions due to their opposing charges.

Ionizing radiation Form of radiation that causes disruption of the electron clouds surrounding atoms, causing the ionization of the atoms.

Ions Electrically charged atoms produced when atoms either gain or lose electrons.

Isoelectric focusing Technique to separate molecules by means of differences in their isoelectric point.

Isotonic Fluid containing the same concentration of dissolved materials as that inside the cell.

Isotype A term referring to one of the five types of constant regions of antibodies.

J chain The protein found on the pentameric form of IgM and the dimeric form of IgA that helps to hold the monomers together.

Jaundice Yellowness of the skin, sclera, mucous membranes, and excretions caused by hyperbilirubinemia.

K antigen Refers to the polysaccharide associated with the capsule found around certain bacteria.

Kaposi's sarcoma A malignancy associated with immunodeficiency in which blood vessels grow in tangled masses that are filled with blood and easily ruptured.

Keratitis Infection of the cornea of the eye.

Killed vaccine Also known as an inactive vaccine. A vaccine composed of virus that is either dead or inactivated.

Kinin A protein that causes vasodilation and attracts phagocytic cells to the site of injury in the process of inflammation.

Koch's postulates Four postulates proposed by Robert Koch in the nineteenth century, used to prove that a particular organism causes a particular disease.

Koplik's spots Red spots with centralized bluish specks that appear on the mucous membranes in the early stages of measles.

Kupffer cell A phagocytic cell that is stationed in the sinusoids of the liver.

Kuru Transmissible spongiform encephalopathy of the human brain. The disease is caused by prions and is associated with cannibalism and with tissue or organ transplantation.

***Lac* operon** A sequence of genes that controls the production of enzymes required to break down the sugar lactose.

Lacrimal apparatus The structures associated with the production and recycling of tears.

Lacrimal gland The gland that produces tears. It is part of the lacrimal apparatus.

Lag phase of bacterial growth The first phase of the bacterial growth curve, in which organisms acclimate to their surroundings; they grow in size but do not increase in number.

Lagging strand Also known as the discontinuous strand. The strand of DNA that is replicated in pieces called Okazaki fragments.

LAL assay The *Limulus* amebocyte lysate assay, which is used to determine whether there is endotoxin contamination.

Langerhans cells Dendritic cells found in the layers of the skin. They are phagocytes that are one type of antigen-presenting cell.

Lantibiotics Antibacterial peptides produced by some Gram-positive bacteria.

Laryngotracheitis Also known as acute croup. A severe infection of the larynx and trachea.

Latency (latent) Pertaining to a virus that has entered a host cell and its nucleic acids have been incorporated into the host's, but no new viruses are produced.

Latent infection A disease in which there are periods of inactivity either before the onset of symptoms or between attacks.

Latent syphilis A stage of syphilis in which there are no symptoms or signs of infection.

LD$_{50}$ Lethal dose 50%. The number of organisms required to kill 50% of the subject population.

Leading strand Also known as the continuous strand. The strand of DNA that is replicated continuously.

Lectin-binding pathway The complement pathway that is activated by the carbohydrate mannose.

Leishmaniasis Infection with the parasite *Leishmania*.

Leprosy A chronic disease caused by *Mycobacterium leprae* which is characterized by granulomatous lesions of the skin, mucous membranes, and central nervous system.

Leukocidin An exotoxin produced by many bacteria that kills white blood cells, including phagocytic cells.

Leukocyte A white blood cell.

Leukocyte endogenous mediator A factor that lowers plasma iron concentration, which limits the availability of iron and thereby inhibits the growth of some pathogens.

Leukotriene A chemical substance released from mast cells that causes prolonged airway constriction, increased dilation and permeability of capillaries, increased secretion of mucus, and stimulation of nerve endings that causes pain and itching.

Light chain of the antibody molecule Two of the four polypeptide chains that make up the antibody molecule. The light chains have a lower molecular weight than the other two chains which are called the heavy chains.

Lipid A A component of the lipopolysaccharide layer of Gram-negative cells, which becomes an endotoxin upon release from the cell.

Lipid carrier cycle In cell wall synthesis, the transport of peptidoglycan precursors (NAG and NAM) through the cell membrane by lipids.

Lipid raft A portion of the plasma membrane of host cells that contain an increased portion of lipid.

Lipocalin A substance that inhibits the scavenging of iron by pathogens, thereby inhibiting their growth.

Lipopolysaccharide layer An outer layer of the cell wall found around Gram-negative cells. It contains endotoxin that is released when the organism dies and this layer falls apart.

Lipoteichoic acid A molecule that is found only in Gram-positive bacteria that penetrates the entire cell wall and attaches to the plasma membrane of the cell.

Live attenuated vaccine Live virus that has been weakened to reduce infectivity, used for some forms of vaccination.

Loaiasis A disease caused by the parasitic *Loa loa* worm and transmitted to humans by the deer fly.

Local infection An infection confined to a specific area of the body.

Lockjaw The early symptom of tetanus, an infection of the nervous system that initially affects the small muscles of the face, preventing them from relaxing.

Log (exponential) phase of bacterial growth The second of the four phases of bacterial growth, in which cells divide at an exponential rate.

Logarithmic decline phase Also known as the death phase. The fourth phase of bacterial growth, in which bacteria die faster than they divide.

Lophotrichous flagella The arrangement in which there are two or more flagella at one or both ends of a bacterial cell.

Lymphadenitis Inflammation of one or more lymph nodes.

Lymphangitis Inflammation or infection of the lymphatic system.

Lymphocyte A form of white blood cell that is involved with the adaptive immune response of the body.

Lymphogranuloma venereum Venereal infection caused by *Chlamydia trachomatis* marked by a primary transient ulcerative lesion in the genital region.

Lymphoma Any neoplastic disorder of the lymphoid tissues.

Lymphopenia A reduced number of lymphocytes resulting from certain forms of infection and also from some kinds of therapy for malignancy or transplantation.

Lyophilization Also referred to as freeze-drying. It is a method of extracting water from a frozen state.

Lysis The bursting of a cell.

Lysosome A small membrane-enclosed organelle seen in eukaryotic cells that contains digestive enzymes.

Lysozyme An antibacterial enzyme found in secretions such as tears and saliva.

Lytic infection cycle Pertaining to a virus, in which the host cell is infected by the virus, produces new viruses, bursts, and releases newly produced virions.

M cells Specialized cells found in the digestive tract. They are used by pathogens to enter the tissues of the body.

M protein A chemical virulence factor found on the bacterial cell surface and also on fimbriae that helps pathogens stay in the host.

Macrophage A highly phagocytic white blood cell.

Macrophage activation A process in which macrophages are stimulated by T_H1 helper cells to eliminate pathogens that proliferate in macrophages.

Macule A discolored spot on the skin that is not raised above the surface.

Maculopapular rash Broad lesions that slope away from a centrally located papule.

Major histocompatability complex (MHC) A group of cell surface proteins that are required for the development of an immune response.

Malaise A vague feeling of discomfort.

Malarial paroxysm The term used to describe the acute phase of infection seen in malaria.

MALT Mucosa-associated lymphoid tissue.

Mannose-binding protein (MBP) Acute-phase protein that binds to mannose sugars found on many bacterial and fungal cell membranes.

Margination The process in which white blood cells traveling in the blood are able to slow down and stop adjacent to the area where the tissue injury has occurred. It is caused by the localized secretion of selectin, which is a sticky molecule on the inner side of the vessels.

Mass spectrometry Analytical technique used to identify amount and type of molecules in a sample via measuring the mass-to-charge ratio and abundance of ions in a gas phase.

Mast cell A white blood cell that releases histamines during an allergic response.

Matrix protein The second layer of viral protein found just inside the envelope of the virus.

Mature vegetation A "mesh" of platelets, fibrin, and inflammatory cells seen in infectious endocarditis.

Maximum growth temperature The highest temperature at which an organism will grow.

Measles Also known as rubeola. A highly contagious viral infection usually seen in children. It involves the respiratory tract and is marked by discrete red papules, which become confluent.

Mechanical vector transmission The passive movement of organisms from a vector to a human or fomite.

Membrane attack complex (MAC) The complex formed during the final stages of the complement pathway. This complex produces a hole in the cell wall of bacteria, leading to their death.

Memory cell A long-lived T or B lymphocyte that is derived as part of the adaptive immune response. It retains a memory of the antigen that it was sensitized to.

Meninges Connective tissue layers that cover the brain and spinal cord.

Meningitis Infection of the meninges (layers) that surround the brain and spinal cord.

Meningoencephalitis Inflammation of the brain and meninges.

Merozoite A trophozoite form of the malaria parasite derived from sporozoites. It is found in red blood cells and hepatocytes during the malaria infection cycle.

Mesophiles Organisms that grow best at temperatures between 25°C and 40°C.

Messenger RNA (mRNA) A type of RNA that results from the transcription of DNA and carries information about the arrangement of amino acids in a protein.

Metabolism The sum of catabolism (the breakdown of organic molecules) and anabolism (the building up of organic molecules).

Metacercariae Encysted form of cercariae seen in the development of a fluke.

Metachromatic granules Polyphosphate granules seen in the cytoplasm of certain bacteria.

Metagenome Term used to refer to genomes from a mixed community of organisms.

MHC restriction Major histocompatability complex restriction associated with the interaction of antigen-processing cells and helper T cells.

Microaerophilic bacteria Organisms that grow best in the presence of small amounts of oxygen.

Microarray analysis A type of biological analysis method that is used in a variety of biotechnology processes.

Microbial antagonism The ability of normal microbial flora to inhibit the growth of pathogens by competing for resources.

Microbial flora Microorganisms that are normally found living in and providing important benefits to the host.

Microbiota Microbial population within a habitat.

Microfilament A protein fiber that makes up part of the cytoskeleton in eukaryotic cells.

Microfilariae Live offspring of tissue nematodes that circulate in the blood and subcutaneous tissues until they are ingested by specific blood-sucking insects.

Microglial cells Resident macrophages found in the central nervous system. There are two forms, ameboid (which travel through developing brain tissue and are also found in damaged brain tissue) and ramified (found in normal brain tissue).

Microtubule A protein tubule that forms the structure of cilia, flagella, and part of the cytoskeleton in eukaryotic cells.

Minimal bactericidal concentration The lowest concentration of an antibacterial agent required to kill a bacterium.

Minimal inhibitory concentration (MIC) The lowest concentration of an antimicrobial agent that prevents growth in the dilution method of determining antibiotic sensitivity.

Minimum growth temperature The lowest temperature at which an organism will grow.

Miracidia The ciliated free-swimming first-stage fluke larvae that have emerged from eggs.

Mitochondria (singular mitochondrion) A membrane-enclosed organelle found in eukaryotic cells that is responsible for the production of ATP.

Moist heat Hot air with a high moisture content. For example, steam.

Mold The filamentous multicellular form of fungi.

Monocyte A nonphagocytic white blood cell found in the blood that will differentiate into a phagocytic macrophage in response to an infection.

Mononuclear phagocytic system Formerly called the reticuloendothelial system. It is a collection of phagocytic cells and tissues that contain phagocytic cells, located throughout the body.

Monotrichous flagellum One of the four types of flagellar arrangement in which there is one flagellum seen on the cell.

Morbidity rate The number of individuals affected by a disease during a set period divided by the total population.

Mordant A substance used during staining that helps a dye bind to the cell.

Mortality rate The number of deaths caused by a disease during a set period divided by the total population.

Motile structures Cellular features used for movement (pseudopods, cilia, and flagella in protozoans).

MRSA Methicillin-resistant *Staphylococcus aureus*.

Mucociliary escalator Mechanism involving ciliated cells that allows materials in the bronchi, trapped in mucus, to be lifted up into the pharynx and subsequently swallowed or spat out.

Mucocutaneous candidiasis A fungal infection of the mucosal tissues caused by the fungus *Candida albicans*.

Mucopurulent vaginal discharge Vaginal discharge containing mucus and pus.

Multiple sclerosis Neurological disease in which autoimmunity causes production of antibody against antigens on the myelin sheath of neurons.

Mumps An acute contagious disease usually seen in children. It is caused by a paramyxovirus and chiefly affects the parotid salivary glands.

Mupirocin An antibiotic that prevents colonization of the nasal passages with *S. aureus*.

Murein Also known as peptidoglycan, a bacterial cell wall component.

Mutagen An agent that causes mutations in DNA.

Mutualism (mutualistic) A form of symbiosis in which two organisms of different species live in a relationship in which both benefit.

Myalgia Muscular pain.

Myasthenia gravis Genetic disorder of infants characterized by apnea, weakness, and fatigue.

Mycelium A mass of long threadlike intertwining structures called hyphae.

Mycolic acid A waxy substance found in the cell wall of certain bacteria, such as the genus *Mycobacterium*.

Mycosis (plural mycoses) A disease caused by fungi.

Myocarditis Inflammation of the muscular walls of the heart.

N-acetylglucosamine (NAG) One of the repeating disaccharides that make up the cell wall of bacteria.

N-acetylmuramic acid (NAM) One of the repeating disaccharides that make up the cell wall of bacteria.

NAD (nicotinamide adenine dinucleotide) A coenzyme that carries hydrogen atoms and electrons.

Naive lymphocyte A lymphocyte that has not found the antigen that binds to its receptor.

Nanocantilever A beam that is immobilized at one end and the other end is free.

Nanotechnology Technologies operating at the nanoscale.

Narrow spectrum antibiotics The range of activity of an antimicrobial agent that attacks only a few kinds of microorganism.

Nationally notifiable disease A disease that must be reported to the Centers for Disease Control and Prevention (CDC).

Natural killer cells (NK cells) Large granular cells found in the peripheral tissue and blood that kill tumor cells, virus-infected cells, bacteria, fungi, and parasites.

Necrotizing fasciitis An infection in which the fascia is destroyed by organisms such as group A streptococci either alone or in a synergistic way with other bacteria.

Necrotizing periodontal disease (NPD) Also known as Vincent's disease or trench mouth. A spectrum of acute inflammatory diseases resulting in the destruction of the soft tissue of the oral cavity.

Negri body A characteristic cytopathology seen in rabies virus infections. These are areas in the cytoplasm that contain masses of viral particles.

Nematode Roundworm.

Neonatal gonorrheal ophthalmia Infection of the eyes of newborn babies from women with gonorrhea. It is caused by infection as the newborn travels down the birth canal.

Nephritis Inflammation of the kidneys.

Neuraminidase A glycoprotein on the envelope of some viruses, which assists in release from the host cell.

Neurotoxin A toxin that acts on the tissues of the nervous system.

Neutropenia A lower than normal number of neutrophil white blood cells.

Neutrophil A phagocytic white blood cell.

Non-cytocidal effect Viral cytopathology that causes a shutdown of host cell function.

Non-gonococcal urethritis The most prevalent sexually transmitted disease. Caused by *Chlamydia trachomatis*.

Non-ionizing radiation A type of radiation, such as ultraviolet radiation, that causes the formation of thymine dimers in DNA.

Nonpolar covalent bond The equal sharing of electrons between two atoms.

Nonseptate fungus A fungus in which the hyphae do not have septa.

Northern blot Analytical tool to detect RNA in a sample.

Nosocomial infection An infection that occurs during a hospital stay.

Nuclear membrane Also called the nuclear envelope. It is the membrane surrounding the nucleus seen in eukaryotic cells.

Nuclear region of bacterial cell Also called a nucleoid. It is the central location where DNA, RNA, and some proteins are found in bacterial cells.

Nucleocapsid The area of the virus in which the capsid and the nucleic acid are found.

Nucleoli (singular nucleolus) Areas in the nucleus of eukaryotic cells where ribosomal RNA is made and ribosomal assembly takes place.

Nucleoplasm The semifluid portion of the cell nucleus of eukaryotic cells that is surrounded by the nuclear membrane.

Nucleoproteins Viral proteins attached to the inner side of the capsid that holds viral nucleic acid in place.

Nucleotide excision The repair mechanism in which enzymes look for distortions in the helical structure of DNA and excise those regions.

Nutrient agar A formulation of media solidified by agar, used for the growth of many types of bacteria.

Nutrient broth A formulation of media in fluid form, used for the growth of many types of bacteria.

Nutrient medium A complex bacterial growth medium in which proteins provide energy, carbon, nitrogen, and other essential nutrients.

O antigen Lipopolysaccharide found on the outer layer of Gram-negative bacteria.

O polysaccharide A type of polysaccharide found on the cell wall of bacteria.

Obligate aerobe A microorganism that cannot grow without oxygen.

Obligate anaerobe A microorganism that cannot grow in the presence of oxygen.

Obligate halophile A microorganism that requires higher than normal concentrations of salt for its growth.

Obligate intracellular parasite The definition of a virus that requires entry into a host cell to reproduce.

Oculoglandular tularemia Purulent conjunctivitis caused by *Francisella tularensis*.

Okazaki fragments Pieces of DNA that are made on the lagging strand of DNA during replication.

Oligonucleotide probe Short sequence of nucleotides that match a specific region of nucleic acid.

Oncogenic virus A virus that causes the host cell to become cancerous.

Oncovirus Also known as an oncogenic virus. A virus that causes the host cell to become cancerous.

Oocyst The encysted or encapsulated stage in the development of any sporozoan.

Open reading frame The starting point for protein synthesis on messenger RNA.

Operator site on DNA A gene in an operon that can bind repressor proteins and inhibit the transcription of the structural genes of the operon.

Operon A sequence of genes that includes both structural and regulatory genes controlling transcription.

Opisthotonos The last stages of tetanus in which the body bends backwards as a result of the relentless contraction of the muscles without relaxation.

Opportunistic infection Infection caused by a microorganism if the immune status of the host is low.

Opportunistic pathogen Resident or transient microorganisms that do not ordinarily cause disease but can do so under certain circumstances.

Opsonin Any molecule that when bound to a microbe enhances the phagocytosis of that microbe by a phagocytic host cell.

Opsonization The process by which microorganisms are rendered more attractive to phagocytes by being coated with antibodies (opsonins) and/or C3b complement proteins.

Optimal growth temperature The temperature at which microorganisms grow best.

Organelle A structure found in the cytoplasm of eukaryotic cells.

Origin of replication The point on the DNA where replication begins.

Osmosis A special type of diffusion in which water "chases" (moves toward) a higher concentration across the plasma membrane.

Oxidation A chemical reaction in which a substance loses one or more electrons.

P pilus A structure that projects from the exterior of a bacterial cell wall and binds to receptors on the epithelial cells of the urinary tract.

P site The area of the ribosome that holds the growing chain of amino acids.

Pancreatitis Inflammation of the pancreas.

Pandemic Global proportions of an epidemic.

Papillomavirus A virus that can cause warts; it is associated with human cervical cancer.

Papule Small elevated lesion of the skin.

Paracortical area of the lymph node An area of the lymph node where T cells are found.

Paragonimiasis Infection with the lung fluke *Paragonimus.*

Paranasal conidiobolomycosis Infection of the submucosa of the paranasal sinuses, resulting in the formation of granulated fibrotic tissue filled with eosinophils.

Parasite-directed endocytosis A unique process in which the microvilli of the epithelial cells surround the organism and escort it into the cell cytoplasm. This process is seen in gonorrhea infection.

Parasitemia Parasites in the blood.

Parasitism A symbiotic relationship in which one organism (the parasite) benefits at the expense of the host.

Parenteral route A portal of entry in which the barrier of the skin is broken, as in a cut, puncture, or surgical procedure.

Passive agglutination Type of agglutination with the antigen first bound to an inert particle to detect an antibody.

Passive immunization A procedure in which an already formed immune product such as antibody is administered to a patient.

Passive transport Movement of materials across the membrane without the expenditure of ATP.

Pasteurization Mild heating to destroy pathogens and other organisms that cause spoilage.

Pathogen An organism capable of causing disease.

Pathogen-associated molecular pattern (PAMP) Molecules associated with groups of pathogens that can be recognized by the innate immune system.

Pathogenic Causing damage in the host and thus disease.

Pathogenicity The ability of a microorganism to cause damage in the host and thus disease.

Pathogenicity islands Sections of the genome that include groups of genes coding for virulence factors that increase the pathogenicity of a microorganism.

PD$_{50}$ Paralytic dose 50%. The number of organisms required to cause paralysis in 50% of a subject population.

Pediculosis A form of lice infection.

Pellicle (1) A thin protein film over the tooth that is the base for the development of a biofilm leading to plaque formation and tooth decay. (2) A strengthened plasma membrane seen in some protozoa.

Pelvic inflammatory disease (PID) An infection of the pelvic cavity in females, caused by any of several organisms.

Penicillin An antibiotic derived from a common blue mold that is used to kill specific bacterium in particular Gram-positive cocci.

Penicillin G A naturally occurring form of penicillin that is taken by injection.

Penicillin V A naturally occurring form of penicillin that is taken by mouth.

Penicillin-binding protein (PBP) Proteins found in the cell walls of bacteria that function in the building of the wall structure. The β-lactam ring of penicillin binds to these proteins.

Pentamidine An antibiotic used in the treatment of *Pneumocystis* infections, leishmaniasis, and African sleeping sickness.

Peptide bond The bond that forms between amino acids in constructing the primary sequence of proteins.

Peptidoglycan The major component of bacterial cell walls.

Peptidyl transferase reaction The enzymatic reaction that links amino acids together by forming the peptide bond.

Perforin An enzyme released by cytotoxic cells that leads to the destruction of target cells.

Period of convalescence One of the five periods that characterizes disease, in which the patient is recuperating.

Period of decline The fourth period used to characterize the disease process, in which the patient is getting better and symptoms are declining. This is the period when secondary infections can occur.

Period of illness The third of the periods that are used to characterize disease, in which the symptoms are greatest. During this period the immune response is functioning at its maximum. This is also the period during which a patient may die.

Periodontitis Inflammation of the gingiva of the oral cavity.

Peripheral tolerance Immunological tolerance that is caused by the inactivation (anergy) in lymphocytes that encounter antigen without co-stimulatory signals.

Periplasmic space The space between the plasma membrane and the outer membrane in Gram-negative bacteria.

Peritrichous flagella This is the arrangement in which flagella are distributed all over the cell.

Permease An enzyme complex involved in the active transport of materials through the cell membrane.

Peroxidase The enzyme used by bacteria to convert hydrogen peroxide to water.

Peroxisome Vesicles in eukaryotic cells that contain the enzymes peroxidase, catalase, and oxidase.

Pertussis A highly contagious respiratory infection caused by *Bordetella pertussis*. Also known as whooping cough.

Petechiae Pinpoint-sized hemorrhages most commonly found in skin folds. They are often seen in rickettsial diseases.

Petechial hemorrhaging Capillary hemorrhage into the skin forming red or purple spots.

Peyer's patches A collection of lymphoid nodules found at the junction between the small and large intestine.

Phage display Technique used to study protein–protein or protein–DNA interactions that uses bacteriophages to display the protein partner.

Phage therapy The use of bacteriophage to treat bacteria-infected hosts.

Phagetyping Method for identifying strains of bacteria by means of bacteriophages that can infect them.

Phagocyte A cell that can carry out phagocytosis.

Phagocytosis Ingestion of materials into cells by means of vacuole formation.

Phagolysosome A structure resulting from the fusion of a phagosome with a lysosome.

Phagosome A vacuole that forms around an organism within the phagocyte that engulfed it.

Pharyngitis Inflammation of the pharynx commonly called a sore throat.

Phase variation A mechanism used by some microorganisms in which the number of pili decreases to prevent the binding of antibody.

Phenol A powerful disinfectant compound used as the standard by which other disinfectants are measured.

Phenol coefficient A numerical expression of the effectiveness of a disinfectant relative to that of phenol.

Phosphoenolpyruvate (PEP) A high-energy molecule used by some bacteria for translocation.

Phospholipid A lipid composed of glycerol, two fatty acids, and a polar head group. It is the basic unit seen in membrane structures.

Phosphorylation The addition of a phosphate group to a molecule, often from ATP, which generally increases the molecule's energy.

Photoautotroph An autotroph that obtains energy from light.

Photoheterotroph A heterotroph that obtains energy from light.

Photoreactivation Also known as light repair. The process of using the enzyme photolyase to unlink thymine dimers in DNA.

Piedra Colonization of the hair shaft by various species of fungi characterized by nodules affixed to the hair.

Pili (singular pilus) Tiny hollow projections used to attach bacteria to surfaces (called attachment pili) or for the transfer of genetic material during conjugation.

Pilin protein Globular proteins that make up the pilus structure.

Pinocytosis The taking in of small molecules by invagination of the cell membrane.

Pityriasis Overgrowth of yeasts of the genus *Malassezia*, which are commonly found on the skin, leads to dermatitis.

Plasma cell Activated B cells that produce antibodies.

Plasmid An extrachromosomal piece of DNA that is small and circular and replicates independently. It can be transferred to another cell.

Plasmid fingerprinting Method for identifying strains of bacteria by analyzing the plasmids they harbor.

Plasmolysis The shrinking of a cell as a result of changes in the osmotic concentration resulting from loss of water in a hypertonic solution.

Pneumonia Inflammation of the lungs.

Pneumonic plague Also known as the Black Death. An infection of the lungs with *Yersinia pestis*, causing a highly contagious form of plague.

Pneumonitis Inflammation of the lungs.

Point mutation Also known as a missense mutation. A change in a single base of the DNA sequence.

Polycistronic mRNA found in prokaryotes, encoding more than one protein.

Polymerase chain reaction (PCR) Technology used to amplify DNA fragments.

Polyribosome A long chain of ribosomes attached at different points along a strand of messenger RNA.

Porins Proteins in the outer layer of Gram-negative bacteria that nonselectively transport polar molecules into the periplasmic space.

Portal of entry A site at which microorganisms enter the body.

Portal of exit A site at which microorganisms can leave the body.

Post-translational modification Changing the structure of proteins so that antibodies do not recognize them.

Precipitation Creation of an insoluble compound.

Prevalence The number of people infected with a particular disease at any given time.

Primary amebic meningoencephalitis Infection of tissues of the nervous system caused by parasites that have ameboid characteristics.

Primary cell line A cell line that must be started from a tissue or organ before each use.

Primary immune response The initial adaptive immune response to an antigen.

Primary immunodeficiency disease A genetic or developmental defect in which T cells or B cells are lacking or nonfunctional.

Primary infection An initial infection in a previously healthy individual.

Primary (obligate) pathogen An organism that can cause disease in healthy hosts.

Primary syphilis Early-stage syphilis infection, characterized by painless ulcerations on the genitals called chancres.

Primase An enzyme that puts the RNA primer on the lagging strand of DNA during replication.

Primer:template junction The area where the RNA primer is located on the strand of DNA. It is required for the replication of DNA.

Prion An infectious protein.

Prion protein scrapie (PrP^{sc}) The abnormally folded, infectious, form of the prion protein.

Prodromal period of disease The second phase of the disease process, in which nonspecific symptoms such as headache and malaise appear.

Prodrug The inactive form of a drug that must be activated enzymatically once in the patient's body.

Products Substances created in a chemical reaction.

Proglottids Reproductive segments seen in cestodes, which contain both male and female gonads.

Prokaryote Microorganism that lacks a cell nucleus and membrane-enclosed organelles. All bacteria are prokaryotes.

Promoter site on DNA The site where RNA polymerase binds to the DNA strand to begin transcription.

Propagated epidemic Disease that involves people-to-people contact and stays in the population for a long time.

Properdin pathway The alternative complement pathway that is activated by contact between lipopolysaccharides and endotoxins on the surface of pathogens and three factors found in the blood (factor B, factor D, and factor P).

Prophage A sequence of DNA from a bacterial virus that is incorporated into the bacterial chromosome.

Prophylactic A medicine or protocol used to prevent infection or disease.

Prospective analytical study A type of epidemiology study in which analysis is ongoing while the disease is occurring.

Prostaglandins Chemical mediators that act as cell regulators and are produced during the inflammatory response. They can stimulate pain and fever responses.

Prostatitis Inflammation of the prostate gland.

Proteasome Large proteins found in both prokaryotes and eukaryotes that degrade unneeded or damaged proteins.

Protein microarrays Miniaturized assay systems that contain small amounts of proteins in a high-density pattern.

Proteomics Discipline considering entire proteomes rather than individual proteins.

Proton motive force A concentration gradient of protons seen during the chemiosmosis step during electron transport.

Protozoan Single-celled, microscopic, animal-like organism.

Provirus The combined nucleic acids of a virus and a host cell during latency.

PrP^c protein The normally folded form of the prion protein.

PrP^{sc} protein The abnormally folded, infectious, form of the prion protein.

Pseudopod Bulge of cytoplasm used as a means of locomotion in some protozoa.

Psychrophiles Bacteria that grow best at very cold temperatures, between 0°C and 15°C.

Psychrotrophs Bacteria that grow best at temperatures between 20°C and 30°C.

Purines The nucleotide bases adenine and guanine.

Pustule A small elevated pus-containing lesion of the skin.

Pyogenic bacteria Pus-forming bacteria.

Pyrimidines The nucleotides thymine and cytosine.

Pyrogens Chemicals that can induce a fever response.

Pyruvate A 3-carbon molecule created by glycolysis.

Pyuria Pus in the urine.

Quaternary ammonium compound (QUATS) Popular detergents containing ammonium cations. They are low-level disinfectants/antiseptics but are odorless, tasteless, and harmless to humans (except at high concentrations).

Quorum sensing Type of decision-making process seen in bacteria that is based on the density of the population of bacteria.

Radiation The energy that is emitted from atomic activities and dispersed at high velocity through matter or space.

Radioimmunoassay Analytical approach where radioisotopes within an antigen or antibody are used to produce a signal for detection.

Radioisotope Radioactive isotope of an element.

Reactants Substances used up in a chemical reaction.

Reading frame The mechanism used to read the DNA-coded sequence for transcription.

Real-time PCR PCR where DNA amplification is measured via a fluorescence signal.

Re-assortment Term used to describe changes in the DNA sequence.

Receptor-mediated endocytosis The process of uptake of materials into the cell through binding to specific receptors on the cell membrane.

Recombinant DNA Genetic material combined from multiple sources by means of genetic engineering.

Recombinant plasmid Plasmid including DNA from two or more sources.

Recombination The combining of DNA from two different cells, resulting in a recombinant DNA molecule.

Redox reaction The oxidation and reduction reactions that move electrons from donor to acceptor molecules.

Reduction A chemical reaction in which a substance gains one or more electrons.

Reduviid A large winged insect that feeds on sleeping hosts in the evening hours.

Re-emerging infectious disease A disease that was thought to be under control through the administration of antibiotics but reappears in the form of drug-resistant disease. A good example is tuberculosis.

Refractory septic shock An irreversible, fatal hypotension caused by septic shock.

Regulatory T cell (T$_{reg}$) A T lymphocyte that regulates the cellular and humoral immune response.

Relapsing fever Disease caused by *Borrelia* species. It is vector transmitted through tick or louse bites and has a poor prognosis, including severe jaundice bleeding and changes in mental status.

Replication fork Location along the double-stranded DNA helix where replication is ongoing.

Replicator sequence An easily opened sequence of A-T pairs in the DNA.

Resident (fixed) macrophages Phagocytic cells that are stationed in specific tissues throughout the body.

Residual body The exocytotic vesicle containing the elements of the destroyed organism. It is seen at the end of the phagocytic process.

Resistance islands Areas of the chromosome in which there is an accumulation of genes associated with resistance to antimicrobial agents.

Reticulate body The larger, more fragile, replicative form of *Chlamydia*.

Reticuloendothelial system Old term for what is now known as the mononuclear phagocytic system.

Retrospective analytical study A type of epidemiological study involving the analysis of data after the episode is over.

Retrovirus An enveloped RNA virus that uses its own reverse transcriptase to transcribe its RNA into DNA in the cytoplasm of the host cell.

Reverse transcriptase An enzyme found in retroviruses that can convert RNA into DNA.

Rheumatic fever Febrile illness caused by infection with Group A hemolytic streptococci.

Rheumatoid arthritis An autoimmune disease that occurs when the immune system attacks the synovial membrane of the synovial joints.

Rhinitis Inflammation of the nasal mucous membranes.

Rhinorrhea Discharge of thin nasal mucous.

Ribavirin A broad-spectrum antiviral antibiotic used to treat severe viral pneumonia caused by respiratory syncytial virus, particularly in high-risk infants. Also used in conjunction with interferon for therapy of hepatitis C infection.

Ribose The form of sugar found in RNA molecules.

Ribosomal RNA The form of RNA that is part of the structure of the ribosomal subunits.

Ribosome The structure in which translation of mRNA into proteins occurs.

Rice stool Also known as rice water stool, a symptom of late stages of cholera infection, in which feces is likened to water from boiled rice.

Ring stage of malaria Stage of malaria in which red blood cells are infected with merozoites, which take on a ring-like shape.

Ringworm A skin lesion caused by fungi that can be found all over the body and is characterized by red margins, numerous scales and reddish itching skin.

River blindness Parasitic infection of the eye spread by the sucking blackfly, which causes corneal ulceration, fibrosis, and blindness.

RNAase H The enzyme that removes the RNA primer from the fragments being made on the lagging strand of DNA during replication.

Rocky Mountain spotted fever Disease caused by *Rickettsia rickettsii*, which is transmitted to humans by ticks.

Rostellum Retractable chitinous hooks that are found on the scolex of tapeworms and are used for attachment.

Rubella Also known as German measles. A viral infection characterized by a short-lived skin rash.

Rubeola Also known as measles or hard measles. Common form of measles.

S pili Fragments of pili intended to bind to and inactivate antibody molecules.

Saint Louis encephalitis Type of viral encephalitis most often seen in humans in the central United States.

Salpingitis Inflammation of the fallopian tubes.

Sanitization Reducing microbial contamination to maintain or improve public health. For example, cleaning, disinfection, and removal of trash from public spaces.

Scalded skin syndrome An infection caused by staphylococci that produces large patches of sloughed skin over the whole body.

Schistosome A term used to describe the reproductive system of some flukes.

Schistosomiasis Infection with the protozoan parasite *Schistosoma*.

Schizogony A reproductive cycle of simple fission followed by sexual reproduction (gametogony).

Scolex The head of the tapeworm used for attachment.

Sebaceous gland Epidermal structure associated with hair follicles that secretes an oily substance called sebum.

Sebum Oily substance secreted by the sebaceous glands.

Secondary immune response The adaptive immune response that occurs when an antigen that has been previously seen is encountered again. This response is quicker and more powerful than the primary adaptive immune response.

Secondary infection Infection that can occur in patients recovering from a primary infection. It can be worse than the primary infection because of the weakened immune response resulting from fighting off the primary infection.

Secondary syphilis Middle-stage syphilis infection, characterized by a skin rash on the extremities.

Secretory piece A protein found on the secretory IgA molecule that attaches it to mucins of the tissues.

Selectin A molecule that is secreted from the epithelial cells of blood vessels that causes the margination of white blood cells at the site of the tissue damage.

Selective medium Medium that encourages the growth of some organisms while inhibiting the growth of others.

Selective permeability The ability to prevent the free passage of certain molecules and ions across the membrane while allowing others to pass through.

Selective toxicity The ability of an antimicrobial agent to kill microbes without causing significant damage to the host.

Semi-continuous cell line A cell line that survives for a moderate period of time before it must be started again from a tissue or organ.

Semi-synthetic penicillin Natural penicillin that has been chemically modified in a laboratory.

Sensitivity Detection limit of a diagnostic test.

Sepsis Presence in the blood of pathogenic microorganisms or their toxins.

Sepsis syndrome Sepsis that causes altered blood flow to organs.

Septa (singular septum) Crosswalls seen in the hyphae of some molds and in bacteria during formation of endospores.

Septic shock A life-threatening hypotensive event caused by endotoxins, in which blood vessels collapse.

Septicemia Rapid multiplication of pathogens in the blood.

Sequence To determine the physical order and placement of bases in a molecule of DNA or RNA.

Serial dilution Stepwise dilution of a concentrated solution repeatedly using the same dilution factor.

Serological Referring to the diagnostic identification of antibodies in the patient serum.

Severe combined immunodeficiency syndrome (SCID) A primary immunodeficiency congenital condition in which there is no T-cell function and no B-cell function.

Severe sepsis Systemic inflammation due to infection.

Sexually transmitted urethritis An infection that presents as dysuria, urethral discharge, or both.

Shiga toxin A dangerous enterotoxin produced by *Shigella* organisms but also found in other Enterobacteriaceae that have acquired the genes for the production of the toxin.

Shigellosis Also known as bacillary dysentery. Gastrointestinal disease caused by several strains of *Shigella* that invade the intestinal tract.

Shingles Sporadic disease caused by reactivation of varicella-zoster virus. It appears mostly in elderly and immunocompromised individuals.

Simple diffusion The net movement of particles from a region of high concentration to a region of lower concentration. It requires no ATP.

Simple stain A single dye used to reveal cell shapes and arrangements.

Single-stranded DNA (ssDNA) virus A virus that possesses only one strand of DNA.

Sinusitis Inflammation of the sinuses.

Sleeping sickness Disease caused by the parasite *Trypanosoma*, also known as African trypanosomiasis. It is marked by intermittent and progressive loss of consciousness.

Slime layer A thin protective structure loosely bound to the cell wall that protects some cells against drying and is sometimes used to bind cells together.

Slow infection A type of persistent infection in which there is a prolonged incubation period before new virions are produced.

Smallpox A formerly worldwide and serious viral infection that was declared eradicated.

Sodium thioglycolate A reducing medium used for growing anaerobic bacteria.

Southern blot Analytical tool to detect DNA in a sample.

Spanish influenza An epidemic of influenza that occurred in 1918 and is believed to have killed 50 million people.

Specialized transduction Type of transduction (transfer of genetic material) in which the DNA being transduced is limited to one or a few genes lying adjacent to the viral insert in the host chromosome that are accidentally included with the viral insert when it is excised from the host chromosome.

Specificity Selectiveness of a diagnostic test.

Sphere of hydration The surrounding of an ion with water molecules. It occurs because of the nonpolar covalent bonding seen in water molecules.

Spirillum (plural spirilla) A spiral-shaped motile bacterium with a rigid cell.

Spirochete A corkscrew-shaped motile bacterium with a flexible cell.

Spirochetemia Spirochetes growing in the blood.

Splenomegaly Enlargement of the spleen.

Spontaneous mutation A mutation that occurs in the absence of any agent known to cause changes in DNA.

Sporadic disease A disease that is limited to a small number of isolated cases, posing no threat to a large population.

Sporotrichosis Fungal skin infection caused by *Sporothrix schenckii* and often transmitted to the body from plants.

Sporozoa A group of protozoan parasites that use both sexual and asexual reproduction during infection.

Sporozoite A malaria trophozoite present in the salivary glands of infected mosquitoes.

Sporulation The formation of a spore.

Stationary phase of bacterial growth The third of the four phases of the bacterial growth curve, in which new cells are produced at the same rate as the old cells die, leaving the number of live cells constant.

Sterilization The killing or removal of all microorganisms in a material or on an object.

Stop codon The last codon of mRNA to be translated at the ribosome, causing the release of the mRNA molecule.

Straintyping Method for identifying strains of bacteria.

Strep throat Serious infection of the pharynx by streptococcal bacteria.

Streptococcal toxic shock syndrome (STSS) Toxic shock syndrome resulting from an infection with *Streptococcus pyogenes*.

Subacute disease A disease that is intermediate between acute and chronic.

Subacute sclerosing panencephalitis A progressive debilitating and fatal brain disorder caused by infection with a mutated measles virus.

Subclinical infection An infection in which there are no apparent symptoms, either because there were insufficient organisms to produce them or because the immune response disposed of the infection before symptoms could appear.

Substrates Substances acted upon by enzymes.

Subunit vaccine A vaccine composed of immunogenic parts of virus derived from genetic engineering and recombinant DNA techniques.

Superantigen Powerful antigens such as bacterial toxins that activate large numbers of T cells, causing a powerful immune response that can lead to toxic shock.

Supercoiling of DNA The characteristic seen in DNA in which coils of the helix are themselves coiled.

Superficial mycosis Fungal infection that does not involve a tissue response.

Superinfection A secondary infection caused by the loss of normal microbial flora, which permits colonization of the body by pathogenic and often antibiotic-resistant microbes.

Superorganism A group of organisms, that functions as a whole unit.

Superoxide dismutase (SOD) An enzyme that converts superoxide to molecular oxygen and hydrogen peroxide.

Suppressor mutations Mutations reversing other previous mutations.

Surfactants Chemicals that reduce the surface tension of solvents. They are very effective for disrupting the plasma membrane of cells.

Swine flu Influenza affecting pigs, or human influenza caused by a related virus.

Symptomatic phase of HIV infection The end stage of the infection in which a variety of infections occur as a result of the loss of the patient's immune defenses.

Symptoms of the disease Characteristics such as headache and nausea that can be observed or felt only by the patient. These are nonmeasurable.

Syncytia Multinucleate masses formed by the fusion of many virally infected cells into one gigantic cell.

Synercid antibiotics Antibiotics containing a pair of antibiotics, quinupristin and dalfopristin.

Syphilis Venereal disease caused by *Treponema pallidum* leading to many structural and cutaneous lesions.

Systemic autoimmunity Autoimmune disease in which the immune response attacks multiple organs.

Systemic infection An infection that occurs throughout the body.

Systemic lupus erythematosus (SLE) An autoimmune disease in which autoantibodies are produced against DNA, RNA, and proteins associated with nucleic acids. It causes damage to small blood vessels, especially in the kidneys.

T cell (T lymphocyte) A thymus-derived lymphocyte involved in the adaptive immune response.

Tachycardia Rapid heartbeat.

Tapeworms Long ribbonlike helminths (worms). They are the largest of the intestinal parasites.

Tegument The viral protein layer located between the capsid and the envelope in complex icosahedral viruses.

Teichoic acid A polymer attached to peptidoglycan in Gram-positive cells.

Tertiary syphilis Late-stage syphilis infection, characterized by localized lesions called gummas, as well as damage to the neurological and cardiovascular systems.

Tetanospasmin The neurotoxic component of tetanus toxin.

Tetanus An acute and often fatal infection caused by the neurotoxin produced by *Clostridium tetani*.

Tetanus toxin A neurotoxin that inhibits the relaxation phase of muscle function.

Tetracycline An antibacterial agent that inhibits protein synthesis.

T_{FH} cells T_{FH} helper cells activate B cells, stimulating them to differentiate into plasma cells that produce antibody.

T_H1 cells T_H1 helper cells activate macrophages to phagocytose and digest ingested microbes and to increase their number of MHC II molecules and cytokine receptors making them better antigen presenting cells.

T_H2 cells T_H2 helper cells promote antiparasitic responses and allergic responses by activating eosinophils and mast cells.

T_H17 cells T_H17 helper cells protect against extracellular bacteria and fungi by stimulating neutrophils.

Thermal death point (TDP) The lowest temperature required to kill all the organisms in a sample in 10 minutes.

Thermal death time (TDT) The shortest length of time needed to kill all organisms at a specific temperature.

Thermophiles Bacteria that grow at high temperatures above 45°C.

Thrombocytopenia A decrease in the number of platelets found in the circulating blood.

Thrombophlebitis Inflammation of veins.

Thrush Milky patches of inflammation on oral mucous membranes. It is a symptom of oral candidiasis caused by *Candida albicans*.

Tincture A solution used to carry other antimicrobial chemicals.

Tinea capitis Also called scalp ringworm. It is a form of ringworm in which hyphae grow in hair follicles, often leaving circular patterns of baldness.

Tinea corporis Ringworm found between the figures, in wrinkles in the palm, and on scaly skin.

Tinea cruris Also called jock itch. It is groin ringworm that occurs in skin folds in the pubic area of the body.

Tinea nigra Condition caused by the fungus *Hortaea werneckii*. Causes brown or black superficial skin lesions found mostly on the palms.

Tinea pedis Also called Athlete's foot. It is a foot infection in which hyphae invade the skin between the toes, causing dry scaly lesions.

Titer The amount or concentration of a substance in a solution.

Toll-like receptors Molecules located on the surface of cells that defend the body. These receptors bind to antigens found on pathogens.

Tonsillitis Inflammation of the tonsils in particular the palatine tonsils.

Topoisomerase An enzyme that breaks the DNA chain, allowing supercoils to relax and strands of the double helix to separate, and then reseals the break.

Toxemia The presence and spread of toxins in the blood.

Toxic shock syndrome (TSS) Potentially fatal condition caused by superantigen exotoxins produced by *Staphylococcus aureus* and *Streptococcus pyogenes*. TSS produced by *Staphylococcus aureus* is often associated with the use of tampons.

Toxoid An exotoxin that has been chemically inactivated but remains antigenic and can therefore be used to immunize against the toxin.

Tracheal toxin The toxin produced by *Bordetella pertussis*. It immobilizes and progressively destroys ciliated cells.

Tracheobronchitis Inflammation of the trachea and the bronchi.

Trachoma A chronic conjunctivitis of the eye that leads to scarring, corneal ulceration, and eventual blindness.

Transcription The process of producing RNA from DNA.

Transcriptomics Discipline considering entire transcriptomes rather than individual mRNAs (transcripts).

Transcytosis The movement of IgA across the epithelial cell barrier.

Transfer RNA Type of RNA that transfers amino acids from the cytoplasm to the ribosome for placement into newly developing peptides.

Transferrin A substance that binds iron.

Transformation A form of transfer of genetic information in which naked DNA is taken up by recipient cells.

Translation The synthesis of protein using the information carried by RNA.

Translational apparatus The large and small subunits of ribosomal RNA and proteins, which come together to form the intact ribosome along with transfer and messenger RNA.

Translocation protein systems Proteins in the cell that move substances out of the cell.

Transmissible spongiform encephalitis (TSE) The neurological disease caused by prions.

Transmitted In regards to a virus that is passed from one host directly to another host.

Transposition The process in which genetic sequences can move from one location to another in the DNA.

Transposon A mobile genetic sequence that contains the genes for transposition as well as one or more genes not related to transposition.

Trematode A fluke.

Trench mouth Also known as necrotizing periodontal disease and Vincent's disease. A spectrum of acute inflammatory diseases resulting in the destruction of the soft tissue of the oral cavity.

Trichinosis Disease caused by eating inadequately prepared meat infected with *Trichinella spiralis*. It is marked by diarrhea, nausea, fever, stiffness, and pain.

Trophozoite The feeding and replicating stage of a protozoan.

Trypanosomiasis A neurological disease caused by the parasite *Trypanosoma*. It is commonly referred to as African sleeping sickness.

Trypomastigote The form of the parasite *Trypanosoma* that is found in the blood.

Tuberculosis Severe infection of the lower respiratory tract, primarily caused by *Mycobacterium tuberculosis*.

Tumor necrosis factor (TNF) A chemical cytokine seen in the inflammatory and immune response to infection.

Type 1 Insulin-dependent diabetes mellitus An autoimmune disease that results when the immune system attacks the insulin producing cells of the pancreas.

Typhoid fever A form of enteric fever caused by *Salmonella enterica* serotype Typhi.

Typhoidal tularemia Tularemia infection with symptoms similar to those of typhoid fever.

Ubiquitin Molecular carrier tags for materials to be taken to the proteasome for recycling.

Ulcer A defect or excavation of the surface of an organ or tissue produced by sloughing off of necrotic inflammatory tissue.

Ulceroglandular tularemia A localized papule that forms at the inoculation site and becomes ulcerated and necrotized; caused by *Francisella tularensis*.

Undulant fever A cycling pattern of symptoms seen in *Brucellosis* infections.

Urethritis Inflammation of the urethra.

Uropathogenic Causing disease in the urinary tract.

Use dilution method A method used to determine whether a chemical substance is bacteriostatic or bacteriocidal.

Vacuolating cytotoxin A circulating protein associated with *Helicobacter pylori* infections.

Vaginal candidiasis A fungal infection of the vaginal area caused by the yeast *Candida albicans*.

Vaginitis Inflammation of the vagina.

Valley fever The disease caused by the infectious form of coccidioidomycosis.

Vancomycin-resistant enterococci (VREs) Enterococcal organisms that are resistant to vancomycin.

Variable region of the antibody molecule The most variable regions of the immunoglobulin molecule made up of variable regions of the heavy and light chain. It contains the antigen-binding site.

Variola major A form of smallpox with a high mortality rate (20% or higher).

Variola minor A form of smallpox with a low mortality rate.

Vasogenic edema A swelling of the brain caused by albumin entering the cerebrospinal fluid.

Vector An organism that transmits a disease-causing organism from one host to another.

Vector transmission Movement of an organism from one host to another. This can be through biological mechanisms such as bites, or through mechanical mechanisms such as shedding from the vector's body.

Vehicle transmission The movement of a disease-causing organism through the use of a nonliving carrier.

Vesicles (1) Membrane-enclosed inclusions in the cytoplasm of cells. (2) Type of skin lesion.

Vincent's disease Also known as necrotizing periodontal disease and trench mouth. A spectrum of acute inflammatory diseases resulting in the destruction of the soft tissue of the oral cavity.

Viral envelope The structure seen around the capsid of some viruses. It is made up of the membrane components obtained when the mature virion leaves the host cell during the process of budding release.

Viremia An infection in which viruses are transported in the blood but do not replicate there.

Virion A mature fully developed viral particle.

Virucidal Term describing a chemical agent that can kill viruses.

Virulence The degree of fitness of the pathogen to overcome the body's defenses and establish themselves.

Virulence factors A structural or physiological characteristic that helps pathogens to establish themselves in the host.

VRSA Vancomycin-resistant *Staphylococcus aureus*.

Wall teichoic acid The form of teichoic acid that extends through only a portion of the Gram-positive cell wall of bacteria as opposed to all the way through.

Wart Small growths characterized by thickening of the skin, which are caused by human papillomaviruses.

Watery diarrhea The most common type of gastrointestinal infection symptom. It develops rapidly and results in frequent voiding.

West Nile meningoencephalitis The most damaging form of West Nile virus infection with mild early symptoms similar to West Nile fever. It progresses to loss of consciousness, near coma, and hyperactive deep tendon reflexes that later diminish.

West Nile virus A virus that causes an emerging infectious disease transmitted by mosquitoes. It causes problems of the central nervous system, including seizures and encephalitis.

Western blot Analytical tool to detect protein in a sample.

Western equine encephalitis Type of encephalitis seen most often in the western United States. It primarily affects horses but can also infect humans.

Whooping cough A highly contagious respiratory infection caused by *Bordetella pertussis*. Also known as pertussis.

Widal reaction Identifying an unknown antibody by using a known antigen.

Wiskott–Aldrich syndrome A disease in which there are defects in the cytoskeleton of eukaryotic cells, causing a predisposition to infection with pyrogenic bacteria.

Wobble hypothesis The theory developed by Francis Crick which states that errors in the third base of a codon can be acceptable because of the degeneracy of the genetic code.

X-linked agammaglobulinemia Primary immunodeficiency disease in which there is an absence of B cells leading to low levels of antibody production, which results in increased bacterial infection.

X-ray crystallography A tool for examining the structure of molecules by examining the diffraction pattern created by beaming X-rays through a crystalline form.

Yeast The single cell form of fungi.

Zone of inhibition A clear area that appears on agar in the disk-diffusion method, indicating where the agent being tested has inhibited the growth of the organism.

Zoonotic disease A disease that can be transmitted from animals to humans.

Zygomatic rhinitis Fungal invasion of the tissue through the arteries causing areas of the mucosa to become grayish black (similar to a blood clot).

Pathogen List

Bacteria				
Latin name	Gram stain	Morphology	Disease	Chapter
Acinetobacter	Gram −		bacteremia	25
Actinomyces species	Gram +		tooth decay	22
Bacillus species	Gram +		ocular infection, meningitis, endocarditis	9, 19, 20
Bacillus anthracis	Gram +		cutaneous anthrax, inhalation anthrax	1, 4, 5, 6, 8, 9, 18, 19, 21, 26
Bacillus cereus	Gram +		food poisoning, ocular infection	5, 22
Bacillus subtilis	Gram +		food poisoning in severely immunocompromised patients	18, 20
Bacteroides species	Gram −		pelvic inflammatory disease	22, 23
Bordetella pertussis	Gram −		pertussis (whooping cough)	4, 5, 6, 8, 21
Borrelia species	Gram −		bacteremia, relapsing fever	6, 25
Borrelia burgdorferi	Gram −		chronic meningitis, Lyme disease	6, 7, 8, 9, 24, 25
Borrelia recurrentis	Gram −		relapsing fever	25
Brucella species	Gram −		bacteremia, brucellosis	6, 8, 25
Burkholderia mallei	Gram −		glanders	8
Burkholderia pseudomallei	Gram −		melioidosis	8
Campylobacter jejuni	Gram −		acute intestinal disease, *Campylobacter* enteritis, dysentery, watery diarrhea	18, 22
Chlamydia species	Gram −		genital tract infections, lymphogranuloma venereum, respiratory and cardiovascular disease, trachoma	4, 19, 23
Chlamydia trachomatis	Gram −		cervicitis, epididymitis, lymphogranuloma venereum, non-gonococcal urethritis, salpingitis, trachoma	5, 6, 7, 18, 23, 26
Chlamydophila pneumoniae	Gram −		atypical pneumonia, chlamydial pneumonia	21

The Bug Parade, with additional pronunciation guide, can be found at **www.garlandscience.com/micro2**

Pathogen List

Latin name	Gram stain	Morphology	Disease	Chapter
Chlamydophila psittaci	Gram –		atypical pneumonia, psittacosis (ornithosis)	6, 8, 21
Clostridium species	Gram +		gastrointestinal infections, botulism, gas gangrene, tetanus	9, 22
Clostridium botulinum	Gram +		botulism, food poisoning	4, 5, 6, 8, 9, 10, 12, 18, 22, 24
Clostridium difficile	Gram +		gastrointestinal infections, superinfections	6, 19, 20, 22
Clostridium perfringens	Gram +		gas gangrene, food poisoning, watery diarrhea	5, 8, 10, 22, 26
Clostridium tetani	Gram +		tetanus (lockjaw)	1, 4, 5, 6, 18, 24
Corynebacterium species	Gram +		catheter bacteremia, intravenous-line bacteremia, meningitis, septic arthritis	7, 9, 25, 26
Corynebacterium diphtheriae	Gram +		diphtheria	2, 4, 5, 6, 8, 11, 12, 21
Coxiella burnetii	Gram –		atypical pneumonia, Q fever	6, 8, 9, 18, 21
Enterobacter species	Gram –		bacteremia, nosocomial pneumonia	7, 21, 25
Enterococcus species	Gram +		dental diseases, diverticular abscesses, endocarditis, infections of the bowel, appendix, and liver	6, 7, 19, 20, 22, 23, 25
Enterococcus faecalis	Gram +		endocarditis	20
Escherichia coli	Gram –		cystitis, meningitis, opportunistic infections, nephritis, urethritis, watery diarrhea	1, 3, 5, 6, 7, 9, 10, 11, 18, 19, 20, 22, 23, 24, 25
Escherichia coli O104:H4	Gram –		bloody diarrhea, hemolytic uremic syndrome	1, 8
Escherichia coli O157:H7	Gram –		hemorrhagic gastroenteritis, hemolytic uremic syndrome, renal failure	1, 5, 6, 8, 9, 20, 22
Francisella tularensis	Gram –		tularaemia	6, 8, 25
Gardnerella vaginalis	Gram variable		bacterial vaginitis	23
Haemophilus ducreyi	Gram –		chancroid	6, 23
Haemophilus influenzae	Gram –		acute purulent meningitis	6, 7, 10, 21, 24
Haemophilus influenzae type B	Gram –		bacteremia, meningitis, pneumonia	24, 25
Helicobacter pylori	Gram –		chronic gastritis, gastric cancer, gastroduodenal ulcer, peptic ulcers	3, 5, 6, 7, 8, 9, 10, 15, 22

The Bug Parade, with additional pronunciation guide, can be found at **www.garlandscience.com/micro2**

Latin name	Gram stain	Morphology	Disease	Chapter
Klebsiella pneumoniae	Gram −		pneumonia	5, 7, 9, 20
Lactobacillus acidophilus	Gram +		dental caries	22, 23
Legionella pneumophila	Gram −		atypical pneumonia, legionellosis (Legionnaires' disease)	6, 8, 21
Leptospira species	Gram −		bacteremia, chronic meningitis	24, 25
Listeria species	Gram +		bacteremia, sepsis, meningitis	20, 25
Listeria monocytogenes	Gram +		listeriosis	4, 6, 18, 24
Mycobacterium species	Gram +		pulmonary disease, skin disease, lymphadenitis	18, 19
Mycobacterium leprae	Gram +		leprosy	4, 6, 7, 9, 10, 15, 16, 19
Mycobacterium tuberculosis	Gram +		chronic meningitis, infections of the central nervous system, tuberculosis	1, 4, 5, 6, 7, 8, 9, 10, 15, 16, 17, 18, 19, 20, 21, 24
Mycoplasma	no Gram reaction		mild pneumonia	2, 4, 9
Mycoplasma pneumoniae	no Gram reaction		*Mycoplasma* pneumonia	5, 21
Neisseria species	Gram −		bacteremia, gonococcal and meningococcal diseases	7, 9, 11, 15, 17, 20, 23
Neisseria gonorrhoeae	Gram −		bacteremia, gonorrhea, neonatorum (neonatal gonorrheal ophthalmia), ophthalmia	4, 5, 6, 7, 9, 17, 18, 20, 23, 25, 26
Neisseria meningitides	Gram −		acute purulent meningitis, bacteremia, meningitis	4, 5, 6, 24, 25
Nocardia	Gram +		nocardiasis	4
Propionibacterium acnes	Gram +		acne, chronic blepharitis, endophthalamitis	7, 26
Proteus species	Gram −		nosocomial urinary tract infections	7, 23
Proteus mirabilis	Gram −		urinary tract infections	5
Pseudomonas species	Gram −		catheter bacteremia, intravenous-line bacteremia	6, 7, 11, 18, 19, 20, 23, 25
Pseudomonas aeruginosa	Gram −		aggressive keratitis, folliculitis, nosocomial infections	4, 6, 10, 19, 21, 23, 24, 26
Rickettsia species	Gram −		vector transmitted diseases	4, 7, 19

Pathogen List

Latin name	Gram stain	Morphology	Disease	Chapter
Rickettsia akari	Gram −		rickettsial pox	25
Rickettsia prowazekii	Gram −		typhus	6, 7, 8, 25
Rickettsia rickettsii	Gram −		Rocky Mountain spotted fever	5, 6, 25
Rickettsia typhi	Gram −		typhus	25
Salmonella species	Gram −		dysentery, food poisoning, gastrointestinal infections	3, 4, 5, 6, 8, 9, 10, 11, 15, 17, 18, 20, 22
Salmonella bongori	Gram −		bacteremia, acute enteritis	22
Salmonella enterica serovar Typhi	Gram −		bacteremia, gastroenteritis, salmonellosis, sepsis, typhoid fever	5, 6, 7, 10, 18, 22, 25
Shigella species	Gram −		shigellosis (bacillary dysentery)	3, 4, 5, 6, 8, 9, 11, 20, 22
Shigella boydii	Gram −		dysentery	22
Shigella dysenteriae	Gram −		bacillary dysentery, bacteremia, food poisoning, gastrointestinal infections	5, 12, 20, 22, 25
Shigella flexneri	Gram −		dysentery	22
Shigella sonnei	Gram −		dysentery	22
Staphylococcus species	Gram +		impetigo	10, 11, 15, 18, 19, 20, 26
Staphylococcus aureus	Gram +		boils (furuncles), carbuncles, catheter bacteremia, endocarditis, folliculitis, food poisoning, intravenous-line bacteremia, nosocomial pneumonia, scalded skin syndrome, skin infections, staphylococcal pneumonia, toxic shock	1, 4, 5, 6, 7, 9, 10, 16, 18, 19, 20, 21, 22, 23, 24, 25, 26
methicillin-resistant *Staphylococcus aureus* (MRSA)	Gram +		antibiotic resistant infections	1, 6, 20, 21, 23
vancomycin-resistant *Staphylococcus aureus* (VRSA)	Gram +		antibiotic resistant nosocomial infections	6, 20

The Bug Parade, with additional pronunciation guide, can be found at **www.garlandscience.com/micro2**

Latin name	Gram stain	Morphology	Disease	Chapter
Staphylococcus epidermidis	Gram +		catheter bacteremia, intravenous-line bacteremia	7, 25, 26
coagulase-negative staphylococci	Gram +		endocarditis, epididymitis, bacteremia, urinary tract and CNS infections	23, 24, 25, 26
Streptococcus species	Gram +		dental diseases, diverticular abscesses, infections of the bowel, appendix, and liver	5, 7, 9, 15, 19, 20, 21, 22, 23, 25, 26
group A streptococci	Gram +		erysipelas, impetigo, necrotizing fasciitis	21, 25, 26
group B streptococci	Gram +		neonatal sepsis, postpartum infections	23, 24
group C streptococci	Gram +		pharyngitis	21
β-hemolytic streptococci	Gram +		bacteremia	25
Streptococcus mutans	Gram +		tooth decay	2, 5, 22
Streptococcus pneumoniae	Gram +		acute otitis media, acute purulent meningitis, bacteremia, meningitis, middle ear infections, pneumonia, sinusitis	4, 5, 6, 7, 9, 10, 11, 20, 21, 24, 25
Streptococcus pyogenes	Gram +		abscesses on the tonsils, acute tonsillitis, endocarditis, erysipelas, necrotizing fasciitis, pharyngitis, pneumonia, scarlet fever, strep throat, toxic shock	4, 5, 7, 9, 10, 19, 21, 26
Streptococcus salivarius	Gram +		tooth decay	22
Streptococcus viridans	Gram +		endocarditis	25
Treponema pallidum	Gram –		chronic meningitis, syphilis	5, 6, 7, 9, 10, 23, 24, 25
Ureaplasma urealyticum	No Gram reaction		non-gonococcal urethritis	23
Vibrio cholerae	Gram –		cholera	5, 6, 8, 9, 12, 20, 22
Vibrio cholerae 0139	Gram –		cholera	8, 20
Vibrio parahaemolyticus	Gram –		food poisoning, watery diarrhea	6, 22
Yersinia enterocolitica	Gram –		gastrointestinal infections	11, 22
Yersinia pestis	Gram –		plague (bubonic plague, pneumonic plague)	1, 5, 6, 8, 15, 19, 25

The Bug Parade, with additional pronunciation guide, can be found at **www.garlandscience.com/micro2**

Pathogen List

Viruses				
Latin name	**Nucleic Acid Type**	**Morphology**	**Disease**	**Chapter**
adenovirus	dsDNA		acute respiratory disease, diarrhea, gastrointestinal infections	12, 13, 22
alphaviruses	+ ssRNA		encephalitis	8
arboviruses	+ ssRNA		hemorrhagic fever, hepatic necrosis, West Nile fever (fever and encephalitis)	6, 8, 24, 25
	+ ssRNA			
	– ssRNA			
arena virus	– ssRNA		zoonotic central nervous system infections	8, 13
Arenaviridae	– ssRNA		Lassa fever	8
avian influenza virus (H5N1)	– ssRNA		avian influenza (bird flu)	8, 13
Bunyaviridae	– ssRNA		hantavirus pulmonary syndrome	8
calicivirus	+ ssRNA		diarrhea, gastrointestinal infections	22
coronavirus	+ ssRNA		acute mild respiratory infections, SARS, MERS	13, 19
cowpox virus	dsDNA		cowpox	13
coxsackievirus A16	+ ssRNA		hand–foot–and–mouth disease	13
cytomegalovirus	dsDNA		acute polyneuritis, mononucleosis syndrome	13, 17, 19, 24, 25
Dengue virus	+ ssRNA		fever, headache, muscle and joint pain	6, 19, 25
eastern equine encephalitis virus	+ ssRNA		encephalitis	24, 25
Ebola virus	– ssRNA		hemorrhagic fever	4, 8, 12, 13, 19, 25
enterovirus	+ ssRNA		diarrhea, gastrointestinal infections, meningitis, persistent enterovirus infection of the immunodeficient	13, 22, 24
Epstein–Barr virus	dsDNA		acute polyneuritis, Burkitt's lymphoma, infectious mononucleosis, lymphoproliferative infections in immunocompromised patients, nasopharyngeal carcinoma	13, 24, 25
flavivirus	+ ssRNA		hemorrhagic fever, yellow fever	6, 8, 25

The Bug Parade, with additional pronunciation guide, can be found at **www.garlandscience.com/micro2**

Latin name	Nucleic Acid Type	Morphology	Disease	Chapter
hantavirus	– ssRNA		hanta fever, hantavirus pulmonary syndrome	6, 8, 21
hepatitis A virus	+ ssRNA		food poisoning, hepatitis A	5, 6, 7, 13, 22
hepatitis B virus (HBV)	– ssRNA (partially ds)		cirrhosis, hepatitis B, hepatocellular carcinoma	5, 6, 13, 20, 22, 23
hepatitis C virus	+ ssRNA		cirrhosis, hepatitis C, hepatocellular carcinoma	5, 6, 8, 13, 15, 19, 22, 23
hepatitis D virus	– ssRNA		cirrhosis, hepatitis D	22
hepatitis E virus	+ ssRNA		liver fibrosis and cirrhosis, hepatitis E	22
hepatitis G virus	+ ssRNA		hepatitis	22
herpesvirus	dsDNA		meningitis	12, 13, 19, 24
herpes simplex virus type 1 (HSV-1)	dsDNA		cold sores (fever blisters), ophthalmic herpes	13, 17, 19, 23, 24, 26
herpes simplex virus type 2 (HSV-2)	dsDNA		genital herpes	5, 13, 17, 19, 23, 24
human immunodeficiency virus (HIV)	+ ssRNA		acquired immune deficiency syndrome (AIDS)	1, 4, 5, 6, 8, 12, 13, 15, 17, 19, 21, 23, 24
human lymphotropic virus	+ ssRNA		lymphoma, T cell leukemia	13
human papillomavirus (HPV)	dsDNA		carcinomas, cervical cancer, papillomas, warts (dermal, genital)	12, 13, 19, 23, 26
influenza virus (influenza A and B)	– ssRNA		influenza	1, 5, 7, 8, 12, 13, 15, 19, 21
Lassa virus	– ssRNA		Lassa fever	8
Marburg virus	– ssRNA		hemorrhagic fever	8, 19, 25
measles virus	– ssRNA		measles, subacute sclerosing panencephalitis	5, 6, 13, 24
Middle East Respiratory Syndrome coronavirus (MERS-CoV)	+ ssRNA		acute respiratory illness, fever, cough, shortness of breath	13, 19
mumps virus	– ssRNA		meningitis, mumps	5, 6, 13, 24
Nipah virus	– ssRNA		central nervous system infections	8

The Bug Parade, with additional pronunciation guide, can be found at **www.garlandscience.com/micro2**

Pathogen List

Latin name	Nucleic Acid Type	Morphology	Disease	Chapter
Norwalk viruses (norovirus)	+ ssRNA		intestinal infections	22
parainfluenza virus	− ssRNA		parainfluenza	21
poliovirus	+ ssRNA		poliomyelitis (abortive, aseptic, paralytic)	5, 6, 7, 12, 13, 24
polyoma virus BK	dsDNA		hemorrhagic cystitis	13
polyoma virus JC	dsDNA		progressive multifocal leukoencephalopathy	13
poxviruses	dsDNA		smallpox	12, 13, 19, 26
pseudorabies virus	dsDNA		Aujeszky's disease	13
rabies virus	− ssRNA		rabies	5, 6, 8, 13, 24
respiratory syncytial virus	− ssRNA		respiratory infections	19, 21
rhabdovirus	− ssRNA		rabies	7, 8
rhinovirus	+ ssRNA		common cold	12, 13, 19, 21
rotavirus	dsRNA		diarrhea, gastrointestinal infections	13, 22
rubella virus	+ ssRNA		progressive rubella panencephalitis, rubella (German measles)	5, 6, 13, 24, 26
rubeola virus	− ssRNA		measles (five–day measles, hard measles)	26
SARS virus (coronavirus)	+ ssRNA		severe acute respiratory syndrome (SARS)	6, 8, 19, 20
St Louis encephalitis virus	+ ssRNA		encephalitis	24, 25
varicella–zoster virus (VZV)	dsDNA		chickenpox, postherpetic neuralgia, shingles	5, 6, 13, 16, 17, 19, 24, 26
variola minor; variola major virus	dsDNA		smallpox	6, 8, 12, 13, 26
Venezuelan encephalitis virus	+ ssRNA		central nervous system infections	13
western equine encephalitis virus (WEE)	+ ssRNA		encephalitis	24, 25
West Nile virus	+ ssRNA		West Nile fever (fever and encephalitis)	6, 8, 24, 25
yellow fever virus	+ ssRNA		hemorrhagic fever, hepatic necrosis	6, 8, 13, 25

The Bug Parade, with additional pronunciation guide, can be found at **www.garlandscience.com/micro2**

Protozoan Parasites, Worms, And Lice		
Latin name	Disease	Chapter
Acanthamoeba species	chronic meningitis, primary amebic meningoencephalitis	19, 21, 24, 26
Ancylostoma duodenale	hookworm infection	14, 22
Ascaris lumbricoides (giant roundworm)	ascariasis	14, 19
Brugia malayi	filariasis (lymphadenitis, lymphadenopathy, elephantiasis)	25
Clonorchis sinensis (Asian liver fluke)	clonorchiasis	14
Cryptosporidium parvum	cryptosporidiosis	8, 17, 22
Entamoeba histolytica	amebiasis	14, 19, 22
Enterobius vermicularis (pin worm)	enterobiasis	14, 19
Giardia duodenalis	gastrointestinal infections, giardiasis, watery diarrhea	22
Giardia intestinales (*G. lamblia*)	giardiasis	5, 6, 14
Leishmania species	cutaneous leishmaniasis (skin lesions)	14, 15, 17, 19, 26
Loa loa	loaiasis	26
Microsporidium species	opportunistic parasite associated with AIDS	17
Naegleria species	primary amebic meningoencephalitis	14, 24
Necator americanus (hookworm)	hookworm disease	14, 22
Onchocerca volvulus	onchocerciasis (river blindness)	19, 26
Paragonimus species (lung fluke)	paragonimiasis	14
Pediculus humanus capitis (head louse)	head lice	26
Pediculus humanus corporis (body louse, clothes louse)	body lice, genital lice (crabs)	26
Plasmodium species	malaria	5, 6, 8, 14, 15, 19
Plasmodium falciparum	malaria	14, 19
Plasmodium vivax	malaria	14
Schistosoma species (flukes)	schistosomiasis	14
Strongyloides stercoralis (roundworm)	gastrointestinal parasitic infection	14
Taenia species	tapeworm	14, 19
Toxoplasma species	toxoplasmosis	17, 19
Toxoplasma gondii	chronic meningitis, toxoplasmosis	6, 14, 24
Trichinella species	trichinosis	6, 14, 18
Trichinella spiralis	chronic meningitis, food poisoning, myalgia	14, 22, 24
Trichomonas vaginalis	trichomoniasis, vaginitis	14, 19, 23
Trichuris trichiura (whipworm)	whipworm infection	14, 19, 22

The Bug Parade, with additional pronunciation guide, can be found at **www.garlandscience.com/micro2**

Pathogen List

Latin name	Disease	Chapter
Trypanosoma brucei brucei, Trypanosoma brucei gambiense, Trypanosoma brucei rhodesiense	trypanosomiasis (sleeping sickness)	6, 14, 19, 24, 25
Trypanosoma cruzi	Chagas' disease, heart and gastrointestinal lesions	6, 14, 19, 25
Wuchereria bancrofti	filariasis (lymphadenitis, lymphadenopathy, elephantiasis)	25

Fungi		
Latin name	Disease	Chapter
Aspergillus species	aspergilliosis	19, 21
Blastomyces dermatitidis	blastomycosis	21
Candida albicans	cutaneous and vaginal candidiasis	15, 17, 19, 23, 26
Coccidioides immitis	coccidioidomycosis	17, 21, 24
Cryptococcus neoformans	cryptococcosis	17, 24
Dermatophytes	dermatophytosis, tinea cruris (jock itch), tinea pedis (athlete's foot)	26
Histoplasma capsulatum	chronic meningitis, histoplasmosis	14, 17, 21, 24
Hortaea werneckii	tinea nigra	14
Pneumocystis (carnii) jirovecii	pneumocystis pneumonia	16, 17, 19, 21

Figure Acknowledgments

Chapter 1
Figure 1.1: Courtesy of Mark Thomas/Science Photo Library.
Figure 1.2: A. Courtesy of Jeroen Rouwkema. B. Courtesy of Moisey.
Figure 1.3: Courtesy of the National Museum of Health and Medicine, Armed Forces Institute of Pathology.
Figure 1.6: Courtesy of Peter Falkner/Science Photo Library.

Chapter 2
Chapter opening quote: From Smith B (1943) A Tree Grows in Brooklyn. Reprinted with permission from HarperCollins Publishers.

Chapter 3
Chapter opening quote: From Margulis L & Sagan D (2000) What Is Life? Reprinted with permission from University of California Press.

Chapter 4
Chapter opening quote: From Thomas L (1974) The Lives of A Cell: Notes of a Biology Watcher. Reprinted with permission of Viking Books, Penguin Random House LLC.
Figure 4.4: B. Courtesy of CDC/Dr. Richard Facklam.
C. Courtesy of CDC/Dr. Mike Miller.
Figure 4.5: B. From O'Connell D (2004) *Nat Rev Microbiol* 2:442. With permission from Macmillan Publishers Ltd.
C. Courtesy of CDC.
Figure 4.6: B. Courtesy of CDC/Janice Haney Carr & Jeff Hageam. C. Courtesy of CDC/Dr. Thomas F Sellers, Emory University.
Figure 4.8: Courtesy of CDC.
Figure 4.9: Courtesy of CDC/Dr. William A Clark.
Figure 4.10: Courtesy of CDC/Dr. George P Kubica.
Figure 4.11: Courtesy of CDC/Larry Stauffer, Oregon State Public Health Laboratory.
Figure 4.17: B. From Ménard R, Dehio C & Sansonetti PJ (1996) *Trends Microbiol* 4:220–226. With permission from Elsevier Limited.
Figure 4.18: From Hossain T, Kappelman MD, Perez-Atayde AR et al. (2003) *J Perinatol* 23:684–687. With permission from Macmillan Publishers Ltd.
Figure 4.19: Courtesy of NIBSC/Science Photo Library.
Figure 4.24: A. From Rothberg KG, Heuser JE, Donzell WC et al. (1992) *Cell* 68:673–682. With permission from Elsevier Limited. B. Courtesy of Dr. Dorothy Bainton.
Figure 4.25: Courtesy of Eye of Science/Science Photo Library.
Figure 4.27: B. Courtesy of Lelio Orci, University of Geneva.

Chapter 5
Figure 5.5: Courtesy of Gillete Corporation/Science Photo Library.
Figure 5.7: Courtesy of Professors P Motta & F Carpino, University "La Sapienza", Rome/Science Photo Library.
Figure 5.8: Courtesy of Professor P Motta, University "La Sapienza", Rome/Science Photo Library.
Figure 5.9: Courtesy of Professor PM Motta et al./Science Photo Library.
Figure 5.10: Courtesy of David Scharf/Science Photo Library.
Figure 5.11: Courtesy of CDC/Dr. David Cox.
Figure 5.15: From James WD, Berger T & Elston D (2006) Andrews' diseases of the skin: Clincal dermatology, 10th ed. Saunders. With permission from Elsevier.

Figure 5.18: Courtesy of the National Center for Immunization and Respiratory Diseases, Division of Viral Diseases.
Figure 5.19: Courtesy of CDC/Dr. Daniel P Perl.
Figure 5.20: Courtesy of CDC/Dr. Edwin P Ewing Jr.

Chapter 6
Figure 6.2: Courtesy of Dr. Gary Settles/Science Photo Library.
Figure 6.13: Courtesy of Biomérieux.
Figure 6.20: From Tortora GJ, Funke BR & Case CL (2007) Microbiology: An Introduction, 9th ed. With permission from Pearson Education, Inc., New York.
Figure 6.21: From Dhuna V, Dhuna K, Singh J et al. (2010) *Adv Biosci Biotech* 1:79–90.
Figure 6.22: Courtesy of CDC.
Figure 6.25: Courtesy of Wladimir Bulgar/Science Photo Library.
Figure 6.28: Courtesy of Alere.
Figure 6.29: Courtesy of Alere.

Chapter 7
Figure 7.1: Adapted from Cho I & Blaser MJ (2012) *Nat Rev Genet* 13:260–270. With permission from Macmillan Publishers Ltd.
Figure 7.2: Adapted from Cho I & Blaser MJ (2012) *Nat Rev Genet* 13:260–270. With permission from Macmillan Publishers Ltd.
Figure 7.6: From Boshoff HIM & Barry CE (2005) *Nat Rev Microbiol* 3:70–80. With permission from Macmillan Publishers Ltd.

Chapter 8
Figure 8.3: Courtesy of CDC/James Gathany.
Figure 8.4: Adapted from Webby R, Hoffmann E & Webster R (2004) *Nat Med* 10:S77–S81. With permission from Macmillan Publishers Ltd.
Figure 8.5: Courtesy of CDC/Dr. Thomas Hooten.
Figure 8.8: From Geisbert TW & Jahrling PB (2004) *Nat Med* 10:S110–S121. With permission from Macmillan Publishers Ltd.
Figure 8.10: Adapted from Towards universal access to diagnosis and treatment of multidrug-resistant and extensively drug-resistant tuberculosis by 2015: WHO Progress Report 2011. Geneva, World Health Organization, 2011 (Map 1, page 15 http://whqlibdoc.who.int/publications/2011/9789241501330_eng.pdf, accessed 04 March 2014). With permission from the publisher.
Figure 8.11: A. Courtesy of Science Photo Library. B. Courtesy of National Museum of Health and Medicine, Armed Forces Institute of Pathology.
Figure 8.15: Courtesy of Stephen Jaffe/AFP/Getty Images.
Figure 8.16: Courtesy of CDC/Dr. LaForce.
Figure 8.17: Courtesy of CDC.
Figure 8.18: Courtesy of CDC.
Figure 8.19: A. Courtesy of CDC/James Hicks. B. Courtesy of CDC/Dr. Lyle Conrad. C. Courtesy of CDC/Dr. Stan Foster.

Chapter 9
Chapter opening quote: From Flemming A (1929) *J Exp Pathol* 10:226–236.
Figure 9.6: A. Courtesy of Dr. Immo Rantala/Science Photo Library. B. Courtesy of Professor Bill Costerton.
Figure 9.7: A. Courtesy of Thomas Deerinck, NCMIR/Science Photo Library. B. Courtesy of Dennis Kunkel Microscopy Inc.

Figure 9.8: A. Courtesy of CDC/NCID/Rob Weyant. C. Courtesy of Eye of Science/Science Photo Library.

Figure 9.9: A. Courtesy of CDC/Dr. David Cox. B. Courtesy of CDC.

Figure 9.10: Courtesy of Kwangshin Kim/Science Photo Library.

Figure 9.13: A. Courtesy of CDC/Dr. William A Clark. B. Courtesy of Dr. Jack Bostrack/Visuals Unlimited, Inc. C. Courtesy of CDC/Dr. William A Clark. D. Courtesy of CDC.

Figure 9.23: Courtesy of Biophoto Associates/Science Photo Library.

Figure 9.24: Courtesy of Dr. Gopal Murti/Science Photo Library.

Figure 9.25: B. Courtesy of CNRI/Science Photo Library.

Chapter 10

Chapter opening quote: From Crichton M (1969) The Andromeda Strain. Reprinted with permission from Random House.

Figure 10.2: A. Courtesy of CDC/Dr. Gavin Hart & Dr. NJ Fiumra. B. Courtesy of CDC.

Figure 10.3: From Special Program for Research and Training in Tropical disease, World Health Organization/Dr. Colin McDougall.

Figure 10.4: Courtesy of Eye of Science/Science Photo Library.

Figure 10.5: Courtesy of Drgnu23.

Figure 10.8: Courtesy of Witmadrid.

Figure 10.9: Courtesy of Navaho.

Figure 10.10: Courtesy of Y tambe.

Figure 10.11: B. Courtesy of Dr. Kari Lounatmaa/Science Photo Library.

Figure 10.14: A. Courtesy of Simon Fraser/Science Photo Library. B. Courtesy of Susan Boyer, Appalachian Farming Systems Research Center, US Department of Agriculture.

Figure 10.16: Courtesy of CDC/Amanda Moore, Todd Parker & Audra Marsh.

Chapter 11

Figure 11.30: Courtesy of Dennis Kunkel Microscopy Inc.

Chapter 12

Figure 12.3: A. Courtesy of Dr. Klaus Boller/Science Photo Library.

Figure 12.4: A. Courtesy of Eye of Science/Science Photo Library.

Figure 12.10: Courtesy of CNRI/Science Photo Library.

Chapter 13

Chapter opening quote: From Crewe M (2012) The Way We Fall. Reprinted with permission from Disney-Hyperion.

Figure 13.2: Courtesy of St Bartholomew's Hospital/Science Photo Library.

Figure 13.8: Courtesy of SOA-AIDS Amsterdam.

Figure 13.9: Courtesy of CDC/Dr. KL Hermann.

Figure 13.10: A. Courtesy of CNRI/Science Photo Library. B. Courtesy of CDC/Dr. Daniel P Perl.

Figure 13.12: Courtesy of CDC/Dr. Herrmann.

Figure 13.13: Courtesy of Kent Wood/Science Photo Library.

Figure 13.14: Courtesy of CDC.

Figure 13.15: Courtesy of CDC/Dr. NJ Flumara & Dr. Gavin Hart.

Chapter 14

Chapter opening quote: From Buckman R (2003) Human Wildlife: The Life That Lives on Us. Reprinted with permission from Johns Hopkins University Press.

Figure 14.1: Courtesy of D Phillips/Science Photo Library.

Figure 14.2: Courtesy of DPDx, CDC.

Figure 14.3: Courtesy of CDC.

Figure 14.4: Courtesy of ER Degginger/Science Photo Library.

Figure 14.5: Courtesy of London School of Hygiene & Tropical Medicine/Science Photo Library.

Figure 14.6: Courtesy of Eye of Science/Science Photo Library.

Figure 14.8: From Taylor TE, Fu WJ, Carr RA et al. (2004) *Nat Med* 10:143–145. With permission from Macmillan Publishers.

Figure 14.9: Courtesy of Martin Dohrn/Science Photo Library.

Figure 14.10: Courtesy of Dr. MA Ansary/Science Photo Library.

Figure 14.11: Courtesy of Gastrolab/Science Photo Library.

Figure 14.13: Courtesy of CDC/Dr. Mae Melvin.

Figure 14.14: Courtesy of CDC.

Figure 14.16: Courtesy of CDC/Dr. Sulzer.

Figure 14.17: Courtesy of N Hamilton, The Aspergillus Website.

Figure 14.18: Courtesy of CDC/Dr. Edwin P Ewing Jr.

Figure 14.19: Courtesy of CDC/Dr. Libero Ajello.

Figure 14.20: Courtesy of CDC/Dr. Lucille K Georg.

Figure 14.21: Courtesy of CDC.

Figure 14.22: Courtesy of www.doctorfungus.org © 2007.

Figure 14.23: Courtesy of www.doctorfungus.org © 2007.

Figure 14.24: Courtesy of CDC/Susan Lindsley, VD.

Chapter 15

Figure 15.1: Courtesy of Power and Syred/Science Photo Library.

Figure 15.2: Courtesy of D Phillips/Science Photo Library.

Figure 15.3: Courtesy of Photo Insolite Realite/Science Photo Library.

Figure 15.7: Courtesy of Anthony Butterworth.

Figure 15.8: A. Courtesy of Biophoto Associates/Science Photo Library. B. Courtesy of Science Photo Library.

Figure 15.10: Courtesy of Professor P Motta, University "La Sapienza" Rome/Science Photo Library.

Figure 15.14: Courtesy of Dr. Dorothy Bainton.

Figure 15.15: A. Courtesy of Dr. Kari Lounatmaa/Science Photo Library. B. Courtesy of Biology Media/Science Photo Library.

Figure 15.17: A. Courtesy of Professor Sucharit Bhakdi, Johannes Gutenberg University Mainz, Germany. B. From Schreiber RD, Morrison DC, Podack ER & Muller-Eberhard HJ (1979) *J Exp Med* 149:870–882. With permission from Rockefeller University Press.

Chapter 16

Figure 16.2: B. Courtesy of N Rooney.

Figure 16.4: Courtesy of David Scharf/Science Photo Library.

Figure 16.5: Courtesy of Dr. Olivier Schwartz, Institute Pastuer/Science Photo Library.

Figure 16.17: Courtesy of Professor Ann Dvorak, Harvard Medical School.

Chapter 17

Chapter opening quote: From Gottlieb MS, Schancker HM, Tan PF et al. (1981) *Morb Mortal Wkly Rep* 30:250–252.

Figure 17.1: A. Courtesy of Hans Gelderblom, Berlin.

Figure 17.3: Courtesy of Domininka Rudnicka, Nathalie Sol-Foulon & Olivier Schwartz, Institut Pasteur.

Figure 17.7: Courtesy of Department of Medical Photography, St Steven's Hospital, London/Science Photo Library.

Figure 17.10: Courtesy of Irene Visintin, Dr. Gil Mor Laboratory, Department of Obstetrics and Gynecology, Yale.

Figure 17.11: Courtesy of Euan Tovey, Woolcock Institute of Medical Research, Australia.

Figure 17.12: Courtesy of Professor Barry Kay, Imperial College London.

Chapter 18

Chapter opening quote: From Brandeis LD (1913) What Publicity Can Do. *Harper's Weekly*, 20 December.

Figure 18.5: Courtesy of Dr. Jack Bostrack/Visuals Unlimited, Inc.

Figure 18.10: Courtesy of CDC/Janice Haney Carr.

Chapter 20
Figure 20.3: Courtesy of CDC/Gilda L Jones.
Figure 20.4: Courtesy of CDC/Dr. Richard Facklam.

Chapter 21
Figure 21.4: Courtesy of Steve Gschmeissner/Science Photo Library.
Figure 21.6: Courtesy of CDC/Dr. Heinz F Eichenwald.
Figure 21.8: Courtesy of Science VU/CDC/Visuals Unlimited, Inc.
Figure 21.9: Courtesy of CDC.
Figure 21.10: Courtesy of AB Dowsett/Science Photo Library.
Figure 21.12: Courtesy of CDC/Dr. George P Kubica.
Figure 21.13: Courtesy of CDC.
Figure 21.15: Courtesy of CDC/Science Photo Library.
Figure 21.16: From Horowitz MA (1984) *Cell* 36:27–33. With permission from Elsevier Limited.
Figure 21.17: B. Courtesy of Dennis Kunkel Microscopy Inc.
Figure 21.18: Courtesy of CDC/Cynthia Goldsmith & Luanne Elliott.
Figure 21.19: Courtesy of CDC/Dr. Russell K Byrnes.
Figure 21.20: Courtesy of CDC/Dr. Liberio Ajello.
Figure 21.21: Courtesy of Michael Abbey/Science Photo Library.
Figure 21.22: Courtesy of CDC/Dr. Liberio Ajello.

Chapter 22
Chapter opening quote: From Schlosser E (2001) Fast Food Nation: The Dark Side of the All-American Meal. Reprinted with permission from Houghton Mifflin Company, NY and Allen Lane, Penguin UK.
Figure 22.1: Courtesy of CDC/Lois Wiggs.
Figure 22.3: Courtesy of David Scharf/Science Photo Library.
Figure 22.4: Courtesy of SCIMAT/Science Photo Library.
Figure 22.5: Adapted from Struthers JK & Westran RP (2003) Clinical Bacteriology. Manson Publishing Ltd.
Figure 22.6: Adapted from Struthers JK & Westran RP (2003) Clinical Bacteriology. Manson Publishing Ltd.
Figure 22.9: Courtesy of CDC.
Figure 22.10: Courtesy of CDC/NIAID.
Figure 22.11: Courtesy of CDC.
Figure 22.12: Courtesy of CDC/Dr. Patricia Fields & Dr. Collette Fitzgerald.
Figure 22.13: Courtesy of AB Dowsett/Science Photo Library.
Figure 22.14: Courtesy of David M Martin/Science Photo Library.
Figure 22.16: Courtesy of Martin M Rotker/Science Photo Library.
Figure 22.17: Courtesy of CDC/Janice Haney Carr.
Figure 22.18: Courtesy of CDC/Dr. Edwin P Ewing Jr.
Figure 22.19: Courtesy of CDC/Dr. Mae Melvin.
Figure 22.20: Courtesy of CDC/Dr. Mae Melvin.

Chapter 23
Figure 23.4: From Struthers JK & Westran RP (2003) Clinical Bacteriology. Manson Publishing Ltd.
Figure 23.7: Courtesy of Dr. Y Boussougan/CNRI/Science Photo Library.
Figure 23.8: Courtesy of Dr. MA Ansary/Science Photo Library.
Figure 23.9: Courtesy of CDC/Susan Lindsley.
Figure 23.10: Courtesy of CDC/M Rein.
Figure 23.11: Courtesy of CDC.
Figure 23.12: Courtesy of CDC/Susan Lindsley.

Figure 23.13: Courtesy of Dr. P Marazzi/Science Photo Library.
Figure 23.14: Courtesy of CNRI/Science Photo Library.
Figure 23.16: Courtesy of CDC/Dr. NJ Flumara, Dr. Gavin Hart.
Figure 23.17: Courtesy of CDC.

Chapter 24
Chapter opening quote: From Brown D (2000) The 'recipe for disaster' that killed 80 and left a £5bn bill. *The Telegraph*, 27 October.
Figure 24.2: Adapted from Struthers JK & Westran RP (2003) Clinical Bacteriology. Manson Publishing Ltd.
Figure 24.3: B. Courtesy of CDC.
Figure 24.5: Courtesy of Alfred Pasieka/Science Photo Library.
Figure 24.6: Courtesy of CDC.
Figure 24.8: Courtesy of CDC/Dr. Makonnen Fekadu.
Figure 24.10: Courtesy of Dr. Frederick Skvara/Visuals Unlimited, Inc.
Figure 24.11: Courtesy of EM Unit, Veterinary Laboratory Agency/Science Photo Library.
Figure 24.12: Courtesy of CDC/Dr. Leanor Haley.
Figure 24.13: Courtesy of Dr. Andrew Bollen & Dr. Walter Finkbeiner.

Chapter 25
Chapter opening quote: From Buckman R (2003) Human Wildlife: The Life That Lives on Us. Reprinted with permission from Johns Hopkins University Press.
Figure 25.4: Courtesy of CDC/Dr. Edwin P Ewing Jr.
Figure 25.6: Courtesy of CDC.
Figure 25.8: Courtesy of CDC/James Gathany.
Figure 25.9: Courtesy of CDC.
Figure 25.10: Courtesy of Dr. MA Ansary/Science Photo Library.
Figure 25.11: Courtesy of Eye of Science/Science Photo Library.
Figure 25.12: Courtesy of CDC.

Chapter 26
Chapter opening quote: From Cowles D & Cowles D (1997) Miracle Victory over the Flesh-Eating Bacteria. Gibbs Smith Inc.
Figure 26.4: Courtesy of S Burdette. In www.antimicrobe.org; Empiric, 2008. Esun Technologies, Pittsburgh PA.
Figure 26.5: Courtesy of Dr. P Marazzi/Science Photo Library.
Figure 26.6: Courtesy of Dr. P Marazzi/Science Photo Library.
Figure 26.7: Courtesy of Science Photo Library.
Figure 26.8: Courtesy of Stevie Grand/Science Photo Library.
Figure 26.9: Courtesy of CDC.
Figure 26.10: Courtesy of CDC/Dr. John Noble Jr.
Figure 26.11: Courtesy of Dr. P Marazzi/Science Photo Library.
Figure 26.12: Courtesy of Dr. P Marazzi/Science Photo Library.
Figure 26.13: Courtesy of CDC/Susan Lindsley.
Figure 26.14: Courtesy of CDC.
Figure 26.15: Courtesy of Science Photo Library.
Figure 26.16: Courtesy of CDC.
Figure 26.17: Courtesy of Gilles San Martin.
Figure 26.18: Courtesy of Western Ophthalmic Hospital/Science Photo Library.
Figure 26.19: Courtesy of Sue Ford/Science Photo Library.
Figure 26.20: Courtesy of CDC/J Pledger.

Index

Page numbers in boldface refer to major discussion of a topic; page numbers followed by F refer to figures, and those followed by T refer to tables.

A

ABC transport systems 204
Abscesses
    brain 596F, 597F, 598
    diverticular 544
    eyelid 656
    skin 645
    tonsils 513
Acanthamoeba
    keratitis 656, 657T
    Legionella reservoir 523
    meningoencephalitis 610, 610F
Acetaldehyde 43
Acetyl-coenzyme A (acetyl-CoA) 40
Aciclovir *see* Acyclovir
Acid-fast stain 51F, **53**, 53F, 519F
Acidophiles 19, 216, 221T
Acids **19–20**
    causing dental caries 541
    denaturation of proteins 438, 438F
Acne 643T, **645–646**
    comedonal 645
    inflammatory 645–646, 645F
    nodular cystic 646
Acquired immune deficiency syndrome
        see AIDS
Acridine 257
Actin 63, 63F, 86
Actinomycetes (*Actinomyces*) 540, 541
Activation energy 34–35, 34F
Active immunity 405–406
Active immunization 309
Active transport **203–204**, 203F
Acute disease 132T, 139
Acute-phase proteins 370
Acute-phase response **370**
Acyclovir 473F, **474**, 587, 652
Adaptive immune response 11, 352, **383–408**
    cells involved 386–388, 386F
    cellular cooperation 403F
    components 384–388
    course **403–404**
    natural and artificial **405–406**
    relationship with innate immunity 406–
        407, 407F
    viral infections 295, 295F
    see also Cellular immune response;
        Humoral immune response;
        Lymphocytes
Adenine (A) 25, 26F
    specific pairing 27, 27F, 237, 237F
    structure 237F
Adenoids, surgical removal 514
Adenosine 27
Adenosine diphosphate (ADP) 27
Adenosine monophosphate (AMP) 27
Adenosine triphosphate *see* ATP
Adenoviruses
    DNA replication 282
    host cell attachment 278, 278F
    incubation period 294T
    intestinal infection 556T
    persistent infections 296T
    recombinant DNA vaccine 310
    structure 274F
    vaccine 308T

Adherence (attachment)
    bacteria 81–83, 81F, 82F
        clinical significance 193–194
        mechanisms 81, 82T
        structures mediating 193–195, 193F,
          194F
    fungi 343
    phagocyte–pathogen 367–368, 368F
    viruses **276–278**
Adhesins 81
ADP (adenosine diphosphate) 27
Aedes aegypti 157, 632
Aedes albopictus 157
Aerobes 219
    obligate 220, 221T
Aerobic growth conditions 112
Aerobic respiration 33, 40
    ATP generation 41
    final electron acceptor 40
Aerotolerant bacteria 220, 221T
African trypanosomiasis *see* Sleeping sickness
Agammaglobulinemia, X-linked 421T
Agar 222
Age
    disease transmission and 102
    susceptibility to infection and 103
    urinary tract infections and 573, 573F
Agglutination test 119, **120**, 120F, 122T
Aging *see* Elderly
AIDS (acquired immune deficiency syndrome)
    103, **414–420**
    development 417F, 418
    emergence 149T
    failure of host defense 11
    incubation period 294T
    Pneumocystis pneumonia 414, 529
    see also Human immunodeficiency virus
AIDS dementia (complex) 419, 606, 606T
Air
    filtration 452
    ultraviolet irradiation 454
Air travel 4, 153, 489
Airborne infection 100
Albendazole 332, 337, 480
Alcoholic fermentation 42, 43, 44F
Alcohols 441, 443T, **445**
Aldehydes **447**
Alkaline phosphatase 118
Alkalophiles 221T
Alkylation, DNA 257
Allergens 426, 426T
Allergic conjunctivitis 429
Allergic reactions 413, **426–429**
    clinical effects 428–429, 428F
    effector mechanisms 427–428
    IgE-mediated 426, 426T, 427–429
    phases 428, 428F
    see also Hypersensitivity
Allergic rhinitis (hay fever) 426, 426T, 429
Allolactose 254, 254F
Allosteric inhibition 37, 37F
    irreversible 37
    regulation of *lac* operon 253–254
Alpha toxin 89T
Alveolar macrophages 362, 362F
Amantadine
    influenza therapy 527, 527T

    mechanism of action 473F, 476
    resistance 165–166, 476, 500
Amastigotes 633–634
Amblyomma americanum 627
Amebiasis **327–328**
    pathogenesis 328
    prevalence 318T, 319, 327
    treatment 328, 479
    see also Entamoeba histolytica
Amebic dysentery 536
Amebic meningoencephalitis, primary (PAM)
    317, **610**
Ameboids 319, 327
American trypanosomiasis *see* Chagas'
    disease
Amino acids 23, 23F
    attachment to tRNA 247F, 248
    bacterial requirements 219T
    sequence 24, 25F
Aminoacyl-tRNA synthetases 248
para-Aminobenzoic acid (PABA) 219T,
    471–472
Aminoglycosides 462T
    mechanism of action **470**, 470F
    resistance **493**
2-Aminopurine 258F
Amoxicillin 463, 463F
AMP (adenosine monophosphate) 27
AmpC β-lactamase **492–493**
AmpG gene 493
Amphitrichous bacteria 198, 198F
Amphotericin B 477, 531
Ampicillin 463, 463F
    resistance 490, 490T, 497T
Anabolism 32, 32F, **43**
Anaerobes **220–221**
    culture methods 220–221, 220F
    facultative 33, 219, 220, 221T
    obligate 220, 221T
    soft tissue infections 644F
Anaerobic growth conditions 112
Anaerobic metabolism 33, 40
    fermentation 42–43
    final electron acceptors 40
Anaphylactic shock 429
Anaphylaxis, systemic 426T, 429
Ancylostoma duodenale 319, 566–567, 566F
Anemia
    acute hemolytic 427
    hookworm infections 567
    malaria 324
Anergy 388
Anidulafungin 477
Animal feeds, antibiotics in **488**
Animals
    bites 301, 604, 644
    pet, infected feces 99, 563
    reservoirs of infection 98, 98T
    see also Zoonotic diseases
Anions 17
Anopheles mosquitoes 322, 323
Anthrax
    bioterrorism incidents (2001) 171, 207
    as bioweapon 5, **172**, 522
    cutaneous 172, **647–648**
    gastrointestinal 172
    on Gruinard Island 435
    inhalational (respiratory) 5, 172, **522–523**

toxin **88,** 89T
vaccine 172
see also Bacillus anthracis
Anti-helminthic drugs **480–481**
Anti-protozoan drugs **478–480**
  resistance to 500
Antibiotic resistance 185, **485–503**
  clinically dangerous **496–499**
  costs 488
  factors promoting development 105,
    486–489
  genes, recombinant plasmids 116
  mechanisms **492–496,** 497T
    drug inactivation 466, 492–493
    efflux pumping 493–494
    metabolic pathway alteration 496
    modification of drug target 495–496
  mutations, selection 486, 486F
  nosocomial infections 7–8
  plasmid-mediated transfer 486–487, 487F
  strategies for overcoming 490–491
  testing **490–491,** 490F, 491F
  timeline **489–490,** 490T
  urinary tract infections 573, 575
Antibiotic resistance islands 487
Antibiotics 11, **459–472**
  animal feeds **488**
  broad spectrum 461
  classes 462T
  defined 459
  development of new **501–503**
  discovery 8, 460, 460F
  effects on normal gut flora 105, 499, 499F
  inappropriate clinical use **487–488,** 499,
    499F
  investigating little-known 502
  microbial sources **460,** 461T
  minimum bactericidal concentration (MBC)
    491
  minimum inhibitory concentration (MIC)
    491, 491F
  narrow spectrum 461
  organ-transplant patients 104
  respiratory system infections 512F
  selective toxicity 64, 465
  spectra of activity **461–462,** 462T
  structure **462–464**
  susceptibility testing **490–491,** 490F
  targets **465–472,** 465F
    bacterial cell wall 187, 466–468
    bacterial metabolism 471–472
    bacterial nucleic acids 470–471
    bacterial plasma membrane 468
    bacterial protein synthesis 64, 469–470
    development of new 501
    dividing cells 228
    modification by bacteria **495–496**
  testing of new **502–503**
Antibodies
  activation of immune cells 401
  allergic reactions 426, 427
  to bacterial capsule 84
  complement activation 372, 373F, 397,
    397F
  diagnostic tests using 113, 119–121
  distribution and function **399–401**
  mechanisms of host protection 397, 397F
  monoclonal **116–118**
  neutralization by 397, 397F
  opsonization 374, 374F, 398
  primary response 401, 401F
  secondary in diagnostic tests 120–121
  secondary response 401, 401F
  see also Humoral immune response;
    Immunoglobulins
Antifungal drugs **476–478**
Antigen(s)
  antibody binding 398

diagnosis of pathogen-specific 113,
    119–121
  epitope 398
  nonself, recognition by Toll-like receptors
    357, 364
Antigen–MHC complex 392
Antigen presentation **391–393**
  activation of naive T cells 393–394, 393F
  dendritic cells 387, 387F
  MHC molecules 391–392, 391F
  superantigens 392
Antigen-presenting cells (APCs) 387, 391
  activation of T cells **393–394,** 393F
  antigen presentation 391F, 392
  response to infection 403–404, 403F
Antigen receptors 387–388, 388F
  gene rearrangements 389
Antigenic drift 296
Antigenic shift 164, 296
Antigenic variation 195, **295–296**
Antimalarials **478–479**
  resistance to 500
Antimicrobial chemical agents 440F, **441–448**
  defined 437
  determination of microbial death 439,
    439F
  evaluation 441–442
  factors affecting efficacy 439–440
  mechanisms of action 443–444, 443T
  selection 442–444, 443T
  targets 437–439
  types 444–448
  see also Antiseptics; Disinfectants
Antimicrobial compounds, naturally produced
    460, 502
Antimicrobial drugs **459–481**
Antiparasitic drugs **478–481**
  resistance to **500**
Antisepsis, terminology 436–437
Antiseptics 11, **441–448**
  defined 436
  evaluation methods 441–442
  factors affecting efficacy 439–440
  first 8
  mechanisms of action 443–444, 443T
  potency 441
  selection 442–444, 443T
  targets **437–439**
  types 444–448
Antiviral drugs **472–476**
  development of new 473, 473F
  mechanisms of action 281, 282, 473–476,
    473F
  resistance **500**
Antiviral proteins (AVPs) 376, 377F
Apoptosis
  induction by cytotoxic T cells 395
  induction by natural killer cells 365
  neutrophils 360
Arachnoid mater 594, 594F
Arboviruses **631–632,** 631T
  central nervous system infections 598T,
    606
  emergence 156
Archaea 59
Armadillos 229
ArrayTube 124F, 125F
Arsenic compounds 480
Artemisinin 479
Arthritis
  hepatitis B 561
  Lyme disease 624–625
Arthroconidium 339–340
Arthus reaction 427
Artificially acquired immunity **405–406**
Ascariasis (*Ascaris* infections) **332**
  pathogenesis 321, 332
  prevalence 318T, 319

Ascaris lumbricoides 331, 331F, **332**
  life cycle 332, 333F
  transmission 321, 322T
Aseptic, defined 436
Asexual reproduction
  fungi 338
  Plasmodium 324
  Toxoplasma 327
  see also Binary fission
Asian influenza pandemic (1957) 164
Aspergillosis 343, **532**
Aspergillus 532, 532F
Aspergillus fumigatus 104
Asthma 426–427, 426T, 429
Astroviruses 556T
Asymptomatic carriers 98
Ataxia telangiectasia 421T
Athlete's foot (tinea pedis) 341, 341F, 654
Atopy 426–427
ATP 20, **27–28**
  production 33, 41
    aerobic respiration 40
    bacteria 201, 201F
    electron transport chain 41F
    fermentation 42
    glycolysis 39
  structure 27, 28F
  utilization
    active transport 203, 203F
    anabolism 43
    efflux pumps 494, 494F
Atrichous bacteria 198
Attachment *see* Adherence
Augmentin® 459, 466, 485, 492
Aum Shinrikyo cult 177
Auto-induction 57
Autoantibodies 425, 425F
Autoclaves 450–451, 450F
Autoimmune disease **424–425**
  causes 424–425, 425F
  mechanisms 425, 425F
  organ-specific 424
  systemic 424
Autotrophy 32
Avian influenza (bird flu) 4–5, **165–166**
  drug resistance 165–166
  potential for human disease 166, 166F
  vaccine 166, 310
  virus (H5N1) 165–166, 166F
Axial filaments, bacterial **195–196,** 195F, 196F
  clinical significance 196, 208T
Axostyle 320F, 329
Azithromycin 469, 491F
  resistance 495
Azole antifungals **477**
AZT (azidothymidine) 473F, **474**
Aztreonam 467

B

B-cell lymphoma 420
B cells (B lymphocytes) 387–388
  activation by T cells 396, **402**
  activation of naive T cells 393F, 394
  antigen presentation 391F
  antigen receptors 397
    role in T-cell activation 394, 402
    structure 387, 388F
    see also Immunoglobulins
  clonal selection 388–389
  co-stimulatory signals 388
  course of infection 403F, 404
  differentiation and maturation 386F,
    388–389, 402
  Epstein–Barr virus infection 630
  humoral response 384, **396–403**
  inherited defects **421–422**
  memory 405
  peripheral lymphoid tissues 390, 390F

survival 389
Bacillary dysentery 536, 548, 549
Bacilli 49
Bacillus, antibiotic production 461T
Bacillus anthracis 172, 172F, 647
    as bioweapon 5, 172
    disease pathogenesis 522–523, 522F
    endospores 207
    toxin **88,** 89T
    see also Anthrax
Bacillus cereus 543F
    food poisoning 539T
    phagocytosis 368F
Bacillus subtilis 449T, 494
Bacitracin 468
Bacteremia 132T, 142, **616–617**
    common sources 616, 617F, 618T
    intravenous line and catheter 616–617
    transient 616
Bacteria **48–54**
    adherence *see under* Adherence
    anatomy **185–209,** 186F
    antibiotic production 460, 461T
    antibiotic resistance *see* Antibiotic
        resistance
    cell structure **59–60,** 59T, 60F
    classification 48–50
    complement-mediated lysis 372F
    encapsulated *see* Encapsulated bacteria
    endospores *see* Endospores
    genetic transfer mechanisms **259–265,**
        265T
    genetics **235–266**
    growth **213–232**
    host relationships 133–134
    identification tests 113, 113F
    intracellular movement **199**
    motility *see* Motility, bacterial
    multicell arrangements 49, 49F
    nuclear region **205,** 205F
    numbers
        measuring **228–229,** 228F, 229F, 230T
        required for infection 83
    pathogenicity and virulence **56–58**
    persistent infections **139–142,** 140T
    plasma membrane *see* Plasma membrane,
        bacterial
    ribosomes *see* Ribosomes, prokaryotic
    secretion **204**
    shapes 48–49, 49F
    size 48, 49F
    staining **50–54**
    structural analysis **501**
    thermal death times 449T
    toxins **87–93**
    see also Prokaryotes
Bacterial cell wall **186–192**
    additional components **188–190**
    as antibiotic target 187, **466–468**
    assembly **187,** 188F
    carbohydrates 20
    during cell division 187, 187F
    clinical significance **191–192,** 208T
    as disinfectant/antiseptic target 437
    endotoxins 91–92, 92F
    Gram-positive *vs.* Gram-negative bacteria
        188–190, 189F, 190T
    protection against host defenses 84
    structural components 186, 186F
    structures inside **199–208**
    structures outside **193–199**
Bacterial endocarditis *see* Infectious
    endocarditis
Bacterial infections
    blood and lymph **620–626**
    central nervous system 596, 597F,
        601–603

eye 657T
    gastrointestinal tract **542–555,** 542F, 543F
    immunocompromised host 104
    lower respiratory tract **516–525**
    recurrent 421
    reproductive system **577–585**
    respiratory system 510, **512,** 512F
    sexually transmitted 577F
    skin and soft tissue 643T, **644–648,** 644F
    specimen collection 229–230, 231T
    treatment 11, **459–472**
    upper respiratory tract **512–515**
    urinary system **573–576**
Bacterial vaginitis 579
Bacteriocidal agents 437, 442
Bacteriocins 105, 134
Bacteriophages 115
    genetic transfer 260–262
    infection cycle 275–276
    phagetyping 124
    therapeutic potential 276, 502
    toxin-encodings genes 275
Bacteriostatic agents 437, 442
Bacteriuria, chronic 573
Bacteroides 543F, 578F
BALT (bronchus-associated lymphoid tissue)
    385
Baltimore, David 280
Barrier defenses **352–356,** 356T
Base pairing, complementary 27, 27F
    DNA replication 240–241, 241F
    DNA structure 237, 237F, 238F
    proofreading by DNA polymerase 241–242,
        242F
    RNA structure 238, 239
    in transcription 246
    in translation 247F, 248, 251, 251F
Bases (alkaline) **19–20**
Bases (nucleotide) 25
    analogs, erroneous insertion 257, 258F
    complementary pairing *see* Base pairing,
        complementary
    deoxynucleotide formation 236, 236F
    DNA 25, 236–237
    excision repair 257–258
    hydrogen bonding between 27, 27F, 237,
        237F
    RNA 25, 238
    structure 26F, 237, 237F
Basophils 359F, 359T, **360–361**
    activation by IgE 400
    allergic reactions 427
Bed sores 105
Benzalkonium chloride 443T, 446F
Beta (β)-lactam antibiotics 463, 463F, 466–467
Beta (β)-lactam ring 462–463, 463F, 464F
Beta (β)-lactamase **492–493**
    action 492, 492F
    AmpC **492–493**
    antibiotics susceptible to 185, 466, 467
    extended spectrum 575
    inhibitors 466, 492
    regulation of activity 493
Betadine 443T, 446
Binary fission
    bacteria 83, 226, 226F
        cell wall assembly 187, 187F
    protozoans 319
Bio-barcode assays 125
Biofilms **57–58,** 58F
    dental 82–83, 83F, 540
    medical equipment 106
    role of slime layer 193
Bioinformatics 124
Biological molecules **20–28**
Biological weapons (bioweapons) **170–177**
    CDC categories 171, 171T

danger of 177
    essential features 170
    history 170
    see also Bioterrorism
Biopsy, tissue 231T
Biosensors 125
Biosynthetic reactions 43
Biotechnology **113–118**
    health-related applications 114F, 115
    history 114F
    industry 115
Bioterrorism 5, **170–178,** 170F
    anthrax letters (2001) 171, 207
    defined 170
    potential agents 171T, 172–177
    probability and effects of an attack 177
    warning signs 177–178, 178T
Bioweapons *see* Biological weapons
Bird flu *see* Avian influenza
Birds
    Chlamydophila psittaci 524
    influenza virus reservoir 163, 165, 166F
    West Nile virus transmission 157
Bismuth sulfate agar 224
Bisphenolics 443, 443F
Bites, animal 301, 604, 644
Bithionol 336
BK polyoma virus 296T
Black Death 54, 148, 174, 620
    see also Plague
Black water fever 324
Blake, William 572
Blastoconidia 339
Blastomyces dermatitidis 530
Blastomycosis **530**
Bleach, chlorine 442, 446
Blood
    circulating pathogens **616–620**
    specimen collection 231T
    virus dissemination via **302,** 302F
Blood agar 224T, 225–226, 225F
Blood and lymph infections **615–637**
    bacterial **620–626**
    extravascular 616
    intravascular 617, **618–619**
    parasitic **633–636**
    rickettsial **626–628**
    viral **629–633**
Blood–brain barrier 594–595
Blood flukes *see* Schistosomes
Body fluids
    hepatitis B transmission 560–561
    HIV transmission 415, 415T
    virus transmission **305**
Boiling water 450T, 451
Boils 86, 645
Bone marrow
    B-cell development and maturation 388
    stem cells 359F, 386F
Bordetella bronchiseptica 53F
Bordetella pertussis 521, 522
    attachment to cilia 64F
    see also Pertussis
Borrelia 625
Borrelia burgdorferi 623–625
    axial filaments 196, 196F
    life cycle 623F, 624
    transmission 623
    see also Lyme disease
Borrelia recurrentis 625
Botulinum toxin **89–90,** 89T, 602–603
    bioterrorism potential 173
    cosmetic use (Botox®) 75
Botulism **602–603**
    adult infectious 173
    as bioweapon **173**
    foodborne 173, 602–603

infant 173, 173F, 593, 603
  inhalational 173
  pathogenesis 602–603
  treatment 603
  wound 173, 603
Bovine spongiform encephalopathy (mad cow disease) 153, 168, 608–609
Bradykinin 515
Brain 594–595
  abscess 596F, 597F, 598
  malaria 324–325, 326F
  microglial cells 362F, 363
  swelling (edema) 595, 595F
  Trypanosoma brucei infection 330
Brandeis, Louis 436
Brilliant green 224
5-Bromouracil 258F
Bronchus-associated lymphoid tissue (BALT) 385
Broth dilution test 491
Brown, David 594
Brucella 622–623
Brucellosis **622 623**
Brugia malayi 635–636
BSE *see* Bovine spongiform encephalopathy
Buboes
  bubonic plague 174, 174F, 621, 621F
  lymphogranuloma venereum 579F
Buckman, Robert 616
Budding
  enveloped viruses **288,** 288F
  yeasts 339, 339F
Burkitt's lymphoma 630, 630F
Burnet, Frank Macfarlane 149–150
Burns 97, **104,** 353

C

C-reactive protein (CRP) 370
C3
  cascade initiated by **374–375,** 374F
  cleavage 372, 373, 373F
  inherited deficiency 375, 375T
C5 374
Cadherin 86–87
Caliciviruses 556T, 557–558, 562
Campylobacter
  central nervous system infection 597F
  enteritis 542, 543F, **553–554**
  sources and transmission 542F
Campylobacter jejuni 552T, 553–554, 553F
Canaliculitis 656
Cancer
  cytotoxic chemotherapy 103–104
  HIV-infected patients 418, 419–420
  viruses and **311**
Candida 642
Candida albicans 9, 342
  cutaneous infections 643T, 653
  immunocompromised host 104, 653
  pathogenesis of disease 343, 344, 589–590, 653
  vaginal infections **589–590**
Candidiasis
  cutaneous 643T, **653**
  mucocutaneous 342, 653
  oral (thrush) 342, 653, 653F
  treatment 590, 653
  vaginal 342, **589–590**
Candle jars 221
Cannibalism 168
Cannulas, as portals of entry 80
CAP protein 254, 254F
Capreomycin resistance 497T
Capsids 272, **273–274,** 273T
  empty 287, 289
  subunit assembly **287**
Capsomeres 273, 273T
  assembly 287

Capsule, bacterial 193, 193F
  clinical significance 193–194
  host defenses **84**
  negative stain **52,** 52F
  protection against host defenses 84F
  see also Encapsulated bacteria
Carbapenems 462T, **467**
Carbohydrates **20–21**
Carbolic acid 8
Carbon (C)
  bacterial requirements **218,** 218T
  covalent bonding 17–18, 18F
  formation of biological molecules 20
Carbon dioxide (CO_2), synthesis 40
Carbon source 113
Carbuncle 645
Carrier molecules 35–36
Carriers, disease 98
Caspofungin 477
Catabolism 32, 32F, **38–43**
Catalase 66, 220
Catalysts 34
Catheters
  associated bacteremia 616–617
  immunocompromised host 103, 104
  nosocomial infections 106
  as portals of entry 80
  urinary, infections related to 572
Cationic detergents 443T
Cations 17
Cats, domestic
  infected feces 99, 563
  Toxoplasma life cycle 326–327
CCR5 chemokine receptor 415
CD4 387, 415
CD4 T cells *see* Helper T cells
CD8 387
CD8 T cells *see* Cytotoxic T lymphocytes
CD40 ligand 394
CDC *see* Centers for Disease Control
Ceftriaxone 464F, 497T
Cefuroxine 464F
Cell culture, viral 310–311
Cell division
  bacterial 83, 226, 226F
    antibiotics targeting 228
    cell wall assembly 187, 187F
    growth curve 227–228, 227F
    snapping 514, 514F
    see also Growth
Cell lines 311
  continuous 311
  primary 311
  semi-continuous 311
Cell wall
  bacterial *see* Bacterial cell wall
  fungal 338, 477
Cells, host *see* Host cells
Cellular immune response 384, **393–396**
  course during infection 403–404, 403F
  inherited defects **422**
  primary 393
Cellulitis (erysipelas) 643T, **645**
Centers for Disease Control (CDC)
  categories of biological weapons 171, 171T
  definition of bioterrorism 170
  nationally notifiable diseases 110, 111F
  universal precautions 106, 107T
Central nervous system (CNS)
  anatomy **594–595,** 594F
  infections **593–611**
    bacterial 596, 597F, **601–603**
    common pathogens 596–597
    fungal 596, **609**
    parasitic **610**
    persistent/slow viral 298, **606–607,** 606T

  prions *see* Prions
  sources and routes 596–598, 596F
  treatment 598–599
  viral 597, 598T, **603–607**
  microglial cells 362F, 363
  virus dissemination **303,** 303T
Central tolerance 389
Cephalexin 464F
Cephalosporins 462T
  chemical modifications 464, 464F
  mechanism of action **466–467**
  resistance 490T, 493
Cephalosporium 461T
Cephalothin 464F
Cephtazidime 464F
Cercariae 335, 336, 337
Cerebrospinal fluid (CSF) 231T, 594, 594F
Cervical cancer 588
Cervicitis 576, 579
  Chlamydia 585
Cestodes (tapeworms) **334–335**
  morphology 320, 321F, 334, 334F
  pathogenesis of disease 334–335
  prevalence 318T
  treatment of disease 335
Chagas' disease (American trypanosomiasis) 329, **633–635**
  pathogenesis 634
  prevalence 318T, 319
  treatment 634–635
Chagoma 634
Chain, Ernst 460
Chancre
  syphilis 580, 580F
  trypanosomiasis 330
Chancroid 578T
Chaperones
  bacterial protein synthesis 469
  viruses 280, 287
Chediak–Higashi syndrome 423, 423T
Chemical agents, antimicrobial *see* Antimicrobial chemical agents
Chemical bonding **16–19**
Chemical requirements, bacterial **218–221,** 218T
Chemiosmosis 40, **41**
Chemistry **15–28**
Chemoautotrophs 32
Chemoheterotrophs 32
Chemokines 358, 378
Chemostat 227–228, 227F
Chemotaxis 367
Chemotherapy, cytotoxic 103–104
Chickenpox (varicella) 383, 643T, **650**
  incubation period 294T
  latency 139, 298, 298F
  vaccine 308T, 383
  see also Varicella-zoster virus
Childbed fever 7
Children
  cryptosporidiosis 564
  enterobiasis 331, 332
  enterovirus infections 558
  gastrointestinal diseases 536, 537, 556T
  giardiasis 563, 564
  hookworms 567
  parainfluenza 516
  respiratory syncytial virus infections 527
  rotavirus infections 557
  whipworms 565–566
  see also Infants
Chitin 338
Chlamydia trachomatis **583–585**
  genital infections 577F, 578–579, **583–585**
  lymphogranuloma venereum 579, 579F
  neonatal conjunctivitis 585, 658
  persistence 140T
  replication cycle **584,** 584F

trachoma 657, 657F, 657T
Chlamydophila pneumoniae 512F, **518**
Chlamydophila psittaci 512F, 524
Chloramines 446
Chloramphenicol
  mechanism of action 469, 470F
  resistance 490T, 497T
Chlorhexidine 443T
Chlorine
  chemistry 16–17, 17F
  disinfectant 442, 446, 448
Chloroquine 478–479
  resistance 500
Chloroxylenol 443T
Chocolate agar 229, 229F
Cholera 90, **552–553**
  emergence 149T
  pandemic 537, 552
  pathogenesis 552–553
  toxin 89T, **90**, 553
  treatment 553
  see also Vibrio cholerae
Cholesterol 23, 60
Chromatin 67
Chromosomes
  bacterial 205
    mutation hot spots 257
    plasmid incorporation 264, 264F
    replication 243, 244F
  eukaryotic 67
  virus-induced breaks 93
Chronic disease 132T, 139
Chronic granulomatous disease 376, 423,
  423T
-Cidal agents 437
Cidofovir 473F, 475
Cilastatin 467
Cilia
  bacterial **63–64,** 63F
    role in infection 64, 64F
  protozoan 319
Ciliates 319
Ciprofloxacin 471
  resistance 495, 497T
Circulating pathogens **616–620**
Circulatory system, human 616F
Cirrhosis of liver 561, 561F, 562
Citric acid cycle *see* Krebs cycle
Citrobacter 512F
Clarithromycin 469, 495
Classification of organisms **48**
Clavulanic acid 492
Climate, disease transmission and 102
Climate change, emerging diseases and 152
Clindamycin 497T
Clinical specimens, collection 229–230, 231T
Clinical trials 118
Clonal deletion 389, 389F
Clonal selection **388–389,** 389F
Cloning 114F, 115
Clonorchiasis **336–337**
Clonorchis sinensis (liver fluke) 322T, 336–337
  host defenses 361F
  life cycle 336, 336F
  morphology 321F
  prevalence 318T
Clostridium 48, 543F
Clostridium botulinum 539T, 597F, **602–603**
  endospores 207, 449T
  optimal growth temperature 215T
  phage encoding of toxin 275
  refrigeration 451
  toxin *see* Botulinum toxin
  see also Botulism
Clostridium difficile 499, 537, 537F
  exogenous infection 543–544, 543F
  nosocomial infections 106, 537
Clostridium perfringens 644F

enzymes 85, 86
food poisoning 539T
gas gangrene **647**
lethality of oxygen to 220, 220F
toxins 85, 89T
Clostridium tetani 597F, **601–602**
  accidental infection 10
  endospores 449T
  morphology 601, 601F
  reservoir of infection 99
  toxin production 89T, 90, 601
  see also Tetanus
Clotrimazole 477
Clue cells 579, 579F
CNS *see* Central nervous system
Co-receptors, virus 277, 279
Co-stimulatory signals 388
  activation of naive T cells 393F, 394
Coagulase 85–86, 85F
Coagulase-negative staphylococci
  bacteremia 617F
  infectious endocarditis 618T
  urinary tract infections 574F
  ventriculo-peritoneal shunt infections 597F
  see also Staphylococcus epidermidis
Coagulation disorders, viral hemorrhagic fever
  160
Cocci 49, 49F
Coccidioides immitis 343–344, 531
Coccidioidomycosis (coccidiomycosis)
  342–343, **531**
  pathogenesis 343–344, 531, 531F
  treatment 531
Coccobacilli 49
Codons 244–245, 245F
  initiation 244, 245F, 247, 250, 250F
  stop 244, 245F, 251
Coenzyme A (CoA) 40
Coenzymes 35–36, 36F
Cofactors 35–36
Cold, common *see* Common cold
Cold sores 271, 303, 303F, 651, 651F
Cold temperatures 38
  bacterial growth requirements 214F, 215
  controlling microbial growth **451**
Coliforms 543F
  brain abscess 597F
  fecal 213, 228
  identification 213
  soft tissue infections 644F
  see also Escherichia coli
Colitis
  hemorrhagic 547
  pseudomembranous 537
Collagenase 86
Colony morphology 113
Color, bacterial colonies 50
Combinatorial chemistry 502
Comedones 645
Comedos 645
Commensalism 133, 134T
Common cold 294T, 295, **515**
  inappropriate use of antibiotics 487
  transmission 100F
  see also Rhinoviruses
Common-source outbreaks 109, 109F
Communicable disease **138–139**
Competitive inhibition 36, 36F
Complement system **372–375**
  activation by antibodies 372, 373F, 397,
    397F
  alternative (properdin) pathway 372–373,
    373F
  C3 to C9 cascade 374–375, 374F
  classical pathway 372, 373F
  defenses against 375
  inherited deficiencies 375, 375T, 421T, 422,
    423F

lectin-binding pathway 373–374
mast cell activation 365
Complementary base pairing *see* Base pairing,
  complementary
Complete blood count (CBC) 358
Compromised host *see* Immunocompromised
  host
Concentration gradients
  active transport against 203, 203F
  proton motive force 41, 201, 201F
  simple diffusion 202
Congenital infections *see* Fetal infections
Conidia 338, 339–340
Conidiobolomycosis, paranasal 342
Conjugated molecules 118, 120, 121
Conjugation **262–264,** 263F, 265T
  pilus 195, 262, 262F, 263F
  transfer of antibiotic resistance 486–487,
    487F
Conjunctiva, normal microbial flora 133T
Conjunctivitis
  allergic 429
  bacterial 656, 657T
  neonatal 585, 658
Constitutive genes 252, 253
Contact lenses 656
Contact transmission **99–100,** 100F
  direct 99
  droplet 100, 100F
  indirect 99–100
Contagious disease **138–139**
Continuous cell lines 311
Convalescence, period of 137, 138F
Corynebacterium 617, 642
Corynebacterium diphtheriae 514
  antibiotic sensitivity 512F, 515
  morphology 514, 514F
  toxin production 88, 89T, 261, 275, 514
  volutin granules 27
  see also Diphtheria
Coughing
  beneficial function 400, 511
  virus transmission 304
  whooping cough 522
Covalent bonds **17–18,** 18F
  nonpolar 17
  polar 17, 17F
Cowles, David L. 642
Cowpox 307
Coxiella burnetii 512F, 524
  endospores 206
Crabs (body lice) 655
Creutzfeldt–Jakob disease (CJD) 608–609
  emergence **168–169,** 168F
  variant (vCJD) 168, 168F, **169,** 608–609
Crewe, Megan 294
Crichton, Michael 214
Crick, Francis 244
Cristae 65, 65F
Croup 516
Cryptococcosis **609,** 609F
Cryptococcus neoformans 609
Cryptosporidiosis 149T, **564–565**
Cryptosporidium 564, 564F
Cryptosporidium parvum 564, 565
Crystal violet 51T, 52
Culture
  microbiological 112–113
  viral **310–311**
Culture media *see* Growth media
CXCR4 chemokine receptor 415
Cyclic AMP (cAMP) 254
Cysticercus 335
Cystitis 573, **575**
Cysts
  Entamoeba 321, 328
  Giardia 563
  nodular cystic acne 646

protozoan 319
Toxoplasma 327
see also Oocysts
Cytocidal effect 93
Cytokines **357–358**
  genetic variation 378
  natural killer cells 365–366, 366F
  T cell 394
  viral hemorrhagic fevers 160
Cytomegalovirus (CMV) **629**
  central nervous system 598T
  congenital infection 629
  neonatal infection 629
Cytopathic effect (CPE) 93, 93F
Cytoplasm **62–66**
  membrane-enclosed structures 65–66
  non-membrane-bound structures 62–64
  role in infection 62
  transport of viral components **279–280,**
    285
Cytosine (C) 25, 26F
  specific pairing 27, 27F, 237, 237F
  structure 237F
Cytoskeleton **63**
  bacterial movement using 199
  role in infection 63, 63F, 86–87
Cytosol 62
Cytotoxic chemotherapy 103–104
Cytotoxic T lymphocytes (CTLs; Tc) (CD8
    T cells) 387, **394–395**
  anti-HIV response 417, 418
  antigen presentation to 391F, 392
  during course of infection 403F, 404
  functions 394–395, 394F
  target-cell killing 395
  viral escape mutants 297
Cytotoxins 88, 89T
  distending (*Campylobacter*) 553
  enterobacterial 544
  vacuolating 554

D

Dacryocystitis 656
Daptomycin 468
Databases 123–124
Death, microbial **439–440**
  factors affecting rate 439–440
  heat sterilization 449, 449T
  rate, determination 439, 439F
Death phase, bacterial growth curve 227F, 228
Decline, period of 137, 138F
Deep sequencing 123
Deer fly 657
Deer mouse 152
Defensins 502
Degerming 436
Dehydration synthesis 19
  carbohydrate formation 21, 21F
  fat formation 21, 22F
  peptide formation 23, 24F
Delavirdine 475
Dendritic cells
  activation of naive T cells 393–394, 393F
  adaptive immunity **386–387,** 387F
  antigen presentation 376F, 387
  development 359F, 386F
  HIV infection 416, 416F
  innate immunity 359T, **363–364**
Dengue fever 631T, **632**
  emergence 152, **157–158**
  pathogenesis 158
Dental caries (tooth decay) 15, **540–541**
  bacterial adherence 81–83, 83F
Dental infections **540–542**
Dental plaque 540, 541F
  formation 82–83, 83F, **540,** 540F
  in gingivitis and periodontitis 541

immunocompromised patients 542
Deoxynucleotides 236, 236F
Deoxyribonucleic acid *see* DNA
Deoxyribose 25, 25F, 236
Dermacentor andersoni 627
Dermacentor variabilis 627
Dermatophagoides pteronyssimus (house
    dust-mite) 426, 426F
Dermatophytes 653–654
Dermatophytosis 643T, **653–654**
Dermis 642F
Detergents 437–438, 446–447
  cationic 443T
Diabetes mellitus, type 1 insulin-dependent
    424, 425, 425F
Diagnostics **112–126**
  growth-based methods 112–113
  new approaches 124–125
  nucleic acid-based 121–124, 122T
  sample collection 112
  serology-based 113, 118, **119–121,** 122T
Diapedesis 360, 360F, 370
Diaper rash 653
Diarrhea 5–6
  amebiasis 328
  Campylobacter 553
  cryptosporidiosis 565
  E. coli 5, 6, 545–547
  giardiasis 564
  shigellosis 549
  traveler's **537**
  viral infections 555
  watery **536**
Didanosine 474
Diethyl carbamazine 636
Diffusion
  facilitated 202–203, 203F
  simple 202
DiGeorge syndrome 421T, 422
Digestive system *see* Gastrointestinal tract
Diiodohydroxyquin 478–479
Dilution tests, antimicrobial **442,** 491
Dimorphism, fungal 340
Diphtheria **514–515**
  antitoxin 515
  pathogenesis 514
  re-emergence 149T
  toxin **88–89,** 89T, 261, 514
  treatment 515
  vaccination 514
Diphtheria, tetanus and pertussis (DTaP)
    vaccine 514, 521
Diphyllobothrium 322T
Diplococci 49, 49F
Dipstick tests 121
Disaccharides 20, 21, 21F
Disease 55, **131–144**
  defined 132
  development 137–142
  duration 139
  etiology 132, **136–137**
  symptoms of 137
  transmission *see* Transmission, disease
  see also Infectious diseases
Disinfectants 11, **441–448**
  bacterial resistance to 441
  defined 436
  evaluation methods 441–442
  factors affecting efficacy 439–440
  mechanisms of action 443–444, 443T
  potency 441
  selection 442–444, 443T
  targets **437–439**
  types 444–448
Disinfection
  by boiling 451
  defined 436
  radiation 453F, 454

terminology 436–437
  see also Disinfectants
Disk diffusion method
  antibiotic susceptibility testing 490–491,
    490F
  disinfectant/antiseptic testing **442,** 442F
Disseminated intravascular coagulation (DIC)
    92, 160
Disulfide bridges 24, 24F
Diverticular abscesses 544
DNA (deoxyribonucleic acid) **236–245**
  bacterial
    as antibiotic target **470–471**
    clinical significance 205–206, 208T
    localization 205
    release using KOH 235
    sequencing 501
  as disinfectant target 439
  eukaryotic cells 67
  genetic code **244–245,** 245F
  microarray analysis 123, 123F
  naked 260
  recombinant **115–116**
  sequence identification **123–124**
  Southern blotting 123
  strands 26–27, 26F
    antiparallel orientation 26–27, 27F, 237,
      238F
    lagging 242, 243F
    leading 242, 243F
  structure 25–27, **236–237**
    building blocks 25, 25F, 236
    double helix 27, 27F, 236, 236F
  supercoiling **239–240**
  viral 67
DNA damage
  mechanisms **257**
  radiation-induced 257, 452, 453F
  repair **257–258**
  see also Mutations
DNA ligase 243
DNA photolyase 258, 259F
DNA polymerase **240–242**
  antiviral drugs targeting 474, 475
  proofreading 240–241, 241F
  replication fork 242–243, 243F
DNA repair **257–258**
  base excision 257–258
  nucleotide excision 258, 258F
  photoreactivation 258, 259F
DNA replication **239–244**
  adding new nucleotides 240–241, 241F
  DNA polymerase 240–242
  erroneous insertion of base analogs 257,
    258F
  errors 241, 245, **256–257**
  initiation 243
  primer:template junction 240, 240F
  proofreading 241–242, 242F
  replication fork 242–243, 243F
  strand unwinding and separation 239–240
  termination 243, 244F
  viruses 281–282
  see also DNA synthesis
DNA synthesis
  antifungals targeting 478
  antivirals targeting 474–475
  lagging strand 242–243, 243F
  leading strand 242, 243F
  viruses 281–282
  see also DNA replication
DNA viruses 272, 272F
  double-stranded DNA (dsDNA) **282**
  replication of genome **281–282**
  single-stranded DNA (ssDNA) **282**
  transcription **282–283**
  transport into nucleus 280
DNase 261

Dogs
    bites 301, 644
    infected feces 99, 563
Double-diffusion precipitation 119F
Doxycycline 470
DRACO **476**
Droplet transmission 100, 100F, 304, 304F
Drugs
    causing autoimmunity 424
    inducing immunosuppression 103–104, 424
    type II hypersensitivity 427
Duodenal ulcers 554, 554F
Dura mater 594, 594F
Dust mites 426, 426F
Dysbiotic communities 136
Dysentery **536**
    amebic 536
    bacillary 536, 548, 549
Dysplasia, cervical 588
Dysuria 574, 575

E

E test 491, 491F
Eastern equine encephalitis (EEE) 598T, 606, 631T
Ebola hemorrhagic fever **632–633**
    2014 outbreak in West Africa 160, 632
    diagnostic testing 160–161
    emergence 149T, 158
    pathogenesis 159–160, 632–633
    treatment 160, 161, 633
Ebola virus 158F
    animal reservoir 159
    as bioweapon 176–177
    vaccine development 160, 177
Ecchymoses, viral hemorrhagic fevers 160
Echinocandins 477
Efflux pumps 203–204, 204F
    antibiotic **493–494**
    classification 493
    mechanism of action 494, 494F
Eflornithine 480
Elderly
    rotavirus infections 557
    susceptibility to infection 103
    tuberculosis 162
    urinary tract infections 573F
    viral infections 307
Electroimmunodiffusion 119
Electron microscopy, viruses 272
Electron transfer 32
Electron transport chain **40,** 41F
    bacterial plasma membrane 201, 201F
    generation of proton motive force 41
    mitochondrial 40, 41F
    prokaryotes 41
Elementary body (EB) 584, 584F
Elephantiasis 635, 635F, 636
ELISA (enzyme-linked immunosorbent assay) **121,** 121F
Embolus 619, 620F
Embryonic stem cells 115
Emerging infectious diseases 9, **148–161, 166–169**
    antibiotic resistance 496–498
    defined 148
    drivers of emergence 148–149
    examples 149T
    interspecies transfer 154
    mechanisms of emergence 151–154
    transition patterns 150–151
    viral infections 154–161
Encapsulated bacteria 83–84
    host defenses **84**
    protection against host defenses 84F
    repeated infections 421

resistance to phagocytosis 84, 84F, 193–194, 369
    see also Capsule, bacterial
Encephalitis 603
    Acanthamoeba 610, 610F
    rabies 604
    viral **606**
Endemic diseases 108
Endocarditis, infectious see Infectious endocarditis
Endocytosis **67–69**
    parasite-directed 582
    receptor-mediated 68, 68F
    role in infection 69
Endoflagella see Axial filaments
Endogenous infections 543, 543F, 544–545
Endoplasmic reticulum (ER) **65–66**
    role in infection 66
    rough (RER) 65
    smooth 65
    viral protein trafficking 286, 286F
Endosomes, virus uncoating 278F, 279
Endospores 53, **206–207**
    clinical significance 207, 208T
    disinfectant/antiseptic resistance 440
    formation (sporogenesis) 207, 207F
    germination 207
    radiation resistance 453
    staining 51T, **53–54,** 54F
    thermal death times 449T
    see also Spores
Endosymbiotic theory 65
Endothelial cells, binding of white cells 360, 360F
Endotoxins **91–92,** 92F
    clinical significance 192
    compared to exotoxins 91T
    complement activation 373, 373F
Energy
    of activation 34–35, 34F
    bacterial production 200–201, 201F
    biological molecules as sources 20
    fueling anabolic reactions 43
    peptide bond formation 248
    storage in ATP 27–28, 28F
Entamoeba dispar 327
Entamoeba histolytica **327–328**
    life cycle 327–328
    transmission 321, 322T
    see also Amebiasis
Enteric fever **536,** 549–550
    see also Typhoid fever
Enterobacter 512T
Enterobacteriaceae 544–552
    cell-surface antigens 544, 544F
    oxygen requirements 33
    virulence factors 544–545
Enterobiasis **331–332**
    pathogenesis 331–332
    treatment 332
Enterobius vermicularis (pinworm) **331–332,** 332F
    prevalence 318T
Enterococcus (enterococci)
    intestinal infections 543F
    nosocomial infections 106, 106F
    urinary tract infections 574F
    vancomycin-resistant (VRE) **498**
Enterotoxins 88, 89T, 544
Enteroviruses
    central nervous system infections 598T, 600, 605
    digestive system infections **558**
    dissemination within host 302F
    incubation period 294T
    persistent infection of immunodeficient 606–607, 606T
Entry, portals of see Portals of entry

Enveloped viruses **274–275**
    attachment 278
    budding **288,** 288F
    complement-mediated lysis 375
    intracellular trafficking 285
    penetration and uncoating **279**
    see also Viral envelope
Environment(s)
    emerging diseases and **151–152**
    reservoirs of infection 99
Environmental Protection Agency (EPA) 115
Enzyme immunoassay (EIA) 119, 121
Enzyme-linked immunosorbent assay (ELISA) **121,** 121F
Enzymes 25, **34–38**
    active site 35, 35F
    allosteric site 37, 37F
    coenzymes and cofactors 35–36
    concentration effects 38
    as disinfectant/antiseptic targets 438, 438F
    factors affecting activity 38
    inhibition 36–37
    pathogenic bacteria **84–87,** 85F
    properties 34–35, 34F
Enzyme-substrate complex 35, 35F
Eosin methylene blue (EMB) agar 224T, 225, 225F
Eosinophils 359F, 359T, **361**
    allergic reactions 428
    parasitic infections 361, 361F
Epidemics 108–109
    common-source outbreaks 109, 109F
    propagated 109, 109F
    viral infections 295
Epidemiological studies **110**
    analytical 110
    descriptive 110
Epidemiology 10, **108–111**
Epidermis 642, 642F
    barrier function 352–353, 353F
    virus transmission via **305**
    see also Skin
Epididymitis 576, 579
Epiglottis **354,** 356T
Episome 630
Epitope 398
Epizootics, influenza 163
Epstein–Barr nuclear antigens (EBNAs) 630
Epstein–Barr virus (EBV) **630–631**
    cancer and 311, 630
    central nervous system infection 598T
    pathogenesis of infection 630
    persistent infections 296T
    treatment of infection 630–631
Ergosterol 23, 60, 338
    antifungals targeting 477
ErmE gene 495–496
Erysipelas 643T, **645**
Erythema migrans 624, 624F
Erythrogenic toxin 89T, 90
Erythromycin 469
    resistance 490T, 495–496, 497T
Eschar 647
Escherich, Theodor 6
Escherichia coli **545–547**
    adherence 81, 81F, 82F, 82T
    antibiotic resistance 493, **498**
    bacteremia 617F
    bacteriocins 134
    conjugation 262–263, 263F
    diarrheagenic 545–547
    doubling time 83
    enteroaggregative (EAEC) 152–153, 546T
    enterohemorrhagic (EHEC) 6, 152, 546T, **547**
    enteroinvasive (EIEC) 546T
    enteropathogenic (EPEC) **546–547,** 546T
    enterotoxigenic (ETEC) 536, 537, **546,** 546T

fimbriae 194
identification 225, 225F
lac operon 253
meningitis 597F, 600
nosocomial infections 106, 106F
nuclear region 205F
O polysaccharides 192
O104:H4 6, 152–153
O157:H7 6, 6F, 542F, 543F
   antibiotic resistance 494
   emergence 152
   outbreak in spinach (2006) 547
   sources and transmission 6, 6F, 7F
O:H terminology 6
opportunistic infections 134
optimal growth temperature 215T
P pili 574
penicillin-binding proteins 466
plasma membrane 200
recombinant DNA technology 118
ribosomes 206
toxins 545, 546T
urinary tract infections 573, 574, 574F, 575
uropathogenic 574
virulence factors 544, 545, 574
Ethambutol 467–468, 520
Ethanol (ethyl alcohol) 442, 443T, 445
Ethidium 257
Ethylene oxide 443T, 447–448
Etiology, disease 132, **136–137**
Eukaryotes 48
cell structure **59–69**, 59T, 60F
cytoplasm and organelles 62–67
plasma membrane 60–62, 61F
Exfoliatins 646
Exocytosis **67–69**, 69F
role in infection 69
viral infection cycle 275, 288
Exogenous infections 542–544, 543F
Exonuclease, DNA proofreading 241–242, 242F
Exotoxins **87–91**, 88F, 89T
compared to endotoxins 91T
vaccines against 90–91
Exponential phase, bacterial growth 227–228, 227F
Extended-spectrum β-lactamase (ESBL) 575
Extravascular infections 616
Eye
barriers to infection 80F, **354**, 354F, 356T
infections 641, **656–658**, 657T
   fungal keratitis 342
   neonatal 447, **658**, 658F
   privileged status 656
   viral entry via **301**, 301F
Eyelid abscesses 656

F

F plasmid
integration into host chromosome 264, 264F
transfer via conjugation 262–263, 263F
F1 protein 621
Factor B 373, 373F
Factor D 373, 373F
Factor P (properdin) 373, 373F
Facultative anaerobes 33
FAD (flavin adenine dinucleotide) 36, 36F
ATP yield 41
electron transport chain 40, 41F
Krebs cycle 40
Famciclovir (Famvir®) 474
Farmer's lung 427
Fasciitis **644**
see also Necrotizing fasciitis
Fastidious bacteria 222, 229
Fatal familial insomnia (FFI) 168, 168F, 608

Fats **21**, 22F
Fatty acids 21, 22F
oxidation by-products 66
saturated 21, 22F
unsaturated 21, 22F
Favus 341
Fecal coliform counts 228
Fecal coliforms 213
Fecal–oral route of contamination 78, 99
amebiasis 328
measuring numbers of bacteria 228, 228F
viruses **306**, 556T
Feedback inhibition 37, 37F
Female reproductive tract
anatomy 577–578, 578F
microbial flora 578F
Females
portals of exit 102F
sexually transmitted infections 576
urinary tract infections 572, 573F, 574
Fermentation 39F, **42–43**, 43F
alcoholic 42, 43, 44F
in diagnostic tests 113
homolactic 42–43, 43F
Fetal infections
cytomegalovirus 629
HIV 306, 415
sexually transmitted infections 576, 581
viruses 306
Fever **371–372**
brucellosis 623
crisis phase 371
malaria 324, 325
pathophysiology 371, 371F
role in host defense 351, 371–372
Fever blisters *see* Cold sores
Fibronectin 517, 517F
Filariasis **635–636**, 635F
Filovirus fevers **632–633**
Filtration **452**
air purification 452
sterilization 452, 452F
testing 228, 228F
Fimbriae 81, **194–195**, 194F
clinical significance 194–195, 208T
Flagella (flagellum)
bacterial 6, **196–198**
   arrangements 198, 198F
   clinical significance 198, 208T
   movement 196, 196F
   structure 196–198, 197F
basal body 196–197, 197F
eukaryotic **64**
filament 196, 197F
hook 196, 197F
protozoans 320, 320F
stain 51T, **53**, 53F
Flagellates 320
Flagellin 196, 197F
Flagyl® *see* Metronidazole
Flavin adenine dinucleotide *see* FAD
Flaviviridae 561, 562, 631T
Flea-borne diseases
endemic typhus 628
plague 620–621
Fleas 101
Fleming, Alexander 8, 186, 460
Flesh-eating disease *see* Necrotizing fasciitis
Florey, Howard 460
Flu *see* Influenza
Fluconazole 477
Flucytosine 478
Fluid mosaic model 61, 61F
Flukes *see* Trematodes
Fluorescent antibodies 118, 120–121
Fluoroquinolone antibiotics 471
Focal infection 132T
Focus of infection 142

Folate agonists 479
Folic acid metabolism 471–472
Folliculitis 340, 643T, **645**
Fomites 100
Food
allergy 426T
irradiation 453
preservation 452
refrigeration and freezing 451
as reservoir of infection 99
as vehicle of transmission 100
Food and Drug Administration (FDA) 115
Food poisoning **537–539**
human story 5–6
infections 538, 539T
intoxications 538–539, 539T
Salmonella quorum sensing 57
Foodborne infections 100
botulism 173, 602–603
Campylobacter enteritis 553
emerging **152–153**
Salmonella gastroenteritis 550
see also Food poisoning
Fore people, New Guinea 168, 608
Forespore 207, 207F
Formaldehyde 443T, 447
Formalin 447
Foscarnet 473F, 475
Fosfomycin 186, 497T
Francisella tularensis 621–622
as bioweapon **176**
see also Tularemia
Frascatoro, Girolamo 148
Freeze-drying **451**
Freezing 38, **451**
Frequency of urination 574, 575
Functional group 20
Fungal cell wall 338
drugs targeting **477**
Fungal infections 338, **340–344**
central nervous system 596, **609**
drug treatment **476–478**
genitourinary system 577F, **589–590**
historical treatment 476
immunocompromised host 104, 528–529
pathogenesis **343–344**
recurrent 421
respiratory system 511, **528–532**
skin 643T, **653–654**
see also Mycoses
Fungemia 616
Fungi **338–340**
adherence 343
antibiotic production 460, 461T
dimorphism 340
growth 338
identification 224
invasion 343–344
mold form 338–339, 338F
size 49F
structure 338
thermal death times 449T
tissue injury 344
yeast form 338–339, 339F
Furuncles (boils) 86, 645
Fusidic acid, resistance 497T
Fusiform bacilli 49
Fusobacterium 543F

G

Gajdusek, D. Carlton 168
β-Galactosidase 253
GALT (gut-associated lymphoid tissue) 385
Gametocytes 323, 327
Gametogony 319
Gamma radiation 452
DNA damage 257

sterilization 453
Ganciclovir 475, 629
Gardnerella vaginalis 577F, 579
Gas gangrene **647,** 647F
    fermentation pathway 43F
    pathogenesis 85, 86, 647
    tissue destruction 220, 220F
Gaseous disinfectants/sterilants **447–448**
GasPak™ jar 220–221, 221F
Gastric adenocarcinoma 554
Gastric juice **355,** 356T
Gastric ulcers 554
Gastritis, *Helicobacter pylori* 554
Gastroenteritis
    Salmonella 542F, **550–551,** 551F
    viral 557–558
Gastrointestinal (digestive system) infections
    535–568
    bacterial **542–555,** 542F, 543F
    clinical symptoms **536**
    defenses against 538F
    endemic **537**
    endogenous 543, 543F, 544–545
    epidemic 537
    epidemiology **536–540**
    exogenous 542–544, 543F
    foodborne 537–539
    nosocomial **537**
    pandemic 537
    parasitic **563–567**
    treatment and management **536–537**
    viral **555–562**
Gastrointestinal tract (digestive system)
    bacterial adherence 81, 81F, 82F, 82T
    defenses against infection 78, 538F
    lymphoid tissue 385
    normal microflora 133, 133T, 135F
        effects of antibiotics 105, 499, 499F
    as portal of entry **77–78,** 77T, 78F
    as portal of exit 78
    viral entry **300**
Gay men *see* Homosexual men
Gel electrophoresis 120, 120F, 122, 124–125
Gender differences
    disease transmission 102
    susceptibility to virus infections 307
    see also Females; Males
Gene(s)
    bacterial transfer **259–265**
    constellations 163
    constitutive 252, 253
    defined 245
    horizontal transfer *see* Horizontal gene
        transfer
    inducible 253
    re-assortment 153, 154, 155F
    transfer methods 265T
Gene expression 236, **245–252**
    DNA viruses 282–283
    induction (activation) 252, **253–254**
    regulation **252–256**
    repression 252, **254–255**
    see also Transcription; Translation
Generation times, bacterial 215, 215T
    factors affecting 226, 229
Genetic code **244–245,** 245F
    degeneracy 244
Genetic engineering 114, 115–116, 116F
Genetic susceptibility to infection **378**
Genetics, microbial **235–266**
    pathogenicity and **265**
Genital herpes 571, **586–588**
    clinical manifestations 587, 587F
    newborn infants 587
    pathogenesis 586–587
    recurrent 587
    transmission 305, 305F, 586
    treatment 587–588

ulcers 578, 578T
Genital ulcers 578, 578T
Genital warts 301F, **588–589,** 588F, 652, 652F
Genitourinary infections **571–591**
    bacterial **577–585**
    fungal **589–590**
    viral **586–589**
    see also Sexually transmitted infections;
        Urinary tract infections
Genitourinary (urogenital) tract
    bacterial adherence 82T
    normal microflora 133T
    portals of entry 77T, **78–79,** 79F
    viral entry via **300–301**
Genomics 124
Genotyping 123
Gentamicin 470
    resistance 493, 497T
Genus 48
Germ theory 5
German measles *see* Rubella
Germination, endospore 207
Gerstmann–Straüssler–Scheinker syndrome
    (GSS) 168, 168F
Ghon complex 520
*Giardia duodenalis (Giardia intestinalis; Giardia
    lamblia)* 320, 320F, **563–564,** 563F
Giardiasis **563–564**
    pathogenesis 320, 563–564
    prevalence 318T
Gilchrist's disease (blastomycosis) **530**
Gingivitis 540, **541**
Globalization **153–154**
Glossina (tsetse fly) 329, 329F
Glucans 338
Glucose
    ATP yield 41
    chemical structure 20–21, 20F
    fermentation 42
    glycolytic breakdown 38–39
    group translocation 204, 204F
    oxidation reaction 33
    as source of energy 253
Glutaraldehyde 443T, 447
Glycerol 21, 22F
Glycocalyx **193–194,** 193F
    clinical significance 193–194, 208T
Glycogen 206
Glycolipids **22,** 23F
Glycolysis 33, **38–39,** 39F
    ATP yield 41
Glycopeptide antibiotics 462T, **467**
Glycoproteins, viral envelope 275
Golgi apparatus **65,** 65F
    viral protein trafficking 286, 286F
Gonococcus *see Neisseria gonorrhoeae*
Gonorrhea **581–583**
    clinical manifestations 578–579, **582–583,**
        583F
    disseminated infection 583
    pathogenesis 582
    treatment 583
    see also Neisseria gonorrhoeae
Gram-negative bacteria 52
    antibiotic targets 466, 467
    cell wall **189–190,** 189F
        clinical significance 191–192
        vs. Gram-positive 189–190, 190T
    compromised host 104
    conjugation 262
    defenses against complement 375
    DNA release 235
    efflux pumps 493, 494
    endotoxins 91–92, 92F
    fimbriae and pili 194–195
    flagella 197–198, 197F
    motility 194–195
Gram-nonreactive bacteria 52

Gram-positive bacteria 52
    antibiotic targets 466
    cell wall **189,** 189F
        assembly 187, 188F
        clinical significance **191**
        vs. Gram-negative 189–190, 190T
    compromised host 104
    conjugation 262–264
    defenses against complement 375
    efflux pumps 493, 494
    endospores 206
    flagella 197, 197F
    identification 225
Gram stain **51–52,** 51T
    procedure 52, 52F
Gram test 235
Gram-variable bacteria 52
Granulocyte–macrophage colony-stimulating
    factor (GM-CSF) 365–366
Granulocytes 359F, 361
Granulomas 140, 141F
Granulysin 395
Granzymes 360, 365, 395
Graves' disease 424
Griffith, Fredrick 260, 261F
Group A streptococcus (*Streptococcus
    pyogenes*)
    bacteremia 617F
    cell wall 84, 191
    chain formation 49, 50F
    childbed fever 7
    erysipelas (cellulitis) 645
    erythrogenic cytotoxin 89T, 90
    identification 225–226, 225F
    impetigo 646
    M proteins 84, 143, 191
    necrotizing fasciitis 86, 86F, 459, 641, 644,
        645F
    pharyngitis (strep throat) 512–513, 513F
    respiratory system infections 512F
    scarlet fever 89T, 90, 513
    skin and soft tissue infections 643T, 644,
        644F
    spreading out 86, 86F
    tonsillar abscess 513
    toxic shock syndrome 143
    virulence factors 512–513, 513F
Group B streptococcus (*Streptococcus
    agalactiae*)
    female reproductive tract 578F, 596
    identification 225F
    meningitis 597F, 600
Group C streptococcus 512F
Group D streptococcus 225F
Group translocation 204, 204F
Growth
    bacteria **213–232**
        antibiotic resistance and 486, 486F
        characteristics **226–229**
        chemical requirements 218–221, 218F
        clinical implications **229–230,** 231T
        measurement 228–229, 229F, 230T
        phases 227–228, 227F
        physical requirements 214–217
        requirements **214–222,** 221T
        role in successful infection 83
    fungi 338
    methods of controlling *see* Microbial
        control methods
Growth-based diagnostics **112–113**
Growth curve, bacterial **227–228,** 227F
Growth factors, bacterial **219,** 219T
Growth media 112, **222–226**
    chemically defined 222–223, 223T
    complex 222, 222T
    differential 112, 224, 224T, 225, 225F
    elective 112
    fastidious bacteria 222

identification of pathogens 224–226
selective 112, 224
selective/differential 225, 225F
Gruber reaction 119, 122T
Gruinard Island 435
Guanine (G) 25, 26F
specific pairing 27, 27F, 237, 237F
structure 237F
Guillain–Barré syndrome 553–554
Gummas 581, 581F
Gut-associated lymphoid tissue (GALT) 385

H

H antigens 544, 544F
E. coli 6, 545
Salmonella 549
Habitat 79, 82
Haemophilus ducreyi 577F
Haemophilus influenzae
growth media 229, 229F
meningitis 597F, 599
respiratory system infections 512F
type B (Hib), vaccine 599
Hairy cell leukemia 376
Hairy leukoplakia 418
Halogens **445–446**
Halophiles 217
extreme 217
facultative 217
obligate 217
Hand washing 6–7, **454**
Hantavirus 302F, 528, 528F
Hantavirus pulmonary syndrome (HPS) 149T,
152, **528**
Haptens 413
Hashimoto's thyroiditis 424
Hay fever (allergic rhinitis) 426, 426T, 429
Head lice 655, 656F
Health care, relevance of microbiology **8–11**
Health care facilities
antibiotic resistance **488**
see also Hospitals
Health care workers
hepatitis B risk 560
MRSA transmission 498
transmission of viruses to 305
Heat **448–451**
denaturation of proteins 438, 438F
dry 448–449
mechanism of antimicrobial activity 448
moist 448, **448–451**, 450T
sterilization 448, 449–451, 449T
see also Pasteurization
Heat capacity 19
Heat-labile toxin 545, 546, 546T
Heat-stable toxin 545, 546, 546T
Heavy chains, immunoglobulin 397–398, 398F
Heavy metals
allosteric inhibition 37
as antimicrobial agents **447**
inhibition of microbial proteins 438, 438F
treatment of protozoan infections 480
Helical viruses 273, 273F
Helicase 240
Helicobacter pylori 78, 552T, **554–555**
exogenous infection 543F
flagellar movements 198
gastritis 554
morphology 554, 554F
peptic ulcers 91, 554, 554F
persistence 140T, 142
portal of entry 78
survival at low pH 216, 217F
transmission 544
treatment of infection 555
urea breath test 31
Helminthic infections **330–337**

digestive system 563, **565–567**
pathogenesis 320–321
prevalence 318T, 319
treatment **480–481**
Helminths (worms) 318, **320–321**
life cycles and transmission 321–322, 322T
morphology 320, 321F
see also Cestodes; Nematodes;
Trematodes
Helper T cells (CD4 T cells) 387, **395–396**
antigen presentation to 391F, 392
B-cell cooperation **402**
during course of infection 403F, 404
differentiation into subtypes 395, 395F
effector
functions 394, 395–396, 395F
production **393–394**, 393F
HIV infection 417, 417F, 418
see also Th1 cells; Th2 cells
Hemagglutinin (HA) 163, 273, 273F
Hematogenous dissemination, viruses 302,
302F
Heme 219T
Hemochromatosis 35
Hemoglobinuria 324
Hemolysins 85
E. coli 545, 546T
Hemolysis 225–226, 225F
alpha 225
beta 225–226
gamma 225
Hemolytic anemia, acute 427
Hemolytic uremic syndrome 152
Hemorrhagic disease, bacterial 6
Hemorrhagic fevers, viral *see* Viral
hemorrhagic fevers
HEPA (high-efficiency particular air) filters 452
Hepadnaviridae 560
Hepatitis 558–559
chronic viral 427
viruses **558–562**, 559T
Hepatitis A 539T, **559–560**, 559T
incubation period 294T, 560
pathogenesis 559–560
vaccine 308T
Hepatitis A virus 559
Hepatitis B 559T, **560–561**
cancer risk 311, 560, 561
chronic carriers 560
hepatitis D co-infection 562
incubation period 294T, 561
pathogenesis 560–561
serum globulin 561
treatment 561
vaccine 308T, 561
Hepatitis B surface antigen (HBsAg) 560
Hepatitis B virus 560, 560F
dissemination within host 302F
persistence 296T
transmission 305, 577F
Hepatitis C 559T, **561**
cancer risk 311, 561
chronic 297
incubation period 294T, 561
liver damage 561, 561F
Hepatitis C virus 561
chronic infection 296T, 297
transmission 577F
Hepatitis D **562**
Hepatitis E **562**
Hepatitis G **562**
Hepatocellular carcinoma 560, 561
Herd immunity **109–110**, 109F, 307
Hermaphrodite trematodes 335
Herpes
genital *see* Genital herpes
incubation period 294T
newborn infants 587

ophthalmic 301, 301F, 652
oral (cold sore) 271, 303, 303F, 651, 651F
Herpes simplex virus (HSV)
central nervous system infections 598T
dissemination 302F, 303, 586
latency 298, 586, 651
persistence 296T
type 1 (HSV-1) 586, **651–652**
type 2 (HSV-2) 577F, **586–588**
pathogenesis 586–587
transmission 305, 305F, 586
see also Genital herpes
Herpes viruses
DNA replication 282
latency 275
portals of entry 301, 301F
structure 274, 274F
Herpes zoster *see* Shingles
Heterotrophy 32, 338
Hexachlorophene 443T, 445, 445F
Hfr cells 264, 264F
High-throughput screening 502
Histamine 360–361, 362
allergic reactions 428
IgE-induced release 400–401
Histatin 354
Histoplasma capsulatum 530, 530F
Histoplasmosis 343, **530–531**
disseminated 343, 343F
History of microbiology 8–9
HIV *see* Human immunodeficiency virus
Homelessness 162
Homolactic fermentation **42–43**, 43F
Homosexual men
AIDS/HIV infection 414, 415F
sexually transmitted diseases 579
Hong Kong influenza pandemic (1968) 164
Hookworms **566–567**, 566F
pathogenesis of disease 320, 321, 566–567
prevalence 318T, 319
treatment 567
Horizontal gene transfer
antibiotic resistance 486–487, 487F
mechanisms 260–264, 265T
Hortaea werneckii 340
Hospital-acquired infections *see* Nosocomial
infections
Hospitals
antibiotic resistance 488
transmission of disease 4, 4F, 7–8, 106
urinary tract infections 572, 573
Host(s)
compromise *see* Immunocompromised
host
definitive 321
establishment in **81–83**
intermediate 321
parasites with multiple 321–322, 322T
portals of entry **76–81**, 76F, 77T
portals of exit **102**, 102F, 103F
susceptibility to virus infection **307**
Host cells
adherence/attachment to *see* Adherence
bacterial movement between 63, 63F
bacterial movement within **199**
receptors for viruses 62, 62F, 276–277
binding mechanisms **277–278**, 278F
structure **59–69**, 59T, 60F
virus-mediated lysis 275, 276F
virus penetration **278–279**
virus release **287–288**, 288F
virus spread between 289, 289F
Host damage **87–94**
bacterial toxins 87–93
direct 87
fungal infections 344
indirect 87
parasitic infections 320–321

virus-induced 93–94
Host defense 10–11, 349
    first line (barriers) **352–356**
    second line 352, **367–378**
    adaptive immune response **383–408**
    compromised *see* Immunocompromised
        host
    defeating **83–87**
    failures **413–430**
    innate immune response **351–379**
    integrated system 406–407, 406F, 407F
    viral infections 312
    see also Immune response
Host–microorganism relationships **132–136,**
        134T
Host–pathogen relationships 48, **54–56**
Hot air ovens 448
Hot temperatures
    bacterial growth requirements 214F, 215
    see also Heat
House dust-mite (*Dermatophagoides
        pteronyssimus*) 426, 426F
Houseflies 101, 101F
HPV *see* Human papillomavirus
HSV *see* Herpes simplex virus
Human Genome Project 114F
Human herpesvirus 4 *see* Epstein–Barr virus
Human herpesvirus 5 *see* Cytomegalovirus
Human immunodeficiency virus (HIV)
        414–420, 577F
    antibodies 121, 419
    attachment to host cells 278, 415
    in body fluids 415, 415T
    cellular targets 415
    co-receptors 279, 415
    cytotoxic T lymphocyte escape mutants
        297
    dissemination within host 302F, 415–416
    entry into target cells 414, 416, 416F
    genome 416
    global spread 153–154
    gp41 and gp120 proteins 414F, 415
    infection
        acute phase 417, 417F
        antiviral drugs 474, 475
        asymptomatic phase 417–418, 417F
        central nervous system 598T
        clinical latency 418
        course **417–418,** 417F
        fetus/newborn baby 306, 415
        immune response **418–419**
        major tissue effects **419–420**
        persistence 296T, 298
        sequence of cellular events 416, 417F
        symptomatic phase *see* AIDS
        tuberculosis co-infection 162, 519
    integration into host DNA 416, 417F
    lymphadenopathy 417, 419
    portals of entry 300, 301
    relationship with host 55
    replication 416, 417F
    transmission 100, 305, **415–416,** 415F
    type 1 (HIV-1) 414
    type 2 (HIV-2) 414
    virion structure 414F
Human lymphotropic virus 311
Human Microbiome Project 134–135
Human papillomavirus (HPV) 577F, **588–589,**
        652
    cancer and 311, 588
    cutaneous infections 643T, 652
    genital infections *see* Genital warts
    incubation period 294T
    pathogenesis of infection 588, 652
    persistence 296T
    reduced immunosurveillance 297, 297F
    transmission 305
    vaccine 308T, 588, 589

see also Warts
Human reservoirs of infection 98
Humoral immune response 384, **396–403**
    course during infection 403–404, 403F
    inherited defects **421–422**
    see also Antibodies; Immunoglobulins
Hyaluronidase 85F, 86
Hybridization **123**
Hybridomas 118, 118F
Hydration, spheres of 19, 19F
Hydrogen atom (H) 18
Hydrogen bonds **18–19,** 18F
Hydrogen ions (H+) 19–20
Hydrogen peroxide (H_2O_2)
    as antiseptic/disinfectant 443T, 446
    bacterial enzymes neutralizing 220
    bacterial production 219–220
    degradation in peroxisomes 66
Hydrogen sulfide (H_2S) 32
Hydrolysis 19
    DNA damage 257
Hydrophilic molecules 22
Hydrophobic molecules 22
Hygiene hypothesis 426–427
Hyperkeratosis, fungal 342
Hypersensitivity **426–429**
    type I 426–427, 426T
    type II 427
    type III 427
    type IV 427
    see also Allergic reactions
Hyperthermophiles 214F, 215, 221T
Hypertonic environments 202, 202F, 217
Hyphae 339, 339F
Hypotonic environments 202, 202F

I

ICAM/ICAM-1 receptors 277, 515
Icosahedral viruses 274, 274F
    complex 274, 274F
    simple 274, 274F
ID_{50} 83, 84F, 306
Illness, period of 137, 138F
Imidazole 477
Imipenem 467, 497T
Immune complexes 427
    infectious endocarditis 619, 620F
Immune response
    adaptive 11, **383–408**
    cellular 384, **393–396**
    course of infection **406–407,** 406F
    failures **413–430**
    humoral 384, **396–403**
    innate 11, **351–379**
    primary 401, 401F
    secondary 401, 401F
    see also Host defense
Immune system 10–11
Immunization 5, 11
    active 309
    passive 309
Immunoassays 121, 122T
Immunoblotting **120–121,** 120F, 122T
Immunocompromised host **102–107**
    aspergillosis 532
    candidiasis 104, 653
    cryptococcosis 609
    cryptosporidiosis 565
    cytomegalovirus infections 629
    deficient phagocytosis 369
    dental plaque 542
    development of antibiotic resistance **488**
    fungal infections 342, 528–529
    nosocomial infections 106
    persistent viral infections 297
    Pneumocystis pneumonia 529
    primary amebic meningoencephalitis 610

toxoplasmosis 327
    tuberculosis 4, 162
    viral infections 307
Immunodeficiencies **414–424**
    caused by infection **414–420**
    drug-induced 424
    primary **421–423,** 421T
        accessory cell defects 422–423, 423T
        B cell defects 421–422
        T cell defects 422
    see also AIDS
Immunofluorescence 119
    indirect 120–121, 120F
Immunoglobulins (Igs) **397–401**
    antigen-binding sites 398, 398F
    constant region 398, 398F
    distribution and function **399–401**
    heavy chains 397–398, 398F
    IgA 399–400, 399F
        dimeric (secretory) 399–400, 399F
        function 399
        J chain 399–400, 399F, 400F
        monomeric (nonsecretory) 399F
        secretory piece 399–400, 399F
        tears 354
        transcytosis 399–400, 400F
    IgD 401
    IgE 400–401
        allergic reactions 426, 426T, 427–429
        basophil/mast cell activation 400–401,
            400F
    IgG 399, 399F
        function 399
        timing of release 401, 401F
    IgM 399, 399F
        timing of release 401, 401F
    isotypes (classes) 398, 399–401
        timing of release 401, 401F
    light chains 397–398, 398F
    structure **397–398,** 398F
    variable regions 398, 398F
    see also Antibodies
Immunological memory 387, **404–406**
Immunologically protected sites 425
Immunosuppressed host *see*
        Immunocompromised host
Immunosuppression
    drug-induced 103–104, 424
    pathogens causing 420
    viral hemorrhagic fevers 159–160, 159F
Impetigo 643T, **646,** 646F
Implants, medical 57–58
Incidence **108,** 108F
Inclusion bodies
    bacterial **206,** 208T
    virus-infected cells 93, 93F
Incubation period 137, 138F
    range in length 137, 138F
    viral infections 294, 294T
Index case 110
Indirect immunofluorescence 120–121, 120F
Infantile paralysis 605
Infants
    botulism 173, 173F, 593, 603
    candidiasis/thrush 653
    E. coli infections 546
    parainfluenza 516
    respiratory syncytial virus infection 527
    rotavirus infections 557
    toxoplasmosis 327
    see also Children; Newborn infants
Infections
    adaptive immune response **403–404,**
            403F
    autoimmune reactions 424
    establishment **81–83**
    genetic susceptibility **378**
    immunodeficiencies caused by **414–420**

number of organisms required **83**
periods of 137, 138F
recurrent, primary immunodeficiencies 421
requirements for 10, 56, **75–94**
scope of **142–143**
stages of a typical acute 406–407, 406F
terminology 132T
transmission *see* Transmission, disease
Infectious diseases **9–11**
　development **137–142**
　diagnosis **112–126**
　duration **139**
　emerging and re-emerging **147–178**
　environment and **151–152**
　etiology 132, **136–137**
　host–pathogen relationship 54–56
　nationally notifiable 110, 111F
　prevention 11
　principles **131–144**
　risk factors 54, 103
　survivors of natural disasters 3
　terminology 132T
　treatment **11**
　trends over time 9, 110F, 150, 150F, 151F
　see also Disease
Infectious dose 50% (ID$_{50}$) 83, 84F, 306
Infectious endocarditis **618–619**
　bacterial sources 618T
　pathogenesis 619, 619F, 620F
　subacute 427
Infectious mononucleosis 294T, 630–631
　see also Epstein–Barr virus
Inflammation **369–370**
　autoimmune disease 424, 425F
　brain 595
　phagocyte migration 370
　role of complement 374, 374F
　vascular changes 369–370
Influenza (flu) 4–5, **525–527**
　acute influenzal syndrome 526
　Asian (1957) 164
　avian *see* Avian influenza
　complications 526–527
　epidemics 526
　epizootics 163
　Hong Kong (1968) 164
　incubation period 294T
　pandemics 163–164, 525
　pathogenesis 526–527
　re-emergence 149T, **163–164**
　Spanish (1918 pandemic) 4, 5F, 163–164, 164F
　swine (2009) 5, 164
　transmission 526
　treatment 527, 527T
　vaccines 5, 273, 308T, 310
Influenza virus 525, 525F
　antigenic shift 164, 296
　antigenic variation 295
　attachment to host cell 278
　dissemination within host 302F
　drug resistance 500
　gene constellations 163
　genetic re-assortment 153
　host cell receptors 62
　release from host cell 288F
　serotypes 525
　structure 273, 273F
Initiation codon 244, 245F, 247
Initiation of translation 250, 250F
Initiator protein, DNA replication 243
Injections 80
Innate immune response 11, **351–379**
　first line of defense 352–356
　second line of defense 352, 367–378
　barrier defenses **352–356**, 356T
　cells involved **358–367**, 359F, 359T
　five mechanisms **367–378**

molecules involved **356–358**
relationship with adaptive immunity 406–407, 407F
viral infections 295F
Inoculation 5, 112
Insect bites 80
Insulin 115F, 116, 425F
Integrase, HIV 414F, 416
Intensive care unit 7, 7F
Interferons (IFNs) **375–376**, 376T
　alpha (IFN-α) 376, 376T
　beta (IFN-β) 376, 376T
　gamma (IFN-γ) 366, 366T, **376**, 376T
　genetic variation 378
　mechanisms of action 376, 377F
　production in influenza 526
　therapeutic use 376, 562
　type I 375
　type II 375
Interleukin(s) 358
Interleukin 1 (IL-1) 371, 371F
Interleukin 2 (IL-2) 366, 366F
Interleukin 6 (IL-6) 370
Interleukin 12 (IL-12) 366, 366F
Intermediate filaments 63
Intestine
　bacterial infections 543F
　dendritic cells 363
　Peyer's patches 385, 385F
　viral entry 300, 300F
Intracellular pathogens (or parasites)
　antigen presentation 391F, 392
　movement within cells 199
　obligate 86, 272
　resistance to host defense 86
Intracranial pressure, increased 595
Intravascular infections 617, **618–619**
Intravenous drug abusers
　AIDS/HIV infection 100, 415, 415F
　viral hepatitis 562
　wound botulism 603
Intravenous line bacteremia 616–617
Invasin 86
Iodine 445–446
Iodophor compounds 443T, 445–446
Iodoquinol 479
Ionic bonds **16–17**, 17F
Ionizing radiation 452–453, 453F
　DNA damage 257
Ions 17
Iron
　bacterial requirements 35, 218T
　binding by transferrins 355
Isoelectric focusing (IEF) 125
Isolation, infected individuals 138–139
Isoniazid 467–468
　resistance 497T, 520
　tuberculosis therapy 520
Isopropanol 443T, 445
Isotonic solutions 202
Isotypes, immunoglobulin 398, 399–401
Itraconazole 477
Ivermectin 480
Ixodes ticks 152F, 623, 623F, 624

J

J chain 399–400, 399F, 400F
Jaundice
　malaria 325
　viral hepatitis 560
JC polyoma virus 296T
Jenner, Edward 5, 307
Junin virus 158F

K

K antigens 544, 544F

Salmonella 549
Kaposi's sarcoma 420, 420F
Keratins 63
Keratitis 656, 657T
　fungal 342
Ketoconazole 477
Kidney
　bacterial infection 575
　macrophages 362F
Kinins 369
Kirby–Bauer test *see* Disk diffusion method
Kissing bugs (reduviid) 329, 633, 634
Klebsiella 543F, 574F
Klebsiella pneumoniae 84, 493
Knowledge-based mathematical models 110
Koch, Robert 5, 136, 172, 507, 510
Koch's postulates 136–137, 137F
Köhler, Georges 117
Koplik's spots 305, 305F, 648–649, 648F
Krebs cycle 39F, **40**, 41
Kupffer cells 362, 362F, 363F
Kuru 168, 168F, 608

L

lac operon 253–254, 253F, 254F
lac repressor protein 253–254, 254F
Lacrimal apparatus 356T
　barrier function **354**, 354F
β-Lactamase *see* Beta (β)-lactamase
Lactic fermentation 42–43
Lactobacillus 134, 578F
Lactobacillus acidophilus 541
Lactoferrin 35
Lactose 253
Lactose operon *see lac* operon
Lag phase, bacterial growth 227, 227F
Lagging strand DNA synthesis 242–243, 243F
Lamina propria, dendritic cells 363
Lamivudine 474
Langerhans cells 363
Lantibiotics 502
Large intestine, normal microflora 133, 133T
Laryngotracheitis 516
Lassa fever 149T
Lassa virus 176
Latency, virus 275, 276F
Latent diseases 132T, 139
Latent viral infections 296, **297–298**, 298F
Latex agglutination 120, 120F
LD$_{50}$ 83, 84F, 306
Lead 37
Leading strand DNA synthesis 242, 243F
Lederberg, Joshua 133
Leeuwenhoek, Antony van 4, 4F
Legionella pneumophila 523, 523F
Legionnaire's disease (legionellosis) 512F, **523–524**
　emergence 149T
　pathogenesis 523–524
　transmission 511
　treatment 524
Leishmania 655
　drug treatment **480**
　pathogenesis of disease 320
Leishmaniasis
　cutaneous **655**, 655F
　prevalence 318T
Leprosy 216, 217F
　see also Mycobacterium leprae
Leptospira interrogans 195F
Lethal dose 50% (LD$_{50}$) 83, 84F, 306
Leukemia, hairy cell 376
Leukocidins 85, 369
Leukocyte adhesion deficiency 423T
Leukocyte endogenous mediator (LEM) 371
Leukocytes *see* White blood cells
Leukotrienes 370
Levofloxacin 471

Lice **655,** 656F
  see also Louse-borne diseases
Life expectancy, trends 150, 151F
Lifestyle, antibiotic resistance and **488–489**
Light chains, immunoglobulin 397–398, 398F
Limulus amebocyte lysate assay (LAL) 92
Lincosamides 462T
Linezolid 470, 497T
Lipid A 92, 192
  as antibiotic target 502
Lipid bilayer *see* Phospholipid bilayer
Lipid carrier cycle 187
Lipid rafts, virus attachment 277
Lipids **21–23**
  plasma membrane 60, 61F
Lipocalin 354
Lipopeptide antibiotics 462T
Lipopolysaccharide (LPS) 92
  clinical significance 192
  complement activation 373, 373F
  layer 189–190
  Toll-like receptor interaction 357
Lipoteichoic acids 189, 189F
  clinical significance 191
Lister, Joseph 8, 441
Listeria monocytogenes
  meningitis 597F
  movement within cells 63
Liver
  Plasmodium life cycle 324
  resident macrophages 362, 362F, 363F
  virus dissemination to 303
Liver fluke *see Clonorchis sinensis*
Loa loa 657, 657F
Loaiasis 657–658, 657F, 657T
Local infection 132T, 142
Lockjaw 602
Log phase, bacterial growth 227–228, 227F
Logarithmic decline (death) phase, bacterial
    growth 227F, 228
Lophotrichous bacteria 198, 198F
Louse-borne diseases
  epidemic typhus 627–628
  relapsing fever 625
Louse infections **655,** 656F
Lower respiratory tract 510, 510F
  ciliated cells 353, 354F
  defenses against infection 511
Lower respiratory tract infections 510F
  bacterial **516–525**
  viral **525–528**
  see also Pneumonia; Respiratory system
    infections
Lung flukes 335–336
Lungs
  dendritic cells 363
  resident macrophages 362–363, 362F
  specimen collection 231T
Lupus erythematosus, systemic (SLE) 424, 425
Lyme disease 615, 616, **623–625**
  emergence 149T, 152
  pathogenesis 624–625, 624F
  transmission 98, 98T, 623
  treatment 625
  vector 152F
  see also Borrelia burgdorferi
Lymph, infections of *see* Blood and lymph
    infections
Lymph nodes
  B cells 390, 390F
  dendritic cells 364
  distribution in body 384F, 385
  resident macrophages 362F
  structure 390, 390F
  T cells 390, 390F
Lymphadenitis
  filariasis 636

sexually transmitted infections 579, 579F
Lymphadenopathy
  HIV 417, 419
  syphilis 580
Lymphatic system 384–385, 384F
  filariasis 635–636
Lymphocytes **387–388**
  anergy 388
  antigen receptors 387–388, 388F
  clonal deletion 389, 389F
  clonal selection 388–389, 389F
  development and maturation 386F, 387,
    388–391
  effector 387
  HIV infection 416, 416F
  lymphoid tissues 390–391
  naive 387
  natural killer cells 365
  survival 389–390
  viral hemorrhagic fevers 159F, 160
  see also B cells; T cells
Lymphogranuloma venereum 578T, 579, 579F
Lymphoid follicles 390, 390F
Lymphoid progenitor cells, common 359F,
    386F
Lymphoid tissues **390–391**
  B cell survival 389
  dendritic cells 364
  distribution in body **384–386,** 384F
  peripheral 390–391
Lymphoma
  Burkitt's 630, 630F
  HIV-related 420
Lyophilization 451
Lysis
  complement-mediated 372, 372F, 374–375
  virus-infected cells 275, 276F, 287
Lysol 437, 442, 445
Lysosomes **66**
  fusion with phagosomes 368
  role in infection 66, 69
Lysozyme 354, 355
Lytic infection cycle 275–276, 276F

M

M cells 363, 385, 385F
  HIV infection 416
  Salmonella-induced ruffling 550, 551F
  Shigella infection 548
  viral entry via 300, 300F
M proteins, streptococcal 84, 191
  toxic shock 143
MacConkey agar 213, 224T
Machupo virus 176
Macrolides 462T, **469,** 470F
Macrophages **361–363**
  activation by T cells 396
  activation of naive T cells 393F, 394
  alveolar 362, 362F
  antigen presentation 391F
  differentiation 361, 361F
  functions 359T, 361
  HIV infection 417–418
  Legionella infection 523, 523F
  phagocytosis 367, 368
  pulmonary interstitial 362, 362F
  resident 362–363, 362F
  Toxoplasma infection 327
  viral hemorrhagic fevers 159F
Macules 643, 644F
Mad cow disease 153, 168, 608–609
Magnetic immunoassays 125
Major histocompatibility complex (MHC)
    391–392
  antigen presentation 391, 391F
  B-cell activation 402
  class I 392

class II 392
  deficiency 421T, 422
  restriction 392
Malaria **323–326**
  Burkitt's lymphoma and 630
  cerebral 325, 326F
  drug resistance 500
  drug treatment **478–479**
  life cycle 323–324, 325F
  pathogenesis 324–326
  prevalence 318–319, 318T
  prophylactic agents 479
  re-emergence 149T
  transmission 322, 323
  treatment 326
  see also Plasmodium
Malarial paroxysm 325
Malassezia 340
Males
  portals of exit 103F
  urinary tract infections 573, 573F, 574
MALT (mucosa-associated lymphoid tissue)
    385
Mannan 338
Mannitol salt agar (MSA) 225, 225F
Mannose 373–374
Mannose-binding protein (MBP) 370, 373–374
Marburg hemorrhagic fever 158, **632–633**
  pathogenesis 159–160, 632–633
  treatment 633
Marburg virus
  as bioweapon 176–177
  relationship with host 55
Margination 360, 360F, 370
Margulis, Lynn 32
Mass spectrometry 125
Mast cells 359T, **364–365**
  activation by IgE 400–401, 400F
  allergic reactions 427, 428–429, 428F
  differentiation 359F, 364
  distribution in body 364, 364F
  mediators 364–365, 365T
Mastoiditis **512,** 512F
Mathematical models, knowledge-based 110
Matrix protein 273, 273F
Maximum growth temperature 215
Measles 643T, **648–649**
  central nervous system 598T
  hard 648
  historical perspective 148
  host susceptibility 307
  incubation period 294T
  Koplik's spots 305, 305F, 648–649, 648F
  pathogenesis 648–649
  reservoirs 98
  treatment 649
  vaccination success 308, 309F
  vaccine 308T
  see also Subacute sclerosing
    panencephalitis
Measles, mumps and rubella (MMR) vaccine
    308, 308T, 648
Measles virus (rubeola virus) 643T, 648
  dissemination within host 302F
  persistence 296T, 606
  portal of entry 77
Mebendazole 332, 480, 566
mecA gene 495, 497
Media, growth *see* Growth media
Mefloquine 478–479
Megasomes 142
Melarsoprol 330, 480
Membrane
  -enclosed cytoplasmic structures **65–66**
  fluid mosaic model 61, 61F
  plasma *see* Plasma membrane
Membrane attack complex (MAC) 374–375,
    374F

inherited deficiencies 375, 375T
lysis of pathogens 372, 372F
Membrane filters 452, 452F
Membrane proteins
bacterial cell 200, 200F
energy production 201, 201F
eukaryotic cell 60–61, 61F
integral 200
peripheral 200
Membrane transport, bacterial **201–204**
active 203–204, 203F
osmosis 201–202, 202F
passive 202–203, 203F
Memory, immunological **404–406**
Memory cells 387, 390, 401, 404–405
Meninges 594, 594F
Meningitis 47, **599–601**
aseptic 600
bacterial 597F, **599–600**
chronic 600, 600T
cryptococcal 609
diagnosis 600
purulent 596F
routes of infection 596, 596F
symptoms 600
syphilis 581
treatment 597F, 598–599, 600
viral **600**
Meningococcus see Neisseria meningitidis
Meningoencephalitis, primary amebic 317, **610**
Mercury 447
Merozoites 324, 326–327
MERS-CoV (Middle East Respiratory Syndrome Coronavirus) 293
Mesophiles 214F, 215, 221T
Messenger RNA (mRNA) 238, 239T
development of new antibiotics 501
genetic code 244–245
pairing with tRNA 248
polycistronic 238, 253
ribosome interactions 247, 248, 250, 250F
synthesis 245–246, 246F
in translation **247**, 251, 251F
Metabolic fingerprinting 113
Metabolic pathways **33–34**
anabolic 43
catabolic 38–43
Metabolism **31–44**
bacterial
alteration, in antibiotic resistance **496,** 496F
as antibiotic target **471–472,** 472F
basic concepts 31–34
component parts 32, 32F
defined 32
Metacercariae 335
Metachromatic granules 206
Metagenome 133
Methicillin 463, 463F
resistance 490, 490T
Methicillin-resistant Staphylococcus aureus see MRSA
Methylene blue 51T
Metronidazole 328, 329, 479
MHC see Major histocompatibility complex
Micafungin 477
Miconazole 477
Microaerophiles 220, 221T
Microarray analysis
development of new antibiotics 501
diagnostic applications 123, 124, 124F, 125F
procedure 123, 123F
protein 125
Microbial antagonism 134
Microbial control methods **435–455,** 440F
chemical 440F, **441–448**

mechanical 440F
physical 440F, **448–454**
terminology 436–437
see also Antiseptics; Disinfectants
Microbial death see Death, microbial
Microbial flora, normal 9, 132–136, 133T
digestive system infections 543F, 544
effects of antibiotics 105, 499, 499F
opportunistic infections 105, 134
protective effects 134
Microbial growth see Growth
Microbiology
chemistry for **15–28**
history 8–9
importance in everyday life 3–8
relevance to health care **8–11**
Microbiome, human 133–136
newborn child 135, 136F
variation 135, 135F
see also Microbial flora, normal
Microbiota, human 132–136
Micrococcus 225F
Microconidia 530, 530F
Microfilaments 63, 63F
Microfilariae 635
Microglial cells 362F, 363
Microhabitat 78
Micromonospora 461T
Microorganisms
numbers of see Numbers of microorganisms
relationship with human host **132–136,** 134T
relative sizes 49F
Microscopes 4, 4F
Microtiter plate 121, 121F
Microtubules 63
arrangement in cilia 63, 63F
Middle East Respiratory Syndrome Coronavirus (MERS-CoV) 293
Milk pasteurization 451
Milstein, César 117
Minerals, bacterial requirements 35
Minimum bactericidal concentration (MBC) 491
Minimum growth temperature 215
Minimum inhibitory concentration (MIC) 442, 491, 491F
Miracidia 335, 337
Mitochondria **65**
chemiosmosis 41
electron transport chain 40, 41F
endosymbiotic theory 65
structure 65, 65F
MMR (measles, mumps and rubella) vaccine 308, 308T, 648
Molds **338–340,** 338F
dimorphic fungi 340
nonseptate 339
septate 339, 339F
Molecular biology 114F
Molecular mimicry 424
Molecules, biological **20–28**
Monobactams 467
Monoclonal antibodies **116–118**
applications 118
production methods 118, 118F
Monocytes 359F, 359T, **361–363**
differentiation into macrophages 361, 361F
Mononuclear phagocyte system 362
Mononucleosis, infectious 294T, 630–631
Monosaccharides 20–21, 20F
Monotrichous bacteria 190, 190F
Montagu, Mary Wortley 307
Morbidity rate **108**
Mordants 52
Mortality rate **108**
Mosquitoes 101
arbovirus transmission 631, 632

dengue transmission 152, 157
filariasis transmission 635
malaria life cycle 323–324, 323F, 325F
malaria transmission 322, 323
West Nile virus transmission 157
Mother-to-child transmission
cytomegalovirus 629
herpes infections 587
HIV 415
sexually transmitted infections 576, 581
viruses 306
Motility, bacterial
axial filament-mediated 195, 195F, 196F
energy efficiency 198
flagellar 196, 196F, 198
intercellular 63, 63F
intracellular **199**
pilus-mediated 194–195
Mouth
infections 540–542, 543F
normal microflora 133, 133T
trench **541–542**
viral infections 305, 305F
mRNA see Messenger RNA
MRSA (methicillin-resistant Staphylococcus aureus) **496–498**
clinical significance 495, 498
mechanism of antibiotic resistance 466, 495
nosocomial infections 7–8, 7F, 106
nosocomial pneumonia 512F
Mucociliary escalator 353, 354F, 511, 511F
Mucosa-associated lymphoid tissue (MALT) 385
Mucous membranes
barriers to infection **353,** 354F, 356T
dendritic cells 363
fungal infections 341–342
lymphoid tissue 385
portals of entry **76–79,** 77T
Mucus, barrier function 353, 354F
Mullis, Kary B. 116
Multiple sclerosis 424, 425
Mumps
central nervous system 598T, 600
incubation period 294T
vaccine 308T
Mumps virus, dissemination within host 303
Mupirocin 191, 497T
murA–F genes 187
Murein see Peptidoglycan
Mutagens 257
Mutations **256–259**
antibiotic resistance 486
causes 256–257
frameshift 256T, 257
generating viral diversity 155F
hot spots 257
point 256, 256T
rates 257
replication errors 241, 245, **256–257**
spontaneous 257
suppressor 257
types 256T
Mutualism 133, 134T
Mutualistic microorganisms 55
Myasthenia gravis 425
Mycelium 339
Mycobacteria
antibiotics effective against 467–468
cell wall 191, 468F
Mycobacterium leprae 137
acid-fast stain 53, 53F
cell wall 191
growth requirements 229
optimal growth temperature 216, 217F
resistance to phagocytosis 369

Mycobacterium tuberculosis 519–521
    acid-fast stain 53, 53F, 519, 519F
    antibiotic susceptibility testing 491
    cell wall 191
    discovery 507
    generation time 215
    growth requirements 229
    new antibiotic targets 501
    optimal growth temperature 215T
    pathogenesis of disease 519–520
    persistence 140, 140T
    protection against host defenses 84, 369
    spread in aerosols 4
    virulence/fitness 162–163
    see also Tuberculosis
Mycolic acid 191
    acid-fast stain 53
    antibiotics targeting 468, 468F
    protection against host defenses 84
Mycoplasma, cell wall 190
Mycoplasma pneumoniae 512F
    adherence 82T
    pneumonia **518–519**
Mycoses **340–343**
    cutaneous 341–342
        secondary 342, 342F
    deep 342–343
    mucocutaneous 341–342
    subcutaneous 342
    superficial 340–341
    see also Fungal infections
Myeloid progenitor cells, common 359F, 386F
Myeloperoxidase deficiency 423T

N

N-acetylglucosamine (NAG) 186, 186F
    peptidoglycan assembly 187, 188F
N-acetylmuramic acid (NAM) 186, 186F
    peptidoglycan assembly 187, 188F
NAD+ (nicotinamide adenine dinucleotide)
    36, 36F
    ATP yield 41
    electron transport chain 40, 41F
    fermentation pathways 42, 43, 44F
    glycolysis 39
    Krebs cycle 40
NADH (reduced nicotinamide adenine
    dinucleotide) 42–43, 219T
Naegleria infections 317, 610
Nail infections, fungal 341, 341F
Naive lymphocytes 387
Nanocantilever arrays 125
Nanotechnology 125
Natural disasters 3
Natural killer cells (NK cells) 359T, **365–367**
    activating and inhibitory receptors
        366–367, 366F
    cytokine production 365–366, 366F
    innate immune response 365–367, 366F
    origins 359F, 365, 386F
    target-cell killing 365
Naturally acquired immunity **405–406**
Nebulizers, transmission of infection 106
Necator americanus 319, 566–567
Necrotizing fasciitis 644, 645F
    case stories 459, 641
    pathogenesis 86, 86F
Necrotizing periodontal disease (NPD)
    541–542
Needle aspiration 231T
Negative stain 51T, **52,** 52F
Negri bodies 93, 93F, 604, 604F
Neisseria
    complement deficiencies 375, 423F
    pilus-mediated immune escape 195
Neisseria gonorrhoeae (gonococcus) 577F,
    581–583

adherence 81, 82T
antibiotic resistance 583
attachment to sperm 64
clinical manifestations 578–579, 582–583
disseminated infection 583
establishment of infection 81, 582
evasion of host defenses 582
fimbriae 194
immunosuppressive effects 420
neonatal ophthalmia 657T, 658, 658F
persistent infections 140T
protection against host defenses 84, 369
two-cell arrangement 49
see also Gonorrhea
Neisseria meningitidis
    adherence 82T
    bacteremia 617F
    meningitis 47, 597F, 599
    vaccines 599
Nematodes (roundworms) **330–334**
    intestinal **330–333,** 331T
    lymphatic system 635–636
    morphology 320, 331
    tissue **333–334**
Neonates *see* Newborn infants
Nephritis 573, **575**
Nervous system
    virus dissemination **303,** 303T
    see also Central nervous system
Neuraminidase 163, 273, 273F
    inhibitors 473F, **475–476**
Neurosyphilis 581, 600
Neurotoxins 88, 89T
Neutralization, antibody-mediated 397, 397F
Neutralizing antibodies, defects in production
    422
Neutropenia **103–104**
    inherited 422
Neutrophiles 221T
Neutrophils **358–360**
    cell killing mechanisms 360
    innate immune response 359–360, 359T
    origins 359, 359F
    passage from blood into tissues 360, 360F
    phagocytosis 367
Nevirapine 473F, 475
Newborn infants
    antimicrobial eye treatment 447, 658
    chlamydial conjunctivitis 585
    cytomegalovirus infection 629
    eye infections **658,** 658F
    gonorrheal ophthalmia 657T, 658, 658F
    herpes infection 587
    HIV infection 415
    humoral immunity 399, 422
    meningitis 600
    microbiome 135, 136F
    sexually transmitted infections 576
    tetanus 601
    transfer of immunoglobulins to 405
Niacin (nicotinic acid) 219T
Nickel allergy 413, 427
Niclosamide 335, 480
Nicotinamide adenine dinucleotide *see* NAD+
Nipah virus 152
Nitric oxide (NO), viral hemorrhagic fevers 160
Nitrofurans 462T
Nitrofurantoin resistance 497T
Nitrogen (N), bacterial requirements **218,**
    218T
Nits (louse eggs) 655
NK cells *see* Natural killer cells
Non-cytocidal effect 93
Non-enveloped viruses
    assembly 287
    attachment 277–278
    intracellular trafficking 285
    penetration and uncoating **278–279,** 278F

Non-gonococcal urethritis (NGU) 579,
    583–585
Non-nucleoside inhibitors 473F, **475**
Normal microbial flora *see* Microbial flora,
    normal
Noroviruses **557–558**
Northern blotting 123
Norwalk virus 557
Nose
    colonization by *Staphylococcus aureus*
        191
    normal microflora 133, 133T
Nosocomial infections 7–8, 7F, **105–107**
    causative organisms 106, 106F
    common sites 106, 106F
    gastrointestinal tract **537**
    intravenous line and catheter bacteremia
        616–617
    pneumonia 516, **517,** 517F, 518
    preventing and controlling 107
    urinary system 572, 573
    wound infections 644
Notifiable diseases, national 110, 111F
Novobiocin resistance 497T
Nuclear membrane 67
Nuclear pores 67, 67F
Nuclear region, bacterial **205,** 205F
Nucleic acids **25–27**
    diagnostic methods based on **121–124,**
        122T
    microbial
        as antibiotic targets **470–471**
        as disinfectant/antiseptic targets 439
    structure 25–27
    synthesis
        antifungals targeting **478**
        antivirals targeting 474–475
        disinfectants targeting 439
    see also DNA; RNA
Nucleocapsid 273T
Nucleoid region *see* Nuclear region
Nucleoli 67
Nucleoplasm 67
Nucleoproteins 273
Nucleoside analogs **474–475**
    mechanism of action 281, 473F
Nucleotide analogs 473F, **475**
Nucleotides 236, 236F
    addition during DNA replication 240–241,
        241F
    excision repair 258, 258F
    linkages between 25–26, 26F
    structure 25, 25F
    see also Bases
Nucleus, cell **66–67**
    role in infection 67
    transport of viral genome into **280,** 281F
    viral DNA synthesis 282
    virus assembly 285
    virus uncoating 278F, 279
Numbers of microorganisms
    disinfectant/antiseptic efficacy and 439
    measurement **228–229,** 228F, 229F, 230T
    required for infection 83
Nutrient agar 222
Nutrient broth 222
Nutrient media 113, 222
Nystatin 477

O

O antigens (O polysaccharides) 192, 544, 544F
    E. coli 6
    Salmonella 549
Obligate intracellular parasites 86
Okazaki fragments 242–243, 243F
Older adults *see* Elderly
Oligonucleotide probe 122–123

Onchocerca volvulus 657, 657T
Onchocerciasis (river blindness) 657, 657T
Oncoviruses (oncogenic viruses) **311**
Onychomycosis 341, 341F
Oocysts
   Cryptosporidium 564
   Plasmodium 323, 323F
   Toxoplasma 327
Open reading frames (ORFs) 247
Operator site, DNA 252–253
Operon 253
Ophthalmia neonatorum 657T, 658, 658F
Opisthotonos 602, 602F
Opportunistic infections **105**
   AIDS/HIV infection 418, 418T
Opportunistic pathogens 9, **55,** 134
Opsonins 372
Opsonization
   bacterial capsule 84
   complement-mediated 374, 374F
   role of antibodies 397, 397F, 398
Organ transplantation **104**
Organelles 48, 62, **65–66**
Organic (biological) molecules **20–28**
Origin of replication 243
Origin of transfer 262, 263F
Ornithosis (psittacosis) **524**
Ortho-phenylphenol 445, 445F
Orthomyxoviruses 525
Oseltamivir (Tamiflu®) 166, 527T
   mechanism of action 473F, 475–476
   resistance 500
Osmosis 201–202
Osmotic lysis 202, 202F
Osmotic pressure **452**
   bacterial growth and **217**
Otitis media 485, **512,** 512F
Outbreaks, common-source 109, 109F
Outer membrane, Gram-negative bacteria
   189–190, 189F
   clinical significance 191–192
Oxazolidonones 462T, 470
Oxidation 32
   DNA 257
   reactions **32–33**
Oxidation–reduction reactions *see* Redox
   reactions
Oxidizing agents **446**
Oxygen (O)
   electron acceptor role 40, 41F
   requirements of bacteria **219–221,** 221T
Ozone 443T, 446

P

P pili 574
Palmitic acid 22F
Pancreatitis, acute 336
Pandemics 109
Pantothenic acid 219T
Papillomas 588, 652
   see also Warts
Papillomaviruses 588
   see also Human papillomavirus
Papules 643, 644F
Para-aminobenzoic acid (PABA) 219T, 471–472
Paracortical areas 390, 390F
Paragonimiasis 318T, **335–336**
Paragonimus 335–336
Parainfluenza **515–516**
Parainfluenza virus 515–516
Paralytic dose 50% (PD$_{50}$) 306
Paranasal conidiobolomycosis 342
Parasite-directed endocytosis 582
Parasitemia 330, 616
Parasites **318–322**
   life cycles and transmission **321–322,**
     322T

   morphology 319–321
   multiple hosts 321–322
   single host 321
Parasitic infections **318–337**
   blood and lymph **633–636**
   central nervous system **610**
   digestive system **563–567**
   drugs for treating **478–481**
   eosinophil responses 361, 361F
   eye 657T
   humoral immunity 401
   pathogenesis 319–321
   prevalence and significance 318–319, 318F
   skin **655**
   see also Helminthic infections; Protozoan
     infections
Parasitism 134, 134T
Parenteral route of entry 77T, **80–81**
PARESIS signs, neurosyphilis 581
Paromycin 469
Passive immunity 405–406
Passive immunization 309
Passive transport **202–203**
Pasteur, Louis 3
Pasteurization 436–437, 448, 451
   batch method 451
   flash method 451
Pathogen-associated molecular patterns
   (PAMPs) 357
Pathogen load 112
Pathogenesis 10
Pathogenicity 9, **55**
   bacterial **56–58**
   genetics and **265**
Pathogenicity islands 56
Pathogens
   accidental 10
   damage to host **87–94**
   defeating host defenses **83–87**
   defined 48, 54
   establishment **81–83**
   media used for identifying **224–226**
   numbers required for infection **83**
   opportunistic 9, **55,** 134
   portals of entry **76–81,** 76F, 77T
   portals of exit **102,** 102F, 103F
   primary (obligate) 9–10, **55**
   relationships with host 48, **54–56**
   requirements for infection 10, 56, **75–94**
   reservoirs **98–99**
   transmissibility **55**
PD$_{50}$ 306
Pediculosis **655,** 656F
Pediculus humanus capitis 655, 656F
Pediculus humanus corporis 627–628, 655
Pellicle, dental 81–82, 83F, 540
Pelvic inflammatory disease (PID) 576, 579
   gonococcal 579, 583, 583F
   non-gonococcal 579, 585
Penicillin(s) 462T
   discovery 8, 9, 460
   mechanism of action 11, **466**
   resistance 185
     mechanisms 466, **495**
     timeline of development 490, 490T
   selective toxicity 465
   semi-synthetic 463–464, 463F
   structure 462–463, 463F
Penicillin-binding proteins (PBPs) 466, **495**
Penicillin G 463, 463F
Penicillin V 463, 463F
Penicillinase 185
Penicillium 460, 461T
Pentamidine 478
Peptic ulcers 31, 78
Peptide bonds 23, 24F
   formation during translation 248, **250**

Peptides 24
Peptidoglycan 11, 21, 186
   antibiotics targeting 466
   assembly **187,** 188F
   components 186, 186F
   endospore formation 207, 207F
   Gram-positive *vs.* Gram-negative bacteria
     189, 189F
Peptidyl transferase reaction 250
Peptococcus 543F
Peptones 222
Peptostreptococcus 543F
Peracetic acid 443T, 446
Perforins 360, 365, 395
Period of convalescence 137, 138F
Period of decline 137, 138F
Period of illness 137, 138F
Periodontal disease
   necrotizing (NPD) **541–542**
   plaque-induced 541
Periodontal infections **540–542**
Periodontitis **541**
Peripheral tolerance 388
Periplasmic space 189F, 190
Peritrichous bacteria 198, 198F
Permeability, selective 199
Permeases 202–203, 203F, 253
Peroxidase 220
Peroxide anion 219–220
Peroxisomes **66**
Persistent infections
   bacteria **139–142,** 140T
   viruses **296–298,** 296T
Perspiration **355,** 356T, 642
Pertussis (whooping cough) **521–522**
   re-emergence 149T
   tracheal toxin 522
   vaccination 521
   see also Bordetella pertussis
Petechiae, viral hemorrhagic fevers 160
Peyer's patches 363, 385, 385F
pH **19–20**
   disinfectant/antiseptic efficacy and 441
   effects on enzymes 38
   requirements of bacteria **216–217,** 221T
   scale 20, 20F
Phage conversion 262
Phage display 118, 119F
Phage therapy 276, 502
Phages *see* Bacteriophages
Phagetyping 124
Phagocytes 367
   inherited defects 421T, 422–423, 423T
   migration to sites of inflammation **370**
   see also Macrophages; Neutrophils
Phagocytosis 68, 68F
   bacterial resistance to 84, 84F, 193–194,
     369
   compromised patients 369
   frustrated 57, 58F
   in host defense **367–369,** 368F
   phases 367–368, 368F
   pseudopod formation 67F, 68, 368
   resistance of pathogens to **369**
   role in infection 69
   role of lysosomes 66
Phagolysosome 368
Phagosome 368
Pharmaceutical industry 113, 116
Pharyngitis **512–513**
   bacteria causing 512F
   diphtheria 514
   sexually transmitted 576
Phase variation 195
Phenol 441, **445,** 445F
Phenol coefficient test **441–442**
Phenol-containing compounds 443T

Phenolic compounds 443T, **445,** 445F
Phosphoenolpyruvate (PEP) 204
Phospholipid bilayer
　bacterial plasma membrane 200, 200F
　eukaryotic intracellular membranes 65, 67,
　　67F
　eukaryotic plasma membrane 60, 61F
　viral envelope 275
Phospholipids **22,** 23F
Phosphorus, bacterial requirements **218,**
　218T
Phosphorylation
　glycolytic pathway 39
　substrate-level 42
Photoautotrophs 32
Photoheterotrophs 32
Photoreactivation 258, 259F
Pia mater 594, 594F
Picornaviruses
　enteroviruses 558
　hepatitis A 559
　rhinoviruses 515
Piedra 340
Pigs *see* Swine
Pili (pilus) **194–195,** 194F
　clinical significance 194–195, 208T
　conjugation (sex) 195, 262, 262F, 263F
　uropathogenic *E. coli* 574
Pilin protein 194
Pine nuts 152
Pinocytosis 67–68, 67F
Pinworm *see Enterobius vermicularis*
Piperazines 481
Pityriasis 340
Plague **620–621**
　as bioweapon **173–175**
　bubonic 54, 174, 174F, 621, 621F
　pneumonic 174, 175, 620, 621
　septicemic 174
　sylvatic 620
　treatment 621
　urban 620
　see also Yersinia pestis
Plantibodies 118
Plaque, dental *see* Dental plaque
Plasma cells 386F, 387, 402, 404
Plasma membrane
　bacterial 60, **199–205**
　　as antibiotic target **468**
　　cell wall assembly 187
　　clinical significance 204, 208T
　　as disinfectant/antiseptic target
　　　437–438, 438F
　　electron transport chain 41
　　energy production 200–201, 201F
　　secretion via 204
　　structure 200, 200F
　　transport across 201–204
　eukaryotic (host) cell **60–61,** 61F
　　role in infection 62, 62F
　　virion budding **288,** 288F
　　virus assembly 285, 286F
　　virus attachment 277, 278–279, 278F
　fluid mosaic model 61, 61F
　fungal 60, 338
　　drugs targeting **477**
　selective permeability 199
　see also Membrane proteins
Plasmids 56, **205,** 205F
　antibiotic resistance 486–487, 487F
　clinical significance 205–206, 265
　dissimilation 265
　fingerprinting 124
　integration into host chromosome 264,
　　264F
　recombinant 115–116, 115F
　role in conjugation 262–263, 263F

spontaneous loss 263
traits encoded by 263T
Plasmodium 318–319, **323–326**
　drug resistance 479
　life cycle 322, 323–324, 325F
　pathogenesis of disease 320, 324–326
　sporozoites 323–324, 323F
　see also Malaria
Plasmodium falciparum 323
　emergence from red blood cells 324F
　pathogenesis of disease 324–325
　transmission 322T
　treatment 326
Plasmodium vivax 323
Plasmolysis 201, 202, 202F
Pleomorphic bacteria 48
Pneumococcus *see Streptococcus*
　　pneumoniae
Pneumocystis (carinii) jiroveci 529, 529F
Pneumocystis pneumonia (PCP) **529–530**
　AIDS patients 414, 529
　pathogenesis 529
　treatment 530
Pneumonia
　atypical 512F, 516–517
　bacterial 512F, **516–518**
　Chlamydophila **518**
　community-acquired 512F, 516, **517**
　complicating influenza 526–527
　Legionella **523–524**
　lobar, stages 517
　Mycoplasma **518–519**
　nosocomial (hospital-acquired) 512F, 516,
　　517, 517F, 518
　Pneumocystis 414, **529–530**
　treatment 518
　walking 518
Pneumonitis 529
Polio (poliomyelitis) **605**
　herd immunity 110
　incubation period 294T, 605
　pathogenesis 605
　prevention 605
　vaccination success 307–308, 309F
　vaccine 308, 308T, 605
Poliovirus **605**
　dissemination within host 302F, 303, 605
　portal of entry 78
　structure 274
Poly-Ig receptor 400F
Polyene antifungals **477**
Polymerase chain reaction (PCR)
　diagnostic use **121–123**
　principle 116, 117F
Polymixin B 468, 497T
Polyoma virus BK 296T
Polyoma virus JC 296T
Polypeptide antibiotics **468**
Polypeptides 24
Polyribosomes (polysomes) 250, 250F
Polysaccharides 20, 21, 21F
Pore-forming toxin 545
Porins 190, 192
Portals of entry **76–81,** 76F, 77T
　viruses **299–301,** 299F
Portals of exit **102,** 102F, 103F
Posaconazole 477
Post-translational modification 195
Potassium clavulanate 492
Potassium hydroxide (KOH), DNA release 235
Poverty 162
Praziquantel 335, 336, 337, 480
Precipitation test 119, 119F, 122T
Pregnant women
　susceptibility to virus infections 307
　urinary tract infections 575
　virus transmission to baby 306
　see also Fetal infections

Prevalence **108,** 108F
Prevention, disease 11
Prevotella 543F
Primaquine 479
Primary immune response 401, 401F
Primary infections 132T, 142
Primase 243
Primaxin 467, 492
Primers
　DNA replication 240, 240F
　lagging strand synthesis 242–243, 243F
Prion diseases 168F, **607–609**
　biology 167–169
　emergence **167–169**
　pathogenesis 607
　pathology 608F
Prion hypothesis **167**
Prions **167–169,** 594, **607–609**
　cellular form (PrP^C) 167, 167F, 607, 607F
　methods of destruction 167
　properties 607
　scrapie form (PrP^{SC}) 167, 167F, 607, 607F
Pristinamycin 469
Procainamide 424
Prodromal period 137, 138F
Prodrugs 474
Products, reaction 33
　concentration effects 38
Proglottids 320, 334–335
Proguanil 479
Prokaryotes 48
　cell structure **59–60,** 59T, 60F
　cytoplasm 62
　ribosomes *see* Ribosomes, prokaryotic
　translation 248–250, 248F, 248T, 249F
　see also Bacteria
Promoter site/region
　control of gene expression 252–253
　initiation of transcription 246, 246F
Properdin (factor P) 373, 373F
Prophage 261
β-Propiolactone 447
Propionibacterium acnes 642, 643T, 644F, 645
Propylene oxide 447
Prospective analytical epidemiological
　　studies 110
Prostaglandins 369
　fever 371, 371F
Prostatitis 573, 574, **575**
Protease, HIV 414F
Protease inhibitors 473F, **475**
Proteasomes **66,** 66F
Protein(s) **23–25**
　3D structure 24, 25F
　　pH effects 216–217
　　temperature effects 216
　denaturation 24, 438, 438F
　　temporary or permanent 444, 444F
　as disinfectant/antiseptic targets 438,
　　438F, 444, 444F
　microarrays 125
　properties 23–24
　structural 24
　synthesis **246–252**
　　as antibiotic target 64, **469–470,** 470F
　　bacteria 206
　　viruses 283, **284–285**
　　see also Translation
　types 24–25
Proteomics 124–125
Proteus, urinary tract infections 574F
Proteus mirabilis 79, 194F
Proton motive force 41, 201, 201F
Protozoan infections **323–330**
　digestive system **563–565**
　pathogenesis 320
　sexually transmitted 577F

significance 318–319, 318T
treatment **478–480**
Protozoans 318, **319–320**
  life cycles and transmission 321–322, 322T
  locomotion 319–320
  morphology 319–320
  size 49F, 319
  thermal death times 449T
Provirus 275, 276F
  integration 281F
PrP^C 167, 167F, 607, 607F
PrP^SC 167, 167F, 607, 607F
Prusiner, Stanley 167
Pseudocyst, *Trypanosoma cruzi* 634
Pseudomembrane, diphtheria 514, 514F
Pseudomembranous colitis 537
Pseudomonas
  antibiotic resistance 493, 494
  bacteremia 617
  in soap and cleaning solutions 98
Pseudomonas aeruginosa
  burn patients 97, 104
  keratitis 656
  nosocomial infections 106, 106F
  nosocomial pneumonia 512F, 516
  optimal growth temperature 215T
  quorum sensing 57
  skin and soft tissue infections 644F, 645
  urinary tract infections 574F
  ventriculo-peritoneal shunt infection 597F
Pseudopodia (pseudopods)
  ameboids 319
  phagocytes 67F, 68, 368
Psittacosis **524**
Psychrophiles 214F, 215, 221T
Psychrotrophs 214F, 215, 221T
Public health measures 11
Puerperal infections 7
Pulmonary interstitial macrophages 362
Purines 25
  bacterial requirements 219T
  structure 26F, 237, 237F
Pus 366
Pustules 643, 644F
Pyogenic bacteria, recurrent infections 421, 423F
Pyrantel pamoate 332
Pyrazinamide 497T, 520
Pyrexia *see* Fever
Pyridoxine (vitamin B$_6$) 219T
Pyrimethamine 479
Pyrimidines 25
  bacterial requirements 219T
  structure 26F, 237, 237F
Pyrogens 371
  endogenous 371
  exogenous 371
Pyruvate
  fermentation 42–43, 43F
  Krebs cycle 40
  metabolic pathways 39, 39F
  production during glycolysis 39
Pyuria 575

Q

Q fever **524**
Quarantine 139
Quaternary ammonium compounds (QUATS) 443T, 446–447, 446F
Quinacrine 477
Quinine 478–479
Quinolone antibiotics 462T, **471**
Quorum sensing **56–57**, 58F

R

Rabies 598T, **603–605**
  incubation period 294T, 604
  pathogenesis 604
  precautions 131
  re-emergence 149T
  reservoir 98, 98T
  sylvatic 604
  treatment 604–605
  urban 604
  vaccine 308T, 605
Rabies virus 604, 604F
  cytopathic effect 93, 93F
  dissemination within host 303, 598, 604
  portal of entry 301, 301F
Radiation **452–454**
  disinfection 453F, 454
  DNA damage 257, 452, 453F
  ionizing 452–453, 453F
  mechanism of microbial killing 439
  non-ionizing 452, 453F
  sterilization 452–453
Radioimmunoassay (RIA) 119, 121
Radioisotope-labeled antibodies 120, 121
Rashes *see* Skin rashes
Re-assortment, genetic 153, 154
Re-emerging infectious diseases 9, **161–165**
  antibiotic resistance **498–499**
  defined 148
  examples 149T
  mechanisms of re-emergence 148–149, 150
Reactants 33
Reactive oxygen species (ROS) 368
Reading frame 245
Real-time polymerase chain reaction (PCR) 122–123
Receptor-mediated endocytosis 68, 68F
  viruses 278F, 279
Receptors, virus 276–277
Recombinant DNA technology **115–116**
  applications 116, 116F
  development of antiviral drugs 473, 473F
  diagnostic methods using **121–124**
Recombinant vaccines 114F, **310**
Recombination, genetic 259
  generating viral diversity 155F
  high frequency of (Hfr) 264, 264F
  mechanisms in bacteria **259–265**
Redox reactions 32–33
  coenzymes 36, 36F
  electron transport chain 40
Reduction reactions **32–33**
Reduviid bugs 329, 633, 634
Reed, Walter 632
Refrigeration 38, **451**
Regulatory T cells (T$_{reg}$ cells) 387
  functions 394, **396**
Relapsing fever **625–626**
  louse-borne 625
  tick-borne 625
Reoviruses 556
  dissemination within host 302F
  portal of entry 300
Replicator sequence 243
Reproductive system infections **576–590**
  bacterial **577–585**
  fungal **589–590**
  portals of entry 79, 79F
  viral **586–589**
  see also Sexually transmitted infections
Reservoirs, pathogen **98–99**
Residual body 368
Resistance islands, antibiotic 497
Respiration **33**
  cellular 33, 39F
Respiratory equipment, transmission of infection 106
Respiratory syncytial virus (RSV) 525, **527–528**

Respiratory system
  anatomy **510,** 510F
  bacterial adherence 82T
  barriers to infection 353, 354F
  ciliated cells 63, 64, 64F
  defenses against infection 511, 511F
  lymphoid tissue 385
  mucociliary escalator 353, 354F, 511, 511F
  normal microflora 133, 133T
  as portal of entry **76–77,** 77T
  viral entry **299–300,** 300F
Respiratory system infections **509–533,** 510F
  bacterial 510, **512,** 512F
  fungal 511, **528–532**
  manifestations 300F
  pathogens causing **510–512**
  transmission 100, 100F, **304,** 304F, 510–511
  viral 510
  see also Lower respiratory tract infections; Upper respiratory tract infections
Restriction enzymes 115F
Reticulate body (RB) 584, 584F
Reticuloendothelial system 362
Retrospective analytical epidemiological studies 110
Retroviruses
  integration **284**
  oncogenicity 311
  transcription **284**
  transport into nucleus 280, 281F
  see also Human immunodeficiency virus
Reverse transcriptase 280, 281F, 284
  HIV 414F, 416
  non-nucleoside inhibitors 475
  nucleoside inhibitors 474
Reverse transcriptase polymerase chain reaction (RT-PCR) 122
Rheumatic fever 424
Rheumatoid arthritis 424, 425
Rhinitis
  allergic *see* Allergic rhinitis
  zygomatic 342
Rhinoviruses **515,** 515F
  antigenic variation 295
  dissemination within host 302F
  host cell receptor 62F, 277, 515
  see also Common cold
Ribavirin 475
Riboflavin (vitamin B$_2$) 219T
Ribonucleic acid *see* RNA
Ribose 25, 25F, 238
Ribosomal RNA (rRNA) 238, 239T
  16S 249, 249F
    gene sequences 123–124
  ribosome subunits 248T, 249, 249F
Ribosomes
  A site 250, 250F, 251, 251F
  attached to endoplasmic reticulum 65
  E site 250, 250F, 251, 251F
  eukaryotic **64**
  P site 250, 250F, 251, 251F
  prokaryotic **206**
    as antibiotic targets 64, **469–470,** 470F
    clinical significance 206, 208T
    linkage to RNA polymerase 248, 248F
    modification in antibiotic resistance **495–496**
  mRNA binding 247, 248
  structure 248–249, 248T, 249F
  in translation 248–250, 248F, 249F
  role in infection 64
  subunit structure 248–249, 248T, 249F
  in translation **248–250,** 248F, 249F
  tRNA binding sites (A, P and E) 250, 250F
Rice stool 90, 553
Rickettsia 137, **626–627**

Rickettsia akari 626T
Rickettsia prowazekii 137, 626T, 627–628
Rickettsia rickettsii 626T, 627
Rickettsia typhi 628
Rickettsial infections **626–628,** 626T
Rickettsial pox 626T
Rifampin (rifampicin)
  mechanism of action 471
  resistance 497T, 520
  tuberculosis therapy 520
Rifamycins 462T, **471**
Rift Valley fever virus 158F
Rimantadine 165–166, 527, 527T
Ring stage 324
Ring test 119, 119F
Ringworm 341, 341F, 654, 654F
Ritonavir 475
River blindness (onchocerciasis) 657, 657T
RNA (ribonucleic acid)
  bacterial
    analysis **501**
    as antibiotic target **470–471**
  as disinfectant target 439
  Northern blotting 123
  primers
    DNA replication 240, 240F
    lagging strand 242–243, 243F
  strand loops 239
  structure 25, 27, **238–239**
  synthesis 245–246, 246F
  types 238, 239T
  see also Messenger RNA; Ribosomal RNA;
    Transfer RNA
RNA polymerase 246, 246F
  as antibiotic target 471
  linkage to ribosome 248, 248F
  virus transcription 282
RNA viruses 272, 272F
  arboviruses 156
  biosynthesis **283–285**
  double-stranded RNA (dsRNA) 284
  emerging infectious diseases 154
  helical 273, 273F
  (-) single-stranded RNA (-ssRNA) 284
  (+) single-stranded RNA (+ssRNA) 284
  transport into nucleus 280, 281F
RNAase H 243
Rocky Mountain spotted fever (RMSF) 626,
    626T, **627**
  rash 627, 627F
Rostellum, tapeworm 321F, 334, 334F
Rotaviruses **556–557,** 556T
  dissemination within host 302F
  pathogenesis of infection 557
  treatment of infection 557
  vaccine 308T
Roundworms *see* Nematodes
rRNA *see* Ribosomal RNA
Rubella 643T, **649**
  congenital 649
  persistent panencephalitis 296T, 606, 606T
  vaccine 308T, 649
Rubeola 648
Rubeola virus *see* Measles virus

S

S pili 195
Sabouraud's agar 224
Safranin 51T, 52, 54
Saint Louis encephalitis 606, 631T
Saliva **354,** 356T, 541
Salmonella **549–552,** 549T
  bacteremia 549
  chronic infection 550
  enteric fever 536, 549–550
  food poisoning 539T, 550
  gastroenteritis 542F, **550–551,** 551F

  motility 549, 549F
  movement within host 86–87, 199
  quorum sensing 57
  serotypes 549
  treatment of infections 552
  virulence factors 544
Salmonella bongori 549
Salmonella enterica 549, 549T
Salmonella enterica serotype Typhi (*Salmonella typhi*) 549T, 551–552
  exogenous infection 543F
  identification medium 224
  pathogenesis of disease 551–552
  persistence 140–142, 140T, 141F
  phenol coefficient 441–442
  portal of entry 80
  treatment of infections 552
  see also Typhoid fever
Salmonella enteritidis 542F, 543F
Salmonella paratyphi 543F
Salmonella typhi see Salmonella enterica serotype Typhi
Salmonella typhimurium 543F
Salpingitis 576
  gonococcal 583, 583F
  non-gonococcal 585
Salt, food preservation 452
Sample collection, clinical 112
Sanitization 436
Saquinavir 473F, 475
SARS (severe acute respiratory syndrome) 153, **154–156**
  host response and treatment 156
  pathogenesis 155–156, 155F
  rapid spread 154–155
SARS coronavirus 156, 156F
Scalded skin syndrome 643T, **646–647**
Scalp ringworm 654, 654F
Scarlet fever 89T, 90, **513**
Schistosomes (*Schistosoma* flukes) 335
  morphology 337, 337F
Schistosomiasis **337**
  pathogenesis 320, 321, **337**
  prevalence 318T
  treatment 337
Schizogony 319
Schlosser, Eric 536
Scolex 334, 334F
Scrapie 167, 168
Seasonal variations, virus transmission 304
Sebaceous glands 642–643
Sebum **355,** 356T, 642–643
Secondary immune response 401, 401F
Secondary infections 132T, 142
Secretion, bacterial **204**
Secretory piece 399–400, 399F
Selectin 360
Selective permeability 199
Selective toxicity 64, 465
Semmelweis, Ignaz 7
Sensitivity, diagnostic test 112
Sentinel cells *see* Mast cells
Sepsis **142–143,** 616, **618**
  severe 143, 618
Sepsis syndrome 143
Septa, hyphal 339, 339F
Septic shock 616, **618**
  acute 143
  refractory 618
Septicemia 132T, 142
Septum, endospore formation 207, 207F
Serial dilution 119
Serological analysis 113, 119, 122T
Serology-based diagnostics 113, 118,
    119–121, 122T
Serum sickness 427
Severe acute respiratory syndrome *see* SARS

Severe combined immunodeficiency
    syndrome (SCID) 421T, 422
Severe sepsis 143, 618
Sexual reproduction
  cestodes 334
  fungi 338
  Plasmodium 323–324
  protozoans 319
Sexual transmission 79
  HIV 415–416, 415F
  Trichomonas vaginalis 328
  viruses 305
Sexually transmitted infections (STIs) 572,
    576–589, 577F
  bacterial **577–585,** 577F
  clinical syndromes **578–579**
  fungal 577F, **589–590**
  localized or systemic 576
  organisms causing 577F
  portal of entry 79
  protozoan 577F
  viral 300–301, 305, 577F, **586–589**
Shiga toxin
  E. coli 89T, 153, 545, 546T, 547
  Shigella 89T, 547, 548T, 549
Shigella **547–549**
  cytoskeletal interactions 63, 63F, 199
  food poisoning 539T
  pathogenesis of disease 548–549, 548F
  sources and transmission 542F, 548
  virulence factors 544
Shigella boydii 547, 548T
Shigella dysenteriae 547, 548, 548T
  adherence 82T
  antibiotic resistance 489
  exogenous infection 543F
  phage encoding of toxin 275
Shigella flexneri 547, 548, 548T
Shigella sonnei 543F, 547, 548, 548T
Shigellosis **547–549**
  pathogenesis 548–549, 548F
  treatment 549
Shingles 643T, **650**
  clinical features 650, 651F
  pathogenesis 139, 298, 298F, 650
  vaccine 650
  see also Varicella-zoster virus
Siderophilic microbes 35
Silver 443T, 447
Silver nitrate 447
Silver sulfadiazine 447
Sin Nombre virus 528, 528F
Sinusitis **512,** 512F
16S rRNA 249, 249F
  gene sequences 123–124
Skin **79–80**
  anatomy 80F, **642–643,** 642F
  barrier function 79–80, 80F, **352–353,**
    353F, 356T
  Langerhans cells 363
  lesion types 643, 644F
  normal microflora 79, 80F, 133
    common organisms 133T, 642
    entry to underlying tissue 353, 353F
  portal of entry 77T, 80–81, 643
  protective mechanisms 642–643
  viral entry via **301,** 301F
  virus transmission via **305**
Skin infections **641–655,** 643T
  bacterial 643T, **644–648,** 644F
  fungal 340, 341, 341F, 643T, **653–654**
  parasitic **655**
  viral 643T, **648–652**
Skin rashes
  chickenpox 650
  Lyme disease 624, 624F
  maculopapular
    endemic typhus 628
    viral infections 305, 305T

Rocky Mountain spotted fever 627, 627F
secondary syphilis 580, 581F
shingles 650, 651F
smallpox 650, 650F
vesicular 305, 305T
viruses causing 305, 305T
Sleeping sickness (African trypanosomiasis)
329–330
pathogenesis 330
significance 318T, 319
treatment 330, 480
Slime layer 193
Slow viral infections 296, **298**, 606–607, 606T
Smallpox (variola) 643T, **649–650**
as bioweapon 148, 170, **175–176**
herd immunity 109, 309
incubation period 294T
inoculation 5, 307
major 649
minor 649
portal of entry 77
signs and symptoms 175, 175F, 650, 650F
vaccination 147, 175–176, 650
herd immunity and 109, 309
history 307
vaccine 308T
Smallpox virus see Variola virus
Smith, Betty 16
Snails 335, 336, 337
Snapping 514, 514F
SNARE proteins 90
Sneezing
beneficial function 400
disease transmission 100F, 304, 304F
Snyder test 15
Soaps 105, 446, 454
Sodium 16, 17, 17F
Sodium chloride (NaCl)
ionic bonding 17, 17F
solubility in water 19, 19F
Sodium dodecyl sulfate polyacrylamide-gel
electrophoresis, two-dimensional
(2D SDS-PAGE) 124–125
Sodium stibogluconate 480
Sodium thioglycolate tubes 220, 220F
Soft tissue infections 643T, **644–648**, 644F
Soil 99
Solubility 19
Southern blotting 123
Spanish flu pandemic (1918) 4, 5F, 163–164,
164F
Species 48
Species barrier, crossing the **154**
Specificity, diagnostic test 112
Specimen collection, clinical 229–230, 231T
Spectinomycin 469
Spectrophotometry 228–229, 229F
Sperm cells 64
Spinal cord 594
Spirilla 49
Spirochetemia 625
Spirochetes 49
axial filaments 195, 195F
Spleen, macrophages 362F
Sporadic diseases 108
Spores
disinfection resistance 445
fungal 338
physical methods of killing 448, 449T, 453
see also Endospores
Sporotrichosis 342
Sporozoans 320
Sporozoites
Plasmodium 323–324, 323F
Toxoplasma 327
Sporulation 53, 206–207, 207F
see also Endospores

Spread of infection see Transmission
Spreading out 86
Sputum samples 229, 231T
Staining **50–54,** 51T
differential 51
negative 50, 51T
simple 50–51
Staphylococci (Staphylococcus species)
coagulase-negative see Coagulase-
negative staphylococci
exotoxins 86
identification 225, 225F
immunosuppression by 420
impetigo 646, 646F
morphology 49, 50F
skin and soft tissue infections 643T
Staphylococcus aureus
appearance 50, 50F
bacteremia 617
cell wall 191
central nervous system infections 597F
exotoxins 90
food poisoning 539T, 542F, 543F
identification 225, 225F
infectious endocarditis 618T
methicillin-resistant see MRSA
nasal colonization 191
normal skin 642
nosocomial infections 106, 106F
nosocomial pneumonia 516
optimal growth temperature 215T
phenol coefficient 441–442
quorum sensing 57
respiratory system infections 512F
scalded skin syndrome 646, 646F
skin and soft tissue infections 643T, 644,
644F, 645
urinary tract infections 574F
vancomycin-resistant (VRSA) 7F, 106, 498
Staphylococcus epidermidis 617, 642
identification 225F
nosocomial infections 106, 106F
Staphylokinase 86
Start codon see Initiation codon
-Static agents 437
Stationary phase, bacterial growth 227F, 228
Steam, pressurized 449–451, 450F, 450T
Stem cells, bone marrow 359F, 386F
Sterilants, chemical 443T
Sterilization 436
filtration 452
gaseous agents 447–448
heat 448, 449–451, 449T
radiation 452–453
Steroids **23**
Stomach, specimen collection 231T
Stop codons 244, 245F, 251
Straintyping 123, 124
Strep throat 512–513, 513F
Streptococcal toxic shock syndrome (STSS)
143
Streptococci (Streptococcus species)
autoimmune reactions 424
bacteremia 617F
bacteriocins 134
brain abscess 597F
capsule 193F
group A see Group A streptococcus
group B see Group B streptococcus
group C 512F
group D 225F
hemolysis 85
identification by hemolysis 225–226, 225F
infectious endocarditis 618T
intestinal infections 543F
kinase enzymes 85, 86
M proteins see M proteins, streptococcal
morphology 49, 49F, 50F

respiratory system infections 511,
512–513, 512F
skin and soft tissue infections 643T
virulence factors 512–513, 513F
Streptococcus agalactiae see Group B
streptococcus
Streptococcus mutans 15, 543F
dental caries 541
dental plaque formation 81–82, 83F, 540,
540F
Streptococcus pneumoniae (pneumococcus)
adherence 81, 82T
alpha hemolysis 225
antibiotic resistance 485, 495
bacteremia 617F
capsule 52F, 83–84, 193
infections caused 512F
meningitis 596F, 597F, 599
morphology 49, 49F
pneumonia 517, 518
portal of entry 80–81
transformation 260
Streptococcus pyogenes see Group A
streptococcus
Streptococcus salivarius 541
Streptococcus sanguinis 543F
Streptogramin 469
Streptokinase 85, 85F, 86
Streptolysin 85
Streptolysin O 226
Streptolysin S 226
Streptomyces, antibiotic production 461T, 469
Streptomycin 469, 490T
Strongyloides stercoralis 331
Styes 656
Subacute disease 132T, 139
Subacute sclerosing panencephalitis (SSPE)
139, 606, 606T
Subarachnoid space 594, 594F
Subclinical infections 142
Substrate-level phosphorylation 42
Substrates 33
concentration effects 38
Sugar
dental caries and 541
food preservation 452
Sulfa drugs 460
antiviral 473
mechanism of action 471–472, 472F
Sulfadoxine–pyrimethamine 500
Sulfamethoxazole 472
Sulfisoxazole 497T
Sulfonamides 462T
mechanism of action 36
resistance 490T
Sulfur, bacterial requirements **218**, 218T
Sunlight 436
Superantigens 90, 392
Superinfections **499,** 499F
hepatitis D 562
influenza 526
Superorganism 132
Superoxide dismutase (SOD) 219–220
Superoxide free radicals 219
Surfactants **446–447,** 446F
mechanism of microbial killing 437–438,
438F
Surgical procedures 80
Svedberg units 249
Swabs 231T
Sweat (perspiration) **355,** 356T, 642
Swine (pigs)
influenza virus incubation 164, 166, 166F
Nipah virus transmission 152
trichinosis 333
Swine flu pandemic (2009) 5, 164
Symmetrel® see Amantadine
Symptoms of disease 137

Syncytia
  virus-infected cells 93, 93F
  virus spread via 289
Synercid 469, 497T, 498
Syphilis (*Treponema pallidum* infection)
  579–581
  cardiovascular 581
  congenital 572, 581
  distribution of lesions 216, 216F
  latent 580–581
  pathogenesis 580
  primary 580
  secondary (disseminated) 580, 581F
  tertiary 580, 581, 581F
  treatment 581
  ulcers 578, 578T, 580, 580F
  see also Treponema pallidum
Systemic infection 132T, 142
Systemic lupus erythematosus (SLE) 424, 425

T

T-cell receptors 388, 388F, 392
T cells (T lymphocytes) 387–388
  activation of B cells **402**
  activation of naive **393–394**, 393F
  clonal selection 388–389
  co-stimulatory signals 388
  differentiation and maturation 386F,
      388–389
  effector
    during course of infection 403F, 404
    functions **394–396**
    production **393–394**, 393F
  immune response 384, **393–396**
  inherited defects **422**
  memory 405
  naive 393–394, 403F, 404
  peripheral lymphoid tissues 390, 390F, 391
  response to infection 403F, 404
  superantigen response 392
  survival 389–390
  type IV hypersensitivity 427
  see also Cytotoxic T lymphocytes; Helper
      T cells; Regulatory T cells
Taenia saginata 321, 334–335
tag O gene 191
Tamiflu® *see* Oseltamivir
Tampons 90
Tapeworms *see* Cestodes
Tears 354, 354F
Tegument 274, 274F
Teichoic acid 189, 189F
  clinical significance 191
  wall 189, 189F, 191
Teicoplanin 467
Telithromycin 497T
Temin, Howard 280
Temperature
  body 371
    see also Fever
  cold, microbial growth 38
  disinfectant/antiseptic efficacy and 440,
      441
  enzyme activity and 38
  high, microbial killing 448, 449T
  maximum growth 215
  minimum growth 215
  optimal growth 215, 215T
  requirements of bacteria **214–216**, 214F,
      221T
  see also Cold temperatures; Heat
Tet genes 494
Tetanospasmin *see* Tetanus toxin
Tetanus **601–602**, 602F
  pathogenesis 601–602
  treatment 602
  vaccination 602
  see also Clostridium tetani

Tetanus immunoglobulin 602
Tetanus toxin 89T, **90**, 601–602
Tetracyclines 462T
  efflux pumps **494**
  mechanism of action **469–470**, 470F
  resistance 490T, 494, 497T
Tetrahydropyrimidines 481
T$_{FH}$ cells 395, 396, 404
T$_H$0 cells 395
T$_H$1 cells 395
  functions 395F, 396
T$_H$2 cells 395
  functions 395F, 396
T$_H$17 cells 395, 395F, 396
Thermal death point (TDP) 449
Thermal death times (TDTs) 449, 449T
Thermophiles 214F, 215, 221T
  extreme (hyperthermophiles) 214F, 215,
      221T
Thiamine (vitamin B$_1$) 219T
Thiosemicarbazones 473, 480
Thomas, Lewis 48
Throat
  normal microflora 133, 133T
  strep 512–513, 513F
Thrush 342, 589, 653, 653F
Thymine (T) 25, 26F
  dimers 257, 258, 259F
  specific pairing 27, 27F, 237, 237F
  structure 237F
Thymus 388
  T-cell development 388
  T-cell survival 389–390
Tick-borne diseases
  Lyme disease 623–625
  relapsing fever 625
  Rocky Mountain spotted fever 627
Ticks 101, 152, 152F
Tincture 445
Tinea capitis 340, 654, 654F
Tinea corporis 341, 341F
Tinea cruris 341, 654
Tinea nigra 340
Tinea pedis 341, 341F, 654
Titer, antibody 119
TLRs *see* Toll-like receptors
Tobramycin 470
Togaviruses 631T
Tolerance (immunological)
  central 389
  peripheral 388
Toll-like receptors (TLRs) **357**, 357T
  basophils 361
  dendritic cells 364
  genetic variations 378
  mast cells 365
  neutrophils 360
Tonsillitis 514
Tonsils
  abscesses 513
  surgical removal 514
Tooth decay *see* Dental caries
Topoisomerase 240, 243, 244F
  as antibiotic target 471
Toxemia 132T, 142
Toxic shock 90, **142–143**
Toxic shock syndrome toxin 90
Toxins **87–93**
  antigen presentation 391F
  encoding by phages 275
  food poisoning 538–539, 539T
  see also Endotoxins; Exotoxins
Toxoids 91
Toxoplasma gondii 326–327
Toxoplasmosis **326–327**
Tracheobronchitis 516
Trachoma 657, 657F, 657T

Traffic accidents 105
Transacetylase 253
Transcription **245–246**
  bubble structure 246, 246F
  DNA viruses **282–283**
  elongation 246, 246F
  initiation 246, 246F
  linkage with translation 248, 248F
  regulation 252–254
  retroviruses **284**
  RNA viruses **283–284**
  termination 246, 246F
Transcriptomics 124
Transcytosis, immunoglobulins 399–400, 400F
Transduction **260–262**, 265T
  generalized 261, 262F
  specialized 261–262
Transfer RNA (tRNA) 238, 239T
  acceptor arm 247, 247F
  anticodon loop 247F, 248
  ribosomal binding sites (A, P and E) 250,
      250F
  structure 247–248, 247F
  in translation **247–248**, 251
Transferrins **355**, 356T
Transformation **260**, 261F, 265T
  competence 260
Translation **246–252**, 251F
  elongation 251
  initiation 250, 250F
  linkage with transcription 248, 248F
  mRNA **247**, 251, 251F
  peptide bond formation **248**, 250
  prokaryotic 206, 248–250, 249F
  ribosomes **248–250**, 248F, 248T, 249F
  termination 251–252
  tRNA function **247–248**, 251, 251F
  viral control **284–285**
  see also Protein(s), synthesis
Translational apparatus 250
Translocation, group 204, 204F
Translocation protein systems 190
Transmembrane efflux pumps *see* Efflux
      pumps
Transmissibility, disease **55**
Transmissible spongiform encephalopathies
      (TSE) 167–169, 607–609
Transmission, disease **98–102**
  environmental change and 151–152
  factors affecting 102
  globalization and **153–154**
  in hospitals 4, 4F, 7–8, 106
  mechanisms 99–101
  parasites **321–322**, 322T
  reservoirs of pathogens 98–99
  viruses 299
Transplantation, organ **104**
Transporters, superfamily of 203–204
Transposition **259–260**
Transposons 256, **259–260**
Travel, international 4, 153, 489
Traveler's diarrhea **537**
Trematodes (flukes) **335–337**
  hermaphrodite 335
  morphology 320, 321F
  see also Schistosomes
Trench mouth **541–542**
Treponema pallidum 577F, **579–581**
  adherence 82T, 83, 83F
  axial filaments 196, 196F
  difficulty of studying 137
  morphology 579, 580F
  optimal growth temperature 216, 216F
  see also Syphilis
Triazoles 477
Tricarboxylic acid cycle (TCA) *see* Krebs cycle
Trichinella spiralis 322T, 333–334, 539T

Trichinosis **333–334**
Trichomonas vaginalis **328–329,** 577F
  life cycle 329
  morphology 320F, 329
  transmission 321, 322T, 328
Trichomoniasis **328–329**
  pathogenesis 320, 329
  treatment 329
Trichophyton 340
Trichophyton rubrum 341F
Trichuris trichiura (whipworm) 331, **565–566**
  morphology 565, 565F
  pathogenesis 321, 565–566
  prevalence 318T, 319
  treatment 566
Triclosan 443T, 445, 445F
Triglyceride lipid, formation 22F
Trimethoprim 497T, 575
Trimethoprim-sulfamethoxazole (TMP-SMX)
  530, 575
Triple sugar iron agar 224T
tRNA *see* Transfer RNA
Trophozoites 319
  Entamoeba histolytica 327–328
  Giardia 563, 563F
  Toxoplasma 326, 327
  Trichomonas 329
Trypanosoma **329–330**
  drug treatment **480**
  transmission 322T, 329, 329F
Trypanosoma brucei 329–330, 329F
Trypanosoma cruzi 319, 633–634, 633F
Trypanosomiasis **329–330**
  African *see* Sleeping sickness
  American *see* Chagas' disease
Trypomastigotes 330, 633, 634
Tryptophan
  operon 255F
  regulation of synthesis 255, 255F
Tsetse fly 329, 329F
Tubercles 520, 520F
Tuberculosis (TB) **161–163**
  contagiousness 4, 4F, 138–139
  directly observed therapy (DOT) 521, 521T
  drug resistant 162, 163F, 498–499, 519
  extensively drug-resistant (XDR-TB) 162,
    163F, 499, 520
  historical aspects 507
  HIV co-infection 162, 519
  mortality 161–162
  multi-drug resistant (MDR-TB) 162, 499,
    520–521
  non-compliance with therapy 162
  pathogenesis 519–520, 520F
  persistent infection 140, 140T
  primary 520
  pulmonary **519–521**
  re-emergence 149T, 162–163
  secondary 520
  signs and symptoms 162, 163
  treatment 468, 520–521
  see also Mycobacterium tuberculosis
Tubulin 63, 64
Tularemia **621–622**
  as bioweapon **176**
  oculoglandular 622
  respiratory 176
  typhoidal 622
  ulceroglandular 176, 622
  see also Francisella tularensis
Tumor necrosis factor (TNF) 358, 366F
Turbidity testing 228–229, 229F
Turner, Glenn 272
Two-dimensional sodium dodecyl
    sulfate polyacrylamide-gel
    electrophoresis (2D SDS-PAGE)
    124–125

Typhoid fever **551–552**
  carriers 535, 551
  clinical symptoms 536
  pathogenesis 551–552
  persistence 140–142, 140T, 141F
  treatment 552
  see also Salmonella enterica serotype
    Typhi
"Typhoid Mary" 535, 551
Typhus 626, 626T
  endemic **628**
  epidemic **627–628**

U

Ubiquitin 66, 66F
Ultraviolet radiation 452, 453–454
  disinfection 454
  DNA damage 257
Unasyn® 492
Undulant fever **622–623**
Universal precautions **106,** 107T
Upper respiratory tract 510, 510F
Upper respiratory tract infections 510F,
  512–516
  bacterial **512–515,** 512F
  overuse of antibiotics 487
  viral **515–516**
Uracil (U) 25, 26F, 27, 238
Urbanization, global 151
Urea breath test 31
Ureaplasma urealyticum 577F
Urease 78, 79
Urethritis **572–573, 575**
  gonococcal 579, 583, 583F
  non-gonococcal (NGU) 579, **583–585**
  sexually transmitted 576, 578–579
Urgency of urination 574, 575
Urinary tract 572, 572F
Urinary tract infections (UTIs) **572–576**
  bacterial **573–576**
  catheter-related 572
  diagnosis 575
  organisms causing 574F
  pathogenesis 105, 573–574
  portal of entry 78–79
  symptoms 574–575
  treatment 575
Urine
  bacterial infections 575, 575F
  barriers to infection **355,** 356T
  specimen collection 231T
Urogenital tract *see* Genitourinary tract
Urticaria, acute (wheal-and-flare) 426T, 428,
  428F
Use dilution method **442**

V

Vaccination 5, 11
  herd immunity 109–110, 307
Vaccines
  against exotoxins 90–91
  inactivated (killed) 309
  live attenuated 308
  recombinant DNA 114F, **310**
  requirements of successful 309, 310T
  subunit 309
  virus **307–311,** 308T
Vagina
  barriers to infection **355**
  microbial flora 577–578, 578F
Vaginal candidiasis 342, **589–590**
Vaginal discharge 579
Vaginitis 579
  bacterial 579
  trichomoniasis 328, 329
Valaciclovir (Valtrex®) 271, 474

Valley fever 531
Vancomycin 467
  resistance 490T, 497T
Vancomycin-resistant enterococci (VRE) **498**
Vancomycin-resistant *Staphylococcus aureus*
  (VRSA) 7F, 106, 498
Varicella *see* Chickenpox
Varicella-zoster virus (VZV) 643T, **650**
  central nervous system 598T
  dissemination within host 302F
  latency 139, 298, 298F
  persistence 296T
  see also Chickenpox; Shingles
Variola *see* Smallpox
Variola virus
  structure 274, 275F
  transmission 650
Vascular permeability, increased 370
Vasodilation, inflammatory response
  369–370
Vectors (disease)
  control 139
  emerging diseases 152, 152F
  parasites 322
  transmission 80, **100–101,** 101F
    biological 101
    mechanical 101
  viral infections 301
Vectors (recombinant) 115–116, 115F
  monoclonal antibody production 118
Vegetations, mature 619, 619F
Vehicle transmission **100,** 101F
Ventriculo-peritoneal shunt infections 597F
Vesicles, skin 643, 644F
Vibrio **552–553**
Vibrio cholerae 543F, **552–553,** 552T
  adherence 82T
  clinical symptoms of infection 536
  enterotoxin 89T, 90, 553
  morphology 552, 552F
  O139 149T
  pathogenesis of disease 552–553
  phage encoding of toxin 275
  pili 194
  treatment of infections 553
  see also Cholera
Vibrio parahaemolyticus 539T
Vincent's disease **541–542**
Viral encephalitis **606**
Viral envelope 273T, **274–275**
  as disinfectant/antiseptic target 438
  glycoproteins 275
  role in infection 62, 69
  see also Enveloped viruses; Non-enveloped
    viruses
Viral hemorrhagic fevers (VHF) **158–161**
  as bioweapons **176–177**
  pathogenesis 159–161, 159F
  treatment 161
Viral infections **293–312**
  acute **295–296,** 295F
  asymptomatic 295
  blood **629–633**
  central nervous system 597, 598T,
    603–607
  chronic 296, **297**
  digestive system **555–562**
  dissemination and transmission **299–306**
  drug resistance **500**
  drug treatment 11, **472–476**
  emerging **134–161**
  genitourinary system **586–589**
  host defense 312
  host susceptibility 307
  incubation periods 294, 294T
  interferon production 375, 376, 376T, 377F
  latent 296, **297–298,** 298F

lower respiratory tract **525–528**
patterns of **294–298**
persistent **296–298**, 296T
recurrent 421, 422
respiratory system 510
sexually transmitted 300–301, 305, 577F
skin 643T, **648–652**
slow 296, **298,** 606–607, 606T
upper respiratory tract **515–516**
vaccine development **307–311,** 308T
Viral particles *see* Virions
Viremia 132T, 142, 616, 629
active 302
primary and secondary 302F
Viridans streptococci 618T
Virions **272,** 273T
assembly **287**
biosynthesis **281–285**
decoy 289
immature 288
release **287–288,** 288F
spread 289, 289F
Virulence 9, 48
bacterial **56–58**
degrees of 83, 84F
virus **306–307**
Virulence factors 10, 56, 76
new antibiotic targets **502**
Viruses **271–292**
animal 272F, 275
antigenic variation **295–296**
assembly **287**
drugs targeting **475–476**
attachment **276–278**
bacterial *see* Bacteriophages
biosynthesis **281–285**
drugs targeting 281, **475–476**
budding **288,** 288F
cancer and **311**
capsid *see* Capsids
chemical inactivation/killing 444
classification 272F
co-receptors 277, 279
complex 274, 275F
culture **310–311**
cytoplasmic transport of components
279–280, 285
dissemination 299, **301–303**
DNA *see* DNA viruses
entry to host cells 278–279
drugs targeting **476**
enveloped *see* Enveloped viruses
genetic diversity, generating 155F
genome
assembly **287**
intracellular trafficking 286
replication **281–282, 283–284**
transport into nucleus **280,** 281F
helical 273, 273F
icosahedral 274, 274F
infection cycle **275–288**
intracellular locations 62, 66, 67
intracellular trafficking **285–286,** 286F
latency 275, 276F

maturation **285–287**
non-enveloped *see* Non-enveloped viruses
number required to cause infection 83
oncogenic **311**
pathogenic effects **93–94,** 93F
penetration **278–279**
plasma membrane interactions 62, 62F,
277, 278–279, 278F
portals of entry **299–301,** 299F
proteins
intracellular trafficking 285–286, 286F
synthesis 283, **284–285**
release from host cell **287–288,** 288F
drugs targeting **475–476**
RNA *see* RNA viruses
shapes 273–274
size 49F
spread **289,** 289F
structure **272–275**
terminology 272, 273T
systemic 301
thermal death times 449T
transmission **304–306**
uncoating **278–279,** 278F
as antiviral drug target **476**
virulence **306–307**
see also Virions; *specific viruses*
Vitamins, bacterial requirements 219T
Volutin 27, 206
Voriconazole 477
VRE (vancomycin-resistant enterococci) **498**
VRSA (vancomycin-resistant *Staphylococcus
aureus*) 7F, 106, 498
Vulvovaginitis, candidal 342, **589–590**

W

Wall teichoic acid 189, 189F, 191
Warts 588, 643T, **652**
dermal 652
genital 301F, **588–589,** 588F, 652, 652F
immunosurveillance 297, 297F
incubation period 294T
pathogenesis 588, 652
treatment 589, 652
see also Human papillomavirus
Water **19**
boiling 450T, 451
covalent bonds 17, 17F
filtration testing 228, 228F
heat capacity 19
hydrogen bonds 18–19, 18F
movement across plasma membrane
201–202, 202F
reactivity 19
as reservoir of infection 99
solubility 19, 19F
as vehicle of transmission 100
Waterborne infections 100
respiratory system 511, 523
Wescodyne® 443T
West Nile fever 157, 598T, 606, 631T
West Nile virus **156–157,** 606
geographical spread 153, 156–157, 157F
susceptibility mutation 157

Western blot 120, 120F
Western equine encephalitis (WEE) 598T, 606,
631T
Wheal-and-flare reaction 426T, 428, 428F
Whipworm *see Trichuris trichiura*
White blood cells
complete blood count (CBC) 358
differential analysis 358
inherited defects 422–423, 423T
innate immune response 358–367, 359F,
359T
leukocidin actions 85
margination and diapedesis 360, 360F
origins and differentiation 359F
Whooping cough *see* Pertussis
Widal reaction 119, 122T
Winter vomiting bug 557
Wiskott–Aldrich syndrome 421T, 422
Wobble hypothesis 244
World Health Organization (WHO) 486
Worms, parasitic *see* Helminths
Wound botulism 173, 603
Wound infections 644
Wounds, as portals of entry 80
Wuchereria bancrofti 635–636

X

X-linked agammaglobulinemia 421T
X-ray crystallography **501**
X-rays 452
Xenopsylla cheopis 620–621, 628

Y

Yeasts **338–340,** 339F
adherence 343
dimorphic fungi 340
Yellow fever 631T, **632**
re-emergence 149T
vaccine 177, 308T
Yellow fever virus 158F
Yersinia enterocolitica 544
Yersinia pestis 9, 620–621
as bioweapon **173–175**
historical perspective 148
life cycle 620–621
portals of entry 81
resistance to phagocytosis 369
see also Plague

Z

Zanamivir 475–476, 527T
Zidovudine (AZT; azidothymidine) 473F, **474**
Ziehl-Neelsen acid-fast stain 51T, **53,** 53F
Zones of inhibition
antibiotic susceptibility testing 490F, 491
discovery of penicillin 460
disinfectants/antiseptics 442, 442F
Zoonotic diseases 98, 98T
emerging 150, 151–152, 153
respiratory system 510
viral 304
Zygomatic rhinitis 342